BRIDGE TO UNITY

Unified Field-Based Science and Spirituality

R. W. Boyer

BRIDGE TO UNITY

Unified Field-Based Science and Spirituality

by R. W. Boyer, Ph.D.

©Copyright 2008 R.W. Boyer

Published by the Institute for Advanced Research
Malibu, California, USA

First Edition 2008
Printed in the United States of America.
Library of Congress Control Number: 2008909226
ISBN: 1-4392-1410-7

TABLE OF CONTENTS

PROLOGUE

*A*ll the research in modern science has led to the doorstep of the ultimate unity of nature. *The next step is direct empirical validation of unity. That requires going beyond sensory experience and reasoning, the basic means of gaining knowledge in modern science. It is most fortunate that we now have systematic means to unify completely our theoretical understanding and empirical experience of nature based on a holistic science of consciousness. Consciousness is the 'lamp at the door,'[1] illuminating both the outer diversified field of knowledge and the inner unified field of knowledge. Through this more inclusive science, we are stepping into a genuine Age of Enlightenment.*

This book is an in-depth analysis and synthesis of the nature of consciousness. The Prologue and Introduction establish the context for Part 1, an examination of ontological levels of *reality*, the nature of space-time, and the relationship of matter and mind to consciousness. It focuses on quantum theory, relativity theory, and their integration in quantum gravity toward unified field theory. Part 2 applies the unified field-based understanding to core issues in cognitive science and neuroscience of levels of mind—human psychoarchitecture—and states of consciousness. It examines how the nature of consciousness is addressed through direct experience in higher stages of human development. Part 3 connects modern objective science with the most ancient subjective science based in the Veda, and introduces research on fundamental consistencies between the structure of individual physiology and the cosmic structure of nature.

What's Been Missing in Modern Science?

The means we use to gain knowledge has far reaching consequences for our life. In our modern civilization the pre-eminent means has been the objective approach of modern science. It is widely recognized that this approach has produced a massive body of reliable knowledge and rigorous standard of validation, allowing us to progress quite a ways out of the shadows of superstition. It has strengthened our ap-

> "It is quite likely that the 21st century will reveal even more wonderful insights than those that we have been blessed with in the 20th. But for this to happen, we shall need powerful new ideas, which will take us in directions significantly different from those currently being pursued. Perhaps what we mainly need is some subtle change in perspective—something that we all have missed..."—*Sir Roger Penrose, mathematician and cosmologist (p. 1045)* [2]

preciation of universal order in nature, and of our incredible, even sacred fortune to witness its ever expanding display in the vast universe that is our home. Applying this knowledge in everyday life, however, our inner experience has remained separate and isolated from the outer world. This separation is so ingrained in our everyday life that we didn't recognize their underlying unity—let alone make use of it. The knowledge gained through modern science is fragmented. It has left everyday experience in our modern civilization fundamentally isolated, devoid of meaning, and unfulfilling. To get right to the heart of the *whole point*, the knowledge we want most still has eluded us: practical knowledge with the power to create a completely unified and fulfilling experience of life, and to alleviate suffering in our civilization. Because that knowledge has not been discovered through modern science, it is imminent cause for pause—to investigate more carefully what has been overlooked and missed.

In this book we will locate what has been overlooked and missed, and systematic means to validate it empirically. We will explore the deep *inner dimension* of nature, excavating the hidden foundation of modern scientific thought and uncovering how it can bridge with religion and spirituality into a universal system of knowledge and experience. In this exploration we will turn the light of science around onto itself to illuminate its holistic basis. This will reveal profound answers to perennial questions about the nature of matter and mind, of space and time, and ultimately of *consciousness itself.* Fortunately what has been overlooked is right behind where we're looking, within our easy grasp, even within the sound of our own voice.

The fragmentation of objective knowledge.
To begin locating what has been overlooked it can be helpful to glimpse below the outer edifice of modern science to its fragmented and jittery foundations. In applying the objective means of gaining knowledge we have used our minds to focus on the concrete, tangible, outer world of ordinary sensory experience. Fundamental to sensory experience is an object of experience, process of experiencing, and experiencer—the *known, process of knowing,* and *knower.* This basic three-fold division sets the direction on the bridge to unity described in this book. While modern science investigates the *outer objective* domain of the known, the *inner subjective* domain associated with the process of knowing and the knower has been conceptually separated from the outer objective domain and almost completely left out of the investigation. As we explore the core issues in this book it will become increasingly clear that through modern science we have investigated the outer objective domain of nature only. This approach has resulted in the seemingly intractable paradoxes, unsatisfying incompleteness, fragmentation, and lack of fundamental grounding in our everyday understanding and experience.

As to how this has come about, unwittingly we have conditioned ourselves into it. The objective means of gaining knowledge entrains attention to the tangible outer surface of experience—the ordinary outer natural world. When attention becomes fixated *out there,* the result is incessant exploration of the gross surface level of objects while deeper, subtler experiences become overlooked, elusive, and rare. Attention shifts from one object to another on the surface, rather than to deeper levels of experience of the objects. This results in shorter attention spans, shorter-term perspectives about values and purposes in life, and more emphasis on immediate sensory gratification. Effort is placed on squeezing more out of surface sensory experiences through increased intensity of sensation, rapid changes, and extreme contrasts—such as is increasingly apparent in the field of entertainment. This pattern of attentional focus is a common habit of everyday experience. It is evident in the rapid pace of surface change in our contemporary society, which reasonably might be characterized as an *attention deficit culture.*

This *objectifying* approach—and resulting picture of the world—is the main feature of the materialistic or physicalist worldview still prominent in modern science. In this picture the *primary locus of experience* is the concrete, sensory, material, physical level of existence. Materialism in this sense doesn't just have the popularized meaning of placing high value on tangible material possessions. It is a much more engrossing relationship to the world in which the surface objective level is understood and experienced to be the most substantial or *real*—indeed only—level of existence. This picture has been more gripping in recent centuries, due to its emphasis in the educational approach associated with modern science. The concentrated reductive focus on the sensory level of empirical experience has been institutionalized in modern civilization. Through this training we have bound ourselves to life on the superficial 'flatland' of material existence.

Consequences of fragmented knowledge.
The fundamental conceptions, assumptions, and values that guide actions reveal themselves in the results of the actions, such as what has been overlooked in house design and construction later shows up in cracks needing repair. When inadequate attention is given to the foundation, major cracks eventually appear.

As the objective approach of gaining knowledge in modern science has become ubiquitous in our civilization, its fragmenting side effects are more apparent. Some now question whether it actually has resulted in significant improvement in the overall quality of human life. Clearly the material wealth and physical comfort of a significant segment of civilization have increased; but whether the level of suffering has been appreciably reduced is far less clear. Empirical research suggests that happiness is correlated positively with material wealth, for the most part, only for people in poverty or with minimal material resources. [3][4] The poor tend to struggle for material sustenance, and the wealthy tend to strive for meaning and purpose—about which modern science almost completely has separated itself from addressing.

Some have expressed further concern that the objective means of gaining knowledge breaks apart experiences of deeper meaning and purpose, leaving shards of mini-theories and facts that cannot contain deep humanistic value associated with the holistic interconnectedness of nature. In this way modern science has contributed to the existential malaise of meaninglessness in modern and postmodern thought, fueling current trends of reactionary hedonism and violence.

"By the 1950s in North America, the United States and Canada attained a standard of living well above that of any previous civilization in history. They were joined soon afterwards by Western Europe, and then other industrialized nations such as Japan... [E]conomists and sociologists predicted that productivity would double by the end of the century. This prediction turned out to be true. They also predicted that by the year 2000 we would be able to enjoy unprecedented affluence by working only half as many hours per day, and turn the rest of our time to leisure pursuits. This of course did not happen... [W]e spend just as much time working as we did before... The pace and stress of daily life have, on the whole, intensified. Our children encounter weapons, drugs, and assaults in their schools. Overt hostility and violence are commonplace..."—*Jonathan Shear, philosopher (pp. 229-230)* [5]

More immediately, applications of the fragmenting objective approach are challenging our fundamental security in extremely grave ways: technologies that break apart the material fabric of nature are placing our basic humanity and even existence at risk. Modern science has not led to development of our inner mental resources to guide balanced use of outer material resources in increasingly powerful technologies. We are at a critical threshold at which the outer focus on matter has gotten ahead of inner development of our minds.

In response to these concerns it is sometimes asserted that the contribution of modern science is descriptive rather than prescriptive—that its purpose is to advance basic knowledge rather than fix the world. This 'value-neutral' perspective also is due to fragmenting assumptions associated with the objectified physicalist worldview. In actual practice even basic research is heavily laden with value judgments about what to study and how to study it. In addition a considerable portion of scientific funding is for technological applications that are thought to be of practical value, but unfortunately often are associated with the highly disintegrating and impractical behavior of waging war. With sincere intent to contribute to societal progress, it is not uncommon for modern scientists to advocate for and forge ahead with the latest innovations, unfortunately without adequate investigation of potential negative effects warranted by their increased power. This clearly reflects aggressive adherence to a value system of pragmatism interpreted within physicalist assumptions and beliefs—as if there were no deeper *reality* and no *subtler* long-term consequences.

While the modern scientific approach has allowed us to progress out of the irrationality of preceding millennia, it has not yet resolved essential concerns of the individual or of civilization. We have become fixated on the outer objective domain, and have not yet developed comprehensive understanding and integrated experience of the holistic, completely unified value of nature.

This is significantly due to the fragmenting reductive physicalist focus in modern science. It is now important to recognize and acknowledge that modern science—as it has been practiced—is powerful enough to disintegrate us, but not powerful enough to *integrate* us.

Consciousness in the Objectified Reductive Physicalist Worldview

The integrative core of the *inner subjective* domain is consciousness. This has been the constitutive issue for psychological science—like matter and energy for physics and life for biology. [6] The practical goals of this field directly concern improving the quality of life and alleviating suffering. Also applying primarily the objective approach, however, it has not established reliable knowledge of the inner subjective domain of mind and consciousness, and it has not achieved its practical goals. This is reflected in the lack of comprehensive theories and the modest benefits of its applications—evident in the continued suffering in our precious world family.

Theories about consciousness in psychological science—and modern science in general—are significantly shaped by the objectified reductive approach and physicalist worldview. In these theories, consciousness is the ability to be *aware of* some object of experience. It is attributed a functional role in attention, intentionality, and the sense of self. It is described as fading out during sleep and coma, restricted by brain damage or malfunctioning, and ceasing when the physical body no longer sustains life. For the most part, in these theories consciousness is an emergent property of complex functional organization in cellular processes of the physical brain. [7,8,9,10]

At more fundamental elemental, atomic, subatomic, and quantum levels, physical systems follow invariant laws of nature that are thought essentially to be a product of random disorder. The basic signifiers of life—intelligence, intentionality, selective attention, the survival instinct—have not been identified at these more fundamental levels of nature, and neither have mind and consciousness. This reflects the physicalist worldview within which the inner subjective domain of mind and consciousness are assumed to fit—but in which nature is fundamentally random and meaningless. While modern science has strengthened belief in the universal orderliness of nature, curiously this belief extends neither to the most fundamental levels of nature identified in modern science (quantum randomness), nor to the inner subjective domain (unreliable subjectivity). Herein are important clues about where the overlooked knowledge can be located.

Commonsense holistic experience. On the other hand, throughout history there have been anecdotal reports and anomalous evidence of subtle, intangible, interconnected phenomena suggestive of a more holistic view of matter, mind, and consciousness as well as a more integrated and fulfilling life experience. Many have reported at least glimpses of being connected with others and with nature through deeper, more expanded conscious experiences that are not easily explained as localized effects of electrochemical activity in the brain. These reports refer to a wide range of subtle but powerful experiences, sometimes difficult to explain, that add deeper meaning and purpose to life—associated with art, religion, spirituality, and also modern science at times.

Even with increasing ability through modern science to explain and manipulate the outer objective domain of nature, the majority of the world's population has not been disabused of beliefs in these subtle, intangible, elusive experiences and phenomena. The beliefs seem to be grounded in a deeply held intuitive sense of something more in life not evident on its material surface—an underlying thread that ties things together and gives deeper meaning to our place and role in the grand play of nature. It might seem reasonable to dismiss such beliefs as due to lack of training in scientific rigor, if it were not that many—again perhaps a majority—of those most revered scientific authorities and key contributors to knowledge throughout history have reported arriving at similar conclusions. It is quite probable that many modern scientists also have an intuitive sense of an underlying *reality* not captured by the objective approach. Indeed most all of us seem to have some inner sense of a deeper, more inclusive *reality* that underpins the physical world; but this inner sense is not developed enough to be of practical value for our everyday life.

Unscientific assumptions in modern science. A key issue of course is how subtle elusive phenomena and experiences of a deeper, more interconnected *reality* can be known accurately and reliably. The objective means of gaining knowledge basically apply rigorous logical reasoning and precise empirical testing, grounded in ordinary sensory experience. Subtle, elusive, highly interconnected, nonlocal phenomena associated with more holistic experiences are virtually unknowable through these means. They are rarely obvious to ordinary gross sensory experience; but without direct empirical experience of them it has been quite difficult even to get beyond their mere possibility to their reasonableness. They are not easily scheduled at the times and conditions required for rigorous objective testing, appearing to disappear in the effort to define them operationally in order to test them. They seem to be outside of the boundaries that frame the ordinary empirical world of modern scientific understanding and experience.

In addition modern science has been understood as basically a methodology of falsification or negation, in which something cannot be proven true but only less likely to be untrue. This methodology, consistent with reductive mentality, has been much more effective for comparing discrete, local, tangible phenomena than capturing deeper holistic, interconnected, nonlocal phenomena and testing them systematically. The knowledge gained has been significantly restricted by the means used to gain and validate it.

It is frequently reasoned that until modern scientific means to test such elusive phenomena are available it is best not to be concerned with them. From this perspective the prudent strategy is not to abandon the approach that has led civilization out of pre-scientific superstitions, but rather wait until modern science has developed means to test them. Without obvious availability or use of the means, however, typically such phenomena are categorized as illusory—or even delusional—and life goes on without addressing them.

But life doesn't just go on. It goes on based on our ordinary surface experience that is fragmented and still fraught with suffering. Frankly, instead of waiting for modern science, most all of us are what are sometimes called *naïve scientists* with respect to experiences not addressed by objective methodology. We commonly make major decisions such as causal attributions about personal relationships, judgments of right and wrong economic and sociopolitical policies, or the chance of avoiding a truck headed our way with little if any systematic objective testing. Much of our daily life is guided by working assumptions about quite important practical matters that have not been objectively verified. We make judgments and take actions based on these assumptions, and then sometimes try to defend them as if they had been objectively verified.

The point here is that this also has occurred with unverified assumptions in modern science. These assumptions, which underpin the rationale for the objective approach of gaining knowledge, are significantly shaped by the objectified reductive physicalist worldview. Not only has the accepted methodology of modern science restricted development of a more comprehensive range of knowledge that addresses the process of knowing and the knower—the inner subjective domain of mind and consciousness—but also unscientific commitment to reductive physicalism has unduly restricted understanding of the methodology of empirical science.

We have been living our lives as if there is nothing more than the physical world, while subtly sensing inside that something more exists. This has greatly contributed to a deep and unfortunate tear in the psychosocial fabric of modern civilization. It has separated subtler holistic experiences many people feel are the most significant aspects of their lives from the more restricted range of knowledge validated through modern scientific methodology. This directly concerns the gap between existential meaninglessness and moral values that separates modern science and spirituality.

One task of this book is to thread through intellectual and emotional gaps in contemporary thought in order to clarify the underlying seamless unity of nature. This integrated understanding leads to systematic means to experience that unity directly, within a rigorous but more inclusive scientific framework. A rational, logically consistent, and more comprehensive alternative is presented that extends modern science beyond the reductive physicalist worldview.

The Overlooked Knowledge and Experience

Fortunately the forefront of modern science has been slowly progressing to the stage where the fragmented shards of understanding of nature are beginning to recollect into a unified state. This is reflected especially in research efforts to build theories of a single completely unified basis of creation—unified field theory. In this progress modern physics has been grappling with the enigma that *matter doesn't have a material basis*. It has arrived at the rational conclusion that the assumption of materialism or physicalism—a main feature of much of the history of modern science—is untenable at fundamental levels of nature.

It is through investigating these fundamental levels of nature that core issues have arisen in quantum and unified field theories about the relationship of the observed and the observer, and objective matter and subjective mind. Historically these issues were prominent in philosophy, associated with the *mind-body problem* and a closed physical system that disallows causal efficacy of mind, and recently with the so-called *hard problem* of consciousness. They also have been recognized to be embedded in the *measurement problem* in the new physics and the *explanatory gap* between brain and mind in neuroscience. They concern subtler, more interconnected or entangled nonlocal levels of nature and their relationship to mind and consciousness. These issues also provide important clues toward locating the overlooked knowledge. They are now on the forefront of scientific investigation, and are major themes to be explored in depth in this book.

Referring back to the knower. The overall direction of this laudable progress is that modern scientists are beginning to consider the importance of looking inside themselves—trying to figure out how to turn attention inward toward investigation of the *knower*, the observer of nature, and to *consciousness itself*. Also fortunately this is leading to a more inclusive interpretation of systematic means of gaining knowledge that recognizes the essential role of consciousness. The cutting edge of modern science is now the empirical investigation of consciousness. An unprecedented phase transition is underway, from matter to mind to consciousness. In other words we are starting to look 'behind where we're looking.'

Unfortunately modern scientific progress toward the ultimate unity of nature has been only in terms of theory and understanding—intellectual wholeness. It hasn't included *direct experience* of that unity. At least in terms of theory the relationship between the outer objective and inner subjective domains is beginning to be considered more explicitly, and we are now poised for a much deeper integration. These developments presage advances in human knowledge far beyond the recognized history of the modern and ancient worlds.

These advances, however, won't come through manipulation of somewhat deeper levels of the outer material surface of nature—such as bioengineering to alter our natural genetic inheritance, nano-implants to build human-machine cyborgs, or colonization of *outer* space. Although in many cases these research initiatives reflect sincere efforts to address major concerns of humankind, they are predicated on a misunderstanding about the fundamental basis of nature. Practical, substantive advances won't come from dismantling nature as if it were merely inert random pieces of matter we can make better through modern scientific reengineering and computational technologies. When we attend only to the world of matter, we begin treating everything as if it were just bits of matter—ourselves included.

Rather than continuing in this increasingly dangerous direction based on untenable physicalist assumptions, the advances will come from a subtler, more profound alignment *with* nature. This is accomplished through scientific application of psychophysical laws in the *inner* subjective domain based on a holistic science of consciousness. This book intends to help debunk the fragmenting reductive physicalist matter-mind-consciousness ontology in modern science and progress to a holistic consciousness-mind-matter ontology. In this newly uncovered understanding of *consciousness-based science*, natural order is found in both the inner subjective and outer objective domains of nature—mind as well as matter. This is a tremendous scientific advance that opens up long-sought prospects for fulfilling our practical goals as individuals and as a civilization. It bridges the gaps between science and spirituality.

An expanded epistemology of knowledge will accompany this advancement, including appreciation of the subjective basis of objective knowledge and the underlying role of consciousness. Chapters 1 and 2 comprising the introduction to this book point to the subjective underpinnings of objective knowledge, and begin to introduce consciousness-based developmental technologies that systematically foster the direct experience of unity.

One way of understanding these developmental technologies in modern scientific terms is that they apply the principle associated with the 3rd law of thermodynamics—reduced disorder through reduced activity. However, this principle is applied to the inner subjective rather than outer objective domain. At this initial stage of discussion, this can be understood to mean systematically settling down mental activity to a more integrated state of globally coherent brain functioning—or even more simply, to establish inner silence.

A rational description of systematic technologies to develop higher states of consciousness is one major theme of this book. As holistic knowledge and experience unfold through these developmental technologies, the outer objective material domain is increasingly recognized to be the least powerful, least fulfilling, and least *real* level of existence. It becomes experienced as the gross outer shell of an infinitely expansive universe far, far more extensive, integrated, meaningful, and fulfilling than modern science has yet envisioned. Many of the puzzles that have been the intense focus and 'heavy lifting' of cutting edge modern science in the 20th Century seem rather like child's play in a sandbox as mind and consciousness are directly developed through these systematic technologies.

> "The science of today is not the science of yesterday. The science of today is the science of total knowledge."—*John Hagelin, theoretical physicist.*[11]

The science of consciousness. This book proposes that the greatest contribution of modern science in formulating theories of the ultimate unity of nature is that it has developed the theoretical basis for linking up with the most ancient continuous knowledge system of *Vedic science* that directly accesses that unity. Only in recent decades has modern science glimpsed deeply enough into nature to be able to link up with this most ancient tradition of knowledge. Like modern science, Vedic science investigates universal laws of nature through systematic empirical means of gaining knowledge that are not dependent on a particular social, cultural, or religious framework or belief system. Unlike modern science, it is a holistic consciousness-based approach that provides both understanding *and* direct empirical validation of the underlying unity of the outer objective and inner subjective domains of nature in the unified field.

Because modern science is just now approaching the doorstep of the ultimate unity of nature in unified field theory, it initially might be difficult to accept that ancient Vedic scientists—Vedic *rishis*—had that knowledge long ago. Ancient records don't seem, at least on their surface, to provide major scientific answers to support the thesis that the ancient records embody advanced knowledge. Archaeological theories have fit these ancient records into a general view of history concluding that all ancient civilizations were at lesser developed stages of knowledge. This book offers a more complete understanding of ancient Vedic science and its power to address the most challenging contemporary issues in modern science.

The general academic understanding has been that ancient Vedic records—estimated to have originated anywhere from 1500-5000 years ago in India—reflect pre-scientific stages of thought. However, this is now under revision as a longer time frame is unfolding. [12][13] Numerous philosophical and religious traditions have emerged over the centuries—including the major traditions of Hinduism and Buddhism—drawn from different interpretations of Vedic records. While its philosophical depth and influence were noted, the practical technologies in Vedic science were not recognized and thus not applied in Indian society or anywhere else. When that knowledge provided little practical value to daily life it fell out of sight and remained hidden for millennia.

The crucial ingredient. As connections between modern science and ancient Vedic science are articulated more thoroughly, appreciation is growing that this ancient science provides the holistic knowledge and experience that has been overlooked in modern science and civilization. The crucial factor in revealing the essential understanding and practical value of Vedic science is the work of the foremost scientist of consciousness, Vedic scientist and educator His Holiness Maharishi Mahesh Yogi. Maharishi has described ancient Vedic science in modern scientific terminology and has built a bridge of both knowledge and experience to connect them.

Veda is referred to by Maharishi as 'total knowledge' [14]—the unity of knower, process of knowing, and known. In modern scientific terms, Vedic science refers to consciousness-based science, or unified field-based science. Maharishi has pointed out that while Vedic science has existed continuously through time, its essential meaning had been lost to human society. Its applied technologies were rendered ineffective and even counterproductive due to severe misinterpretation, most evident in the unfortunate conditions prevailing even in India—the *'Land of the Veda.'*

A deeper investigation into the Veda brings to light Maharishi's inestimable accomplishment of recollecting the disparate parts of Vedic literature and reestablishing its unified structure and practical application. Over the past 50 years Maharishi has revitalized the ancient science of Veda in terms of the modern science of consciousness, *Maharishi Vedic Science and Technology,* and has focused on empirical validation of its technologies. [15][16]

Study in any approach to knowledge that is sufficiently penetrating will eventually encounter issues of universal significance. From a legalistic perspective, approaches to knowledge are not classified by their major topics of study but rather by their stated purpose, overall intent, and predominant methodologies. Although similar in some ways to other traditions of knowledge—including religious and spiritual traditions—Maharishi Vedic Science and Technology has consistently investigated laws of nature using systematic means of validation in a scientific framework.

In a nutshell, for the most part modern science has been concerned with the object of knowledge or known, religion with the process of knowing—in the sense of 'binding back' to the source of all, to the Godhead—and spirituality with the knower. A universal knowledge system—a *mature* science that addresses *total knowledge*—unifies all three in one subject. In testing and applying its developmental technologies systematically, Maharishi has demonstrated the power of Vedic science to bridge our modern Age of Science into an Age of Enlightenment. The deep significance for modern science, religion, and spirituality of this bridge will unfold as we proceed.

A question may arise whether the strategy in this book of emphasizing Vedic science is sufficiently broad or eclectic enough to get to a universal knowledge system. The strategy is not intended to take away from the invaluable contributions of other approaches. With utmost respect for these contributions and reverence for the profound wisdom of their contributors—in religion, spirituality, and also science—a careful investigation of Maharishi Vedic Science and Technology reveals its unique contribution of the integration of theoretical knowledge and practical applications, science and technology, reason and experience. This book brings out how its consciousness-based technologies have the power to unify the current fragmented state of knowledge and experience, and thus mend the damaging tear in the fabric of our disintegrated civilization.

Overview of Book Contents

Describing the bridge of knowledge and experience Maharishi has built from modern science's current fragmented reductive physicalist view to holistic Vedic science—and showing how it resolves perennial questions—are key objectives of this book. Many planks need to be put in place in order to traverse this bridge. Planks are laid out generally through summarizing major findings, theories, and limitations in modern science, and then 'connecting the dots' into holistic Maharishi Vedic Science and Technology. Topics addressed in this way include:

- Objective and subjective means of gaining knowledge
- Progress in quantum theory, relativity theory, quantum *reality,* and the nature of space
- Quantum gravity and information space underlying conventional space-time
- Gross relative and subtle relative structure of space and time, and the arrow of time
- Unified field theory and the consciousness-mind-body connection
- Human psychoarchitecture, attention, memory, and the emergence of mind
- Free will, determinism, downward and upward causation, and karma
- The top-down structure of nature, and implications for a science of ethics and morality
- Enlightenment and the full range of human development
- Consciousness-based technologies for mental, physical, and societal health
- Collective consciousness and the developmental basis for world peace
- Introduction to Vedic language and the structure of Veda and Vedic literature
- Direct correspondences between the structure of the individual and the cosmos

Chapters 1 and 2 comprising the Introduction establish an expanded rational framework for understanding scientific means of gaining knowledge. Part I of the book —Physics Unbound: *From Here to Infinity*—summarizes progress toward a completely unified field theory. Part II of the book—Psychology Unbound: *From Infinity to Here*—unites body, mind, and consciousness in the unified field. Part III of the book—Introduction to Veda and Vedic Literature: *Here is Infinity*—introduces research that tangibly demonstrates the universality of individual life.

The attempt is to include enough planks on the bridge to unity such that any leaps are less likely to result in falling off the bridge. A way to look at this is that the start of the bridge is the physicalist worldview--the known—and the bridge is the process of knowing that links to the unified experience of known, process of knowing, and knower. Another way to look at it is that the path across the bridge starts at the ordinary waking state of consciousness and the goal is full development in unity consciousness. However, the bridge is not crossed via intellectual understanding alone. Understanding and experience complement each other; deeper experience validates deeper understanding, and eventually they merge together. This book is intended to help clarify the *essential nature of consciousness itself*—the overlooked knowledge—and encourage its direct validation through *direct experience of consciousness itself*—the overlooked experience.

Maharishi emphasizes the principle of "knowing by being." [17] Knowledge has to be validated by direct experience—by being the knowledge—not just intellectually understanding it. In that light the book also provides a logically consistent explanation of the processes and stages of development through higher states of consciousness. It describes systematic scientific means to develop wise, healthy, and permanently fulfilled individuals as the basis for eliminating suffering in our civilization. It identifies the essential *active ingredient* to achieve these most practical goals as the direct unconditional experience of what consciousness is. In Maharishi Vedic Science and Technology *direct experience of consciousness itself* is the essence of spirituality. [18]

This book characterizes knowledge and experience of consciousness as beyond words and concepts. The term *knowledge* has different meanings, from its common usage as contextual facts or relative truths to universal and timeless truth. It is not just that knowledge is relative to its context, or that language is inadequate for describing experience—though these are useful points to appreciate. It also is that the conceptualizing mind cannot fully grasp consciousness itself. Consciousness is a field of being, knowing, direct experience—or whatever you wish to call it—that is beyond all contexts, conceptions, and descriptions. Every *thing* can be said to be relative to every other *thing*. Consciousness is beyond *things*, beyond *thing-ness*, beyond the conceptualizing mind. Thus any description or labels for it inevitably will be incomplete and misrepresentative. It seems useful to recognize this as we nonetheless use concepts and words to point beyond them.

Maharishi Vedic Science and Technology recognizes that our minds and the universe we observe with them share the same source and the same laws of nature. This allows knowledge to be unfolded directly—educed from within—through *systematic subjective means of gaining knowledge*. The holistic consciousness-based educational system of Maharishi Vedic Science and Technology begins with unity, and then sequentially unfolds the parts of knowledge within that unity—identifying wholeness in every part. An essential point is that the holistic approach emphasizes how the parts of nature sequentially emerge from the wholeness of nature, rather than the reductive approach of wholeness emerging from combining the parts. Wholeness is the basis of the parts: eternity is the basis of time; infinity is the basis of space; immortality is the basis of mortality. This subtle change in perspective is fundamental to a more inclusive and logically consistent science.

A key principle is that "knowledge is different in different states of consciousness." [19] Our state of consciousness determines our view of the world and understanding of the causes of fulfillment and suffering. Three basic points about consciousness are presented:

1. There is an unchanging, unmanifest, unified field of consciousness that underlies and permeates all changing, manifest, diversified nature.

2. The individual is capable of developing higher states of consciousness based on full enlivenment in the individual of the unified field of consciousness.

3. Maharishi Vedic Science and Technology provides systematic, natural, consciousness-based technologies to develop higher states of consciousness and live full enlightenment permanently.

This meaning of consciousness dramatically contrasts with physicalist theories in modern science. In the holistic approach of ancient Vedic science and Maharishi Vedic Science and Technology, the physical brain and body don't produce consciousness but rather just the opposite. Consciousness creates mind and body—the consciousness-mind-body ontology. Mind and body can be said to localize consciousness into a *state of consciousness* in the individual—to be discussed at length in the book.

The primary consciousness-based developmental technology in Maharishi Vedic Science and Technology is the *Transcendental Meditation* ™ (TM) technique, which Maharishi has taught throughout the world for the past 50 years. As described in this book, this systematic method of effortless transcending can be viewed as the most important technological contribution in the history of knowledge development. Its simplicity and subtlety reflect an integrated understanding of the natural functioning of the human mind and body. It is a simple, effective, repeatable means to go from the fragmented experience of the ordinary waking state of consciousness to the underlying, transcendent, unified state of consciousness itself. It provides an experiential bridge from ordinary thinking to unity. As Maharishi states, "Transcending thought is infinitely more valuable than thinking" (p. 444). [18]

In revitalizing ancient Vedic science Maharishi also has revived its extensive system of natural medicine, *Maharishi Vedic Healthcare*. This is a revolutionary contribution to the integration of mental and physical health care and disease prevention based on a holistic understanding of the consciousness-mind-body connection. This aspect of Maharishi Vedic Science and Technology, as well as other disciplines totaling 40 areas of the Veda and Vedic literature, will be introduced toward the end of the book after many of the planks of understanding have been put in place.

Author's Note. One objective of this book is to communicate how Maharishi Vedic Science and Technology can be appreciated rationally as a consistent system of scientific knowledge, compatible with modern science but subtler, more comprehensive and more integrated. Because readers will be highly intelligent and probing—perhaps also confident in their approach to knowledge and critically discerning of other approaches—this is a major challenge.

The extent of this challenge emerged in reactions even to the initial sentence in the initial draft of the initial chapter. To avoid complications that might overshadow the overall picture, a reasonably safe starting point was thought to be a general definition of the term *knowledge* based on dictionary definitions: 'Knowledge may be defined as that which is understood with clarity and certainty.' The reaction from the first reader was, 'You're just trying to sell TM, right?' A second reader said that *knowledge* has nothing to do with *psychological* impressions of clarity and certainty; it must be true, and if it is not true, then it is not knowledge. It relates to universal truths of empirical propositions such as, 'All swans are white;' or, 'My hand has five fingers.' A third reader said that humans can be sure of something but can't really *know* anything, referencing a statement by St. Thomas Aquinas that knowledge belongs to God. A fourth reader said that there is no knowledge in the phenomenal world—it is Maya. The only knowledge is *eternal Being*.

To whatever degree this challenge is met in the book, it is in its wholeness and may not be obvious in any particular point. The emphasis is on the 'forest rather than the trees.' In other words *the whole is the point;* or alternatively *the point is the whole*—to be unfolded as we proceed. Many topics and details are either omitted or covered in a cursory manner, especially the details of modern knowledge development and its numerous contributors. The focus is on the unity that has been lost in reductive detail. It may take some reading through the book to establish enough of the planks to feel comfortable with this strategy, especially because it covers regions that could be a bit foggy—fairly common when crossing bridges. Readers may find it more satisfying to get to the unifying thread and overall import, and later micro-analyze it to the mind's *content*. Hopefully you initially will forgive its many trespasses. At the outset I acknowledge there may be assertions and descriptions that seem awkward or unnecessary, and points important to you not adequately addressed. I extend an apology wherever inadequate sensitivity to your perspective is present, and ask for your patient focus on the global sense of what is attempting to be communicated. Your feedback to help with points in need of further clarification will be gratefully received.

As author of this book my only claim is to be a student of Vedic science whose fortune is to have the greatest of teachers. In that respect a distinction can be made between the authority of the book and the author. That which is accurate and authoritative is due to Maharishi and the Vedic tradition. Whatever may be inaccurate and non-authoritative is attributable to the author.

Progress on the path across the bridge to unity is quite interesting, and often a lot of fun. In the unique case of the ultimate unity, however, the path cannot be more fulfilling than the goal. It is said that in *that* goal every moment is *infinitely fulfilling*.

Prologue Notes

[1] Maharishi Mahesh Yogi (2003). Maharishi's Global News Conference, November 19.

[2] Penrose, R. (2005). *The road to reality: A complete guide to the laws of the universe*. New York: Alfred A. Knopf.

[3] Brickman, P., Coats, D, & Janoff-Bolman, R. J. (1978). Lottery Winners and Accident Victims: Is Happiness Relative? *Journal of Personality and Social Psychology, 36*, 917-927.

[4] Diener, E. (1984). Subjective Well-Being. *Psychological Bulletin, 95*, 542-575.

[5] Shear, J. (2006). *The experience of meditation*. London: Paragon Press.

[6] Miller, G. A., Trends and debates in cognitive psychology, *Cognition*, 10 (1981), 215-225.

[7] Damasio, A. (1999). *The feeling of what happens: Body and emotion in the making of consciousness*. New York: Harcourt, Inc.

[8] Baars. , B. J. (1988). *A cognitive theory of consciousness*. Cambridge: Cambridge University Press.

[9] Chalmers, D. J. (1996). *The conscious mind: In search of a fundamental theory*. New York: Oxford University Press.

[10] Churchland, P. S. (2002). *Brain-wise: Studies in neurophilosophy*. Cambridge, MA: MIT Press.

[11] Hagelin, J. (2004). Maharishi's Global News Conference, October 6.

[12] Feuerstein, G., Kak, S., & Frawley, D. (2001). *In search of the cradle of civilization: New light on ancient India*. Quest Books.

[13] Sidharth, B. G. (1999). *The celestial key to the Vedas: Discovering the origins of the world's oldest civilization*. Rochester, VT: Inner Traditions.

[14] *Maharishi Vedic University: Introduction*. (1994). Holland: Maharishi Vedic University Press.

[15] *Scientific research on Maharishi's Transcendental Meditation and TM- Sidhi program: Collected papers, Vols. 1-5* (Various Eds.) Fairfield, IA: MUM Press.

[16] Nader, T. (2000). *Human physiology: Expression of Veda and the Vedic Literature*. The Netherlands: Maharishi Vedic University.

[17] Maharishi Mahesh Yogi (2005). Maharishi's Global News Conference, February 2.

[18] Maharishi Mahesh Yogi (1967). *Maharishi Mahesh Yogi on the Bhagavad-Gita: A new translation and commentary, chapters 1 to 6*. London: Penguin Books.

[19] Maharishi Mahesh Yogi (1972). *Science of Creative Intelligence: Knowledge and experience*. [Syllabus of videotaped course]. Los Angeles: MIU Press.

INTRODUCTION
Objective and Subjective Means of Gaining Knowledge

Chapters 1 and 2 comprising the introduction to this book place science and religion in a simplifying and unifying context of means of gaining knowledge. Chapter 1 brings attention to the subjective underpinnings of objective knowledge and the fundamental subjective nature of the known, the process of knowing, and the knower. It points out that the outer-directed, objective approach of modern science has overlooked the inner knower of knowledge.

Chapter 2 considers how knowledge systems are products of different means of gaining knowledge—different processes of knowing or ways of using the mind. The subjective means of gaining knowledge in ancient Vedic science and Maharishi Vedic Science and Technology is introduced. It extends principles of systematic empirical investigation into the underlying subjective processes in the mind. It emphasizes investigating beyond or transcending the thinking mind in order to develop the intra-subjective and inter-subjective consistency necessary for universal knowledge and experience.

The purpose of these two introductory chapters is to establish the rationality and scientific nature of systematic subjective as well as objective means of gaining knowledge. This provides the basis for later chapters that extend beyond the current range of knowledge gained through the objective means in modern science. The discussion of the means of gaining knowledge unfolds from an initial basic level, in order to assist readers who may not have given much consideration to these issues, as well as to offer what may be a different perspective for many who have.

In the interest of not 'getting lost in the trees' on the first step, it is perhaps useful at the outset to acknowledge that knowledge can mean different things—ranging from knowledge that turns out to be untrue, relatively true (contextual), or absolutely true (beyond contexts). As described in these chapters, the condition or state of the knower has a lot to do with what is meant by knowledge.

Chapter 1

The Subjective Basis of Objectivity

This first chapter of the Introduction begins with a brief discussion of the means of gaining knowledge and their application in the two most general knowledge systems—religion and science. We will summarize the objective approach of gaining knowledge in modern science and begin to uncover its subjective underpinnings. The overall main point of this chapter is that knowledge fundamentally depends on the knower, and subjective consistency in the knower is the practical basis for reliable, accurate knowledge.

Experience and Reason

Experience and reason are commonly considered to be the basic means of gaining knowledge. As an initial point of reference we can say that experience relates to impressions the mind receives of the world through the five senses of perception; and reason relates to the underlying cognitive processes of rational thinking and understanding. One simple way to represent this is in terms of the outer objective world and the inner subjective levels of sensory experience and underlying reasoning about it.

Sensory World

Sensory Experience

Reason

Experience and reason interact—curving back onto each other—in gaining and validating knowledge. We experience the world and inductively reason about it to discover, uncover, unfold, or educe knowledge. For example, from observing the Sun traverse the sky, people reasoned that the Sun moves around the Earth. More detailed observations of the stars, eventually including telescopic observations, led to the reasoning that the Earth rotates on its axis and revolves around the Sun. Also we deductively reason about how the world is, and then try to verify it through experience. People reasoned that Earth might be spherical, and then set out to verify it. As knowledge validated through experience accumulates, a general body of knowledge or knowledge system develops. The two most recognized general knowledge systems that have developed are religion and, more recently, modern science.

For much of recorded history religion was the most prominent general knowledge system. Religion typically focuses on connection to a Supreme Being, essence, or creator of the universe and on divine laws of nature that guide life. It also can be understood to be a system of knowledge about how to live a good or virtuous life. In part religious knowledge unfolds through participation in rituals, charitable service, and various forms of prayer, contemplation and related practices—which emphasize experience—as well as through study of sacred texts, sermons, and counsel from religious leaders—which tend to emphasize reason.

Religious practice also can develop deep appreciation and acceptance of principles that go beyond ordinary experience and reason. Insight, wonder, awe, devotion, religious inspiration and revelation can foster belief and faith in religious principles. Belief and faith are associated with trust in the truth of some knowledge not based only on tangible empirical evidence and reason.

Ministers, priests, clerics, rabbis, historical religious authority figures, and other religious leaders sometimes help extend belief and faith into areas not directly experienced or clearly understood by their followers. Some religious figures are believed not only to have deep experience and understanding of religious principles but also direct access to truth. This relates to a third means of gaining knowledge underlying ordinary sensory experience and reason, which can be associated generally with the concept of *intuition*.

Sensory World

Sensory Experience

Reason

Intuition

Intuition

The term *intuition* relates to cognition or apprehension of knowledge not mediated by ordinary experience and reason. The implication is that accurate knowledge can be gained directly through *deeper levels of mind*—not requiring ordinary sensory and logical analyses.

As it is sometimes used the term relates to expert judgment. A track coach's hunch that certain athletes will become Olympic stars, a chess master's insight into a complex set of moves, or the fluid finger movements of a concert pianist deeply absorbed in a musical interpretation are sometimes referred to as intuitive because such quick judgments and actions are suggestive of an unmediated mental process. [1] However, these skills are based largely on building information processing functions that simplify ordinary sensory-perceptual and cognitive processes. With extensive practice the processes become unitized into an almost automatic sequence, allowing quick and frequently accurate judgments that satisfy some of the criteria for intuition. Ordinary levels of experience and reason can function efficiently in this mode of processing under consistent conditions; but deeper levels of mind are only minimally developed and utilized.

A more profound form of intuition is said to involve enlivenment of deeper, more integrated levels of *feeling* in the mind. Some individuals are believed to possess innate intuitive ability of this type. Others are believed to unfold it using techniques such as religious or spiritual practices to unlock deeper levels of mind—from the *inside*.

These means of gaining knowledge—sensory experience, reason, and intuition—have been relied upon throughout recorded history. For example intuition was emphasized more by Plato, whereas sensory experience and reason were emphasized more by his student Aristotle. Also intuition has been emphasized in spirituality and religion, and ordinary sensory experience and reason emphasized more in modern science.

"Aristotle (384-322 B.C.)...was the son of a physician and had marked predilection for natural history, and a distinct dislike for mathematics. Plato, who was considered the 'father of mathematicians', was his teacher. Early in his career, Aristotle reacted strongly against the mathematical philosophy of his teacher and began to build up his own system, which had a strongly biological bias and character. Psycho-logically, Aristotle was a typical extrovert, who projects all his internal processes on the outside world and objectifies them: so his reaction against Plato, the typical introvert, for whom 'reality' was all inside, was a natural and rather inevitable consequence. The struggle between these two giants was typical of the two *extreme* tendencies which we find in practically all of us, as they represent two most diverse and yet fundamental psycho-logical tendencies... In science, the extreme extroverts have introduced what might be called gross empiricism... The extreme introverts, on the other hand, originated what might be called the 'idealistic philosophies'... We should not overlook the fact that both these tendencies...account in a large degree for many bitter fights in science and life."—*Count Alfred Korzybski, neurolinguist (p. 87)* [2]

These means of gaining knowledge reflect different perspectives about the source of knowledge and the relationship between the knower and the sensory world of the known. The different perspectives will be developed further and reconciled later in the book; and then a much deeper meaning of intuition will be discussed. The purpose at this point is simply to acknowledge intuition as a potential means of gaining knowledge in addition to ordinary experience and reason.

The great challenge of supplementing ordinary experience and reason with intuition as means of gaining knowledge is the public validation of intuitions. Intuitions frequently involve abstract concepts and subtle experiences that, although at times accompanied by deep conviction, can be difficult to verify by other people applying only the means of ordinary experience and reason. For example it has been quite difficult to validate the efficacy of faith in God. Given there is no God, Godhead or ultimate unity of creation, this faith is irrational. Given there is however—which now can be interpreted as generally consistent with unified field theory, to be discussed later—then this faith can be viewed as a deep intuitive connection to that truth, even if not validated by ordinary sensory experience and reason.

In the increasing objectification of the means of gaining knowledge, about 400 years ago knowledge gained through systematic observations of the outer natural world accumulated that prominently challenged the accuracy of some religious knowledge and beliefs. Interest developed in systematic methods to explore, explain, and gain control over the outer natural world. These methods produced knowledge that contrasted with some religious beliefs based on subjective, intuitive-like faith. As these religious beliefs came under careful scrutiny, some discerning people began to deemphasize intuition as well as faith—and over time also religion—as reliable means of gaining knowledge. Modern science emerged from within the general framework of religion, emphasizing objective knowledge grounded in ordinary outer sensory experience rather than subjective faith in the inner intuitions reflected in scriptures and testimony of religious authorities.

The systematic methodology of objective modern science since has been increasingly successful for explaining phenomena of ordinary gross empirical experience. Many long-held beliefs once attributed to supernatural or 'other worldly' causes have been shown through modern scientific research to be explicable within the context of natural laws in the ordinary world of experience. Relatively few reliably reported phenomena have consistently defied modern scientific description and explanation, and for the most part these exceptions have not yet been viewed as essential for scientific progress. However, this historical picture is now undergoing major transformation, and will be placed in a much more expansive and integrating framework in this book.

Objective Means of Gaining Knowledge

Science generally refers to systematized study to gain accurate, reliable knowledge of the natural world. As will become clearer as we proceed, modern science has examined only the outer objective domain of the natural world. Different connotations of objective and subjective will be used in this book, especially in this introduction. However, the overall discussion proceeds toward the use of *objective* as meaning specifically the gross outer material domain, and *subjective* as meaning the subtle inner mental domain. As the book unfolds we will identify more precisely what these domains are. Also we will consider how principles of natural law and the fundamental orderliness of nature apply to both inner subjective as well as outer objective domains.

Modern science has focused on rigorous application of ordinary experience and reason as the means of gaining knowledge. In this approach experience relates to controlled observation and quantification of natural phenomena; and reason is exemplified in the use of formal mathematical logic in describing, analyzing, and predicting natural phenomena.

To summarize briefly the generally accepted methodological principles of modern science, scientific observations provide tangible data, evidence, or facts used to develop logical theories describing the data and accurately predicting how natural phenomena behave. These theories must be testable—usually meaning subject to disconfirmation—based on data collected by applying modern scientific methodology. If the data support the theory, the theory remains a viable explanation of the data; if not, the theory is modified or discarded. Modern science is characterized as a *self-correcting* process, continually improving theories to make them more accurate and reliable in accounting for data from careful observation of the natural world. An important hallmark of modern science is mathematical formalisms of theories that allow precise calculable predictions of natural phenomena.

Although naturalistic, correlational, and quasi-experimental research methods are commonly used in modern science, the most rigorous and formal method is the *experimental method*. A practical understanding of this method is reflected in the circular, self-correcting format used in many scientific research reports. In the *introduction* section a testable hypothesis that predicts the results is deduced from a general theory—associated with *construct validity*. In the *method* section the procedures and measures for testing the hypothesis through experimental manipulation and control of relevant variables are described—*internal validity*. In the *results* section the empirical findings are analyzed, frequently using statistical methods to help determine whether they logically and reasonably can be said to support the hypothesis—*statistical conclusion validity*. In the *discussion* section the results are referred back to the general theory in a self-correcting process, and their range of application in nature is considered—*external validity*.

Scientific experiments are designed to identify *causal relationships* between things in nature. One thing—*independent* variable—is isolated, and its influence on some other thing—*dependent* variable—is tested under controlled conditions. The causal inference is that consistent changes in a second thing must have been caused by a change in the first thing if nothing else changed. The strength of the causal inference depends on how carefully the experiment controls the relevant variables (internal validity); and its generality or range of application depends on the variety and naturalness of the testing conditions (external validity). This approach encounters severe limitations as scientific understanding goes beyond the core assumption of the independence of objects from each other and from the observer—fundamental implications of quantum theory to be discussed at length in upcoming chapters.

The method section of a scientific research report is a detailed description of how carefully an experiment controls the relevant variables and thus how strong of a causal inference, if any, can be made based on the experiment. It is written in a manner that allows other scientists to evaluate critically the validity of the research and replicate it if it is deemed necessary. This critical evaluation process, which helps build general agreement among scientists about the accuracy of the theories and their empirical support, is sometimes called *consensual validation*.

Consensual validation relates to public agreement. For example a group of people can observe and publicly agree that a clock tower indicates it is five o'clock. This is in contrast to a private subjective experience that is known only to the experiencer—such as choosing a time in one's own mind. Scientific observations are public in the sense that, in principle, they can be shared by anyone. This is a key feature of the objective means of gaining knowledge.

Consensual validation can be considered an open, democratic process in the sense that anyone can validate results through replication of research presented in public forums such as publications and conferences. Also theories supported by the majority of scientists in an area of research receive stronger acceptance. On the other hand there are recognized experts and authorities whose informed opinions receive stronger acceptance than others. In this sense consensual validation is in part an appeal to authority. Rather than appealing to the authority of individuals believed to have direct intuitive access to truth—as in religion—the appeal is to the authority of recognized experts in the scientific community who critically examine the theories and data. In this sense modern science is certainly not a democratic process. However, it does rely heavily on the empirical experiences that investigators who build the general consensus commonly share.

Further, some widely acknowledged theories are not products of the tedious work of experimentation and replication. They are based on logical arguments, theoretical insights, and intuitive-like judgments of highly respected and revered scientific authorities. In part due to the complexity of the subject matter in the behavioral and social sciences—especially human behavior—some theories in these fields can become broadly accepted and popular on the basis of anecdotal evidence and logical arguments of recognized authorities, with little experimental validation. Examples include Freudian and other theories of personality and psychopathology, humanistic theories such as Maslow's theory of self-actualization, Darwin's theories of emotional expression in humans and animals, Mendel's genetic theories, Galton's theory of inherited characteristics, theories of gender ability equality, and theories of the origin of language and civilization.

As scientific investigation progresses beyond directly observable phenomena, which it clearly has in the past century, this situation is increasingly attributable as well to the physical and biological sciences—discussed in later chapters. Modern science has some similarity to religion with respect to the role of authority in establishing generally accepted or orthodox theoretical knowledge. Also both science and religion reflect strong beliefs about what is accurate and inaccurate, verified and unverified, right and wrong knowledge—as well as how to develop it.

Eventually a scientific hypothesis associated with a theory may accumulate sufficient experimental validation and consensus to be identified by the scientific community as a law of nature. There are no established criteria in modern science about the amount of validation or consensus necessary for a theory to be accepted as identifying a law of nature. Accepted laws and theories remain subject to further scrutiny, which potentially could limit them to a specific range of contexts, or possibly invalidate them.

According to modern science, theories cannot be proven true. Rather their *truth-value* increases—that is, their falsity more likely decreases—with rigorous testing and consensual validation or agreement between scientists about the theories and the supporting evidence. To a significant degree, progress in gaining knowledge through modern science is a matter of strengthening the acceptance of theories in the minds of a group of scientists informed about the area of research. It involves a set of assumptions about nature and a systematic empirical approach to test them that are shared by a community of like-minded individuals—with a similar range of experiences.

> "The whole of science is nothing more than the refinement of everyday thinking... In contrast to psychology, physics treats directly only sense experiences and the 'understanding' of their connection. But even the concept of the 'real external world' of everyday thinking rests exclusively on sense impressions..."—*Albert Einstein, theoretical physicist.*" (p. 290) [3]

Subjective Underpinnings of Objective Knowledge

As noted earlier, it is a common view that knowledge is gained by experiencing something through the senses and then reasoning about it to come to an understanding. It is useful to recognize, however, that sensory experience is shaped by underlying assumptions imposed upon the process of observing by prior belief, learning, and reasoning—as well as by the functional structures built into the organism's cognitive and sensory-perceptual systems. An important area of study in cognitive science and neuroscience is the functional influence of conceptually driven *top-down* processes on the data driven *bottom-up* processes associated with sensory input.

Students of modern science typically learn about its methodology and content without studying its top-down conceptual and philosophical bases. Indeed many career scientists may not have spent much time considering the epistemological underpinnings of the work in their research specialization. This is the case even though these underpinnings can contribute fundamentally to the sensory perceptions associated with their empirical experiences in experimental work. Empirical experience, of course, isn't merely naïve sensory reception of outer *objective reality*.

"Modern science has spearheaded one assault after another on evidence gathered from our rudimentary perceptions, showing that they often yield a clouded conception of the world we inhabit."—*Brian Greene, theoretical physicist (p. 5)* [4]

Although rarely referenced in science textbooks or taught in standard courses, philosophers of science generally recognize that there are presuppositions and assumptions underlying scientific theories and facts—sometimes called *pretheoretical assumptions*. These assumptions form the philosophical basis of modern science. They shape the interpretation of facts and theories, as well as the methodology and content of modern science. These assumptions generally have not been tested using modern scientific objective methods. As working assumptions they have achieved acceptance largely on the basis of reason and intuitive-like belief or faith.

Examples of such assumptions that are prominent in modern science include: the universe originated from random events without purpose or design, there is universal order through strict cause-effect relationships, everything in creation exists within the framework of relativistic space-time, the manifest universe is fundamentally symmetrical, there is an objective world independent of the observer, the universe can be described by a logically consistent mathematical framework, the reductionistic objective approach is the most efficacious means of gaining knowledge, and modern scientific knowledge has the highest truth-value or is the most accurate knowledge achieved thus far in the evolution of humankind. The contextual limitations of these assumptions will become more evident as we proceed.

"Everybody agrees that the workings of nature exhibit striking regularities. The orbits of the planets, for example, are described by simple geometrical shapes, and their motions display distinct mathematical rhythms... On the basis of such experiences, scientists use inductive reasoning to argue that these regularities are lawlike... [I]nductive reasoning has no absolute security. Just because the sun has risen every day of your life, there is no guarantee that it will therefore rise tomorrow. The belief that it will—that there are indeed dependable regularities of nature—is an act of faith, but one which is indispensable to the progress of science."—*Paul Davies, theoretical physicist and cosmologist (p. 81)* [5]

Paradigms. Pretheoretical assumptions or presuppositions relate to the concept of a *paradigm*. This term refers to a model, pattern, or set of rules a group of people use for building a body of knowledge. Related terms include schema, mental frame, cognitive set, stereotype, contextualism, ethnocentrism, zeitgeist, global perspective, world hypothesis, and worldview. In the broader sociological sense of the term as used here, a paradigm can be understood to be a view of the world with accompanying means to develop it.

Major paradigms in modern science include such examples as Newtonian classical physics and quantum physics, germ theory and evolutionary theory in biology, and behaviorism and the information processing approach in psychological science. A widely accepted aspect of the general paradigm of objective modern science, the doctrine of materialism or physicalism, is now being superseded by progress in modern physics in developing quantum and unified field theories—summarized in Chapters 3-8. Implications of this progress with respect to the means of gaining knowledge will be discussed in Chapter 9.

> "Materialism, of course, is the belief that only the physical is ontologically valid and that, going even further, nothing that is not physical—of which mind and consciousness are the paramount examples—can even exist in the sense of being a measurable, real entity. (This approach runs into problems long before minds and consciousness enter the picture: time and space are only two of the seemingly real quantities that are difficult to subsume under the materialist umbrella.)"—*Jeffrey M. Schwartz, research neuropsychiatrist, and Sharon Begley, science writer/editor (p.28)* [6]

> "The real problem surrounding consciousness studies is not consciousness itself, after all, consciousness works. The problem is our tenacious adherence to materialism, a paradigm that has failed to generate even the simplest working models of cognitive processes, such as memory, intelligence, perception, let alone consciousness. What are at the roots of our underlying loyalty to materialism...[H]ow can we like something so much that works so little?—*Morey Kitzman, research psychologist* [7]

Paradigms can be viewed as conceptually driven top-down belief structures that exert a strong influence on ordinary sensory experience and on reason. They affect the perception and interpretation of research data or facts, as well as the reasoning processes involved in theory development. Scientific facts are *theory-laden*, in that the overall theoretical framework gives the facts meaning and significance. In part what is and what is not observed is a product of the theory that specifies the observation and measurement procedures.

> "It's the theory which decides what we can observe." —*Albert Einstein, theoretical physicist (p. 23)* [8]

Likewise scientific theories can be said to be *paradigm-laden*, in that the paradigm significantly shapes what theories are conceived, proposed, and tested. An example of how paradigms limit theories can be seen in the conception popular in early modern science that the universe is a cosmic machine built from indivisible atoms underlying all physical objects and actions. This billiard ball conception bred numerous linear mechanistic models that dominated modern science for many years, for the most part giving way to field, systems, network, and non-linear models.

> "Rather than propose a new theory or unearth a new fact, often the most important contribution a scientist can make is to discover a new way of seeing old theories or facts... [A] change of vision can, at its best, achieve something loftier than a theory. It can usher in a whole climate of thinking... What we are talking about is not a flip to an equivalent view, but, in extreme cases, a transfiguration"—*Richard Dawkins, biologist (p. xi)* [9]

Modern science has been described as progressing through periods of normal science when a paradigm is firmly established, and revolutionary periods when there is a major shift to a new paradigm. [10] During so-called periods of normal science, anomalies that don't fit the generally accepted or orthodox paradigm tend to be de-emphasized or disregarded. Gradually as more anomalies are encountered, their significance is confronted and carefully investigated by application of generally accepted methodologies. A revolutionary period may ensue, through which the old paradigm is recognized as inadequate and a new paradigm enthusiastically emerges that accounts for data supporting the old paradigm and some anomalies. This is evident in what might be seen as the religious-like zeal in the quantum paradigm, as well as super-symmetry and string theories.

Such paradigm shifts can occur based on new facts, new theories, new methods to test the theories, and new transformative ways of viewing or understanding the theories and facts. Perhaps the most cited example of a broad scientific paradigm shift is the transition from Newtonian classical physics to quantum physics. Newtonian classical physics dominated until the 20th Century when serious anomalies—such as the inability to explain magnetism and the nature of light—led to the formulation of quantum physics and quantum field theory. As will be described at length in this book, modern science can be said to be undergoing its most significant and encompassing paradigm shift: from materialistic object-based science to consciousness-based science.

> "There is no logical way to the discovery of... elemental laws. There is only the way of intuition, which is helped by a feeling for the order lying behind the appearance."—*Albert Einstein, theoretical physicist* [11]

Paradigms can be viewed as products of many factors including psychological, biological, and sociological influences involved with training in a particular knowledge system. They are relevant not only in the social enterprise of modern science but also in other knowledge systems and sub-systems—such as medical arts, politics, education, law, literature, business, and technology. Different religions also can be understood to be different paradigms for viewing God, humans, the world, and their relationships. In large part paradigms are identifiable by their underlying pretheoretical assumptions and intuitive-like beliefs—significantly influenced by the typical range of experience and reason the paradigm reflects.

> "Science is often an adversarial process in which advocates and opponents of a new theory or finding compete to determine whether it gets included in the knowledge base that the larger society can use and benefit from. Even though scientific methodology is designed to be objective, it is frequently the case that different laboratories produce inconsistent empirical evidence. Moreover, the meaning of the same evidence may be interpreted differently to support highly contrasting theoretical perspectives... To insure that the consensus-building process is done objectively and in the best interest of the nation, the National Institutes of Science and Health and other government agencies have evolved peer-review processes...typically comprised of eminent researchers in the field who ideally have no vested interest in the outcome of the review. Yet, even with these safeguards...there are many non-objective factors at work in moving knowledge from the laboratory to scientific consensus to social policy. These may include pre-existing convictions and assumptions, conflicting value systems, different reward systems, and different languages..."—*David Orme-Johnson, Charles N. Alexander, and Mark Hawkins, research psychologists (pp. 383-414)* [12]

Once a paradigm is generally accepted, it can be quite resistant to change. Paradigms not only aid in seeing new information that fits the paradigm but also filter out or constrain recognition and acceptance of information that doesn't fit. This phenomenon is sometimes referred to as paradigm blindness. A simple analogy is that a person wearing blue glasses sees everything in shades of blue. An example of paradigm blindness in business was the difficulty of Swiss watchmakers to accept the digital watch—somewhat ironically developed by their own researchers. The Swiss watchmakers' paradigm, which defined a watch as having bearings, springs, gears, and so on, apparently restricted the acceptance of the digital watch. In 1967 when the digital watch was presented to manufacturers at annual watch conferences—without patent protection—it was immediately appreciated by other international companies that quickly manufactured and marketed it. Within 10 years the Swiss watchmaker's share of the world watch market reportedly plummeted from an estimated 65 percent to less than 10 percent. [13]

> "I know that most men, including those at ease with problems of the highest complexity, can seldom accept even the simplest and most obvious truth if it be such as would oblige them to admit the falsity of conclusions which they have delighted in explaining to colleagues, which they have proudly taught to others, and which they have woven, thread by thread, into the fabric of their lives."—*Leo Tolstoy, author* [14]

"I had the distinct privilege of cofounding and coediting four journals in the field of clinical neuropsychology... I learned a few things as a result... One was that I could eventually determine potential reviewers (thankfully, very few) who would almost certainly recommend against publication of a particular manuscript, as well as those who would almost certainly see considerable merit sufficient for publication... What both had in common: a view of scientific information and its dissemination that was influenced by extrascientific values... In a word, results and conclusions that coincided with their notions of social (or any other form of) advocacy or ideology were acceptable; those that did not, were routinely seen as not worthy of publication."—*Byron P. Rourke, clinical neuropsychologist (pp. 38-39)* [15]

"We see how strongly matters of scientific fashion can influence the directions of theoretical scientific research, despite the traditional protestations from scientists of the objectivity of their subject. Nevertheless, it should be made clear that the apparent lack of objectivity is not the fault of Nature herself. There is an objective physical world out there... The apparent subjectivity that we see in the strong influences of fashion...are simply features of our gropings for this understanding, where social pressures, funding pressures, and (understandable) human weaknesses and limitations play important parts in the somewhat chaotic and often mutually inconsistent pictures that we are presently confronted with."—*Sir Roger Penrose, mathematician and cosmologist (p. 1024)* [16]

Paradigms, as well as paradigm blindness and shifts, reflect important dynamics of knowledge systems. They relate to how people maintain a relatively stable view of the world as a cognitive and affective platform for functioning in daily life. There are significant benefits and costs of adherence to a particular paradigm, related to the emotional attachment people have to it. It can be quite difficult to relinquish a paradigm that may have been the habitual way of viewing the world perhaps for an entire career, associated with a habitual pattern of how the mind is used to gain knowledge. The process of surrendering a paradigm can evoke emotions of fear, resistance, and defensiveness—in that it may imply that investment of time, energy, money, and reputation in the old paradigm was wasted or even was wrong.

The emotional attachment adherents have to their paradigms is evident in controversies contrasting the general knowledge systems of modern science and religion. Many modern scientists fear the loss of rational objectivity which they see as a persistent feature of religion and as contributing to the aggressive emotionalism and dogmatism that have fueled persecution and terrible destruction throughout human history. On the other hand many religious people deeply fear 'going against God's Will,' viewing secular modern science as extremely dangerous in its manipulation of nature not guided by an ethical or moral code that honors its sanctity and natural divine order—such as for example in atom and gene splitting. In addition sociopolitical and especially economic factors are increasingly recognized as major emotional influences in the overall paradigm of modern science—and religion as well.

Whether scientific, religious, or otherwise, knowledge systems apply experience and reason as means of gaining knowledge. They also incorporate pretheoretical assumptions associated with intuitive-like beliefs, and are subject to paradigm blindness, paradigm shifts, and other significant emotional influences. These issues point to the important influence of the underlying *subjective* processes in all knowledge systems.

Objectivity is a Special Case of Subjectivity

The concept of *objectivity* has been central to modern science. Scientific objectivity can be described in terms of basic principles applied to gain objective knowledge. The following three principles are core aspects of the modern scientific concept of objectivity: 1) belief in a natural world independent of the observer; 2) methodological rigor or precision of experience and reason;

and 3) consensual validation. These principles fundamentally are based in subjectivity. They assume a separate world that *I* can sense, that *you* can also sense, and about which *we* can agree. What this world is understood to be depends significantly on what *I* can sense and understand about it, and what *you* and *I* seem to agree about it—based on our current level of sensibilities.

Independent world. The prethcoretical assumption of an objective world existing separate from subjectivity—a mind-independent *reality* that can be observed objectively, not influenced by the subjectivity of the observer [17]—has been fundamental to scientific investigation as it is classically and popularly known. It is as if the world were enveloped in a glass bubble that the investigator peers into from somewhere outside. [18] Modern scientists have focused intently to maintain this assumption. The assumption seems to fit well with common ordinary experience, and also has received strong support by the success of research on a macroscopic scale of time and distance. As microscopic and ultramicroscopic scales are probed, however, this form of the assumption of the independence of observer and observed is no longer viable.

> "Present day physics does not explain the process of observation—really the most important concern in physics—because if we can't observe the state of something then we certainly don't know anything about it. The process of observation is a mystery in present day quantum mechanics."—*Eugene Wigner, theoretical physicist (p. 134)* [19]

> "Observations not only disturb what has to be measured, they produce it... [W]e compel (the electron) to assume a definite position... We ourselves produce the results of measurement."—*Pascal Jordan, theoretical physicist (p. 136)* [20]

Briefly the orthodox view of quantum theory states that an elementary matter particle is no particular *thing* prior to being observed. It is characterized mathematically as a *probability wave function* in a state of potentiality, containing infinite possible characteristics. As theorized, only when there is an observation does the wave function somehow collapse into a concrete measured object or form with specific dynamic characteristics. For one example light can be described in terms of either a particle or a wave. The characteristic that is observed is determined by the way an observer chooses to observe it—the experimental set-up. It can be observed as particles in the photoelectric effect, or indirectly as waves in a light diffraction experiment.

> "Consciousness should have long since been the topic of reasoned scientific study, and yet it has been largely ignored... We must recognize that objective reality is a flawed conception... Now the observer, consciousness, something self-like or mind-like, becomes a provable part of a richer reality than physics or any science has ever dared to envision."—*Evan Harris Walker, theoretical physicist* (p. 137) [20]

For another example a scientist measures the amplitude of an acoustic signal using a dB meter and observes that it indicates 80 dB. The dB meter is a measuring instrument for making precise observations of acoustic energy, but can be said not in itself to be a knowing entity—to be unfolded in later chapters. It might be said that the observation occurred when the sound waves contacted the dB meter, or when the meter indicator pointed to 80 dB, or when the light reflected off the indicator into the eye of the observer, or when the retina was activated, or when the optic nerve was activated, or when the visual cortex was activated, and so on. All of these events are describable theoretically in terms of quantum probability wave functions. The quantum wave function descriptions could continue indefinitely, without ever observing any of the characteristics of the measuring device or event. No particular breaking point in this chain appears to be favored by quantum theory. No matter how far this chain of wave functions is traced, all that is present prior to entry of a conscious observer is abstract quantum wave functions. At some point a conscious observer needs to observe the phenomenon. Until then it can be said that there is no observation, no observer, and no relationship between the observed and observer.[21] These fundamental distinctions seem somehow to be embedded in and emerge due to the conscious experience of the observer.

For a long while in the objectified approach of modern science the concept of an *observation*—and even more so an *observer*—was overlooked; but now there is increasing recognition of the significance of the observer and process of observing.[22] The observer interacts with an object to observe it, and this interaction in some sense *creates* what is observed. What the objective world is observed to be is influenced by the observer's choices of how, when, and where to make an observation. It is important to recognize that it is based as well on what the observer is *capable of observing*. As fundamental time and distance scales are probed it is increasingly evident that objectivity is not independent of subjectivity. The core principle of the independence of observer and observed which appears to apply at the gross macroscopic level of creation is an *inadequate* description of more fundamental levels. A more comprehensive understanding is that objectivity is a product of subjectivity. The objective world might be said, for ordinary practical purposes, to exist apart from any particular observer. But what appears to exist depends on how it is observed—and also *who* observes it. Not only is just beauty in the 'eye of the beholder,' but also *objective reality* is to some extent in the '*mind's eye of the beholder*.' These topics will be discussed in great depth later. These discussions will summarize important progress in physics beyond orthodox quantum theory as it is described here, toward deeper understanding of the relationship between objective *reality* and subjective mind.

Methodological rigor. A major contribution to scientific objectivity is methodological rigor, or precision in applying experience and reason. Scientists try to be as careful as possible in making observations and analyzing them logically. Precision is increased by use of accurate measuring devices and protocols; careful recording, categorization, and classification of observed phenomena; use of standardized terminology and symbology, formal application of descriptive and inferential statistics, mathematical modeling in theory construction; and other safeguards to avoid errors.

However, hopefully it will be recognized and acknowledged—upon a little self-reflection—that all of these procedures are based on processes of observing or knowing within the subjectivity of the observer. Sensory, perceptual, affective, cognitive, memorial, as well as intuitive processes through which knowledge is gained—and the knowledge itself—fundamentally involve the *subjectivity* of the knower. Knowledge is structured and *resides* in the subjectivity of the knower.

Methodological rigor and precision are important for increased accuracy and consistency. But they don't eliminate the subjective nature of the underlying mental processes that are the basis of objective observation. No matter how precise a measurement, a conscious observer or subjective *knower* is needed to *know* it. The fact that high degrees of orderliness and precision can be identified through methodological rigor attests to the fundamental consistency of underlying subjective processes in the knower.

Consensual validation. Consensual validation concerns the distinction between public and private. Commonly, *public* is associated with objectivity and *private* is associated with subjectivity. Scientific theories and facts are placed in the public domain where they can be scrutinized and a consensus about them can be built. This is basically a matter of an individual scientist's private experience and reason, or *intra*-subjectivity, being expressed publicly; and other scientists in turn indicating publicly that it matches their own individual intra-subjective processes.

> "By first-person events we mean the lived *experience* associated with cognitive and mental events... [T]he process being studied (vision, pain, memory, imagination, etc.) appears as relevant and manifest for a 'self' or 'subject' that can provide an account; they have a 'subjective' side... In contrast, third-person descriptions...refer to properties of world events without a direct manifestation in the experiential-mental sphere... Such 'objective' descriptions do have a subjective-social dimension, but this dimension is hidden within the social practices of science."—*Francesco J. Varela, cognitive neurobiologist, and Jonathan Shear, philosopher (p. 1)* [23]

A public observation necessarily is also a private one. In other terms *second-person* and *third-person* observations fundamentally are also *first-person* observations. Making something public adds to its objectivity by making agreement about it *between* people possible. But it doesn't remove the fundamental subjective basis of the agreements *within* each person. The contribution to objectivity of consensual validation is from intra-subjective consistency *across* individuals. This is sometimes called *inter*-subjective agreement, and it remains fundamentally based in subjectivity. Objective consistency across observers, or inter-subjective agreement, also attests to the fundamental consistency of underlying intra-subjective processes within each observer.

> "The study of consciousness in modern science is hampered by deeply ingrained, dualist presuppositions about the nature of consciousness. In particular, conscious experiences are thought to be private and subjective, contrasting with physical phenomena which are public and objective... Phenomena can be objective in the sense of intersubjective, investigators can be objective in the sense of truthful or dispassionate, and procedures can be objective in being well-specified, but observed phenomena cannot be objective in the sense of being observer-free. Phenomena are repeatable in the sense that they are judged by a community of observers to be tokens of the same type. Stripped of its dualistic trappings the empirical method becomes *if you carry out these procedures you will observe or experience these results*—which applies as much to a science of consciousness as it does to physics." —*Max Velmans, research psychologist* (pp. 299-300) [24]

Objective *reality*, methodological rigor, and consensual validation or inter-subjective agreement are all based in subjectivity. Objective knowledge is a type of subjective knowledge. It is a sub-set or special case of subjective knowledge that satisfies criteria for objectivity. Basically these criteria can be simplified into one thing: *subjective consistency*. The objective means of gaining knowledge depends on *intra-subjective* (first-person) and *inter-subjective* (second and third-person) consistency. Importantly the apparent relatively high degree of inter-subjective agreement associated with the objective approach of modern science attests to relatively high intra-subjective *consistency* in individual scientists. [25]

Subjective Consistency

Modern scientists frequently associate *objectivity* with unbiased, reliable knowledge and *subjectivity* with biased, unreliable knowledge. The scientific method was designed in part to reduce the inconsistency attributed to subjectivity. But the subjectivity of the scientist or knower would seem to require relatively high consistency for any precision of experience and reason, as well as for inter-subjective agreement, replication of observations, or anything else involved in gaining reliable knowledge. [26]

For example the scientist's subjective mental processes are quite consistent in conducting experimental research. If measurements appear to be inconsistent, the scientist typically first checks for a malfunction in the device or other environmental factors due to the high level of confidence in the stability, consistency, and reliability of his or her own intra-subjectivity.

Being objective cannot mean being free from all subjectivity. The concept of objectivity free from all subjectivity is like saying we could have a *mind-independent mind*—an experience without an experiencer— which doesn't seem to make sense at least with respect to ordinary individual minds. Rather, to be objective more fundamentally means to minimize distortion and increase orderliness in our subjective minds, as well as in our measurement devices and procedures.

Identifying consistent patterns or orderliness is basic to gaining reliable knowledge. People seem continually to seek consistency and reduction of uncertainty, for survival if for no other reason. A relatively high degree of intra-subjective consistency or inner stability is a prerequisite for identifying order and consistency in the outer natural world. The pretheoretical assumption fundamental to science that the universe is orderly would seem to apply also to the subjective mind if it is part of this same universe. Further it would hardly seem possible to know about or function in an objective world if subjective minds were not at least relatively orderly, stable, reliable, and consistent—both intra-subjectively and inter-subjectively.

Sources of subjective inconsistency. There are of course influences that reduce the consistency of subjective mental processes. Errors in perception, affect, reason, judgment, memory, and intuition—whether unintentional or affected by intentional bias—happen. In the same way that a measuring device can malfunction, physiological and psychological processes also can malfunction due to fatigue, stress, disease, or other disorderly influences that interfere with rational decision making and reliable perception, and that reduce subjective consistency.

For the most part the intact human information processing system appears to be sensitive enough to detect much of its own malfunctioning—within its ordinary range of experience and reason. [27] It usually can take the malfunctioning it detects into consideration such that a complete reevaluation and revamping of its entire knowledge system is not constantly undertaken—although regularly updated. Even though experience and reason can become somewhat inconsistent, a relatively high degree of core consistency and stability is necessary—both inter-subjectively or publicly and, more fundamentally, intra-subjectively or privately—in order to do science.

There are several important sources of subjective *inconsistency* in addition to physiological and psychological malfunctioning. One major source of *inter-subjective* inconsistency is associated with enculturation, especially related to language. Language differences can result in lack of agreement between people even when the underlying *intra-subjectivity* may be quite consistent across individuals but difficult to appreciate due to language barriers. As the various forms of language barrier are bridged, the underlying intra-subjective consistency across individuals is recognized and higher inter-subjective agreement can be achieved.

Another source of *inter-subjective* inconsistency is developmental differences, such as perceptual-cognitive development—as modeled in Piaget's [28] well-known theory. Perceptual-cognitive development has a major impact on the cognitive lens through which the world is perceived and understood. At different stages of perceptual-cognitive development, people experience and reason quite differently.

For a basic example a preoperational child, typically between two and seven years of age, will agree that two containers of the same size and amount of water have equal amounts of water in them. However when the water from one container is poured into a tall thin container, the preoperational child frequently appears to infer that the taller container now has more water in it than the shorter container. Only the dimension of height appears to be considered, ignoring width due to inability to assess both dimensions simultaneously. A more developed concrete operational child, usually above seven years of age, reasons that the water is higher in the taller container due to less width, and correctly infers that the amounts of water are still about equal. The two inferences don't simply differ. The inference of the concrete operational child is more comprehensive and accurate. As the younger child matures, his or her intra-subjectivity hopefully will develop to the level where inter-subjective agreement with the older, more developed child occurs. What was initially experienced and reasoned as incorrect, impossible, or even mystical from one perspective eventually becomes obvious as a more developed stage unfolds. [29]

Individuals in the highest stage in Piaget's model, the stage of formal operations, have the ability to engage in the abstract reasoning required for scientific thinking. Such individuals are capable of comprehending the effects of relative differences due to enculturation and similar factors, and frequently can communicate across those differences to achieve higher levels of inter-subjective agreement. In addition there are levels of abstract reasoning ability, which also affect the degree of inter-subjective agreement that can be achieved. But even among individuals and groups with highly developed abstract reasoning and compatible communication systems, fundamental paradigmatic differences still occur. Perhaps the most significant contributions to subjective *inconsistency* are developmental differences in the deeper levels of mind underlying the formation of the differing paradigms of knowledge and worldviews that have emerged in human history.

> "[N]aive empiricism—that science simply and innocently reports to us the unshakable givens of experience—is...an extreme and untenable view. It is the myth of the given... We do not, for example, perceive a tree. What we actually see, what is given in our experience, is simply a bunch of colored patches. On this, empiricists, rationalists, and Idealists all agree. The traditional empiricist then attempts to ground all knowledge in these sensory "givens'—*the colored patches...* [T]hese objective features are differentiated, conceptualized, organized, and given much of their actual form and content by conceptual structures... These *interior structures* include not only deeply background cultural contexts, intersubjective linguistic structures, and consensus ethical norms, but also most of the specific conceptual tools that scientists use as they analyze their objective data, tools such as logic, statistical displays, and all forms of mathematics... *None of these structures can be seen or found anywhere in the exterior, empirical, sensory world.* They are all, all of them, subjective and intersubjective occasions, interior occasions... And nobody has ever found any way whatsoever to reduce this knowledge to colored patches... Empirical science depends upon these interior domains (subjective and intersubjective) for its own objective operation. But because they *cannot be accessed* by...objective and sensorimotor methods, empirical science, in its more brutish forms, has simply rejected these interiors altogether, interiors which not only allow its own operations, but also contain the within of the Kosmos... This self-obliterating reductionism is not genuine science..."—*Ken Wilber, philosopher (pp. 145-148)* [30]

Developmental theories that propose higher levels of cognitive, moral, and self development beyond abstract reasoning ability describe integrative perspectives as a consistent feature of higher levels.[29 30 31 32 33] Individuals thought to be very highly developed—indicated by their ability to function at levels which incorporate and also go beyond all earlier developmental levels—have a greater tendency to express more integrative worldviews in the direction of a perennial philosophy and toward a universal knowledge system.[34 35] This suggests that higher levels of consensus regarding universal knowledge can be achieved when the investigators' minds are more fully developed and provide more consistent, integrated understanding and experience of the natural world.

The objective methods of modern science emerged as distinct means of gaining knowledge in reaction to the unreliability of some intuitive-like religious knowledge. This is not to say that all knowledge based on intuition is inaccurate, but rather that the deep levels of the mind involved in intuition were not sufficiently accurate and reliable across the population of religious practitioners to establish broad inter-subjective agreement. The same is the case across scientific practitioners in order to connect systematically with intuitive knowledge from religion to build a universal consensus across both science and religion. The difficulty has been insufficient development of the mind to reduce and eliminate the unreliability of deeper levels of individual subjectivity—including intuition—across religious and scientific practitioners. Systematic reliable means to develop deep levels of the mind to establish very high subjective consistency and accuracy within and between individuals has been overlooked and missed. When this is provided the gap between the two major knowledge systems of religion and science can be bridged.

> "Try and penetrate with our limited means the secrets of nature and you will find that, behind all the discernable laws and connections, there remains something subtle, intangible and inexplicable. Veneration for this force beyond anything that we can comprehend is my religion."—*Albert Einstein, theoretical physicist (p. 384-385)* [36]

In the objective approach of modern science the object of knowing—the known—is commonly identified as the outer natural world, as well as knowledge about it. The process of knowing at least includes the inner sensory-perceptual and cognitive-intellectual processes through which the outer natural world is experienced and understood—and also to some degree intuitive processes associated with deeper inner feelings. For the most part, however, the *knower* of knowledge has not been identified in modern science.

Modern scientists generally assume that the knower is some aspect of conscious mind, but have identified neither mind nor consciousness as to what and where they are—and in some cases even whether they exist. Modern scientists have gained considerable knowledge of the outer objective material world, but don't know *who* knows it. Because of its outer-directed reductive physicalist focus, it has not discovered systematic reliable means to turn inward to investigate the knower of knowledge and systematically develop deeper levels of subjective consistency. Without knowing the knower, knowledge is baseless. [37] As succinctly stated by His Holiness Maharishi Mahesh Yogi:

> "Through its objective approach, modern science reveals that which is perceived, the object. The subject, the perceiver, remains separate from it. Modern science investigates the field of the known, but it does not touch at all the field of the knower and the spontaneous process of knowing." (p. 1) [38]

Chapter 1 Notes

[1] Kahnemann, D. A. (2003). Perspectives on Judgment and Choice: Mapping Bounded Rationality. *American Psychologist*, 58:9, 697-720.

[2] Korzybski, A.(1994, 5th Ed.: 1933 1st Ed.). *Science and sanity.* Englewood, New Jersey: Institute of General Semantics.

[3] Einstein, A. (1965). *Ideas and opinions.* New York: Bonanza Books.

[4] Greene, B. (2004). *The fabric of the cosmos: Space, time, and the texture of reality.* New York: Alfred A. Knopf.

[5] Davies, P. (1992). *The mind of God: The scientific basis for a rational world.* New York: Simon & Schuster.

[6] Schwartz, J. M., & Begley, S. (2002). *The mind and the brain: Neuroplasticity and the power of mental force.* New York: HarperCollins Publishers, Inc.

[7] Kitzman, M. (2002). The Hard Problem of Materialism: 'How Can We Like Something So Much, That Works So Little.' Abstract for Poster Session, Thursday, April 11. Toward a Science of Consciousness 'Tucson 2002,' April 8-12.

[8] Joseph, R. (2002). The Myth of the Big Bang: Cosmic Organic Clouds & Creation Science. In Joseph, R. (Ed.). *Neuro Theology: Brain, science, spirituality, religious experience.* San Jose, CA: University Press.

[9] Dawkins, R. (1989). *The selfish gene.* Oxford: Oxford University Press.

[10] Kuhn, T. S. (1970). *The structure of scientific revolutions,* 2nd Edition, Enlarged. Chicago, IL: The University of Chicago Press.

[11] Einstein, A. <http://www.brainyquote.com/quotes/quotes/a/alberteins138241.html>

[12] Orme-Johnson, D. W., Alexander, C. N., & Hawkins, M. A. (2005). Critique of the National Research Council's Report on Meditation. In Schmidt-Wilk, J., Orme-Johnson, D. W., Alexander, V. K., & Schneider, R. H. (Eds). Applications of Maharishi Vedic Science: Honoring the Life Work of Charles N. Alexander, Ph.D. Special issue of the *Journal of Social Behavior and Personality,* Vol. 17, No. 1. Select Press, 383-414.

[13] Barker, J. A. (1990). *Discovering the future: The business of paradigms,* 2nd Edition. Goodyear Rubber and Tire Company, Reg. No. 3393.

[14] Goldstein, S., (2002, Winter). Bohmian Mechanics. *The Stanford Encyclopedia of Philosophy.* Edward N. Zalta (Ed.). <http://plato.stanford.edu/archives/win2002/entries/qm-bohm/>

[15] Rourke, B. P. (2008). Neuropsychology as a (Psycho) Social Science: Implications for Research and Clinical Practice. *Canadian Psychology,* Vol. 49, No. 1, 35-41.

[16] Penrose, R. (2005). *The road to reality: A complete guide to the laws of the universe.* New York: Alfred A.

Knopf.

[17] Warner, R. Facing Ourselves: Incorrigibility and the Mind-Body Problem. In Shear, J., Ed. (2000). *Explaining consciousness: The hard problem*. Cambridge, MA: The MIT Press, pp. 133-147.

[18] Charleston, D. E., (1984, Spring). Personal communication, Norman, OK.

[19] Kleinschnitz, K. & Muehlman J. M. (1987, January). A Special Seminar at Maharishi International University: Perspectives on Consciousness in Physics, with Special Addresses by Professors Eugene Wigner and John Hagelin. In Chandler, K. & Wells, G. (Eds.). *Modern Science and Vedic Science*, Vol. 1, No. 1, Fairfield, IA: Maharishi International University, p. 134.

[20] Walker, E. H. (2000). *The physics of consciousness*. Cambridge, MA: Perseus Books.

[21] Farwell, L. (2000). *How consciousness commands matter: The new scientific revolution and the evidence that anything is possible*. Iowa: Sunstar Publishing.

[22] Hodgson, D. The Easy Problems Ain't So Easy. In Shear, J., Ed. (2000). *Explaining consciousness: The hard problem*. Cambridge, MA: The MIT Press, pp. 125-131.

[23] Varela, F. J. & Shear, J. First-Person Methodologies: What, Why, How? In Varela, F. J. and Shear, J., (Eds.) (2000). *The view from within: First-person approaches to the study of consciousness*. Bowling Green, OH: Imprint Academic Philosophy Center, Bowling Green University, pp. 1-14.

[24] Velmans, M. Intersubjective Science. In Varela. F. and Shear, J., (Eds.). (2000). *The view from within: First-person approaches to the study of consciousness*. Bowling Green, OH: Imprint Academic Philosophy Center, Bowling Green University, pp. 299-306.

[25] Hut, P. & Shepard, R. N. Turning 'The Hard Problem' Upside Down and Sideways. In Shear, J, (Ed.) (2000). *Explaining consciousness: The hard problem*. Cambridge, MA: The MIT Press, pp. 305-322.

[26] Damasio, A. (1999). *The feeling of what happens: Body and emotion in the making of consciousness*. New York: Harcourt, Inc.

[27] Chalmers, D. J. (1996). *The conscious mind: In search of a fundamental theory*. New York: Oxford University Press.

[28] Piaget, J. (1972). Intellectual Evolution From Adolescence to Adulthood. *Human Development, 15,* 1-12.

[29] Alexander, C. N. (1982). Ego Development, Personality and Behavioral Change in Inmates Practicing the Transcendental Meditation Technique or Participating in Other Programs: A Cross-Sectional and Longitudinal Study. Doctoral dissertation, Harvard University. *Dissertation Abstracts International,* 43 (2), 539B.

[30] Wilber, K. (1998). *The marriage of sense and soul: Integrating science and religion*. New York: Random House.

[31] Pascual-Leone, J. (1990). Reflections on Life-Span Intelligence, Consciousness, and Ego Development. In Alexander, C. N. & Langer, E. J. (Eds.). *Higher stages of human development. Perspectives on adult growth*. New York: Oxford University Press.

[32] Kohlberg, L & Ryncarz, R. A. (1990). Beyond Justice Reasoning: Moral Development and Consideration of a Seventh Stage. In Alexander, C. N. & Langer, E. J. (Eds.). *Higher stages of human development: Perspective on adult growth*. New York: Oxford University Press, pp. 191-207.

[33] Miller, M. E. & Cook-Greuter, S. R. (Eds.) (1990). *Transcendence and mature thought in adulthood*. Lenham, Maryland: Rowland and Littlefield Publishers, Inc.

[34] Maslow, A. (1976). *The farther reaches of human nature*. New York: Penguin.

[35] Huxley, A. (1945). *The perennial philosophy*. New York: Harper.

[36] Isaacson, W. (2007). *Einstein: His life and universe*. New York: Simon and Schuster.

[37] Maharishi Mahesh Yogi (1972). *The science of creative intelligence: Knowledge and experience* (Videotaped course). Los Angeles: Maharishi International University Press.

[38] Maharishi Mahesh Yogi (1987). Maharishi on modern science and Vedic science. *Modern Science and Vedic Science,* 1, 1, January, 1-3.

Chapter 2

Developing Subjective Consistency and Coherence

In this second and final chapter of the Introduction, we will discuss knowledge systems as products of different means of gaining knowledge—different ways of using the mind. The subjective means of gaining knowledge in Maharishi Vedic Science and Technology, based on ancient Vedic science, is introduced as a systematic technology to develop subjective consistency and coherence. The overall main point of this chapter is that transcending the mind to contact its underlying basis develops the intra-subjective and inter-subjective consistency necessary for a universal system of knowledge and experience.

The educational or developmental emphasis associated with the objective approach in modern science has been on increasing precision of experience and reason for increased methodological rigor. In this approach attention is outer-directed, toward empirical observation and experimental manipulation of the outer natural world. Ordinary sensory experience and logical reason have been emphasized over deeper levels of mind associated with feeling and intuition.

By focusing the inquiry into knowledge primarily using ordinary sensory experience and abstract thinking or reason, most everything relevant to deeper processes of knowing and the knower in the inner subjective domain has been overlooked and practically unexamined. The product of this outer focus has been an approach to gain knowledge that is more tangible, rational, less emotional, highly conceptual, public, impersonal, amoral, skeptical, concerned with external control of nature, and that defines *reality* largely in terms of the gross physical level of nature experienced through ordinary sensory perception.

This objective—outer directed, objectifying, or *object-referral*—approach has raised the standard for validation of knowledge and has established more accurate and reliable knowledge of the ordinary physical level of nature. In this effort, however, it has tended to direct attention away from deeper levels of the subjective mind, as well as away from ultimate questions about human existence generally thought not amenable to scientific investigation. It also has tended to devalue—and sometimes categorically reject—what many religious people consider to be the deep inner spiritual level of life.

In contrast, religion tends to emphasize emotions, feelings, and intuition more than senses and intellectual reasoning. At least in theory—considerably less so in practice—it also de-emphasizes the material level of nature. Generally religious approaches to gain knowledge have been more private, personal, emotional, moralistic, concerned with self-control, more reliant on intuitive-like feelings and faith, and place more emphasis on what is believed to be the deep inner spiritual level of life.

Religion has recognized the importance of deep intuitive levels of the mind and the historical bias against them in terms of belief and faith associated with the objective approach in modern science. However, the diversity of fundamental beliefs and faiths in different religions indicates that inter-subjective agreement and consensus have not been broadly achieved across religious populations. Certain individuals may have developed universal levels of understanding and experience that go beyond specific religious doctrines, and they may have attracted large groups of adherents that develop into relatively long-standing traditions. But the various religious groups in general have not effectively applied reliable means to develop fully the deep levels of mind in their practitioners to establish a universal consensus across religious groups and traditions. - Religions frequently emphasize the importance of deep feeling and intuition, but have not been successful in developing their reliability and consistency across the broad population of religious practitioners. Religious knowledge thus has remained largely on the level of faith, based on the intuitions of historical authority figures rather than on development of the intuitive ability to experience fundamental truths directly. Outer emotionalism, rather than deep inner intuitive feelings and experience, has become more prominent among many religious groups. Increasingly faith has been relied upon, even *blind faith* absent of deep inner experience—in lieu of deeper experiences that directly confirm the intuitions and revelations referenced in historical scriptures and testimony of religious and spiritual authorities. Faith is quite helpful when it points to and leads to deeper truth, but that truth needs direct empirical validation for oneself.

In the absence of widespread deep inner development, much of the focus in the diverse religions has been on more surface doctrinal distinctions. Unfortunately this has had the general effect of over-emphasizing differences between religions and fragmenting them into numerous sects, contributing to the rigidity and zealotry that has had effects opposite of the intended harmonizing and unifying values of religion. The fragmenting, object-oriented, materialistic pattern—with fragmented interpretations of doctrines and scriptures—characterizes religion nearly as much as modern science. Fragmentation of understanding and experience is due to the state of functioning of the mind, regardless of whether the general worldview is religious or secular.

> "Newton's theory was that...we have the sensation of light when...corpuscles [particles] strike the retinas of our eyes. Huyghen's theory was that light consists of minute waves of trembling in an all-pervading ether... Scientists have to leave it at that, and wait for the future, in the hope of attaining some wider vision which reconciles both... We should apply these same principles to the questions in which there is variance between science and religion... It may be that we are more interested in one set of doctrines than in the other. But, if we have any sense of perspective and of the history of thought, we shall wait and refrain from mutual anathemas... The clash is a sign that there are wider truths and finer perspectives within which a reconciliation of a deeper religion and a more subtle science will be found... Science is concerned with the general conditions which are observed to regulate physical phenomena; whereas religion is wholly wrapped up in the contemplation of moral and aesthetic values. On the one side there is the law of gravitation, and on the other the contemplation of the beauty of holiness. What one side sees, the other misses; and vice versa... A clash of doctrines is not a disaster—it is an opportunity."—*Alfred North Whitehead, mathematician and clergyman (pp. 184-186)* [1]

Fortunately in the 20th Century the forefront of modern science progressed to the point where it was confronted with fundamental questions about nature that in the past had been mainly the purview of religion or philosophy. It has progressed past the pretheoretical assumption of the independence of objectivity and subjectivity into the intangible realm of the abstract unified field. It now is starting to address more carefully core questions about the relationship between the underlying unified basis of nature and the conscious mind that can know it—the *knower*.

Modern science views physical creation as structured in layers, from the concrete level of macroscopic objects to more abstract microscopic molecular, atomic, nuclear, and sub-nuclear levels. As will be summarized in Chapters 3-12, increasing unification has been uncovered at more fundamental levels. The entire universe is now theorized to be fluctuations of abstract quantum fields, and mathematical theories are being developed to unify, at least conceptually, these abstract fields into a single completely unified field. This underlying unified field has been proposed to be the *source of everything*. Development of viable theories that are powerful enough to unify conceptually the vast diversity of nature into a single completely unified field is a wonderful achievement. It encourages further recognition that the same orderliness that structures the outer natural world also structures the inner world of human subjectivity, and that it is possible to develop sufficient intra-subjective and inter-subjective consistency and coherence to achieve a universal knowledge system. It is significant that the general concept of a single underlying source of everything in creation—apart from its specific descriptors and cultural icons—has been perhaps the most widely accepted intuitive belief across religions. Unified field theory has brought modern science to the doorstep of a unified view of nature that religion long ago intuited.

The Inner Laboratory of the Mind

As deeper and more integrated levels of nature are investigated, deeper and more integrative levels of understanding and experience are needed. This requires correspondingly more refined, integrated functioning of the physiological and psychological processes of knowing in the knower through which knowledge is gained. Knowledge requires a *knower*, and depends on the state or condition of the *knower*.

In the centuries preceding the British empiricists of the 1800's the predominant paradigm in early modern science began to view the subjective mind as a mechanistic system that could be investigated systematically. The perspective that the natural world is a cosmic machine eventually included the human body—e.g., Descartes[2]—and later the human mind—e.g., Locke. [3] This set the stage for a positivistic, materialistic view that since has been a primary feature of modern scientific thought. The mind-body problem—and the gap between objectivity and subjectivity— was thought to be resolved by a view of mind as nothing other than the mechanical functioning of the physical body. Unfortunately a presumption that also developed was that mechanistic descriptions of nature inevitably were incompatible with spiritual experience.

> "When Descartes said 'I think therefore I am,' he put Descartes before the source."—*J. W. Jarvis, Vedic research* [4]

Once this mechanistic perspective emerged as part of the physicalist worldview associated with the Age of Reason—naively also called an age of enlightenment—systematic attempts to investigate subjective mental processes began in modern science. Early researchers developed evidence-based experimental approaches relating objective physical stimuli with subjective sensory experience. For example the psychophysical research by Fechner, [5] Weber, [6] and Helmholtz[7] provided precise mathematical formulae relating physical intensity of stimuli with *perceived* magnitude, demonstrating the feasibility of a systematic approach to the study of subjective mental processes. This significantly contributed to the founding of psychology as a specialized scientific discipline in the latter half of the 19th Century.

Black box mentality. In this emerging scientific psychology initial efforts also were made to systematize the investigation of subjective experiences deeper than sensory processes, such as through the trained introspection research of Wundt, [8] Titchner, [9] and James.[10] This research did not result in sufficient intra-subjective or inter-subjective consistency to approximate the rigorous standards of the physical sciences. These efforts gave way to the reactionary paradigm of *behaviorism*, which attempted to eliminate all forms of *mentalism* in favor of a strict positivistic focus on observable behavior only. Unobservable constructs such as attention, mind, and consciousness were conceptually placed in a so-called *black box*, considered outside the range of scientific inquiry. [11] The rationale for the black box approach was scientific rigor, following assumptions in the physical sciences. But more fundamentally it was a product of objectification related to inconsistency in deep levels of subjective experience. It is quite descriptive of the lack of clear experience of mind and consciousness—of darkness in the deep inner levels of the mind.

Consistent with paradigm blindness, when something is deemed untestable by the modern scientific method it is sometimes viewed as unknowable, unworthy of study, non-existent, or even delusional. These emotional reactions to the challenge of the systematic investigation of mental processes were fueled by the pretheoretical assumptions of objectivity and physicalism. It was as if the assumption of universal order in modern science somehow didn't apply to the inner subjective domain; and the consistent integrated functioning of the mind of the scientist was not critical to the knowledge about the world the scientist was able to gain. There was little consideration of the *knower* behind the objects of knowledge or known—it was an unillumined black box.

The closest thing to investigation of the *knower* in modern science since the black box approach has been investigation of the neurophysiological machinery of experience and reason—associated with the process of knowing. In earlier years of modern science everything in front of the eyes was considered objective and everything behind was subjective. In recent years processes behind sensory receptors in the nervous system and brain have been investigated using objective methods. A considerable body of objective knowledge has accumulated about the structure and function of the human nervous system and brain. This neurophysiological and neuropsychological research extends into cellular, molecular, electrochemical, and now quantum theories of brain functioning—to be discussed in later chapters.

The field of scientific psychology advanced beyond behaviorism to deeper interior levels of the mind in the second half of the 20th Century in its study of attention, cognition, and memory from a disembodied functional perspective. This was based in part on development of the computer as a mechanistic model of human information processing *inside* the black box. An important aspect of this research is the use of software programs to simulate perceptual, cognitive, and affective aspects of human information processing as models of psychological processes. There also has been a major effort to integrate structural and functional research on brain physiology with computer simulations of psychological processes to create models of how the subjective mind inside the black box might work. Progress in this work has strengthened the fervor in some investigators of the physicalist belief that the mind is nothing other than neural function. Prominent examples of these functional and structural research strategies will be discussed in Chapters 13-18.

These strategies *objectify* mental processes by turning the means of gaining knowledge—the human information processing system—into an object of knowledge and studying it from the outside third-person perspective. They don't investigate the subjective experiential component of information processing from the *inside*—using a direct, systematic, first-person experiential approach. Low dopamine levels in the brain don't equate with the subjective experience of depression. The speed of a button press in a reaction time task is not equivalent to the subjective experience of deciding that two letters are different. Computer modeling of emotional behavior doesn't address the experience of emotion, the *feeling* or *felt sense* of emotion, nor does any computational algorithm.

The experiential component of mental processes cannot be examined directly using only objective, reductive, third-person means of investigation. The subjective mind and consciousness are not knowable through outer objective means. In other words first-person experience cannot be explained adequately from a third-person perspective. [13] A systematic inside approach is needed to investigate deep subjective experience, in the *inner laboratory* of the investigator's mind.

"It is fine for us scientists to bemoan the fact that consciousness is an entirely personal and private affair and that it is not amenable to the third-person observations that are commonplace in physics and in other branches of the life sciences. We must face the fact, however, that this is the situation and turn the hurdle into a virtue. Above all, we must not fall in the trap of attempting to study consciousness exclusively from an external vantage point based on the fear that the internal vantage point is hopelessly flawed. The study of consciousness requires both internal and external views."—*Antonio Damasio, clinical neuroscientist* (p.82) [12]

[T]he distinction between objectivity and subjectivity…requires reference to, and corroboration in terms of, perspectives outside of one's own subjectivity. What is required then is some combination of objective and subjective approaches… Such combinations are of course standard fare in psychophysiological studies… Yet, useful and exciting as such studies can be, they suffer from a significant asymmetry. For while their objective side employs sophisticated scientific methodologies, capable of isolating and evaluating variables completely outside the ken of ordinary sense perception, their subjective side typically uses mere everyday sorts of introspection, capable of isolating only ordinary internal phenomena such as sense perception, imaging and verbal thought… The need for systematic first-person methodologies here is thus starkly apparent… In contrast to the introspective methods usually relied on in modern Western treatments of consciousness, the Eastern approaches…appear to be cross-culturally congruent, despite great differences of metaphysical frameworks and social milieu. Thus examining them and their effects in the context of modern scientific methodologies and criteria may well prove useful to us in our own task of developing a significant science of consciousness."—*Jonathan Shear, philosopher, and Ron Jevning, research neuroscientist* (pp. 189-190) [14]

The Systematic Subjective Approach of Vedic Science

Science is not just an objective description of nature outside us. Nature includes our own nature—and it can be said in some sense that all the laws of nature exist within us and shape our body, mind, and experience. While modern science associated largely with Western traditions of knowledge has focused on objective means of gaining knowledge to investigate the outer natural world from the outer third-person perspective, Eastern traditions have long explored the inner natural world of the mind from the inside—through direct first-person research.

The most ancient continuous knowledge system is the tradition based on the *Veda*. The term *Veda* refers to *total knowledge*. [15] This ancient tradition emphasizes empirical investigation of the mind and consciousness using *systematic subjective means of gaining knowledge*. Ancient Vedic science provides extensive descriptions of the unified field of nature that modern science has glimpsed intellectually and is attempting to model mathematically. This is suggestive of an ultimate integration of objectivity and subjectivity at the most fundamental level of nature. In the holistic approach of ancient Vedic science, physical and psychological laws of nature are cut from the same cloth and are ultimately consistent with each other across the mind-body spectrum—to be discussed at length in later chapters.

Ancient Vedic science is a comprehensive, integrated knowledge system. For many centuries it remained in obscurity as irrelevant to daily life, due to misinterpretation by Western and Eastern scholars without the needed experiential development and empirical experience to interpret it. It was classified as mythological, pre-scientific, and largely only of historical significance. These investigators did not follow through with sufficient empirical research to validate Vedic knowledge in the inner laboratory of their own minds using the systematic subjective technologies contained in ancient Vedic science. Maharishi has pointed out:

> "This tragedy is the fate of a path of knowledge based on direct experience when the means to that
> direct experience has been lost. Past attempts to interpret the *Vedas*, whose basic subject matter
> is the recorded experience of evolution through…(higher) states of consciousness, must obviously
> have been hopeless in the absence of any personal knowledge of these…states." (pp. 312-313) [16]

During the past 50 years this ancient knowledge system has been revived and made understandable in modern scientific language by Maharishi as *Maharishi Vedic Science and Technology*. It is identified as a science because it is based on systematic, reliable, and verifiable empirical means of gaining knowledge. However, it emphasizes systematic and replicable subjective means of gaining knowledge to develop the mind directly—in addition to the objective approach that focuses only on outer-directed experience and reason within ordinary developmental limitations.

Religion attempts to develop ultimate knowledge largely through faith, prayer, and religious works, based on intuition and historical revelation. Modern science attempts to gain and validate knowledge primarily through reason and ordinary sensory experience. Ancient Vedic science uncovers knowledge through intuition, reason, and direct empirical experience of all the levels of the mind and its transcendental basis in consciousness itself. Maharishi describes the similarities and differences between modern science and ancient Vedic science:

> "Science is universal. The terms 'eastern science' and 'western science' simply denote different approaches to the object of inquiry, different approaches to knowing and to living the reality of life... Western science is the investigation of reality through a purely objective approach, while eastern science is the subjective investigation into the nature of consciousness, the field of pure subjectivity, in order to systematically gain the knowledge of the whole creation and live fulfillment in life... Western science rejects any trace of subjectivity on the path of investigation because the observer's state of awareness differs according to the condition of his nervous system and when this changing subjective element is involved in perception, perception will never by reliable... The subjective approach to knowledge aspires to create a state of consciousness that does not change. On the basis of this steady, non-changing value of consciousness...the observer's perception and evaluation of reality is always reliable... Through the eastern approach to knowledge it is possible for every man to be a scientist, a knower of truth, by developing that non-changing state of consciousness as a permanent reality." (pp. 122-123) [17]

Maharishi Vedic Science and Technology extends the fundamental knowledge of the lawfulness, orderliness, and coherence of nature from the outer natural world to be inclusive of the inner natural world of the subjective mind. It also extends the fundamental requirement of the empirical testability of theories to the deep levels of the inner subjective domain. Knowledge of mind and of consciousness is verifiable systematically through direct empirical experience—in addition to indirect objective methodologies. It can be described as direct *experiential* science, not just indirect *experimental* science.

Importantly, in Maharishi Vedic Science and Technology the outer-directed objective approach of modern science is understood to be based on the modes of mental functioning and corresponding means of gaining knowledge associated with the ordinary waking state of consciousness. The ordinary waking state is associated in adulthood with abstract reasoning ability, the stage of formal operational thought considered to be the necessary developmental basis for scientific thinking—referred to in Chapter 1.

"Vedic science goes from inside out; modern science goes from outside in."—Hement Gupta, Ayurvedic physician [18]

The ordinary waking state is characterized by the experience of *being aware of* some thing, some object of experience. This common subjective experience is a representational, reflective, or *object-referral* mode of knowing in which there is a separate object of experience, process of experience, and experiencer—referred to in the Prologue. This is reflected for example in the distinctions of subject, object, and predicate in many language grammars. It is the basis for the common definition of consciousness as being *aware of* a separate object of experience. It also directly relates to the pretheoretical assumption of the independence of observer and observed that has been prominent in modern science. Maharishi Vedic Science and Technology identifies this type of separation of objectivity and subjectivity as characteristic of the ordinary waking state of consciousness that produces fundamentally fragmented knowledge and experience of nature. It is a core feature of modern scientific realism, and also modern scientific naturalism. In Maharishi

Vedic Science and Technology the ordinary waking state provides a fragmented conditional experience of *reality* that can be understood to be quite unreal and unnatural with respect to the full range of states of human consciousness and a universal knowledge system.

The outer-directed focus common in the ordinary waking state conditions the mind and brain to function only in active mental states, typically engaging the mind with the senses in outer-directed, object-based sensory experience. This type of mental experience can become so entrained that the only experiences of alertness are in active thinking, memory, and perception—fixated on the outer objects of experience—with virtually no clear experience of the deeper underlying process of knowing and knower. The mind in the waking state is habitually active in thought, memory, and perception, which fosters belief that the active state of the mind is the only possible natural experience of alertness. Experiences of settling down mental activity are associated with sleep and loss of alertness. Alertness is associated only with active *awareness of* objects of experience.

In the objective means of gaining knowledge the positivistic focus is almost entirely on outer tangible observables. In the 20th Century scientific research progressed far beyond tangible, directly observable sensory phenomena. This has required increasing consideration of the processes of knowing deeper than the ordinary sensory levels—especially the reasoning aspect of the process of knowing. However, this is still an active mental state in which the knower is not clearly experienced but only a vague, virtually unnoticeable background. Thinking about something, whether concrete or abstract, whether matter, energy, nothing, the unified field, or God—as well as introspecting about oneself or being mindful of some experience—keeps the knower in the mental activity of ordinary waking. In this habitual pattern, the underlying state of *pure alertness* is rarely attained or understood—it is experientially an unillumined black box

Direct experience of the knower. Maharishi Vedic Science and Technology investigates the knower directly—*from the inside*. Rather than attention going outward to sense a separate outer object of knowledge, in this systematic subjective approach there is an inward direction to experience the object of knowledge at deeper levels of mind. In this inward direction of attention the object of experience is the mental activity of thought. In the delicate and subtle process of experiencing deeper levels of thought, the mind relaxes its ordinary outer-directed focus and naturally settles down to lesser excited interior states while alertness or wakefulness is maintained.

> "In the Eastern tradition...the seeds of seer-seen non-duality...matured into a variety...of impressive philosophical species which have been attractive to many westerners because they seem so exotic in relation to our own—and because they bear at least the promise of fruits that we Westerners lack but still crave... By no means do all of these systems assert the non-duality of subject and object, but...[of] the most influential...none...completely denies the dualistic relative world that we are familiar with and presuppose as commonsense: the world as a collection of discrete objects, interacting casually in space and time. Their claim is rather that there is another, non-dual way of experiencing the world, and that this other mode of experience is actually more veridical and superior to the dualistic mode we usually take for granted... It is not that these claims are not empirical, but if they are true, they are grounded on evidence not readily available. This is the source of difficulty in evaluating them... The non-dual experience...cannot be attained or even understood conceptually... [T]his is because our usual conceptual knowledge is dualistic... [I]t is knowledge *about* something, which a subject *has;* and such knowledge must discriminate one thing from another in order to assert some *attribute* about some *thing*... [T]he problem with philosophy is that its attempt to grasp non-duality conceptually is inherently dualistic and thus self-defeating. Indeed, the very impetus to philosophy may be seen as a reaction to the split between the subject and the object: philosophy originated in the need of the alienated subject to understand itself and its relation to the objective world it finds itself in. But, according to the "nondualist" systems"...philosophy cannot grasp the source from which it springs and so must yield to praxis: the intellectual attempt to grasp non-duality conceptually must give way to various meditative techniques which, it is claimed, promote the immediate experience of non-duality."— *David Loy, philosopher and professor of international studies* (pp.3-5) [19]

"But one of the most interesting things about the scientific method is that nothing in its says that it must be applied only to sensory domains or to sensory experience alone... [T]here is evidence seen by the *eye of flesh* (e.g., of the sensorimotor world), *mental empiricism* (including logic, mathematics, semiotics, phenomenology, and hermeneutics), and *spiritual empiricism*...spiritual experiences... [T]he experiential evidence in each of these modes is actually quite *public* or shared, because each of them can be trained or educated with the help of a teacher... In all of these ways and more, empiricism in the broadest sense is the surest way to anchor the objective component of truth and demand for evidence... On the other hand, empiricism has also historically been given an extremely narrow meaning, not of experience in general, but of sensory experience alone... But sensory experience is only one of several different but equally legitimate types of experience, which is precisely why mathematics—seen only inwardly, with the mind's eye—is still considered scientific (in fact, is usually considered extremely scientific!)."—*Ken Wilber, philosopher (pp. 152-153)* [20]

As the object of knowledge and process of knowing settle down to their least excited state—the ground state of the mind—the conceptual boundaries of ordinary active thinking are transcended, analogous to a wave settling back into the ocean. The knower then directly experiences the basis of his or her mental activity, without any other object of knowledge or process of knowing. The separate trinity of known, process of knowing, and knower characteristic of ordinary waking experience is transcended.

Through this natural settling process—involving less excitation and effort—the individual mind settles down to an underlying ground state of consciousness itself, unmixed with mental activity. This process can be likened to the process of decreasing entropy through reducing temperature associated with reduced activity—related to the principle of the 3rd law of thermodynamics (referred to in the Prologue). However, this principle is applied to the inner subjective domain rather than the outer objective domain. The common subjective experience that the mind is more orderly and coherent when it is in a more settled and calm but alert state is systematically extended to its most coherent, settled, and alert ground state.

The least excited or ground state of the mind is empirically neither a state of sleep nor hypnotic trance. It is pure alertness, awareness, or wakefulness with no boundaries of thought or mental activity—a natural, simple, settled state of inner silence. It is the simplest form of awareness, the direct experience of pure subjectivity, the direct experience of the knower without any separate object of knowing. A reliable technology for transcending mental activity, drawn from ancient Vedic science, has been taught by Maharishi since the 1950s in a systematic effortless procedure known as the *Transcendental Meditation*™ technique. The subtle process of softer and softer thinking and effortless transcending is described more thoroughly in Chapter 19.

The phrase *direct experience* as used here doesn't refer to subjective states associated with impressions the mind receives from the senses or memory. In the simplest state of consciousness all mental activity including perception, cognition, affect, memory, and intuition is transcended. That *direct experience* is silent unbounded consciousness, outside of the experience of boundaries in time and space, outside of the conceptual and contextual limitations of the individual mind. It is beyond the division of object and subject, beyond reflective and self-reflective thought.

If it is to be described in terms of subjective experience with some form or content, it can be described as the *direct experience of consciousness itself*, beyond any object of experience other than itself. As we unfold deeper levels of nature in deeper levels of experience, it becomes clearer that the concept of empirical experience not only applies to familiar, ordinary sensory experience; it also includes subtler, more refined and holistic levels of sensory experience and thinking, [20] as well as direct transcendental 'experience' beyond sensory and intellectual functioning.

Consciousness itself is described as the unified state of the known, process of knowing, and knower—a *self-referral* state in which the known and process of knowing settle back into, or completely refer back to, their basis in the knower. But it is direct *self-referral* consciousness, not *self-*

reflective thought. It is called *transcendental consciousness* because it transcends or goes beyond mental activity, beyond all objects and processes of knowing and boundaries of mental content—while alertness, wakefulness, awareness or consciousness itself remains.

That simplest state is sometimes called *pure consciousness*, which suggests it is content-less or content-free. In other terms it can be directly experienced independent of any qualitative experiential content or *qualia*.[21] This may be difficult to comprehend if the state has not been *directly experienced*, because being conscious is ordinarily and habitually associated with some content of awareness—*awareness of* something. In the context of ordinary waking experience, typically consciousness is taken to be an intentional state that inherently has some object or content. The state being described here can be said to be a non-conceptual experience, beyond content and mental intention, beyond being *aware of* some thing other than itself. If it is thought of as necessarily having content, then it can be said that the content is nothing other than itself—conscious only of itself, self-referral consciousness. If it is thought of as necessarily containing a content of intentionality at least in some form, then it can be said to intend only to itself, with no other separate object of intent—that is, completely self-referral. In the following descriptions of the state of pure consciousness, Maharishi also refers to it as *pure Being*:

> 'When people learn to meditate effectively and settle down completely, one often hears comments like 'Oh, that's what was being talked about. Very simple. And nothing like anything I'd imagined!" —Jonathan Shear, philosopher, and Ron Jevning, research neuroscientist (p. 195)[22]

> "The state of Being is one of pure consciousness, completely out of the field of relativity; there is...no trace of sensory activity, no trace of mental activity. There is no trinity of thinker, thinking process and thought; doer, process of doing and action; experiencer, process of experiencing and object of experience." (p. 394)[23]

> "Underneath the subtlest layer of all that exists in the relative field is the abstract, absolute field of pure Being which is unmanifested and transcendental... Experience shows that Being is the essential, basic nature of the mind; but, since It commonly remains in tune with the senses projecting outwards toward the manifested realms of creation, the mind misses or fails to appreciate its own essential nature, just as the eyes are unable to see themselves. Everything but the eyes themselves can be seen through the eyes. Similarly, everything is based on the essential nature of the mind...and yet, while the mind is engaged in the projected field of manifested diversity, Being is not appreciated by the mind, although It is the very basis and essential constituent." (pp. 23-25)[24]

The clarity of experience of this simplest state of wakefulness is influenced by the condition of the mind and body associated with and accompanying it at the time. Initially it might be described as a brief period involving no clear thoughts, but not sleep. It also might be described as a vague but comfortable sense of inner quiescence, or a deep inner silence or stillness, or a moment of inner bliss, or a clear period of no thought, or perhaps pure unbounded non-conceptual alertness. Through repeated transcending the clarity of this simplest ground state of inner silence increases. Here are two anecdotal reports from TM practitioners describing their experience of pure consciousness:

> "I distinctly recall the day of instruction, my first clear experience of transcending. Following the instructions of the teacher, without knowing what to expect, I began to drift down into deeper and deeper levels of relaxation as if I were sinking into my chair. Then

for some time, perhaps a minute or a few minutes, I experienced a silent, inner state of no thoughts, just pure awareness and nothing else; then again I became aware of my surroundings. It left me with a deep sense of ease, inner renewal, and happiness." (pp. 338-339) [25]

"As I spontaneously become aware of more fundamental and abstract levels of the object of attention...during meditation, the rigid boundaries of the object begin to fade. As the object becomes more and more unlocalized and the focus of attention continues to spread, comprehension becomes more and more unbounded. When the faintest impulse...dissolves and there is no localized content to experience, my awareness is completely unbounded. I am left with the experience of a pure, abstract, universal field of consciousness, unlocalized by specific content or activity of the mind—just the Self wide awake within its own unbounded nature." [26]

It is quite helpful to understand attributions about, and descriptions of, transcendental consciousness in the developmental context of higher states of consciousness beyond the ordinary waking state—to be discussed in detail in Chapters 19-21. Comparisons between this meaning of consciousness and physicalist models based on functional and structural analyses of the brain will be addressed in Chapters 13-18. The subtle philosophical issues arising from the notion of a non-conceptual experience influenced by the state of the mind and body also will be discussed, as well as important practical distinctions between the Transcendental Meditation technique and other approaches such as phenomenological introspection, contemplation, and concentration.

The difficulty of identifying the nature of consciousness in modern science has been due to lack of systematic reliable means to isolate consciousness from the mental activity of ordinary waking experience. Based on personal descriptions and theoretical vantage points, many modern investigators seem at best to have only a foggy sense of distinction between mind and consciousness. Presumably this is due to lack of systematic experience of the deepest levels of mind and underlying transcendent nature of consciousness itself. The Transcendental Meditation technique reliably provides these experiences, and thus has been invaluable for experimental research on states of consciousness. A large body of research on the psychophysiological, physiological, and behavioral correlates of transcendental consciousness has accumulated during the past 40 years in refereed scientific journals. [27] This research is based largely on findings from regular TM practitioners, who report consistent and clear experiences of transcendental consciousness.

Transcendental consciousness has been distinguished from the three ordinary states of waking, dreaming, and deep sleep using psychophysiological and phenomenological criteria. [28] [29] The waking state is characterized by experience of some separate object, the dream state basically by experience of illusory objects, the deep sleep state basically by virtually no experience, and transcendental consciousness by awareness without a separate object of experience—self-referral consciousness. This experimental research supports ancient Vedic references on the transcendent state as a distinct *fourth state of consciousness*, identified in Vedic literature by various terms including, for example, *Atma*, *samadhi*, *turiya chetana*, *para chetana*, and *parame vyoman*.

Although the self-referral state of transcendental consciousness is a natural experience, it can be difficult to understand within the limitations of the ordinary, active, object-referral waking state. Without clear direct experience of the state, the *thought* of the state and *direct experience* of the state frequently are hard to distinguish. This concerns the general difficulty of comprehending stages of development outside the current range of experience, a phenomenon related to paradigm blindness. The non-representational transcendent state is not reported to be a common experience among large groups of modern scientists—or religionists. It thus has not been fully consensually validated according to many contributors to the contemporary general consensus. However, it has received considerable experimental validation in rigorous studies, using regular TM practitioners as subjects. Also it has been described extensively in various ways in historical artistic, religious, and spiritual literature.

Chapters 19-21 include numerous anecdotal reports of states of consciousness beyond the ordinary three states of waking, dreaming, and deep sleep—which may be helpful to glean a better sense of these experiences. While these reports have appeared across cultures and time periods, there also have been those who assert that such experiences are illogical or impossible on various grounds—apparently due to not yet having clear and distinct experiences of them.

"Part of the reason why people have difficulty considering the concept of pure conscious-ness is that they have trouble *imagining* what such a state would be like. When one tries to imagine it, one inevitably brings to mind some image, such as of white light or a dark void, or a soft, diffused feeling. Such activities of imagination necessarily fail to represent pure consciousness because pure consciousness is neither light nor dark nor a feeling or quality of any kind... Hume encountered difficulty trying to imagine what pure consciousness is in the appendix to the *Treatise* [p. 634], where he asks the reader to imagine a life reduced to the level of an oyster, with only one feeling such as hunger or thirst. He then surmised that if this last feeling (or content) of awareness were removed, that nothing more would be required to make him non-existent... Hume's *Treatise* [p. 252]...asserts that the idea of a simple, continuous identity cannot be based on impressions of direct experience because there *are no impressions* of sense or emotion that could give the idea of a simple, unchangeable identity of the self or consciousness... He simply did not know *how* to gain the experience of the simple underlying field of pure consciousness... [C]onstructivists... hold that *all* experience is a product of prior anticipation or categories of interpretation, and that there can be, therefore, no unchanging, universal "core" experience... John Dewey [1934, *A common faith*] argued essentially the same point... Katz [*Mysticism and philosophical analysis,* 1978], for example, states: "There are NO pure (i.e., unmediated) experiences... That is to say, all experience is processed through, organized by, and makes itself available to us in extremely complex epistemological ways... This 'mediated' aspect of our experience seems an inescapable feature of any epistemological inquiry"... Kuhn, Hesse, Feyerabend, and Hanson, for example, hold that even sensory experience is always embedded in a web of constructed belief... [But a] fundamentally different kind of aware-ness that is not conditioned or complex constitutes pure consciousness, the simplest state of human consciousness." [30]

Transcendental consciousness: The basis of subjective consistency and coherence. The systematic subjective means of gaining knowledge can be related to principles in the objec-tive approach of modern science. The process of transcending mental activity can be understood in the context of the experimental method of isolating an independent variable in order to gain knowledge of its nature and effects. In the systematic subjective approach the independent vari-able to isolate or disembed from other confounding variables is consciousness in its pure state, un-mixed with the ordinary mental activity of feeling, thinking, perceiving, and sensing. In the inner laboratory of the mind, however, this process does not involve control of other variables—such as through mental control or concentration—but rather natural and effortless settling down of mental activity to its least excited ground state of inner silence.

> "Plausibility is one thing, objective fact quite another. Here, real research, following normal scientific protocols, is clearly necessary. And this is especially important when the topic involves the subjective experience of people who have invested their time, and often their hearts, in meditation procedures hoping to gain significant, even life-changing, experiences... Fortunately, however, a growing body of meditation-related research...has been accumulating... This research began, naturally enough, by evaluating some of the traditional claims of physiological correlates of the pure consciousness experience... One striking cross-cultural claim...is that it is accompanied not only by significant reduction of metabolic activity but complete cessation of the normal respiratory activity... As the *Yoga Sutras*...puts it, the processes of 'taking in the breath and exhaling...do not appear in...a reposeful mind', that is, in a mind in the pure consciousness state of *samadhi*...(Patanjali, 1977, p. 81)... This rather surprising physiological claim has now received support from contemporary scientific studies showing high correlations between periods of complete respiratory suspension and reported episodes of experiences of pure consciousness in subjects practicing the TM technique."—*Jonathan Shear, philosopher, and Ron Jevning, research neuroscientist* (pp. 195-196) [31]

In both objective and subjective approaches, knowledge can be understood to be gained through eliminating confounds, mistakes, errors, and biases. The systematic subjective approach conceives of subjective bias as mental distortions or errors, comparable to measurement, calculation, or logic errors in the objective approach of modern science. Consistent with the modern scientific assumption of the lawfulness of nature, distortions are understood as stress in the mind and body to be eliminated in order to gain more coherent brain functioning and more accurate knowledge.

> "Those sold on the program of science believe that the approach to explaining consciousness is to...isolate the causal influences with respect to consciousness and model them."—*Valerie Gray Hardcastle, philosopher* (p. 62) [32]

From this perspective scientific investigation not only involves objective research but quite reasonably also subjective research in consciousness—systematic, empirical, experiential research in consciousness—with its resultant refinement and development of mental and physiological functioning. This research includes systematically settling down the active mind to its least excited state and uncovering or isolating pure consciousness that constantly underlies and ordinarily is embedded or mixed with mental activity. This natural effortless settling process simultaneously provides deep rest to the body. This is said to activate natural healing mechanisms in mind and body to eliminate the internal noise and incoherence due to accumulated stress that produces distortion and bias.

The stress-reducing effects of regular TM practice—as well as associated enhancement of mental and behavioral performance with increased orderliness in brain functioning—are well documented and will be summarized briefly in later chapters. Elimination of the subjective distortions that are due to stress fosters the capability for more accurate internal models or paradigms of both the objective and subjective levels of nature to be unfolded. Through repeated application of systematic developmental technologies in the inner laboratory of the minds of investigators, there is increasing orderliness and natural attunement with the fundamental orderliness that exists throughout nature. It is a negentropic influence that reestablishes increasingly orderly functioning in both physiological and psychological health—a *recoherence* effect. [33]

In the advancement of systematic knowledge, internal cognitive models or theories develop that increasingly approximate the orderliness of nature. The accuracy of these models depends on objective observation and, more fundamentally, subjective empirical experience based on how orderly, consistent, and coherent the mind of the investigator functions. As the mind of the investigator, observer, or knower is more in tune with the fundamental orderliness of nature—when the mind has less distortion and bias—then the observer is able to understand and experience nature more accurately and universally. This systematic *subjective* experiential method, corresponding to the systematic *objective* experimental method, fosters increasingly integrated understanding and experience of the laws of nature.

Both modern science and Maharishi Vedic Science and Technology can be viewed as *self-correcting* processes—referred to in Chapter 1. But in modern science the referential loop from theory to empirical validation and back to reevaluation of the theory is not experienced as extending deeper into the underlying basis of reason and experience in the pure consciousness of the knower. It remains on the outer levels of mental activity—with the deeper, underlying inner levels of mind remaining an experiential black box. In Maharishi Vedic Science and Technology the transcending process takes the mind to the underlying source of thought in consciousness itself, to pure subjectivity, unbounded awareness beyond all contextual limitations—self-referral consciousness. This involves much deeper 'self-correcting' processes, the result of natural healing mechanisms activated through deep rest, that eliminate deep-rooted distortions and stress in body and mind. It establishes a silent, inner, unrestricted, unbounded, direct experience of pure consciousness that provides a non-changing platform of inner stability or fundamental grounding.

In this more expanded and integrated understanding of objectivity and subjectivity, inner development of the knower and the process of knowing are fundamental to gain, unfold, actualize, or educe accurate and reliable scientific knowledge. The inability of the majority of modern investigators of reliable and accurate knowledge—whether in science or religion—to develop an integrated, universal understanding and experience of nature is due basically to incomplete development of the inner subjective domain. It is due to the deeper levels of the mind and state of consciousness being too inconsistent, coarse, or insufficiently refined—an experiential black box. This results in more disintegrated inner experiences and correspondingly fragmented worldviews.

The modern scientific framework relies on inter-subjective agreement or general consensus to protect against the inconsistency of individual subjectivity in reason and experience. But the general consensus is determined by—and limited by—the level of intra-subjective consistency developed in those who contribute to it. Practically the entire enterprise of modern science is based on a view of the world according to the common level of development in the ordinary waking state of consciousness. Investigators are so engrossed in this view that there is virtually no appreciation of this particular state-dependent limitation, and virtually no systematic pursuit of higher levels of development that would expand and accelerate the growth of deeper, more holistic knowledge and experience of nature. Indeed, adherents of this worldview have been attempting to construct a completely closed causal nexus of the physical universe, which closes off deeper views of nature—the inadequacies of which will be discussed systematically in the next part of the book.

The subjective means of gaining knowledge in Maharishi Vedic Science and Technology connects progress in the systematic scientific understanding of nature to development of the mind. This can be understood in terms of increasing subjective consistency and coherence through the elimination of distortions in mind and body. This increased first-person intra-subjective consistency and coherence is the platform for third-person inter-subjective consistency. It is the basis for a universal, consensually validated or publicly agreed upon universal knowledge system.

Maharishi Vedic Science and Technology recognizes that the most comprehensive advances in knowledge ultimately depend on the degree of refinement and integrated development of mind and body—reflecting the state of consciousness that knowers have generally been able to achieve. It recognizes that, as intuited by religious authorities and recorded in most religious texts, how life is lived influences the functioning of body and mind, and thereby the beliefs, paradigms, and worldviews—all interactive with and limited by the state of consciousness of the investigator. Not only is beauty and objective *reality* in the eye of the beholder but also the worldview is a product of the state of consciousness of the beholder.

It is in this important sense that the systematic subjective methodology in Vedic science can be said to establish a practical basis for a science of ethical and moral behavior—in the context of an integrated holistic approach to health. In this understanding, morality relates to the effects on ourselves and others of our thoughts, speech, and action. It means for example that what we take in to our digestive, respiratory, or sensory systems through interactions with the environment either refines or coarsens our mind and body. This influences the functioning of our mind and body, and it directly relates to our ability to unfold universal knowledge and experience.

Maharishi Vedic Science and Technology provides systematic scientific technologies designed to develop *spontaneous right action* in accord with the totality of the laws of nature in permanent higher states of consciousness. It is fostered naturally through regular contact with the simplest ground state of awareness—transcendental consciousness. This will be an important topic later in the book once many of the planks on the bridge to unity are in place.

The worthy aspiration to be *objective*, in the sense of being able to observe nature in an unbiased, accurate, and reliable manner, is more fundamentally understood as establishing consistent and coherent subjectivity. It means attunement of mind and body with the inherent orderliness of nature—attunement with *natural law*. This non-changing basis of consistent subjectivity is said to develop when the individual mind and body achieve a stress-free, coherent style of functioning. When permanently established it is identified as the state of *enlightenment*.

The state of enlightenment, referred to in Maharishi Vedic Science and Technology as the *fifth state of consciousness*, is described as permanent non-changing transcendental consciousness (fourth state) simultaneously coexisting with the changing states of waking, dreaming, and deep sleep (ordinary three states). This meaning of enlightenment can be found in general terms in many religious and spiritual traditions. It is quite different from how it was used in the intellectual 'age of enlightenment' of the 19th Century, more accurately described as an age focused on intellectual reasoning. Enlightenment is not just an intellectual process. Fortunately, the Ages of Reason and Science are now advancing to a new era worthy of the name Age of Enlightenment.

Systematic unbiased observation of nature can be said to begin in its full sense in the state of enlightenment. In a fundamental sense it is on the platform of permanently non-changing consciousness that unbiased observations of nature can be made. In that state the investigator's most fundamental background experience does not change with changes in the objects and processes of observation—the known and the process of knowing. Only then are confounds eliminated that are a product of the consciousness of the observer or knower seeming to be identified with and to co-vary with the objects of experience. Only then is knowledge based on a consistent, coherent, undistorted *knower*. In this natural higher state the 24-hour routine of life is spontaneously experienced on the inner platform of unbounded consciousness underlying ordinary sensory experience, reason, and feeling. Through repeated direct experience of transcendental consciousness, the duality of observer and observed characterizing ordinary waking is permeated by an underlying unity. This cultures and attunes the individual to live spontaneously the fundamental unity of nature, the basis for a stable, consistent, and universal knowledge system. Without this development the deepest part of the mind capable of serving as the means to gain universal knowledge is not fully activated and doesn't yield consistent, undistorted, holistic experiences of the inner and outer natural world. There remains a fundamental gap between understanding and experience.

It can take considerable time to develop the mathematical and technical background in the modern scientific laboratory to verify empirically for oneself complicated theories and evidence about physical phenomena. For example it may take years to understand and demonstrate the phenomenon of superconductivity—rather than to accept it from a textbook as a matter of faith. In a comparable manner it can take time to verify empirically using systematic subjective methodology the accuracy of descriptions of higher states of consciousness by direct first-person experience in the inner laboratory of one's own mind.

As will be discussed later in the book, Maharishi Vedic Science and Technology describes in detail a developmental progression starting from the object-referral of the ordinary waking state through higher self-referral states of consciousness. The object-referral state, frequently associated with the physicalist worldview, is a type of *flatland* monism in which the objective level of existence is experienced and understood to be the only *real* level of existence—sometimes called *materialist monism*. In this state there is an underlying fragmentation of known, process of knowing, and knower. Because of lack of clarity of experience of the deeper hidden levels of mind and consciousness—and to be consistent with the ontology of physicalism—*reality* is associated only with the ordinary outer objective world of the known. With repeated experience of the self-referral state in which the trinity of known, process of knowing, and knower is transcended, experience of the knower as pure consciousness develops. That direct experience eventually stabilizes in the ultimate unity, oneness, singularity, monism, or non-dualism of everything in terms of the universal level of the knower in the highest state of enlightenment—*unity consciousness*. Eventually the entire structure of the cosmos in its infinite totality can be said to unfold within one's own consciousness. Adding the ground state of consciousness to the simple model of levels from Chapter 1, we now have:

> "Does Spirit exist?... The technically correct answer is: Take up the injunction, perform the experiment, gather the data (the experiences), and check them with a community of the similarly adequate... If you want to *know* this, you must *do* this... Thus: take up the injunction or paradigm of meditation; practice and polish that cognitive tool until awareness learns to discern the incredibly subtle phenomena of spiritual data; check your observations with others who have done so, much as mathematicians will check their interior proofs with others who have completed the injunctions; and thus confirm or reject your results. And in the verification of that transcendental data, the existence of Spirit will become radically clear—at least as clear as rocks are to the eye of the flesh and geometry is to the eye of mind."—*Ken Wilber, philosopher (pp. 172-173)* [34]

Sensory World

Sensory Experience

Reason

Intuition

Consciousness

A comprehensive delineation of levels of mind and their relationship to matter—which profoundly addresses the mind-body problem—will be described in Chapters 10-18. Then the various functions of subjective levels will be described in some detail.

The Veda: The Unified Field of Subjectivity and Objectivity

In modern science the laws of nature that structure orderly relationships are inferred based on consistent experience and understanding of their effects. Whatever natural laws exist, they are abstract principles of order that underlie their concrete manifest expressions. A primary goal of science is to identify and understand fully these underlying invariant laws of nature.

"[B]ecause most individuals do not have the personal experience of the inner levels of life, an intellectual understanding of the ultimate reality has been obscure and superficial. For this reason modern scholars who have tried to analyze ancient records from a materialistic angle have...usually classified them as primitive, unintelligible and mythical. With the rise of modern science our society has experienced a diminishing interest in understanding the spiritual nature of the ultimate reality. For the most part world religions today have degraded to...dogmatic and distorted commentaries. Another great tragedy of modern times is the excessive reductionism, fragmentation and hyperspecialization of knowledge that has resulted in the massive accumulation of details regarding the material nature of our universe... [H]idden within the detailed information in various disciplines of modern science are the primal patterns of intelligence and consciousness—the same fundamental mechanisms understood by ancient cultures thousands of years ago." —*Nirmal Pugh and Derek Pugh, Vedic Researchers* (p. 12) [35]

In Maharishi Vedic Science and Technology the total collection of the abstract underlying laws of nature that govern the manifestation and evolution of the universe—the unified field of total knowledge—is the Veda. The Veda does not refer only to the words and texts of an ancient tradition of literature, which are conceptual representations that can only point to its experiential *reality*. Modern scholars and researchers who have described the Veda as a collection of pre-scientific mythological stories have interpreted it in disastrous and trivializing ways. As described later in the book, *Veda* refers to the transcendental unified field in which the abstract laws of nature reside. It is described as the *blueprint of nature*, the *Constitution of the Universe*, the total organizing power of the laws of nature that structure and govern phenomenal existence. [36] [37] In the following quote Maharishi explains further:

"The self-interacting dynamics of this Unified Field constitutes the most basic level of Nature's dynamics, and is governed by its own set of fundamental laws. Just as the constitution of a nation represents the most fundamental level of national law and the basis of all the laws governing the nation, the laws governing the self-interacting dynamics of the Unified Field represent the most fundamental level of Natural Law and the basis of all known Laws of Nature... The laws governing the self-interacting dynamics of the Unified Field can therefore be called the *Constitution of the Universe*—the eternal, non-changing basis of Natural Law and the ultimate source of the order and harmony displayed throughout creation." (pp. 78-79) [38]

Maharishi emphasizes that the Veda—the unified field, the field of total knowledge—is not known through book learning:

"Veda is not a thing to be studied in the books...[but] in the experience of transcendental consciousness... It is much easier to be than to know... Book learning is a waste of time in the sense that you can't get total knowledge through reading... In comparison to That [total knowledge in transcendental consciousness], it's a waste of life." [39]

In ancient Vedic science, described in Maharishi Vedic Science and Technology, the ultimate unified field of all the laws of nature is referred to constantly. The 'nature' of the unified field is described as the coexistence of the opposite values of infinite silence and infinite dynamism. Maharishi explains:

"Life is silence and dynamism together. Dynamism is promoted by the silent dynamism in the silence. Silent dynamism is the promoter of all dynamism; dynamism in silence is the promoter of all silence. In one is the promoting of the other. When dynamism overbalances silence, then life becomes uncomfortable... So Transcendental Meditation [is] to investigate the value of silence—silence brought to human awareness, to balance the exaggeration of dynamism. If this program of bringing silence to dynamism…[is applied, then there will be] balance in human society... Governments only driven by the power of dynamism have to wake up to the power of silence... Silence supporting dynamism, dynamism supporting silence—this is the functioning of the Constitution of the Universe. This is the functioning of Natural Law." [40]

The Veda is said to be open to empirical validation through direct experience in one's own pure consciousness. It is also corroborated through logical reasoning, consensual validation, ancient Vedic records which give detailed descriptions of the authentic knowledge and experience—as well as through measurement of empirical effects using objective methods. As in modern science, the core principles of rigorous empirical investigation and consensual validation continue to hold in Maharishi Vedic Science and Technology. In this subjective approach of gaining knowledge, however, systematic empirical investigation not only includes outer sensory experience but also inner *direct experience* of deep levels of mind and of consciousness itself. Generally accepted knowledge in modern science, which for the most part is based only on ordinary experience and reason, is not invalidated. Rather its contextual limitations become appreciated as the investigator disembeds from all contextual limitations through transcending them. With time, that experience unfolds comprehensive knowledge which is deeply integrating and directly fosters optimal brain functioning, improved health, and fulfillment in daily life.

Individual paradigms of knowledge eventually merge into universal knowledge as individual subjectivity is established permanently in unbounded pure consciousness. Systematic development of individual subjectivity progresses until the complete orderly structure of natural law is enlivened in the individual through the elimination of subjective distortion, which provides the foundation for universal knowledge and experience.

In Maharishi Vedic Science and Technology transcendental consciousness is described as a unified field of consciousness at the source not only of the individual mind but also of all the laws of nature expressed in phenomenal creation. Maharishi describes transcendental consciousness as:

"…a field of all possibilities where all creative potentialities exist together, infinitely correlated but as yet unexpressed. It is a state of perfect order, the matrix from which all the laws of nature emerge, the source of creative intelligence." (p. 123) [41]

Precise descriptions of the structure and dynamics of the unbounded field of natural law can be found throughout ancient Vedic literature. These descriptions describe the unified field of nature as directly *experienceable* at the core of one's own inner self. Previously appearing to be at variance with modern scientific accounts of nature, ancient Vedic science has been strikingly supported by contemporary formulations in modern physics which provide similar descriptions of an infinitely dynamic, self-interacting, unified field at the basis of physical existence. [42 43] The most parsimonious explanation for the correspondence between modern scientific and ancient Vedic accounts is that these two traditions of knowledge ultimately converge on the same unified field. Logically there is only one completely unified field. It is extensively described in ancient Vedic science, and recently has been glimpsed from the objective perspective of modern science. [44]

A fundamental tenet of Maharishi Vedic Science and Technology is that the unified field of nature is a field of consciousness that can be directly experienced as the source of both objectivity and subjectivity. This sharply contrasts with the prevailing physicalist worldview in the mainstream of modern science. In the physicalist view consciousness is a property of a living brain at cellular and molecular levels and may not exist at more fundamental levels of nature. Maharishi Vedic Science and Technology systematically fosters growth to higher states of consciousness, through which the gap of understanding and experience leading to these divergent views is bridged.

Building planks of understanding in order to bridge theories of the unified field of nature and transcendental consciousness—the *unified field of consciousness*—is the focus of the next several chapters. A tremendous amount of theorizing and experimental investigation has gone into building up to unified field theory in modern science. Only a small portion of major threads of this progress can be covered in this book, but for some readers even this amount of detail may seem tedious. The intent is to provide enough planks on the bridge described in this book for the knowledge in Maharishi Vedic Science and Technology that goes beyond modern science to be evaluated rationally by readers as more advanced, comprehensive, practical, and needed.

In the following quote Maharishi summarizes many of the key points in the two chapters of this Introduction:

"Vedic Science is a complete science, which extends and fulfills the objective approach of modern science by incorporating the knower and the process of knowing into the field of investigation. It provides a complete and comprehensive knowledge of the unified field of all the laws of nature, which can be described as the unified state of the knower, the known, and the process of knowing. Vedic Science also describes the sequential mechanics through which this three-in-one structure of the unified field gives rise to the infinite range and diversity of natural law displayed in the universe.

The knowledge of the unified field has been discovered by modern science during just the last few years, but complete knowledge of the unified field has always been available in the Vedic literature. Today quantum physics has glimpsed the details of the unified field and is locating its three-in-one structure. This is precisely the three-in-one structure of the self-referral state of consciousness.

Credit must be given to modern science because its objective approach has now uncovered the reality of pure subjectivity. The world of scientists should know, however, that the objective approach comes to an end there. Many more discoveries may be made on the surface, relative levels of existence, but the goal of physics has been reached.

Today the most advanced level of modern science needs a complete approach to investigation, which includes the two other values of knowledge—the value of the knower and the value of the process of knowing. All three values are uncovered in their totality in Vedic Science.

If progress is to continue, a shift is required from the science of only one category to a total science. Vedic Science is that total science. It uncovers the knowledge of the total potential of natural law in its completeness and brings human awareness in tune with those fine creative impulses that are engaged in transforming the field of intelligence into the field of matter.

The approach of Vedic Science is such that the knowledge gained enlivens that most fundamental value of consciousness from where all thoughts and actions emerge. Therefore, the very methodology of gaining knowledge through Vedic Science is such that as one gains the knowledge of natural law on the intellectual level, one begins to live that natural law in daily life in a most spontaneous way. This is the basis of the practical application of Vedic Science.

Vedic Science is applied through the Technology of the Unified Field. We speak of the unified field in connection with Vedic Science because of the similarity between what has been discovered by physics and what exists in the self-referral state of human consciousness. The Technology of the Unified Field is a purely scientific procedure for the total development of the human psyche, the total development of the human race. This is the time when objective, science-based progress in the world is being enriched by the possibility of total development of human life on earth, and that is the reason why we anticipate the creation of a unified field based ideal civilization." (pp. 1-3) [45]

Chapter 2 Notes

[1] Whitehead, A.N. (1925). *Science and the modern world*. New York: The Free Press.

[2] Descartes, R. (1972). *The philosophical works of Descartes* (2 vols.). Cambridge, England: Cambridge University Press.

[3] Locke, J. (1959). *An essay concerning human understanding*. Fraser, A C. (Ed.). New York: Dover Publications. (Original work published 1690).

[4] Jarvis, J. (1974, Spring). Personal communication, Los Angeles, CA.

[5] Fechner, G. T. (1860-1966). *Elements of psychophysics* (Vol. 1). New York: Holt, Rinehart & Winston.

[6] Weber, E. H. (1978). *The sense of touch*. Ross, H. E. (Trans.). New York: Academic Press (Original work published 1834).

[7] Helmholtz, H. (1925). *Treatise on physiological optics,* Vol. 3. Southell, J. P. C. (Trans. from the 3rd German Ed., 1910). Benjamin Backus, U. Pennsylvania.

[8] Wundt, W. (1912). *An introduction to psychology.* London: George Allen.

[9] Titchner, E. B. (1913). *A text-book of psychology.* New York: Macmillan.

[10] James, W. (1890). *The principles of psychology.* New York: Holt.

[11] Leahey, T. H. (1980). *A history of psychology: Main currents in psychological thought.* Englewood Cliffs, New Jersey: Prentice-Hall, Inc.

[12] Damasio, A. (1999). *The feeling of what happens: Body and emotion in the making of consciousness.* New York: Harcourt, Inc.

[13] Velmans, M. (2000). Intersubjective Science. In F. Varela & J. Shear (Eds). *The view from within: First-person approaches to the study of consciousness.* Bowling Green, OH: Imprint Academic Philosophy Center, Bowling Green University, pp. 299-306.

[14] Shear, J. & Jevning, R. (2000). Pure Consciousness: Scientific Explorations of Meditation Techniques. In Varela, F. & J. Shear (Eds). *The view from within: First-person approaches to the study of consciousness.* Bowling Green, OH: Imprint Academic, pp. 189-209.

[15] Maharishi Mahesh Yogi (1986). *Life supported by natural law.* Washington DC: Age of Enlightenment Press.

[16] *Maharishi International University Catalogue, 1974/5.* Los Angeles: MIU Press.

[17] Maharishi Mahesh Yogi (1997). *Maharishi speaks to educators: Master over natural law.* India: Age of Enlightenment Publications (Printers).

[18] Gupta, H. (2001, Spring). Personal communication, Caraka Samhita course on Sharirstanam, Maharishi University of Management.

[19] Loy, D. (1997). *Non-duality: Study in comparative philosophy.* Atlantic Highlands, New Jersey: Humanities Press International, Inc.

[20] Wilber, K. (1998). *The marriage of sense and soul: Integrating science and religion.* New York: Random House.

[21] Shear, J. (2000). The Hard Problem: Closing the Empirical Gap. In Shear, J. (Ed.). *Explaining consciousness: The hard problem.* Cambridge, MA: The MIT Press, pp. 359-375.

[22] Shear, J. & Jevning, R. (2000). Pure consciousness: Scientific Explorations of Meditation Techniques. In Varela, F. & Shear, J. (Eds.). *The view from within: First-person approaches to the study of consciousness.* Bowling Green, OH: Imprint Academic, pp. 190-209.

[23] Maharishi Mahesh Yogi (1967). *Maharishi Mahesh Yogi on the Bhagavad-Gita: A new translation and commentary, chapters 1 to 6.* London: Penguin Books.

[24] Maharishi Mahesh Yogi (1963). *Science of Being and Art of Living.* Washington, DC: Age of Enlightenment Press.

[25] Alexander, C. N. & Boyer, R. W. (1989). Seven States of Consciousness: Unfolding the Full Potential of

the Cosmic Psyche in Individual Life Through Maharishi's Vedic Psychology. *Modern Science and Vedic Science, 2* (4), 325-371.

[26] Orme-Johnson, D. W. & Haynes, C. T. (1981). EEG Phase Coherence, Pure Consciousness, Creativity, and TM-Sidhi Experiences. *International Journal of Neuroscience, 13*, 211-217.

[27] *Scientific research on Maharishi's Transcendental Mediation and TM-Sidhi programme—Collected papers, Vols. 1-5* (Various Eds.) Fairfield, IA: Maharishi University of Management Press.

[28] Alexander, C. N., Cranson, R. W., Boyer, R. W., & Orme-Johnson, D. W. (1986). Transcendental Consciousness: A Fourth State of Consciousness Beyond Sleep, Dreaming, and Waking. In Gackenbach, J. (Ed.). *Sleep and dreams: A sourcebook.* New York: Garland Publishing, Inc., pp. 282-315.

[29] Mason, L.I. (1995). *Electrophysiological correlates of higher states of consciousness during sleep.* Doctoral dissertation, Maharishi International University, Fairfield, IA. Printed 1998 by UMI Dissertation Services, Ann Arbor, MI. *Dissertation Abstracts International,* Vol. 56-10B, p. 5797.

[30] Alexander, C. N., Chandler, K., & Boyer, R. W. (1989). *Experience and understanding of pure consciousness in the Vedic Science of Maharishi Mahesh Yogi.* (Unpublished manuscript). Maharishi International University, Fairfield, IA.

[31] Shear, J. & Jevning, R. (2000). Pure Consciousness: Scientific Explorations of Meditation Techniques. In F. Varela & J. Shear (Eds). *The view from within: First-person approaches to the study of consciousness.* Bowling Green, OH: Imprint Academic. pp. 190-209.]

[32] Hardcastle, V. G. (2000). The Why of Consciousness: A Non-Issue for Materialists. In J. Shear (Ed). *Explaining consciousness: The hard problem.* Cambridge, MA: The MIT Press, pp. 61-68.

[33] Hebert, R. (2008, April). Personal communication. Fairfield, IA:

[34] Wilber, K. (1998). *The marriage of sense and soul: Integrating science and religion.* New York: Random House.

[35] Pugh, N. D. & Pugh, D. C. (1999). *Unveiling creation: Eight is the key.* Fairfield, IA: Sunstar Publishing, Ltd.

[36] *Inaugurating Maharishi Vedic University* (1996). India: Age of Enlightenment Publications.

[37] Nader, T. (2000). *Human physiology: Expression of Veda and Vedic Literature,* 4th Edition. Vlodrop, The Netherlands: Maharishi Vedic University.

[38] Maharishi Mahesh Yogi (1996). *Maharishi's Absolute Theory of Defence: Sovereignty in Invincibility.* India: Age of Enlightenment Publications.

[39] Maharishi Mahesh Yogi (2003). Maharishi's Global News Conference, October 29.

[40] Maharishi Mahesh Yogi. (2005). Maharishi's Global News Conference, May 17.

[41] *Creating an Ideal Society: A global undertaking* (1976). Rheinweiler, W. Germany: Maharishi European Research University Press.

[42] Hagelin, J. (1987). Is Consciousness the Unified Field? A Field Theorist's Perspective. *Modern Science and Vedic Science,* 1, 1, January, 29-87.

[43] Alexander, C. N., Boyer, R., & Alexander, V. (1987). Higher States of Consciousness in the Vedic Psychology of Maharishi Mahesh Yogi: A Theoretical Introduction and Research Review, *Modern Science and Vedic Science,* 1, 1, January, 1987, 89-126.

[44] Hagelin, J. S. Restructuring Physics from its Foundation in Light of Maharishi's Vedic Science. *Modern Science and Vedic Science,* 3, 1, 1989, 3-72.

[45] Maharishi Mahesh Yogi. (1987). Maharishi on Modern Science and Vedic Science. *Modern Science and Vedic Science,* 1, 1, January, 1-3.

Part I

PHYSICS UNBOUND:
FROM HERE TO INFINITY

Part I of the book links modern scientific theories of matter, energy, space, and time to the unified field as described in modern science and in Maharishi Vedic Science and Technology. The progress of integrating relativity and quantum theories toward unified field theory is discussed, including recent developments in theories of quantum gravity, the holographic model associated with quantum information space, and the concept of nonlocal quantum mind as underpinning physical matter.

The challenge of integrating the process of observing and the observer into the theorized causally closed physical universe is emphasized, and a logically consistent alternative is outlined in the expanded framework of gross relative, subtle relative, and transcendental levels of nature described in terms of an inner dimension based on ancient Vedic science. Specific connections between levels of nature in modern science and in Maharishi Vedic Science and Technology are identified.

Chapter 3

Unifying the Objective World?

In this chapter, we will summarize key steps of progress in the reductive search in modern science for the essential constituent of matter—from atoms to elementary particles, quanta, and perhaps strings and branes—toward ultimate unification in a theory of everything. Examples are provided of theories on the forefront of modern physics that challenge commonsense views of our everyday world. The intent is to foster more familiarity with the abstract theories to be explored in this part of the book. The overall main point of this chapter is that the two primary theories—relativity theory and quantum theory—need to be integrated in order to establish a unified field theory, a theory of everything.

The pretheoretical assumption that the natural world can be investigated objectively—because it exists independent of the subjectivity of the observer—has been quite useful in the general knowledge system of modern science for investigating macroscopic and microscopic scales of the natural world. This pretheoretical assumption fundamentally divides existence into outer objective nature and inner subjective experience of it. As modern science has probed deeper into matter and is glimpsing more unified levels of nature, the fragmenting effect of this assumption has become increasingly clear. A more integrated understanding and experience of subjectivity and objectivity is needed.

Although the relationship of subjectivity and objectivity has not been an active concern for many in the larger community of modern scientists, in the form of the mind-body problem it has been one of the most persistent issues in the history of human thought. Throughout the 20th Century, investigators working on the cutting edge of theoretical physics have been confronted with this issue in the context of the *measurement problem* associated with quantum theory. This has contributed to the modern emergence of research on consciousness, which again is working on the issue, associated with attempts to think through the *'hard problem'* of the experiential nature of consciousness [1] [2]—to be discussed in later chapters.

The measurement problem, the 'hard problem' of consciousness, and the mind-body problem all contain a similar *explanatory gap* between matter and mind that is a central issue for modern science to address in order to explain the workings of the natural world. In addressing the gap between matter and mind, contemporary theories go beyond the assumption of materialism or physicalism, a key assumption that separated the understanding of consciousness in modern science from that in ancient Vedic science and Maharishi Vedic Science and Technology. This issue—introduced at the end of Chapter 2—is the crux of the matter in this and the following several chapters. These chapters review important aspects of the progress toward unified field theory, which requires unification of objectivity and subjectivity, and which represent key planks on the bridge from the physicalist view of consciousness to Maharishi Vedic Science and Technology.

One objective of these chapters is to make the progress beyond physicalism more comprehensible to non-physicists (and perhaps some physicists too). Because this knowledge is highly conceptual and abstract, the review needs to consider deeply key issues in quantum theory, relativity theory, and related models leading to unified field theory. It may take some time to think through this material and grasp the abstract ideas. Given the thoughtful contemporary knowledge in theoretical physics and mathematics, forbearance on the part of experts in these fields is requested, especially with respect to the introductory aspects of this review. In Chapter 9, the progressive review and its implications are referred back to issues from Chapters 1-2 about the means of gaining knowledge, hopefully then bringing a more integrating perspective to readers already familiar with recent progress on quantum physics, quantum mind, and unified field theories.

However, those familiar with this cutting edge research also may find interesting and illuminating the perspective from which progress toward unified field theory is examined, in order to understand the limitations of quantum and relativity theories. It will be helpful for important discussions later that address the integration of objectivity and subjectivity. The chapters in this part of the book lay down many of the planks on the bridge to unity. These chapters demonstrate how the process of knowing and the knower have not been addressed directly in much of the progress in quantum physics, and how they are crucial for a coherent understanding of even physical *reality*.

To contrast the different perspectives for heuristic purposes, implications of a strict interpretation of the materialistic doctrine—much of which is still commonly expressed today—are first presented. In Chapters 4-8, developments in quantum theory and relativity theory that go beyond this view are discussed. Chapters 10-12 identify specific links between quantum and unified field theories in modern science and Maharishi Vedic Science and Technology.

Nothing *Matters*?

A prominent objective in modern science has been to go as far as possible using precise logic and ordinary sensory experience to explain the outer objective, independently existing world. Scientific rigor has been conceptualized in terms of avoiding descriptions of nature that default to vague subjective concepts associated with the process of observing and observer. If the analysis doesn't get too subtle, the surface material levels of nature can be described quite well in this approach, even *almost* to the point of closure as exemplified in the belief that the physical universe is a closed causal nexus. The human mind is creative enough to take this approach a long way.

Unfortunately it is the long way. It has taken a relatively long time to gain an understanding of matter sufficient enough to go beyond it rationally, using the basic means of ordinary sensory experience and abstract reasoning. In this process, some investigators become enthralled with intellectual rigorism, puzzle solving, wonder, cynicism, and sometimes existential empathy and compassion. At some point, the reductive objectifying intellectual mind paints itself into a corner and finds *nothing*. Eventually the approach is recognized to be fragmented, outside of oneself, incomplete, and unfulfilling. Hopefully not too far into the investigator's lifespan, the investigator starts to look *inside*—to investigate what has been overlooked.

[F]or the purpose of constructing the picture of the external world, we have used the greatly simplifying device of cutting our own personality out... In particular, and most importantly, this is the reason why the scientific world-view contains of itself no ethical values, no aesthetical values, not a word about our own ultimate scope or destination, and no God... Science cannot tell us a word about why music delights us, of why and how an old song can move us to tears... Science is reticent too when it is a question of the great Unity—the One of Parmenides—of which we all somehow form part, to which we belong. The most popular name for it in our time is God... Science is, very usually, branded as being atheistic... If its world-picture does not even contain blue, yellow, bitter, sweet...if personality is cut out of it by agreement, how should it contain the most sublime idea that presents itself to human mind?—*Erwin Schrödinger, theoretical physicist (p. 208)* [3]

A strict interpretation of the materialistic doctrine implies that something came from nothing, that literally *nothing* begat matter—that material creation is an effect without a cause. Accordingly the big bang thought to initiate creation an estimated 13.7 billion years ago was a spontaneous event that just happened, without initial conditions, inherent nature, inevitability, purpose, design, or precedent of any kind. [4] Consistent with this view, there is no non-material existence outside of the space-time continuum emerging from the big bang, no a priori potential to manifest, no underlying deep *reality* from which the big bang emerged—*nothing*.

In contrast, according to the doctrine of determinism, after the big bang material creation has followed orderly cause-effect patterns according to invariant natural laws. Apparently at the first instant of time, literally *nothing* spontaneously formed an incredible symmetry that also spontaneously underwent sequential symmetry breaking. At the onset of the big bang, instantaneously and without any initial inevitability, invariant natural laws of space, time, energy, and matter emerged that immediately began to structure and then maintain the orderly functioning of the entire cosmos throughout all subsequent history.

Spontaneously and randomly over billions of years, but according to these invariant laws, energy and matter congealed into stars and planets and eventually into complex chemical compounds, molecules, and cells that somehow developed into primitive living organisms. Through random mutation and natural selection, these organisms developed the capability for complex non-random behaviors associated with life, including the survival instinct and intelligent, life-sustaining, negentropic behavior. Eventually human organisms evolved with sufficiently complex brains for the emergence of consciousness. In this view, consciousness is generally thought to be a by-product—or epiphenomenon, or even non-existent misperception—of neural complexity at the cellular and molecular levels of neurophysiological functioning. But also it is recognized to be central to the means by which the cosmos is being systematically investigated and at least partially comprehended.

Consistent with this view, the emergence of higher-order, intelligent, conscious behavior does not mean that human organisms, as individuals or as a species, are evolving toward some teleological endpoint other than meaningless survival. Eventually all living things succumb to the thermodynamic principle of entropy. In this view, humans have no more intrinsic value than—and are just different from—other animals, insects, vegetables, and minerals; intrinsic values have no place in this view. Religious or spiritual experiences, life after death, and related futile imaginings many people feel give additional meaning and value to human existence are basically considered to be by-products of abnormal electrochemical activity in the brain. They may have, however, an adaptive role relevant to meaningless survival, such as to reduce psychological angst. Death is the consistent empirical outcome when survival systems succumb to entropy, at which time consciousness and all experiences cease for the individual being.

"The reigning belief that the thoughts we think and the choices we make reflect the deterministic working of neurons and, ultimately, subatomic particles seemed to me to have subverted mankind's sense of morality. The view that people are mere machines and that the mind is just another (not particularly special) manifestation of a clockwork physical universe had infiltrated all our thinking..."—*Jeffrey Schwartz, research neuropsychiatrist, and Sharon Begley, science writer/editor (p. 258)* [2]

This view of life and of nature has contributed to existential nihilism and cultural relativism. It is directly associated with prominent modern beliefs, especially among intelligentsia, that there are neither absolute objective nor subjective grounds for ethical and moral behavior.

"Indirectly, driven by popular misunderstanding rather than a fealty to Einstein's thinking, *relativity* became associated with a new *relativism* in morality and art and politics... In both science and moral philosophy, Einstein was driven by a quest for certainty and deterministic laws. If his theory of relativity produced ripples that unsettled the realms of morality and culture, this was caused not by what Einstein believed but by how he was popularly interpreted... Whatever the causes of the new relativism and modernism, the untethering of the world from its classical moorings would soon produce some unnerving reverberations and reactions."—*Walter Isaacson, journalist and biographer (pp. 278-280)* [5]

[I]n the humanities—in art, literature, and philosophy—the growing awareness of groundlessness has taken form not through a confrontation with objectivism but rather nihilism, skepticism, and extreme relativism. Indeed, this concern with nihilism is typical of late-twentieth-century life. Its visible manifestations are the increasing fragmentation of life, the revival of and continuing adherence to a variety of religious and political dogmatisms, and a pervasive yet intangible feeling of anxiety... Indeed, our scientific culture has only just begun to consider the possibility of pragmatic and progressive approaches to experience that would enable us to learn to transform our deep-seated and emotional grasping after a ground. Without such a pragmatic approach to the transformation of experience in everyday life—especially within our developing scientific culture—human existence will remain confined to the undecidable choice between objectivism and nihilism"—*Francisco Varela, cognitive neurobiologist, Evan Thompson, neurophilosopher, and Eleanor Rosch, cognitive psychologist (pp. 238-244)* [6]

"The world into which we are born is brutal and cruel, and at the same time of divine beauty. Which element we think outweighs the other, whether meaninglessness or meaning, is a matter of temperament. If meaninglessness were absolutely preponderant, the meaningfulness of life would vanish to an increasing degree with each step in our development. But that is—or seems to me—not the case. Probably, as in all metaphysical questions, both are true: Life is—or has—meaning and meaninglessness. I cherish the anxious hope that meaning will preponderate and win the battle."—*Carl Jung, psychiatrist (pp. 358-359)* [7]

The physicalist worldview reflects thoughtful beliefs, theories, and evidence about nature. However, in Maharishi Vedic Science and Technology it is a product of the reductive approach common in modern science, limitations of the ordinary waking state of consciousness, and the lack of development of deeper, more integrating levels of mind through direct experience of consciousness itself. It unfortunately has restricted itself to an outside, indirect, object-based, 'third person' view, which has led to major negative consequences. A fragmented, localized model has been imposed on the natural world by the observer's limited range of ordinary waking experience. As Maharishi explains:

"Being objective in its approach, modern science brings only intellectual understanding about the functioning of the laws of nature. It does not penetrate into the life of the scientist. It does not integrate his personality. He can do some little jugglery here and there in the field of creation, converting this into that and that into this, but he himself is open to all kinds of destructive values because the modern approach to the investigation of natural law does not and cannot enable the scientist to imbibe knowledge and live it in daily life." (pp. 122-123) [8]

Maharishi further points out that the current fragmented approach in modern science and the educational system based on it:

"...Inspires the seeker of knowledge to focus on isolated areas for many years... Reveals that in the pursuit of knowledge whatever knowledge one gains, that knowledge itself reveals that there is yet more knowledge to be known... This makes the path of knowledge endlessly long, and fulfillment out of reach. Permanent lack of fulfillment is the natural outcome of modern science-based education—it is so sad... [It] Searches for knowledge outside of the Self; thus the knower runs after knowledge and exhausts himself without reaching the goal of complete knowledge... Failing to bring fulfillment, it is destined to be surpassed by a more complete science... [It] Creates an imperfect man, who makes mistakes and creates problems and suffering for himself and others... [It] Creates a society characterized by problems, crime, stress, sickness, unhappiness, and disharmony... [It] Bestows upon the superpowers the ability to destroy life on earth..." (pp. 192-195) [9]

What's the *Matter*?

An initial step toward an expanded and integrated view beyond physicalism is to recognize the progress in modern science over the past century in understanding the basis of matter. Over the past few centuries, the field of physics has applied the objective means of gaining knowledge to focus intently on understanding what matter is—and also energy, space, and time. Early physical theory, associated with Newtonian physics, mainly dealt with macroscopic and microscopic levels of nature. It conceptualizes nature in terms of atomic particles and forces—matter and fields. Newtonian physics identified two basic fields as accounting for almost all change in the natural world: electromagnetism and gravity. Einstein greatly expanded aspects of this model with relativistic theories of gravity as a space-time continuum, extending it to the *ultramacroscopic* cosmology of the universe. Newtonian and relativistic theories together are generally identified as *classical physics* to distinguish them from the *new physics,* which developed throughout the 20th Century.

The new physics, closely associated with quantum theory, additionally focuses on *ultramicroscopic* levels of nature. It conceptualizes matter and force fields in terms of quanta, which have attributes of both particles and waves. In recent decades, unified field theories also have developed that focus on both the ultramicroscopic and ultramacroscopic levels of nature and their ultimate common underlying basis. These quite abstract theories are attempting to identify a single underlying field thought to be the essence of matter, energy, space, and time—a single source of the entire cosmos. This chapter introduces some of the findings in this progression, as well as some of the challenges the findings present to the classical perspective.

Atoms are the matter. Early forms of materialistic theory conceptualized matter as microscopic atoms, thought to be the "uncuttable constituents" of all objects. [9] Eventually atoms were conceptualized as composed of even tinier subatomic particles such as protons, neutrons, and electrons surrounded by comparatively vast areas of empty space. About 99.999999999999% of the atom's size was described as empty space, similar on a microscopic scale to the relative distances between the Sun and planets on the macroscopic astrophysical scale of our solar system.

It was later theorized that the comparatively vast areas of space between subatomic particles are suffused with invisible energy or force fields distributed in space, such as electricity, magnetism, and gravity. This led to modern scientific explanations for many phenomena previously thought to be examples of mysterious *action-at-a-distance*. At the beginning of the 20th Century, Einstein's special and general theories of relativity took a major step further. Einstein proposed the fundamental equivalence of matter, force fields, and energy, and also of space and time.

In these classical theories, causal influences between objects are mediated by an unbroken chain of measurable physical events. Objects are localized in space and time, with the forces that influence them generally decreasing with distance but never falling to zero. Although the range of effects can be thought of as sort of unbounded, the significant influences in determining causal effects were believed to be only local influences.

Subsequent research led to theories that subatomic particles are constituted of even more elementary particles (e.g., gluons, quarks). In particle-scattering experiments, 150 or more subatomic particles were identified as theoretically likely to exist, some with inexplicable features. Sometimes particle pairs emerged that were mirror matches with opposite electrical charge, called a particle and its *antiparticle* partner (e.g., negatively charged electrons and positively charged positrons). All known particles are now theorized to have antiparticle partners, and are classified into three groups according to the concept of mass. [10] However, antiparticles have been detected only in laboratory experiments; no examples have been found as occurring naturally in the world.

Quanta are the matter. As the layers of creation have been further unpeeled, all forms of matter have been conceptualized to be expressions of fundamental *quantum fields*. Elementary particles are described as excitations, fluctuations, vibrations, or waves of these abstract fields. Fields are the substrates that fluctuate in waves. These abstract quantum fields are described as extending throughout the universe. In some sense, they are thought to have boundaries or discrete properties; but at the same time they also are thought to be unbounded. The concept of a quantum attempts to embody the discreteness of particle-like packets of energy, while at the same time recognizing their nature as extended energy fields. The simultaneous local point particle and nonlocal field wave qualities have been challenging to comprehend and model.

Quantum fields are depicted as fluctuating only at certain potential energy states, which form *quanta*, discrete wave packets with particle properties. A particle is conceptualized as a point in the quantum field that moves through the field in a stable wave pattern. Matter particles are thought to have no material existence other than being stable states of fluctuation—*physical particle states*—of quantum fields.

These quantized, digitized packets or wavelets of energy are multiples of what is called the *Planck scale*, the fundamental unit of space and time in physical nature. This scale is named after Max Planck, who took a first major step toward quantum theory in 1899. Planck attempted to derive empirically a universal unit of measurement in nature. It was calculated based on what were understood to be the universal constants of the speed of light, Newton's gravitational constant, and Planck's constant, derived from experimental measurement. The Planck scale is the incredibly tiny distance scale of 10^{-33} cm (a millionth of a billionth of a billionth of a billionth of a centimeter), the incredibly brief time scale of 10^{-43} seconds (the time it takes light to move across the Planck length), and the incredibly powerful energy scale of 10^{19} GeV (gigaelectron volts).

In addition to localized matter particles, quantum fields are also theorized to be the basis of the spatially extended forces such as electromagnetism and gravity. The types of fluctuations determine the quantum field's role as either a particle or a force. Stable states of fluctuation produce the *particle* quality of the field and are associated with the propagation of quantized energy packets through the field, such as an electron matter particle. Transient fluctuations of the field are associated with the *force* quality of the field, transferring quantized energy and momentum between stable matter particles propagating through the field, such as the electromagnetic force.

The transient fluctuations—sometimes called force carrier particles, exchange particles, messenger particles, or virtual particles—mediate the exchange of energy between matter particles. They are conceptualized as coming in and out of existence so incredibly rapidly that their existence is described as *virtual*. In quantum field theory, a force is an effect on a particle due to the presence of another particle. This effect is depicted as being mediated by a virtual exchange particle that passes between interacting matter particles.

In addition to fluctuating in a stable pattern associated with discrete matter particles and a transient pattern associated with spatially extended virtual particle forces, the quantum field is also described as capable of being in a least excited or ground state. This is called the *vacuum state*. However, the vacuum state does not correspond to a state of absolute zero energy with no fluctuations or vibrations at all. The most fundamental quantum field is depicted as continuously exhibiting *zero point energy* or *zero point motion*—called *quantum* or *vacuum fluctuations*—whether it is in its particle, force, or vacuum state. These ultramicroscopic fluctuations are theorized to be an inherent dynamic quality of the quantum field.

Experimental observations such as the *Lamb Shift* and the *Casimir Effect*, for example, are interpreted as providing evidence for quantum vacuum fluctuations. The Lamb Shift is the finding that the observed energy states of electrons around an atomic nucleus are greater than predicted. The difference is theorized to be due to the energy contribution of the virtual quantum fluctuations. The Casimir Effect refers to the finding that there is an attractive force between two metal plates even after everything in between them that could account for it—such as light, air, or other forces—has been removed. As two objects move closer to or farther from each other, the measured energy density decreases or increases due to the theorized quantum fluctuations in the volume of space between them.

All particles and forces are theorized to be excitations of four quantum forces. These four forces are 1) the *electromagnetic force* (the basis of electricity, magnetism, and the laws of chemistry), 2) the *weak nuclear force* (which is responsible for radioactivity and produces the energy emitted from the Sun and other stars, 3) the *strong nuclear force* (which holds together atomic nuclei and elementary particles in them) and 4) the *gravitational force*.

Quantum fields are the matter. A primary focus of the new physics is to identify the unity of these fundamental quantum forces and how they dynamically emerge from their underlying basis. This research has been fostered by the recognition that diverse phenomena in nature exhibit more integrated patterns or symmetries at smaller, more fundamental space-time scales, associated with the principle of spontaneous symmetry breaking. At the onset of the theorized big bang, the emerging universe broke from its undifferentiated state through spontaneous symmetry breaking in a sequence of stages of increasing diversity. This occurred automatically as the extremely high levels of energy distributed and temperature began to drop. One approach toward understanding the unity of the fundamental quantum fields is to reflect backward in time to events immediately following the theorized big bang, and attempt to reconstruct how this sequential symmetry breaking may have taken place. In other words, this means attempting the difficult conceptual task of putting the cosmic egg back together—possibly back into literally *nothing*.

The principle of symmetry has been important for developing quantum field theories that describe how—at about 10^{-16} cm—the electromagnetic force and weak forces unify into the *electroweak* quantum field. This level of unification is called electroweak unification, and it relates to what is called the *Standard Model*. There is also some evidence that at even higher energy and smaller time and distance scales—about 10^{-27} cm—the electroweak field unifies with the strong quantum energy force, called *grand unification* or *strong-electroweak* unification. This level of unification is thought to bring everything down to two fundamental quantum fields: the strong-electroweak field (the level of grand unification, of electromagnetism and the weak and strong nuclear forces) and the gravitational field.

The principle of symmetry has facilitated the development of theories that unify quantum fields in the same *spin* type. In this context, spin is an important mathematical concept characterized as a discrete amount of angular momentum that determines some of the properties of matter and force particles. It does not have a direct physical interpretation yet, but is sometimes likened to rotational movement analogous to the external spin of a top. Particles are classified into five spin types (0, ½, 1, 3/2, and 2) in half-units of Planck's constant.

Whole number or integral types (0, + or - 1, + or -2) are the force carrier or virtual particles. Called *bosons*, they have the statistical property of unifying or collecting together in the same position and momentum or physical state. Bosons relate to coherence phenomena such as laser light; they are not discrete particles and cannot be distinguished completely from each other.

Half-integral spin types (+ or -1/2, 1 or -3/2) are the matter particles. Called *fermions*, they have the property of diversifying, and cannot occupy the same energy state. Fermions comprise the various patterns of matter particles that create the vast diversity of behavior throughout creation. They are characterized as discrete particles that can be clearly distinguished from each other, in contrast to bosons.

All particles are either fermions or bosons. Fermions are the matter particles, and they interact via boson force carrier particles that mediate the fundamental forces. All interactions are described in terms of the exchange of these quantized energy particles. Fermions and bosons both interact through the gravitational force. Heavier fermions, called *baryons* (such as the proton), interact through the strong, weak, and electromagnetic forces. Lighter fermions (*leptons* such as the electron) interact through the electromagnetic and weak forces. These details are included here as the basis for additional discussions in later chapters that relate these mathematically derived theories to Maharishi Vedic Science and Technology.

Discovery of the principle of super-symmetry in the 1970s aided development of theories attempting to unify bosons and fermions. This was a major step toward a unified field theory that unites the strong-electroweak field with the gravitational field. However, this principle requires that super-symmetric partners—sometimes called *sparticles*—with identical properties exist for all the particles and antiparticles. To verify this theory, super-symmetric partners for all the known particles need to be found in nature—such as the photino as the theorized partner to the photon, the gluino for the gluon, or the gravitino for the graviton. Each particle is thought to have a super-symmetric partner with a spin either ½ larger or ½ smaller. Although there may be some indirect evidence, hypothesized super-symmetric partners have not been found in nature. They are sometimes related to *dark matter*, or the *hidden sector*—because they are not visible. Dark matter was proposed in part to construct theories applying the principle of super-symmetry, and to help explain how galaxies hold together. Dark matter is different from *dark energy*. Dark energy was proposed more recently to explain empirical findings that the universe is expanding at an increasing rate—discussed shortly, and again in Chapter 12.

Some theories suggest that the hypothesized super-symmetric partners probably don't interact with the known particles and antiparticles in the lower energy microscopic or macroscopic levels. This might explain the difficulty in detecting them, if they do exist. For various reasons the hypothesized lightest super-symmetric sparticle, the photino, is expected to be the easiest to find. Although it is unlikely to be detected using particle detectors, indirect evidence might be found of its effects through particle interactions with missing energy—interactions that don't add up to predicted amounts. According to most super-symmetric theories, photino sparticles don't decay over time, so they would be abundant in the universe today and contribute significantly to its overall mass-density and structure.

Einstein's general theory of relativity implies that the universe either expands continually or expands and collapses. This depends on the average mass-density, and thus the overall strength of gravity, in the universe. The estimated total mass-density adding together all visible objects such as stars, planets, and nebulae fits predictions of an expanding universe. However, recent *inflationary* models predict the universe might be on the brink of starting to shrink or collapse. This requires much higher mass-density for gravity to pull the universe back together. The additional mass-density might come from the theorized dark matter and dark energy that are undetected so far. The additional mass-density required for the universe to collapse approximates the total mass-density of photinos predicted by some super-symmetric theories. If this were correct, it would also help solve the problem that the speed of rotation of stars within galaxies suggests that the mass of galaxies may be much higher than expected from the visible objects in the universe. Related issues of cosmological dynamics will be referred to again in Chapter 12.

Perhaps the most prominent concern in the new physics over the past few decades is how to combine the three forces—electromagnetic, weak, and strong nuclear forces—with or into the one force of gravity. One major approach is based on the principle of super-symmetry. Extensive research has been underway to develop super-symmetric theories that unify the strong-electroweak field with the gravitational field, hypothesized to occur at the Planck scale. Called *super-unification*, it is a momentous undertaking with many theoretical and experimental difficulties.

The model of spontaneous symmetry breaking describes the unified field as progressively breaking into the four fundamental quantum fields. At the scale of 10^{-33} cm, the most fundamental field is thought to break from its ultimate unified ground or vacuum state into the gravitational and strong-electroweak fields. At about 10^{-27} cm, the vacuum state of the strong-electroweak field breaks or differentiates into the strong and electroweak fields, and at about 10^{-16} cm, the vacuum state of the electroweak fields breaks into the weak and electromagnetic fields.[11]

Unifying the strong-electroweak field with the gravitational field is particularly significant because the gravitational field is identified as the field of space-time itself. The theory would unify all objects made of matter particles and force particles. There are believed to be initial indirect indicators that super-symmetric Higgs bosons and possibly even dark matter and dark energy may exist in nature. This would support the theory of a single underlying unified field. But again, to date no super-symmetric sparticles have been found, and integration of relativity theory and quantum theory has not yet been accomplished.

Relativity theory mostly concerns very large scale cosmological phenomena such as the motion of planets and galaxies and the overall shape of the universe. It is based on a conception of space-time as a smooth four-dimensional geometry that, for the most part, is gently curved and molded under the influence of the gravitational fields of objects with mass. This contrasts radically with very small scale ultramicroscopic phenomena, a major focus of quantum theory. Quantum field theory conceptualizes ultramicroscopic quantum or vacuum fluctuations that produce a constantly changing, chaotic, choppy, or grainy texture to space at the level of super-unification, the Planck scale. As

"For many years, physicists found that the central obstacle to realizing a unified theory was the fundamental conflict between the two major breakthroughs of twentieth-century physics: general relativity and quantum mechanics... However...whenever the theories are used in conjunction...out of their combined mathematics pops an *infinite* probability. That doesn't mean a probability so high that you should put all your money on it because it's a shoo-in. Probabilities bigger than 100 percent are meaningless. Calculations that produce an infinite probability simply show that the combined equations of general relativity and quantum mechanics have gone haywire. "—*Brian Greene, theoretical physicist (p. 16)* [12]

a simple analogy, at ordinary viewing distances photographs look smooth and clear, but become increasingly granular at very small scales, such as under a microscope.

It has been quite difficult to integrate these two prominent and most successful theories—the relativistic space-time model of gravity and the model from super-symmetric quantum field theory—into a theory of *quantum gravity* needed for a completely unified field theory. In terms of spin states, this means to connect the hypothesized spin 2 gravity field with the other spin-type fields, to connect the space-time continuum itself with the particle and force fluctuations of the other hypothesized super-symmetric quantum fields. The mathematical calculations repeatedly produced infinite values, and it has been a quite arduous task to find the rationale to cancel out the meaningless infinities to obtain mathematically meaningful results.

Strings and branes are the matter. One of the major directions toward a theory that unifies all matter and force particles and their hypothesized super-symmetric partners is a prominent new approach called *string theory.* Combining super-symmetry and the theory of strings into super-string theory (frequently just called string theory) has resulted in what some investigators believe is the best candidate for a unified field theory that includes quantum gravity—although there still is considerable debate about it. Upcoming chapters include discussions of this and other theories of quantum gravity.

String theory basically replaces the concept of a dimensionless point particle used in both classical and quantum theories with a one-dimensional string. The concept of a particle is of a point in space with no internal structure—no extension in space—and conceptually only the capability of movement through space. A string has extension in space, which allows for an internal structure and thus the additional potential for internal movement or fluctuation. This allows mathematically for more complex, higher-order patterns of fluctuation which adds explanatory power to the theory. The higher-order fluctuations within the string are significant at the ultramicroscopic scale; otherwise, strings have much the same mathematical properties as dimensionless point particles in quantum field theory and in classical physics.

In this theory, strings may be the ultimate 'uncuttable constituents.' The size of a one-dimensional string is theorized to approximate the incredibly small Planck length of 10-33 cm. The energy levels associated with their higher-order, internal vibrations are thought to be in the incredibly high range of the Planck energy (10 to the 19th GeV). There is only one type of string, but it can vibrate in many different ways. The different vibratory patterns of the string are thought to produce all the various particles that make up the entire universe. The precise patterns of fluctuation determine the different matter and force particles, anti-particles, and their super-symmetric partners. One of the patterns of fluctuation that is allowed by the extended spatial nature of strings matches the properties of the hypothesized super-symmetric graviton. This would link string theory with quantum gravity, if super-symmetric gravitons were to be found in nature. The mathematical properties of strings result in precisely canceling out the meaningless infinities that prevented a completely consistent finite theory of quantum gravity.

Using mathematical analyses, predictions of string theory can be compared to properties of the known particles and forces. But coming up with sufficiently detailed predictions to do this is beyond current computational abilities. From this ongoing research, however, amazing new conceptions of nature are emerging. One amazing feature of string theories is that mathematically they require *dimensions* in addition to the ordinary four dimensions of the space-time continuum. The additional mathematical dimensions—as many as six or seven—are conceptualized as spatial dimensions that are curled up in the internal structure of the string within the Planck scale, called *space-time compactification.* The classical four dimensions of space-time observed in nature at the larger microscopic and macroscopic scales are thought to be the non-compactified or unfurled dimensions that make up our ordinary world of sensory experience. The higher-order dimensions in string theory are mathematical dimensions in conceptual, imaginary, or mathematical space. They are not necessarily comparable to the ordinary physical space-time dimensions.

String theory is so complicated that its exact mathematical equations have not been able to be determined. Approximate equations yield many perspectives or models, but there are indicators of a smaller set of consistent models (Types I, IIA, IIB, Heterotic-O, Heterotic-E, and 11-D Supergravity). Recent advances pull together these different models into an overriding framework called *M-theory*. [10]

M-theory involves *11 dimensions*, including time, the ordinary three spatial dimensions, and seven compactified dimensions in mathematical space.[10][13] In addition to one-dimensional strings, however, other mathematical objects including two, three, and higher dimensional objects called *branes* (short for membranes) are posited. A brane is a mathematical object with length in p-directions, called a p-brane (a p=1 brane would be a one-dimensional string, a p=2 brane would be a surface such as a membrane, etc.). Higher-order p-branes might be curled up into complicated, convoluted geometric shapes. If branes exist in the natural world, calculations show that they would be massively heavy, and thus require energy levels existing only at the earliest stages of the big bang. This means that they would have very little effect on most of the larger scale, lower energy activity in today's universe.

Understanding *M*-theory and developing exact equations for it are major concerns in theoretical physics in this early part of the 21st Century. Importantly, the theory is said to provide a logical integrating framework that encompasses much of the progress in the past century in the new physics—including quantum field theory and super-symmetry. There is considerable enthusiasm and hope among some groups of physicists and mathematicians that *M*-theory will soon lead to the long-sought completely unified field theory—the theory of everything (the big T.O.E.). However, there remains considerable debate as to whether string theory is the appropriate direction in order to gain an accurate understanding of quantum gravity and the ultimate underlying unity of nature. String theory is not yet a viable theory of fundamental unification, and there is major concern whether super-symmetry upon which it is based even exists in the natural world. Though mathematically compelling, to date the actual evidence is considered to be less than scant. [13]

To summarize, material objects are conceptualized as built of atoms, which are composed of elementary particles and forces, which are quantum-wave energy fields, which might be composed of geometric patterns called strings, branes, or other similar mathematical objects. The overall picture is of probing smaller and smaller time and distance scales and getting down to the size of the Planck scale. This ultramicroscopic scale is so incredibly tiny as to be *almost* like a dimensionless point in space. For purposes of building mathematical theories of this fundamental ultramicroscopic level of nature, it is conceptualized that compactified near this tiniest size are mathematical objects or geometric patterns such as strings and branes. Through different patterns of fluctuation, these abstract, conceptual, geometrical patterns are theorized to generate all the objects in our ordinary objective four-dimensional world.

These terms, however, give the impression that the geometric patterns are physical objects. They are mathematical 'objects' described in terms of spatiotemporal metaphors in order to help develop and explain the theories. At the incredibly tiny Planck scale, it is thought that ultimate limits to space-time are being reached. Theories about phenomena at this hypothesized limit of space-time are challenged by the issue of the meaning of ontologically *real* physical existence and how it differs from what exists only as mathematical 'objects' in conceptual space—assuming this distinction is meaningful. In this issue are clear glimpses of the historical mind-body problem.

If branes and strings exist as some form of objects on some level of the *real* natural world, they would be understood in terms of fundamental curvatures *of* space-time itself not made of matter. Modern physics has taken the search for the essence of matter beyond all forms of matter—beyond elementary matter and force particles, beyond invisible non-local quantum fields, and beyond space and time—to a theorized *non-physical* basis of physical creation. In other words, fundamentally there is no matter. This suggests that there is much more to the story than just ordinary physical *reality*.

Implications That Challenge Common Sensory Experience

Relativistic space-time. Even from the perspective of relativistic physics (not considering quantum theory), scientific theories about the universe challenge ordinary sensory experience. For example, Einstein's Special Theory of Relativity is sometimes interpreted to mean that no object can travel faster than the speed of light (about 300,000 kilometers or 186,000 miles per second). As noted earlier, this is thought to set an ultimate limit on any object traveling from one event to another in order to influence it, including *any* form of information.

Sometimes this is interpreted to mean that for events separated by vast cosmic distances, cause and effect doesn't have meaning because no information could possibly travel fast enough (faster than light-speed) to influence, measure, or verify the causal relationships. Also, two events could happen so close together in time—even over relatively short distances—that information from one event still could not travel fast enough to affect the other, within an extremely short period of time. The range of possible cause-effect relationships within the speed of light is called the *light cone*. It is related to the general notion of *Einstein locality*, which places a limit on causal relationships to interactions within the time and distance frame of the speed of light.

It is quite challenging to think we can never know what—if anything—is happening at *this instant* in some part of the universe outside of our light cone; but apparently we cannot if relativity theory is the last word on it. According to the theory, there isn't an absolute '*this instant*' in a far away place or anywhere else—time and space are relative. For example, we cannot know for certain that the Sun is still in the sky *right now*. The Sun is about 55 million kilometers away, so it takes sunlight about eight minutes to travel to Earth. What could be happening on the Sun right now, if it is there right now, is outside the range of our future light cone. Thus we can say that the Sun appeared to be there about eight minutes ago; but we have no way to confirm—nor according to the theory will we ever have a way to confirm—that it is there right now, or even that there is a *there right now*. After eight minutes we are usually able to see the effects of what very likely took place, and can infer that the Sun was continuing to radiate light eight minutes ago. But we still won't know if it is *there right now*. In relativity theory the concept of *there right now* is undefined.

Further relativity theory states that two observers who are traveling with respect to each other will measure different times between the same two events, due to their relative motion Photon particles, which move at the speed of light, are thought not to be subject to time—it is *as if* they exist in timelessness. If it were possible for some object to approach the speed of light, it also would approach timelessness, from the perspective of an observer.

Thus theoretically if you left Earth in an unbelievably fast space ship traveling near the speed of light, time would slow down for you with respect to family members on Earth. Although time would not seem different to you, it would look like the family members you waved goodbye to as you zipped away—if this were possible to do—were waving back very slowly. Upon returning to Earth a few years later from your perspective, you might be shocked to find that your family members *actually* aged a lot more than you, because time would have been much slower for you from their perspective. Again, according to relativity theory there is no underlying absolute time or absolute space. Time and space have quantitative meaning only in relation to objects moving relative to each other. In relativity theory, however, space and time as one variable—space-time—is an absolute value.

Quantum gravity. Quantum field theory stretches the credulity of ordinary commonsense even further. An example is the concept of *quantum gravity.* The hypothesized vacuum fluctuations at the Planck scale, as energy, have gravitational effects. The energy is so incredibly high at this very highly interactive level of nature that the fluctuations are thought directly to affect themselves—they are described as *self-interacting.* An example of a self-interacting field is that the gravitational field of the Earth changes the gravitational field of the Moon, which in turn feeds back to change the Earth's gravitational field, and so on. These self-interacting effects make it very difficult to measure gravity when even as few as three gravitational masses are involved—let alone entire galaxies, all of which interact with each other within their light cones.

These self-interacting dynamics are thought to change the ultramicroscopic structure of the space-time continuum, creating self-interacting quantum field fluctuations called *space-time foam.* These quantum or vacuum state fluctuations are theorized to be an inherent feature of the constant dynamism in nature at the level of the Planck scale. They are sometimes likened to a soupy froth of virtual particles randomly being created and destroyed at an incredible rate (about 10^{44} per cubic cm/sec) at the Planck scale.

An important feature of hypothesized space-time foam is that the self-interacting, superposed quantum fluctuations theoretically could allow the possibility of a sort of action-at-a-distance effect not limited by the known dynamics of the particles and forces at lower-energy microscopic and macroscopic scales. At this ultramicroscopic level, motion is not described in the ordinary sense as objects traveling along a continuous path, but rather in terms of jumping into and out of the space-time foam—to be discussed in more detail in the next chapter.

Accordingly it is thought to be possible—even if a very low, almost non-existent probability—for a classical macroscopic physical object on one side of a classical physical wall spontaneously to appear on the other side of the wall, *without* having traveled through the wall. This type of 'travel,' called *quantum mechanical tunneling,* is thought to be a common quantum process associated with very short distance processes such as nuclear reactions in radioactive decay. On a macroscopic level, it can be described as a process of decomposing at one classical macroscopic four-dimensional space-time position, and then recomposing somewhere else.

Based on this theory, it has been speculated that it may be possible to 'travel' anywhere in the unbounded quantum field. This quantum 'traveling' would be via *wormholes* or quantum mechanical tunnels that create instantaneous connections between regions of space—even to other regions of the universe (that according to relativity theory are undefined and we cannot ever know are *there right now*). This 'travel' would somehow involve going beyond the boundaries of ordinary space-time, and thus it would not be limited by the restrictions of space-time in the four-dimensional world of ordinary sensory experience. It wouldn't be travel in the ordinary sense, but more like appearing at one location, then decomposing all the information from that location and recomposing it at another location. This is the basis for recent speculations about the scientific possibility of *teleportation.* At this stage of research, models are being developed to describe how the complete information about a physical object necessary to *port* it could be extracted from the object—which would decompose the object. However, it is not yet understood how the information could be directed to a new location, as well as how to recompose it ('travel' even more dissipating than jet lag).

The theoretical possibility based on quantum theory of *instantaneous* 'porting' contrasts with the notion in relativity theory of space-time and the limitations of light-speed. Consistent with many peoples' intuitive sense, it seems to imply that there indeed may be a *'there right now,'* as opposed to the notion in relativity theory that space-time outside the light cone is undefined and can never be known. This reflects the major challenge of how to reconcile quantum theory and relativity theory—to be discussed further in later chapters.

The holy universe? Another important theoretical oddity relevant to the clash of the classical relativistic space-time model and super-symmetric quantum gravity is *black holes*. This phenomenon was originally predicted based on relativity theory.

If the mass of a star becomes very densely packed and concentrated beyond a critical value, the resulting curvature or warping of space-time is theorized to produce a strong gravitational pull on anything in its vicinity. The gravitational pull would be so strong that any object or information—including light—which came within the range of what is called its *event horizon* would not be able to escape. It is called a horizon because it is a boundary at which no light can escape; nothing that might be in the black hole can be seen from outside of the horizon. Inside the event horizon is a hidden region, but no information about it is available to outside observers because no information inside can escape. Objects passing near the black hole but outside its event horizon would be pulled off course by the gravitational force of the black hole, and then continue along their modified path. But objects that move into the event horizon could not escape, and inevitably would be pulled deeper toward the center of the black hole and its increasingly powerful and destructive gravitational pull.

Although we could never see a black hole because light cannot escape from it, there is some indirect cosmological evidence that black holes may exist in the natural world. Some of this evidence comes from observations of unusual patterns from light-emitting stars that are likely to be the product of the gravitational influence of nearby event horizons of black holes. As the dust and gases of the outer layers of stars approach a black hole, the black hole's gravitational force is thought to speed them up to near light-speed, creating friction which produces visible light—and X rays—just outside the event horizon. These effects can be observed telescopically, or measured using other means. Growing evidence of this type suggests that all galaxies might have a black hole in their center, and that black holes may be a core feature of the creation and evolution of a galaxy. It is thought that the center of our own Milky Way galaxy may have a black hole as much as two and a half million times the mass of the Sun; and many times more massive black holes might exist in other parts of the universe.

As referred to earlier, applying the relativistic theory of gravity to the entire universe suggests that the overall fabric of relative space-time changes—expanding and contracting, stretching and shrinking. Telescopic observations that all galaxies appear to be rapidly moving away from each other provide evidence that the universe is expanding. If this expansion continues uninterruptedly, then we have what is sometimes called an *open* universe. Other recent theories suggest that this type of universe, in which all objects are moving away from each other, eventually will dissipate everything into a dark cold emptiness. If the energy of expansion dissipates and massive objects slow down, however, at some point the force of gravity could reverse the expansive direction of movement and shrink the entire universe. This is hypothesized to result in a *closed* universe. A closed universe possibly could cycle between expansion and contraction—big bangs and big crunches—over vast epochs of time.

Conceiving of space-time as expanding and shrinking might bring up the question of what is outside the edge of the universe. Where does it *expand to*, and what remains where it *shrinks from*? Such questions are thought to reflect incomplete understanding of relativistic space-time. According to the theory, there is nothing *out there* beyond space-time—no residual quality of space or time or existence that was there before or after space-time expands or shrinks—it is undefined in the theory. (But this suggests that space is not infinite, or at least not the whole story.)

One notion based on relativity theory is that if you were to travel as far as possible in one direction, ultimately somehow you would arrive back at the starting point. In this model, it would be as if the universe were a gigantic sphere (although other shapes have been proposed also), but with nothing outside of the sphere. Ultimately inside and outside are complementary and go together as a whole—without anything *outside* of the wholeness of the cosmic egg. This is sometimes likened to a *Mobius strip* (a shape that is similar to the common symbol for infinity), in which movement in one direction eventually leads all the way around and back to the original starting point.

Some contemporary cosmological theories of a closed universe propose that as the fabric of space-time shrinks, the universe with all its massive content would be increasingly compressed. If all the galaxies in the universe were to become severely compressed, the density and temperature would rise and disintegrate everything into a thick hot soup of elementary particles and forces. If this process of contracting space-time were to begin today and continue for another 14 billion years or so—the length of time by some estimates since the big bang—in theory everything could by squeezed or crunched into a smaller and smaller region of space. Eventually it would be scrunched to the size of a basketball, soccer ball, baseball, tennis ball, golf ball, and possibly an ultramicroscopic point—like the core of a black hole.

At this point the description of a black hole and the source of the big bang may sound quite similar to the description of an elementary particle, or a string or brane. Indeed some theories suggest the potential equivalence of these theorized states. The fundamental substrate of matter particles and the origin of the big bang could be the same thing: ultramicroscopic curvatures of space-time such as particles, strings, or branes. According to string theory, however, this compressed point of space-time would not be a point with no spatial extension. Rather it would be the smallest size, about the Planck length, like a string or brane. It is theorized that such a smallest size may have been the origin of the big bang. This smallest size of space, however, is not conceived to be a point, string, or brane floating somewhere in cosmic space, because space-time would not exist outside of it. The big bang could not have occurred at some point in space-time, because space-time *would not have existed* within which the big bang could happen. It is difficult to describe such theorized phenomena that in themselves seem to contain contradictions.

It also has been speculated that the positive and negative charges of many elementary particles might suggest that there is an opposite of the black hole—a *white hole*—that would radiate energy rather than capture it. Much like opposite charges, if a black hole were connected somehow to a white hole, it has been speculated that the energy falling into the black hole could be expelled through the white hole. Because the connection between them somehow would be beyond space and time, it might result in an instantaneous nonlocal transfer of energy—such as teleportation via a wormhole—to even a distant (relativistically undefined) region of space-time. Such opposite pairs could be the weave of a local-nonlocal fabric of the space-time continuum.

On the other hand, some recent perspectives question whether black holes exist. One emerging alternative theory suggests that stellar masses may not shrink all the way down to the ultramicroscopic scale of space-time or to an infinitely dense singularity. Rather, they might arrive at some sort of equilibrium on a macroscopic scale, producing a pool of undifferentiated plasmic mass called a *gravistar*. This model has been proposed to avoid potential difficulties of ultramicroscopic compressed black holes with the contradiction of *infinite quantities* of density and mass.

Adding possibilities from quantum and super-string theories to the relativistic theory of black holes, the universe becomes a much more vast and dramatic place than meets the ordinary eye. It has been recently theorized that many universes, some virtual and some lasting a relatively long time from our human perspective, are created and disappear into black holes—possibly reappearing in an undefined elsewhere through *wormholes* in other sectors of the universe. It also has been theorized that distinct self-contained universes within universes may exist, with no means of interacting with each other. These distinct universes might go through private cycles of creation and dissolution, big bangs and crunches, similar to virtual particles in space-time foam.

Some contemporary theories even propose that our universe with its billions and billions of galaxies, each with billions of stars, is but one of an uncountable multitude of universes, some of which could possibly have different laws of nature. The total collection of all possible universes is sometimes referred to as the *Big Universe*, or the *multiverse*. [10] [14] [15] [16] In the multiverse, many universes such as our *special universe* might be created and destroyed continuously through big bangs and crunches. What may be special about our universe is that the particular combination of natural laws that emerged in it happened to lead to life—especially human life with our incredible capability and fortune of at least partially comprehending it. Of course our universe might not be so special either.

Contemporary *many-worlds* quantum theory takes it even further, speculating that each possible outcome of each event in nature produces a parallel world. Each of those possible outcomes also would produce an additional parallel world, and so on in an unbelievably mind-boggling expansion of potentialities that are thought to have some sense of actuality to them. [17] At this point it might seem that the frontiers of modern physics have entered territory expected to be encountered in the realm of mysticism—or at least science fiction. But these speculations are based on established theories consistent with the large body of reputable modern scientific evidence.

It is reasonable, however, to wonder how all this remains within a strict framework of

> "*The* overarching lesson that has emerged from scientific inquiry over the last century is that human experience is often a misleading guide to the true nature of reality. Lying just beneath the surface of the everyday is a world we'd hardly recognize. Followers of the occult, devotees of astrology, and those who hold to religious principles that speak to a reality beyond experience have, from widely varying perspectives, long since arrived at a similar conclusion. But that's not what I have in mind. I'm referring to the work of ingenious innovators and tireless researchers—the men and women of science—who have peeled back layer after layer of the cosmic onion, enigma by enigma, and revealed a universe that is at once surprising, unfamiliar, exciting, elegant, and thoroughly unlike what anyone ever expected." —Brian Greene, theoretical physicist (p. 5) [12]

objective scientific methodology. As will be discussed in later chapters, fortunately it doesn't. It rather can be viewed as representing awkward but constructive progress beyond the rigid limitations of objective physicalist thinking and methodology. This doesn't mean that modern science is becoming *sci-fi*, however, but rather what might be called *psy-phy*. The *psyche* or mind is starting to be considered as fundamental in developing a coherent theory of even the physical universe. Theories in the new physics are starting to reflect expanded notions of space that connect with the more abstract level of the psyche or mind and mental or conceptual space, sometimes called *quantum reality* or *quantum mind*—to be unfolded in upcoming chapters. It is a wonderful advance, none too soon, that reflects long-sought progress on the mind-body problem and the integration of subjectivity and objectivity, of psychology with physics and physiology. But it still can be quite perplexing within the physicalist worldview and ordinary sensory experience.

In the past century we have progressed toward theories of the ultimate underlying unity of nature at its most fundamental level. However, at this stage what has resulted appears to be just the opposite: an infinite diversity of universes. In trying to put the cosmic egg back together, infinities of cosmic eggs have been conceived. As we approach what is thought to be the ultimate unity of the entire physical world, it seems to shatter into infinite diversity. This is due to not integrating the objective and subjective domains of nature. It concerns how this basic gap relates to the fundamental particle-wave, point-field, and even deeper point-infinity *duality* of nature—discussed in the next chapters. As we will see, a core challenge to modern physics and related fields is to integrate infinitesimally small particles with cosmologically large, even infinite fields—and also the ordinary sensory world with the conceptual mathematical world of the mind. These fascinating and important topics are embedded in the *measurement problem* in quantum theory and the related mystery of the *collapse of the quantum wave function*—the focus of Chapters 4 and 5.

> "Science in the first quarter of the twentieth century had not only eliminated materialism as a possible foundation for objective truth, but seemed to have discredited the very idea of objective truth in science. But if the community of scientists has renounced the idea of objective truth in favor of the pragmatic idea that 'what is true for us is what works for us', then every group becomes licensed to do the same, and the hope evaporates that science might provide objective criteria for resolving contentious social issues... This philosophical shift has had profound social and intellectual ramifications. But the physicists who initiated this mischief were generally too interested in practical developments in their own field to get involved in these philosophical issues. Thus they failed to broadcast an important fact: already by mid-century, a further development in physics had occurred that provides an effective antidote to both the 'materialism' of the modern era, and the 'relativism' and 'social constructionism' of the post-modern period."—*Henry Stapp, theoretical physicist (p. 142)* [18]

As will be increasingly revealed, the two major theories in modern physics—relativity theory and quantum theory—can be related to fundamental principles associated with the duality of unifying and diversifying tendencies throughout nature. The unifying principle can be thought of as associated more with relativity theory, tangibly expressed in the law of gravity. The diversifying principle can be thought of in some ways as associated more with the quantum principle in quantum theory, tangibly expressed in the fundamental discreteness of nature. These tendencies also are exhibited in the duality of bosons and fermions, as well as particle and wave, which ultimately need to be unified in a theory of the unified field. They also relate to innumerable fundamental dualities such as part/whole, individuality/universality, nature/nurture, free will/determinism, nationalism/internationalism, and so on that reflect the need to transcend the ordinary intellect in order to reconcile fully the phenomenal opposites in nature. This is the direction of the upcoming chapters on theories of the physical universe. In a later part of the book, this duality will be associated with mind and body, and eventually with phenomenal universality and individuality as the indivisible 'nature' of consciousness itself.

Chapter 3 Notes

[1] Chalmers, D. J. (1995), *The conscious mind: In search of a fundamental theory.* New York: Oxford University Press.

[2] Schwartz, J. M. & Begley, S. (2002). *The mind & the brain: Neuroplasticity and the power of mental force.* New York: ReganBooks.

[3] Schrödinger, E. (1956). *What is life? and other scientific essays.* New York: Doubleday Anchor Books.

[4] Dossey, L. (1989). *Recovering the soul: A scientific and spiritual search.* New York: Bantam Books.

[5] Isaacson, W. (2007). *Einstein: His life and universe.* New York: Simon and Schuster.

[6] Varela, F., J., Thompson, E., & Rosch, E. (1993). *The embodied mind: Cognitive science and human experience.* Cambridge, MA: The MIT Press.

[7] Jung, C. G. (1963). *Memories, dreams, reflections.* Winston, R. & C. (Trans.). New York: Random House.

[8] Maharishi Mahesh Yogi (1997). *Maharishi speaks to educators: Mastery over Natural Law,* Vol. 4. India: Age of Enlightenment Publications (Printers).

[9] *Maharishi Vedic University: Introduction* (1994).. Holland: Maharishi Vedic University Press.

[10] Greene, B. (2000). *The elegant universe*. New York: Vintage Books.

[11] Hagelin, J. S. (1987). Is Consciousness the Unified Field? A Field Theorist's Perspective. *Modern Science and Vedic Science*, 1, 1, (1987), 29-87.

[12] Greene, B. (2004). *The fabric of the cosmos: Space, time, and the texture of reality*. New York: Alfred A. Knopf.

[13] Woit, P. (2006). *Not even wrong: The failure of string theory and the search for unity in physical law*. New York: Basic Books.

[14] Smolin, L. (2001). *Three roads to quantum gravity*. New York: Basic Books.

[15] Hawking, S. (2001). *The universe in a nutshell*. New York: Bantam Books.

[16] Penrose, R. (2005). *The road to reality: A complete guide to the laws of the universe*. New York: Alfred A. Knopf

[17] Everett, H. (1957). *Rev of Mod. Physics*, 29, 454-462.

[18] Stapp, H. P. (2007). *Mindful universe: Quantum mechanics and the participating observer*. Berlin Heidelberg: Springer-Verlag.

Chapter 4

Perspectives on the *Matter*

In this chapter, we will review fundamental perspectives on quantum theory, including the uncertainty principle and quantum indeterminism, the measurement problem, the quantum wave function and its theorized instantaneous collapse via conscious observation, the experimental finding of nonlocality, and the question of superluminal causal interaction. The overall main point of this chapter is that a deeper exploration of the nature of space and levels of reality is needed in order to address the challenging paradoxes in quantum theory.

Material objects can be studied from many different levels and perspectives. For example, humans can be investigated in terms of the world system, international relations, nation states, regional cultures, communities, families, individual behavior, the organism as a unit, organ systems, cells, molecules, chemicals, atoms, elementary particles and forces, quantum fields, strings, branes, space-time curvatures, non-material mathematical 'objects,' abstract information fields, and so on. At either extremely small or large time and distance scales, the investigations become increasingly abstract to ordinary sensory experience and commonsense.

Investigating the essence of matter using the objective approach has primarily involved a reductive strategy of probing and measuring smaller and smaller time and distance scales, and higher and higher energy and temperature states, in pursuit of the most fundamental constituent of matter—briefly summarized in the past chapter. There are limits to the measurement process using objective means involving theorized independent external probes. These limits are determined in part by the size and nature of the probes, and the feasibility of attaining the necessary energy levels for probing these underlying fundamental layers of the physical world. Also, measurement is limited by the probing process itself: at extremely small scales, probing and measurement processes significantly alter the objects being observed.

At some point, obtaining tangible empirical evidence *directly* through the ordinary senses becomes impossible. *Indirect* evidence is then required, based on precise theories and mathematical models. Analyses of theorized events occurring at time and distance scales smaller than are perceivable through the ordinary senses, as well as presumed events outside of the observer's light cone, must rely on indirect evidence based on logical theories and corresponding models. They increasingly depend on mental conceptions of what is being measured, as well as what *measurement* means, and decreasingly depend on ordinary sensory experience. The subtle workings of reason—the intellectual function of the process of knowing—are necessarily relied upon more than ordinary sensory processes. It becomes a bit more obvious that what is perceived in an observation depends on the mind and consciousness of the observer.

Inevitably assumptions embedded in the concepts of an object, a probe, the resulting evidence from their interaction, as well as of an observer or investigator, all must be considered. In investigating an ultimate completely unified basis of nature, it is also inevitable that at some point there will be *no difference* between the objects, the probes, the processes of investigating, and the investigator. At that ultimate level there would be no difference between the measurement, the process of measuring, and the measurer—known, process of knowing, and knower.

Macroscopic and Microscopic Perspectives

Analogous to the introductory comments about levels of *subjectivity* in Chapters 1 and 2, the following levels associated with the concept of *objectivity* can be identified from the discussion in Chapter 3:

Ultramacroscopic levels	~ cosmic expanse to Infinity
Macroscopic levels	$\sim10^{-3}$ to ~cosmic expanse
Microscopic levels	$\sim10^{-4}$ to $\sim10^{-8}$
Ultramicroscopic levels	$\sim10^{-9}$ to $\sim10^{-33}$ cm (Planck length)
Unified field level	~Infinitesimal point to Infinity

The resolving power of our ordinary senses for *direct* sensory observation and measurement is comparatively quite limited and on a vastly larger time and distance scale than the level of the Planck scale (10^{-33} cm, 10^{-43} sec). Direct sensory observation has been extended with the aid of equipment such as electron microscopes and high frequency oscilloscopes to about 10^{-8} cm (a hundredth of a millionth of a centimeter, still larger than an atomic nucleus). The wavelength of visible light is in the range of .00005 cm, too wide to analyze anything smaller than a cell. Probes that are smaller than this range are required to investigate smaller scales, and thus cannot provide a basis for direct observation and inspection using the ordinary unaided senses.

At these smaller scales, *indirect* methods of probing and measuring are required. The results or basic data, both direct and indirect—such as the movement of a meter or pointer on a measuring device—are phenomena at the macroscopic level of experience observed via the ordinary senses. Indirect methods involve measuring phenomena at these larger scales that are predicted based on detailed models of processes theorized to occur at the smaller unobservable scales. These methods are quite analogous to indirect methods to investigate information processing in the 'black box' of the mind—referred to in Chapter 2.

One prominent indirect method is described as colliding elementary particles and then measuring visible traces of the outcomes of the collisions that are believed to take place at the smaller unobserved scales. This *particle-scattering* method is described as employing particle accelerators that build up high-energy streams of particles used to bombard other particles. The target particles may be destroyed, generating other particles from the collision, or may be scattered at different angles. The results of this method are different patterns that leave observable traces recorded using particle detectors and computers. From the thickness, length, and curvature of the traces, many properties of the resulting charged particles can be calculated, as well as non-charged particles. The traces that have properties fitting the predictions provide support for the theories of the unobserved elementary matter and force particle interactions.

Attaining the levels of energy needed to probe ultramicroscopic scales require large particle accelerators. The largest particle accelerators are at the Fermilab in Batavia, Illinois in the U. S. and the Large Hadron Collider (LHC) at CERN near Geneva, Switzerland (which with its new 2008 remodeling is nearly 27 km around). These installations have been or are being expanded to increase their technological limit of probing nature from about 10^{-18} cm to about 10^{-19} cm. However, the energy levels needed to probe the hypothesized smallest scales of space-time are so immense that they are impossible to reach using this technology. The scales where grand unification (10^{-27} cm) and super-unification (10^{-33} cm, the Planck scale) are theorized to occur are far beyond the range of even these most powerful particle accelerators.

An important alternative indirect method is to search for measurable remnants of the hypothesized extremely high temperature states associated with big bang theory that still may be circulating around the cosmos. This involves making predictions based on models of processes believed to have occurred near the time of the theorized big bang, and then looking for evidence still remaining in the cosmos to support the predictions. An example is the search for neutrinos—which might originate from hypothesized super-symmetric dark matter—using particle detectors shielded in deep underground placements, so far unsuccessful.

Such cosmological methods have been associated primarily with research on gravity and the structure of the cosmos, while particle accelerator methods have been used for quantum field and super-symmetry research. In recent years, cosmological methods have been increasingly prominent in both areas of research, due in part to the limitations of particle accelerator technology.

Investigations in these and other areas of the new physics have obtained evidence that has required radically different perspectives about nature than the classical physics perspective. Large teams of modern scientists (expending large amounts of mental and financial resources) have been busy sorting out the implications of these new perspectives, and this process has been triggering major reevaluations of fundamental assumptions and beliefs. An explanation of all phenomena in nature—a theory of everything—needs to span both the infinitesimally small and the infinitely large, and even go beyond both to an ultimate unity that at the same time is smaller than the smallest and bigger than the biggest (Katha Upanishad 1.2.20).[1] This is the direction we are headed on the bridge to unity, and quantum theory represents important planks on the bridge.

The Quantum Perspective

The historical evolution of modern physics might suggest that the reductive investigation of deeper layers of nature could lead to an infinite regress—taking lots of time and money without ever getting to a verifiable conclusive understanding of the essence of matter and energy. This potential concern is thought to be abated in several ways.

For example the hypothesized quantum fluctuations at the Planck scale—described in Chapter 3—where ordinary space and time end can be understood as establishing an ultimate limit to probing and measuring matter. Also, the extended spatial nature of hypothesized strings and branes at the Planck length is thought to be the smallest possible size of a material object, including any possible measuring probe. More generally, the potential concern is thought to be resolved due to core principles in quantum theory. Especially, it involves measurement limitations associated with the Heisenberg uncertainty principle and a related fundamental *indeterminism* in the very heart of nature.

The measurement problem and quantum indeterminism. A general sense of some of the issues related to measurement limitations might be gained by the example of trying to measure a river. The coordinates of a river marking its position and path can be depicted reliably on a map; but the map is far from precise. A river is a constantly flowing, changing process—water is splashing around, evaporating, banks are eroding, and so on, all at the same time. At some level of precision it becomes difficult even to define what and where the river is. Would the flood plains surrounding the seasonal flow of water, the moisture in the mud along the riverbanks, the atoms in the mud, the force fields surrounding the atoms, and so on all be included? At what *exact* point would a raindrop count as part of the river, or an evaporating drop no longer be counted?

Also, at every scale of measurement there is a relationship between the object being measured and the means used to measure it. If we tried to be very precise about the position of each molecule that makes up the river by measuring it with precision instruments—such as sloshing through the river with an electron microscope in tow—the act of measuring obviously would change the *exact* position and path of the river. This concern is more severe at smaller scales.

In addition, although some processes of change are relatively slow, such as for example decay of an ancient fossil, still change is present and limits precise measurement. A fossil appears to be a static concrete object that can be measured precisely. But at tinier scales it also involves many constantly changing processes that limit how precise measurements of it can be and that will be affected by attempting to measure it exactly. Its stability is a conception associated with the macroscopic level of perception. This applies not only to the object being measured, but also the instruments and probes used to make the measurements. At some level, everything is changing. For example it is estimated that over the course of this year about 99% of the atomic matter that makes up our own bodies is being exchanged with matter from the environment, and an estimated 70% of synaptic connections in the brain may change almost daily. Even our own brain and body as measurement tools are constantly changing.

The unpredictability or unspecifiability in these kinds of measurement problems is sometimes called *classical ignorance*. It is associated with a common meaning of randomness or chance. A simple example is the chance or probabilities involved in shuffling a deck of cards. The unpredictability in card shuffling is due to the complexity of calculating and measuring all the factors that determine where the cards end up. It is a causally determinate process that yields an unpredictable probabilistic outcome due to the complexity of tracking all the events involved. Even if practically quite difficult, however, *in principle* it is possible to trace every macroscopic change from each card's starting position in the deck to predict where it will end up after shuffling. This is the case because it appears that on the macroscopic level of analysis each change is linked to the next in an identifiable continuous physical chain of cause and effect.

At the macroscopic scale, it is usually obvious what and where a playing card is and how to count it (unless a magician is at work). It is also relatively simple to figure out the probabilities of a particular card appearing on top of the deck, even if we are not sure which particular card will actually end up there. The reason that our predictions of the card positions are probabilistic—rather than exact—is that we have not measured all the causal factors and traced all the steps involved.

However, different principles are thought to emerge as objects are investigated from the quantum perspective. At this deeper and finer-scaled level, both the discrete particle attributes of objects and the less discrete wave or field attributes need to be considered. It then becomes much more difficult to specify exactly what the object is and where and when it can be found.

There is thought to be an ultimate limit to precision, built into the measurement of any physical object in nature, that becomes prominent at quite tiny time and distance scales. This is related to another kind of uncertainty in quantum theory in addition to classical ignorance. This additional kind of uncertainty may crucially affect measurements of change at quite tiny time and distance scales: an inherent indeterminacy or randomness in the very fabric of nature that prevents a complete deterministic specification of events. This kind of uncertainty or indeterminism is sometimes called *quantum ignorance*.

On the classical macroscopic level where cards are played, we may be uncertain about which particular card ends up on top of the deck after shuffling. But we know precisely the probabilities involved, and we could also predict the particular card that ended on top if we knew the initial positions of the cards and carefully tracked every step. But even though the system is deterministic, it still could be that the best we can do is to make probabilistic predictions. We might not know the initial positions—sometimes called *initial conditions* or *boundary conditions*. Also, we might never be able to measure and calculate all the steps and causal influences involved.

From the quantum perspective, we also can know precisely the probabilities involved in a quantum event, even if we cannot predict exact outcomes. But both *in principle* and *in practice*, we cannot determine every event because nature itself is thought to have an indeterminate random component. In some interpretations of quantum theory, it isn't that there are hidden causal factors that determine quantum events, but that quantum events are *fundamentally* random.

It is a major challenge in modern science to *determine* whether quantum ignorance is ultimately some perhaps particularly difficult form of unspecifiability—classical ignorance—or quantum events are ultimately random and indeterminate—quantum ignorance. Further, this question forever may be impossible to answer due to the uncertainty principle, which places an ultimate limit on the precision of simultaneous measurement of certain complementary physical quantities (discussed below). Even if there were hidden deterministic processes functioning on an extremely tiny scale—perhaps even beyond our conventional notions of space and time—we might never be able to measure them precisely enough according to the framework of what is possible given our conventional classical methodologies.

The Heisenberg uncertainty principle. Quantum uncertainty or ignorance is directly related to the Heisenberg uncertainty principle. This principle applies to measurement of quantities or attributes that are complementary to each other. They sometimes also are called conjugate quantities, or *dynamic attributes*—such as position and momentum, time and energy, or the shape of a field and its rate of change.

The uncertainty principle states that the values of these complementary attributes cannot be known exactly at the same time. Their complementarity means that attempting to measure one of them exactly will change the other. Extremely precise measurement of position necessarily means less precision in simultaneous measurement of momentum, and vice versa. As measurement of position approaches exactitude, the uncertainty in the measurement of momentum would approach infinitude, according to the principle. Likewise, if we could measure the exact energy state of an object, it would take an infinite amount of time to do so.

This uncertainty relation is thought to apply to all dynamic attributes of objects in the physical world. It becomes crucial when the level of precision involves microscopic scales in which wave properties of objects also need to be considered. The uncertainty principle is related to the nature of wave mechanics at microscopic and smaller scales. In ordinary measurement of macroscopic objects, the principle applies but is not generally a concern. Measurements of a river with a mileage chart, a house with a measuring tape, or a small aircraft part with a micrometer are not precise enough for the principle to be a significant factor. However, it becomes quite significant at microscopic and smaller scales, such as reflecting light off of an elementary particle to measure it.

Specifically the uncertainty principle states that the product of the measurement of complementary attributes—position and momentum, energy and time, as well as all others— cannot be more precise than Planck's constant (6.08×10^{-22} MeV-seconds). Planck's constant—the hypothesized size of a quantum and smallest possible size of any material object—is calculated to be the ultimate limit of precision in measurement. Any physical change would require at least a quantum-size or Planck-size change or jump, and couldn't possibly be more precise than this tiny amount. This means that, according to the uncertainty principle, there is a small but unavoidable degree of imprecision built into measurement of everything in the physical world. Directly related to this measurement limitation, most interpretations of quantum theory hold that there is a fundamental randomness in the functioning of nature itself, which produces quantum ignorance or uncertainty. These are core issues in the mathematical formalism of quantum theory. Basically quantum indeterminism, fundamental randomness, and the uncertainty principle can be viewed as different ways of referring to the same phenomenon. [2]

Quantum theory was formulated in part to help explain the reason that atoms don't break down instantly, as well as how light interacts with atoms, which classical physics couldn't explain. Since its development, quantum theory has withstood all experimental tests of its predictions.

There appears to be no experimental results for which its principles have not been able to account. It is the most successful, precise, and durable theory in modern science.

At first quantum theory generated great enthusiasm as a key to deeper knowledge of *reality*. But many physicists have since concerned themselves with its practical applications and the increased control over nature it has afforded, rather than its significance as a window into a deeper, sub-phenomenal, sub-classical *reality*. This is in part due to the difficulty that the mathematical model in quantum theory does not produce predictions of the type that come from our ordinary everyday experience of the *real* classical macroscopic world.

The quantum wave function. Quantum theory describes objects in terms of wave functions—specifically the mathematical *Schrödinger equation*. Much of the theorizing about the world in physics since the development of quantum theory has been trying to work out the mysteries that seem to derive from this wave function model of nature. This research has led to the related key issue of how to interpret the mathematical wave function in terms of the concrete classical objects of our ordinary sensory experience. This issue was embedded in classical theories, but had not been recognized. This was in part due to the lower precision of measurement at the larger time and distance scales typical in classical physics. As precision greatly increased, and also the need arose to account for both particle and wave attributes of objects, the issue emerged as central to quantum theory.

As generally conceptualized in quantum theory, physical objects that appear in ordinary sensory experience to be at particular locations and times are described more fundamentally as wave packets of *potential matter*, not localized and concrete in the way they appear. Objects are conceptualized as having both particle properties and wave-field properties.

The effects of the particle nature of elementary objects can be easily demonstrated, such as when an electron creates a blip of light on a TV screen. The wave nature of elementary particles can also be demonstrated experimentally, such as in the famous double-slit experiment. This experiment is described as involving photons of light passing through a small slit in a filter and hitting a sheet of film on the other side of the filter. As single photons of light passing through the slit are recorded on the film, a pattern builds up on the film that shows the photons travel directly through the slit in pretty much a straight line (there is some diffusion of light, however) to hit the film. There is a clearly defined bright area across from the slit where the photons hit the film, and no indication of the light beam of photons hitting other areas of the film. But the situation is dramatically different when the filter has two slits that are near each other. Under this experimental set-up, bands of bright and dim areas build up on the film, showing interference effects associated with waves.

It can be demonstrated in this way that photons are not just straight-moving particles of light, but also have wave properties. Some kind of wave property of even a single photon, sort of like a cloud of energy potential, somehow passes through both slits at the same time. The observed interference patterns are thought to be due to the wave nature of light, and these patterns cannot be explained by a theory that describes light only as discrete particles.

Photons of light energy thus appear to have both particle and wave properties. When observed or measured, light demonstrates its particle nature. But depending on the experimental set-up, the evidence from patterns of the particles show that wave properties also can be inferred. All quantum objects exhibit particle and wave properties. Whenever there is a measurement, only particles are detected; but the patterns of particles show that wave mechanics are also present.

The uncertainty principle, however, may forever hide from experimental validation the answer to the question whether elementary objects are particles or waves. As particles, conceptually they are just points in space that have no size, *no extension* or volume in space. As waves, in some sense they have *infinite extension*. It is theorized that an infinite number of these abstract waves of different wavelength combine to give the appearance of boundaries, the wave packet or particle. But how the unbounded infinity as quantum waves appears in terms of the discrete boundaries of a quantized particle is a conceptual mystery.

To model the particle and wave properties mathematically, a material object made of elementary particles is conceptualized to be an infinite number of discrete quantum waves that extend throughout the entire quantum field. The mathematical quantum wave (also represented by the Greek letter *psi*) reflects all possible shapes or states of the field. The quantum waves that make up the quantum wave function are conceptualized as all coexisting and overlapping each other, called *superposition*.

A material object is represented mathematically to be a superposition of all possible shapes of the quantum field. This field is thought of in quantum theory as a mathematical field of possibilities of dynamic attributes of objects such as position and motion *when* they are measured. It is a field of possibilities that can be used to predict accurately dynamic attributes of actual objects.

It is not, however, that all the overlapping superpositions of quantum waves that mathematically describe an object have equal distribution of potential throughout the entire abstract field. The quantum wave function reflects the *probability amplitude* of an object to be at a specific time and location associated with where and when the object is measured or observed.

The quantum wave function—called a potential energy state or tendency to exist—is described mathematically in terms of a *probability distribution* that follows the laws of wave patterns in nature. Their mathematical properties are basically the same as ordinary waves of energy, such as sound waves or ocean waves. But they are generally thought of as waves of probability or possibility, rather than *real* waves of energy and matter such as ocean or sound waves. Thus they have been described as *probability waves* or *proxy waves,* in contrast to *real* physical waves.

As will be discussed in the next chapter, in most early interpretations of quantum theory the wave function was not thought of as physical in the sense of having energy or mass. It was thought to be only a conception in mathematical possibility space, used to model attributes of objects associated with patterns of movement in actual physical space. The ontological nature of quantum waves has turned out to be a major issue that will be discussed at length in upcoming chapters.

Probability distributions are the result of a large number of individual measurements made on individual events that are as similar to each other as possible. The final probability is an accumulation of the set of frequencies based on individual events. As a probability distribution, the wave function represents the non-negative percentages of likelihood of an object appearing anywhere. The quantum wave function represents a discrete quantum wave that is spread out through the entire quantum field. Each object, before it is measured as having particular dynamic attributes, is thought in some at least mathematical sense to exist throughout the entire quantum field. Thus there is theoretically at least some—even if in most locations an extremely low—probability of the object appearing *anywhere* in the field when measured physically.

Because of the inherent dynamism and uncertainty of objects from the perspective of quantum theory, the ordinary descriptors of macroscopic objects—such as location, speed, direction of motion, and causal relations—cannot be applied in the same way as in classical physics. This uncertainty in measuring time and distance in quantum theory means that classical cause-effect relationships requiring knowledge of temporal sequence cannot be specified precisely. Even further, according to quantum theory, objects don't ever touch or interact with each other directly, either spatially or temporally. The classical notion of an unbroken seamless chain of physical causation in time and space is inadequate in quantum theory.

This means that the concept of an object moving along a *path* from one location through all the intermediate points in space to another location does not completely apply. It is not that there is a perfectly smooth, seamless motion of objects along a path from one place to another that can be measured precisely (referred to in Chapter 3)—as is typically envisioned based on ordinary sensory experience and classical physics. Rather, motion in quantum theory can be explained as involving a classical sensory object appearing to be at one location and then appearing at another location, but the classical object cannot be precisely located or measured at each intervening step. If it were possible to observe an object at such a very minute scale, it is assumed to be as if the object would appear in one position, then jump to the next, and so on.

In quantum theory, the distance between two points has only a statistical meaning based on repeated measurements—not a definite, exact answer. Distance measurements at the macroscopic level are basically thought to be averaging across the hypothesized random space-time fluctuations of the quantum field. At macroscopic scales, the fluctuations are thought for practical purposes to cancel each other out. Calculations based on quantum theory yield basically the same results as the classical physics model with most common events at the macroscopic level.

Instead of specifying a precise location, time, and path of a material object, however, a physical object can be described *precisely* by specifying the probabilities of the object appearing at particular locations, times, and paths. The overall set of possibilities—the quantum wave function—can be precisely described in terms of probabilities as it changes through time. The quantum wave function is precisely deterministic through time. But the exact dynamic state of an object—position and momentum or energy at each point in time—cannot be exactly specified.

In quantum theory, the object is not definable as being in an exact state in which all of its dynamic (conjugate) attributes can be measured simultaneously. The object is in some sense a set of possibilities, a wave or cloud of probabilities or tendencies to exist in particular states. It is an overall quantum wave state or process that can be given a precise probabilistic description. But it is not exactly identifiable at the same time in terms of position and momentum or energy and time, until it is measured on a classical level associated with ordinary gross sensory experience.

Collapse of the wave function. The more abstract mathematical way of representing objects in quantum theory involves challenging puzzles about the nature of *reality*. Especially challenging is how to bridge from the representation of objects in abstract mathematical space—*probability* or *possibility space*—to actual objects in ordinary physical space and time. This relates to fundamental issues about how we model—reason—and how we measure—experience—objective *reality*. For the first time in modern physics there was acknowledgement of the necessity to account for the process of observing in objective measurements. Quantum theory remained within the framework of Newtonian classical concepts of space and time, but now needed to deal with the measurement process and the role of the observation and the observer. Quite reasonably this led to the *measurement problem*. This problem can be associated with the gap between mathematical theory and phenomenal experience of objective *reality*, between reason and ordinary sensory experience—and also between mind and matter.

> "Although there is still controversy...most physicists agree that probability is deeply woven into the fabric of quantum reality. Whereas human intuition, and its embodiment in classical physics, envision a reality in which things are always definitely one way or another, quantum mechanics describes a reality in which things sometimes hover in a haze of being partly one way and partly another. Things become definite only when a suitable observation forces them to relinquish quantum possibilities and settle on a specific outcome. The outcome that's realized, though, cannot be predicted—we can predict only the odds that things will turn out one way or another."—*Brian Greene, theoretical physicist (p. 11)* [3]

The description of objects as probability amplitude distributions is based on a mathematical model rather than on what we sense in the ordinary objective world. To get from the mathematical model of a quantum wave function to a classical material object measured as concrete at the macroscopic level of ordinary sensory experience—to get from the potential to the actual—it is posited that there must be a *collapse* of the wave function (also called state-vector reduction).

This collapse is related to the process of measurement or observation of the object. It is theorized to take place *instantaneously* in the observation process when the object or event is measured. This hypothesized instantaneous collapse of dynamic attributes of a quantum object is also associated with the concept of a *quantum jump*. The role of the process of observing and observer in collapsing the quantum wave function is implied in some interpretations of quantum theory, but there is no resolution as to how (or even whether) this collapse occurs. It is assumed that immediately after the measurement or observation, the quantum wave function in terms of its mathematical description in the Schrödinger equation takes over again until another measurement is taken, but there is no way to know for sure according to quantum theory. When the process of observing and the observer are incorporated into the mathematical model of the quantum wave function, they are incorporated as additional quantum wave functions that add more complexity (recall the dB measurement example from Chapter 1). But they don't provide means to account for how the theorized collapse of the wave function takes place.

Experimental measurements can be thought of as controlled, intentional circumstances or contexts that attempt to formalize and make more precise the process of observing. They are special cases of what is a relatively constant process of observing, significantly defined by the observer's worldview. Quantum theory doesn't address what an observation or measurement is, when it takes place, or what is happening in the ordinary world between measurements. It even can be said to bring into question, or at least suggest is ultimately undefinable and unknowable, whether there is an ordinary classical world between measurements—that is, when no one is observing it.

Quantum theory is quite accurate in predicting the probabilistic results of measurements. But it doesn't address what the natural world is like when it is not being measured, and it doesn't predict the specific classical outcomes that result from measurements. Quantum theory doesn't describe how the abstract *mathematical potentiality* links up with the concrete *material actuality* of ordinary objective *reality*.

Schrödinger's cat paradox. One example of the fundamental indeterminacy or quantum uncertainty related to the wave function model in quantum theory is the phenomenon of particle decay. Some types of atomic substances decay spontaneously by emitting energy in the form of an elementary particle, such as a photon. The average number of particles that is emitted over a given period of time can be predicted exactly. But there seems to be no way to predict when a particular individual atom will decay, or in what direction the decaying particle will be emitted—apparently due to the uncertainty of quantum ignorance or indeterminism.

According to the orthodox interpretation of quantum theory (to be discussed in the next chapter), the uncertainty is unavoidable—because the quantum wave function provides the most complete description of *reality* that is possible. No other information is available or exists. The only way to resolve the uncertainty is to make an observation, a measurement, which is thought to collapse the quantum wave function, resulting in a discrete certain outcome.

A well-known thought experiment to illustrate the problems of quantum uncertainty is the *Schrödinger's cat* paradox. One way to describe the paradox is that a cat is inside a steel box and there is no way to look inside the box. The box is connected to a Geiger counter with a small piece of radioactive material that has a 50% chance of one of its atoms decaying over the course of an hour. If the particle decay occurs, the Geiger counter will record it and then trigger the release of a lethal gas into the box. If no atom decays during the hour, the lethal gas would not be triggered, and the cat presumably would still be alive.[4]

A quantum wave function can be used to represent this system, indicating that the cat is alive and not alive with equal probability when not observed. Experimental findings of interference effects (such as light waves in the double slit experiment described above) support the theory that there is a probability wave function of possible states associated with quantum phenomena.

From the standpoint of the observer outside the box, according to the quantum wave function the cat is both alive and not alive with some probability—not one or the other. It might seem reasonable to assume that the cat is not in a smeared out state covering both possibilities; but there is no way to know whether it is alive or not until it is observed. Until then, the cat *in some sense* is both alive and not alive at some probability. The issue is how to reconcile the quantum wave probability function model with the apparent actual specific objective state of the world—in this thought experiment, the discrete macroscopic state of the cat being alive or not.

Of course it seems possible that many other things might have happened to the cat during the hour. It might have expired on its own, such as from a brain aneurysm, or tunneled spontaneously through the wall of the box to chase laboratory mice (obviously in another lab), or ported through a wormhole to another (undefined) part of the universe. There are many possibilities that could have occurred in the hour. The measurement problem in quantum theory is associated with how observing the cat changes things. How does the observation change the quantum wave function—in which the cat is *in some sense* both alive and not alive at some probability—to the specific macroscopic state in which the cat is in one state or the other? Although the commonsense classical view would assume that the cat is in one state or the other all the time, we cannot say that this is the *real* situation, because there is no way to know it. Thus the state of the cat in some sense *is* uncertain for us, and more accurately described as both alive and not at some probability.

The two perspectives—probabilistic quantum model and deterministic classical model—are not easily reconciled. But neither adequately accounts for the world. Even more perplexing, getting from the quantum model to the classical model *seems to depend on the process of measurement by a conscious observer* in collapsing the quantum wave function. This certainly would not be expected to be crucial in an objective world that is independent of the subjectivity of the observer. The uncertainty principle and quantum theory appear to identify an unavoidable uncertainty or indeterminacy in the measurement of nature. This is thought to bring into question the strict cause-effect determinism of the classical physics perspective. Also, positing a collapse of the wave function is sometimes interpreted as introducing the subjective process of observing and observer into the causal chain of events. This brings into question in the first place the fundamental assumption in classical science of an independent objective world apart from the observer.

These issues have deeply challenged even the most eminent modern scientists. Theoretical physicist Albert Einstein, for one, argued vigorously that quantum theory is *incomplete*. He believed that there must be unknown information or processes involved—sometimes called *hidden variables*. Eventually when known, these hidden variables will allow us to get beyond the probabilistic uncertainty or ignorance of quantum theory to a completely deterministic and objective account of nature. As Einstein asserted:

> "The belief in an external world independent of the perceiving subject is the basis of all natural science." (p. 201) [4]

Reflecting on the issue of quantum ignorance or uncertainty, Einstein commented:

"I cannot believe that God plays dice with the universe." (p. 199) [4]

To the notion of a collapse of the wave function upon observation, Einstein said:

"I cannot imagine that a mouse could drastically change the universe by merely looking at it." (p. 199-201) [4]

Bell's Theorem: A Crucial Test of Quantum Theory?

After years of debate, a crucial experimental test was developed that is based on *Bell's theorem* (also called *Bell's inequality*). [4] [5] Bell's theorem can be interpreted as containing several key assumptions about nature, including that objects have an objective existence independent of the observer, and that matter or any form of information cannot travel faster than the speed of light—associated with Einstein locality and the light cone. It also involves the assumption that nature can be known through logic and empirical tests—reason and experience—according to deterministic laws of cause and effect.

Experiments based on Bell's theorem might be interpreted as testing whether there is an inherent indeterminate, probabilistic component fundamental to nature or there are as yet unknown hidden variables that can explain the indeterminacy. In other words, this crucial test of quantum theory might be viewed as addressing whether quantum ignorance, or some form of classical ignorance, is involved in quantum uncertainty.[4]

Tests of Bell's theorem were conducted using actual experiments with a design similar to the famous thought experiment called the *EPR paradox* of Einstein, Podolsky, and Rosen.[6] The EPR paradox was perhaps Einstein's (and colleagues) strongest argument in an extensive debate with Neils Bohr, the father of quantum theory, about the nature of quantum *reality*. Both of them agreed that quantum theory correctly represents the results of all known experiments, a monumental accomplishment. But the question was whether it is a *complete* theory of *reality*, accounting for the mechanics of nature whether measured or not. Einstein claimed that the EPR paradox demonstrated logically that there must be unobservable hidden variables not addressed by quantum theory. The issue was long debated by physicists, many of whom eventually gave up on trying to resolve it and even on addressing the fundamental issue of the nature of *reality*. Years later, Bell developed the theoretical basis for crucial experimental tests relevant to the issue.

One example of the experimental set-up involves directing a beam of atoms into a laser light beam. The atoms absorb a photon and then give off a mirror twin pair of photons, which travel in opposite directions. The twin photons each then are aimed at a series of filters that allow only some of the photons to pass through them. The probability of a photon passing through a filter depends on the angle at which the filter is set. If one of the twin photons passes through a filter in a certain state, such as with spin up, then according to the laws of physics based on the above fundamental assumptions, this will have consequences on its twin photon if it passes through a corresponding filter—it will be spin down—with a probability determined by the filter's angle with respect to the first twin photon's filter. The angles of the filters are changed at a rapid enough rate to prevent light particles from the filters reaching each other to influence the result. Mathematical predictions based on the two models—the quantum model and the hidden variables deterministic model—then can be compared empirically.

If the behavior of the twin photons correlate highly, then it could be viewed that the second twin photon somehow instantaneously gets information—as if telepathically—from the first twin photon to calculate the correct probability of passing or not passing through its corresponding filter. But according to Einstein locality, it is not possible for the photons to transmit any information to each other if the information would have to travel faster than the speed of light between the twin photons. An alternative explanation is that the twin photons share some hidden information before they separate, which each carries as they go in opposite directions.

The question is whether it is possible for the twin photons to carry information that allows them to correlate highly with each other under any possible filtering conditions. Bell's theorem calculates that there is a limitation on what information can be carried by the separated twin photons, and that this limitation can be predicted and measured. This is thought to provide a means to compare any possible hidden variables or pre-planned information explanation with the probabilistic predictions of quantum theory.

When the predictions based on Bells' theorem were compared to the predictions of quantum theory in actual experiments, the outcomes *supported* quantum theory. [7][8][9] This is an example of the phenomenon called *quantum entanglement*—specifically phase entanglement. It refers to the highly correlated behavior of particles after they interact and separate, even when the speed of light would have disallowed them from exchanging information with each other.

It turns out, however, that the results of these crucial experiments are *not* thought of as an ultimate test of the hidden variables interpretation of quantum theory—a test of quantum versus classical ignorance. Rather, the results are understood to be a definitive test of *Einstein locality*. Bell emphasized that the actual results show that any understanding of quantum theory—and importantly any understanding of the natural world—must recognize a fundamental *nonlocality* in the fabric of nature.

The hidden variables interpretation of quantum theory favored by Einstein (EPR Paradox) was based in his strong belief in complete objectivity and in complete determinism, reflected in his assertion that God does not play dice. The test of Bell's theorem may not have addressed whether nature is completely objective and deterministic, but it is strong evidence that the fabric of nature is nonlocal. [4][10][11][12] One way of putting this is that whether God plays dice is yet to be determined, but God clearly moves in mysterious ways and is not just a local phenomenon. This momentous finding, reluctantly acknowledged by Bell and others, indicates that one of the most fundamental beliefs about the nature of *reality*—the belief that matter and energy interact only locally within the speed of light—is untenable. As theoretical physicist and string theorist Brian Greene states:

> "...Einstein, Podolsky, and Rosen were proven by experiment—not by theory, not by pondering, but by nature—to be wrong... But where could they have gone wrong? Well, remember that the Einstein, Podolsky, and Rosen argument hangs on one central assumption: if at a given moment you can determine a feature of an object by an experiment done on another, spatially distinct object, then the first object must have had this feature all along... More precisely, since nothing goes faster than the speed of light, if your measurement on one object were somehow to cause a change in the other—for example, to cause the other to take on an identical spinning motion about a chosen axis—there would have to be a delay before this could happen, a delay at least as long as the time it would take light to traverse the distance between the two objects... We are forced to conclude that the assumption made by Einstein, Podolsky, and Rosen, no matter how reasonable it seems, cannot be how our quantum universe works." (p. 113) [3]

Nonlocal communication? The classical model of causal interaction in modern science has been that all action is mediated by a continuous chain of mechanistic physical events. All causal events in nature are thought to be local interactions. Even today many modern scientists summarily reject suggestions of nonlocal action–at-a-distance influences not mediated by the four fundamental forces. Although these forces are now understood to be considerably more abstract than originally conceptualized in modern science, they are still thought to function as local influences that diminish with increasing distance.

The strongest limitation to these forces is believed to be the speed of light; movement that is faster than the speed of light—*superluminal* action—has been thought to be impossible. In dramatic contrast, tests of Bell's theorem demonstrated that nonlocal effects are instantaneous or at least not limited to the speed of light, not mediated by the known forces of nature, not diminishing with distance, and also a common feature of everyday life. [4]

What may be disconcerting for many modern scientists about nonlocality is that it opens up the possibility of 'action-at-a-distance' not mediated by the four known fundamental forces. It can be quite challenging to comprehend how distant objects can highly correlate with each other when there is no material thing—not even any relativistic space-time gravitational field—to connect them. Nonetheless, based on Bell's theorem and its definitive experimental tests, anecdotal reports of subtle nonlocal action-at-a-distance that have been so persistent in non-scientific contexts throughout history can no longer be rejected, neither on theoretical nor empirical grounds. Embedded in the experimental finding of nonlocality in quantum theory is a fundamental challenge to classical physical causality limited by the speed of light, or Einstein locality.

However, it still may be possible to accept nonlocality but not accept the possibility of nonlocal *communication* of information that implies a subtler form of causality suggested by anecdotal reports of action-at-a-distance. Many modern scientists believe that no one as yet has demonstrated nonlocal communication of information within the framework of objective methodology. Several important logical concerns argue against it, even if it were possible based on nonlocality. One concern is that the high correlations in the experimental tests of Bell's theorem demonstrating nonlocality are extremely sensitive and delicate. It is thought that if a communication message were added in some way to this delicate phenomenon, it would disrupt the correlation. Also, according to some interpretations of relativity theory, superluminal motion would imply the highly counterintuitive possibility of backward causality—going backward in time to change something that in our classical world has already happened—frequently assumed to be impossible.

In addition it is thought that nonlocal communication is impossible due to quantum indeterminism, the theorized random nature of quantum events. This is related to *Eberhard's impossibility proof*. This proof holds that if quantum theory is correct, communication signals cannot be based on the build-up of individual information messages at the quantum level. Even if there is nonlocality and even if individual messages were able to travel faster than light, it is thought that they cannot be the basis for patterns of information that serve as communication signals because each individual message is random with respect to each other. According to this proof, superluminal communication of signals that convey meaningful information is possible only if quantum uncertainty turns out to be some form of classical uncertainty.[4] However, progress in quantum information science in the past decade may eventually provide approaches to quantum error correction that could challenge some of these and related perspectives.[13]

At this point in reviewing quantum theory, there is the classical perspective of *reality* tied to ordinary experience, and the more inclusive quantum perspective tied to mathematical probabilities that predict aspects of nature the classical perspective cannot predict. From either perspective—and whether either or both are correct—it is necessary to accept that nature is nonlocal.

But there remains the key question whether nonlocal communication or information transfer is possible. This is directly associated with whether quantum uncertainty is ultimately a form of deterministic classical ignorance. It also is relevant to Einstein locality and the concept of the light cone associated with relativity theory. As discussed in Chapter 3, major issues remain in order to reconcile quantum theory with relativity theory and build a consistent theory of *quantum gravity* needed for a unified field theory.

Importantly, however, a comprehensive model of *reality* also needs to account for the process of observing and the observer. As significant as the crucial test of Bell's theorem is, in some ways even more significant is the importance of reflecting back on the process of observing and ultimately on what constitutes the observer—the knower. As will be unfolded in the next chapters, explanations of nature that don't adequately address these issues, including quantum theory, are fundamentally incomplete. If at least some levels of the natural world are nonlocal, then it could be that the mind is nonlocal also. This represents a potential basis for resolving the question of nonlocal communication—to be discussed in the next chapters.

Sensory Experience and Reason as Ontological Levels of *Reality*

The mysteries generated by quantum theory will be recognized eventually as an artifact of contextual limitations of the assumptions that define them. As these and related assumptions are reevaluated, the collapse of the wave function can also be de-mystified. The Schrödinger's cat paradox, and a well-known elaboration called *Wigner's friend,* will be discussed briefly to bring out some of the issues that the apparent paradox presents, and also to suggest an approach to resolve the issues that will be elaborated in the following chapters. A fundamentally different understanding of ontological levels of *reality* is needed to resolve the paradox.

To get from the quantum wave probability function representing the uncertain state of the cat to the discrete state that it is either alive or not—the measurement problem—it is believed that an observer must observe the cat. Orthodox quantum theory states that the conscious awareness of the observer somehow is coincident with the probability wave function collapse into a discrete state of the cat. The classical perspective assumes that the cat exists in a discrete state whether observed or not. This classical perspective—sometimes associated with the philosophy of *materialistic realism*—isn't able to account for the probability wave function and the success quantum theory has had with it. On the other hand, quantum theory cannot account for the theorized instantaneous collapse of the wave function at the time of the observation into the classically observed state of the cat. If it is instantaneous, no causal explanations even seem possible.

To add *Wigner's friend* into this paradox, suppose we ask the friend to observe the state of the cat and tell us the results. If the collapse of the wave function is via a conscious observer making an observation, then it might be asked which observer causes the collapse. Is it that our friend and the cat remained in uncollapsed probabilistic uncertainty until we observed our friend and also received the report from the friend about the cat, at which point the collapse of the wave function of the cat instantaneously occurred? If so, how come our friend's observation of the cat was insufficient to collapse the wave function?

One proposed solution to this is that there is only one consciousness in the universe. This is an alternative to materialistic monism, sometimes called *monistic idealism*. In this view, any observation by a conscious observer collapses the quantum wave function, and the paradox arises only due to assuming that our friend's consciousness and our consciousness are not one and the same. This is thought to resolve the question of what observation causes the collapse, because all conscious observations are the same and any one would collapse the wave function. This theoretical position is summarized by physicist Amit Goswami:

> "Quantum collapse is a process of choosing and recognizing by a conscious observer; there is ultimately only one observer." (p. 88) [14]

However, this proposed resolution adds another complicating distinction, namely between unitary universal consciousness and specific intentional individual awareness. In this view, it is not consciousness per se that collapses the probability wave function, but the intentional awareness or conscious attention of any particular individual observer. The distinction between consciousness and individual intentional conscious awareness is quite important and will be discussed at length in later chapters on levels of mind; it is core to understanding the nature of consciousness. But in this context, it places the power to collapse the wave function into intentional awareness of any particular observer, rather than unitary universal consciousness. Our friend's conscious observation would collapse the wave function, apparently even for us; but we wouldn't know the cat's state until he told us, so our knowledge of the cat's state would still be indefinite.

This also brings up the important issue of what actually constitutes an individual intentional awareness. What degree or quality of consciousness is required to collapse the wave function? Considering the reasonable assumption that, not only we and our friend are conscious beings but also the cat is a conscious being, couldn't the cat collapse its own wave function? If the cat's awareness could collapse the wave function, then when could the wave function not be collapsed inasmuch as it is said to collapse spontaneously and instantaneously upon any conscious observation? Wouldn't the cat be aware of its own existence whenever it was awake, and thus the wave function would have collapsed in the chamber even when we could not have looked inside it?

This brings up the additional issue of at what point is the awareness of the cat no longer capable of collapsing the wave function when the gas is released and the cat expires, but we have not yet observed it. Would the cat then return to an indeterminate state, with equal probability of being alive or not, awaiting another observer to collapse it after the cat lost consciousness and no longer could collapse the probability wave function? It also might be asked how the radioactive material collapses into a discrete particle, and how the Geiger counter functions as a classical measuring device, when there are no observers observing them to collapse their wave functions.

The *objectifying* attribution of a subjective phenomenon of the collapse of a mathematical model of the world into the objective world is the issue in these hypothetical examples. The resolution of the Schrödinger's cat paradox and Wigner's friend elaboration requires delineation of different ontological levels of *reality*, in order to avoid the initial pattern of *objectification* from which the paradox arises. The collapse of the probability wave function can be viewed as having very little to do with the outer objective domain, but rather primarily with the process of observing in the inner subjective domain of the mind of the observer.

In the act of observing and measuring, it is the observer's knowledge of the cat that 'collapses' from the probability of the cat being alive or not to the actuality of it being one or the other—not the cat. At one point we can say that what is *real* about the state of the cat *from the observer's perspective* is that the cat is alive or not at some probability. After observing it, what is *real* about the state of the cat *from the observer's new perspective* is that the cat is in only one of the two states. It is the observer's inner state of knowledge that collapses upon observation. This 'collapse' would occur with the observation, whether or not the particle was emitted in the experiment.

The understanding that the collapse of the wave function is a collapse of the uncertain knowledge state of the observer might be more obvious if we replace the cat in the box with Wigner's friend—

> "When a quantum state collapses, it's not because anything is happening physically, it's simply because this little piece of the world called a person has come across some knowledge, and he updates his knowledge... So the quantum state that's being changed is just the person's knowledge of the world, it's not something existent in the world in and of itself."—*Christopher Fuchs, theoretical physicist (p. 42)* [15]

and of course also replace the lethal gas, let's say with laughing gas. This doesn't change the situation from our perspective as outside observers. We still will be uncertain about the state of the friend in the chamber until we observe him. Then, our state of knowledge will collapse from the quantum wave function model from our perspective to the discrete state of knowing that the friend is laughing or not. However, it is quite reasonable to assume that the friend in the chamber will not be uncertain about his own state. (It is also possible that the friend will be in a state of intense laughter without the particle decay, but still we would be uncertain about it until we observed him.)

The friend would know his own state, and we as observers outside the chamber would at the same time only have the capability of knowing the uncollapsed probabilities of his state, until we observed or measured it. In this example, the friend would be simultaneously an object in a discrete state and a quantum object in a potential state. Both would have their *reality*, according to the perspective of the observers—our friend in the chamber, or us outside.

However, if any conscious observation instantaneously and automatically collapses the probability wave function, and the friend in the box is conscious and thus experiences his own state classically, then presumably his state would be collapsed for all observers, including us outside even though we have no way to observe it. According to the logic of the situation as described, our friend would be in a classical state even though we would have no way to know otherwise until we observed it. Until we observed our friend, he would be in an uncertain state from our view.

From another angle, let's suppose this time that the chamber included a clock, set up to stop when the cat expired. But we as outside observers would have no access to the clock until opening the chamber and directly observing the cat and the clock. The collapse of the wave function into a discrete state of the cat being alive or not theoretically would occur upon our observation. But the recording device might well have recorded an earlier time of expiration of the cat. When did the collapse of the wave function occur? Would the clock not stop until we observed it, and then instantaneously record a prior stop time upon our observation of it? Would the clock also not be a classical measuring device that could record anything but probabilities until observed? Although perhaps a bit tedious, hopefully these examples are useful for pointing to the understanding of the 'collapse' of the wave function as a change of state of knowledge in the mind of the observer, for the most part not causally related to the state of the cat. This is important in establishing the basis for a resolution of not only Schrödinger's cat paradox but fundamental paradoxes about the nature of *reality* that quantum theory seems to pose.

Different interpretations of the theorized collapse of the wave function are discussed in the next chapter. Some may consider the approach discussed here—that it is a collapse of the knowledge state of the observer—to be associated with the orthodox interpretation of quantum theory. [25] However, this knowledge state approach to the collapse will turn out to be quite different than interpretations associated with orthodox quantum theory. The early orthodox interpretations of quantum theory, from which the Schrödinger's cat paradox appears to have originated, theorize that the wave function collapses upon observation by a conscious observer. Through time this interpretation has morphed somewhat into an interpretation that sounds more like the description here that the collapse is a 'collapse' of the knowledge state of the observer not causal with respect to the state of the object—cat, friend, clock, or whatever. This is exemplified in the following explanation.

The probabilistic knowledge associated with the quantum wave function is sometimes now explained to mean that if we were to take 32 boxes with a cat in each one and wait for five half-lives of the radioactive material (in our example, five hours), then upon opening the boxes we will find on the average only one live cat. Even though it is said that the wave function collapses upon observation—that is, it doesn't collapse until the box is opened and an observation is made—it is now explained that the observations will show on the average that one cat survived. In this view, what *really* happened inside the box when we couldn't observe it are 'classical' questions that are not appropriate because the answer to them cannot be known. However, it is interesting how this view is more toward separating the collapse of the wave function upon observation from the state of the cat. In other terms, there is less objectification of the mathematical wave function onto the classical world, associating it a little more with the subjective state of the knower—while still unfortunately not making the leap to recognize fully that the state of the cat is an objective phenomenon that is not due to the collapse of the wave function upon conscious observation.

But does the perspective of wave function collapse as a 'collapse' to a discrete state of knowledge in the observer somehow ignore the issue of quantum indeterminism? In these examples, quantum indeterminism is associated with the random emission of an elementary particle. Although this contributes to the uncertainty associated with the quantum wave function, it can be viewed as a separate issue with respect to the part of the quantum wave function said to collapse upon observation. Even without the probabilistic uncertainty established by the experimental set-up of the particle emission, we would still be uncertain about the state of our friend or the cat in the chamber prior to our observation. The collapse of the wave function due to the observation would collapse the state of knowledge within the observer's subjective experience, regardless of whether the particle is emitted or was even part of the experiment. The collapse due to observation does not causally affect the particle emission, and vice versa. The observation can be said to collapse the wave function in the sense that it changes the observer's knowledge, but not necessarily in the sense of the actual state of the cat as alive or not, or in-between.

Of course these examples are thought experiments. In the *real* world, the issue is thought to be that in the case of the probabilistic wave function of an elementary particle there is no other possible observer, such as the friend, clock, or cat in the box. Under these conditions, it is thought that there is no way to know anything more than the probabilities known to the outside observer. All observations are from an outside perspective, and the discrete state of the world cannot be known apart from making an observation—and thus there is no way to know whether there is a discrete state of the world apart from the observation. In quantum theory, there is no way to write a wave function that includes the observer in the system.

But also, it is important to recognize that without a classical observer that can make a conscious observation, there would be no box, no radioactive material, no Geiger counter, no laboratory—and no mathematical quantum wave function models either. It seems reasonable to assume that a classical world, required for observers to have nervous systems and brains upon which consciousness is believed to depend, would also have to exist in order for probability wave function models to be conceived. This would seem to be the case even though the probability wave functions cannot account for the classical world by themselves, so to speak, because they are just random processes that are mathematical possibilities and are not classical actualities. From another angle, there would seem to be no way to ask either a classical or quantum question without a classical observer. In quantum theory, a classical conscious observer is needed to collapse the wave function; but getting to a conscious observer first requires a classical world—a problem of circular reasoning. This is suggestive that if the model of the collapse of the quantum probability wave function is correct, consciousness cannot be a product of brain processes in the classical world.

A more comprehensive and integrated understanding of the observed, process of observing, and observer that involves levels of *reality* is needed to address these paradoxes. There is some level of *reality* associated with the object as discrete and independent, as a probability wave, and as the inner subjective knowledge of the object's state. The collapse of the wave function is associated with the inner state of knowledge of the observer, and becomes paradoxical when the observer's state of knowledge is *objectified* onto the wave and particle attributes of the object. The quantum wave function is a mathematical *reality* based on reasoning, and the discrete state of the object is a sensory *reality*. The inability to reconcile these two levels of *reality* leads to the measurement problem and the paradox of the instantaneous collapse of the quantum wave function by the conscious mind of the observer. These points will be elaborated in the following chapters, which explore different interpretations of quantum theory. A major underlying theme is progress toward the ontological existence of quantum reality, as well as mental reality. Perhaps it is reasonable at this point to propose that a rational and logically consistent understanding of physical *reality* requires an understanding of subjectivity and conscious mind. One might eventually go so far as to propose that physics is a part of a more inclusive science of mind.

Various interpretations of quantum theory have developed that concern the issue of whether the stamp of *reality* is best attached to the ordinary world we observe—sensory experience—or theoretical descriptions of it in the mathematical framework of quantum theory—reason—which we don't directly observe with the senses. Are classical descriptions of what we ordinarily see as objective *reality* more or less *real* than quantum descriptions that on some level are more accurate and comprehensive? How do our models of *reality* relate to the process of observing and the observer? Are there degrees of *reality* that depend on the level observed, or that depend on the level of the observer? Major interpretations of quantum theory, which reflect different perspectives on these questions, are the focus of Chapter 5.

Chapter 4 Notes

[1] Nader, T. (2000). *Human physiology: Expression of Veda and the Vedic Literature*. The Netherlands: Maharishi Vedic University.

[2] Klauber, R., (2004, May). Personal communication, Fairfield, IA.

[3] Greene, B. (2004). *The fabric of the cosmos: Space, time, and the texture of reality*. New York: Alfred A. Knopf.

[4] Herbert, N. (1985). *Quantum reality: Beyond the new physics*. New York: Anchor Books.

[5] Walker, E. W. (2000). *The physics of consciousness*. Cambridge, MA: Perseus Books.

[6] Einstein, A., Podolsky, B., & Rosen, N. (1935). Can Quantum-Mechanical Description of Physical Reality be Considered Complete? *Phys. Rev. 47*, 777-780.

[7] Clauser, J., Horn, M., Shimony, A., & Holt, R. (1969). *Phys. Rev. Lett., 26*, 880-884.

[8] Aspect, A., Grangier, P., & Roger, G. (1982). Experimental Realization of Einstein-Podolsky-Rosen-Bohm Gedankenexperiment: A New Violation of Bell's Inequalities. *Phys. Rev. Lett. 49*, 91.

[9] Aspect, A., Dalibard, J., & Roger, G. (1982). Experimental Test of Bell's Inequalities Using Time-Varying Analyzers. *Phys. Rev. Lett. 49*, 1804.

[10] Bell, J. S. (1964). On the Einstein-Podolsky-Rosen Paradox. *Physics*: 195-200.

[11] Bell, J. S. (1987). *Speakable and unspeakable in quantum mechanics*. Cambridge: Cambridge University Press.

[12] Goldstein, S., Bohmian Mechanics. (2002) *The Stanford Encyclopedia of Philosophy* (Winter Edition), Edward N. Zalta (Ed.), URL = <http//plato.stanford.edu/archives/win2002/entries/qm-bohm/>

[13] Nielsen, M. A. (2003). Simple Rules for a Complex Quantum World. *Scientific American*, Special Edition, May.

[14] Goswami, A. (1993). *The self-aware universe: How consciousness creates the material world*. New York: G. E. Putnam's Sons.

[15] Folger, T. (2001). Quantum Shmantum. *Discover*, September.

Chapter 5

Perspectives on Quantum Theory

In this chapter we will consider some of the major interpretations of quantum theory as an investigation into the foundations of reality. Included are the Copenhagen or orthodox, duplex, consciousness-creates-reality, quantum logic, many-worlds (many mind-worlds), continuous spontaneous localization, consistent histories, decoherence, and neorealist interpretations of quantum theory. We will identify where the interpretations don't adequately address the process of observing and the observer, and how a more expanded approach of levels of reality begins to address the measurement problem and the related issue of the relationship of matter to mind. This is based on a more holistic conception of space or existence. The overall main point of this chapter is that interpretations of quantum theory reflect the explanatory gap between matter and mind—the mind-body problem—and need an expanded ontology of nature that is inclusive of gross, subtle, and transcendental levels of reality.

The quantum wave function and its collapse into a discrete classical state in the measurement process have been core concepts in quantum theory. Implicit in these concepts is nonlocality, for which the test of Bell's theorem explicitly provided strong evidence as a fundamental feature of nature. These issues have prompted major reevaluations of fundamental beliefs in modern science. The nature of space and time, relativity, determinism, and even *objective reality* are being considered somewhat more holistically in this research. This is reflected in interpretations of quantum theory briefly described in this chapter, as well as theories of quantum gravity in the next chapter. These interpretations generally can be looked upon as attempts to characterize that elusive phenomenon of *reality*. They are products of not recognizing the ontological *reality* of mind and consciousness and their relationship to matter.

The interpretations of quantum theory described in this chapter don't exhaust the array of interpretations, but do include the major ones. All of them are generally consistent with the accepted evidence, but additional evidence that discriminates between them has yet to be produced in the general framework of modern physics. The interpretations are outlined here in simplified summaries that give no sense of their formal mathematical bases. In these summaries, the interpretations are reminiscent of historical attempts to address the *mind-body problem*. They have similarities to philosophical concepts such as realism which denies the fundamental *reality* of mind, idealism which denies the fundamental *reality* of matter, and forms of psychophysical dualism and parallelism which accepted both as *real* but didn't explain how they interact.

One task in this major area is to identify what attributes of objects can be understood to exist objectively and independently—in the objects themselves—and what attributes are relational consequences of the act or process of measuring them. In Newtonian classical physics the entire universe contains *ordinary* or *classical* objects that have static or structural attributes—such as spin, charge, and mass—as well as dynamic attributes—such as position and momentum—that are intrinsic to the objects. These attributes are viewed as independent of observation and measurement. This divides nature into two fundamental parts that modern science has not yet been able to integrate: the objective world and subjective experience of it.

In most interpretations of quantum theory, on the other hand, *quantum objects* have intrinsic static attributes that are the same whether measured or not, but their dynamic attributes are a product of interaction with the particular measurement circumstances. This is a core issue in the *reality* crisis arising from the *explanatory gap* between ordinary classical experience and quantum theory (refer to Chapter 4). It is part of the contemporary effort to address the nature of objective physical *reality* and its relationship to subjective mental *reality*. More generally, it concerns what *reality* is—which is changing with emerging recognition of deeper *levels of reality*.

"An essential aspect...of things in physics is that they lay claim, at a certain time, to an existence independent of one another, provided these "objects" are situated in different parts of space... Unless one makes this kind of assumption about the independence of the existence of (the "being-thus") of objects which are far apart from one another in space—which stems in the first place from everyday thinking—physical thinking in the familiar sense would not be possible. It is also hard to see any way of formulating and testing the laws of physics unless one makes a clear distinction of this kind. This principle has been carried to extremes in the field theory by localizing the elementary objects on which it is based and which exist independently of each other, as well as the elementary laws, which have been postulated for it, in the infinitely small (four-dimensional) elements of space."—*Albert Einstein, theoretical physicist (p. 154)* [1]

"[T]he realist naturally thinks that there is a distinction between our ideas or concepts and that which they represent, namely, the world. The ultimate court of appeal for judging the validity of our representations is this independent world... The idealist, on the other hand, quickly points out that we have no access to such an independent world except through our representations. We cannot stand outside of ourselves to behold the degree of fit that our representations might have with the world. In fact, we simply have no idea of what the outside world is except that it is the presumed object of our representations. Taking this point to the extreme, the idealist argues that the very idea of a world independent of representations is itself only another of our representations..."—*Francisco Varela, cognitive neurobiologist, Evan Thompson, neurophilosopher, and Eleanor Rosch, cognitive psychologist (p. 161)* [2]

At what point in the measurement process do objects change from *possibilities* to *actualities?* Where is the dividing line between quantum *reality* and the ordinary classical *reality* that is the outcome of measurements we ordinarily observe? Are ordinary physical objects more or less *real* than quantum objects? Attempts to address these questions, described in this chapter, reflect growing appreciation of the fundamental role of mind and consciousness in determining the nature of *reality*. Progress in quantum theory reflects movement toward recognizing the ontological *reality* of the quantum level. It brings us closer to unified field theory, from which key planks to connect modern science to Maharishi Vedic Science and Technology can be put in place.

However, many physicists seem to have opted out of the *reality* game. While the success of quantum theory suggests it addresses something *real* about nature, *reality* questions are frequently considered impractical because it is believed that *real* progress cannot be made on them scientifically. Perhaps most of these scientists consider themselves to be pragmatists, viewing quantum theory as a calculation tool or recipe to connect measured physical phenomena with mathematical representations, useful for solving technical problems with *real* practical applications. Others go further, considering *reality* questions a complete waste of time because they are unanswerable. Among this category are strict positivists, who feel that science is only about phenomena that are empirically verifiable through the ordinary senses. Firmly nestled in an implicit physicalist worldview, they frequently are not concerned about interpretations of the nature of the quantum world underlying the classical world, and consider it vain philosophical speculation. Other positivists believe that although *reality* questions may be unanswerable, they present useful puzzles that serve as means to advance scientific theories.[3]

There is, however, appreciation among some physicists of the significance of quantum theories for understanding the nature of *reality*. These scientists generally recognize that the foundations upon which the theoretical scaffolding or edifice of modern science and technology is built are not nearly as solid as might be expected given its widespread acceptance in modern civilization.[4] They appreciate that the theoretical and pretheoretical foundations of modern science are fragmented and on shaky or jittery ground, but also frequently tend to be hesitant to examine them. This is in part due to the apparent complexity of the issues and the inevitability of having to address fundamental questions about the nature of subjectivity that many feel are best avoided. These issues directly involve the role of the process of knowing and knower, mind and consciousness. They require modern science to go beyond the fragmenting objectification of *reality* familiar to the objective means of gaining knowledge. We at least now know enough about these issues to recognize that interpretations of quantum theory have to address them directly. They are fundamental to the measurement problem and the collapse of the quantum wave function. This chapter exemplifies progress in the new physics toward understanding the necessity of addressing mind and consciousness, the process of observing and observer, in theories of objective *reality*.

Brief Overview of Major Interpretations of Quantum Theory

It still may be the case that the majority of physicists who express interest in *reality* adhere to some version of what is called the *orthodox* interpretation of quantum theory. As will be seen, this interpretation has had close affiliation with pragmatic and positivistic views. Most of those who believe in this interpretation generally have favored the *Copenhagen* interpretation associated with Neils Bohr and Werner Heisenberg, while some others connect more with the related but more integrative interpretation stemming from the work of von Neumann.[5]

Copenhagen interpretation. Historically the Copenhagen orthodox interpretation of quantum theory emphasized the phenomena of ordinary sensory experience as the only actual *reality*. It asserts that there is no deep underlying quantum *reality*, and that quantum theory is not a representation of such a *reality*. Quantum theory is a probabilistic model of the relationships between the actual objects and processes of ordinary observed *reality* and the quantum concepts used to understand them; it is a way of explaining the results of experiments. As theoretical physicist Neils Bohr put it: "There is no quantum world... [T]here is only an abstract quantum description." (p. 22)[5]

Reality consists of the phenomena of ordinary experience, created by observation, and there is no deeper underlying quantum *reality*. At the quantum level, no additional explanation of the underlying mechanics of nature is needed because there are no underlying mechanics. At this level, quantum indeterminism takes over. Quantum theory is a mathematical tool for calculating relationships between abstract quantum possibilities and the concrete *real* phenomena of ordinary sensory experience. It is not an explanation of quantum *reality*, because no further explanation is needed or even possible. This might be termed a type of *phenomenal realism*, in that it emphasizes ordinary classical phenomenal experience as the only *reality*. In this sense it also can be associated with a type of *materialistic monism*, or *scientific realism* or *naturalism*, based on ordinary sensory experience only. But it is not classical realism in the sense that all static and dynamic attributes of objects inhere in the objects themselves independent of measuring them. In the Copenhagen framework, dynamic attributes of objects are a product of the process of measurement or observation and do not inhere in the objects prior to observation. They become phenomenally *real* objects only through the process of observation. As we will see, however, in recent years orthodox interpretations of quantum theory have been progressing toward a type of *idealism,* as well as different types of *dualism,* depending on what is emphasized in the interpretations. [67]

According to the original orthodox Copenhagen interpretation, quantum theory is a complete theory. It emphasizes quantum uncertainty associated with the uncertainty principle and quantum ignorance. It provides statistical or probabilistic results not because there are yet unknown hidden variables that will eventually allow a completely deterministic account—as Einstein believed—but rather because nature is fundamentally indeterminate and random. The quantum wave function represents completely the physical state of a single quantum object. Quantum uncertainty is not thought of as *caused* by the limit of precision of Planck's constant—the smallest amount of change possible—or by quantum indeterminacy or randomness. There are no specific underlying dynamic attributes to be certain or uncertain about prior to measurement. They emerge in the interaction with a measurement device in the measurement process, and they don't exist outside of the interaction.

Each experiment or measurement must be taken as a whole, and the dynamic attributes of quantum objects are a product of the entire interaction. The dynamic attributes can be measured precisely. But the free choice of attribute to measure ensures that the value of its complementary or conjugate attribute will be completely uncertain in the undivided wholeness of that particular experimental set-up, consistent with the uncertainty principle.

This reflects a major advance toward acknowledging the process of observing or knowing as important to physical *reality*—due to the role of the measurement or observation process, in contrast to the classical physical view. In this interpretation, it isn't quite that quantum objects such as electrons or photons don't exist at all. Rather, they don't have dynamic attributes—position, momentum, and so on—until measured. An elementary particle's dynamic attributes come about through the relationship between it and the probe or measuring device, and change as this relationship changes. According to Bohr: "Isolated material particles are abstractions, their properties being definable and observable only through their interaction with other systems." (p. 161) [5] As theoretical physicist Werner Heisenberg stated it: "Atoms are not things." (p. 162) [5]

From this perspective, it is fantasy to think that an individual electron is heading toward a TV screen to create a particular blip on the screen. Before the observation or measurement that is recorded as a blip, the electron as a quantum object has no *reality* other than its theoretical possibility of existence. If we have to imagine it as existing and traveling somewhere prior to measurement, it would be as if it were traveling everywhere in all directions simultaneously.

In this view, the process of observation somehow creates what quantum objects appear to be like as classical objects by the attributes measured and the corresponding measuring device. Measuring devices are *real* classical ordinary objects, and thus are not just quantum wave functions. They are not undergirded by more fundamental independent quantum objects in space and time, which rather are just mathematical probabilities. Commenting on this interpretation of quantum theory as proposed by Bohr and Heisenberg, theoretical physicist Henry Stapp explains:

"Their theoretical structure did not extend down and anchor itself on fundamental microscopic space-time realities. Instead it turned back and anchored itself in the concrete sense realities that form the basis of social life. This radical concept, called the Copenhagen interpretation, was bitterly challenged at first but became during the thirties [1930's] the orthodox interpretation of quantum theory, nominally accepted by almost all textbooks and practical workers in the field." (pp. 162-163) [5]

A central concern with this interpretation is that it divides nature into two disjointed sections: discrete observed phenomena identified as *real,* and indefinite *virtual* quantum probability tendencies to exist that are not discrete or *real.* But it cannot make specific non-probabilistic predictions about the *real* phenomenal world. The division into *virtual* quantum 'objects' and *real* classical measuring devices is subject to the concern brought up in Chapter 1 in the example of the chain of wave functions. If a measuring device became the object of measurement, it would be represented by a quantum wave function, and then the device used to measure it would be considered a classical object. In turn this second measuring device could also become an object of measurement and then would be represented as a quantum wave function, and so on in an infinite regress. The measurement problem and the problem of the collapse of the wave function—how to get from *virtual* probabilities to discrete *actualities*—are inherent in this view.

In addition, those parts of nature that are not being measured—the overall experimental arrangement including measuring device, act of measurement, and also measurer—cannot be examined because it is assumed they have a special status as *real* objects for which quantum principles don't apply. Thus much of the world is placed inside measurement acts and devices not accessible to being examined quantum mechanically. How they got into this classical state is unclear.

Also, it is not clear how and where to draw the line between *real* phenomena—classical objects that apparently exist whether being observed or not, such as cats or friends—and quantum 'objects' that become actual discrete phenomena only in the process of observation. Some adherents to this interpretation go so far as to say that nothing is *real* until it is observed (similar to the 'tree falls in the forest and no one hears' paradox). This implies that much of what becomes the universe is indefinite and virtual—as if waiting for a measurement to be actualized into *reality.*

Moreover, it impels the question of what constitutes an observation or measurement. Some say that the act of making a record—a film recording, needle mark on a scroll, or any automated scoring system—counts as an observation. This suggests that a collapse to a discrete outcome can occur with simply a measuring device as the observer. What it is about the measuring device that compels the collapse of the wave function is not addressed. It only asserts that the collapse occurs, and gives no justification that there is a collapse that occurs instantaneously at the time of a measurement, or how it might occur. It also doesn't address what is in the measuring device that makes it a *measurement* device, or what initiates or intends the measurement act that then compels the collapse to occur, as well as what or who establishes the experimental conditions that at some point constitute the measurement act. If any event can be said to be measured or observed whether or not a conscious observer is involved, then what distinguishes an observation from events not observed? How come all events don't involve a collapse of the wave function, whether or not a conscious observer with intention to observe is involved? A classical measuring device first requires a classical observer to create it and give it the intent or purpose to measure.

Viewing it in the context of *levels of reality,* this orthodox interpretation of quantum theory focuses attention on the ordinary sensory level and away from the inner process of observing and observer. In typical positivistic style, it tucks away mind and consciousness—as well as all the related core issues—in the inaccessible black box. However, it does reflect some acknowledgement of the process of observing or knowing in creating phenomenal *reality.* It embodies a more explicit level of recognition than classical physics that the conscious mind—in the act of measurement—is involved in phenomenal experience as the ultimate observer of *reality,* even if the conscious mind or observer is considered unexaminable. This can be seen as a major step of progress toward recognizing the ontological *reality* of mind and consciousness and their fundamental importance in the epistemology of modern scientific theories that attempt to explain *objective reality.*

Duplex interpretation. An offshoot of the Copenhagen interpretation of quantum theory is the *duplex* model associated with Werner Heisenberg, [5] in which *reality* consists of both potentialities and actualities. In this interpretation, which might be described as a bit closer to a type of dualism, the unmeasured world is *real* in the sense that wave functions describe *real unrealized* possibilities. They are *sort of real* tendencies to exist, simultaneous potentials that change in accord with precise laws represented by the quantum wave function. The act of measurement makes them *real* discrete events in time and space generally associated with ordinary sensory experience. It can be viewed as another small step toward quantum *reality.*

However, quantum *reality* doesn't exist in the same sense attributed to actual ordinary phenomena in classical time and space. Nothing actually happens in quantum *reality*, other than changing tendencies to exist. Quantum 'objects' follow precise probabilistic laws that somehow result in phenomenal experiences of the *real* world, but only in the context of a particular observation or measurement. When not being measured, quantum 'objects' have a virtual, semi-*real* existence as possibilities, called *potentia.*

The duplex interpretation emphasizes the measurement act as the bridge between the *semi-real* possibilities or potentia and the actual *real* phenomena of experience. These two aspects are accepted literally as how nature is. The world not being measured is *actually* a superposition of potentia containing many possibilities. Such potentia don't exist in the collapsed ordinary world realized into actualities through the process of measurement. The measured world is made of *real* phenomena, including *real* objects and *real* measurement acts. Heisenberg noted:

> "In the experiments about atomic events we have to do with things and facts, with the phenomena that are just as *real* as any phenomena in daily life. But the atoms or the elementary particles themselves are not as *real*; they form a world of potentialities or possibilities rather than one of things or facts." (p. 26) [5]

This interpretation of quantum theory has many of the same unresolved issues as the Copenhagen interpretation. But it does go another bit further toward attributing *reality* to quantum phenomena. It also recognizes the role of the process of observing or measurement—which involves mind and consciousness—as the process that relates *real* potentia to *real* objects in nature.

Consciousness-creates-reality interpretation. In part due to dissatisfaction with the division of classical objects and quantum potentialities in the Copenhagen interpretation, an alternative approach developed that is associated primarily with John von Neumann. [8] It is different enough from the Copenhagen interpretation that it might better be classified as non-orthodox.

This view is more directly reminiscent of idealism than the Copenhagen view, but epistemologically it is a type of dualistic interactionism. It conceives of the entirety of *reality*—not just the act of measurement and its discrete results—as an undivided whole, called *quantum wholeness.* This is accomplished by identifying both quantum objects and classical objects such as measuring devices—*all* aspects of the entire measurement arrangement—as quantum wave functions. If mathematical quantum probabilities can't seem to make a logical quantum jump into classically deterministic actualities, then perhaps classical *reality* can be placed in quantum *reality.*

This conclusion developed in an attempt to define more clearly the point at which the collapse of the quantum wave function takes place. Von Neumann logically considered the potential points from the quantum object to the observer's consciousness where the knowledge of the discrete result as classical objects would ultimately be observed. In carefully evaluating each link in the chain of events of the measurement process to identify where the collapse could take place, he found no particular point that is favored by quantum theory. The quantum jump or collapse could be placed anywhere in the chain of quantum wave functions, but could not be eliminated.

In this interpretation, every process in this chain is a quantum process. A major challenge is to explain how, in the collapse to a discrete result, individualized and qualitatively different events are produced from indistinguishable random quantum processes. Von Neumann reluctantly concluded that the only location where the necessary collapse or transition would have to take place was where the chain ultimately stopped—namely, the conscious mind of the observer. The only place where the processes presumably were no longer just quantum probabilities—by virtue of at least being observed or appearing to be *real* classical phenomena—was where events in the environment and in the brain somehow become a discrete experience in the conscious mind of the observer. The collapse must take place somewhere in order for actual discrete objects to be experienced, and there was no other logical place to assign it. This identifies the crucial factor in the measurement problem and collapse of the wave function as the black box of the conscious mind. As theoretical physicist Eugene Wigner noted:

> "It is not possible to formulate the laws of quantum mechanics in a fully consistent way without reference to the consciousness... It will remain remarkable in whatever way our future concepts may develop, that the very study of the external world led to the conclusion that the content of the consciousness is an ultimate reality." (pp. 25-26) [5]

This interpretation of quantum theory has similarities to Feynman's *multiple histories* or *sum-over-histories* mathematical model. This model calculates a quantum object—the quantum wave function—as occupying all possible paths of motion at the same time in superposition. In von Neumann's all-quantum *reality* interpretation, the act of measurement to obtain a single discrete result collapses almost all of the possibilities. But the quantum object always maintains some possibilities, presumably at least to the degree of Planck's constant and the Heisenberg uncertainty principle. All objects remain quantum objects even during and after measurement. In the measurement act, the possible values of the dynamic attribute being measured become squeezed down into a phenomenally experienced discrete result. But again, consistent with the uncertainty principle, by contracting the value of an attribute to a discrete state, simultaneously the value of its complementary attribute becomes infinite. [5] In what has been called *von Neumann's proof,* [8] he showed mathematically that if electrons are ordinary classical objects with intrinsic dynamic attributes whether the attributes are observable or not, their behavior must show differences from the probabilistic predictions of quantum theory. If quantum theory is correct, the theory that *reality* is made of ordinary classical objects is mathematically impossible, according to this proof. [5]

A necessary assumption of the view that there is only quantum *reality*—no *real* classical objects—still would seem to be that quantum objects and the collapse of the wave function in the act of measurement are part of the causal chain. We need to get to apparent physical causation at some point. The only place where such a change could take place is theorized to be the unexamined realm of the conscious mind of the observer. But this is also comprised of nothing but indistinguishable quantum wave functions, if everything is quantum *reality*. We seem to have a paradox that while everything is quantum *reality,* the conscious mind of the observer needs to function classically in order to collapse the wave function. We still seem to have some kind of division between quantum and classical *reality*. Assuming that consciousness is something different from just potentia or proxy waves at least in that it has some power to collapse the quantum wave function in a manner that other proxy waves cannot, then it also impels the question of what constitutes conscious awareness and a conscious observer. Can a computer develop into a conscious observer, can a bacterium, mouse, cat, or does it require a human mind? Prior to the evolution of a level of consciousness competent to serve as an observer—such as possibly human organisms with language and self-reflection according to some definitions of consciousness—did the universe exist only as quantum wave functions or proxy waves with nothing definite taking place? How could, for example, Darwin's evolutionary stages that supposedly led to the development of conscious beings take place in a world inhabited only by random indistinguishable proxy waves because there were no creatures developed enough to have consciousness (according to some popular definitions)? Relevant to this as well as the other orthodox interpretations, how could a wave function collapse due to conscious observation before conscious observers existed?

This interpretation of quantum theory certainly is more direct in acknowledging the *reality* of conscious mind, more explicitly identifying it as the crucial locus where wave function collapse occurs. Conscious mind is identified as the only site of wave function collapse. But at the same time, the ontological nature of the conscious mind of the observer remains unexamined. The conscious mind is viewed as a separate and necessary part of nature (Process 1) in its influence in the collapse of a separate physical part of nature that functions according to quantum mechanics (Process 2); these two processes interact, but are fundamentally separate aspects of nature.

Further, no distinction is made in this interpretation—and the others as well—between mind and consciousness. It thus might be identified more appropriately as a *mind-creates-reality* interpretation of quantum theory. It has the same concern mentioned before that where there is no conscious observer there would seem to be no concrete phenomenal world. This might be resolved if consciousness were associated with some universal observer that is not limited to individual observers or individual minds—such as might be conceived in a non-individualized *monistic idealism* (see Chapter 4). But then wouldn't it create the collapse of the wave function all the time and everywhere, if there is an actual collapse of some kind? In this case, how could there be an uncollapsed quantum *reality,* or how could consciousness sometimes collapse the wave functions and sometimes not? Addressed in another way, if all things are quantum wave functions, quantum potentia, how do we *ever* get a classical observer? The only thing theorized to have the power to collapse the wave function is the conscious observer, and classical brains are theorized to be necessary for consciousness; but some wave function has to be collapsed first in order to get any classical object, including a conscious observer who can collapse the wave function.

Such questions remain unanswered in this and the previous interpretations, due to not understanding the ontological nature of mind and consciousness. They are indicative of the fragmented, incomplete approach of attempting to build an objective view of the world focusing only on the observed—the outer objective domain—and not also directly addressing the inner subjective domain associated with the process of observing and observer, mind and consciousness. The consciousness-creates-reality' and Copenhagen interpretations certainly reflect acknowledgement of the role of mind and consciousness in attempting to explain *reality*, but like most other interpretations, don't address them directly—this is brought out more in the next two chapters.

Quantum logic interpretation. Another interpretation, also drawn from the work of von Neumann, concerns *quantum logic*.[9] This approach asserts that because quantum objects have logically incompatible dynamic attributes—complementary or conjugate quantities—a different logic is needed for them. The new basis of quantum logic was proposed for understanding how dynamic attributes combine in ways that seem illogical from the perspective of classical logic.

Classical objects follow *Boolean* linear logic; but quantum objects conform to a non-Boolean, non-linear logic associated with wave mechanics. The dynamic attributes of quantum objects are different from classical objects in that they are not all measurable at the same time. The complementary attributes follow a logic that is not based on the familiar classical relationship between part and whole. They don't have discrete parts that combine in a single way, as do discrete events in classical *reality*. They follow less discrete wave mechanics that don't have specific parts and can be divided up arbitrarily in many different ways according to how they are analyzed or measured. Specifically these dynamic attributes of quantum objects don't follow the classical distributive law—that is, mathematically, they are *non-commutative*. These are additional ways of saying that the dynamic attributes are complementary attributes and cannot be measured at the same time within the framework of classical Boolean logic. However, embedded in quantum logic seem to be principles of sequence, directionality, and even hierarchical structure, which are more explicitly brought out in interpretations of quantum theory associated with the principle of decoherence—to be discussed shortly.

In quantum logic, quantum objects generally are sort of *real* objects with intrinsic dynamic attributes, but these attributes must combine in non-classical ways. This approach addresses the measurement problem by placing it in the non-classical framework of quantum logic. It is another angle on the same issues, but it doesn't give an adequate basis to explain how classical logic seems to work most of the time in our ordinary experience if nature is fundamentally non-classical. It makes an important contribution to quantum theory, but is not a comprehensive interpretation. There are also several variants of this approach, beyond the degree of detail in this book.

Many-worlds (mind-worlds) interpretation. Initially developed in the 1950s in the work of Hugh Everett, [10] the many-worlds interpretation takes *idealism* to a completely different level. Whereas the Copenhagen interpretation didn't directly acknowledge quantum *reality* at all, the view of the many-worlds interpretation is that uncountable numbers of quantum *realities* are generated in every measurement act.

This interpretation is generally consistent with von Neumann's all-quantum *reality*, in which everything is represented by proxy waves. However, it posits that the possibilities *don't ever collapse* but rather become alternate parallel worlds that continue as separate *realities*, each on their own without any ability for them to interact with each other. This approach, sometimes taken to mean that a multitude of actual *real* parallel worlds is created, is perhaps better understood from a heuristic perspective as a *many mind-worlds* interpretation. Rather than actual physical parallel worlds, which would violate the law of conservation of energy, there is one very big world with many parallel mind-worlds in it. Of course this implies that all the parallel mind-worlds other than the one the observer appears to be in don't have physical energy; and it raises the question of how the observer's mind-world happens to have it, but apparently not for other observers.

> "[T]he mind should not multiply entities beyond necessity. What can be done with fewer...is done in vain with more."—*William of Occam, 14th Century philosopher* [11]

Quantum theory states that a wave function represents all possible values of a quantum object. In the many-worlds—or many mind-worlds—interpretation, the wave function does not collapse upon observation. The entire world at all of its levels remains in a superposed state of all possibilities. Rather than collapsing into a discrete state, each possibility branches off into its own parallel mind-world. Each of these parallel mind-worlds presumably is somewhat like the world you are experiencing now, but it is based on a different possible outcome of each probabilistic event, and also presumably it would not have any actual physical objects with mass and energy in it. Superposition is taken to be at *the level of the mind*, creating an infinity of simultaneous worlds, each with its own mind associated with one of the possible outcomes. Each particular observer is in one of these worlds, and is unable to perceive any of the others.

In this no-collapse interpretation, for example, shuffling a deck of cards creates an entirely new mind-world from each possible outcome. This is sometimes interpreted as suggesting that there is a vast multitude of copies or virtual copies of you that get created every moment for each possible outcome. Each mind-world—theoretically containing its very own possible virtual copy of you—is associated with a different result of possible outcomes of each event. But you are forever unable to know of any of the other uncountable versions of you that exist and multiply.

Incredibly this interpretation of quantum theory has had some popular acceptance among physicists. Apparently this is because it is thought of as resolving the measurement problem without giving special power to the measurement process in collapsing the wave function. This interpretation can be said to incorporate the observer in the sense that an observer is placed in each parallel world. This is said to eliminate the necessity for the collapse of the wave function caused by the subjective consciousness of the observer upon observation. In lieu of a collapse, each copy or version of observer has the ability to experience only one outcome. The observer cannot experience the unlimited number of parallel mind-worlds that simultaneously and instantaneously spring into existence as a consequence of each possible outcome. It posits the 'creation' of a separate mind-world for every possible outcome, rather than collapsing the wave function into a discrete objective outcome across observers. It hides the concept of a collapse of the wave function by virtue of positing the creation of separate individual mind-worlds that cannot interact.

It avoids the issue of the role of the process of observing by associating it with just one of the mind-worlds—the only one experienced by the observer—without the observer having a causal role in this process. How the world that is phenomenally experienced is associated with one of the possible outcomes of superposed quantum waves, and how this process occurs based on quantum randomness, are not addressed.[12] Also not addressed is how the individual observer appears to observe his or her own mind-world as a discrete macroscopic world, inasmuch as it is presumably just part of a quantum wave or proxy wave system that has not collapsed at all.

Further this interpretation does not define the branching or partitioning process, other than to assume that it automatically and instantaneously happens whenever there are specific possible outcomes of an observation. It provides no explanation or mechanism for how the mind-worlds arise—they just magically appear instantaneously whenever there are possible outcomes of an event without any mediating processes. Also, it doesn't explain how the one mind-world the observer thinks he or she is in—what might be described as an at least apparently classical *reality* for the particular version of observer—appears to be highly correlated or deterministically continuous with successive worlds in which the observer resided. The question of how the continuity or sense of continuous identity, the sense of '*I*,' occurs in the unending procession of partitioning of independent mind-worlds is particularly relevant. Perhaps in the other worlds there is no sense of continuity. But then how do *I* end up in the special world that at least seems to have it? The underlying continuity of the sense of *I* will be discussed in the next section of the book.

This interpretation divides *reality* into *my* phenomenally experienced world and all other worlds. The branching process, as well as all the other worlds, are inaccessible due to *my* limitations. In this sense, it is similar to orthodox interpretations, in that it is another unexamined black box attempt at a solution. The issues concerning a collapse are hidden in the branching of worlds and association of the observer with one of the worlds, and then the collapse is said not to occur. It objectifies a mathematical concept onto the ordinary sensory world in a quite inadequate way.

Continuous spontaneous localization interpretation. This interpretation attempts to improve upon the orthodox interpretations of quantum theory by describing the collapse of the wave function in a way that is not clouded by the vagaries of the subjective notion of measurement. In this interpretation, the collapse of the wave function is a *real* physical process that occurs spontaneously as objects move through time.

The wave function appears to evolve deterministically, but contains a fundamentally probabilistic stochastic perturbation (quantum indeterminism). The perturbation is insignificant in very small systems but produces a collapse of the wave function for large systems into localized position and other dynamic attributes typical of ordinary macroscopic objects. When the small stochastic perturbation associated with each quantum particle adds together across large objects, it becomes significant enough to cause a collapse into a discrete localized macroscopic state on its own, regardless of whether it is observed. This is theorized to occur spontaneously and continuously in all large systems as the wave function of a quantum system changes through time.

This interpretation reestablishes some discreteness and definiteness to the phenomenal world without relying on the measurement process and its related subjective underpinnings, arguing for a tolerable level of indefiniteness while maintaining the principle of quantum indeterminism fundamental to the orthodox interpretation. It identifies both the uncollapsed and collapsed wave functions as *real* states of processes in nature, and gives no causal role to the process of observing and the observer with respect to collapse of the wave function. [13]

In this interpretation, quantum wave collapse into classical objects takes place through interaction with the classical environment, without involvement of a conscious observer. This importantly suggests that both quantum and classical objects are *real*, and moreover that they causally interact. It clearly implies that the quantum wave is a *real* wave of some kind that is not just a mathematical description of nature as posited in orthodox quantum theory interpretations. The quantum wave has ontological existence as a *real* wave. This reflects a major step beyond the orthodox interpretation toward the ontological existence of quantum *reality*. The next interpretations go even further in this direction. These interpretations are progressing toward the quantum level as increasingly *real*, and toward wave function collapse as an *objective reduction* that can occur independent of measurement and the conscious mind of an observer.

Consistent histories interpretation. Another variant of no-collapse interpretations of quantum theory is the *consistent histories* approach, associated with the concept of *decoherence*. It also attempts to account for the consistency between a measurement that has an infinite number of possible outcomes and the phenomenally definite outcome without positing a collapse of the wave function via the process of observation by a subjective conscious observer. Decoherence is an alternative description of quantum mechanical principles that does not rely on the concept of measurement dependent on an observer. It is said to be an observer independent means to assign values to quantum processes. [14]

The concept of decoherence concerns interactions between an object and its environment, which involve a multitude of influences. Many influences don't substantially change the object, but do limit its range of possibilities. Objects don't exist in isolation; they exist in some environmental context. In interacting with its environment, an object imprints itself on the environment, and this imprint influences the object through time. These interactions set limits on the object, spontaneously narrowing down quantum possibilities into an allowable consistent set of states.

The wave-like nature of objects moving in space and time is exhibited in the pattern of quantum wave interference effects. The wave interference effects are more prominent when the wave pattern is coherent and not disrupted by environmental influences. But in interaction with complex environments—classical environments in the *real* natural world—the environmental influences disrupt the coherence. This *decoherent* effect interferes with and suppresses the quantum interference pattern. This provides an explanation for the classical results of observations as discrete classical particles rather than waves, without positing an instantaneous collapse of the quantum wave function via conscious observation. Greene explains:

"Once environmental decoherence blurs a wave function, the exotic nature of quantum probabilities melts into the more familiar probabilities of day-to-day life... If a quantum calculation reveals that a cat, sitting in a closed box...has a 50 percent chance of being alive...decoherence suggests that the cat will *not* be in some absurd mixed state of being dead and alive... [D]ecoherence suggests that long before you open the box, the environment has already completed billions of observations that, in almost no time at all, turned all mysterious quantum probabilities into their less mysterious classical counterparts... Decoherence forces much of the weirdness of quantum physics to 'leak' from large objects since, bit by bit, the quantum weirdness is carried away by the innumerable impinging particles from the environment." (pp. 210-211) [14]

Greene further explains:

"The wave function of a grain of dust floating in your living room, bombarded by jittering air molecules, will decohere in about a billionth of a billionth of a billionth of a billionth (10^{-36}) of a second. If the grain of dust is kept in a perfect vacuum chamber and subject only to interactions with sunlight, its wave function will decohere a bit more slowly... And if the grain of dust is floating in the darkest depths of empty space and subject only to interactions with the relic microwave photons from the big bang, its wave function will decohere in about a millionth of a second... For larger objects, decoherence happens faster still. It is no wonder that, even though ours is a quantum universe, the world around us looks like it does." (514) [14]

As the concept of decoherence is sometimes applied in a *consistent histories* framework (see next chapter on loop quantum gravity), it not only emphasizes the *context dependence* of objects but also their *consistent histories*. It is not just that a quantum object, interacting with its environment, decoheres to produce the appearance of a classical particle-like object. The perceived consistency of an object from the perspective of an observer also needs to be considered—without the observer's consciousness collapsing quantum wave possibilities into particle-like appearances. Consistency through time incorporates the perspective of the observer as part of the decoherence process. The contextual environment serves as a selection process that automatically narrows quantum possibilities, and from the perspective of an observer these processes are consistent through time. There is a consistent process of change associated with and central to the nature of an object appearing as discrete in time and space. As a simple analogy, a sentence is constructed in a consistent series of words with the first word setting a context that narrows down possibilities for the next word. For example, if the first word is 'the,' then the next word likely would be consistent with it, such as 'tree' rather than the unlikely possibility of 'however.' For another simple analogy, in the game of baseball, for a player on first base the next consistent location is second base, depending on the context or frame of reference in the game.

The many mind-worlds interpretation discussed earlier needs to explain how the discrete results that the observer experiences are consistent with the questions initially asked about the world—such as in the attempt to measure it—instead of an infinity of other possible questions. It doesn't explain how the world that is experienced has a logical relation to the world in which questions requiring the observations or measurements are posed. This is another way of stating the concern that the many mind-worlds interpretation does not address the apparent continuity across possible worlds that is a core aspect of phenomenal experience (as well as of science). Decoherence in a consistent histories framework at least is a beginning for addressing this issue. It refers to abstract logical principles relating questions or measurements to definite answers or outcomes. Sets of questions are identified as decoherent if specific answers to them are not superpositions of answers to other questions. Theoretical physicist and quantum gravity theorist Lee Smolin explains:

> "This approach lets you specify a series of questions about the history of the universe. Assuming only that the questions are consistent with one another, in the sense that the answer to one will not preclude our asking another, this approach tells us how to compute the probabilities of the different possible answers." (p. 43) [15]

In the interpretation of quantum theory based on consistent-histories decoherence, possibilities are narrowed down due to consistent contextual histories through time. The phenomenal outcome of measurements is a product of initial measurement questions about the world and the environmental context that objects being measured interact with through time. Phenomenal outcomes are dependent upon, and limited to be consistent with, the initial questions being posed.

In this context-dependent view, there is one world with many different perspectives or minds inside it—related to the many mind-worlds approach. But the world we get depends on the questions about it we ask—such as the measurement choices—and their historical environmental contexts through time. According to this interpretation, there is not a physical collapse of the wave function at the time of measurement by conscious observation. Rather, definite answers emerge from the initial questions in a highly-context dependent manner involving natural processes of decoherence. Smolin states:

> "It is almost as if the questions bring reality into being. If one does not first ask for a history of the world that includes the question of whether dinosaurs roamed the Earth a hundred million years ago, one may not get a description in which the notion of dinosaurs—or any big 'classical objects'—has any meaning." (p. 44) [15]

In this interpretation there is one world with many consistent decoherent histories of it that are experienced or brought into being independently by many different observer-minds. Still, in this interpretation *reality* remains on the level of perhaps an infinite number of possible consistent histories due to an infinite number of possible observers. Smolin again states:

> "Each history is incompatible with the others, in the sense that they cannot be experienced together by observers like ourselves. But each is, according to formalism, equally real... Either quantum mechanics is wrong when applied to the whole universe, or it is incomplete in that it must be supplemented by a theory of which set of questions corresponds to reality..." (p. 44) [15]

According to this quite abstract approach, worlds with inconsistent histories are eliminated but there are still an infinite number of possible observers who view the world within their own exclusive consistent perspective. There remains an unlimited number of equally *real* phenomenal experiences of the one world. It doesn't offer a basis for determining a set of questions about the one *real* world that corresponds to definite consistent answers. It also doesn't account for how different observers seem to agree on the *real* world. The principle of decoherence is not sufficient to narrow down the phenomenal perspectives of the one world to definite consensually validated macroscopic outcomes. It eliminates inconsistent histories, but doesn't explain how consistent histories are narrowed down to specific phenomenal outcomes. This interpretation seems to attribute ontological *reality* to quantum phenomena, in that the wave collapse is objective and independent of a conscious observer. It also recognizes the thorny issue of how these objective processes relate to the subjective observer. It attempts to do this by accounting for the infinite quantum possibilities in terms of an unlimited number of observer perspectives. In this way it relates to the many mind-worlds interpretation, placing the possibilities into individual minds (or placing individual minds into each possible individual mind-world). It thus has some of the same difficulties with respect to how a consistent history for a particular observer is built, as well as how histories can be consistent across observers as the basis of consensual validation. Also, it doesn't address the issue of how conscious minds might exist at a quantum mechanical level of nature.

Further, a major implication of this interpretation is that the basis for consistent histories must be in the initial conditions of the universe. The entire history of the universe that gives particular consistent answers has to be consistent with the initial conditions. Further still, the questions embedded in the initial conditions might seem to imply some kind of observer capable of conceiving questions that somehow are consistent with the subsequent history of the universe. These are quite amazing and noteworthy implications about the nature of the initial universe, whatever the unexamined observer in the interpretation might end up being.

Decoherence is an important principle that clarifies to some degree how infinite initial possibilities are narrowed into definite outcomes through context dependence. In addition to its relevance to the ontological existence of both classical and quantum levels, there are at least two other important implications. It implies an overall asymmetric direction to time—sometimes referred to as the *arrow of time*—fundamental to how infinite possibilities manifest as definite phenomena through consistent contextual histories. Also, its emphasis on context dependence ultimately gets back to the *initial conditions* inherent in the structure of nature that shape the entire universe—to be discussed further in later chapters. It seems to imply that the initial conditions must include all possible consistent outcomes or answers to questions about the universe, which suggests that the initial conditions include an immense amount of inherent order, an implication also of unified field theory. How this is consistent with *fundamental* randomness seems relevant.

The context-dependent principle of decoherence differs significantly from conceptions of quantum wave collapse in orthodox interpretations of quantum theory. If the wave function can interact intimately with the physical environment, then it certainly is not just a non-physical mathematical object in conceptual space, not just a mathematical function, and not just virtual or semi-*real* potentia. The notion that a quantum object is shaped into a classical object through interaction with the classical environment implies that the quantum 'object' is something *real* and involved all along in causal interaction with its environment. This is a quite different notion than the orthodox interpretation which holds that the quantum level is just random virtual proxy waves or potentia. The *wave* part of the quantum is increasingly *real,* not just the particle part.

In addition, this interpretation reflects growing recognition that the transformation of a quantum 'object' into a classically observed phenomenal object does not depend on the process of observation by a conscious observer that is theorized to collapse the wave function. It is rather an objective collapse, or *objective reduction*, that does not necessarily involve a conscious observer. It thus may be helpful to make a distinction between the probability wave as some type of indefinite amorphous but *real* quantum 'object' that interacts with the physical environment, and the probability wave *function* as a mathematical equation in non-physical conceptual space or an observer's mind—an important distinction made in Chapter 4 that will lead to resolution of the Schrödinger's cat paradox. On the other hand, there is also growing recognition that a conscious observer with a perspective or frame of reference is intimately involved in a consistent history.

The ontological *reality* of quantum phenomena is taken much further in the next interpretation of quantum theory, in which the quantum wave is a *real* wave that is inherently nonlocal. In this view, quantum uncertainty and indefinite possibilities involve deterministic processes. We will return to these points after the overview of quantum theory interpretations. We will then make distinctions between the particle level of nature, an underlying wave level of nature, and an even more abstract mental level of nature which relates to the mathematical conception of the quantum wave function—all being ontologically *real* levels of nature.

In the following quote, mathematician and cosmologist Sir Roger Penrose points to some of the same issues the above overview of quantum theory interpretations has been uncovering:

"In fact, almost all the 'conventional' interpretations of quantum mechanics ultimately depend upon the presence of a 'perceiving being', and therefore seem to require that we know what a perceiving being actually is! We recall that the Copenhagen interpretation... takes the wavefunction not to be an objectively real physical entity but, in effect, to be something whose existence is 'in the observer's mind'. Moreover, at least in one of its manifestations, this interpretation requires that a measurement be an 'observation,' which presumably means something ultimately observed by a conscious being—although at a more practical level of applicability, the measurement is something carried out by a 'classical' measuring apparatus. This dependence upon a classical apparatus is only a stopgap, however, since any actual piece of apparatus is still made of quantum constituents, and it would not actually behave classically—even approximately—if it adhered to the standard quantum...evolution... The issue of environmental decoherence...also provides us with merely a stopgap position, since the inaccessibility of the information 'lost in the environment' doesn't mean that it *actually* is lost, in an objective sense. But for the loss to be subjective, we are again thrown back on the issue of 'subjectively perceived—by whom?' which returns us to the conscious-observer question... The many-worlds description of reality is manifestly dependent upon having a proper understanding of what constitutes a 'conscious observer', since each perceived 'reality' is associated with an 'observer state', so we do not know what reality states (i.e., worlds) are allowed until we know what observer states are allowed. Put another way, the behavior of the seemingly objective world that is actually perceived depends upon how one's consciousness threads its way through the myriads of quantum-superposed alternatives. In the absence of an adequate theory of conscious observers, the many-worlds interpretation must necessarily remain fundamentally incomplete... The consistent histories approach...is also explicitly dependent upon some notion of what 'an observer' might be... As far as I can make out, the only interpretations that do *not* necessarily depend upon some notion of 'conscious observer' are that of de-Broglie-Bohm...and most of those...require some fundamental change in the rules of quantum mechanics... I am an adherent of this last view... Accordingly, there is no need to invoke any conscious observer in order to achieve the reduction of the quantum state...when a measurement takes place." (pp. 1031-1032) [16]

The final interpretation of quantum theory to be outlined here can be classified as a no-collapse approach in the specific sense that quantum wave collapse is not due to the conscious mind of an observer. But it also incorporates principles of decoherence to explain the unpredictability and probabilism of motion contributing to wave-particle effects that are described as accounting for the collapse of the wave function. It is related to the interpretation mentioned in the Penrose quote above. It is associated with the work of mathematician and theoretical physicist David Bohm. It has opened entirely new ground in physics, and is an important plank in the direction of a systematic understanding of mind as more fundamental than matter. It does this by positing a *sub-quantum reality.*

"[G]rowing dissatisfaction with several of the other prevailing views in physics caused Bohm to become increasingly troubled with Bohr's interpretation of quantum theory… [W]riting a textbook…he still wasn't satisfied…and sent copies of the book to both Bohr and Einstein… [In] a six-month series of spirited conversations, Einstein…admitted he was still every bit as dissatisfied with the theory as was Bohm… Bohr and his followers   claimed that quantum theory was complete and it was not possible to arrive at any clearer understanding… This was the same as saying there was no deeper reality beyond the subatomic landscape, no further answers to be found… Inspired by his interactions with Einstein, he [Bohm] accepted the validity of his misgivings about quantum physics and decided there had to be an alternative view… He began by assuming that particles such as electrons *do* exist in the absence of observers. He also assumed that there was a deeper reality beneath Bohr's inviolable wall… [B]y proposing the existence of a new kind of field on this subquantum level he was able to explain the findings of quantum physics as well as Bohr could. Bohm called his proposed new field the *quantum potential* and theorized that, like gravity, it pervaded all of space. However, unlike gravitational fields, magnetic fields, and so on, its influence did not diminish with distance." —*Michael Talbot, science writer (p. 39)* [17]

Neorealist interpretation. Von Neumann's proof that ordinary objective *reality* cannot exist if quantum theory is correct had the effect of placing much of the work on the nature of *reality* as investigated by theoretical physicists on hold for several decades. In the 1950s, however, an interpretation began to emerge that seemed to defy the proof. It provided a mathematical model of elementary particles as *real* classical objects with intrinsic dynamic attributes that matched the probabilistic predictions of quantum theory. This interpretation can be described as a deterministic realization of the hidden variables approach to quantum theory favored by Einstein, in part coming out of conversations between Einstein and Bohm, its principle developer. The approach is sometimes called *neorealism* because of its renewed emphasis on ordinary classical deterministic *reality*, but it is much deeper. It was largely ignored for years, but is now becoming prominent. [18]

Neorealist nonlocal interpretations of quantum theory other than Bohmian mechanics are not covered in this book. As an example of one of them, in the *transactional* interpretation [19] the collapse of the wave function does not occur instantaneously. Rather it occurs in a space-time interval during which actual physical waves moving forward and backward in time allow the communication of information necessary, for example, for the matched behavior of twin photons in quantum entanglement found in the experimental tests of Bell's theorem.

The neorealist interpretation associated with Bohm is described as a deterministic reformulation of quantum theory within an objective perspective that doesn't invoke the subjectivity of the observer in the collapse of the wave function. [18] Classified as a no-collapse model, it is a mathematical theory of the motion of particles in which the path of a *real* elementary particle is guided by a *real* nonlocal wave. It is thus sometimes described as a resolution to the dilemma of wave-particle duality. [1]

In this interpretation, related to the mathematical framework of Bohmian mechanics, atoms are *real* classical objects, for the most part made of *real* classical particles. The underlying waves are also *real,* but are a subtler nonlocal level of *reality*. In this sense, the interpretation has aspects of dualism rather than just classical realism (although it also emphasizes quantum wholeness and the underlying unity of matter and mind). It even might be classified as a type of Cartesian dualism in the sense that it involves a field of mind in addition to and beyond ordinary matter, but importantly not in the Cartesian dualistic sense that mind is dimensionless and not spatially extended with respect to the notion of ordinary classical space. Rather, the underlying wave field is nonlocal and functions at a level of nature that is beyond the conventional level of ordinary space-time—indeed, even beyond the quantum level to an underlying, more fundamental, deterministic, causally efficacious *sub-quantum* field, beyond the classical model of material particles.

This interpretation has been referred to as the most successful interpretation of non-relativistic quantum theory in that it provides a reasonable solution to the measurement problem and also a model of quantum processes. [1] However, it accomplishes these things through an approach that is considered quite radical from the perspective of the orthodox interpretation, in that it can be interpreted as positing additional ontological levels of nature underlying ordinary physical existence. Both a wave function and a function for the positions of particles are used to describe a quantum mechanical system, with both waves and particles considered to be *real* processes occurring objectively in nature. The particle aspect is local, and the wave aspect is nonlocal.

In direct contrast to the orthodox interpretation, it holds that quantum objects are not all the same but are intrinsically different, and that their dynamic attributes exist independent of the observer in the objects themselves—hidden when not classically measured. It posits that the wave function gives a probabilistic description of a group of quantum objects, but is an incomplete description of an individual quantum object; and that a single quantum object is unpredictable due to *classical* ignorance or uncertainty (deterministic but unspecifiable). In this interpretation, quantum theory is incomplete, and eventually there will be deterministic explanations for quantum indeterminism and for the measurement process that can be said to be independent of the observer, in contrast to orthodox interpretations of instantaneous wave function collapse. In other words, it sides with (a further elaboration of) Einstein rather than Bohr.

In this theory *reality* is the same whether measured or not. Electrons are particles when measured, as well as when not being measured. Their dynamic attributes of motion are directed by a *real* but very abstract *guiding wave, pilot wave,* or *psi wave,* also called the *quantum potential*. To match the behavior of objects according to quantum probability predictions, this quantum potential or psi wave must be connected to every particle in the universe, must be classically invisible, must be superluminal, and must be a common aspect of nature. [5] A vastly different landscape of nature is proposed, which has causal relations and determinism but is nonlocal in its character.

Recall from the discussion of Bell's theorem in Chapter 4 that experiments have demonstrated properties associated with nonlocality as necessary to any description of nature, whether classical or quantum. In the previous discussion, nonlocality was described as *unmediated*. In the neorealist interpretation, nonlocality is *mediated* by the psi wave or quantum potential, an underlying field characterized by quantum wholeness and nonlocal interconnectivity.

Starting with the initial random configuration, this interpretation models change in a system by the quantum potential or psi wave that yields results empirically equivalent to quantum mechanical accounts. Interference effects found in the double slit experiments, for example, are due to the psi wave or guiding wave that guides the path of particles through specific slits. The wave function represents actual nonlocal waves that pilot the motion of actual local particles.

The notion of the collapse of the wave function is accounted for in terms of decoherence and of the destruction of interference patterns of the waves when another system in the environment—measurement probe of some kind—impacts the system. The collapse of the wave function is not based on the influence of an observer, other than the environmental or contextual disturbances that might come from the measuring equipment in the measurement process. The wave function itself does not collapse in the same sense that it does in other interpretations of instantaneous collapse upon measurement—the wave aspect is the guiding wave or psi wave, and it doesn't collapse into a particle. The psi wave functions according to wave mechanics, directing the classical elementary particle such as an electron or photon along its path. The particle does not have a wave nature; it is the wave nature of the psi wave that produces the appearance of wave patterns such as interference effects (although it can be said to be made up of wave patterns). The actual path of a particle is fundamentally determined by the psi wave, influenced by the myriad of mechanical influences—made of numerous electric, magnetic, and gravitational effects of the known fundamental forces—that comprise the entire experimental arrangement and measurement process. This guiding influence is somewhat analogous to the way a relatively small steering wheel directs a large powerful truck, a radio signal directs the path of a distant orbiter in space, or the Moon's gravitation influences the tides. However, the psi wave is a much, much subtler and sensitive guiding influence than these classical examples, and it does not drop off with distance in the same way that they do. This can be understood to be a more abstract theory of the nature of space and time in which there is a subtler nonlocal level of *reality* underlying ordinary local conventional space-time.

This interpretation accepts that the world is a seamless and inseparable whole—quantum wholeness, as in the orthodox interpretation. It further adds that the act of observation dissolves the separation of observed and process of observing. Bohm emphasizes that:

> "One is led to a new notion of unbroken wholeness which denies the classical analyzability of the world into separately and independently existing parts... The inseparable quantum interconnectedness of the whole universe is the fundamental reality..." (p. 18) [5]

Physicist Nick Herbert comments:

> "Undoubtedly we are all connected in unremarkable ways, but close connections carry the most weight. Quantum wholeness, on the other hand, is a fundamentally new kind of togetherness, undiminished by spatial and temporal separation. No casual hookup, this new quantum thing, but a true mingling of distant beings that reaches across the galaxy as forcefully as it reaches across the garden." (p. 19) [5]

This neorealist interpretation addresses the apparent fundamental indeterminism in quantum theory by offering a deterministic account of quantum probabilities as arising from averaging over initial random unknown conditions—rather than from an inherent quantum indeterminism. The path of an elementary particle is entirely deterministic, at all times definite in all its dynamic attributes such as position and momentum. Its path is a combination of the influence of the quantum potential or psi wave and the myriad of other deterministic contextual influences—specified and unspecified—that make up the whole experimental arrangement or environment. Together these constantly changing influences produce a fuzzy, jittery, complex path of motion that is unpredictable with exact precision. Hence in this interpretation the uncertainty can be said to be due to a type of classical ignorance rather than quantum ignorance.

The complexity of influences and the extreme sensitivity of the psi wave that lead to unpredictability are similar to a chaotic system, as described in *chaos theory*. It also can be directly related to context-dependent decoherence described earlier. The entire experimental set-up determines the path of the particle. Because the psi wave doesn't drop off with distance and is highly interconnected due to quantum wholeness, the influences that make up the entire experimental arrangement must include the totality of the whole extended nonlocal environment—in effect, about everything and everywhere in the universe. All of these influences together are so complex that they are *unfathomable*, and thus the exact path of the particle is not precisely predictable. It is in this sense that the experimental arrangement and act of measurement must be taken as an undivided contextual whole, impossible to be taken as separate parts. [5]

Also it is in the sense that all components of the experimental arrangement—the measuring device, the probe, and so on—influence the process of measurement that they can be said to *create* the discrete actual *reality*. Any change in the experimental arrangement would alter these influences, including the subtle psi wave, and thus would *create* a different outcome. This gives an explanation for the process of observation creating the outcome objectively, but within a framework of causal determinism and realism—and without giving the subjectivity of the observer a role in the sense referred to in orthodox interpretations of quantum theory of a collapse of the wave function via the conscious attention of the observer.[5]

It further provides an explanation within the framework of determinism and realism for quantum wholeness in the sense of nonlocality exemplified by phase entanglement—in which distant objects influence each other superluminally. The experimental verification of nonlocality associated with Bell's theorem did not—and could not within its ontological limitations—explain the mechanism through which the phase entanglement might be mediated. This interpretation provides such a mechanism: the psi wave or quantum potential. The phase-entangled twin particles don't influence each other directly; they are both directed by the extremely subtle guiding influence of the pilot or psi wave.

Until recently this neorealist interpretation was viewed by orthodox quantum physicists as naïve and outmoded thinking, because it was thought to represent a reemergence of the historical concept of ordinary classical *reality*. However, it has generated renewed interest as it has been examined more carefully. The interpretation provides formal mathematical explanations of quantum wholeness and nonlocality through the psi wave. It also gives a logical alternative account of the collapse of the wave function and of quantum uncertainty, and it achieves these within a deterministic framework. But it necessitates the addition of a subtler sub-phenomenal level of *reality* that is not the gross phenomenal world of ordinary sensory experience on which modern science has fixated. In this interpretation of quantum theory can be seen even more the disembedding of the concept of classical physical *reality* from the notion that the conscious mind of the observer creates it via the measurement process. But it doesn't eliminate mind; as we will see, it adds another ontological level and even places mind as a nonlocal level underlying all of classical physical existence.

However, does this theory of the quantum potential or psi wave violate Eberhard's impossibility proof (refer to Chapter 4), which holds that even if there is nonlocality it would still be impossible for *information* to be sent between objects superluminally? In order for the psi wave to be a *signal* transporting information superluminally, it must be able to be modulated in a systematic way that results in the particle changing its pattern of motion. But the approach describes the psi wave as extremely sensitive, such that any change in the system would seem to produce an unpredictable effect that would interfere with any form of message transmission.

In this quantum theory interpretation, the psi wave does carry information, in that it sensitively reflects the totality of the experimental arrangement. It represents 'active information,'[13] through which it deterministically directs the motion of a particle, but not through the strength of the forces involved. Whether it is possible to change the content of this information in a subtle enough manner to redirect a particle systematically and predictably remains to be answered. Certainly the path of particles are influenced by changes in the fundamental forces on the gross physical level. But is it possible to guide the path of particles from the deeper level of information associated directly with the psi wave, such as through an intentional message or information?

It is in trying to account for this possibility that this neorealist interpretation also brings mind into it. Bohm has speculated that the nonlocal mind may operate with extreme subtlety in a way that would allow such sensitive but systematic information transmission. He has proposed this as a general framework for the way the universe operates—how mind influences matter.[5] At this level, nature may function in terms of nonlocal, highly interconnected processes rather than the discrete local activity characteristic of the grosser physical level of particles associated with ordinary classical *reality*. In other words, the psi wave may be related to extremely subtle action on an underlying, abstract, nonlocal level of mind. [20] Importantly this finally begins to bring into the natural world and its causal chain the possibility of the causal efficacy of conscious minds, which has been impossible to address in modern science within the reductive physicalist worldview—to be extensively unfolded in upcoming chapters.

Later elaborations of this interpretation have expanded the concept of an elementary particle of matter and its underlying psi wave to the concept of a subtle, nonlocal, non-linear field that constantly enfolds and unfolds, and in which any local region of space is interconnected nonlocally with the entire universe. In this conception, nature is viewed as a single wholeness or totality enfolded in each individual region of space—akin to holographic models to be discussed in Chapters 6-7. In the following quote, Bohm and his colleague B. J. Hiley summarize how their model of undivided wholeness associated with the concept of the *implicate order* relates to physical and mental phenomena:

> "One may then ask what is the relationship between the physical and the mental processes? The answer that we propose it that there are not two processes. Rather, it is suggested that both are essentially the same. This means that that which we experience as mind, in its movement through various levels of subtlety, will, in a natural way ultimately move the body by reaching the level of the quantum potential and of the 'dance' of the particles. There is no unbridgeable gap or barrier between any of these levels. Rather, at each stage some kind of information is the bridge. This implies that the quantum potential acting on atomic particles, for example, represents only one stage in the process... It is thus implied that in some sense a rudimentary mind-like quality is present even at the level of particle physics, and that as we go to subtler levels, this mind-like quality becomes stronger and more developed." (pp. 385-386) [13]

The view emerging from this work can be understood in terms of two fundamental levels that make up the one wholeness, both causally determinate, interconnected, and interacting with each other but with different properties associated with different levels of subtlety. The surface classical level—*explicate order*—is the familiar world of mechanical particle interactions mediated by the four fundamental forces of nature within Einstein locality. The subtler level—*implicate order*—is a highly interconnected, enfolded *reality* of nonlocal interactions mediated by the psi wave that is more of the nature of mind than matter, and which permeates and shapes apparent movement in the explicate order. In specifying two levels, however, it would be inappropriate to identify this theory as dualistic. Rather it emphasizes the causal seamlessness between these two levels, and it also emphasizes quantum wholeness, in which the more expressed classical physical level—the explicate order—is a grosser manifestation of the subtler, more encompassing implicate order. In this sense it is a non-dual or monistic account of nature. In the following quotes, Bohm outlines how the explicate order is embedded in the implicate order, both emerging sequentially from the one wholeness or universal flux:

> "Our proposal to start with the implicate order as basic, then, means that what is primary, independently existent, and universal has to be expressed in terms of the implicate order. So we are suggesting that it is the implicate order that is autonomously active while...the explicate order flows out of a law of the implicate order, so that it is secondary, derivative, and appropriate only in certain limited contexts. Or, to put it another way, the relationships constituting the fundamental law are between the enfolded structures that interweave and interpenetrate each other, throughout the whole of space, rather than between the

abstracted and separated forms that are manifest to the senses (and to our instruments)." (p. 235) [22]

"[T]here is a universal flux that cannot be defined explicitly but which can be known only implicitly, as indicated by the explicitly definable forms and shapes, some stable and some unstable, that can be abstracted from the universal flux. In this flow, mind and matter are not separate substances. Rather, they are different aspects of one whole and unbroken movement. In this way, we are able to look on all aspects of existence as not divided from each other, and thus we can bring to an end the fragmentation implicit in the current attitude toward the atomic point of view, which leads us to divide everything from everything in a thoroughgoing way." (p. 14) [22]

The neorealist interpretation based on Bohmian mechanics represents a significant step toward acknowledgement of the *reality* of the subtle level of nonlocal mind and attempting to conceptualize its place and causal role in the natural world. In this interpretation, mind can be associated generally with the psi wave or quantum potential—a subtler, nonlocal, more holistic level that undergirds and permeates the explicate classical world. The *reality* of the level of mind as a deeper, more holistic and interconnected domain of existence underlying ordinary classical *reality* is clearly emerging in this approach. This model provides for the first time in modern science a logically consistent approach to the conscious mind as having a causally efficacious role in nature. The level of particulate matter is a manifestation from the level of mind, and thus not ultimately separate from it. In this interpretation, mind is nonlocal, and not just in the physical head as a product only of gross neural activity. This is in a constructive direction that will develop into an important plank of understanding to link modern science to Maharishi Vedic Science and Technology. It is moving toward enumerating three fundamental levels of *reality*, corresponding to gross, subtle, and transcendental (unified field) levels in Maharishi Vedic Science and Technology, while at the same time positing an ultimately seamless unity or oneness or super-implicate order. Three phenomenal levels or parts of nature within one ultimate wholeness is basic to ancient Vedic Science. It is reflected in this interpretation of quantum theory in terms of the explicate, implicate, and super-implicate or universal flux levels of the wholeness of nature.

"According to standard quantum mechanics, when we perform a measurement and find a particle to be here, we cause its probability wave to change: the previous range of potential outcomes is reduced to the one actual result our measurement finds... In the standard approach, the collapse happens instantaneously across the whole universe... In standard quantum mechanics, then, it is this instantaneous change in probability waves that is responsible for the faster-than-light influence... Nevertheless, there is a hitch. *After more than seven decades, no one understands how or even whether the collapse of a probability wave really happens.*"—Brian Greene, theoretical physicist (p. 19) [14]

Summary of Key Issues in Interpretations of Quantum Theory

It may be helpful at this stage to summarize some of the key issues in the interpretations of quantum theory presented in this chapter. One key issue is whether nature is fundamentally random and indeterminate (quantum ignorance), or completely deterministic even if not fully specifiable due to inability to establish initial conditions or calculate all the determining environmental influences (classical ignorance). Orthodox interpretations are on the side of fundamental indeterminism, and at least Bohmian neorealism is on the side of determinism.

Another issue is whether there is a collapse of the quantum wave function. Orthodox interpretations view the collapse as a product of measurement or observation that is spontaneous, instantaneous, unmediated, and cannot be analyzed. Some non-orthodox interpretations are no-collapse models. Neorealistic Bohmian mechanics accounts for the notion of the collapse of the wave function by theorizing that some aspects of the concept of wave collapse are a product of contextual interference effects and the measuring devices and processes, and other aspects of the wave function are associated with the psi wave. Neither collapse instantaneously due to the measurement process. In other words, the orthodox interpretations of quantum theory are fundamentally incomplete, which is consistent with Einstein's intuitions.

Another related issue is the degree to which objects are independent of the process of observing. This is sometimes described in terms of which attributes of objects are intrinsic to the objects and which are products of the process of observing involving the interaction of object and measurement. Orthodox interpretations posit static attributes are intrinsic and dynamic attributes are a product of the measurement process. The neorealist interpretation, for the most part, holds that both static and dynamic objects are intrinsic to the objects.

An additional major issue is whether there is a nonlocal force field not mediated by the four known fundamental forces that is a level of *reality* underlying ordinary local classical *reality*. Orthodox interpretations posit either that ordinary phenomenally experienced *reality* is the only *reality*, or that there is an additional semi-*real* quantum *reality*, or that quantum *reality* is the only *reality*. None of these posit an underlying nonlocal force field (except possibly the unified field, to be discussed later). The Bohmian neorealist interpretation posits such a *real*, underlying, subtle field or psi wave that is causally efficacious in guiding the behavior of grosser matter particles in ordinary space and time.

Closely related to this is the issue whether superluminal communication is possible. Whereas orthodox interpretations generally are on the side that it is not possible (except in the sense of tunneling), the neorealist interpretation suggests that it is. The basis for nonlocal relationships—experimental tests of Bell's theorem—is strong both theoretically and empirically. It requires that physical *reality*, not just the phenomena of ordinary experience, must have this property. It seems reasonable to wonder whether the neorealist approach that a subtle superluminal field mediates nonlocality is in fact less radical than the orthodox approach. The orthodox approach is associated with inexplicable quantum jumps, tunneling, and teleportation in which objects disappear and reappear anywhere in a magical, unmediated, instantaneous manner in a relativistically undefined space and time, and that such jumps might be directed experimentally as implied in research to develop teleportation while their instantaneity seems to preclude any causal mechanics.

An overriding issue has been emerging that contrasts the view of the world in terms of locality—captured in the concepts of Einstein locality and the light cone in relativity theory—with the more entangled, integrative view of nonlocality, sometimes associated with the concept of *holism* in quantum theory. Bohr's undivided wholeness of the experimental arrangement, von Neumann's quantum wholeness or all-quantum *reality*, Bell's quantum nonlocality, and Bohm's quantum potential or psi wave represent emerging major steps toward a more holistic and interconnected understanding of the universe than the local model in classical physics including Einstein relativity theory. Prior to the mechanistic view that emerged in modern science a couple hundred years ago, nature was conceived in a more holistic way. The more recent generally accepted and objectified view in modern science has been that holistic conceptions accepting superluminal communication are too radical. [5] But cutting edge understanding, as summarized in this chapter on developments in quantum theory, has progressed far beyond this perspective.

Both quantum theory and relativity theory can be associated to some degree with an emerging emphasis on holism. However, the underlying principle in relativity theory that all things are related to each other can be associated more with a fundamental unifying value in nature, whereas the underlying principles of quantum discreteness in quantum theory expresses more the diversifying value in nature. This fundamental duality of unifying and diversifying values will later be shown to characterize the most fundamental nature of the unified field, and also the nature of subjective experience, the nature of the discriminating intellect, as well as the 'nature' of consciousness itself.

The concept of holism cuts to the core of the objective means of gaining knowledge in modern science. It brings into question whether there is such a *thing* as an *independent object* in the first place. Referring to the finding of phase entanglement, Greene states,

> "We used to think that a basic property of space is that it separates and distinguishes one object from another. But we now see that quantum mechanics radically challenges this view. Two things can be separated by an enormous amount of space and yet not have a fully independent existence... Space, even a huge amount of space, does not weaken their quantum mechanical interdependence." (p. 122) [14]

Historically modern scientific methodology hinged on the pretheoretical assumption of the independence of objects in nature (refer to Einstein's quote near the beginning of this chapter). From a holistic perspective, however, all things are connected and therefore nothing is completely independent, as exemplified in nonlocality and quantum wholeness. Holism implies that the concepts of independence of objects and objective *reality* represent a conditional or contextual *reality* that has a limited domain of application, associated with ordinary objective experience.

This more advanced view of objective *reality* as a conditional *reality* is somewhat reflected in the relational view of relativity theory. But relativity theory sets specific limits on the way things are related: they are related only through local influences mediated by the four fundamental force fields of nature within the light cone—Einstein locality. Herein is a major concern relevant to interpretations of quantum theory, orthodox as well as non-orthodox, and also relativity theory. In the quantum wave function and its collapse throughout the infinite quantum field, quantum theory implies nonlocality and holism. For the most part, relativity theory as proposed by Einstein implies locality. These two theories have not yet been reconciled.

On the other hand, whereas in relativity theory time and space are interdependent, in quantum theory they are treated differently—with one time coordinate and many space coordinates. The work toward integrating these two theories involves the framework of quantum fields. For the most part, quantum theory is based on the non-relativistic classical framework in which space and time are conceived as a non-changing backdrop upon which objects move. In this view, space and time constitute a non-changing frame of reference in which 'motion' theoretically *could be* instantaneous and unmitigated by distance (on the face of it, inconsistent with Einstein locality). The way motion is quantified differs, depending on whether it is the observer or the object observed that is thought to be moving—both yielding correct formulations of the laws of nature within their own perspective.

However, at near the speed of light these non-relativistic formulations have been shown to be less accurate than relativistic formulations that don't distinguish between the perspective of the observer and the object observed. In the relativistic framework, the speed of light does not change with the perspective of the observer. Interpretations of quantum theory based on non-relativistic physics are generally thought not to be viable until they can be extended into the relativistic space-time framework. On the other hand, the classical relativistic framework—generally associated with a local space-time conception or Einstein locality—cannot deal with the fundamental nonlocality demonstrated experimentally in tests of Bell's theorem.

A major challenge of the unified field theories emerging over the past few decades is to integrate the quantum framework with the classical framework of general relativity. In these theories, change is relativistic and local within the ordinary physical universe, or unmediated and instantaneous—such as quantum jumps and wormhole tunneling in quantum theory. Bohmian mechanics points to the theoretical possibility of an additional underlying nonlocal field theorized to relate to the concept of mind. Such a nonlocal field would need to exist in between the ordinary classical world and the undifferentiated unified field—hinted at in the earlier Bohm quote in this chapter and to be discussed more in the next few chapters.

The locality-holism controversy embodies the same core issue discussed in Chapter 3 in the context of building a unified field theory. Contemporary work in theoretical physics and mathematics has focused on the challenge of unifying the strong-electroweak force with the gravitational force. This is directly relevant to how to reconcile quantum theory and relativity theory—a huge task in theories of *quantum gravity*. Major approaches to quantum gravity are the focus of Chapter 6. In these approaches, issues relevant to the process of knowing and the knower are increasingly recognized as core issues, but still remain implicit and unexamined. The need for a more encompassing framework becomes increasingly obvious.

Embedded in these issues, and especially clear with respect to the contrast of nonlocal holism and relativistic Einstein locality, is a central concern that modern science has not dealt with directly and which is an important theme in this book: limitations of the objective means of gaining knowledge. To what degree is it appropriate to do an *objectified* science in terms of independent objects when ultimately there are no independent objects? Is it meaningful to base science solely on a particular means of gaining knowledge—the objective approach—when its basic assumption of independence of objects is at best conditional? Modern science has finally achieved an understanding of nature that recognizes the necessity of addressing the conscious mind even for an adequate model of physical *reality*. As enough planks are in place to go beyond the objectified physicalist framework, the immense practical advantage of the subjective means of gaining knowledge (introduced in Chapter 2) for comprehensive, integrated knowledge and experience of nature will become more evident.

In Maharishi Vedic Science and Technology, the fragmented view of *reality* that has characterized modern science and modern civilization is a consequence of the developmental limitations of the ordinary waking state of consciousness. In the non-relativistic classical framework, the observer and the observed are independent of each other. In the relativistic classical framework, the observer and the observed are relative and dependent, but the concept of the observer remains unexamined. Theories of quantum gravity need to incorporate fully relativity theory into quantum theory. However, even then the process of observing and observer also need to be incorporated into a holistic theory. A framework that integrates relativistic space-time and nonlocality with the process of observing and observer in an expanded, integrated model is necessary, involving ontologically *real* levels of nature beyond the physical. It can be quite helpful to approach these issues from the angle of an expanded understanding of the nature of space. Levels of existence that incorporate more abstract notions of space than the relativistic notion of space-time as defined by Einstein locality and light-speed are clearly unfolding in quantum gravity theories.

Recognizing the Process of Knowing in Determining Reality

To overview our progress on the bridge to unity so far, we can say that the ordinary world of physical objects constitutes a constantly changing system of energy patterns or processes that have a particle level of *reality* and also an underlying level of *reality* that can be characterized more in terms of waves. These constantly changing particle and wave processes appear as concrete objects at the macroscopic level due to the functioning and resolving power of the ordinary senses, which can be identified as the outermost part of the process of observing or knowing.

The ordinary senses structure the appearance of static, objective, macroscopic form—contributing to objectification—from these changing particle and wave properties. This is analogous to seeing a constantly flowing river as a discrete locatable object or as a constantly changing process or flow, depending on where attention is placed at the time. The ordinary senses *transduce* patterns of waves and particles into subjectively meaningful sensations and perceptions with qualities such as odor, taste, shape and color, texture, and sound. Because humans generally have reliable macroscopic sensory systems that function similarly, there is relatively high agreement or consensus across observers about gross *objective reality*, how the ordinary natural world appears.

The subtler wave properties of objects permeate the particle properties, and their effects can be inferred on the microscopic and macroscopic levels under certain conditions. Both particle and wave properties of objects can be thought of as *relatively real*. Neither are ordinarily experienced through the senses at the macroscopic level, but experimental conditions can make them apparent—at least indirectly, and in the sense of theoretically reasonable.

The particle properties are associated with classical *reality* and have more limitations than the underlying, more holistic level of *reality* that is better characterized in terms of wave properties. Groups of material particles can exhibit wave patterns that appear to follow the same statistical patterns as waves associated with the subtler wave level of *reality*. But these classical wave patterns are built of particles; and as ordinary particles observed at local microscopic and macroscopic levels, they don't exhibit quantum wholeness or phase entanglement in their interactions.

The underlying wave nature of objects can be more accurately represented by the mathematical model of superposed waves. The mathematical probability distribution captures this wave nature, which the ordinary senses don't capture. When an idealized mathematical model of quantum wave functions is superimposed on the particle and wave properties of objects—coupled with the presumption of independent objectivity—it compels the assumption that a collapse of the wave function is needed to get from the mathematical model to the localized independently existing objects we experience. The inner mental 'collapse' or updating of the knowledge state associated with the probability function is *objectified* into the outer objective world, typical of the reductive physicalist worldview. This directly leads to paradoxes in quantum descriptions of nature.

Physical objects are experienced as discrete independent objects by virtue of how the ordinary senses function in the process of observing. When the inner cognitive or intellectual functions impose a mathematical model onto ordinary sensory-perceptual experience, it seems to require a collapse of the mathematical probability function into the concrete objects. The collapse of the mathematical wave function can be said to be *real* on the level of mathematical conceptions in the mind—as a reduction of possibilities and probabilities into specific states of knowledge. There is a *real* 'collapse' of the state of knowledge from a number of possibilities to one mathematical outcome upon observation and measurement. But this 'collapse' of the quantum wave function is concerned more with the process of knowing than with the objects known: it is not a collapse of a quantum 'object' into the classical material world. Distinctions need to be made between the particle, the wave, and the mathematical functions describing them.

Classical *reality* is associated with the ordinary senses, which is the level of nature structured in terms of discrete particles. Quantum *reality* is associated with waves, a subtler level of nature that involves less discrete, more fluid wave mechanics. Conceptual or *mental* reality is even subtler and more abstract than the quantum wave level. It is at this level that there is a *real* collapse of the *quantum probability wave function as a mathematical model* into a discrete state of knowledge. But the classical particle and quantum wave levels don't collapse in the same way. The dynamics of the inner levels of subjectivity—the level of thinking and conceptualization—in the process of observing or knowing will be discussed in detail in the next part of the book.

It may be helpful, as an example, to consider the many mind-worlds interpretation of quantum theory in the picture we are now sketching. It deals with the contrast between the inner conceptual domain and outer concrete sensory domain of ordinary experience by assuming the *reality* of the conceptual possibilities. It posits that every event in some way may *actually* generate new worlds—at least mind-worlds. This reflects an awkward, fragmented attempt to make a conceptual shift in emphasis from the concrete discrete limitations of the objective domain of matter to recognition of the subtler, less limited, highly interconnected and nonlocal domain of the mind and conceptual *reality*. But due to its incomplete examination of mind and consciousness, it contains untenable components that do not integrate into a coherent theory of nature.

Core issues in quantum theory can be viewed in terms of the relationship between the 'body'— the concrete objective level of matter and energy that makes up ordinary sensory experience—and the mind—the abstract subjective level of conceptual models and mathematical logic. Embedded in the collapse of the quantum probability wave function, particles and waves, locality and nonlocality, and similar fundamental contrasts is the issue of how to relate these two realms. This is the measurement problem and, more fundamentally, the mind-body problem.

"I am aware that there will still be many readers who find difficulty with assigning any kind of actual existence to mathematical structures. Let me make the request of such readers that they merely broaden their notion of what the term 'existence' can mean to them. The mathematical forms of Plato's world clearly do not have the same kind of existence as do ordinary physical objects such as tables and chairs... Objective mathematical notions must be thought of as timeless entities and are not to be regarded as being conjured into existence at the moment that they are first humanly perceived... Those designs were already 'in existence' since the beginning of time, in the potential timeless sense that they would necessarily be revealed precisely in the form that we perceive them today, no matter at what time or in what location some perceiving being might have chosen to examine them... Thus, mathematical existence is different from physical existence but also from an existence that is assigned by our mental perceptions. Yet there is a deep and mysterious connection with each of those other two forms of existence: the physical and the mental... I have schematically indicated all of these three forms of existence—the physical, the mental, and the Platonic mathematical—as entities belonging to three separate 'worlds'... There may be a sense in which the three worlds are not separate at all, but merely reflect, individually, aspects of a deeper truth about the world as a whole of which we have little conception at the present time."—*Sir Roger Penrose, mathematician and cosmologist (. pp. 17-23)* [16]

The mysteries associated with quantum theory are primarily products of the *objectified* means of gaining knowledge in modern science. In recent decades, modern science has gone beyond direct sensory evidence and has had to rely more on intellectual processes of reasoning associated with mathematical modeling as key means to advance scientific theory. There is a lack of integration of the abstract level of reason with the concrete level of ordinary sensory experience. It is difficult to develop integrated knowledge when there is a large gap associated with the key means of gaining knowledge used. The concrete level of ordinary sensory experience contrasts with the more abstract level of mathematical conceptualization and reason. This separateness or gap contributes to the disintegrating effect of the objective means of gaining knowledge. Modern science has not applied educational procedures to develop the mind—especially the ability to experience subtle levels of objects in nature and their transcendental unified basis—in order to integrate the ordinary sensory and conceptual reasoning levels of the process of knowing. In the following quote, Maharishi explains the importance of progressive refinement of the ability of perception:

> "When only the surface value of perception is open to our awareness, then the boundaries of the object are rigid and well-defined—the only qualities that are perceived are those which distinguish the object from the rest of the environment. However, when the unbounded awareness becomes established on the level of the conscious mind—then the perception naturally begins to appreciate deeper values of the object, until perception is so refined that the finest relative is capable of being spontaneously perceived on the gross, surface level... As long as our perception is not refined, our vision falls only on the yellow, and we speak of the yellow rose. However, the possibility exists to refine our perception (to rise to a higher state of consciousness) so that the deeper reality of the situation will be cognized..." [23]

The challenge of understanding quantum processes, reconciling them with classical processes, and also of establishing a consistent theory of quantum gravity that builds into a theory of a completely unified field is a direct result of incomplete development of our minds. When our mental capabilities are in use such that all the means of gaining knowledge are applied, then holistic and completely integrated knowledge will unfold naturally. Hopefully soon it will be recognized that the more direct and comprehensive approach of ancient Vedic science already accomplished this integration much more profoundly, practically, and efficiently long ago.

In addressing the measurement problem and the mind-body problem there is increasing appreciation that there are not two but rather *three* fundamental areas, levels, or domains: 1) matter and energy (the known), 2) the process of observation associated with the subtle level of the mind (the process of knowing), and 3) consciousness itself (the knower). This trinity—referred to in the Prologue and throughout this book—is a universal theme that not only implicitly runs through modern scientific theories about the nature of *reality* but also in various forms through most religions and other worldviews striving for universal knowledge.

As will be discussed at length in upcoming chapters, Maharishi Vedic Science and Technology relates this trinity to the three-in-one self-interacting dynamics of the process of creation within the Veda—within the unified field. It describes in detail how the three-in-one self-interacting dynamics become expressed in three major domains or levels of nature—gross, subtle, and transcendental. These levels reflect fundamentally different and more abstract conceptions of space or existence, necessary in order to develop a holistic understanding of the natural world and its ultimate underlying unity.

In this context it might be a little easier to conceptualize space broadly—more like the concept of *existence*. The concept of space is frequently associated with Einstein locality, and it may be difficult to think of space in a more abstract sense, even though it is frequently used in a general way—such as in conceptual space or mathematical possibility space. This is one important focus of the next chapter.

A key term for space in ancient Vedic terminology is *Akasha*. Grammatically this word is associated with the ability to shine, view, or reveal. For heuristic purposes, it can be interpreted as containing three fundamental parts (A, ka, sha), which generally can be related to the three levels of transcendental, subtle, and gross space, existence, or *reality*. At this point in the discussion, these three domains can be likened to the level of space that is particle-like with locality (gross), quantum wave-like with locality and nonlocality (subtle), and point-infinity or singularity-infinity (transcendental; wholeness in every part, ultimate unity). In Maharishi Vedic Science, from the standpoint of the diversified point value of nature, nonlocal wholeness is embedded in every local point; from the standpoint of wholeness or unity, local points are embedded in the wholeness. Both simultaneously are embedded in the nature of the unified field—the Veda. These points will be unfolded extensively in later chapters,

Enlivenment of deeper processes of knowing will be discussed in detail when we consider how the subjective means of gaining knowledge in Maharishi Vedic Science and Technology integrates knowledge and experience in the development of higher states of consciousness. But first, we need to lay down several planks toward clarifying classical and quantum *realities*, next discussing the major perspectives of quantum gravity. First we need to be clearer about the two levels of nature—gross and subtle—that are related generally to classical physical *reality* and nonclassical, non-physical quantum *reality,* and also associated with the levels of the physical body and non-physical mind. Then in later chapters these two levels can be related to their underlying transcendental source in the unified field. Chapter 6 addresses the developing understanding of an ontologically *real* level of existence and underlying mental *reality* in terms of more abstract concepts of space.

> "...[I]n order to solve the mind-body problem we need, at a minimum, a new conception of space. We need a conceptual breakthrough in the way we think about the medium in which material objects exist, and hence in our conception of material objects themselves. That is the region in which our ignorance is focused: not in the details of neurophysiologic activity but, more fundamentally, in how space is structured or constituted. That which we refer to when we use the word 'space' has a nature that is quite different from how we standardly conceive it to be; so different, indeed, that it is capable of 'containing' the non-spatial (as we now conceive it) phenomenon of consciousness."–*Colin McGinn, philosopher (p. 103)* [24]

Chapter 5 Notes

[1] Dickson, M. W. (1998). *Quantum chance and non-locality*. Cambridge: Cambridge University Press.

[2] Varela, F. J., Thompson, E., & Rosch, E. (1993). *The embodied mind: Cognitive science and human experience*. Cambridge, MA: The MIT Press.

[3] Hawking, S. (2001). *The universe in a nutshell*. New York: Bantam Books.

[4] Hut, P. & Shepard, R. N. Turning "the Hard Problem" Upside Down and Sideways. In Shear, J., Ed. (2000). *Explaining consciousness—The hard problem*. Cambridge, MA: The MIT Press, pp. 305-322.

[5] Herbert, N. (1985). *Quantum reality: Beyond the new physics*. New York: Anchor Books.

[6] Velmans, M. The Relation of Consciousness to the Material World. In Shear, J., Ed. (2000). *Explaining consciousness—The hard problem*. Cambridge, MA: The MIT Press, pp. 325-336.

[7] Stapp, H. P. The Hard Problem: A Quantum Approach. In Shear, J., Ed. (2000). *Explaining consciousness—The hard problem*. Cambridge, MA: The MIT Press, pp. 197-215.

[8] von Neumann, J., (1932). *Mathematische Grundlagen der Quantenmechanik*. Berlin: Springer Verlag. Trans. R. T. Beyer (1955). *Mathematical foundations of quantum mechanics*. Princeton: Princeton University Press.

[9] Birkhoff, G. & von Neumann, J. (1936). The Logic of Quantum Mechanics. *Annals of Mathematics, 37,* 823-843.

[10] Everett, H. (1957). "Relative State" Formulation of Quantum Mechanics. *Review of Modern Physics, 29,* July, 454-462.

[11] Pigliucci, M. Neuro-Theology, a Rather Skeptical Perspective. In Joseph, R. (2002). *NeuroTheology: Brain science, spirituality, religious experience*. San Jose, CA: University Press, pp. 269-272.

[12] Goswami, A. (1993). *The self-aware universe: How consciousness creates the material world.* New York: G. E. Putnam's Sons.

[13] Bohm, D & Hiley, B. J. (1993). *The undivided universe.* London: Routledge.

[14] Greene, B. (2004). *The fabric of the cosmos: Space, time, and the texture of reality.* New York: Alfred A. Knopf.

[15] Smolin, L. (2001). *Three roads to quantum gravity.* New York: Basic Books.

[16] Penrose, R. (2005). *The road to reality. A complete guide to the laws of the universe.* New York: Alfred A. Knopf.

[17] Goldstein, S. (2002). Bohmian Mechanics. *The Stanford Encyclopedia of Philosophy (Winter Edition),* Edward N. Zalta (Ed.), URL = <http//plato.stanford.edu/archives/win2002/entries/qm-bohm/>

[18] Talbot, M. (1991). *The holographic universe.* New York: HarperCollins Publishers, Inc.

[19] Cramer, J. G.(1980). The Transactional Interpretation of Quantum Mechanics. *Physics Review,* D22, 362.

[20] Goldstein, S. (1998). Quantum Theory Without Observers, *Physics Today,* March, 42-46; April, 38-42.

[21] Eccles, J. C. (1986). Do Mental Events Cause Neural Events Analogously to the Probability Fields of Quantum Mechanics? *Proceedings of the Royal Society of London,* B227; 411-428.

[22] Bohm, D. (1980). *Wholeness and the implicate order.* London: Routledge & Kegan Paul.

[23] Maharishi Mahesh Yogi. (1972). *Science of Creative Intelligence: Knowledge and experience* (Videotaped course), Lesson 23. Los Angeles: MIU Press.

[24] McGinn, C. Consciousness and Space. In Shear, J., Ed. (2000). *Explaining consciousness—The hard problem.* Cambridge, MA: The MIT Press, pp. 97-108.

Chapter 6

Perspectives on Quantum Gravity

In this chapter, we will examine three major approaches to quantum gravity—string theory, quantum loop gravity, black hole thermodynamics—and attempts to integrate them using the holographic principle. Reflecting the increasing abstraction of concepts of space, existence, or reality, we will consider the emerging recognition of information or mental space underlying physical space, associated with the concept of the quantum mind. The importance of the process of observing and the observer, and how modern science has progressed to the stage that it is necessary to address them directly in pursuit of a coherent theory of physical reality, will be emphasized. The overall main point of this chapter is that quantum gravity theories are progressing toward recognizing an underlying nonlocal field of quantum reality, information space, or quantum mind.

Theories of quantum gravity attempt to account for the tiniest part as well as the vast expanse of space and time—in a nutshell the whole universe. Before taking a glimpse at three major theories in this challenging area of research, it may be helpful to introduce briefly some of the perspectives about space and time contributing to the theories. The perspectives reflect the contrast between physical space (matter, body) and mathematical, conceptual, information, or mental space (mind) that underlies efforts in modern science to develop accurate models of the *real* physical world. Formal mathematical logic has been the primary means for building the conceptual models, and the ordinary senses have been the primary means for their empirical verification. Not fully integrating these two means of gaining knowledge has resulted in the explanatory gap between matter and mind, the measurement problem, the gap between ordinary classical objects and their quantum descriptions, the gap between physical space and mathematical or conceptual space, and also the historical mind-body problem.

A key issue in this chapter is the nature of space. As will be discussed, recent quantum gravity theories propose an underlying field of information space that generates ordinary four-dimensional space-time. These theories relate this field to the concept of a sub-phenomenal quantum *reality*, and even a quantum field of pure information underlying physical matter. The Cartesian notion of mind was that it is non-spatial, at least in the sense that it doesn't exist in the domain of ordinary space and time. These emerging notions of a field of *quantum mind* contain some sense of spatial extension, but not in the sense of ordinary space-time. As will be described, this reflects constructive progress toward recognizing an ontological level of nature underlying ordinary space-time that is more mind-like than matter-like.

Relativistic descriptions of ordinary space as something that can be compressed and stretched, curved and shaped, enfolded and unfurled, might seem to imply that it is an incredibly fine rubbery *medium* or *aether* of some kind. This conception of space is now associated with a Planck-size quantized texture, directly related to the speed of light and Einstein locality that limits all objects generated from and moving within it. In ancient Vedic science, this level of ordinary space is generated from a nonlocal level of existence, a subtler medium—the subtle level of space—that has fewer limitations and is more closely associated with what could be called mental space. The expanded framework of a subtler level of space, existence, or *reality* will turn out to be central for addressing the mind-body problem.

Through much of recorded history a common intuition has been that the *real* world is composed of three dimensions of physical space and one dimension of time. In the Newtonian classical framework, the three spatial dimensions represent an invisible, smooth, continuous, unbounded substrate or background upon which *real* objects interact with each other. Space is conceived as a separate non-changing basis or reference for measuring motion or change. Space is assumed to be infinite, and time is assumed to be eternal.

In this classical *non-relativistic* framework there are basically two types of objects: localized material objects and the more diffuse energy fields that surround them in space. Material objects—built of corpuscular things such as atoms—are highly discrete and localized. Energy fields such as the electromagnetic and gravitational force fields are more spread out, but drop off with distance and thus also are localized to some extent (within the light cone). Unless gravity is something different than space, this seems fundamentally inconsistent with the assumed infinity of space and eternity of time. Indeed through much of modern science the gravitational force and its relationship to space has been a mystery. In the vast majority of instances involving measurement of ordinary matter, however, the non-relativistic continuous background framework worked fine for quantifying motion or change; but more careful analysis revealed its limitations, leading to Einstein's relativistic framework of space-time that identifies gravity as the curvature of space-time.

In the classical *relativistic* framework, space is generally thought of as smooth and continuous, but not a non-changing background or substrate. It is not thought of as an underlying, separate, fixed reference or preferred perspective upon which to measure change. Space and time, as well as energy and mass, are interchangeable. All change in nature is conceptualized in terms of changing relationships between objects. Curiously, in this framework physical objects are thought of as localized and independent of each other, but at the same time are understood in relation to each other, and according to a frame of reference or observer perspective. It is perplexing to understand how objects can be independent from each other but exist only in relation to each other—as well as in relation to a frame of reference or observer perspective, to be discussed more in this chapter.

Quantum theory developed within the Newtonian non-relativistic framework in which space is conceived as a continuous background. But its core principle of discrete, discontinuous, quantized units contrasts with the Newtonian smooth continuous notion of space. This core principle has not been able to be reconciled with the conception of gravity as the curvature of space-time in relativity theory. The model of space in quantum theory is that it is discontinuous, rather than smooth and continuous, but also that it is background as in the Newtonian model. Quantum field theory attempts to integrate the quantum model and the model in relativity theory. To build a unified field theory, the gravitational field needs to be unified with the three other quantum fields and with the theory of space-time in general relativity. This is the arena of *quantum gravity*.

In addition, both quantum theory and relativity theory need to incorporate fully the experimental finding of nonlocality. Quantum theory continues to wrestle with particle-wave duality and similar issues that concern how to integrate locality and nonlocality. On the other hand, nonlocality is also at least quite difficult to fit into the classical relativistic framework based on Einstein locality and the speed of light. The work of trying to resolve these issues has led to more elaborate mathematical models, including some that posit additional dimensions in an attempt to bridge the gap between relativistic space-time theory and non-relativistic quantum theory, the two fundamental and most successful theories in modern science.

The circularity of relativistic conceptions of space-time. The classical framework of three spatial dimensions relates to linear Euclidean geometry. Einstein's General Theory of Relativity is a generalization into a non-Euclidean geometry in which gravitational forces are accounted for in terms of space curving, bending, twisting, and folding back onto itself. Einstein showed that the gravitational force could be modeled in terms of the curvature of space-time. He had hoped that the other fundamental force fields also could be modeled as curvatures of space-time—a non-Euclidean, non-linear, relational space—which could serve as the basis for a unified field theory.

For a long time space was thought to be a subtle intangible substance—sometimes referred to as the *aether* (mentioned above). This idea largely had been abandoned in modern science due to Einstein's arguments against it, and because no such substance was detected. It was found that light does not change its speed relative to an observer moving in either the same or opposite direction in empty space, which would not be expected to be the case if light existed in a medium such as an aether-like field. However, if the essence of light is more abstract than photon-based light, as an inherent property of an even subtler medium or aether that underlies and permeates ordinary space-time, then the apparent absolute speed of light in ordinary space-time noted above might still make sense as a limitation of this subtler medium—to be discussed later.

Even in the relativistic framework it sounds as if gravitational space is an incredibly flexible, elastic substance or abstract field of *something* with specific limiting properties, reminiscent of the concept of aether. But the relativistic conception of space was thought to be more abstract, in that space is described as existing *only* in terms of object relations—not an underlying background field or substance. However, it isn't at all obvious how the relativistic framework of space as existing only in terms of relationships between objects fits with the concept of physical space as a force field of gravity in a similar sense as the strong, weak, and electromagnetic force fields. To say both that space exists only in terms of relationships with no independent existence and that it is a gravitational force field with specific properties at least sounds contradictory. If space has only a relational existence, and objects are fundamentally nothing other than curvatures of space-time, then it suggests mass and energy that are interchangeable with space-time also would have only relational existence. What does a theory of relationships between objects mean when there are no fundamental objects? What would the relationships be between? It would seem that objects of some kind are needed in order to have relationships between them.

This impels the question of what an object is. Physical objects certainly would not be able to exist prior to space; but it might be concluded from relativity theory that space doesn't exist prior to and independent of objects. Also, more than one object is required in order to have a relationship between them. How do at least two objects get created, apparently existing in some space-time-like manifold, from which their relationship *generates* the space-time manifold? This might not be such an issue if space had no properties at all; but it is thought to be the field of gravity that has specific properties—that is, properties that define Einstein locality. As generally recognized, this circular 'chicken and egg' type of paradox, in which the primacy of either space or relationships that create it seems irresolvable, will be circled back to later. It will be discussed in connection with the concept of abstract information rather than material objects, the important principle of *duality*, the conditional limitations of Einstein locality as fundamental to the concept of space, and the unexamined concept of a frame of reference or observer perspective.

The concept of space as 'something' that doesn't just have a relational nature is again becoming recognized as a result of quantum theory. For example, the theorized phenomenon of vacuum fluctuations or quantum jitters fundamental to quantum theory, as well as quantum entanglement associated with nonlocality, suggest a quite different notion of space than that it is ultimately empty or that it has only a relational nature. As Greene notes:

"The concepts of empty space and of nothingness take on a whole new meaning when quantum uncertainty takes the stage. Indeed, since 1905, when Einstein did away with the luminiferous aether, the idea that space is filled with invisible substances has waged a vigorous comeback... [K]ey developments in modern physics have reinstituted various forms of an aetherlike entity, none of which set an absolute standard for motion like the original luminiferous aether, but all of which thoroughly challenge the naïve conception of what it means for spacetime to be empty. Moreover...*the* most basic role that space plays in a classical universe—as the medium that separates one object from another, as the intervening stuff that allows us to declare definitively that one object is distinct and independent from another—is thoroughly challenged by startling quantum connections." (p. 76) [1]

"[I]f one applies the rules of quantum theory to the currently accepted general theory of relativity, one finds that the gravitational field is also constituted of such 'wave-particle' modes, each having a minimum 'zero-point' energy. As we keep on adding excitations corresponding to shorter and shorter wavelengths to the gravitational field, we come to a certain length at which the measurement of space and time becomes totally undefinable... When this length is estimated it turns out to be about 10^{-33} cm [Planck scale]. If one computes the amount of energy that would be in one cubic centimeter of space, with this shortest possible wavelength, it turns out to be very far beyond the total energy of all the matter in the known universe... What is implied by this proposal is that what we call empty space contains an immense background of energy, and that matter as we know it is a small, 'quantized' wavelike excitation on top of this background... [F]urther developments in physics may make it possible to probe the above-described background in a more direct way... In this connection it may be said that space, which has so much energy, is full rather than empty. The two opposing notions of space as empty and space as full have indeed continually alternated with each other in the development of philosophical and physical ideas. Thus, in Ancient Greece, the School of Parmenides and Zeno held that space is a plenum. This view was opposed by Democritus, who was perhaps the first seriously to propose a world view that conceived of space as emptiness (i.e., the void) in which material particles (e.g., atoms) are free to move. Modern science has generally favored this latter atomistic view, and yet, during the nineteenth century, the former view was also seriously entertained, through the hypothesis of an ether that fills all space. Matter, thought of as consisting of special recurrent stable and separable forms in the ether (such as ripples or vortices), would be transmitted through this plenum as if the latter were empty... It is being suggested...that what we perceive through the senses as empty space is actually the plenum, which is the ground for the existence of everything..."—*David Bohm, mathematician and theoretical physicist (pp. 241-243)* [2]

"[L]et us imagine what would happen if our eyes were able to see gravitational effects, rather than the narrow band of wave energy in the eletromagnetic field we call visible light. The universe would appear quite different from the way it appears to our eyes now. The Earth would no longer be apparently limited in its dimensions by the physical structure on which we live. We could see the Earth extend far beyond its apparent dimensions and fade gradually into infinity; we would also see that the Sun and the planets physically touch us even though extremely lightly. This simplified example indicates how the perception of fullness or void can be a phenomenon that depends on the perceptual or computational instrumentation being used. The statement that between the Earth and Sun space is essentially empty is true only in a relative sense. Since the force of gravity is operating, there must be a field-energy exchange of particles (gravitons). Space is not, therefore, empty even from that relative perspective.—*Thomas J. Routt, computer scientist (pp. 201-202* [3]

The manner in which the ancient concept of space as an aether has been interpreted in modern science exemplifies a general point. It is that in important cases, such as this one, modern thinkers have attributed more crude and restricted meanings to the ancient concepts that missed important subtleties. Some of these ancient concepts seem to have been *objectified* in modern science in a manner that overlooked their original subtlety; and only recently have we arrived at a refined enough level of understanding to consider that the ancient ideas may have had much more to them than have been credited to them. This will become more obvious in discussing details of ancient Vedic science in later chapters, especially with respect to the fundamental elements and to the nature of mind and consciousness. Even Einstein, after his arguments discredited the classical notion of the aether, supported its revival in a different form:

> "Space without ether is unthinkable; for in such space there not only would be no propagation of light, but also no possibility of existence for standards of space and time...nor therefore spacetime intervals in the physical sense. But this ether may not be thought of as endowed with the qualities of ponderable media, as consisting of parts which may be tracked through time..." (p. 318) [4]

Mathematical versus physical space. Mathematical models in geometry that concern theories of space typically start with the concept of a dimensionless point. Any two points can be connected to form a line, but there are an infinite number of points between any two of them— always an infinite regression of points. How to make a point *in the first place*, how to distinguish the point from infinity or from nothing, and how dimensionless points add together to create one, two, or three-dimensional objects are for the most part unexamined.

As mental conceptions, mathematical theories make certain initial assumptions or axioms— sometimes called an *undefined quantity*. An intuitive-like intellectual concept such as a dimensionless point or straight line is assumed in conceptual or mathematical space, and this is the point of departure for models in geometry. How such initial assumptions are formed *to begin with* necessarily involves consideration of the underlying subjective process of knowing in the knower, which are virtually never addressed in mathematics. The concepts are projections of the knowers, but are focused on as conceptual objects with no consideration of their source—the knower—and the dynamics of their formation—the process of knowing. This contributes to objectification, and the fundamental subjective basis of objectivity remains unattended to and unexamined.

There are geometric models in mathematics that generalize non-Euclidean concepts to higher-order dimensions, especially the field of *topology* which focuses on relationships between objects. It concerns geometric objects or spaces that can be transformed—twisted, compressed, and so on—and how these transformations affect the relationship between the inside and outside of objects, which can be applied to the concept of distance in space. In this approach, geometric objects such as circles or cubes are mathematically transformable and interchangeable in higher-order topological space. Such conceptualizations of higher-order dimensions are thought to be helpful in modeling the complex processes from which quantum force and particle fields arise— such as in M-theory, for example (refer to Chapter 3).

Landscape and still life paintings are examples of how a three-dimensional perspective can appear to be represented on a two-dimensional space. Stereographs and holographs are also examples of how three spatial dimensions can be enfolded into two dimensions. Higher-order mathematical dimensions can be enfolded into fewer dimensions in complex but analogous ways. As mentioned in the brief discussion of quantum logic in Chapter 5, quantum processes are thought to require geometries that are non-linear—associated with non-commutative mathematical logic. It is thought that any model that attempts to account **for non**locality would require a notion of space that is not a linear Euclidean geometry. As a simple analogy from theoretical physicist F. David Peat for such a non-linear geometry:

> "Take...a drop of ink placed on a tangle of string. When the string is unwound, the ink dots become very far apart. Measured along the string itself, the distance between the dots is several feet, but across the tangle they were only a few millimeters apart. The two points are both close together and far apart. To put it another way, measured with respect to a linear unfolded order, the points are a great distance from each other. But with respect to the tangled enfolded order, they are in contact." (pp. 133-134) [5]

Bohm's distinction of explicate and implicate orders, introduced in Chapter 5, might be pictured in this way, as an analogy. In this distinction, the implicate order can be thought of as a higher-order enfolded space, and the explicate order as ordinary unfolded four-dimensional space-time. In Bohm's theory, however, both explicate and implicate orders can be understood to be ontologically *real* levels of existence—not explicate being *real* and implicate being only mathematical or imaginary space.

Envisioning how the classical four-dimensional model of space and time relates to higher-order mathematical geometries—such as the eleven-dimensional model of M-theory—can be quite challenging. On the one hand, models of higher dimensions in *conceptual, mathematical* space might be thought of as distinguishable from *physical space*. Conceptual space also is associated with terms such as configuration space, field space, imaginary space, hyperspace, superspace, unconventional space, and mental space, among others. On the other hand, the higher-order dimensions as mathematically defined constructs are commonly conceptualized in terms of spatial dimensions; and frequently there is at least a tacit assumption that they may represent higher-order dimensions either of ordinary physical space and time or possibly even a sub-phenomenal level of existence or *reality*. In describing progress in interpretations of quantum theory in Chapter 5, the quantum wave level was increasingly attributed ontologically *real* status.

Some recent models envision physical gravitational forces extending into imaginary mathematical space and back into physical space, which certainly brings into question the distinction between these two kinds of space—to be discussed further in the next chapter. [6] In addition, theories such as inflationary big bang theory—to be discussed in Chapter 12—propose finite expansion of the field of space itself outside of ordinary space-time that is not limited by the speed of light, but not instantaneous expansion. This also suggests a fundamentally different notion of space and of motion than the notion associated with Einstein locality, relativistic space-time, and its equivalence with the gravitational field. New (or ancient) notions of space as a subtler field, medium, or aether are now emerging in these cutting edge theories.

Is the observer an object in space? Topological theories involve non-linear models that emphasize the relational nature of mathematical logic, like the relational nature of space-time in relativity theory. Relativity theory associates relationships with the concept of a frame of reference or observer perspective—sometimes called *observer dependence*. When applied to the universe as a space-time continuum, it is assumed that all observers are necessarily inside the universe and unable to observe the entire universe. All relative observer perspectives are thought to be limited by, and dependent upon, their particular relationship to other objects in the universe. Importantly this impacts the fundamental assumption of the independence of observer and observed, making it relativistic also. Although it incorporates the observer into the framework, it also contains the tacit assumption that observers are ordinary independent **objects.** If so, then it is not at all obvious how relationships said to depend on the perspective of an observer—observer dependence—can occur prior to observers in any ordinary sense of what an observer is.

A *perspective* seems to presuppose an observer—some kind of conscious knowing entity that can sense, think, remember, and so forth. Such a conscious knowing entity presumably would have an evolved physical brain, preceded by an extensive physical evolutionary history. Brains exist in space and time. Prior to evolved brains associated with observers that have perspectives, there would be the curious situation of numerous objects with numerous relationships but no observer perspectives because ordinary observers didn't yet exist. The notion of a frame of reference is *objectified*—and the subjective observer ordinarily needed to have an observer perspective is, practically speaking, completely overlooked and unexamined.

This of course depends on the definition of an observer—as well as the process of observing. Implicit in relativity theory is an unexamined conception of mind and consciousness of the observer important in the concept of an observer perspective or frame of reference. Eventually this will be seen to be a key issue in characterizing the limitations of relativity theory—to be discussed in this and the next chapter.

Is space fundamentally discrete? Whereas in both the non-relativistic and relativistic classical frameworks space-time is continuous and smooth, quantum theory proposes the notion of the discontinuity and quantization of space. Ultimate limits to the investigation of smaller and smaller scales of physical space and time—associated with the uncertainty principle—clashed with the mathematical model of a continuous space of dimensionless points, and with relativity theory. What has been emerging in the quantum paradigm is a model in which space-time itself is thought to be discrete and quantized at the Planck scale. On macroscopic and ultramacroscopic scales space-time appears to be continuous, but at the ultramicroscopic Planck scale its ultimate quantized nature is theorized to be revealed.

According to the quantum paradigm, discrete quantum fluctuations are the essential fabric of space-time. Space-time is envisioned as inherently enfolded into quantized units—such as of space-time foam. All events in nature involving motion or transfer of energy in any form are thought to be composed of a finite number of smallest possible quantized units of space-time. All objects that make up the universe—whether elementary particles or possibly black holes—are viewed as composed of non-linear topologies of discrete units of enfolded space-time at the Planck scale.[7] In other terms, space-time itself is thought to be digital rather than analog—what might be called the ultimate fragmented or *sandbox* view of nature. This view will eventually be seen as an incomplete and inadequate articulation of the nature of space.

The three major theoretical approaches to quantum gravity examined in this chapter—string theory, loop quantum gravity theory, black hole thermodynamics—all posit that at the basis of the entire physical universe is the smallest quantized time and distance scale. These important new theories also generally are consistent with the understanding that this hypothesized fundamental unit of nature is the Planck size. The Planck size is thought to be the smallest possible structure of nature itself, the ultimate limit of space and time, beyond which space and time can be said to be completely meaningless. It is calculated using known constants derived from fundamental principles—Planck's constant, the speed of light, and the gravitational constant from Newton's laws of gravitation. It also is directly related to the uncertainty principle. There is the Planck time, the fastest change; the Planck length, the shortest distance; the Planck energy, the basic unit of energy; and the Planck temperature, the hottest anything can be. According to this perspective, summarized by theoretical physicist and loop quantum gravity theorist Lee Smolin, the world we live in today—the four-dimensional space-time emerging after the big bang—is "incredibly big, slow, and cold…" (p. 63) [8]

String Theory

As discussed in Chapter 4, measurement precision is limited by the size of the probes used: smaller probes allow increased precision of measurement. For example, shorter wavelengths of light, which contain higher energy, allow increased precision or finer-grained probing and measurement. However, this relationship is thought to be fundamentally different at the range of the Planck scale, the ultramicroscopic scale of hypothesized strings and branes. Whereas at larger scales the wavelengths get shorter with increased energy, at near the Planck scale strings and branes are found mathematically to *grow* in size when energy is added, thus preventing the possibility of creating anything smaller—including a smaller probe.[9] This appears to derive from the uncertainty principle, placing an ultimate limit on measurement because there can be nothing available with which to probe lengths smaller than the Planck length (10^{-33} cm). Even the concept of a metric of time and space is thought to break down at the Planck scale—at least our ordinary conventional concept.

This provides an approach to reconcile the Newtonian classical framework with the measurement limitations associated with hypothesized quantum fluctuations. [8] The problem of trying to reconcile the smooth, continuous, classical model with the frenetic quantum model may be due to mistakenly applying the mathematical concept of a *dimensionless point* used in the classical framework to the ultramicroscopic level associated with the theory of the quantum. When the concept of a point that has no extension in space is used as a model for quantum processes, the calculations necessarily have a probabilistic component because at this level the processes may not be point processes but rather *Planck-size* processes.

In string theory, space-time is *not* an idealized smooth geometry of infinitesimal points with no spatial extension. Space-time is *quantized* into the smallest possible quantity—hypothesized to be the ultimate uncuttable or irreducible constituent of nature. This ultimate quantum unit is thought to be the basis of all objects in space-time in the unfurled universe, as well as the core of black holes and the ultramicroscopic level of the vacuum state of space-time foam. Greene states:

> "…[S]uperstring theory starts off by proposing a new answer to an old question: what are the smallest, indivisible constituents of matter? For many decades, the conventional answer has been that matter is composed of particles—electrons and quarks—that can be modeled as dots that are indivisible and that have no size and no internal structure… Superstring theory…does not deny the key role played by electrons, quarks, and the other particle species revealed by experiment, but it does claim that these particles are not dots. Instead, according to superstring theory, every particle is composed of a tiny filament of energy, some hundred billion billion times smaller than a single atomic nucleus…which is shaped like a little string… Going from dots to strings-so-small-they-look-like-dots might not seem like a terribly significant change in perspective. But it is. From such humble beginnings, superstring theory combines general relativity and quantum mechanics into a single, consistent theory, banishing the pernicious infinite probabilities afflicting previously attempted unions." (p. 17-18) [1]

In a major step toward unification of all matter and force fields in a unified field theory, string theory proposes that all particles that make up the entire universe are tiny strings that vibrate at different frequencies and patterns. More intense vibrations involve more energy, and are thus more massive. Greene explains:

> "From $E=mc^2$, we know that mass and energy are interchangeable... But Einstein's equation works perfectly well in reverse...and that's the direction in which string theory uses Einstein's equation. The *mass* of a particle in string theory is nothing but the *energy* of its vibrating string... [A] massless particle like a photon or a graviton corresponds to a string executing the most placid and gentle vibrational pattern that it possibly can." (p. 354) [1]

> "Since even a placidly vibrating string has *some* amount of energy, you might wonder how it's possible for a string vibrational pattern to yield a massless particle. The answer, once again, has to do with quantum uncertainty. No matter how placid a string is, quantum uncertainty implies that it has a minimal amount of jitter and jiggle. And, through the weirdness of quantum mechanics, these uncertainty-induced jitters have *negative* energy. When this is combined with the positive energy from the most gentle of ordinary string vibrations, the total mass/energy is zero." (p. 528) [1]

A key issue in string theory and its further development in M-theory is how strings and higher-dimensional branes produce the fundamental particles and forces that comprise all material objects in nature. Another related issue is how the higher-order dimensions remain compactified or *enfolded* while the ordinary four dimensions of space and time are *unfolded* into our familiar universe. These issues concern how changes in the geometric patterns of interacting strings or branes could produce different masses, charges, and spin types of elementary particles. One promising group of higher-order, non-linear geometric patterns is called *Calabi-Yau shapes,* which are used to model compactified (enfolded) dimensions in the Planck scale range, the theorized smallest possible size of an object in ordinary space and time.

However, the theories of strings and branes still imply a Newtonian-type background or substrate of space and time within which strings and branes vibrate. One way of thinking about this is that although strings are hypothetical objects that involve higher mathematical dimensions, the theory draws parallels to—if not attempts to represent—processes that require a background of physical space-time.

For example, strings and branes in the compactified dimensions of the Planck scale are conceptualized as changing from one spatial geometry to another, such as from four-dimensional to three-dimensional space, or seven-dimensional to six-dimensional space, through a process in which the fabric of space tears and reconnects into a different Calabi-Yau shape. These compactified processes near about the level of the Planck scale are hypothesized to determine matter and force particles in the non-compactified four-dimensional world. When this mathematical theory is applied to physical space and time, space and time as a substrate of these dynamic tearing and reconnecting processes is implied.

Without a background of space and time, in what do the strings vibrate when the string vibrations themselves are thought somehow to produce space-time? Space-time needs to be a product of the theory, rather than be dependent on it. Again, Greene explains:

> "[O]ur present formulation of string theory presupposes the existence of space and time within which strings (and the other ingredients found in M-theory) move about and vibrate... Finding the correct mathematical apparatus for formulating string theory without recourse to a pre-existing notion of space and time is one of the most important issues facing string theorists. An understanding of how space and time emerge would take us

a huge step closer to answering the crucial question of which geometrical form actually *does* emerge." (pp. 378-380) [8]

String theory shares with quantum theories in general the non-relativistic framework of space and time, sometimes called a *background dependent* framework—in contrast to the General Theory of Relativity that is said to be background *independent*. Indeed, relativity theory is sometimes interpreted to imply that there is no background at all from which space-time could be dependent or independent. However, in relativity theory space-time can be said to be *observer dependent*, which can be thought of as relating to a background in some sense—to be discussed later. As a background *dependent* model, string theory doesn't yet integrate the background *independent* relativistic framework. Although string theory can be understood as a *theory of everything* in the sense of describing a basis of all physical objects, it doesn't integrate the relativistic space-time continuum. In this way, it is not a theory of a unified field. It is strongly believed by most physicists that a unified field theory needs to integrate fully the space-time continuum so that there is no background of space-time. String theory is a consistent model of quantum gravity within a non-relativistic framework; but a fully relativistic string theory has not emerged yet. [7 8]

String theory has other challenges that make it not yet a developed candidate for a unified field theory. There are numerous possible consistent string theories. M-theory represents an integration of some of them; but its predictions are not yet precisely calculable from well-defined principles in order to eliminate other consistent theories. Also, string theory requires the existence of super-symmetric partners for fermions and bosons that have not been found experimentally.

According to string theories, a large portion of the mass of the universe is in a hypothesized hidden sector of dark matter. Although major efforts to find evidence for this are underway, the evidence is scant, and some might say non-existent. In order for these theories to be a viable approach to quantum gravity, evidence for the graviton and its super-symmetric partner, associated with the gravitational force field, also need to be found—as well as the super-symmetric sparticle partners of all the other particles. Gravitons would interact with all the super-symmetric partners of all the forms of physical energy, and also with each other, and a theory of quantum gravity needs to be able to predict the results of such interactions more precisely than these string theories are able to do. [6] In addition, evidence needs to be found for the hypothesized higher dimensions within which the strings are conceived as vibrating, as well as the specific forms of the hypothesized higher-order dimensions, which to date have not been found. How the higher-dimensional *reality* ends up appearing as the ordinary four-dimensional world we are familiar with is another important unresolved issue, containing several complicated problems.

Another issue is that string theory appears to be at variance with recent experimental findings supporting the notion that most of the energy in the universe is in the form of a positive cosmological constant [8]—to be discussed in Chapter 12. Further, the more developed versions of string theory include mathematical objects in addition to strings, such as branes. There is yet to be an

> "Supersymmetry was initially devised as a sort of addition to certain physical theories known to not correspond to reality, but which were being studied for their interesting mathematical properties. Essentially, physicists were playing a mathematical game... Later, after the mathematical ideas and techniques were developed, theorists began to see their applicability to particle physics. While of this writing (December 2003), there is zero direct evidence that SUSY [supersymmetry] is a property of the universe, there is strong suspicion that it might be, as evidenced by the approximately 10,000 scientific papers written on the subject... SUSY may yet prove to be an interesting, yet ultimately wrong, idea."—*Don Lincoln, particle physicist (p. 395)* [10]

explanation of whether these other geometric entities are composed of strings, or vice versa; or whether some other entities underlie them.

An interesting new direction in M-theory explores a possibly more fundamental field underlying strings, from which a coherent background of string vibrations emerges that produces classical space and time in our familiar four-dimensional world. This relates to the concept of a *zero-brane* in M-theory.[8] Zero-branes have no spatial extent, like the classical notion of a point particle, except that they are attached to strings, function differently than classical point particles, and thus don't produce the same mathematical problems associated with quantum fluctuations that plagued classical point-particle theory and that string theory resolved. Greene points out in the following quote that zero-branes might be placed:

> "...possibly in an era that existed before the big bang or the pre-big bang (if we can use temporal terms, for lack of any other linguistic framework)... [A] *zero-brane*...may give us a glimpse of the spaceless and timeless realm... [W]hereas strings show us that conventional notions of space and time cease to have relevance below the Planck scale, the zero-branes give essentially the same conclusion but also provide a tiny window on the new unconventional framework that takes over. Studies with these zero-branes indicate that ordinary geometry is replaced by something known as non-commutative geometry... In this geometrical framework, the conventional notions of space and of distance between points melt away, leaving us in a vastly different conceptual landscape... [I]t gives us a hint of what the more complete framework for incorporating space and time may involve... Already, through studies in M-theory, we have seen glimpses of a strange new domain of the universe lurking beneath the Planck length, possibly one in which there is no notion of time and space." (pp. 379-387) [8]

This glimpse at a more fundamental level that is beyond *conventional* space and time is taken further in the next theory of quantum gravity, called loop quantum gravity. This theory identifies conceptual or information space as underlying and generating physical space. Later this will be associated with the notion of *quantum mind*. As we proceed further in the book, connections between the notion of an ontological level of quantum *reality* or quantum mind and the subtle relative domain of nature in Maharishi Vedic Science and Technology will be articulated further.

In the context of uncovering aspects of nature not addressed in string theory, it may be useful to point out that in addition to not integrating completely the relativistic space-time continuum, string theory also doesn't directly address fundamental issues related to the process of observing and the observer. A unified field theory needs to describe how everything in creation emerges from the unified field, without anything other than or outside of itself—including the process of knowing and the knower. This is a crucial issue for string theory, as well as any potential unified field theory, to address.

In quantum theory, to which string theory fundamentally relates, the observer is outside the system and there is no way to include the observer in it—no way to include the observer in the quantum wave function, as noted in Chapter 4. This clearly calls for a more fundamental framework than quantum theory in order to formulate a completely unified field theory. This is the direction we are headed, for which the planks of understanding are being laid.

The two other major approaches to quantum gravity draw more from cosmological research and relativity theory. Loop quantum gravity attempts to build a model of quantum gravity in a relativistic framework that attempts to incorporate at least some notions relevant to an actual observer. It offers an alternative to strings through the concept of a network of quantum gravity loops, while also maintaining the conception that space-time is ultimately quantized at the Planck scale.

Loop Quantum Gravity

Newtonian physics conceptualized an object as an independent entity and focused on the *state* of the object at a fixed space and time, rather than change in the object through time. Quantum theory, and to some degree relativity theory, incorporates a more dynamic characterization of an object as a *process*, focusing on change through time and considering at least some of the history of the object. The concept of a *system* is a more expanded characterization, emphasizing relationships between processes through time. The state of an object is a momentary snapshot of an ongoing process, within the larger context of a system of ongoing processes.

An even more expanded approach is that processes and states emerge from increasingly limiting perspectives that focus on parts of a *holistic* system. This emphasizes the wholeness of the system, toward an important change from the reductive approach of explaining higher-order complex phenomena by lower-order, less complex phenomena—referred to in the Prologue. This will be shown later to be a key approach for developing a more comprehensive understanding of space and time, as well as a completely unified field theory. It reflects the holistic approach of Maharishi Vedic Science and Technology toward which these theories are progressing.

Loop quantum gravity characterizes the universe as a complex interrelated system of constantly changing causal events or processes. It represents a formal attempt to unite quantum theory with the background independent—but observer dependent—framework of relativity theory. It states that the universe is a closed system, all observers are in the system, and the speed of light sets a fundamental limit that defines the range of causal effects—the past and future light cones—for a particular observer in the cosmological system. Along with these relativistic principles, it adds the quantum perspective of space as discrete and having the tiniest possible geometric area and volume—the Planck scale.

In the systemic approach of loop quantum gravity, motion or change is primary, and space-time is defined in terms of change. The universe is not made up of objects *in* time and space, but rather an incredibly large number of bottom-line information *processes* or *events,* which can be thought of as the smallest units of change. Matter is composed of atoms, and the universe is composed more abstractly of histories of elementary information events. [7]

Meaningful events are events that are causally related to each other—that is, occurring within the range of influence of the past or future light cones. The universe is a system of causal relations between events defined by interacting light cones—the *causal structure*—which is in constant dynamic flux. The interacting light cones are pulled in the direction of massive objects—the curvature of space-time constituting the gravitational force. [7] This causal structure is a constantly changing network of the smallest possible events or bits of information change that underlie and generate macroscopic phenomena we experience as the classical four-dimensional world.

When the system under consideration is the entire universe, it seems logical to assume that all observers must be inside the universe and cannot view its entirety. This is a key assumption used in loop quantum gravity in an attempt to place the observer inside the system. In this system of dynamic causal structure, loop quantum gravity emphasizes the context dependence of events with respect to observer perspectives. It attempts to *relativize* quantum theory by emphasizing the relational, context-dependent nature of events tied to observer perspectives inside the cosmological system.

Context dependence is related to consistent histories of events associated with particular observers, extending the principle of decoherence described in Chapter 5. In this view, there is one universe, with many overlapping perspectives based on innumerable different observers and their consistent decoherent histories. This is thought to explain how dynamic change in the quantum universe produces the phenomenal classical *reality* of definite decoherent outcomes that appear consistent across observers, without the subjectivity of the observer taking a role in collapsing the wave function in the act of observation. It associates superpositions of quantum events with the context dependence of an unlimited number of observer perspectives or minds.

Instead of each quantum superposition associated with its own mind-world, however, there are many separate observer perspectives of the same world. There is one universe, composed of many different quantum perspectives based on separate observers. Taking the concept of context dependence a bit further, it refers to the entire past history of the observer as defining the current momentary perspective of the observer in the one universe. In the one universe is an unlimited set of separate but overlapping perspectives, defined by the experience of each observer based on the observer's particular location, time, and history in the universe.[7]

This draws from the mathematical approach of *topos theory*, which allows for many quantum descriptions to be derived from one mathematical formalism according to different contexts.[6] The descriptions of the universe are different for each observer, but they have to be logically consistent or compatible with each other in a decoherent manner if the same questions are asked about it across observers and their answers are to match—which allows for intersubjective consensual validation.

In this manner the paradox of superpositions in the quantum wave function associated with quantum descriptions of *reality* compared to classical *reality* is thought to be resolved—by making the superpositions a consequence of the different overlapping perspectives of an unlimited number of observers. It describes superpositions of histories that are independent of each other, each associated with a different observer's perspective. However, it allows agreed-upon outcomes or observational answers when there are the same initial questions or conditions and similar histories. In other words, it rather amazingly seems to place the apparent consistency across individual observers as originating in the initial conditions of the universe. Unaddressed is what is meant by an observer and the unique observer perspective, as well as how each unique independent perspective can actually overlap with many other observer perspectives for an agreed-upon view of the world based on consistent initial questions about it.

Further, core aspects of the subjectivity of the observer still remain *outside* the relativistic system. The observer still experiences a unitary state from outside the relativistic system. Smolin points out:

> "The quantum description is always the description of some part of the universe by an observer who remains outside it... If you observe a system that includes me, you may see me in a superposition of states. But I do not describe myself in such terms, because in this kind of theory no observer ever describes themselves. Rather than trying to make sense of metaphysical statements about their being many universes—many realities—within one solution to the theory of quantum cosmology, we are constructing a pluralistic version of quantum cosmology in which there is one universe. That universe has, however, many different mathematical descriptions, each corresponding to what a different observer can see when they look around them. Each is incomplete, because no observer can see the whole universe. Each observer, for example, excludes themselves from the world they describe. But when two observers ask the same questions, they must agree... One universe, seen by many observers, rather than many universes, seen by one mythical observer outside the universe." (pp. 47-48)[7]

Although this attempts to place the individual observer into a relativistic quantum system, core aspects of the observer still remain fundamentally separate. The system of the universe is thought to be background independent with respect to space and time, and to be context dependent and relativistic. But the observer's *subjective experience* remains outside the observed system—tacitly continuing to be a phenomenal background of some kind, or at least in the background, and not an explicit part of the theory. The frame of reference and observer perspective are core concepts in relativity theory. But they are highly objectified concepts, referring to any position in space-time that establishes conceptually a view or reference point of the universe. Presumably an actual observer—a conscious mind—is involved somehow, at some point. Examples to illustrate the relativism of space and time typically do place an actual observer into the observer perspective or frame of reference. For example, implicit in the concept of past and future light cones is an observer that establishes the frame of reference. But this use of *observer dependence* is a quite objectified notion of a frame of reference—only implicitly involving a conscious subjective mind. It is recognized that an observer is needed to view the universe, but the nature of an actual conscious observer that establishes a frame of reference remains unexamined.

In the orthodox interpretation of quantum theory, a conscious observer 'creates' the quantum wave function collapse upon observation into a discrete classical state. The conscious mind of the observer has a central role in the collapse to a discrete state of the object observed. The world can be said to be observer dependent in the sense that the conscious observer is the only object with the power to collapse the quantum wave function. But in quantum theory there is no way to incorporate the observer into the quantum wave function representation of nature.

In the attempt to *relativize* quantum theory in loop quantum gravity theory—making it observer dependent—the relativistic frame of reference becomes a partial consistent history of the universe from a particular observer's perspective. [6] In this theory, the observer perspective is *slightly* more explicit as involving an actual conscious observer. But any ordinary meaning of a conscious observer—such as the ability to self-observe or have some sense of individual self—is not included. These aspects of an observer still are *somewhere* outside the observed cosmological system. Crucial aspects of the observer are excluded from the observed world on the one hand; but on the other hand, they are assumed to be in it and thus only able to observe part of it. The inner conscious experience of the observer still is not integrated into the system being observed; and can be thought of as a background in some unspecified sense.

Further the process of observing—how the observer's separate perspective is formed—is not addressed. The theory is *observer dependent* in terms of the relational context, but it does not incorporate fully the observer or the process of observing. This is inconsistent with the assumption that there is no underlying background of any type or any thing outside of the closed relativistic system. Something still remains outside of the observed system—namely, the observer and the ability of the observer to self-observe as well as the unitary experience of observing even a part of the entire system. The notion that the theory is observer dependent inside the relativistic system, and that the observer somehow observes at least aspects of the system from the outside—independent of it—remains a fundamental quandary.

Loop quantum gravity thus can be viewed as another black box model with respect to the mind and consciousness of the observer, even though it purports to place the observer into the overall system of the universe as specific separate event histories and observer perspectives. In this theory, the observer is an unusually limited observer who cannot self-observe or self-reflect—which is a core feature of definitions of consciousness in most theories of mind, including the physicalist paradigm. The subjectivity of the observer is still not incorporated into the theorized closed causal nexus attempting to be built and assuming to include everything in it.

In ancient Vedic science and Maharishi Vedic Science and Technology, the 'nature' of consciousness is self-referral. This is not part of the observer perspective in loop quantum gravity theory. The objectified focus on the object of experience, the known—while leaving out consideration of the process of knowing and the knower—is evident in this attempt to describe a closed causal universe. The attempt to integrate the major approaches to quantum gravity discussed in this chapter by the use of the holographic model does not resolve these issues either. This will be brought out in Chapter 7, but first let's consider other key aspects of loop quantum gravity.

Quantum chromodynamics. In loop quantum gravity, the background independent, observer dependent framework of relativistic space-time is associated with findings from the important phenomenon called *superconductivity*, in which strange properties emerge when particular metals drop below a critical temperature near absolute zero. At this temperature, electrical resistance falls to zero—apparently no resistance at all.

Electric fields in metals at ordinary temperature ranges exhibit continuous field lines. But in superconductors, the magnetic field lines are discrete. The idea that electric field lines might also exhibit discreteness has been useful in describing how elementary particles called *quarks* function in protons, neutrons, and other particles. Quarks are theorized to have the property of never being able to be separated completely in order to move on their own outside of the particle in which they are embedded. The force holding the quarks together is a connecting charge that functions somewhat like electric and magnetic field lines. In quarks there are three kinds of charge, each either positive or negative, which results in six different types of quarks. The property associated with these forces is called *color* charge, and the area of physics in which it is studied is *quantum chromodynamics (QCD)*.[7]

As described in QCD theory, when quarks are very close to each other in a particle they are pretty much able to move freely in a way to suggest that the force or charge holding them together is weak. But if any two quarks move farther apart, the force keeping them together rises to a fixed discrete value that doesn't fall off when they attempt to be separated. This is in contrast to the electric force, as well as the other known forces, that decrease with distance. It is as if the quarks are always tethered to each other, such that even as they move farther apart a fixed amount of force holds them together. It is not known of what this connecting force with such unusual properties is made; but it appears to behave like a magnetic field line in a superconductor, and is sometimes envisioned as a string or quality of 'stringiness.'[7]

It is theorized that space itself may function like a superconductor, in which the discrete lines of force are color charges of quarks. In this theory, the color force field could be thought of as the most fundamental level, associated with QCD. This approach attempts to describe how empty space functions like a superconductor in a field of randomly oscillating vacuum fluctuations. A second way, related to string theory, is that the charge as a string or connection between quarks is the more fundamental level, and the field is only an approximation of how the more fundamental strings behave. These two alternatives may be just different ways of describing the same phenomenon—the *hypothesis of duality*—a possible way to integrate string and field concepts of space.[7]

A fundamental difficulty in general of the theory of fields is that it leads to infinite quantities. In field theory, the strength of a field increases closer to a particle. But a particle is conceived as a dimensionless point, so the field would increase in strength at closer and closer distances until it theoretically reaches infinite strength. This is an unacceptable result in physically meaningful calculations, which require only finite values of energy consistent with the laws of conservation.

In loop quantum gravity, this concern about field theory is eliminated by extricating the field from the notion of a fixed continuous background. The field is defined relativistically—only in terms of its relationships—as a network of relationships that change and evolve dynamically. A simple analogy of a system that is defined by the relationships within the system is a graph, which is a set of points connected by its edges, embedded in a fixed substrate. In the relativistic background *independent* framework of loop quantum gravity, space-time is thought to emerge from a network of interacting loops that has no underlying fixed substrate. It can be likened to the concept of a *fractal* built up of changing loop networks that define space. [7] There is no underlying substrate or background framework; it is background independent with respect to space and time. Space and time emerge only due to relationships that build the network.

Spin networks and a non-physical, pure geometry of information space. In the theory of loop quantum gravity, space is *created* from topological relationships occurring in a dynamically evolving network of intersecting loops—called a *spin network*. Smolin describes the theory as a quantum version of Einstein's theory of gravity, and further elaborates:

> "…[T]he theory depended only on the relationships of the loops to one another—on how they knot, link and kink…based on the idea of a discrete line of force, as in a magnetic field in a superconductor. Translated into the loop picture of the gravitational field, this turns out to imply that the area of any surface comes in discrete multiples of simple units. The smallest of these units is about the Planck area, which is the square of the Planck length. This means that all surfaces are discrete, made of parts each of which carries a finite amount of area. The same is true of volume… The discreteness of space is associated with the mathematical concept that areas and volumes can only come in integer values, multiples of certain fixed units, and that the most fundamental unit cannot be zero… A spin network is simply a graph…whose edges are labeled by integers. These integers come from the values that the angular momentum of a particle are allowed to have in quantum theory, which are equal to an integer times half of Planck's constant… The integers on each edge of a network correspond to units of area carried by that edge. Rather than carrying a certain amount of electric or magnetic flux, the lines of a spin network carry units of area. The nodes of the spin networks also have a single meaning: they correspond to quantized units of volume. The volume contained in a simple spin network, when measured in Planck units, is basically equal to the number of nodes of the network… A very large spin network can represent a quantum geometry that looks smooth and continuous when viewed on a scale much larger than the Planck length… In the spin network picture, space only seems continuous—it is actually made up of building blocks which are the nodes and edges of the spin network… The spin networks do not live in space; their structure generates space. And they are nothing but a structure of relations, governed by how the edges are tied together at the nodes." (pp. 130-138) [7]

This model of a constantly changing network of intersecting loops also draws from Penrose's *twistor theory*. [7] In twistor theory, the fundamental geometry is not dimensionless points but an expansion of the concept of a line of infinite extent called a twistor. A spin network of dynamic nonlocal processes associated with twistors generates its own curved space-time in a manner that produces quantum particles of localized matter. The spin network in loop quantum gravity has similarities with space-time foam, and is sometimes called *spin foam*. [7] It also sounds like string theory, but an important difference is that the loops are not *in* space but are said to *generate* it.

Spin networks are described as a *pure geometry*, from which physical space is generated. [7] But how can these spin networks—envisioned as structural networks of nodes and edges with minimal area and volume in *mathematical, functional, conceptual* space—generate actual *physical* space? There appears to be a sort of backward quantum jump from physical space to mathematical, conceptual space—to functional structure without physical form.

If the basis of space is a *non-physical* pure geometry, then an explanation is needed for how this fits into relativistic space-time with its defining features of Einstein locality and a closed causal nexus that doesn't envision anything outside it. Also, given that the spin network is different from physical space, what are the causal relationships extending across physical, structural space and non-physical, functional, conceptual space? This is basically the same issue that was not adequately addressed historically in dualistic mind-body theories (also discussed in Chapter 4 as a core issue in quantum theory). In the current form, it is the relationship between the structure of Planck-size quantum bits of physical space-time and the functional concept of abstract non-physical bits of information space—akin to the "it from bit" model that the physical world has an underlying abstract informational structure.[11][12]

In loop quantum gravity theory, the source of physical space-time—the pure geometry of conceptual space—implies *something* outside of the closed relativistic physical universe. The pure geometry of information space is outside of physical space and is said to be its underlying generator. We also discussed earlier in this section on loop quantum gravity that there is something else outside of the closed causal network of the relativistic physical universe: namely, the observer's ability to observe and self-observe. Core aspects of the observer's subjectivity remain outside of the relativistic system. Could it be that these two *outside* things or processes—the pure geometry of information space, and what might be called the mental space of the subjectivity of the observer—relate directly to each other? Could it be that these theories are beginning to identify the foggy outlines of a deeper level of ontological existence underlying conventional notions of physical space-time that can be associated with a nonlocal mental *reality*?

The emerging ontology of a quite abstract information space beyond physical space is reminiscent of Bohm's implicate order. Could it also be that this deeper level relates to observer dependence—and to the level of mind, as Bohm theorized? In this more mind-like level of *reality,* a distinguishing feature would need to be that it somehow simultaneously incorporates locality and nonlocality. For this, it would need to go beyond the conventional notion of space-time characterized by Einstein locality.

Moving from physical, relativistic, quantized space-time to quantized information space is an important step closer to recognizing mental space that underlies physical space—a major plank on the bridge to Vedic science. These issues, including the role of the observer perspective, again will be referred to after summarizing a related major approach to quantum gravity, black hole thermodynamics. In this third approach to quantum gravity, the concept of observer perspective or frame of reference can be said to relate even more explicitly to an actual conscious observer.

Black Hole Thermodynamics

This third approach to quantum gravity focuses on the extreme dynamics of black holes as a window into the nature of space-time. It draws upon thermodynamics and information theory—especially the concept of entropy, directly related to the concept of order. It posits a quantized structure of space-time at the Planck scale, as well as quantum ignorance and the inherent randomness of nature. It also emphasizes observer dependent limitations or reference frames, an important but unexamined feature of relativity theory associated with the causal light cone.

A key concept in this approach is the horizon of a black hole, defined by the mechanics of light waves interacting with the incredibly strong gravitational forces forming the black hole. The pull of gravity due to the density of mass in the black hole prevents light—as well as anything else, including any physically-based or quantized form of information—from leaving it. This establishes separate perspectives for observers inside compared to observers outside of the event horizon of the black hole. Observers outside the black hole would not see anything inside it—only a black area or hidden region. However, observers drawn into the black hole would not experience the event horizon as would outside observers, because they would be able to see inside the black hole. One interesting implication of this approach seems to be that even abstract information is subject to gravitational forces—which ties it directly to mass and energy and suggests it has a conventional physical-like character in a manner that doesn't seem to be implied in a pure information space that underlies and generates conventional space-time including the gravitational force.

Hidden regions and entropy. A hidden region is established when there are events from which an observer will not be able to receive information because the information is outside of the observer's light cone. Interiors of black holes are one type of hidden region for outside observers. Hidden regions may also be created if the rate of expansion of the universe increases such that light signals cannot be received by some observers, such as in inflationary big bang theory.

The role of the observer's perspective is evident in hypothetical descriptions of what an actual observer might experience in a spaceship near a black hole. Einstein's Special and General Theories of Relativity depend on whether the observer is undergoing constant inertial motion or accelerating motion. In the General Theory of Relativity, an observer cannot tell the difference between accelerating motion and being in a gravitational field—Einstein's *equivalence principle*. An observer in a spaceship floating near a black hole in inertial motion would see just empty space around. But as soon as the rockets of the spaceship are turned on and the spaceship accelerates, theoretically the situation would instantaneously and dramatically change. Smolin hypothesizes what will be experienced by an observer in the accelerating spaceship:

> "First she will experience the normal effect of acceleration, which is to make her feel heavy, just as though she were all of a sudden in a gravitational field... [A]s soon as she accelerates, our observer's particle detectors will begin to register, in spite of the fact that, according to a normal observer who is not accelerating, the space through which she is traveling is empty... Our accelerating observer sees herself as traveling through a region filled with particles... Even more remarkable is what she will see if she looks at her thermometer. Before she began accelerating it read zero, because temperature is a measure of the energy in random motion, and in empty space there is nothing to give a non-zero temperature. Now the thermometer registers a temperature, even though all that has changed is her acceleration. " (p. 81) [7]

Accelerating observers will experience being surrounded by a hot gas of photons, with their instruments registering non-zero temperature. Apparently even observing the same region of space, however, an inertial observer will see it as empty space and the instruments will not register.

In order for the instruments in the accelerating spaceship to register a non-zero temperature, a source for the heat energy must be present. It is theorized that the heat energy comes from the rocket engines, because it is present only when the spaceship is accelerating due to firing of the rockets. [33] According to the uncertainty principle, electric and magnetic fields are complementary, which means that precise measurement of one of the fields necessarily means inability to measure the other field. Both of the fields, even in empty space, cannot contain zero energy. There is zero point motion or intrinsic random motion even in a field with zero temperature—the hypothesized vacuum fluctuations of the quantum field. Although these fluctuations cannot be detected by instruments at rest because they carry no energy, they can be detected by accelerating instruments because the acceleration provides the needed energy source. This raises the temperature of the thermometer on the accelerating spaceship. [7]

As to how the energy of the rockets becomes the random energy of heat recorded as temperature, Smolin explains:

"It turns out to have to do with...non-local correlations between quantum systems... The photons that make up the vacuum electric and magnetic fields come in pairs that are correlated... What is more, each photon detected by our accelerating observer's thermometer is correlated with one that is beyond her horizon. This means that part of the information she would need if she wanted to give a complete description of each photon she sees is inaccessible to her, because it resides in a photon that is in her hidden region. As a result, what she observes is intrinsically random. As with the atoms in a gas, there is no way for her to predict exactly how the photons she observes are moving. The result is that the motion she sees is random. But random motion is, by definition, heat. So the photons she sees are hot!" (p. 84) [7]

Entropy is a measure of the randomness in a system, which also can be understood to be a related to information. Smolin further states that according to *Unruh's law*:

"Accelerating observers see themselves as embedded in a gas of hot photons at a temperature proportional to their acceleration." (p. 86) [7]

He also states that according to *Bekenstein's law*:

"...[With] every horizon that forms a boundary separating an observer from a region which is hidden from them, there is associated an entropy which measures the amount of information which is hidden behind it. This entropy is always proportional to the area of the horizon... These two laws are the basis for our understanding of quantum black holes..." (pp. 86-87) [7]

If a particle is trapped in a black hole and cannot escape, the entropy *outside* the black hole would decrease because there are fewer particles and thus less information. Correspondingly, the entropy *inside* the black hole would increase. Because the area of the horizon of the black hole is proportional to the entropy, the area would also increase. The opposite situation, in which the horizon of a black hole decreases, is associated with the prediction that black holes emit radiation—the heat of the photons carries the entropy that shrinks the horizon. [7][13]

This means that the entropy outside the black hole is in terms of missing information due to fewer matter particles; but the entropy inside the black hole is in terms of the *area* of the horizon and thus the geometry of space and time. The amount of missing information associated with the boundary of a hidden region turns out to be one quarter of the area of the horizon, in units of Planck's constant. Smolin elaborates on the implications of this point:

"If a surface can be seen as a kind of channel through which information flows from one region of space to another, then the area of the surface is a measure of its capacity to transmit information... No matter what is on the other side of the boundary, trapped in the hidden region, it can contain the answer to only a finite number of yes/no questions per unit area of the boundary... If the world really were continuous, then every volume of space would contain an infinite amount of information to specify the position of even one electron. This is because the position is given by a real number, and most real numbers require an infinite number of digits to describe them... In practice, the greatest amount of information that may be stored behind a horizon is huge—10^{66} bits of information per square centimeter. No actual experiment so far comes close to probing this limit. But if we want to describe nature on the Planck scale, we shall certainly run into this limitation, as it allows us to talk about only one bit of information for every four Planck areas... This new limitation tells us there is an absolute bound to the information available to us about what is contained on the other side of a horizon. It is known as *Bekenstein's bound...*" (pp. 103-105) [7]

This is the basis for the model of quantum gravity in black hole thermodynamics, which directly links the concept of bits of non-physical information and bits of physical space-time in a formal mathematical relationship—the Bekenstein's bound. In this model, physical space must be quantized because the smallest possible area of space has an inherent mathematical limit to the amount of information it can contain. In this way black hole thermodynamics arrives at the same conclusion as loop quantum gravity, as well as a similar conception in string theory: space-time is quantized. But it may be said to go a bit further, positing that the area and volume of the quantized units of space-time directly corresponds to bits of information in conceptual space. It also attempts to place the observer in the system, and to connect the limitations of an observer perspective to quantitative limits of information.

The finding that the amount of information that a region of space can contain is directly proportional to its surface area provides a direction for how string theory, loop quantum gravity, and black hole thermodynamics can be linked together. In a formal mathematical relation, the holographic principle precisely connects the quantum geometry of bits of space-time with bits of abstract information. The holographic principle suggests that all information transfer can be viewed in terms of going through channels that encode the information and that have limits to the amount of information that can be transmitted. [7] As an example of this principle, information from a volume of space-time, such as a three-dimensional object, can be encoded in a two-dimensional area—analogous to a computer screen or a hologram. [15] Smolin suggests implications of this point:

> "...[T]he analogy between the history of the universe and the flow of information in a computer is the most rational, scientific analogy... What is mystical is the picture of the world as existing in an eternal three-dimensional space, extending in all directions as far as the mind can imagine... When we look out, we are looking back in time through the history of the universe, and after not too long we come to the big bang... When we imagine we are seeing into an infinite three-dimensional space, we are falling for a fallacy in which we substitute what we actually see for an intellectual construct. It is not only a mystical vision; it is wrong—*Lee Smolin, theoretical physicist and loop quantum gravity theorist* (pp. 64-65) [7]

> "If we continue to miniaturize computers more and more, we shall eventually be building them purely out of the quantum geometry in space... Imagine we can then build a computer memory out of nothing but the spin network states that describe the quantum geometry of space... [T]he most efficient memory we could construct out of the quantum geometry of space is achieved by constructing a surface and putting one bit of memory in every region 2 Planck lengths on a side. Once we have done this, building the memory into the third dimension will not help." (p. 172) [7]

This holographic model of the limits of information flow is thought to provide some insight into how from an abstract mathematical space-time geometry our ordinary world appears to be projected. Smolin describes the holographic principle as having strong and weak versions. [7] In the strong version, all the information that is ever available can be encoded in an area of a screen boundary. If this is the case, then it is contended that *reality* might consist of *only* the screen and there are no separate objects existing somewhere outside of the screen that is the source of the information on the screen. If an observer couldn't tell the difference between a separate object and the activity on the screen because there is no additional information available, then perhaps the screen is the only thing that actually exists—a Turing test notion of *reality*.

One problem with the strong version of the holographic principle is thought to be that it still describes the world as a world of things or objects [7]—ideas that loop quantum gravity theory suggests must be gone beyond for an abstract pure geometry of information space. In contrast, the weak version of the holographic principle is held to be a dynamic conception of the world as interacting processes that is more abstract than theories of ordinary physical space which don't recognize an underlying information space that generates it. Once again, Smolin conjectures:

"In such a world, nothing exists except processes by which information is conveyed from one part of the world to another. And the area of a screen—indeed, the area of any surface in space—is really nothing but the capacity of that surface as a channel for information. So, according to the weak holographic principle, space is nothing but a way of talking about all the different channels of communication that allow information to pass from observer to observer. And geometry, as measured in terms of area and volume, is nothing but a measure of the capacity of these screens to transmit information... This more radical version of the holographic principle...relies on the idea that the universe cannot be described from the point of view of an observer who exists somehow outside of it. Instead there are many partial viewpoints, where observers may receive information from their pasts. According to the holographic principle, geometrical quantities such as the areas of surfaces have their origins in measuring the flow of information to observers inside the universe... Thus, it is not enough to say that the world is a hologram. The world must be a network of holograms, each of which contains coded within it information about the relationships between the others. In short, the holographic principle is the ultimate realization of the notion that the world is a network of relationships. Those relationships are revealed by this new principle to involve nothing but information. Any element in this network is nothing but a partial realization of the relationships between the other elements. In the end, perhaps, the history of a universe is nothing but the flow of information." (pp. 177-178) [7]

"[A]ny mode of knowing can be collapsed and confined merely to surfaces, to exteriors...crushed by the flatland madness sweeping the modern world... The systems sciences...reduced all of them to nothing but interwoven ITs in a dynamical system of network processes...hobbled and chained to the bed of exterior processes and empirical ITs. This was holism, but merely an exterior holism that perfectly gutted the interiors... The shackles were no longer atomistic; the shackles...were now holistically interwoven chains of degradation...subtly embodying and even extending the reductionistic nightmare...reason still trapped in flatland. They were...another twist on flatland holism, material monism...sliding chains of material marks—in other words, sliding chains of ITs... [T]here is nothing under the surface; there is only the surface... [A]gainst its dominant and domineering mood...we wish to introduce the within, the deep, the interiors of the Kosmos, the contours of the Divine."—*Ken Wilber, philosopher (pp. 133-136)* [15]

The quantum gravity and holographic theories demonstrate progress in understanding the holistic interconnectedness of nature beyond the physical into a pure geometry of information space. The theories reflect increasingly abstract approaches to the concept of an *object*. The focus changes from objects to relationships between objects, then relationships between processes, then relationships between elements of information change or flow—defining an element as a partial realization of the information between all the other elements. Matter is reduced to fundamental quantized units of physical space-time, then to pure geometry in conceptual space, and further to very abstract flows of non-physical bits of information. The primary locus or attribution of *reality* shifts from the domain of matter to the more abstract functional domain of information, while retaining the quantum concept of the smallest unit—now not in the form of space-time bits but rather information bits. But in the typical objectifying pattern, the observer is still presumed to be a classical object, even while 'objects' are progressively understood more abstractly.

In this progression, the concept of space is getting closer to the philosophy of idealism, which holds that matter is fundamentally a product of mind. Matter doesn't exist as independent objects in the physical world, but rather as a projection of non-physical information space. This reflects a continuing trend that is moving the orthodox interpretation of quantum theory from what might be called a phenomenal realism toward a quantum idealism. It contains further acknowledgement of the ontological *reality* of information space—at least a bit closer to *mental reality*.

Implicit in the notion of a purely abstract information space is that it is not based on any particular individual mind that has a relative perspective within the system. Individual minds, which can know information, are likened to two-dimensional holograms in a disjointed network. But there still is no explicit consideration of subjective minds of actual individual observers—apparently in the continuing attempt to remain 'objective.' There is recognition of an abstract underlying field of information that generates physical space-time. This underlying abstract information field is composed of surfaces or areas through which information flows, associated with different partial histories of events related to different perspectives. But the perspectives are not yet explicitly associated with actual conscious observers or minds. Rather they are just parts of an abstract information field, partial event histories possibly comprising a network of holograms—discontinuous and discrete. There is yet nothing explicit in the models with the power to experience and enjoy the information as meaningful—still remaining an objectified view of nature.

Of particular interest is how the concept of the observer perspective is dealt with in the progressive abstractions toward a non-material field of information space. According to the weak holographic principle, in the universe there is nothing but representations transmitted from one partial set of events in the history of the universe to other parts of the universe—also described as screens (or screen processes since at this level of nature in the model there are no objects such as screens). What was an observer's perspective is now a partial history of events or information flow. Somehow this is thought to remove the possibility that the universe can be viewed from an outside perspective—by placing all events as partial histories or perspectives in the universe. But there still is at least a tacit distinction between representations and partial event histories that send and receive them. What makes up this distinction? How do partial event histories that are nothing but relationships between each other send and receive representations? Who or what is observing the representations or screen-like processes? These issues are yet to be considered.

Use of the image of a computer memory chip as an analogy to describe the holographic principle further could be misleading. It is clear that the processes or elements of information flow—the partial event histories—are not the type of *objects* or *things* that have material substrates *in* physical space-time, with which we ordinarily associate computer chips or screens. They are abstract functional processes in information space. If they are made of *something*, they would be made of abstract information flows in conceptual space—possibly *thoughts* in mental space, or *mindstuff*, but not *matterstuff*. This is an important distinction that needs to be explained in models based on the holographic principle. This is especially the case for models such as the holographic model described above that proposes the possibility of building computational systems or computers out of information bits of conceptual space that generate physical space.

Inasmuch as the holographic model as described above is moving toward an ontologically *real* information space closer to a *mental reality*, it may be useful to clarify a point about 'screens' referred to in the description. It possibly might be tempting to associate the 'screen' in this model

128

with the philosophical concept of *screen of the mind*—sometimes called the *Cartesian theatre*. However, the 'screen' described in the holographic model is a channel of information flow, such as the area of an event horizon. It is a type of barrier that establishes a transmission limit for information. Although the idea can be somewhat helpful toward distinguishing the physical and mental domains, it is important not to equate it with the much more abstract concept of screen of the mind. As discussed in later chapters, there are several layers of existence between the subjective screen of the mind and the flatland-like two-dimensional barrier of a holographic screen *reality*. Unpacking the different layers of the mental domain—associated with the process of observing or knowing—is the focus of Chapter 11.

The theoretical work toward an integrated theory of quantum gravity—including holographic versions—still doesn't consider an actual observer. The notion of an observer is hidden in the concept of partial realizations of relationships of information flow, or partial event histories. The observer remains central but implicit, somehow existing as a disembodied abstract functional observer perspective or frame of reference, but at the same time mysteriously remaining outside of the relational system. It is appropriate that these quantum gravity and holographic models continue not to capture the perspective of the observer. The reason is not just that the objective approach attempts to progress as far as possible without reverting to ill-defined subjective concepts; indeed the models have the problem of the implicit use of subjective concepts without defining them. The more significant reason is that the observer or knower cannot be accounted for within these theories. The mind and consciousness of the observer repeatedly are placed in some version of a black box because they fundamentally don't fit inside the limited theoretical framework of a closed physical universe. They are subtler levels of nature that underlie and permeate the physical. This is a focus of the next two chapters, which outline contextual limitations of the physical domain and begin to specify more clearly its relationship to the subtler mental domain of nature. They address core issues on quantum *reality* and the emerging framework of *quantum mind*.

Chapter 6 Notes

[1] Greene, B. (2004). *The fabric of the cosmos: Space, time, and the texture of reality*. New York: Alfred A. Knopf.

[2] Bohm, D. (1980). *Wholeness and the implicate order*. London: Routledge & Kegan Paul.

[3] Routt, T. J. (2005). *Quantum computing: The Vedic fabric of the digital universe*. Fairfield, IA: 1st World Publishing.

[4] Isaacson, W. (2007). Einstein: His life and universe. New York: Simon and Schuster.

[5] Peat, F. D. (1990). *Einstein's moon: Bell's theorem and the curious quest for quantum reality*. Contemporary Books, Inc.

[6] Hawking, S. (2001). *The universe in a nutshell*. New York: Bantam Books.

[7] Smolin, L. (2001). *Three roads to quantum gravity*. New York: Basic Books.

[8] Greene, B. (1999). *The elegant universe: Superstrings, hidden dimensions, and the quest for the ultimate theory*. New York: Vintage Books.

[9] Smolin, L. (2003). Loop Quantum Gravity. In Brockman, J. (Ed.), *The new humanists: Science at the edge*. New York: Barnes & Noble Books, pp. 329-352.

[10] Lincoln, D. (2004). *Understanding the universe: From quarks to the cosmos*. Singapore: World Scientific Publishing Co., Pte. Ltd.

[11] Wheeler, J. A. (1990). *A journey into gravity and spacetime*. Scientific American Library.

[12] Chalmers, D. J. (1996). *The conscious mind: In search of a fundamental theory*. New York: Oxford University Press.

[13] Hawking, S. & Penrose, R. (1996). *The nature of space and time*. Princeton: Princeton University Press.

[14] t'Hooft, G. (1993). Dimensional Reduction in Quantum Gravity. In Ali, A., Ellis, J., & Randjbar-Daemi, S. (Eds.) *Salamfest*. Singapore: World Scientific, pp. 284-296.

[15] Wilber, K. (1998). *The marriage of sense and soul: Integrating science and religion*. New York: Random House.

Chapter 7

A Closed Physical Universe?

In this chapter, we will examine the overall model of a closed physical universe that modern science has unsuccessfully attempted to build. We will identify major problems and the incompleteness of the model, and the necessity of incorporating an underlying information field of mental space or quantum mind. Key issues in this chapter include Einstein locality and the light cone, the concepts of background independence and observer dependence, and fundamental entropy and order. The overall main point of this chapter is that a more holistic and comprehensive model is needed for a coherent description of nature, and even of the physical universe. The past few chapters have focused on cutting edge theories in quantum physics, which may be somewhat taxing for many readers. Hopefully the underlying theme of the relationship of objectivity and subjectivity, and matter and mind, runs clear enough through these chapters that you will bear with it as planks are laid down in order to proceed to unified field theory—and then to the ultimate unity of the individual and the cosmos.

The progress in modern physics described in this part of the book is toward increasingly abstract conceptions of matter and energy, and space and time—within an objectified reductive approach. This approach applies the objective scientific method to validate theories in a manner that is significantly framed by the physicalist worldview. It has its empirical basis in a broad range of ordinary sensory experiences and its theoretical basis in logical reasoning, generally within the experiential *reality* of the ordinary waking state of consciousness. This state is characterized by a highly restricted and localized subjective experience that supports the belief that mind and consciousness are products only of neural activity in the physical brain.

As exemplified by the holographic model and theories of quantum gravity, however, progress is leading beyond the view of the world as essentially *physical reality*. Contemporary models have progressed from a predominantly structural focus on nature as matter toward a functional focus that is beginning to recognize a more abstract conceptual information space toward an underlying nonlocal *mental reality* or *quantum mind*. This is an important aspect of the revolutionary paradigm shift underway—from the matter-mind-consciousness ontology to the consciousness-mind-matter ontology—that is more significant than any previous transition in modern science.

However, it has developed slowly, building up to the current phase transition over the past hundred years or so. This is because it doesn't have a coherent direction, due to lack of systematic understanding about the relationship of matter to mind and consciousness. More significantly it doesn't yet utilize systematic technologies that directly develop the minds of the investigators to experience the fundamental holistic nature of this relationship. It remains an objectified, fragmented approach—perhaps expected upon reflection to have produced a disintegrated view of nature, as well as contributing significantly to our disintegrated civilization.

At the current stage of progress on the bridge to unity described in this book, the limitations of the objective material domain as the only level of existence are coming into view. Also, the significance of the subjective mental domain is starting to be recognized—though descriptions of it still remain quite foggy. This chapter provides an outline of the material domain and summarizes some of the key points discussed in the past few chapters. It offers additional examples of how contemporary theories extend beyond the material domain, and points to challenges that require even more integrated understanding of nature that includes additional levels of space or existence.

> "[W]ithout...opening into a new physics, we shall be stuck within the strait-jacket of an entirely computational physics, or of a computational cum random physics. Within that strait-jacket, there can be no scientific role for intentionality and subjective experience. By breaking loose from it, we have at least the potentiality of such a role... Many who might agree with this would argue that there can be no role for such things within *any* scientific picture. To those who argue this way, I can only ask that they be patient... I believe that there is already an indication, within the mysterious developments of quantum mechanics, that the conceptions of mentality are a little closer to our understandings of the physical universe than they had been before."—*Sir Roger Penrose, mathematician and cosmologist (. 420)*[1]

Outline of the Material Domain

At this stage in the discussion the objective material domain generally can be described as an *almost* closed, background independent, observer dependent cosmological system. The system is mostly deterministic and quantized into four fundamental forces through which relatively independent physical objects interact in local causal relationships according to the limitations of the speed of light—but the cosmological system also is thought to be fundamentally random. It is modeled primarily by the theory of the relativistic space-time continuum, attempting to be combined with the theory of the discrete quantum discontinuum.

The mathematical consistency and precision of these theories make them a compelling view generally consistent with the range of ordinary sensory experience and logical reasoning. For some this view is so compelling that it is thought to be the only domain that exists, the *entire* universe. Fortunately conceptions that go beyond it are emerging in pursuit of a more coherent and unified understanding. This doesn't mean that the physicalist worldview is incorrect, but rather incomplete and conditional. The physical universe is perhaps better characterized as a *quasi-closed* domain underlain by more fundamental, expansive and interconnected levels.

As noted in the Prologue, the view of a closed fundamentally random physical universe also unfortunately has come to have quite dangerous applications in modern civilization. This is due to increasingly powerful, destructive applications based on partial knowledge and experience that places matter as more fundamental than mind and random disorder more fundamental than order. Because of the emphasis on only the objective means of gaining knowledge, the power of objective technology to manipulate nature has advanced more rapidly than the power of our subjective minds to apply it for healthy integrated societal progress.

Conceptualizing nature as objective and separate from our inner nature, there is a fundamental disconnect between our own inner selves and the outer natural world. Applications of this fragmented physicalist worldview are contributing to a civilization that has been tearing itself apart and is in danger of dismantling itself or blowing itself up. This situation ironically involves good intentions and sincere efforts in the rigorous search for reliable knowledge that can be applied to help society. It is crucial that our modern civilization soon get beyond this closed, fragmented understanding and experience of nature.[2]

Closed universe? A central pillar of the closed relativistic framework in which the physical domain is held to be the only domain of existence is the speed of light, thought to be an absolute value in nature. It is a key factor in how the principle of causality is applied, associated with the light cone that defines the range of cause-effect relationships as limited by how fast light photons can travel between objects. According to the general understanding in this framework, objects outside the light cone cannot influence each other—indeed, we can never know at any moment if there are objects outside our light cone. All of the forces involved in motion or change—the four fundamental particle-force fields that comprise the physical universe—are believed not to propagate faster than the speed of light. Not surprisingly, a major challenge to this view is the experimental finding of nonlocality and apparently superluminal relationships.

In the conceptual framework of a causally closed relativistic universe characterized by Einstein locality, mass is thought of as resistance to being accelerated—it is not a measure of the density or size of an object. Mass increases as an object increases in speed. An object approaching the speed of light gains mass and theoretically would become infinitely massive at the speed of light, generally considered impossible. Thus no physical object with mass can ever travel the speed of light. Moreover, at speeds faster than the speed of light mass would have an imaginary mathematical value proportional to the square root of -1, also considered impossible. There is no evidence for theorized particles—called *tachyons*—that move faster than light, and even their theoretical acceptance is diminishing. Relativity theory mathematically doesn't rule out the possibility, but commonsense might seem to do so in that pushing a tachyon in one direction theoretically would result in it moving in the opposite direction. There is a theoretical possibility of 'objects' that travel superluminally, but to date no such 'objects' have been detected in a manner to receive consensual validation within modern science.

However, Einstein's relativity theory doesn't exactly state that nothing can travel faster than the speed of light. It can be interpreted to imply that objects traveling slower than the speed of light can slow down or speed up to almost the speed of light, and objects traveling faster than the speed of light can slow down almost to the speed of light or speed up towards infinitely fast. But objects must stay on one or the other side of the speed of light for their entire existence. The speed of light thus can be described as a kind of barrier that cannot be crossed, especially by objects with physical mass.

It generally is assumed that gravitational waves—as well as all *massless objects*—travel at the speed of light. Because mass is resistance to being accelerated and massless objects would have no resistance, they also are conceived as traveling at light-speed. The speed of light is a barrier in nature fundamental to the concepts of mass, inertia, and motion in the gravitational field. From a more expanded perspective, however, it may reflect a basic textural quality of the medium or aether of gross conventional space-time that limits the speed of all massive and massless *quantized* energy. Light photons have physical energy that interacts with gravitation and the other known fields. If an object of some kind were able to travel faster than the speed of light, it would not have any mass as it commonly has been defined, and it would not have any known form of interaction with the fundamental particle-force fields measurable so far. Objects that might have these attributes are undefined in the framework of a closed relativistic universe.

The speed of light is an absolute value in the sense that all ordinary observers under the same relative conditions—such as in the same medium or substrate, and the same relative velocities—will measure the speed as being the same. But the speed of light varies according to the medium through which it travels. Historically the evidence indicated it travels fastest in so-called empty space—although some recent studies suggest that some aspects of light might travel faster under certain specialized conditions, even though still having some speed limit. Recent theories propose that different colors or wavelengths of light may travel at slightly different speeds. These and other specialized experimental tests based on tiny but measurable differences possibly could require modification of the theory of the speed of light. Evidence for these quite minute measurable effects may eventually support a view of the universe that is much more open than the closed universe depicted in modern physical theories. This is not to suggest that the speed of light is violated or that it is incorrect, but rather that it is a conditional understanding within a more comprehensive and inclusive context.

> "Some of the effects predicted by the [loop quantum gravity] theory appear to be in conflict with one of the principles of Einstein's *special* theory of relativity...the constancy of the speed of light... [T]he theory predicts that the speed of light has a small dependence on energy. Photons of higher energy travel slightly slower than low-energy photons... [T]he principle of relativity is preserved, but Einstein's special theory of relativity requires modification so as to allow photons to have a speed that depends on energy. The most shocking thing I have learned in the last year [about 2002] is that this is a real possibility."—*Lee Smolin, theoretical physicist (pp. 339-341)* [3]

The nonlocal psi wave theory in Bohmian mechanics is a possible example of the more inclusive context. The psi wave would have to interact with the known particle-force fields to influence them causally, and also would have to do so in a manner that is so subtle as to involve amounts of energy at least smaller than can be detected according to the limitations of the uncertainty principle. On the other hand, the theory can be understood to posit that information is transmitted faster than the speed of light in the nonlocal field of the psi wave.

In Einstein's General Theory of Relativity both massive objects and energy produce curvatures of space-time. Because energy has been identified as having a positive value, and gravity is an attractive force, it bends light pathways toward each other. This is important in predicting whether space and time have a beginning and an end. If light rays in a light cone were to be traced back in time, they would be found to bend toward each other more strongly; and the boundary of the light cone eventually would contract to zero in a finite amount of time, indicating that the light cone—and thus time—has a beginning. [4]

At least in his early career Einstein seemed to believe that the universe is infinite and eternal. However, the General Theory of Relativity appears to be mathematically consistent with the hypothesis that the universe does begin, and may end. [4] Einstein also apparently thought that massive stars would settle into a final steady state, but the current mathematical evidence predicts that they contract into black holes. As described by theoretical physicist and cosmologist Stephen Hawking, in the General Theory of Relativity time and space:

> "...are defined by measurements within the universe, such as the number of vibrations of a quartz in a clock or the length of a ruler. It is quite conceivable that time defined in this way, within the universe, should have a minimum or maximum value—in other words, a beginning and an end. It would make no sense to ask what happened before the beginning or after the end, because such times would not be defined." (pp. 178-181) [4]

It sort of makes sense that space can be curved, but what about time? [5] In a way the phrase 'the curvature of space-time' seems to be a misnomer—at least it isn't obvious how time can be *curved*. Motion in space can be forward or backward, but time certainly seems to be asymmetrical in the forward direction. This is related to the important concept of the *arrow of time*—to be discussed later.

When we move out of physical space and time into conceptual or imaginary space and time, however, such commonsense limitations can be bypassed, at least conceptually. This line of conceptualizing offers considerably more flexibility for purposes of theory building. As discussed in earlier chapters, the effort to integrate relativity theory with quantum theory uses mathematical constructs such as imaginary space and time that don't have direct interpretations in terms of ordinary physical *reality*. Again, Hawking states:

"To describe how quantum theory shapes time and space, it is helpful to introduce the idea of imaginary time... One can think of ordinary numbers such as 1, 2, -3.5, and so on as corresponding to positions on a line stretching from left to right: zero in the middle, positive real numbers on the right, and negative real numbers on the left... Imaginary numbers can then be represented as corresponding to positions on a vertical line: zero is again in the middle, positive imaginary numbers plotted upward, and negative imaginary numbers plotted downward. Thus imaginary numbers can be thought of as a new kind of number at right angles to ordinary real numbers. Because they are a mathematical construct, they don't need a physical realization; one can't have an imaginary number of oranges or an imaginary credit card bill... One might think this means that imaginary numbers are just a mathematical game having nothing to do with the real world. From the viewpoint of positivist philosophy, however, one cannot determine what is real. All one can do is find which mathematical models describe the universe we live in. It turns out that a mathematical model involving imaginary time predicts not only effects we have already observed but also effects we have not been able to measure yet nevertheless believe in for other reasons. So what is real and what is imaginary? Is the distinction just in our minds?... Einstein's classical (i.e., nonquantum) general theory of relativity combined real time and the three dimensions of space into four-dimensional space-time. But the real time direction was distinguished from the three spatial directions; the world line or history of an observer always increased in the real time direction (that is, time always moved from past to future), but it could increase or *decrease* in any of the three spatial directions. In other words, one could reverse direction in space, but not in time... On the other hand, because imaginary time is at right angles to real time, it behaves like a fourth spatial direction. It can therefore have a much richer range of possibilities than the railroad track of ordinary real time, which can only have a beginning or an end or go around in circles. It is in this imaginary sense that time has a shape... The history of the universe in real time determines its history in imaginary time, and vice versa, but the two kinds of history can be very different. In particular, the universe need have no beginning or end in imaginary time. Imaginary time behaves just like another direction in space. Thus, the histories of the universe in imaginary time can be thought of as curved surfaces... At least in imaginary time, they appear to be closed surfaces." (p. 59) [4]

As exemplified in this quote, mathematical models employing conceptual tools of imaginary space and time are going beyond the view of the universe as a four-dimensional space-time system, such as string theory and M-theory with six or seven additional dimensions. Some of these models not only blur the distinction between physical and conceptual, mathematical, or imaginary space, but they go much further and attribute to conceptual, mathematical, imaginary space and time the ability to interact with physical space and time; and further, as discussed in Chapter 6, to *generate* it. The melting of the conceptual distinction between *real* and *imaginary* space and time is evident in the following quote from Hawking:

"Up to recently it was thought that the six or seven extra dimensions would all be curled up very small... However there has recently been the suggestion that one or more of the extra dimensions might be comparatively large or even infinite... Large extra dimensions are an exciting new development in our search for the ultimate model or theory. They would imply that we live in a brane world, a four-dimensional surface or brane in a higher-dimensional space-time... Matter and nongravitational forces like the electric force would be confined to the brane. Thus everything not involving gravity would behave as it would in four dimensions... On the other hand, gravity in the form of curved space would permeate the whole bulk of the higher-dimensional space-time. This would mean that gravity would behave differently from other forces we experience, because gravity would spread out in the extra dimensions..." (pp. 59-85) [4]

This reflects an attempt to conceptualize a subtler, higher-dimensional and possibly infinite space while maintaining the relationship of the notion of space to the gravitational field. It exemplifies model building using an imaginary or conceptual space and time that is freed up from at least some of the constraints of ordinary physical space-time, but still having interactions with the gravitational field, or at least some form of gravitational-like attraction. From a positivistic perspective, it doesn't matter whether imaginary space-time *really* is different from ordinary space-time because it is believed that *reality* cannot *really* be known. The models, however, are generally consistent with relativity and quantum theories, as well as with other established theories of ordinary *real* phenomena in nature.

When the models have a high degree of mathematical consistency and predictive success, there is a tendency to attribute a status to the constructs in them as perhaps actually existing in nature—rather than being just abstract mathematical constructs. The direction of M-theory and similar models is toward higher mathematical dimensions that are increasingly described as *actually* interacting with our ordinary four-dimensional physical world—as an underlying sub-phenomenal *reality*. With the possibility of extra dimensions or levels of *reality*, Einstein's belief that space and time, at least in some meaning, are infinite may turn out to be accurate after all.

Theories of quantum gravity go further to propose that ordinary four-dimensional space-time emerges from information space that is more fundamental and underlies our ordinary world. This conceptual space or field of information flow is more abstract than ordinary space-time; it cannot be found in ordinary physical space-time. Its ontological *reality* is not based in *physical reality,* and it does not emerge from material existence. It is more associated with mathematical space and the concept of information—which implies meaning and intelligence—and thus might be thought of as more closely associated with a subtler field of *mind* than matter.

These models reflect initial attempts to conceptualize connections and interactions between conventional physical space and unconventional information space. They can be viewed as anticipating progress toward a causal connection between mind and matter, beginning to bridge the mind-body gap and implying a notion of causality that applies to both ordinary physical and some form of extraordinary non-physical existence.

In these ways contemporary models are increasingly recognizing *mental reality*, not only as a functionally existent abstract *reality*, but also importantly as the basis of *physical reality*. This is clearly progressing toward a dualistic view of nature—at least in that there is local physical *reality* and an underlying nonlocal mental *reality* of some kind—akin to Bohm's explicate and implicate orders. On the other hand, it is starting to bridge the dualistic gap between materialistic monism and idealism by suggesting that material *reality* is built from an underlying information space—or mental *reality*—that can causally interact in some yet to be defined way with physical *reality* and the gravitational force. But the models also include the notion of underlying non-dualistic unity.

However, in proposing higher dimensions beyond four-dimensional space-time, or a spin network of pure geometry, or an abstract field of information flow underlying our *real* ordinary world, there is still no direct examination of mind and consciousness. On the one hand, there is increasing recognition of a level of existence beyond *physical reality* and that somehow generates it. It is described as a more abstract conceptual or functional *reality* of information flow that is non-linear, extra-dimensional in the sense that it is not subject to the limitations of ordinary physical space-time, and may have components that are infinite. On the other hand, mind and consciousness continue to be placed outside of it—*somewhere*—or at least not accounted for in it. In quantum theory, the observer remains outside of the system and it is not known how to include the observer in it—even though loop quantum gravity and holographic theories partially attempt to do so.

> "Our conclusion is that attempts to *embed* consciousness in space and time are doomed to failure, just as equivalent attempts to *embed* motion in space only. Yes, motion does take place in space, but it also partakes in time. Similarly, consciousness certainly takes place in space and time, but in addition seems to require an additional aspect of reality...in order for us to give a proper description of its relation with the world as described in physics."—*Piet Hut, astrophysicist, and Roger N. Shepard, cognitive and evolutionary psychologist (p. 319)* [6]

There is clear progress toward an underlying non-material information field that is somewhat more mind-like, and that generates the entire discrete measured classical world. But the models continue to reflect the pretheoretical assumption that the ordinary objective world is independent and external to the conscious mind of the scientific investigators, while also asserting that in a causally closed relativistic universe the observer has to be inside it. [7] It is assumed that the mind of an actual conscious experiencer or observer is in the brain, and that the mind is not a field of existence that extends beyond individual brains. This is how many investigators seem to interpret their own experiences of mind and consciousness—to be only in their heads. Accordingly, because mind and consciousness are products of the physical brain, they don't need to be addressed in models concerned with these much smaller and more fundamental levels of nature. Rather it is thought that they are appropriately dealt with in research concerning grosser electrochemical, molecular, and cellular levels. It is due to this pretheoretical assumption—associated with the physicalist view that matter creates mind and consciousness—that mind and consciousness are repeatedly placed in a black box in these models, because the models cannot account for them. It appears that for most investigators the subjective experience of subtler levels of mind and consciousness as underlying ordinary thinking and perception is both a conceptual and *experiential* black box—a foggy region or void, with virtually no clear experiences of these deeper levels.

> "Our quest for the fabric of reality has brought us from religion to science. But that science, when asked to show us reality, has caused us to look into a mirror to see what we are. It is the last place that science would have chosen to search, yet now we must look there into that image we see within ourselves. Now we must find out what the observer is and find out what threads consciousness weaves in creating the quantum mind"—*Evan Harris Walker, theoretical physicist (p. 138)* [8]

In probing theoretically the more fundamental layers of nature at the extreme basis of matter, there is inexorable progress toward concepts and corresponding language reflective of an underlying abstract information space—toward *mental reality*—from which physical *reality* emerges. Although still not explicit, these models are increasingly incorporating concepts ordinarily associated with mind and consciousness. In the phase transition from matter to mind to consciousness, theories in which mind and consciousness are products of brain functioning and exist just in the physical brain are beginning to give way to these more expanded models. The brain is one kind of instantiation of a more abstract information processing function, as is a computer—both of which are physical. But as deeper levels of physical structure are probed, the models have gone into functional concepts of information flow and non-physical networks of pure geometry beyond matter, and into underlying, ontologically *real* nonlocal fields of existence.

An abstract field of information modeled as a higher-dimensional conceptual space, a pure geometry, or a functional information space that underlies and generates four-dimensional physical space-time clearly goes beyond models of any object—including the brain—as just a highly localized physical structure. The brain and mind are no longer *just* in the head, because brains, heads, and other ordinary objects are no longer just localized physical matter but rather are also more abstract nonlocal processes. In the words of mathematician C. J. S. Clarke, "Mind breaks out of the skull." (p. 174)[9] It is important to appreciate that nonlocality is not just a conceptual or functional nonlocality in some imaginary mathematical space, but an experimentally verified property of *real* everyday existence.

> "...[S]omething vital has been left out of almost all the modern efforts to understand our mental life—something that counts as a first principle, without which everything is bound to be incomplete and off base... This missing element is the mind's *nonlocal* nature... If nonlocal mind is a reality, the world becomes a place of interaction and connection, not one of isolation and disjunction. And if humanity really *believed* that nonlocal mind were real, an entirely new foundation for ethical and moral behavior would enter, which would hold at least the possibility of a radical departure from the insane ways human beings and nation-states have chronically behaved toward each other. And, further, the entire existential premise of human life might shift toward the moral and the ethical, toward the spiritual and the holy." —*Larry Dossey, physician and science writer (pp. 1-7)* [10]

These issues will be discussed further in Chapters 13-18 on models of the brain and mind, some of which extend into the quantum level and deeper. The point here is that contemporary models in physical science extend past physical matter, including the brain. They are beginning to reflect more expanded conceptions of the fundamental basis of the physical universe, exemplified by such terms as *quantum mind*. The concept of *quantum mind* abstracts the concept of the mind beyond the confines of the brain and into a nonlocal mental domain that somehow underlies and also generates the material domain. However, the term *quantum mind* is a misnomer, in that the theorized subtle nonlocal field of mind may not be made of quantized information bits—*qubits*. It may not be quantized into Planck-size bits of either space-time or information, but rather may be even subtler and more holistic—to be discussed in Chapter 8.

> "I propose, then, that "mental force" is a force of nature generated by volitional effort... One should not, needless to say, posit the existence of a new force of nature lightly. The known forces—gravity, electromagnetism, and the strong and weak forces...do a pretty good job of explaining a dizzying range of natural phenomena... But mental force, its name notwithstanding, is not strictly analogous to the four known forces. Instead, I am using *force* to imply the ability to affect matter. The matter in question is the brain. Mental force affects the brain by altering the wave functions of the atoms that make up the brain's ions, neurotransmitters, and synaptic vesicles... Mental force is the causal bridge between conscious effort and the observed metabolic and neuronal changes"— *Jeffrey Schwartz, research neuropsychiatrist, and Sharon Begley, science writer/editor (p. 318-320)* [11]

Still, these models continue to be vague about how they account for more holistic concepts such as nonlocality, higher and possibly infinite dimensions, purely abstract information networks, let alone an all-encompassing unified field. These concepts particularly challenge the reductive approach that focuses on local properties and the ultimate discreteness of nature. This inevitably brings up the conundrum of nonlocality in locality, whereas locality as a limitation of nonlocality would seem to make more sense. The issue of infinite regress will continue to present a dilemma for the reductive approach, until a completely holistic approach is also recognized and applied.

Although much evidence appears to support the conclusion that the universe is a closed system that is ultimately quantized, some contemporary models imply that more fundamentally it is not closed. The universe appears closed only if the explanatory gap between mental experience and physical existence is overlooked, and mind and consciousness are not explicitly included. Contemporary models are positing the existence of an underlying field of some kind that is nonlocal and possibly infinite. This is not captured by focusing on the ultimate quantized structure of space-time in a closed universe. Amidst the challenges of modeling a completely causally closed physical universe, the foggy outlines of an underlying mental space that is outside and beyond it is starting to appear. The advancing models are thus in a constructive direction, but still have a ways to go to describe the level of the mind, and even further to get to the all-encompassing completely unified field that incorporates fully the process of knowing and the knower, mind and consciousness. To go further, the concept of space needs to be expanded beyond our notions of Einstein locality, the gravitational field, and discontinuous quantum bits of space-time in order to accommodate nonlocality and a non-physical information space, as well as mental space.

> "Consciousness is the next big anomaly to call for a revision in how we conceive space—just as other revisions were called for by earlier anomalies. And the revision is likely to be large-scale... [T]o represent consciousness as it is in itself—neat, as it were—we would need to let go of the spatial skeleton of our thought... We can form thoughts about consciousness states, but we cannot articulate the natural constitution of what we are thinking about. It is the spatial bias of our thinking that stands in our way (along with perhaps other impediments). And without a more adequate articulation of consciousness we are not going to be in a position to come up with the unifying theory that must link consciousness to the world of matter in space. We are not going to discover what space must be like such that consciousness can have its origin in that sphere. Clearly, the space of perception and action is no place to find the roots of consciousness! In that sense of 'space' consciousness is not spatial; but we seem unable to develop a new conception of space that can overcome the impossibility of finding a place for consciousness in it—*Colin McGinn, philosopher (p. 103)* [12]

Background independent and observer dependent? The concept of background independence generally means that there is a not an underlying field or substrate of space-time that serves as a non-changing basis or reference for measuring motion or change. Whether there is ultimately a non-changing basis or absolute reference—which the unified field would seem to be—will be discussed in Chapters 10 -11. But in positing more fundamental levels than our ordinary four-dimensional physical world, possibly with extra dimensions, contemporary theories are at least implying a substrate of some kind. The models positing that information space generates physical space clearly imply that the physical level of space would be background *dependent*, at least in some important sense.

The notion of observer dependence also is relevant to the issue of a background. Observer dependence similarly implies some form of background or substrate. The universe is observer dependent at least in the sense that an observer is needed in order to observe it and know that it exists; there is at least an experiential substrate of the universe we observe—namely *us*, the observers. A closed cosmological system needs to account fully for this kind of observer dependence. Quantum theory places the observer somewhere *outside* the probabilistic quantum descriptions of nature; but in relativity theory as it is frequently interpreted, the observer must be *inside* the system—a fundamental quandary. This quandary is a product of the physicalist assumption that mind and consciousness are only in the brain; but already we know that the brain, and all matter, is fundamentally nonlocal.

When the concepts of observer and observer perspective begin to be examined in contemporary models of a closed universe, inconsistencies become more apparent. Relativity theory is said to be observer dependent in the sense that the outcome of measurements of space and time depend on the perspective of the observer—specifically the relative motion of an observer. This is associated with the concept of *Lorentz Invariance*, which states that regardless of the speed between two objects—such as observer and observed—the laws of nature that describe them will be the same. It is thought to be meaningless to determine which is moving, because motion has meaning only in relation to each other.

This is drawn from Einstein's work on general relativity which is interpreted as being demonstrated, for example, by the phenomenon that two observers who are close to each other cannot distinguish between the effects of gravity and the effects of accelerated motion—the *principle of equivalence*. This places the frame of reference as central to understanding the relativity of space-time. However, to have an *actual* frame of reference that an observer experiences, an observer is needed—as well as some way to observe, a process of observing. In this sense relativity theory can be seen to contain implicitly a psychological theory of mind that assumes local limits to human sensory perception. This will be discussed in later chapters with respect to the limitations of the ordinary waking state of consciousness.

The concepts of the event horizon and of the light cone also depend on the observer perspective. A fundamental issue that is not examined in theories that incorporate these concepts is what constitutes an observer and an observer perspective, which are implicit in the concept of a *frame of reference*. This is comparable to the illusive issue in quantum theory of what constitutes a *measurement* or *observation* (refer to Chapters 4-5).

Fortunately the theory of the causal light cone can be said to provide a clear basis for locating the observer. The light cone specifies the range of possible causal influences in an observer's causal world. By definition the observer is located at the beginning of the observer's future light cone, as well as at the end of the observer's past light cone. Thus the observer exists *in between* these personal past and future light cones. If in typical reductive fashion smaller and smaller scales of the transition between past and future light cones of an observer are examined—getting down to the essential core—what is the essence and basis of the observer perspective and frame of reference, presumably embedded in which can be located the observer?

According to descriptions of the holographic model summarized in the past chapter that attempts to integrate quantum gravity theories, this essential core is a partial realization of relationships between events that make up the universe—a Planck-size area or channel of information flow. It involves a separate collection of historical events, somewhat analogous to a physical memory chip—but beyond material structure. It is individualized in that it is distinguished from other channels or partial histories. How it becomes individualized, as well as features that might relate more clearly this separate channel to ordinary meanings of an observer perspective—and also an observer—are not addressed. There appears to be no explanation of whether this information channel is able to know any information, or even to know itself—frequently associated with self-reflection and with conscious attention and a observer perspective with respect to a *real* ordinary observer. If referred to at all in these quantum gravity theories, these aspects of an observer are placed somewhere outside the cosmological system—of which the observer only has a partial perspective. But in the closed universe, there is no outside; further, there is no place or role for a causally efficacious observer inside either.

It is at least quite difficult in these theories and models to include in the observer perspective the typical features associated with the concept of an actual *real* conscious observer. According to the physicalist worldview, an observer's mind and consciousness which are usually associated with having a perspective are products of electrochemical and molecular activity in an individual brain. Brains exist and function at a much larger time and distance scale than are being referred to in these quantum gravity and holographic models.

Again, however, in these models there is no longer just physical space but also some kind of information space that generates physical space—a few planks closer to the concept of an underlying level or field of quantum mind. At the same time the models don't articulate any ordinary sense of an observer as having an individual mind and consciousness.

> "All through the physical world runs that unknown content, which must surely be the stuff of our consciousness. Here is a hint of aspects deep within the world of physics, and yet unattainable by the methods of physics. And, moreover, we have found that where science has progressed the farthest, the mind has but regained from nature that which the mind has put into nature."—*Sir Arthur Eddington, astrophysicist (p. 276)* [13]
>
> "...[T]he stuff of the world is mind-stuff."—*Sir Arthur Eddington, astrophysicist (p. 200-201)* [14]

Perhaps mind and consciousness are not necessary in order to have an observer perspective. Then what is it that delineates a particular set of historical events into a perspective or frame of reference? How is the entire universe divided into events or frames of reference or perspectives? What is the basis for dividing them into individualized partial historical perspectives or mind-worlds with continuity through time at a fundamental level of nature where actual observers with perspectives don't exist because brains don't exist at these levels?

The inability to integrate mind and consciousness—the process of observing and the observer—is due to the objectification of mind and consciousness as emergent products of the physical domain or as epiphenomena of the brain. Unexamined physicalist presuppositions of an observer perspective fail to carry over into deeper and more inclusive models of non-material, nonlocal information space or the subtle implicate order.

Moreover it seems reasonable to expect that individualized channels of information flow would serve some functional meaning and role in the models, related to the core concept of observer dependence. When the models propose a non-material information space that generates physical space, the concept of information as having functional meaning seems relevant.

In the attempt to remain objective and avoid vague subjective concepts, the concept of information also has been *objectified*. Information is sometimes conceptualized in a manner similar to the concepts of matter or energy, sort of like a substance or structure with certain attributes but not associated with inherent meaning or intention. Matter and energy are objective phenomena, whereas meaning and intention are typically considered subjective. Although communication theory posits a definition of information that is without regard to meaning, more common definitions of information carry the implication of intentional meaning. Information is commonly related to communication of meaningful knowledge between informer and informed.

The concept of information as meaningless abstract structures seems problematic when placed in a functional network of bits or qubits of information space that is beyond physical structure. In this sense it carries meaning—for example, yes/no answers to questions. This implies intention, awareness of significance, informer, and informed. Associated with the causal structure made of interacting causal light cones, it also involves the concept of cause, and thus is associated with systematic selective transfer of impulses of energy from one to another part of the system, in which both the concepts of intention and meaning seem quite relevant.

But apparently the functional ability to act on the flow of information—as in sending and receiving meaningful information and causally affecting events intentionally—also would not be necessary in the objectified type of observer perspective or frame of reference assumed in these models of information space. Intentional agents with the means to take action by having a brain and body would not exist at these tiniest of time and distance scales dealt with in the models.

If there were no agent in the system that receives, sends, and acts on the information, then the flow of information would not be functionally *informative*.[7,15] It would be merely flow of bits of abstract *something* devoid of information in the sense of meaning—because there aren't any knowing entities to inform. If the concept of the observer has any relationship to ordinary meanings such as the ability to act, function, reflect, be aware of, and so on—which the concept of the light cone would reasonably seem to imply—then the system couldn't be observer dependent because there are no observers to depend on with perspectives. Observer dependence then would seem to contain only the concept of a separate history or channel—the notion of a *part* or *bit* of the entire cosmological system—not the observer as a conscious mind or as a causal agent. The closed system apparently has no agent that either experiences or causes the flow or change at fundamental levels of nature, and has no *meaningful* information in it at fundamental levels.

On the other hand the quantum gravity models that attempt to incorporate the observer into the closed causal network do contain at least to a limited degree the concept of an observer as something that has a mind—associated also with interpretations of quantum theory such as the many mind-worlds interpretation. What would notions of an individualized consistent historical perspective through time, an observer perspective, or a mind-world mean when there are no observers with conscious minds? Hopefully it is clear that the concept of observer perspective is objectified and unexamined in these models of a causally closed physical universe.

This also brings up the issue of what *causes* change in this kind of model of a closed universe. Is the universe a network of constantly changing processes or events that fundamentally are random—with no conscious observer and no agent, either individually or universally—and in which change or information flow isn't informative or functional and is fundamentally meaningless and devoid of any kind of purpose? How is this consistent with the concept of bits of *information flow* in conceptual space causally generating physical space-time? Much further, how is this consistent with our somewhat orderly ordinary world of experience in which we as conscious observers at least believe we are agents with causal power? How does the closed chain of cause and effect somehow unlink itself and insert a causally efficacious conscious mind at some level of neural complexity? Is our consciousness epiphenomenal and our direct inner experience of causal power just a misperception—some kind of delusional state that afflicts the human species?

Deterministic and localized? The designation of a law of nature commonly refers to an absolute, invariant, deterministic, mathematically describable pattern of change that is thought to apply throughout nature under similar conditions. However, it is sometimes used in a less absolute manner, such as in the classical laws of thermodynamics based on the statistical probabilities of a large collection of events. As statistical averages, the 'laws' of thermodynamics describe what is most probable. But the laws are not absolute—there is inherently a small probability of the laws being 'violated.' In this sense, they are sometimes identified as principles rather than laws. But these 'principles' simply may reflect the practical impossibility of identifying all the innumerable causally determinant influences in physical processes, which necessarily results in a statistically probabilistic analysis.

The 2nd 'law' of thermodynamics states that collections of particles tend over time toward increasing disorder—increasing entropy. However, it is possible for some systems to have memories and other components that help at least preserve if not evolve to increasing states of order, such as a DNA molecule, or a human observer. Systems that increase or preserve order are sometimes called *negentropic systems*, usually associated with biological organisms. It is thought generally that non-biological processes at more fundamental levels than where life emerges don't exhibit negentropic behaviors that imply intelligence, intentionality, and functionality. Also, however, any negentropic process the creates a less entropic state—so to speak, extracting order from the environment—would have to result in increasing disorder or entropy in the environment taken as a whole, in order to comply with the 2nd law of thermodynamics.

If there is no intelligent ordering agent for the entire universe, then the cosmological system would have to contain in itself processes that drive change in the system. The system must originally have the lowest entropy or highest order, which then inevitably dissipates due to the 2nd law of thermodynamics—of course for no purpose or intent. These processes would function on much more fundamental time and distance scales than the levels at which negentropic behaviors indicative of intelligence and associated with functional meaning are thought to emerge. How purpose and meaning, as well as life—and also the initial state of the highest orderliness—would emerge in a fundamentally random closed system is at the least quite a challenge to explain. This will be discussed in a biopsychological context in Chapters 13-19.

Apparently what primarily drives the closed cosmological system is the dynamic random interplay of the initial kinetic energy of the spontaneous big bang and the gravitational attractive force that holds it together—active within the framework of thermodynamic principles of dissipation and increasing entropy through thermodynamic radiation. From a reductive perspective that the universe began from an infinitely dense singularity that spontaneously exploded in a hot big bang, the evolution of the universe is a cooling down process. These issues will be discussed again in Chapter 12.

As briefly mentioned in Chapter 3, this cooling process might continue until everything in creation dissipates into a dark cold emptiness. In this model the universe would have had a beginning in the spontaneous big bang, but would dissipate into a kind of amorphous fog in which it remains forever. Another possibility is that gravity could contract the universe back into another infinitely dense singularity that disappears into *nothing*—in this case, somehow beginning and ending in literally *nothing*. Still another possibility is that *nothing* again randomly and spontaneously explodes, expands, and then contracts in an eternally repeating pattern (which is beginning to sound like *something* is going on).

It has been calculated that the histories of the universe at least in imaginary time appear to be closed surfaces. If this holds not just for imaginary time but also for *real* physical time, then according to Hawking:

> "...[T]he universe began in a big bang, a point where the whole universe, and everything in it, was scrunched up into a single point of infinite density. At this point, Einstein's general theory of relativity would have broken down, so it cannot be used to predict in what manner the universe began... [T]he reason general relativity broke down near the big bang is that it did not incorporate the uncertainty principle; the random element of quantum theory that Einstein had objected to on the grounds that God does not play dice. However, all the evidence is that God is quite a gambler... [The] universe would be entirely self-contained; it wouldn't need anything outside to wind up the clockwork and set it going. Instead, everything in the universe would be determined by the laws of

science and by rolls of the dice within the universe. This may sound presumptuous, but it is what I and many other scientists believe." (pp. 79-85) [4]

This view suggests that the universe runs automatically, initiated by a fundamental random spontaneity, and that as a whole it is not negentropic and most fundamentally not orderly. Negentropic behavior would emerge only much later in its evolution — at much larger and cooler time and distance scales where molecular and cellular processes begin, associated with the spontaneous formation of life and intelligent behavior.

In a big bang universe, however, there needs to be an explanation for the impetus of the initial big bang, because the laws that drive the expansion and contraction of the universe would have had to come into play only after the big bang. [12] 'Prior' to the big bang, no forces of nature would exist, and no inherent tendencies to manifest in any way—such as to explode—would be present in the infinitely dense singularity, including any principles of order. On the other hand, if the infinitely dense singularity is thought of somehow as spontaneously generating heat, and the expansion subsequent to the big bang was due to the dissipation of heat energy, then it would seem that thermodynamic principles somehow would need to be present and actively functioning initially.

> "You might ask yourself what were the conditions of the universe prior to the Big Bang. You might also ask yourself another question. If there was no space, then where *was* the quantum singularity? Was there some sort of "other" space of which we know nothing?... While it makes sense to ponder such questions, there is sufficient mystery surrounding the conditions of the universe in the tiniest moments following the Big Bang, that I believe that trying to quantify the nature of the pre-expansion universe to be essentially pointless."—Don Lincoln, particle physicist (p. 478) [16]

In addition, if in a closed cosmological system the unified field of nature is at the Planck scale and there is nothing below or outside of it, then nature must be fundamentally random at this level, having no inherent principles of order or of anything else. This wouldn't seem to be a feature of the hypothesized spin network of quantum gravity (described in Chapter 6), said to be composed of fundamental units of information flow. The concept of Planck-size units of *random information flow* that depends on the perspective of an observer at levels where no observers exist, and from which physical space-time is generated, doesn't seem to be a particularly coherent explanation.

It further might be questioned how the theory of a hot big bang relates to the notion that an underlying information space generates physical space-time. The concept of information, by itself, doesn't carry with it the physical property of being hot; and also is not typically associated with high entropy but rather with order. As well, it would seem that at the initial big bang when the universe is in its hottest state, it would involve *high* entropy. If the initial universe were hot and with *high* entropy, it isn't obvious how thereafter its evolution has been toward *increasing* entropy according to the 2nd 'law' of thermodynamics—sometimes called the *entropy problem*. How could the initial big bang simultaneously have high entropy *and* high order, which then dissipates as the universe expands and cools toward *increasing* entropy? According to the 3rd 'law' of thermodynamics, heat is associated with high entropy, and cold with low entropy or order. To be consistent with the 2nd law of thermodynamics, the initial state of the big bang must have been a state of the highest order and lowest entropy. But according to the model, the universe has been cooling down since the big bang. If the universe emerging at the onset of the big bang is the state of lowest entropy, then this also would seem to distinguish the initial infinitely dense singularity from the singularity of a black hole, which is sometimes described as having the highest amount of entropy. [17]

Moreover it is not clear how the universe can be initiated by a spontaneous big bang based on randomness, because this would suggest that both randomness as well as something that initiates the big bang are inherent in the infinitely dense singularity. The uncertainty principle cannot be attributed to the infinitely dense singularity when there is nothing definable or extant in it. According to the overall model, it also would not be possible to specify initial conditions 'before' the big bang, because there is nothing 'before.' Initial conditions would need to mean the conditions present once the big bang *initially began*. Then it must have been instantaneously random and indeterminate—the only initial condition must have been just randomness—but also the hottest possible, a state of low entropy, and instantaneously generate orderly natural laws.

If nature were *fundamentally* random, any outcome would have equal possibility, making any consistency of context through time incredibly unlikely. But 'when' the big bang 'began,' an orderly temporal sequence of initial and subsequent stages also began—even apparently prior to any physical existence being generated from the spin network of information flow. It appears not to be the case that any possibility unfolds without any relation to the previous one, as would be the case if it were *completely random*. One event manifests in a consistent and orderly manner from the previous event, constrained to some degree by the context of the previous event—related to the principles of the 2nd law of thermodynamics, decoherence, and the arrow of time.

> "[A] random sequence is one that can't be algorithmically compressed. But...you cannot know whether or not a shorter program exists for generating the sequence. You can never tell whether you have discovered all the tricks to shorten the description. So you can't in general prove that a sequence is random, although you could disprove it by actually finding a compression. This result is all the more curious since it can be proved that almost all digit strings are random. It is just that you cannot know precisely which!... It is fascinating to speculate that apparently random events in nature may not be random at all according to this definition. We cannot be sure, for example, that the indeterminism of quantum mechanics may not be like this... It certainly *appears* random... Could there exist a more elaborate sort of "cosmic code," an algorithm that would generate the results of quantum events in the physical world and hence expose quantum indeterminism as an illusion? Might there be a "message" in this code that contains some profound secrets of the universe?... [Q]uantum indeterminism offers a window for God to act in the universe, to manipulate at the atomic level by "loading the quantum dice," without violating the laws of classical (i.e., nonquantum) physics. In this way God's purposes could be imprinted on a malleable cosmos without upsetting the physicists too much."—*Paul Davies, theoretical physicist and cosmologist (pp. 131-132)* [17]

Even though the initial conditions cannot be identified due to the uncertainty principle, it suggests that principles of order are at least instantaneously involved at the initial moment of the big bang. According to the principle of decoherence, the initial conditions need to contain the possibilities that become actualized through consistent context dependent change. The actuality of one event in one moment would need to be included as a possibility in the prior event. These points also suggest that somehow order would have been characteristic of the initial conditions that emerged instantaneously at the 'initiation' of the big bang. Then wouldn't order at least have to have been a possibility or potential state before the big bang?

> "We take for granted that there is a direction to the way things unfold in time. Eggs break, but they don't unbreak; candles melt, but they don't unmelt; memories are of the past, never of the future, people age, but they don't unage. These asymmetries govern our lives; the distinction between forward and backward in time is a prevailing element of experiential reality. If forward and backward in time exhibited the same symmetry we witness between left and right, or back and forth, the world would be unrecognizable... But where does time's asymmetry come from? What is responsible for this most basic of all time's properties?... It turns out that the known and accepted laws of physics show no such asymmetry... Nothing in the equations of fundamental physics shows any sign of treating one direction in time differently than the other, and that is totally at odds with everything we experience."—*Brian Greene, theoretical physicist (p. 13)* [18]

"[I]f the universe started out in a thoroughly disordered, high-entropy state, further cosmic evolution would merely maintain the disorder... Even though particular symmetries have been lost through cosmic phase transitions, the overall entropy of the universe has steadily increased. In the beginning, therefore, the universe must have been highly ordered."—*Brian Greene, theoretical physicist and string theorist (p. 271)* [18]

There are other challenges associated with the concepts of the big bang and the causally closed physical universe. For example, the *horizon problem* concerns how regions of space outside each other's causal light cone appear to be homogeneous. Big bang theory predicts a chaotic and lumpy or not homogeneous universe, with no means for it to be homogenized due to no causal connection between them. Also, there is the *flatness problem*, associated with the calculation of *omega*, the ratio of kinetic and gravitational energy in the universe. This concerns how there is such a close balance of these opposing forces in the universe, and whether the universe is permanently expanding or follows a cycle of expansion and contraction. The *age problem* concerns reconciling differences between estimates of the age of the universe ranging from 7–20 billion years depending on the value of the *Hubble constant* with estimates of the age of the oldest stars from 13-17 billion years and of fundamental elements from 12-16 billion years. Presumably the universe is not younger than objects such as particles and stellar masses that occupy it, but at this point the value of the Hubble constant needed for a reasonable temporal sequence seems to be arbitrary. Inflationary big bang theory offers explanations for some of these problems—to be discussed in Chapter 12—but the theory is far from being well-established.

In addition the *monopole problem* has to do with big bang theory and grand unified theory predicting the existence of magnetic monopoles—magnetic objects with either only a south or a north pole. Some inflationary universe theories suggest that there may be many completely independent universes, each one with a monopole. However, monopoles have not been found in nature, [19] and there may be no way to test theories about their possible existence.

There is also what has been called the *space problem*. This problem concerns how apparently non-spatial mind and consciousness can be constructed out of a purely spatial physical universe. On the other hand, there is the problem of how space began at the big bang from an apparently *non-spatial* infinitely dense singularity of nothing. [12]

In addition, it might be pointed out that the theory of multiple independent universes in different stages of evolution doesn't seem consistent with the model of a closed causal structure of relativistic space-time. Simultaneity in space—different locations outside the closed causal structure—as well as sequentiality of different stages of universes, would be undefined outside of the relativistic causal structure, and suggestive of some more fundamental substrate or background. Instantaneous tunneling to remote regions of the universe outside the light cone, and even to other universes, is also suggestive of a notion of space and time as a substrate beyond the background independent relativistic space-time causal structure of a closed universe.

Another major issue is the *antimatter problem*. Standard big bang and super-symmetry theories predict equal amounts of matter and antimatter in the universe. The universe appears to have quite a bit of matter, but antimatter has not been found as a naturally occurring phenomenon.

Also research on *chirality*—handedness—indicates that there is a 'violation' of parity invariance in the universe. Big bang and super-symmetric theories—as well as relativity theory—imply symmetry of direction throughout the universe—no favored spatial direction. However, the evidence with respect to weak force interactions demonstrates that nature seems to be asymmetrical. [19]

All of these issues and problems exemplify challenges that remain to be addressed in order to build a coherent model of a causally closed physical universe. Certainly one of the most significant issues in both quantum theory and relativity theory—the two primary bases of models of quantum gravity and a causally closed universe—is the experimental finding of nonlocality. The evidence indicates that the universe is not just localized, but rather both localized and nonlocalized. Nonlocality is implicit in quantum theory—as reflected in the quantum wave function spread throughout the unbounded quantum field, as well as the notion of the instantaneous collapse of the wave function throughout the entire quantum field. The theory attempts to capture both the discrete particle or quantized attribute of nature, as well as the unbounded field or wave attribute of nature. Interpretations of quantum theory struggle with the contrasting attributes of nonlocality and a fundamentally fragmented local atomistic discreteness.

The contrast becomes most challenging in trying to reconcile the models of an infinitely dense *nothing* and an infinite unified field of *everything*—to be discussed in Chapter 10. It is quite challenging to encompass these opposing attributes in a single theory. Because of the strong reductive pattern of thinking, the discreteness attribute—associated with the quantum principle—has been emphasized. This has resulted in models of the ultimate discreteness of nature at the Planck scale, as well as the models of black holes and an infinitely dense singularity bringing everything down to *nothing*. To get out of this reductive endpoint, the model of a *hot* big bang is proposed as a means to explain the expansion or unfolding of nature to its 'present big, cold, slow' appearance. But this picture clearly is incomplete.

Relativity theory also implies holism in the sense of the interconnectedness of nature, even going so far as to identify everything as fundamentally interdependent and even interchangeable. Everything is just changing relationships. But the theory developed within the assumption of locality and the fundamental independence of objects. In relativity theory, space and time don't exist *on their own*—they don't constitute a background or absolute frame of reference. They depend on the existence of objects that have relationships and their relative frames of reference or observer perspectives. The focus on relationships between objects becomes challenging when there are only processes that don't have independently definable objects—as well as no definable subjects or observers that can have perspectives or reference frames either.

The contrast of locality and holism in this relativistic framework is between objects that exist independently *and* relational processes that don't have independent existence. Inasmuch as the theory captures the fundamental principle that all things are relative—that all things are related to each other in the universe—it is a powerful and widely applicable conception of nature, because all things that exist in the universe must be related to some degree and on some level. The theory works in part because subjects and objects are implicit in the concept of observer perspective or frame of reference—but also are unexamined in it. Questions emerge when the theory defines locality in terms of the speed of light, delimiting the range and nature of the relationships. The finding of nonlocality cannot be reconciled easily with the view that all things in nature includes only things limited by the speed of light, because it indicates that there may very well be causal effects—or at least highly correlated relationships—outside of it.

The concepts of causality and determinism also need to be developed further in order to account for this expanded nonlocal system

> "[R]elativity and quantum theory agree, in that they both imply the need to look on the world as an *undivided whole*, in which all parts of the universe, including the observer and his instruments, merge and unify in one totality. In this totality, the atomistic form of insight is a simplification and an abstraction, valid only in some limited context."—*David Bohm, theoretical physicist (pp. 13-14)*[20]

of relationships. However, the concept of causality doesn't necessarily require locality—and neither does the general concept of relativity if it is not defined entirely in terms of Einstein's theories of relativity. Both the general concept of causality and the concept of relativity are likely to survive in nonlocal models of the universe. It is Einstein locality that is fundamentally challenged by the experimental validation of nonlocality. Determinism and relativity aren't necessarily challenged.

Philosopher Michael Dickson makes this point:

"...[N]either special or general relativity is about how matter interacts with matter and hence neither puts much restriction on *how* matter can interact with matter. Of course, one can introduce forces into the theory—one can define force fields, or potential fields, of various kinds, and in this case, the only restriction from relativity is that the symmetries of space-time be respected at least observationally. However, there is no apparent reason that these fields cannot be 'nonlocal.'"... What? Do we not hear talk, in relativity theory, about 'causal connectibility' and the like? Is relativity not exactly about causal relations?... The claim that only time-like related events are 'casually connectible' is better read as the claim that particles cannot move faster than light... Only if causal connections must be mediated by particles may we conclude that causal connections can exist only between time-like separated events. Even if we *do* wish to suppose that causal connections are always mediated by particles (quite a substantive assumption—and no part of relativity theory *per se*), in fact relativity theory is compatible with particles moving at superluminal velocities... Relativity theory simply does not rule out superluminal causes." (pp. 176-178) [21]

In addition to uncoupling the concept of relativity from Einstein locality defined by the speed of light, progress in the fundamental issue of *probabilism* needs to be made in order to loosen the closed grip of the presumption of a local, relativistic closed causal nexus of the physical universe. As has been discussed, different circumstances can result in indeterminate probabilistic outcomes even based on underlying deterministic processes. Unspecifiability and unpredictability don't necessarily indicate quantum ignorance or inherent randomness. There may be inability to know the present state of a system (quantum wholeness) due to inability to know all the determining factors because of measurement limitations. These limitations may include the uncertainty principle, complexity as in non-linear systems, calculation limits and inability to solve the prediction equations, inadequacy of predictive theories, or impossibility of knowing initial conditions. Both in practice and in principle, the state of a deterministic system—even the system of the entire universe—might not be completely predictable, even given the history of its present, prior, and initial states. Again quoting Dickson:

"After all, it may be that a theory is completely deterministic at the level of the universe, but due to interaction, indeterministic at the level of any subsystem of the universe." (p. 21) [21]

It is not necessarily the case that the statistically probabilistic results of measurement lead to the conclusion that the universe is fundamentally random—that God plays dice. A better question might be whether God *chooses* to play dice; and if so, when and where.

"Should we conclude that the universe is a product of design? The new physics and the new cosmology hold out a tantalizing promise: that we might be able to explain how all the physical structures in the universe have come to exist, automatically, as a result of natural processes. We should then no longer have need for a Creator in the traditional sense. Nevertheless, though science may explain the world, we still have to explain science. The laws which enable the universe to come into being spontaneously seem themselves to be the product of exceedingly ingenious design. If physics is the product of design, the universe must have a purpose, and the evidence of modern physics suggests strongly to me that the purpose includes us."—*Paul Davies, theoretical physicist and cosmologist (p. 243)* [22]

On the other hand non-linear systems are thought to be able to generate order spontaneously, from processes that appear to be fundamentally random. These so-called *self-organizing* systems are not describable in terms of ordinary linear models of causality.[23] This is due to their exponentially complicating feedback loops and self-interacting properties, in which all parts of the system affect all the other parts. Even very few initial requirements are necessary for such non-linear systems to generate order. However, there at least *some initial requirements*, which still leaves open the issue of from whence they came—from God, from their own nature, from chance, or from some combination or integration of these. If the universe is fundamentally random, it would seem inconsistent to have even a few initial non-random features (namely three, refer to Chapters 11-12). On the other hand, a fundamentally highly-ordered universe could employ random processes at certain functional stages.

> "…[A]ccording to Newton's absolute space and absolute time, everyone's freeze-frame picture of the universe at a given moment contains exactly the same events; everyone's now is the same now, and so everyone's now-list for a given moment is identical… Two observers in relative motion have nows—single moments in time, from each other's perspective—that are different; their nows slice through at different angles… Observers moving relative to each other have different conceptions of what exists at a given moment, and hence they have different conceptions of reality."—*Brian Greene, theoretical physicist (pp. 133-134)*[18]

It is particularly interesting that the nonlocality of quantum theory and the relational notion in relativity theory—while both reflecting a more interconnected holistic understanding of the universe—result in a fundamental contrast with respect to a cosmological scale of space and time. In relativity theory, there is no absolute space and time apart from the observer's frame of reference, no present moment or universal 'now.' What is *really* happening 'now' is relative; literally one observer's 'now' is another observer's 'future.' On the other hand, there does seem to be a simultaneous 'now' from the perspective that the collapse of the wave function occurs instantaneously throughout the unbounded quantum field—fundamental to orthodox quantum theory. Although in orthodox quantum theory the wave function is not thought of as a material thing—probability waves are different from physical waves such as ocean waves—the quantum entanglement tests of Bell's theorem imply something closer to simultaneity than the constraints of Einstein locality and the light cone allow. Also quantum theory envisions instantaneous quantum mechanical tunneling and 'porting' to regions of space even beyond the light cone, while in relativity theory such regions are undefined and unknowable at least until they come into the range of the light cone of a particular observer. If both theories are valid—as voluminous evidence indicates—then it would seem to be the case that the notion of instantaneity or universal 'now' indeed may be accurate, may be *real*. Within that, the notion of the relativity of space and time and the corresponding limitations of the light cone and Einstein locality also would be *real*—at least conditionally. This will be discussed in Chapter 12.

> "It is instructive at this point to contrast the key features of relativistic and quantum theories… (R)elativity theory requires continuity, strict causality (or determinism) and locality. On the other hand, quantum theory requires non-continuity, non-causality and non-locality… It is therefore hardly surprising that these two theories have never been unified in a consistent way. Rather, it seems most likely that such a unification is not actually possible. What is very probably needed instead is a qualitatively new theory, from which both relativity and quantum theory are to be derived as abstractions, approximations and limiting cases… The best place to begin is with what they have basically in common. This is undivided wholeness. Though each comes to such wholeness in a different way, it is clear that it is this to which they are both fundamentally pointing."—*David Bohm, theoretical physicist (P. 223)*[20]

> " I regard consciousness as fundamental. I regard matter as derived from consciousness."—*Max Planck, theoretical physicist (p. i)*[24]

Eventually fundamental constructs in relativity and quantum theories will sift down to the principle of locality, the principle of relationship, and the principle of nonlocality. The challenges brought up in this chapter point to the incompleteness of models of a closed, local, physical universe in accounting for these principles. The emphasis on reductionism reflects a tendency for premature closure. There is more to the universe—as soon will be discussed, a *whole* lot more.

In order to describe this expanded nonlocal universe, a more holistic perspective is needed. The perspective needs explicitly to account for how the concepts of locality, relationship, and nonlocality connect with mind and consciousness, as well as how they all merge from and into the all-inclusive completely unified field. It may be that consciousness is *not just inside* the theorized closed causal nexus of the physical universe.[25] Rather, the quasi-closed physical universe may be a conditional limitation of a local physical *reality* within a nonlocal mental *reality*, again a conditional limitation of the unified field itself. This view is much closer to Maharishi Vedic Science and Technology, and is the direction we are headed on the bridge to unity.

The connection of nonlocality, relationship, and locality to the knower, process of knowing, and known is a key issue in Chapter 10. In subsequent chapters the challenges related to causality—and determinism versus indeterminism—will be referred to again in the context of discussions of upward and downward causation, emergence theory, natural law, free will, and karma. Later in the book all of these concepts will find their place in a comprehensive understanding of *levels of reality* related to states of consciousness, exemplifying further Maharishi's profound point that 'knowledge is different in different states of consciousness.'[26] Planks that allow us to progress further toward understanding the relationship between the model of the material domain of nature and the emerging model of an underlying information space, mental domain or quantum mind are the focus of the next chapter.

Chapter 7 Notes

[1] Penrose, R. (1994). *Shadows of the mind: In search of the missing science of consciousness*. New York: Oxford University Press.

[2] Katz, V. (2003, Summer). Personal communication, Fairfield, IA.

[3] Smolin, L. (2003). Loop Quantum Gravity. In Brockman J. (Ed.). *The new humanists: Science at the Edge*. New York: Barnes & Noble Books.

[4] Hawking, S. (2001). *The universe in a nutshell*. New York: Bantam Books.

[5] Hensley, P. (Summer, 2003). Personal communication, Fairfield, IA.

[6] Hut, P. & Shepard, R. N. (2000). Turning 'the Hard Problem" Upside Down and Sideways. In Shear, J. (Ed.). *Explaining consciousness—The hard problem*. Cambridge, MA: The MIT Press, pp. 305-322.

[7] Edelman, G. M. & Tononi, G. (2000). *A universe of consciousness: How matter becomes imagination*. New York: Basic Books.

[8] Walker, E. W. (2000). *The physics of consciousness*. Cambridge, MA: Perseus Books.

[9] Clarke, C. J. S. (2000). The Nonlocality of Mind. In Shear, J. (Ed.) (2000). *Explaining consciousness—The hard problem*. Cambridge, MA: The MIT Press, pp. 165-175.

[10] Dossey, L. (1989). *Recovering the soul: A scientific and spiritual search*. New York: Bantam Books.

[11] Schwartz, J. M. & Begley, S. (2002). *The mind and the brain: Neuroplasticity and the power of mental force*. New York: HarperCollins Publishers, Inc.

[12] McGinn, C. (2000). Consciousness and Space. In Shear, J. (Ed.). *Explaining consciousness—The hard problem*. Cambridge, MA: The MIT Press, pp. 97-108.

[13] Eddington, A. (1974). *The nature of the physical world*. Ann Arbor, MI: The University of Michigan Press.

[14] Eddington, A. (1920). *Space, time and gravitation: An outline of the general relativity theory* (Reprint, New York: Harper and Row, 1959).

[15] Seager, W. (2000). Consciousness, Information, and Panpsychism. In Shear, J. (Ed.). *Explaining consciousness—The hard problem.* Cambridge, MA: The MIT Press, pp. 269-286.

[16] Lincoln, D. (2004). *Understanding the universe: From quarks to the cosmos.* Singapore: World Scientific Publishing Co., Pte. Ltd.

[17] Davies, P. (1992). *The mind of God: The scientific basis for a rational world.* New York: Simon & Schuster.

[18] Greene, B. (2004). *The fabric of the cosmos: Space, time, and the texture of reality.* New York: Alfred A. Knopf.

[19] Penrose, R. (2005). *The road to reality: A complete guide to the laws of the universe.* New York: Alfred A. Knopf.

[20] Bohm, D. (1980). *Wholeness and the implicate order.* New York: Routledge Classics.

[21] Dickson, M. W. (1998). *Quantum chance and non-locality.* Cambridge: Cambridge University Press.

[22] Davies, P. (1984). *Superforce: The search for a grand unified theory of nature.* New York: Simon & Schuster, Inc.

[23] Strogatz, S. (2003). *Sync: The Emerging science of spontaneous order.* New York: Hyperon.

[24] Klein, D. B. (1984). *The concept of consciousness: A survey.* Lincoln, NB: University of Nebraska Press.

[25] Chalmers, D. J. (1996). *The conscious mind: In search of a fundamental theory.* New York: Oxford University Press.

[26] *Maharishi International University Catalogue, 1974/75.* Los Angeles: MIU Press.

Chapter 8

What *Matters?*

In this chapter, we will explore more explicitly the relationship between the quantum model of physical reality, the subtler mental reality, and the underlying unified field. This will be discussed in terms of the gross, subtle, and transcendental levels of nature. Key topics include the anthropic principle, the organizing agency of nature, and the quantization and materialization of space or existence into physical matter. The overall main point of this chapter is that matter emerges from an underlying nonlocal field that in turn emerges from the indivisible wholeness or totality of the unified field, the source of everything in nature including all objectivity and subjectivity.

Does God *Matter?*

Some of the recent attempts to understand the agency that determines the structure and function of the universe make reference to the *anthropic principle*. This principle is based on appreciation of the amazing degree of fine-tuning and delicate balance manifested in the dynamic structure of nature. A good example is the *omega* number, the balance of attractive and repulsive forces involved in the creation of a star. If the gravitational force varied as little as one in 10^{40}, it is hypothesized that stars could not have formed and human life as we know it would not exist. [1]

There are infinite mathematical possibilities in the universe from the standpoint of quantum theory, and also relativity theory. Comparable to the question of what collapses the quantum wave function discussed in Chapter 5, what is the organizing power or agency that narrows down these possibilities to get the outcomes we observe in nature—especially the particular outcome of an extremely subtle balance of forces in the universe that has shaped us as observers of it?

The anthropic principle. The fundamental question of what structures the universe that we experience, as well as what structures us, relates to the *anthropic principle*. Roughly this principle can be understood to mean that the universe has to be as it is because otherwise we wouldn't exist

> "It seems that the universe we live in is very special. For a universe to exist for billions of years and contain the ingredients for life, certain special conditions must be satisfied: the masses of the elementary particles and the strengths of the fundamental forces must be tuned to values very close to the ones actually we observe. If these parameters are outside certain narrow limits, the universe will be inhospitable to life. This raises a legitimate scientific question: given that there seem to be more than one possible consistent set of laws, why is it that the laws of nature are such that the parameters fall within the narrow ranges needed for life? We may call this the *anthropic question.*"—Lee Smolin, theoretical physicist (p. 242) [2]

to observe it. [3] This is sometimes viewed as implying that the universe may have been designed for human observers. At least it recognizes that the incredible orderliness exhibited throughout the universe suggests that there is some overriding framework that organizes the laws of nature in order to materialize creation in the way that it appears. This could perhaps come from some outside ordering agent, such as an all-powerful God, or possibly from some principle of order inherent in the laws of nature (God the Father, or Mother Nature?). These can be associated with the strong and weak versions of the anthropic principle.

The strong version proposes that it is rational to consider God as the ordering agent if science cannot find an overriding and unifying principle of order in the laws of nature. This default position is based on the inability of science to explain the source of order in creation. One difficulty of this version is that it is believed not to be verifiable, because there always might be more knowledge to be gained that

> "If we want to stick to our principle that there is nothing outside the universe, then we must reject any mode of explanation in which order is imposed on the universe by an outside agency."—*Lee Smolin, theoretical physicist* " (p. 191) [2]

would identify the ultimate source of order in the laws of nature themselves. Unfortunately it also embodies a common conception of agency that separates the laws of nature from God, as if only one or the other is possible. The weak version of the anthropic principle proposes that because there are many universes it is not surprising that at least one of them would have the special feature of intelligent human observers—to be considered further in Chapter 12 associated with inflationary big bang theory. Hawking explains:

> "The weak anthropic principle amounts to an explanation of which of the various possible... parts of the universe we could inhabit... One would feel happier about the Anthropic principle, of course, if one could show that a number of different initial configurations for the universe are likely to have evolved to produce a universe like the one we observe. This would imply that the initial state of the part of the universe that we inhabit did not have to be chosen with great care." (p. 86) [3]

One approach consistent with the weak version of the anthropic principle proposes many different universes in different phases of evolution. Smolin describes this possibility:

> "The big bang is then not the origin of all that exists, but only a kind of phase transition by which a new region of space and time was created, in a phase different than the one from which it came, and then cooled and expanded. In such a scenario there could be many big bangs, leading to many universes... It is possible that the process creates universes in random phases... These universes will have different dimensions and geometries, and they will also have different sets of elementary particles which interact according to different sets of laws. If there are adjustable parameters, it is possible that they are set at random each time a new universe is created." (pp. 198-199) [2]

In this way of thinking the initial state of the universe would not have to be chosen with great care. If a very large number of universes are created, then eventually one of them would be expected to be compatible with the formation of human life as we know it. It has been proposed that the odds are improved if universes are born—such as in black holes—through a process similar to natural selection, called *"cosmological natural selection."* (p. 199) [2]

This line of reasoning suggests the possibility that sub-cellular, sub-molecular, inorganic levels of nature might exhibit processes previously attributable only to living organisms. It exemplifies a direction in developing theories that attribute biological qualities even to the universe as a whole, which is beginning to blur the distinction between living and non-living systems. It suggests that there might be properties characteristic of life that function at more fundamental levels of nature than the level at which intelligent biological organisms are thought to emerge. By attributing qualities associated with life to fundamental levels of nature thought to involve only non-living processes, it can be seen as another plank toward a more expanded and integrated view of nature—to be discussed more in later chapters.

Although the theory of natural selection deals with biological systems, it still is based on the concept of randomness and chance, not some outside agent that creates order in the system.

Applied to non-living processes underlying biological systems, the theory that the systems are fundamentally random can be maintained. The universe as fundamentally random and indeterminate is a core aspect of orthodox quantum theory. It also is an important concept related to the belief that the universe is closed in the sense that there is no outside ordering agent.

Even then, however, the likelihood of our *special* universe—involving such things as the appearance of three-dimensional space extending into infinity—is extremely low. (Of course, we could say that even an extremely rare random event would occur an infinity of instances in an eternity of time.) The theory of loop quantum gravity, which applies one approach to quantify the smallest unit of space, provides a basis for calculating the likelihood of a spin network generating something like a smooth three-dimensional space. Smolin estimates this likelihood:

> "Each node of a spin network graph corresponds to a volume of roughly the Planck length on each side. There are then 10^{99} nodes inside every cubic centimetre. The universe is at least 10^{27} centimetres in size, so it contains at least 10^{180} nodes. The question of how probable it is that space looks like an almost flat Euclidean three-dimensional space all the way up to cosmological scales can then be posed as follows: how probable it is that a spin network with 10^{180} nodes would represent such a flat Euclidean geometry? The answer is, exceedingly improbable!" (p. 203) [2]

This is suggestive of a deep correlation between our objective universe and our subjective ability to experience it and develop a high degree of inter-subjective agreement about it. It also reflects the continuing challenge to integrate subjective and objective, possibilities and actualities, wave and particle, mind and body, as well as many other fundamental dualities that run through modern scientific conceptions of nature. How this orderly objective-subjective relationship is structured will be discussed in more detail in Chapters 10-12. Also, its relationship to the concept of God or Godhead will be discussed in later chapters.

In a closed cosmological system, order is not caused by an outside ordering agent, but nonetheless the system somehow becomes at least almost universally orderly, deterministic, and causal. One reason it cannot be caused by an outside agent is that, in the model of a closed causal universe, there is no outside. But in this framework, neither can it be caused by an inside ordering agent. The cause cannot be identified as being inherent in the infinitely dense singularity, in that it is said that nothing is definable or inherent 'there.' It also cannot be caused by an inside ordering agent if it is fundamentally random and indeterminate. How then do we get to a causal deterministic universe when there are initially no inside or outside causal agents and no pre-existing deterministic patterns?

This seems to suggest the contradiction that principles of order are built into the initial conditions, possibly even inherent in some way as possibilities—and maybe even probabilities— in the infinitely dense singularity. Is it possible that the origin of the universe somehow contains initial possibilities related to decoherent context dependence and simultaneously no initial conditions, and to contain high entropy and indeterminate randomness while at the same time low entropy and deterministic causal order as laws of nature? Could the universe be a product of both a deterministic God and fundamental indeterminism in nature? What would the concept of nature mean if there was no fundamental orderliness in it, just randomness? Once again, perplexing inconsistencies arise that suggest a more expanded and integrated framework is needed.

Does Anything *Matter*?

Contemporary theories discussed in this and the past few chapters are beginning to develop somewhat more expanded conceptions of the universe beyond a completely closed and background independent—but observer dependent—system comprised only of the physical domain. These theories are beginning to define an additional ontological level of *reality*, which at least includes a subtle substrate of information space out of which the material world 'matters' or materializes that is in the direction of an abstract nonlocal field sometimes described as quantum mind. Also, unified field theories suggest even another additional level—the unified field itself.

The concept of quantum mind—at least in the sense of higher-order information or mental space—is directly related to the concept of observer dependence and an underlying background of the observer. The subjective domain is starting to be uncovered as the underlying basis of the objective domain. But a great challenge in this is to explain how mind can be a nonlocal field, when our everyday experience convinces many of us that the mind is localized in the brain. Eventually these issues and levels will come together on the bridge to unity in describing the 'nature' of the unified field and its phenomenal subtle and gross levels of existence.

> "[W]hen you get down to the Planck length (the length of a string) and Planck time (the time it would take light to travel the length of a string)..."going smaller" ceases to have meaning once you reach the size of the *smallest* constituent of the cosmos. For zero-sized point particles this introduces no constraint, but since strings have size, it does. If string theory is correct, the usual concepts of space and time, the framework within which all of our daily experiences take place, simply don't apply on scales finer than the Planck scale... As for what concepts take over, there is as yet no consensus. One possibility...is that the fabric of space on the Planck scale resembles a lattice or grid, with the 'space' between the grid lines being outside the bounds of physical reality... Another possibility is that space and time do not abruptly cease to have meaning on extremely small scales, but instead morph into other, more fundamental concepts. Shrinking smaller than the Planck scale would be off limits not because you run into a fundamental grid, but because the concepts of space and time segue into notions for which "shrinking smaller" is...meaningless... [A]lthough you can divide regions of space and durations of time in half and half again on everyday scales, as you pass the Planck scale they undergo a transformation that renders such division meaningless... Many string theorists, including me, strongly suspect that something along these lines actually happens, but to go further we need to figure out the more fundamental concepts into which space and time transform."—*Brian Greene, theoretical physicist (pp. 350-351)* [4]

> "[T]he current attempt to understand our 'universe' as if it were self-existent and independent of the sea of cosmic energy can work at best in some limited way... Moreover, it must be remembered that even this vast sea of cosmic energy takes into account only what happens on a scale larger than the critical length of 10-33 cm [the Planck length]... But this length is only a certain kind of limit on the applicability of ordinary notions of space and time. To suppose that there is nothing beyond this limit at all would indeed be quite arbitrary. Rather, it is very possible that beyond it lies a further domain, or set of domains, of the nature of which we have as yet little or no idea."—*David Bohm, theoretical physicist (p. 244)* [5]

Smaller and smaller? In the more expanded context that is emerging in cutting edge science, theories that emphasize quantization can be considered basically corpuscular models that involve smaller time and distance scales than their atomistic predecessors. In these theories, strings, loops, branes, qubits of information flow— however the smallest entity, process, or event is envisioned—embody some notion of a membrane or boundary that delimits them. This inevitably brings up the issue of infinite regress. The Planck scale is thought to be the smallest possible size of an object and the smallest possible boundary between objects. But at least theoretically, the boundaries of the object could be even thinner and thinner. On the other hand, there needs to be some level at which discontinuous quanta merge into indivisible continuity or wholeness if there is a *completely* unified field—because it would be a field beyond all gaps and boundaries, beyond all differences, completely unitary and one with itself. If there are boundaries and discreteness of some kind, then it would not be completely unified. The quantum principle related to the Planck scale ultimately cannot be fundamental if there is a completely unified field. Whether this transition to complete unity takes place at the Planck scale or at some even subtler level that somehow underlies the Planck scale is quite significant. The resolution to this issue requires an expanded conception beyond relativistic and quantum notions of space-time, which will be unfolded as we proceed through this chapter.

> "Despite the great success that atomic theory has so far enjoyed, ultimately it will have to be abandoned in favor of the assumption of continuous matter."—*Max Planck, theoretical physicist* [6]

Major contemporary models usually posit that the unification of all of nature is at the level where the fundamental forces merge into a single field at the Planck scale—the hypothesized level of super-unification. This also has been theorized to be the field of quantum gravity that is fundamentally discrete. This provides a model of a unified basis for all material objects, but it doesn't provide a model of a *completely* unified field.

If the unified field were to be completely undifferentiated in its unity, it would have to underlie any discrete quantized field, including the field of quantum gravity and the Planck scale. Abstracting matter into an extra-dimensional conceptual or information space, described as a pure geometry of discrete information qubits from which ordinary space-time is generated—kind of a 'minddust' theory—is more abstract and thus a step closer, but still does not yet describe a *completely* unified field.

According to quantum field theory, space-time comes to an end at the Planck scale and the concept of something smaller is meaningless—or at least undefined given the conditional limitations of the theory. In this theoretical framework, there is either motion within the limitations of the speed of light or there are quantum jumps or tunneling that instantaneously port objects between (relativistically undefined) regions of ordinary space-time without traveling in between.

On the other hand, for example, the interpretation of quantum theory in Bohmian mechanics proposes a subtle field associated with the psi wave through which nonlocal effects are mediated, the implicate order. This subtle field would underlie quantum force fields including the gravitational field, but also would not be the level of the undifferentiated completely unified field. There has long been the concept in modern science of phenomenally ordinary physical *reality,* and now progress is being made toward unified field theory, but it is quite challenging to link these together. The subtle, non-material, nonlocal level of nature fills in the missing link, an aethereal field in which motion is faster than the speed of light but not instantaneous.

> "It is possible to propose a deeper theory of the individual quantum process which is not relativistically invariant... In other words, we say that underlying the level in which relativity is valid there is a subrelativistic level in which it is not valid even though relativity is recovered in a suitable statistical approximation as well as in the large scale manifest world... Although there is no inherent limitation to the speed of transmission of impulses in this subrelativistic level, it is quite possible that quantum nonlocal connections might be propagated, not at infinite speeds, but at speeds very much greater than that of light... As the atomic free path quantum indeterminacy or randomness is the first sign of a 'subcontinuous' domain in which the laws of continuous matter would break down at the quantum level, so the free path in our trajectories would be the first sign of a subquantum domain in which the laws of quantum theory would break down... The next sign of a breakdown of the quantum theory would be the discovery of some yet smaller dimension whose role

might be analogous to the dimension of an atom in the atomic explanation of continuous matter [the classical microscopic level]. We do not as yet know what this dimension is, but it seems reasonable to propose that it could be of the order of the Planck length, where, in any case, we can expect that our current ideas of space-time and quantum theory might well break down."—*David Bohm and B. J. Hiley, theoretical physicists (pp 347-348)* [7]

Deterministic causal interactions in this subtle field would not be the mechanistic classical causality of ordinary relativistic space-time. At this point in the discussion, it can be associated with the concept of quantum mind which is characterized by nonlocality, non-instantaneity, and causal determinism—but not Einstein locality, the speed of light, and classical particle interactions.

Contemporary quantum gravity theories attempting to characterize some kind of geometric processes within the Planck scale rely on the mathematical concept of non-linear, higher-order dimensional space—such as eleven-dimensional M-theory which proposes 10 spatial dimensions plus time. It is interesting that analogies used to envision these higher-order dimensions typically are based on a notion of the limits of precision of measurement of ordinary four-dimensional space-time. For example one analogy is of a rope that from a distance looks like a one-dimensional string but closer inspection reveals the possibility of movement around the circumference, analogous to a hidden 'higher-order' dimension. Likewise Calabi-Yau shapes with multiple folds creating additional surfaces are used to envision higher-order dimensions (refer to Chapter 6).

In these theories it is sometimes suggested that coordinates in addition to the ordinary three axes of space and one axis of time would be needed to identify the specific place and time of an object or event. As a simplistic example, in order to identify a particular pro ballplayer's signature written on a basketball covered by signatures located in a hi-rise apartment building, one might need to know the answers to seven questions, such as the street name, address number on the street, apartment number that specified the floor of the building, the time the ball will be in the office, and technically also the longitude and latitude of the signature on the ball's surface. This is analogous to a six-dimensional space plus time. However, if the information were precise enough, the spatial location of the signature would be completely specified by the ordinary three global space coordinates (altitude, latitude, longitude) used in the global positioning system. The additional two spatial coordinates are needed only because the coordinate system in the example—street address and so on—was not precise enough. The three-axis global positioning system accurate to a centimeter, along with time, would be sufficient to specify the position of the signature.

This is analogous to the types of examples used to depict higher-order spatial dimensions, such as in with Calabi-Yau shapes. In the typical examples, however, the information with respect to the 'extra' dimensions might be specifiable within the ordinary three dimensions if it were precise enough. The assumption that space and time lose meaning 'below' or at smaller scales than the Planck scale adds to the impression that additional coordinates are needed to specify the exact location in analogies used to depict extra spatial dimensions.

When it comes to the physical world, 'extra' dimensions also could relate to the lack of precision of information in ordinary three-dimensional space. The extra dimensions would be 'extra' only if they did not exist anywhere—at any degree of precision—in ordinary four dimensional space-time. As has been discussed in Chapter 6, string theory and M-theory assume a classical background of space and time, and are not background independent. It seems that all positions on such a notion of non-linear space might be specifiable in three-dimensional space with a sufficiently precise coordinate system—if somehow not subject to the uncertainty principle.

Based on a holistic concept of space as infinity and time as eternity, no extra dimensions beyond the familiar three space and one time dimensions would be necessary even to account for phenomena that are subtler than the Planck scale. This is not to say that the mathematical higher dimensions don't have meaning. They indeed may be pointing to additional degrees of freedom that relate to additional principles and structures in nature. In this sense, they would be associated with higher dimensions—but not necessarily require spatial dimensions in addition to the familiar three. The challenge may be to understand the local-nonlocal contrast, rather than to posit additional spatial dimensions. The extra dimensions may relate, for example, to the degree of interconnectedness of subtle phenomena. Higher dimensionality may refer to distinctions other than space and time as classically conceived in terms of infinite extent in space and time The qualities of space-time, such as the speed of objects within it and the relationships between fundamental forces within it (like classical gravity), may characterize the aethereal fabric rather than mean that there are additional space-time dimensions.

Simultaneously smaller and bigger? Theories attempting to go beyond localized matter particles and into a nonlocal *reality* subtler than the Planck scale are clearly moving toward a very abstract field of some kind theorized to underlie and generate ordinary four-dimensional space-time. The distinguishing feature of this subtler field may not be its dimensionality in terms of conventional space and time, but rather its nonlocal property—its fundamental interconnectedness. This underlying field would exist somehow in between the localized material domain and the completely unified field that is *infinitely* correlated or interconnected. It would be unified with respect to the material domain and all matter, but not in the sense of being completely unified and one with itself.

This nonlocal field underlying physical matter and ordinary or conventional space-time would need to be *subtler* than any aspect of the material domain. Subtlety refers to being more refined, less coarse, and able to permeate grosser levels. This underlying field might therefore be *both smaller*—less grainy—*and bigger*—more extensive and nonlocal—than the material or physical domain associated with the four fundamental particle force fields. This suggests that, even though the Planck scale is incredibly small, there could be an even more abstract field, possibly at an even smaller—or in some sense higher frequency—time and distance scale that has fundamentally different properties than the physical domain of nature associated with ordinary conventional space-time and Einstein locality. This field would be hidden not so much due to it involving higher-order spatial dimensions, but rather because the medium or 'aether' it is composed of is finer-grained (and due to its nonlocal property, simultaneously larger) than local space-time.

Another way of saying this is that there are levels of four-dimensional space-time within infinite eternity. The ordinary localized notion of space-time is the gross level, and there may be another level that is subtler and that is characterized by nonlocal properties. In this more abstract expanded framework, space and time are disembedded from the notion of Einstein locality and the speed of light, while retaining the concept of four dimensions. Likewise, the notion of space-time is disembedded from the gravitational field associated with Einstein locality. That is, the concept of relativity—in the sense that all things are relative to all else—is disembedded from Einstein's Special and General Theories of Relativity. In this expanded perspective, space-time could be infinite and eternal, with different levels or domains within it that are characterized by specific limitations of the mediums or aethers—somewhat analogous to air being subtler than water and earth, for example. However, air, water, and earth are all gross matter, whereas subtle space would permeate gross space, like gross space permeates air, water, and earth. This will be discussed more in Chapter 11.

From an additional angle, gross space-time has similarities to the notion of a three-dimensional *braneworld* (with time being the fourth dimension) as proposed in cosmological implications of M-theory. In this view, a brane, medium, or aether could expand to encompass the entire universe of ordinary space-time, subject to gravity as we now conceive it—alluded to in Chapter 3. However, ordinary gravity would not extend into the subtler level of nature underlying gross space-time. There likely would be a principle associated with an attractive force like gravity in the underlying subtle domain; but it would not be characterized in the manner of the gravitational force associated with Einstein locality and the speed of light. The subtler level would be distinctly nonlocal rather than local—and thus not subject to our familiar notion of a gravitational field and dissipation of its force with the square of the distance.

The Planck length is directly related to the speed of light and is the distance light travels (10^{-33} cm.) in the Planck time (10^{-43} sec.)—it is defined with respect to the speed of light. The four fundamental forces that mediate change in our ordinary world are subject to this limitation. However, the experimental finding of nonlocality cannot be accounted for within this limitation. It suggests a nonlocal level of nature beyond the material domain that does not function within the limitation of the speed of light and the related gravitational force. This additional level would constitute a nonlocal, non-quantized fabric of existence that is more fundamental, interactive, unrestricted, and subtle. It would permeate the grosser ultramicroscopic and microscopic levels of nature, as well as the macroscopic and ultramacroscopic levels of ordinary objective sensory experience. The texture or fabric of this subtler aether would be that it has the property of nonlocal interconnectivity that simultaneously incorporates small and large, discreteness and continuity, embedded in an infinite background—rather than the local connectivity of conventional physical space-time and Planck-size quanta.

Where would such a subtle level be if not somehow underneath the physical domain, and somehow underlying and permeating the Planck scale? Given that the overall theories and findings about the physical universe are correct in modern science, it certainly seems that this subtle field would need to be outside of the range of conventional time and space limited by Einstein locality. If the texture of this nonlocal field were essentially larger than the Planck scale, then it seems unavoidable that it would be mediated by the four known fundamental particle-forces and be subject to ordinary gravity—like everything else at this grosser level. At least it would need to interact with the gravitational field, and it would seem that at least indirect evidence of it could be found within the framework of gravity and Einstein locality. But already we know from Bell's nonlocality that some aspects of even so-called *physical reality* are not accounted for within Einstein locality.

It thus may be helpful to distinguish 1) the structure of the gross material, physical domain, 2) its intellectualized delineation into parts as the basis for measurement, and 3) an ultimate undifferentiated continuum (refer to Chapter 4 and 5). Measurements of space and time—and measurable objects in conventional space and time—are quantized or discrete Planck-size bits; but perhaps there are underlying levels that are not quantized in this way. It would still be the case that there is no smaller scale of measurement of matter than the Planck scale using any independent probe built of matter. Measurement limitations at this limit of the scale of matter would prevent any further objective physical investigation of matter.

From this perspective the Planck length is the smallest curvature of space-time from which quantized material objects are constructed. This may be the physical size that reflects the most fundamental *quantization* of a more abstract, subtler, underlying field, imposing on the field the quantum principle as applied to matter—the fundamental units of which are the Planck length, mass, charge, time and temperature. In this view microscopic and macroscopic objects at the

Planck scale and larger exhibit both particle and wave properties. But the overall direction of cutting edge theories is pointing to the possibility of even subtler processes beyond quantized matter and particle-forces. Planck-scale quantization may be the limitation or concretization of an underlying non-quantized field, rather than additional spatial dimensions that are compactified at the Planck scale as hypothesized in string theories. In Chapter 12 we will discuss this important issue with respect to cosmology.

The issue here is whether there is a meaning to space, existence, or *reality* that is subtler than the notions embedded in the framework of quantum gravity, which involve relativistic space-time defined by all motion being limited by the speed of light, and the Planck scale as the ultimate smallest size of the quantum unit. Of course the notion of an 'object' would fundamentally change in this subtler, more inclusive framework. At larger scales, objects are *material* in that they are built of matter and force particles—quantized energy. At these *materialistic* levels of nature, particle properties generally are prominent. At 'smaller' scales, objects or processes may not be quantized. It might be helpful to distinguish objects with quantized particle properties as *material*, and 'objects' with nonlocal wave-like properties as *non-material*. Material objects exhibit individualized particle properties and underlying wave properties, and non-material 'objects' might exhibit individualized but nonlocal wave properties. Again the distinction of explicate and implicate orders in Bohmian mechanics is helpful (refer to Chapter 5).

The issue of gross versus subtle levels of space or existence is important for bridging the explanatory gap between matter and mind. Core to this issue is the conception that space-time is only material, physical space associated with Einstein locality, ignoring a subtler nonlocal domain of existence—such as mental space. The gap between matter and mind has been made more challenging to bridge due to the Cartesian dualistic conception of matter as having extension in space, and mind as non-spatial or not having extension in space. What is being described here is that mind has a kind of extension in *non-conventional* space—a subtler field, medium, or aether. The nature of this subtle field is much more abstract than ordinary conventional, gross space defined by Einstein locality and gravity, and by Planck-size quantization. Its nature would be characterized by nonlocal wave mechanics. But both levels would be within infinite space.

Mind as associated with information space, or mental space, is not confined to or does not exist only in the material brain or head. The brain doesn't even exist just in the head, and neither are material objects just made of material particles with no subtler underpinnings. The head, the brain, and all other material things are underpinned by nonlocal processes. In ancient Vedic science, the mind exists at an even subtler, expanded, nonlocal level—a much finer, non-quantized (in the sense associated with the Planck scale) level of existence.

A way of understanding abstract geometric objects in string theory, M-theory, or loop quantum gravity theory—such as strings, branes, or loops—is that these geometric objects may represent attempts to describe the mechanics of quantization *into* material particles. They can be viewed as mathematical objects used to model quantization, rather than using the idealized mathematical conception of a point with no spatial extension. Models of these geometric objects can be viewed as attempts to understand how the underlying nonlocal field *becomes* quantized. The great difficulty of developing a consistent theory of quantum gravity may be because there are some aspects of space-time or existence that are *not* quantized and that relate to levels of *reality* beyond local relativistic conventional space-time. In other words, conventional or gross space-time ends at the Planck scale of nature, but non-conventional or subtle space-time continues—even to infinity.

Reductively the Planck scale is theorized in string theory to be the level of nature where space-time is *compactified*. In string theory the classical macroscopic and microscopic world is where the four dimensions of space-time are unfolded, and spatial dimensions near the ultramicroscopic Planck scale are enfolded or compactified. But the *opposite* view may be more appropriate: quantization at the Planck scale is the limiting of an even more abstract, underlying, unfurled, nonlocal field into discrete, localized enfolded particles. To use the term compactification in a somewhat different way, *quantization into discrete Planck-size particles is the compactification.*

In string theory the additional mathematical dimensions at about the Planck scale are conceptualized as being compactified. But there may be a more abstract level that permeates the quantum level of the Planck scale and that is not characterized by particle properties. In this view some properties of space associated with the gravitational field are quantized. These are expressed in the process of sequential symmetry breaking when material quantization takes place and nature begins to limit itself into matter built of particles—the field of quantum gravity. This may occur, for example, at the level of symmetry breaking in which the underlying field manifests into the gravitational and strong particle force fields—hypothesized to occur at the Planck scale, the level of the breaking of hypothesized super-symmetry. Material objects built of particles are subject to the limitations that characterize the field of quantum gravity and that exist at the level of the Planck scale and larger. This includes all *material* objects in nature; but it would not include subtler *non-material* processes that are not quantized matter. These subtler processes would be nonlocal processes not subject to the limitations of the gravitational field as characterized by Einstein locality.

Much of the difficulty of envisioning an underlying nonlocal level of nature is due to the strong reductive tendency in modern science. It is not surprising given this tendency that models of nature would propose a smallest unit of space-time and a closed cosmological causal structure as ultimate, and that nonlocal phenomena and more holistic conceptions of nature would be particularly difficult to articulate.

In addition, the habit of trying to understand the material world based on *idealized* mathematical concepts may add to the difficulty. This approach has led to various hypothetical constructs to account for the lack of fit between the gross material, quantized level and the more abstract non-material, non-quantized level more closely associated with idealized mathematical constructs such as the dimensionless point. In this context the indeterminacy associated with hypothesized quantum fluctuations at the Planck scale in part may be due to applying an idealized dimensionless point mathematical model to the level of quantized matter (refer to the discussion on string theory in Chapter 6).

> "My guess...is that the whole issue of 'vacuum fluctuations' will need to be radically overhauled when we have a better quantum theory of gravity and, indeed, of QFT [Quantum Field Theory]."—*Sir Roger Penrose, mathematician and cosmologist (p. 1012)*[8]

Superimposed on the particle and wave properties of objects are idealized mathematical concepts based on the model of a particle being a dimensionless point with no extension in space. In order to relate this mathematical model with the particle and wave properties, it is conceived that these mathematical points fluctuate randomly—quantum vacuum fluctuations—and thus limit measurement precision at the Planck scale. It is characterized as a process of a dimensionless point randomly fluctuating within the Planck scale, which limits precision of measurement to an average proportional to ½ of Planck's constant. At least one source of the indeterminacy of nature may be directly related to the mathematical model of *dimensionless* points superimposed onto quantized Planck-size processes—as well as onto wave mechanics. A dimensionless point is an even subtler and more abstract concept than either particle or wave aspects of the quantum. It is an intellectual division superimposed on an infinite conceptual space. This is a much more abstract notion of the *quantum principle* than the quantized level of the Planck scale. A dimensionless point is not quantized in the sense that quanta have Planck-scale size extensions of space-time.

Particle and wave motion. The classical mathematical model works well on the macroscopic level, with statistical averaging, because this level of measurement precision is not fine-grained enough for quantum processes to have much effect in calculations. This mathematical model, based on a dimensionless point, also works well at microscopic and ultramicroscopic scales associated with quanta, but necessitates the addition of hypothesized random quantum fluctuations. At the level of the Planck scale, however, quanta with spatial extension such as a string, brane, or loop may be more appropriate for mathematical modeling than a dimensionless point, [9] though this might make the mathematics much more challenging. The quantum wave function mathematical model reflects an attempt to conceptualize both the particle level and the more holistic self-interacting property of nature associated with wave mechanics. Quantum and classical mechanics would apply at the Planck scale and larger, but perhaps not at the level of the underlying non-quantized field. In this underlying field, non-quantized, highly interconnected, nonlocal wave-field mechanics may be a more appropriate modeling framework. But in this more abstract field, there would be no possibility of measurement using a probe built of matter particles.

Some sense of the mechanics of motion at these different levels might be captured by a simple analogy. At the level of matter and quantum mechanics, motion is described in terms of an object moving through a field analogous to a surfer riding on a wave—discrete particle motion. At the subtler level of the underlying field without quantized objects, motion might be described in terms of impulses moving through the field analogous to the wave upon which the surfer is riding. The surfer moves through the ocean, but the underlying water doesn't move along with the surfer. In wave mechanics the water fluctuates in a manner to propagate an impulse through the field, while staying in the same general location with respect to the surfer and not moving laterally along with the surfer. Motion of independent objects in the gross domain of nature might be found to be more apparent motion at a subtler level. The principle of causality could still apply in this subtler level of existence, but not the classical notion of material causality and Einstein locality. Causality could be mediated by something subtler than particles colliding with each other within the speed of light, in order to explain exchanges of energy.

This could be likened to different levels of touch causing a change in direction of a car, such as due to two cars colliding, or due to feeling vibration of a rough road and changing lanes via a mental intention, or even due to feeling 'touched' by remembering a loving smile from your spouse and via mental intention turning the steering wheel of the car to return home. Decreasing amounts of physical energy expenditure are involved in causing—in the sense of initiating or guiding—the change of direction of the car in these examples. The concept of causality can be understood in terms of both the physical implementation of change and the subtler guiding mental intention that can be implemented in a physical change, with varying degrees of expenditure of energy.

The psi wave in Bohmian mechanics might be an example of non-material causality that involves no detectable physical energy change within the limitation of the uncertainty principle. This is somewhat analogous to chaos theory in which small perturbations result in large effects. Microscopic changes can cause macroscopic change. At even smaller scales perhaps a virtually undetectable change guides macroscopic change. An undetectable change on the level where all change is subject to the limitations of the uncertainty principle could be transduced into a large macroscopic change on the material level. In other words, *matterstuff* could be guided or causally influenced by very subtle information waves of *mindstuff* generally undetectable in matterstuff.

This subtler nonlocal field would be in some ways a less differentiated field underlying the four known quantized matter and force particle fields. But it still would have properties of individuality or localization, analogous to an individual wave as an individualized part of the ocean. At this non-material, non-quantized level the abstract wave-field mathematical model may be used that does not necessarily involve random fluctuations and related concepts such as space-time foam. Processes may be describable stochastically at all the levels, at macroscopic and microscopic scales in terms of the wave-like flow of groups of particles and on smaller scales in terms of superposed waves in a nonlocal field. The waves would be very abstract energy that is considerably more powerful than Planck-size quanta, but still not infinite energy associated with the underlying unified field that is *infinitely self-interacting*—to be discussed in Chapter 10.

What is emerging is a view of nature in which the material domain associated with ordinary classical objects is a *quasi-closed* and limited system within a quantum gravity-like aethereal field. The limitations are associated with the quantization of energy into matter particles—defining properties of the gravitational field, which include particle motion localized within the speed of light. This field of quantum gravity is unified in the sense that all objects associated with quantization into particles manifest in it. The gross particle field relates to the ordinary senses, which are capable of experiencing the material level of nature on a macroscopic scale. Action within this level of nature is, for the most part, in terms of objects of sense that are phenomenally independent and subject to the limitations described by relativity and quantum theories.

To summarize, the view unfolding here is that the gross relative domain of nature is built out of an aether-like substance or medium of gross space-time with specific textural properties. These properties are defined by Einstein locality and the speed of light according to relativity theory, and Planck-size quantization according to quantum theory and the uncertainty principle that is directly related to the speed of light. All objective material movement by objects made of this medium are characterized by the limitations of Einstein locality and Planck-size particle dynamics.

The particle attributes of objects follow classical and quantum mechanics at the Planck scale and larger. Underlying the particle attributes of objects at all levels of measurement may be their nonlocal wave properties, at least in part modeled mathematically as probability wave functions. The assumption of Einstein locality may be underlain by the much more interconnected nonlocal field properties of nature, which under specialized experimental conditions manifest as a more fundamental level than localized particles.

The smallest curvature of physical space-time—the smallest size from which material objects are built—is the Planck length, defined with respect to the speed of light. This may be the scale at which the most fundamental material quantization or compactification takes place, which theories of strings, branes, and loops are attempting to model. The even more abstract, idealized mathematical concept of a dimensionless point has been applied in classical and quantum mechanics to model motion of matter at grosser levels and larger time and distance scales. It works well enough on macroscopic levels associated with ordinary sensory experience due to statistical averaging, but becomes problematic when measurement precision approaches the ultramicroscopic Planck scale. Even subtler than quantized fields associated with the material domain may be a nonlocal, highly interacting, non-material, non-quantized field. This conception is more abstract than in classical and quantum theories.

Orthodox quantum theories hold that motion and change are mediated by the four fundamental forces limited by the speed of light, but also hypothesize instantaneous *porting* between regions of space-time through wormholes or similar nonlocal, non-physical connections. However, what is being outlined here is another, intermediate type of motion associated with a subtle nonlocal level of nature. This level would have far fewer limitations than the material level. It still would have some limitations—not the completely undifferentiated level. It would be both finer and less localized at the same time. In this intermediate level 'objects' would exist, move, and casually interact; but they would be impulses of information flow or intelligence—energy impulses more like mindstuff, or we might say thought impulses—not matter built of Planck-size quanta.

As noted in Chapter 6, it is interesting that the concept of motion much faster than the speed of light but not instantaneous has been proposed in inflationary big bang theory—to be discussed again in Chapter 12. In this theory the very early universe underwent a brief but vast expansion much faster than the speed of light. Space, time, gravity, and causal interaction retain meaning in this theory, but apparently different from the conventional meanings associated with Einstein locality. It represents an example of hypothetical motion that is not instantaneous and yet much faster than light-speed.

The subtle domain of existence being described here would have the property of further differentiating or *compactifying* into quantized elementary particles associated with the four fundamental force fields comprising the gross material domain. This subtle aether would permeate the grosser field within which all material objects *manifesting from and in it* are subject to the limitations that define gravity and Einstein locality. However, the subtler field would also have nonlocal properties not subject to these limitations, such as that it would be frictionless—analogous to information flow not involving thermodynamic heat. It would be capable of dynamic activity such as wave-like motion, but subtler than the four fundamental forces and at least virtually undetectable with respect to them. It would involve causality, but not in the sense of material causality with influences between objects via billiard ball-like particle mechanics. It would be 'relativistic,' but not in the sense of limited by the speed of light, Einstein locality, and conventional relativistic notions of gravity. However, it would have the capability of influencing causally the gross material domain of matter that it permeates and which is generated from it.

This view is a few planks closer to the description of levels of nature in ancient Vedic science, which generally distinguishes gross, subtle, and transcendental levels. In this classification, the quantized gravitational field is associated with the *gross* level and the corresponding gross level of material creation composed of matter and force particles. It is underlain by a subtler level that contains the information necessary to structure quantized matter and force particles. The material domain *matters*—that is, materializes or manifests—from this more abstract underlying, nonlocal field. This subtler field in turn manifests from the completely unified field—associated with the consciousness-mind-body ontology. In Chapter 11, the subtle levels of nature that underlie gross material existence—including the level of mind—will be outlined in ancient Vedic science and Maharishi Vedic Science and Technology. After that discussion we will consider in some detail how the conception of matter reflected in quantum field theory directly fits into the gross, subtle, and transcendental levels of space, existence, or *reality* in the more comprehensive framework of ancient Vedic science. This will be an important topic toward the end of Chapter 12.

"God is what mind becomes when it has passed beyond the scale of our comprehension. God may be either a world-soul or a collection of world-souls. So I am thinking that atoms and humans and God may have minds that differ in degree but not in kind."—*Freeman Dyson, physicist and mathematician* [10]

Holism and Reductionism

A helpful strategy in building a more holistic understanding of levels of nature is to disembed from the reductive approach that brings everything down through smaller and smaller scales to nothing. The reductive approach involves starting with ordinary sensory experience and then analyzing material objects to their most fundamental constituents. In so doing we now are arriving at an intellectual understanding of *reality* beyond the material, physical domain. But still, the means of gaining knowledge is objectified—outside oneself—and maintains its strong habit of reductionism. This makes holistic conceptions quite challenging, and also has rendered the mind-body problem, the relationship of subtle mind and gross body, unanswerable for centuries.

"First we need to turn round physics, so that we could see the local Newtonian picture as a specially disintegrated case of the fundamentally global reality... Second we need to turn round our whole approach by putting mind first. We would be in a position to understand how it was that mind could actually do something in the cosmos... We have to start exploring how we can talk about mind in terms of a quantum picture which takes seriously the fundamental place of self-observation; of the quantum logic of actual observables being itself determined by the current situation. Only then will we be able to make a genuine bridge between physics and psychology."—*C. J. S. Clarke, mathematician (pp. 174-175)* [11]

In the reductive physicalist perspective, it is logically impossible to explain how the whole comes from the parts, is greater than the parts, and has top-down causal control over the parts. Although a profound tool for probing into nature intellectually, the reductive perspective fundamentally has things upside down—or at least outside in. Instead of the universe narrowing down to an infinitesimal point such as in some theories of black holes, it may be just the opposite: there are deeper, subtler, increasingly expansive nonlocal levels of nature. In other words it is helpful to clarify the principle that the whole is greater than the parts with the more profound perspective that the parts come out of the whole, rather than the whole coming out of a collection of parts.

Later in the book we will discuss an even more advanced holistic perspective that the whole is contained in each part. This suggests that at the completely unified field level of nature every point in the universe is infinite—an ultimate completely continuous field beyond space and time, beyond all limitations. Chapter 6 included a quote pointing out that an underlying continuum to space would mean that every volume of space would contain an infinite amount of information to specify the position of even one electron. At the ultimate level of the unified field, every point *would indeed* be infinitely self-interacting—containing infinite potential information and energy.

In the description of levels of nature introduced in this chapter the material domain represents a quasi-closed, relatively *real*, classical, local, four-dimensional universe. But a subtler—and relatively more *real*—underlying level that is not nearly as limited is emerging in foggy outline. In ordinary waking consciousness the gross, most tangible, objective level is taken to be the most *real* level, when it may be the least *real* in the sense that it is the least inclusive *reality*. This will be discussed in the chapters on higher states of consciousness, in connection with the concept of *Maya* and the *relative reality* of both matter and mind.

The holistic approach of Maharishi Vedic Science and Technology begins with wholeness or unity, not with the gross, tangible, fundamentally fragmented level of ordinary waking experience. In 'analyzing the parts' of unity, limitations are superimposed on it that are relative and conditional, and that can be considered relatively *unreal* compared to the ultimate unified *reality*.

If the ultimate unity is to be analyzed into parts, its first conceptual delineation can be in terms of infinity of points in the infinite wholeness, each point containing the whole—infinitely self-interacting. The intellectual delineation of indivisible wholeness into parts can be described as a duality of part and whole, with each part remaining the whole. In ancient Vedic science this is the unmanifest level of Veda in which the abstract laws of nature are structured, and from which the subtle and gross levels of phenomenal creation sequentially emerge. Nature manifests as matter through sequential limitations that superimpose phenomenal divisions onto the undivided wholeness or indivisible unity.

"What distinguishes the explicate order is that what is thus derived is a set of recurrent and relatively stable elements that are *outside* of each other. This set of elements (e.g., fields and particles) then provide the explanation of that domain of experience in which the mechanistic order yields an adequate treatment. In the prevailing mechanistic approach, however, these elements, assumed to be separately and independently existent, are taken as constituting the basic reality. The task of science is then to start from such parts and to derive all wholes through abstraction, explaining them as the results of interactions of the parts. On the contrary, when one works in terms of the implicate order, one begins with the undivided wholeness of the universe, and the task of science is to derive the parts through abstraction from the whole."—*David Bohm, mathematician and theoretical physicist (pp. 226-227)*[5]

From a reductive perspective, we could say that embedded in gross is subtle, and embedded in subtle is the unified field. From a holistic perspective, gross is a phenomenal limitation of subtle, and subtle is a phenomenal limitation of the unified field. There is nothing outside of the ultimate unified field. There may be individual big bangs, such as with respect to specific black holes. But with respect to the entirety of existence—the multiverse—it could be said that the big bang would have to be not an explosion but rather an implosion—or perhaps better still, a condensation—because everything that resulted from the big bang remains inside the unified field. Both reductive and holistic perspectives are needed to get some sense of these ultimate dynamics of nature that involve both point value of nothing and infinite value of everything simultaneously.

The big bang—or whatever mechanics of nature result in phenomenal manifestation or materialization—would be a concretization of the infinite into finite values. It would not create time and space, but rather be a phenomenal limitation of eternity and infinity. Without this holistic perspective, it is quite difficult to bridge the explanatory gap and comprehend how the material domain and relativistic conventional notions of space-time relate to the even subtler level of information space or mental reality mind and the transcendent level of the unified field.

Reductively, space relates to the measurement of distance and time relates to the measurement of duration. This perspective is associated with the intellect or discriminating mind—sometimes called Buddhi in ancient Vedic science. From the holistic perspective, space refers to infinity and time to eternity. When the reductive perspective is primary and dominates, the wholeness of life is lost to experience. In Vedic science this is called Pragya aparadh, referred to as the mistake of the intellect. [12] The process of evolution and development to the highest state of consciousness involves reestablishing wholeness or unity as the natural primary experience. This will be discussed at length in later chapters.

What would a domain or plane of existence that is subtler than the ordinary objective level of conventional space-time be like? From a reductive perspective, it is hard even to imagine, because space-time is thought to collapse at the Planck scale and the notion of subtler in terms of smaller scales of time and distance is at least undefined, and frequently is held to be impossible. Within a holistic perspective that the most fundamental level of space or existence is infinity and time is eternity, however, subtle levels of existence can be conceived a bit more easily—as being more like infinity and eternity than the ordinary gross level of space-time. The conceptions of space and time as infinite, eternal purely abstract existence without any limitations can be thought of as conditioned by the manifestation of forms and objects. These forms and objects would be fluctuations of pure existence that appear to take on degrees of limitation, at the subtlest levels only in terms of abstract shapes, which might be considered thought forms or vibrations existing in a more abstract field of existence. These abstract forms of information materialize, becoming more concrete and tangible, as additional limitations of the qualities of localization, discreteness, and material substance are placed upon them. The difference in levels of nature might then be better conceived in terms of mediums or aethers with progressively limiting properties, each level completely encompassing within it and permeating the more limited grosser levels.

From this holistic top-down perspective, no new dimensions of space and time would need to be added in order to account for nonlocality. The subtle and gross aspects of the senses are sufficient for substances and forms to be perceived at any level of manifestation. What makes for subtle or gross domains of nature is not necessarily any fundamentally new spatial dimensions, in that they are conceptual limitations of infinity and eternity. Rather, phenomenal manifestation involves limitations that allow the perception of increasingly concrete form and substance in the purely abstract field of infinite eternity.

One way to conceive of this continuity is in terms of smaller and smaller time and distance scales, with the addition that smaller and smaller also means bigger and bigger at the same time— that is, more permeating and increasingly unbounded. This is already an important principle in quantum theory; but the reductive, more concrete, localized, quantized aspect of it overshadows the more abstract nonlocal aspect. The levels of space can be thought of in terms of different mediums, aether-like substances, or fields that become more limited, from the infinite eternal to the familiar conventional space-time of the gross material domain on which modern science has fixated. The defining features of gross conventional space-time relate to the speed of light and Einstein locality. The levels of purely abstract transcendental, minimally limited subtle, and highly limited and localized gross levels are laid out systematically in Vedic literature and will be described in more detail in Chapter 11.

> "There is enough space in a point for an infinity of universes. There is no lack of capacity. Self-limitation is the only problem. But you cannot run away from yourself. However far you go, you come back to yourself and the need of understanding this point, which is as nothing and yet the source of everything."—*Sri Nisargadatta Maharaj (p. 337)* [13]

It also might be worth directly pointing out that the Vedic understanding of gross, subtle, and transcendental levels of space or existence provides a basis for reconciling the contrasting views of space in relativity and quantum theories described in Chapter 3. According to Einstein's relativistic space-time theory, motion is limited to the speed of light, and the notion of time and place existing *right now* outside of the light cone is undefined and can never be known. According to the nature of space in Newtonian classical and non-relativistic quantum theories, however, even instantaneous quantum mechanical tunneling anywhere in the quantum field is theorized to be possible. In the more expanded understanding of levels of existence, the first notion of space relates to the gross relative domain of nature, the gross aether. The second notion of space relates to the subtle relative domain, the subtle aether, in which motion is not limited by the speed of light but still not instantaneous. The third notion of space relates to the infinitely self-interacting unified field itself, beyond the relative concepts of distance and duration, time and space.

Before addressing the concept of the 'nature' of the infinitely self-interacting unified field, however, in Chapter 9 we will develop further the discussion of the means of gaining knowledge from the Introduction. We will review content from Chapter 1 and link it to intuitive-based assumptions in modern science. Following this discussion we will proceed in Chapter 10 to unified field theory in modern science and begin to connect it to the unified field of consciousness in Maharishi Vedic Science and Technology. Once these planks are in place, Chapter 11 will change from the reductive approach to the holistic approach. We will explore the sequential process of phenomenal manifestation of levels of subjective and objective creation within the unified field. This reflects a remarkably integrated and expansive scope of knowledge beyond modern scientific knowledge, based on direct empirical experience of the transcendental, subtle, and gross levels of nature. Then sufficient planks will have been laid down to be able to discuss more specifically the concrete quantized structure of the material domain and relate it to the levels of existence in ancient Vedic science. Sequential symmetry breaking into the fundamental particle-force fields that materialize the apparently independent objects of our ordinary objective world, and how they might correlate with the structure of the gross material domain in ancient Vedic science, will be a major theme in Chapter 12.

Chapter 8 Notes

[1] Carter, B. (1974) Large Number of Coincidences and the Anthropic Principle in Cosmology. In Longair, M. S. . *Confirmation of cosmological theories with observation*. Holland: Reidel.

[2] Smolin, L. (2001) *Three roads to quantum gravity*. New York: Basic Books.

[3] Hawking, S. (2001) *The universe in a nutshell*. New York: Bantam Books.

[4] Greene, B. (2004) *The fabric of the cosmos: Space, time, and the texture of reality*. New York: Alfred A. Knopf.

[5] Bohm, D. (1980). *Wholeness and the implicate order*. New York: Routledge Classics.

[6] Isaacson, W. (2007). *Einstein: His life and universe*. New York: Simon & Schuster, p. 95.

[7] Bohm, D. & Hiley, B. J. (1993). *The undivided universe*. London: Routledge.

[8] Penrose, R. (2005). *The road to reality: A complete guide to the laws of the universe*. New York: Alfred A. Knopf.

[9] Greene, B. (1999). *The elegant universe: Superstrings, hidden dimensions, and the quest for the ultimate theory*. New York: Vintage Books.

[10] Dyson, F. (2000), *Progress in religion: A talk by Freeman Dyson*, May 16, Acceptance talk for Templeton Prize for Progress in Religion, Washington National Cathedral, as quoted in URL = <www.edge.org/3rd_culture/dyson_progress_index.html>

[11] Clarke, C. J. S. (2000) The Nonlocality of Mind. In Shear, J. (Ed.) *Explaining consciousness—The hard problem*. Cambridge, MA: The MIT Press, pp. 165-175.

[12] Nader, T. (2000). *Human physiology: Expression of Veda and Vedic Literature*, 4th Edition. Vlodrop, The Netherlands: Maharishi Vedic University.

[13] Nisargadatta Maharaj (1973). *I am That*. Durham, NC: Acorn Press.

Chapter 9

The Subjective Sense of Objective Reality

In this chapter, progress in modern science beyond the physicalist worldview described in the past few chapters will be referred back to the discussion in the Introduction about the means of gaining knowledge. Starting to place this progress into a developmental context, we will identify a general relationship between the emerging understanding of objective, subjective, and transcendental domains of reality and levels of subjective experience associated with the means of gaining knowledge—sensory experience, reasoning, intuition, and direct experience of consciousness itself. This chapter is intended to help integrate various planks of the discussion so far, in preparation for exploring the 'nature' of the unified field. The overall main point of this chapter is that knowledge and experience are based on the state of development of the knower.

As noted in the Prologue and in Chapter 1, modern science has relied on ordinary sensory experience and logical reasoning as the basic means of gaining knowledge. As deeper levels of nature are probed, it becomes harder and harder to validate theories directly through ordinary sensory experience. Macroscopic validation of microscopic or ultramicroscopic events can be done only when effects are large enough to be measurable directly on macroscopic scales, or using indirect methods that rely on logical inferences linking unobserved phenomena to observed effects.

At the hypothesized limit of the most fundamental ultramicroscopic scales, investigation goes beyond the material domain into abstract information space underlying conventional space-time. Mathematical models of these levels of nature are not direct representations, and the concepts in the models may or may not have counterparts in them. The mathematical models are such highly idealized abstract conceptions that it is helpful to use metaphors that provide a more tangible sense of how they might relate to ordinary experience. The primary significance of the models is in their mathematical consistency—the metaphors are heuristic tools. However, by invoking some degree of sensory imagery, the metaphors contribute to a more tangible sense of *objective reality* associated with the abstract models. On the other hand, there indeed may be a very abstract sub-phenomenal *reality* that the mathematical models and metaphors are attempting to characterize—if still only in foggy outline.

> "Perception is used for objects which are in contact with sense-organs; inference is used when only the characteristic marks are known; valid testimony is used for knowledge of those things that are beyond the perception of the senses and beyond the logical analysis of the mind."—*Theos Bernard, Indian philosophy scholar (p. 69)* [1]

Objective and Subjective Perspectives

At this vantage point on the bridge to unity, it seems appropriate to consider some of the implications of the contemporary models described in the past few chapters with respect to the discussion in the Introduction about the means of gaining knowledge and the corresponding sense of *objective reality*. In the fundamental distinction of known, process of knowing, and knower, the *known* generally refers to the objects in the outer objective world. It also refers to the knowledge about them. But knowledge about the objects resides somewhere in the inner subjective domain associated with the process of knowing in the knower. The markings in this book, for example, contain knowledge by virtue of the symbolic and semantic value attributed to them. The knowledge doesn't primarily reside in the book, but rather in the reader, observer, or knower who is able to attribute meaning to the markings—for whom they can be meaningful and *informative*.

In the past few chapters, one key issue has been the independent nature of objects in the outer objective world. Let's now consider contributions to the sense of *objective reality* associated with these phenomenal objects. This is particularly relevant, based on contemporary theories that posit these objects have no essential material or physical existence. According to quantum and unified field theories, material objects are particle-wave processes of quantum fields, abstract potentialities or tendencies to exist, which ultimately emerge as excitations of an underlying completely abstract field beyond energy and matter—a field of nothing, or everything, or both. By now it is clear that this goes beyond the physicalist worldview.

To briefly summarize points in the discussion of the means of gaining knowledge from Chapter 1, the sense of *objective reality* is contributed to by the consistency within an observer of ordinary sensory experience, reasoning, and also intuition—major components of intra-subjective consistency. A key additional contribution is consistency across observers, or inter-subjective agreement. Inter-subjective agreement can be based on consistency of sensory experience, reasoning, intuition and intuitive-like beliefs, or all of these processes of knowing taken together.

A major contribution to the sense of independent objects existing in *objective reality* comes from the attributes and qualities of objects as perceived through ordinary sensory experience. The intact sensory system—which seems for the most part to be quite similar across human observers—reliably produces consistent sensations that at least appear to be elicited by outer independent objects. It is a common assumption of naïve realism associated with ordinary sensory experience that the attributes, properties, or qualities of objects are present in the objects themselves.

Whether ultimately there are outer independent objects with such inherent attributes, the sensory system seems to be highly consistent within and even across most human observers—which supports the notion. This part of the process of knowing is a consistent *macroscopic observing system*. This system is described as inputting certain physical properties of outer objects such as elements, chemical compounds, wavelengths, and so on. As this input is processed further, higher-order top-down perceptual and other cognitive and affective processes attribute qualitative meanings to the objects, associated with the paradigm or worldview of the observer or knower.

For example, hopefully many human observers can agree that this book contains symbols that translate into useful knowledge or information when they share language and have broadly agreed upon general views of the world. For other human observers, possibly such as some aboriginal peoples with little or no experience with this cultural context, the book might be experienced to be very thin sheets of wood with dark marks on them, or perhaps just potentially good kindling. To non-human macroscopic observers such as an elephant or a mouse, the book might be simply an object to walk over or eat. The book may not even be observed at all as a distinct object in the environment by microscopic observers such as amoebae or bacteria. Certain physical properties seem to remain the same; but the sense of their *objective reality* as objects with specific meanings differs according to the observer.

At scales outside the range of even the aided ordinary human senses, obviously the *objective reality* of objects cannot be based on ordinary sensory experience. In such cases the sense of *objective reality* relies on indirect observation based more on the deeper reasoning aspect of the process of knowing. As long as reasoning appears to be reliable and consistent within and between observers, it provides an important basis for inter-subjective agreement, contributing to a fuller, stronger sense of *objective reality*. The more developed, consistent, and reliable the reasoning processes appear to be across observers, the deeper into nature the agreed-upon 'objective' investigations can indirectly extend. Beyond the level of the ordinary senses aided by tools such as a microscope, however, the empirical basis of *objective reality* is no longer primarily the *sensory* aspect but rather the *reasoning* aspect of the process of knowing.

A basic contribution to *objective reality* is the assumption that objects exist independent of the observer. It seems quite reasonable to assume that objects exist independent at least of any *particular individual* observer. There might be no sensory basis for establishing an object's existence, if it could not be observed directly. But a relatively high degree of confidence in its independent existence can be established indirectly through consistent reasoning and consensus across observers. Elementary particles have not been observed directly by humans, for example, but a fairly high degree of certainty of their existence has been established based on indirect evidence, reasoning, and consensus.

"[W]e say that we can easily see that our fingers are different from one another... The fact is, "difference" is a mental concept that we superimpose on certain raw sensations...we construct it, impose it, interpret it; we never actually perceive it. In other words, much of what we take to be perceptions are actually conceptions, mental and not empirical... Thus, when many empiricists demand sensory evidence, they are actually demanding mental interpretations without realizing it... Everything we see is the product of mind... The "difference" between your fingers might be a mental construct, but the fingers themselves in some sense preexist your conceptualization of them; they are not totally or merely a product of mental constructions... A diamond will cut a piece of glass, no matter what cultural words or concepts we use for "diamond," "cut," and "glass," and no amount of cultural constructivism will change that simple objective fact... So it is one thing to point out the partial but crucial role that interpretation plays in our perception of the world... But to go to extremes and deny any moment of objective truth at all (and any form of correspondence theory or serviceable representation) is simply to render the discussion unintelligible."—*Ken Wilber, philosopher (pp. 121-123)*[2]

Some attributes of objects thus seem to exist independent of particular observers and are not projections only of an individual observer's mind. Like-minded observers seem to agree on many of these attributes. On the other hand, it also seems reasonable to conclude that some qualities of objects depend on the observer's perspective. They do not independently inhere in the object, but rather are products of the contextual interaction of the object and the observer—such as possibly dynamic qualities of objects, but at least specific qualia such as an object's flavor or color. Like-minded observers seem to be able to have a high level of agreement about many of these attributes and qualities too.

These perspectives of *objective reality* and its subjective underpinnings reflect core issues that are prominent in the new physics. They also are core aspects of debates in cognitive science and neuroscience to be discussed in the next part of the book. These perspectives likewise have direct counterparts related to philosophical, ethical, and even sociopolitical issues in the social sciences and humanities.

The *objective reality* of outer objects can be said to be significantly a product of the perspective attributed to them through the process of knowing by the knower. With apparent inter-subjective agreement across observers, the objects take on a more specific, detailed, concrete, tangible sense of being in outer *objective reality*. Their phenomenal status in *objective reality* as independent objects is significantly due to the consistency within and across observers—intra-subjective and inter-subjective consistency—of the qualities observers apply to them.

Inter-subjective agreement is the product of the contribution of both the consistent qualities of outer objective objects apparently independent of individual observers, as well as the corresponding consistency associated with the subjective make-up of sensory and cognitive functions within and across observers. That is, the objective and subjective environments may match each other across individuals. There seems to be orderly principles and laws on both outer objective and inner subjective levels of nature. This will be discussed further in later chapters.

The objective physical realist perspective that objects exist independent of the observer, and the subjective idealist perspective that objects depend upon the context and interpretation by an observer both have relative validity. These perspectives interact—and curve back on each other— in building a relative view of the relative world and *objective reality*.

> "[T]he development of mathematics may seem to diverge from what it had been set up to achieve, namely simply to reflect physical behaviour. Yet, in many instances, this drive for mathematical consistency and elegance takes us to mathematical structures and concepts which turn out to mirror the physical world in much deeper and more broad-ranging ways than those that we started with. It is as though Nature herself is guided by the same kind of criteria of consistency and elegance as those that guide human mathematical thought."—*Sir Roger Penrose, mathematician and cosmologist (p. 60)* [3]

A Unified Perspective

At grosser levels of nature, objects and observers are sensed as distinct and independent of each other. From the quantum perspective, it is reasoned that they interact as quantum wavelets, associated with superposed probability waves. This perhaps results in less strong of a sense of distinctiveness and independent existence, and more a sense of the participative influence of the observer in creating what is observed. From a unified perspective, neither the object nor the mind of the observer has independent existence. In a unified field-based perspective, objects ultimately have no independent existence at all. Further, not only the object but also the observing system and the observer are all nothing other than excitations of the unified field. The independent existence of objects can be understood to be of a *phenomenal* nature, associated with fluctuations or vibrations comprising the object as sensed by an observer, and corresponding vibrations comprising the observing system and the observer—all ultimately nothing other than the underlying unified field fluctuating within itself.

Whether taking the perspective of independent objectivity or the perspective of subjectivity—historically associated respectively with realism and idealism—eventually we go beyond both and arrive at an underlying unity. If a completely unified field is to be known, the objective and subjective domains also need to be unified—all are essentially the unified field itself.

Indeed, as will be discussed in the chapter on higher states of consciousness, the degree of independent existence or *objective reality* of the *observer*—and what the observer identifies herself or himself to be—is also based on the perspective of the observer. It is most fundamentally based on the state of consciousness of the observer. The assumption of the independence of objects is underlain and accompanied by the experience of individuality in the observer. With respect to the unified field, ultimately there is no separation of observed and observer, no fundamental distinction or separation of objectivity and subjectivity.

The reductive approach of investigating smaller and smaller scales of matter to its essential substrate can result in a kind of *tunnel vision* that gives the impression the unified field exists only at the basis of creation. It is not that this field would exist only as an underlying field at the basis of creation—as if it were only at the bottom of the ocean of creation. Going deeper into nature is a conceptual way of *locating* the unified field. The unified field would not only underlie but also *permeate* every level of every object, including the surface level. Similar to the way in which space permeates material objects, or wave-field properties of quantum fields coexist with and permeate particle properties, the unified field would permeate all particle and wave-field properties—as well as all subjectivity. All objectivity and subjectivity are phenomenal *realities* of the unified field.

As an analogy, every level of the ocean is made up of water and is just different expressions of water. The waves on the surface of the water can become objectified and dominate experience, sometimes even to the point that the water itself—which is also on the surface as well as throughout the entire ocean—is hardly even noticed. When we look at the ocean we may become so engrossed in the waves on its surface that we don't think much about the waves as water. This analogy breaks down, however, when we think of getting out of the ocean of water such as flying in the air above it. There is no *getting out of* the all-encompassing unified field.

Somehow *inside* or *within* the unified field must be both the objective domain of matter and the subjective domain of mind. At some stage in the process of manifestation there must be a delineation of observed and observer that contributes to the subjective experience associated with *objective reality* that outer objects exist independent of individual observers. The unified field needs not only to appear to diversify into independent objects, but also into independent observers in order to construct the phenomenal experience of *objective reality*.

Models of *objective reality* based on the concept of observer-object independence need to be able to account for the independence with respect to an individual observer as well as across observers, while also recognizing the ultimate dependence of all observers and objects with respect to their ultimate unified basis. The process of manifestation and materialization leading to the phenomenal experience of the independence of observer and observed will be discussed in Chapter 11 when the relationships between the levels of mind and matter emerging within the unified field are described in ancient Vedic science and Maharishi Vedic Science and Technology.

At this stage in the discussion, however, it can be said that matter is phenomenal at least in the sense that it requires mind to experience it. Whatever properties of objects exist independent of the individual observer, at least they are dependent on the mind of the observer in order to be observed. Material objects appear to be relatively independent of each other, as well as relatively independent of individual observers. But to be experienced they are dependent upon the mind of the observer. This is not to confirm the idealistic philosophy that all objects are in the minds of each individual observer and don't in at least some sense exist outside of individual observers. Rather it is to suggest that matter is at least experientially dependent on the mind, and both matter and mind are dependent on the unified field as phenomenal limitations of it.

The assumption of the independent existence of observed and observer is so ingrained in the physicalist view of the world that efforts to take into account the process of observing and the observer place them into the world of independent objects, when it is rather more accurately just the opposite. The objective universe, subjective process of observing it, and observer of it are all ultimately within the unified field. Objectivity is a subset or special case of the subjective process of observing, which in turn is a special case of the observer, a special case of consciousness—and a special case of the unified field of nature as the source of everything. This is not to discount or invalidate objective realism or mentalistic idealism, but rather to place both in a holistic context.

Has Modern Science Become Immaterial?

In going beyond physical objects as comprising a completely independent *objective reality* toward the ultimate unification of nature, the question may arise whether modern science is moving out of an empirically based methodology into a modern version of natural philosophy—and thus is no longer practicing science. However, the basic means of gaining knowledge in modern science is not being abandoned. Rather, in its progress a less rigid, less bound, and more comprehensive conception of the empirical world, *objective reality,* and scientific epistemology is developing. Emphasis on different aspects of the means of gaining knowledge changes in pursuit of more comprehensive knowledge; and correspondingly, the objects of knowing are understood at more abstract levels. Modern science relies on direct sensory observation as much as possible, but with progress is also now relying more on deeper aspects of the means of gaining knowledge.[4] This progress also needs to incorporate more abstract conceptions of what the knower or observer is—which largely have not been examined in modern science.

This progress, however, does suggest that the model of the experimental method—empirical comparisons of hypotheses with precise control of independent variables—is not as prominent in some areas of modern scientific research. The basic approach of putting theories to empirical test still holds and is pursued as is possible. But increasingly, evaluation of theories relies more on reasoning, especially formal mathematical logic. Increasingly, theories are evaluated based on mathematical consistency than on direct comparisons through experimental manipulation and ordinary sensory observation such as in the experimental laboratory setting.

> "[D]uring the twentieth century with the advent of quantum theory (and to a lesser extent, relativity) the mathematics became much more highly developed and the physical interpretation much more abstract and indirect, as well as much less clear. As a result there was a constantly increasing focus on the mathematics, while the physical ideas were given less and less importance... This view has become the common one among most of the modern theoretical physicists who now regard the equations as providing their most immediate contact with nature (the experiments only confirming or refuting the correctness of this contact)."—*David Bohm and B. J. Hiley, theoretical physicists (p. 320)*[5]

Further, in some advanced areas of research the classical approach of establishing lawful cause-effect relationships via experimental tests no longer primarily characterizes the modern scientific endeavor. Theories are being developed and evaluated sometimes with little empirical evidence in the form of either direct or indirect measurements. There is in some cases even little concern whether the theories can be associated with measurable phenomena in the natural world, but rather whether the models are logically consistent. Quantum theory, for example, concerns phenomena that don't seem to fit at all with the classical framework of local causes and effects.

This can be viewed as reflecting a progression of increasing abstraction, from matter to the potential matter of quantum theory, from physics to metaphysics, from the geometry of space-time to abstract pure geometry or pre-geometry, from mathematics to metamathematics, from formal logic to meta-logic, and so on. It further is sometimes thought that eventually the level of logical consistency and mathematical precision will be so high as to discriminate sufficiently between alternative theories and support only a single unique theory of the natural world. However, the hope and intent still remains that the theories will lead to some form of empirical verification.[6]

> "After all, if there is one unique theory it does not need experiments to verify it—all that is needed is to show that it is mathematically consistent... [I]f we accept the assumption that there is one unique theory, then it will pay to concentrate on the problem of testing that theory for mathematical consistency rather than on developing experimental tests for it."—*Lee Smolin, theoretical physicist (p.195)*[6]

These changes of emphasis in the means of gaining knowledge can be likened to the shift beyond the interpretation based on naïve sensory experience that the Sun moves across the sky to the more accurate understanding that Earth rotates, based on deeper reasoning and empirical experience. In this research milieu there is *objective reality* in the sense of a superposition of probability waves appearing as a classically observable object that are somewhat distinct from the probability waves comprising a group of like-minded observers, all having no ultimate independent existence. Only in the less expanded, less holistic, and more fragmented classical context is there a completely independent objective world and independent objective means to validate it.

It may be useful here to reiterate that the quantum perspective does account for phenomena more successfully and completely than the Newtonian classical perspective. Even though it may not be commonsensical—indeed it may seem nonsensical—within the familiar perspective of ordinary sensory experience and general popular understanding, in important ways it is a more comprehensive and accurate view of the world. In this sense it is relatively more *real*. It reflects progress toward a more comprehensive and coherent paradigm and knowledge system, involving increasing refinement, subtlety, and abstraction. Many of the interpretations of quantum theory discussed in Chapter 5 reflect the challenge of accounting for experimental findings which reflect more abstract nonlocal properties of nature that are viewed from the restricted materialistic understanding of mind and consciousness as macroscopic processes in local physical *reality* only.

Frequently there are significant emotional reactions to such changes in epistemological emphases and corresponding conceptual shifts toward increasing subtlety of understanding. Even more than the quantum perspective, the unified field perspective deeply challenges long-established assumptions in popular commonsense perspectives as well as classical scientific perspectives about the natural world. As will be discussed in later chapters, unified field-based science has significant psychological and sociological implications that greatly impact the physicalist worldview and, fortunately, how our modern civilization will behave.

As fundamental assumptions become relatively more accurate and universal representations of nature they become more abstract and holistic, and thus more difficult to test via objective means in order to validate or falsify them. (Of course something that is true *cannot* be falsified.) When a broad consensus is developed that represents considerable accuracy and fit with nature, the assumptions can become firmly established as a paradigm or worldview. This then can become the primary basis upon which the truth-value of other phenomena or alternative explanations of them are judged. It can so strongly shape the reasoning and perception of adherents of the paradigm that they can categorically reject alternative views. This can be said to be helpful for inaccurate views, but unhelpful for more refined, accurate, and comprehensive views. It can be said that bias is good as long as it is to truth.

The macroscopic observing system of the human senses can be viewed as a relatively simple mechanistic system compared to deeper, higher-order cognitive and affective information processing systems. It generally seems to produce consistent results, and it is easier to identify malfunctions compared to the deeper, higher-order processes unless the higher-order processes are significantly dysfunctional. There are many more possibilities for interpretation of sensory input on the deeper inner levels of perception, cognition, and feeling. Even with considerable consistency on deeper levels, the huge range of possible patterns of reasoning and feeling sometimes allow irrational and superstitious patterns of association to develop. More careful logic and experience through scientific rigor eliminates many of these irrational patterns. This process is facilitated by eliminating distortions and biases in measurement, in mathematical calculations, and also in logical reasoning. Consistency of deep levels of thought require increased refinement and subtlety.

Even more significant, however, consistency is directly facilitated by eliminating distortions and increasing orderly coherent functioning in the body, brain, and mind of the investigators. This underscores the value of systematic means to enliven more clarity at subtler levels of thinking and reason that interact with the even deeper levels of feeling. As the observer's mind itself is recognized to be the investigative probe—as well as to be directly investigated—and it is developed to function in a more refined, reliable, and orderly manner, then the observer or knower is able naturally to penetrate deeper into nature, including his or her own nature. When a large enough group of contributors to knowledge advancement have the degree of consistency of subtler, deeper levels of thought and feeling necessary to agree on fundamental assumptions, then significant progress can be made in building consensus toward a universal knowledge system. But full agreement won't be reached until the totality of knowledge is directly experienced (and even then will have culturally specific flavors and tones).

The means of gaining knowledge in modern science—primarily ordinary sensory experience and abstract reasoning—are fairly widespread in the adult population of our modern civilization and significantly contribute to the general meaning of *commonsense*. However, the cutting edge of modern science sometimes involves uncommonly precise abstract reasoning. The high degree of specialized training and technical language required to function in this research milieu also make it quite difficult for the general population—even the highly intelligent population—to comprehend the ideas and build a deeper, fuller sense of their *reality*. It is useful to recognize the significant challenge these factors add to the process of building general consensus on a societal scale. This is particularly the case when the individuals who function on the cutting edge of modern science themselves aren't accustomed to direct experience of transcendental consciousness and higher states. Higher levels of subjective development establish the subjective consistency needed for a universal system of knowledge and experience deeper than the intellectual level.

Intuitive-Based Beliefs in Modern Science

Hopefully the point made in Chapter 1 that experience and reason are influenced by underlying assumptions and beliefs is now also obvious. Experience and reason intimately interact with fundamental presumptions or presuppositions about the reasonableness or plausibility of certain possibilities over others. These presuppositions relate to deep inner beliefs in things that are true, even if beyond the information provided through the senses. The beliefs even underpin logic and mathematical reasoning,

> "All sorts of plausibility considerations play a role in shaping our theories, over and above the role played by empirical evidence, in all sorts of domains. Consider, for example, our acceptance of the theory of evolution, as opposed to the theory that the world was created fifty years ago with memories and fossil record intact... Empirical evidence is not all that we have to go on in theory formation; there are also principles of plausibility, simplicity, and aesthetics, among other considerations."—*David Chalmers, philosopher (p. 216)*[7]

which involve forms of thought that don't directly present themselves on any sensory level of empirical experience but nonetheless seem to have a *reality* beyond our own thinking. Not only do these forms of thought seem valid based on logical consistency, but their validity also is contributed to by a deeper intuitive sense of their beauty and self-evident correctness.

Although principles of logic and mathematics cannot be heard, touched, seen, tasted or smelled, they are accompanied by a deep sense of validity—as if they are true and valid apart from our individual belief in them. In a similar way that our senses suggest that independent objects exist in nature outside of our individuality, abstract principles of logic and mathematics also seem to have an independent empirical *reality*. But this level of *reality* nowhere presents itself to our sensory empirical experience and cannot be proven through gross sensory observation. They are thought to be proven by their logical consistency via that aspect of the process of knowing we call reason.

> "...[S]ensory experience is only one of several different but equally legitimate types of experience, which is precisely why mathematics—seen only inwardly, with the mind's eye—is still considered scientific (in fact, is usually considered extremely scientific!)."—*Ken Wilber, philosopher (p. 153)*[2]

In the following quote, Penrose argues for the 'existence' of mathematical forms beyond any particular individual mind. The concept of 'objectivity' refers not to a sensory-based *reality* but a universal thought level of *reality* that is not a projection of individual minds. It is held to be an external standard for determining the validity of individual patterns of logical thinking.

"But what is mathematical proof? A proof, in mathematics, is an impeccable argument, using only the methods of pure logical reasoning, which enables one to infer the validity of a given mathematical assertion from the pre-established validity of other mathematical assertions, or from some particular primitive assertions—the *axioms*—whose validity is taken to be self-evident... [W]e must be careful...whether to trust the 'axioms' as being, in any sense, actually *true*... But what does 'true' mean, in this context?... Plato made it clear that the mathematical propositions—the things that could be regarded as unassailably true—referred not to actual physical objects... He envisaged that these ideal entities inhabited a different world, distinct from the physical world. Today, we might refer to this world as the *Platonic world of mathematical forms*... [T]here is something important to be gained in regarding mathematical structures as having a reality of their own... The precision, reliability, and consistency that are required by our scientific theories demand something beyond any one of our individual (untrustworthy) minds. In mathematics, we find a far greater robustness than can be located in any particular mind. Does this not point to something outside ourselves, with a reality that lies beyond what each individual can achieve?... Nevertheless, one might still take the alternative view that the mathematical world has no independent existence, and consists merely of certain ideas which have been distilled from our various minds and which have been found to be totally trustworthy and are agreed by all... Do we mean...'agreed by those who are in their right minds', or 'agreed by all those who have a Ph.D. in mathematics'...and who have a right to venture an 'authoritative' opinion? There seems to be a danger of circularity here; for to judge whether or not someone is 'in his or her right mind' requires some external standard. So also does the meaning of 'authoritative', unless some standard of an unscientific nature such as 'majority opinion' were to be adopted (and it should be made clear that majority opinion, no matter how important it may be for democratic government, should in no way be used as the criterion for scientific acceptability)... Platonic existence, as I see it, refers to the existence of an objective external standard that is not dependent upon our individual opinions nor upon our particular culture. Such 'existence' could also refer to things other than mathematics, such as to morality or aesthetics...but I am here concerned just with mathematical objectivity, which seems to be a much clearer issue... To my way of thinking, Platonic existence is simply a matter of objectivity and, accordingly, should certainly not be viewed as something 'mystical' or 'unscientific'..." (pp. 110-115) [3]

If the way to access these mathematical forms is through logical reasoning in the mind—not through sensory experience of things outside the mind—must they be outside of the mind? Perhaps these mathematical principles or Platonic forms are not external to individual minds, but are deeper within all of them. They may be a consistent deeper level of 'existence' in all of our individual minds, but just beyond our ordinary range of variable subjective experience. In other words, perhaps the deepest structures of all our individual minds participate in the same laws of nature that structure logical reasoning, as well as that structure the entire objective universe. Our individual minds seem to have the inherent ability to access levels of nature that are beyond our variable individual subjectivity. We have the inherent potential to tap into and experience directly non-changing universal values of nature that have shaped our individual minds and all existence. Penrose points to this recognition in stating that: "Nature is potentially present within all of us, and is revealed in our very faculties of conscious comprehension and sensitivity, at whatever level they may be operating. (p. 420) [8]

But the bottom line of our ability to know, the processes of gaining knowledge within us, is not intellectual reason and not mathematical logic. The senses and body cannot reveal the mind and intellect, because mind and intellect are deeper and subtler and more inclusive; they can only re-veal their expressions in outer behavior. Also in the same way, mind and intellect cannot reveal

> "Physicists who are trying to understand nature may work in many different fields and by many different methods... But the final harvest will always be a sheaf of mathematical formulae. They will never describe nature itself...never put us into contact with reality"—*Sir James Jeans, physicist, astronomer, and mathematician (p. 8)* [9]

consciousness; they can only reveal images and conceptions of consciousness, because conscious-ness is transcendental to even the subtlest activity of mind and intellect.

The senses and the deeper level of logical reasoning (refer to Chapter 1) are not the only bases for our sense of *reality* and what we hold to be true or *real*. The sense of what is accurate, true, or *real* is based also on deep inner feelings of how things must, ought, or at least seem to be. These deep inner feelings go beyond ordinary sensory experience and beyond the formal and informal rules of logical reasoning. They involve a fundamental inner sense about how the world is struc-tured that is meaningful and pleasing on a deeper underlying level of feeling in the mind, a *felt sense* associated with the concept of intuition.

Like levels of objective creation associated with various scales of time and distance and in-creasing abstraction, there are levels of mathematical objects—such as whole numbers, integers, rational numbers, real numbers including irrational numbers, imaginary numbers, and complex numbers. [10] Mathematical logic not only extends into even more abstract conceptions of pre-ge-ometry or metamathematics but also is underpinned by deeper levels of mind associated with in-tuition and intuitive-like feelings.

The fundamental presuppositions or pretheoretical assumptions in mathematics, or any knowledge system, are not completely verifiable through the senses or through logical reasoning associated with mathematical proofs we assume to be self-evidently true. More abstract inclusive forms of logical reasoning and meta-logic emerge in the pursuit of more comprehensive under-standing. But even further these quite abstract forms of thought draw from a deeper sense of the validity of fundamental presuppositions associated with direct apprehension of deeper structures of natural law—associated with an inner intuitive sense of inevitability or truth.

Axioms, undefined quantities, simplicity, elegance, symmetry, balance, self-consistency, uni-versal order, the comprehensibility of nature, and the inevitability of the laws of nature are ex-amples of abstract pretheoretical assumptions and presuppositions associated with inner feelings shared by many scientific investigators about how the world is. These relate to intuitive-based beliefs that significantly shape the direction of theories and empirical research to test them. The most fundamental of these intuitive-based beliefs or felt senses even go beyond the aspect of the intellect that discriminates what is logical and illogical.

Gödel's incompleteness theorem demonstrates that mathematical systems complex enough to include simple arithmetic can be consistent or complete, but not both at the same time. Consis-tency is an important principle in all of mathematics. Relational principles can be devised within a mathematical system that are consistent with other principles in the system. But it has been shown that mathematics taken as a whole contains paradoxes. Mathematics assumes consistency, but is fundamentally incomplete. The sense of consistency comes from outside or beyond the particular rules of a mathematical system themselves—in a deeper and more transcendent level of mind or subjectivity. This transcendent level, which even extends beyond the logical discriminating intellect, is open to direct experience and increasing degrees of clarity.

> "There is a common misconception that Gödel's theorem tells us that there are 'unprovable mathematical propositions', and that this implies that there are regions of the 'Platonic world' of mathematical truths... that are in principle inaccessible to us. This is very far from the conclusion that we should be drawing from Gödel's theorem. What Gödel actually tells us is that whatever rules of proof we have laid down beforehand, if we already accept that those rules are trustworthy (i.e., that they do not allow us to derive falsehoods) and are not too limited, then we are provided with a new means of access to certain mathematical truths that those particular rules are not powerful enough to derive."—*Sir Roger Penrose, mathematician and cosmologist (p. 377)* [3]

> "Gödel's argument does not argue in favour of there being inaccessible mathematical truths. What it does argue for, on the other hand, is that human insight lies beyond formal argument and beyond computable procedures. Moreover, it argues powerfully for the very existence of the Platonic mathematical world. Mathematical truth is not determined arbitrarily by the rules of some 'man-made' formal system, but has an absolute nature, and lies beyond any such system of specifiable rules... We do not just 'calculate', in order to form these perceptions, but something else is profoundly involved—something that would be impossible without the very conscious awareness that is, after all, what the world of perceptions is all about."—*Sir Roger Penrose, mathematician and cosmologist (p. 418-419)* [8]

The inner intuitive sense underlies the deep sense of consistency that is fundamental to logic and mathematics. Although rarely expressed explicitly in these fields, logical reasoning fundamentally relates to a deep feeling of correctness, rightness, or accuracy. What is it that informs us of the accuracy of some logical relation, such as $2 + 2 = 4$? It is a deep inner feeling of commitment or conviction that it *makes sense, it is right.* This conviction is strengthened by the consensual validation of teachers and other historical authorities important in our knowledge training, as well as consistent experience in successfully applying the principles empirically. But at a deep experiential level it is a subtle tacit feeling of almost unquestionable clarity and certainty that something is self-evident, self-consistent, inherently reasonable, sensible, even necessarily true—that our subjective intuitions and logic reflect actual inherent principles and structures, laws of nature, even absolute truths, or at least relatively absolute truths.

The core of feeling that results in a firm sense of logical consistency or of universal harmony, for example, draws from a deeper inner level of experience that goes beyond the discriminating intellect and the consistent principles of logic the intellect applies. The sense of logical consistency or belief in the comprehensibility of nature comes from an intuitive-like level of feeling that underlies the intellectual aspect of the process of knowing. However, this deep level of feeling is underlain by even deeper levels, such as a sense of *knowingness*; and more fundamentally, a sense that *I know who I am*, or direct apprehension that *I am the knower*. Again more fundamentally, it is underlain by the sense that *I exist*; and even again more fundamentally, awareness, wakefulness, or *consciousness itself*. These will be important issues in discussions on levels of the process of knowing and models of the mind in the next part of the book: Psychology Unbound.

The most fundamental inner intuitive sense can apprehend the simultaneous coexistence of opposites, beyond the intellectual sense of logical consistency, beyond all dualities and pairs of opposites to a coherent unity of even what appears to be logically opposite values. In so doing, it integrates the nature of the conceptualizing mind and discriminating intellect with the fundamental holistic nature of consciousness itself—the simultaneous coexistence of the opposites of infinite dynamism and infinite silence of *self-referral* consciousness—discussed in the next chapter.

> "It seems to me that, if one perseveres with the principle of sufficient reason and demands a rational explanation for nature, then we have no choice but to seek that explanation in something beyond or outside the physical world—in something metaphysical—because, as we have seen, a contingent physical universe cannot contain within itself an explanation for itself."—*Paul Davies, theoretical physicist and cosmologist (p. 171)* [11]

The different levels of the inner sense of rightness or intuitive-like sense of self-evident truth can have distortions and biases, and thus may be only relatively and conditionally accurate. Systematic subjective technologies that naturally allow the mind to settle down, transcend, and disembed from these deeper inner levels of feeling directly facilitate growth in their coherence and clarity. This results in more integrated, accurate, and profound appreciation of the beauty and orderliness of nature. Transcending the intuitive-like presuppositions—which automatically occurs in transcending thought, including conceptual meaning—fosters higher levels of integration. Through this development *objective reality* and *subjective reality* become increasingly consistent. Without the individual mind settling down and transcending its conceptual limitations, clarity and consistency at deeper levels of the mind is restricted, and may progress only marginally throughout an entire life span or even sociological paradigmatic era. In the natural growth to higher states of consciousness the concrete gross level is experienced as less *real,* and eventually the transcendental level can be said to be the most *real.*

A very important intuitive-based belief is the sense of *inevitability* of lawful relationships in nature—such as that the relative strengths and masses of fundamental forces will demonstrate orderly patterns. This relates to a deep feeling that nature cannot do anything other than exhibit orderly principles such as symmetry, simplicity, or the law of least action. The intuitive-based sense of inevitability is clearly evident, for example, in Einstein's commitment to his fundamental belief in the independence of objects, and also to the deterministic principle expressed in his famous comment that God does not play dice. These beliefs fueled his persistent efforts to demonstrate that quantum theory is incomplete, even in the face of strong opposition. These are, of course, among the most significant issues in modern science.

Belief in the independence of objects fundamentally concerns the relationship between *objective reality* and *subjective reality*—matter and mind. Implicit in it is not only the issue of the independent existence of objects with respect to an individual observer—such as *you* or *me*—but also with respect to a universal or cosmic observer—such as *God.* It is interesting that Einstein both strongly asserted the independence of objects as a primary assumption essential to do science, and also formulated what might be viewed as the contrasting framework of relativity theory which is fundamentally observer dependent and relational. What he seemed to want most, however, is an understanding of the mind of God, as reflected in his statement:

> "I want to know how God created this world. I am not interested in this or that phenomenon, in the spectrum of this or that element. I want to know His thoughts, the rest are details." [12]

This suggests he wanted to know the source of natural law, but didn't know how to experience it directly. It also is interesting that at some point Einstein seemed to believe that the universe ultimately may be infinite and eternal, when the model of the relativistic space-time continuum is now thought to lead to the conclusion mathematically that it is a closed system.[13] These persistent convictions or intuitive-like assumptions—associated with hidden variables that will eventually lead to a completely deterministic understanding of nature—strongly suggest that he believed there is more to the universe than accounted for by relativity and quantum theories, and that even more comprehensive, invariant, *absolute* laws of nature are *knowable by the human mind.*

Symmetry, invariance, and the laws of conservation.   One of the most challenging intuitive beliefs in modern science is in *universal order*—that the universe is fundamentally causal and deterministic. As commented on in the Prologue, it is curious that while historically a central task of science has been to gain knowledge of lawful cause-effect relationships in nature, in prominent contemporary theories determinism is neither extended to the most fundamental levels—due to quantum uncertainty, quantum ignorance—nor to the subjective domain, in the sense that it is frequently said to be fundamentally untrustworthy and unreliable (for example, refer to Penrose quotes earlier in this chapter and in Chapter 1). Perhaps this is where objective science gets too close to personal subjectivity, in that this can be related directly to whether *you* and *I* have free will. It isn't obvious, however, how free will is consistent with either of the two choices reflected in the positions that either God does or doesn't play dice[14]—to be discussed in a later chapter.

Belief in universal order leads to principles of symmetry that are considered by many modern scientists to be perhaps the most important design components in the universe. Principles of symmetry have broad consensual validation in the arts also as of fundamental aesthetic significance. As believed by some, these principles may be inherent both in the human mind as well as in the structure of nature. In other words, the orderliness of natural law may encompass both the objective and subjective domains.

> "It is widely believed among scientists that beauty is a reliable guide to truth, and many advances in theoretical physics have been made by the theorist demanding mathematical elegance of a new theory. Sometimes, where laboratory tests are difficult, these aesthetic criteria are considered even more important than experiment."—*Paul Davies, theoretical physicist and cosmologist (p. 175)*[11]

> "A perfect sphere is highly symmetric, since any rotation about its center...leaves it looking exactly the same. A cube is less symmetric, since only rotations in units of 90 degrees about axes that pass through the center of its faces...leave it looking the same... The symmetry underlying the known laws of physics are closely related to these, but zero in on a more abstract question: what manipulations—once again real or imagined—can be performed on you or on the environment that will have absolutely no effect of the laws that explain the physical phenomenal you observe?... Over the last few decades, physicists have elevated symmetry principles to the highest rung on the explanatory ladder... Physicists also believe these theories are on the right track because, in some hard-to-describe way, they feel right, and the ideas of symmetry are essential to this feeling."—*Brian Greene, theoretical physicist and string theorist (pp. 220-225)*[15]

Symmetry not only has deep appeal as an elegant basis for organizing the universe, but it also has specific mathematical interpretations that are central to the most comprehensive theories developed so far in modern science. Symmetry translates directly into principles of invariance reflected in fundamental laws of nature, especially the laws of conservation. The invariance is in equations that don't change in describing processes under different conditions.

The mathematical principle of symmetry is now guiding much of the theoretical as well as experimental research in particle physics.[4] The principles of symmetry are so respected that, for example, they are the core of super-symmetric string theories which posit an entire collection of partners to the known particles—theoretically doubling the number of fundamental particles in the universe. None of these theorized particle partners or sparticles have yet been found, but are strongly believed to exist by some groups of researchers because they are predicted based on principles of symmetry. Elaborate, time-consuming, and expensive research efforts to find them attest to the extent of belief in them—when a more unified set of beliefs might lead to even more productive research strategies.

"I heave the basketball; I know it sails in a parabola, exhibiting perfect symmetry, which is interrupted by the basket. Its funny, but it is always interrupted by the basket."—Michael Jordan, professional basketball player [16]

For every conservation law there is a symmetry associated with an invariance principle. According to *Noether's theorem*, any conserved quantity gives rise to a symmetry that is mathematically derived in its complementary or conjugate quantity.[16] This theorem applies to what are called external invariance principles. Invariance principles are classified into two types: *external* and *internal*. This distinction can be related to the contrast between classical and quantum *reality*, as well as physical and conceptual or mathematical *reality*—discussed in prior chapters.

External invariance principles are associated with events that take place and are measurable in ordinary conventional space-time. Major external invariance principles relate to conservation of energy associated with time symmetry; conservation of momentum (motion in a straight line) associated with position (space) symmetry; conservation of angular momentum (motion about a turning point) associated with rotational symmetry; and conservation of electric charge associated with symmetry of positive and negative charge, called *local gauge symmetry*.

For example, in analyzing the energy picture of a wave function related to the conservation of energy, symmetry is found in its complementary quantity of time. This is symmetry between *past* and *future*—called time invariance. This symmetry refers to the laws of nature being the same in different positions of straight lines in space.

For conservation of charge, local gauge symmetry of the electromagnetic field means that exchanging *positive* and *negative* charges in a system could change things locally but the overall charge in the system will be the same. Quantum theory and the general theory of relativity, for example, are local gauge invariant theories. Nothing can be done that would cause the observer to have a preferred point of reference that yields different laws of nature, in local gauge invariance. All four fundamental force fields exhibit local gauge invariant external symmetry.

Internal invariance principles relate to symmetry in conceptual or mathematical space, coming from quantum and super-symmetric string theories. They may not be as fundamental as quantities associated with external invariance, in the sense that they don't necessarily relate to fundamental conservation laws. But on the other hand, they may be more fundamental, and just not yet understood. They were developed from applying the mathematics of group theory to the behavior of particles. Particles follow patterns of behavior that can be classified into groups not dependent on space, time, or electric charge. They include such concepts as flavor, colour, spin, lepton family number, baryon family number, as well as others. They are not continuous measurable quantities; by convention they are labeled using integer values such as 1, 2, and so on, but also sometimes fractions such as ½ or ⅓. These internal quantities, also called *quantum numbers*, relate to reliable patterns of behavior in experimental findings, but the processes underlying them are not yet understood.

For example, baryons such as protons and neutrons are heavy particles associated with the nucleus of the atom. They interact with all the four fundamental forces, but baryon number is conserved only with the strong force and not necessarily conserved with the other three forces.

Lepton number is conserved in electroweak interactions, but not necessarily in any other interactions. There are also internal gauge symmetries associated with some particles.

These principles of symmetry and invariance reflect the general pattern of finding increased order at finer-grained levels of nature. Importantly the assumption of order is applied not just to the ordinary objective sensory world, but now also even to the abstract subjective level of mathematical space and *quantum reality* associated with internal symmetries and quantum numbers. Symmetry principles applied to mathematical space and quantum *reality* reflect the relatively high subjective consistency of beliefs within and between modern scientific investigators working on these theories that go far beyond ordinary macroscopic sensory experience.

This is again suggestive of a deep correspondence between *objective reality* and our subjective ability to experience and agree on it. When coherent and reliable enough, the sensory, reasoning, and intuitive aspects of the subjective process of knowing may directly correspond to the structure of the natural world. This reflects additional recognition that the same principles of order may govern both objective and subjective domains.[12]

Levels of Subjective and Objective Reality

Intuitive-based assumptions, presuppositions, and beliefs are tested by what *seems* to be true logically and by what *seems* to be true according to the ordinary senses. The assumptions, the logical reasoning, and the sensory experiences interact with each other to build more accurate models of the world in the self-correcting process of empirical science. These processes of knowing are significantly influenced by the level of nature that is the primary locus of experience of individual observers and their like-minded consensually validating cohorts.

In the ordinary waking state of consciousness the observed and the observer generally are experienced as separate and independent. At deeper levels of experience a more holistic view of nature involving higher levels of interaction, entanglement, and interdependence becomes more prominent. At the level of the unified field the oneness or unity of nature predominates. As will be discussed in later chapters, these various levels of subjective experience have a general correspondence with higher states of consciousness.

Hopefully emerging from the discussions in this chapter and in the Introduction is increasing recognition of the top-down influence of the means of gaining knowledge, associated with the fundamental principle in Maharishi Vedic Science and Technology that 'knowledge is structured in consciousness.'[17] Sensory means of gaining knowledge are shaped by underlying reasoning, which in turn is shaped by intuitive-like beliefs, which in turn is shaped by the level or state of consciousness of the individual knower. In some non-trivial sense *objective reality* as experienced through the senses is a projection of the state of the observer and the collective state of other observers who contribute to consensual validation. These levels interact with each other in the process of more accurately attuning the individual experience of nature with the universal orderliness of natural laws. This universal level is beyond individual minds, but individual minds and the objects they experience are both structured from the same universal laws—which is the basis for intra-subjective and inter-subjective consistency—and can be directly apprehended at the deepest underlying level of individual minds in the transcendent unified field.

The rigorous and disciplined focus in modern science on objectivity was in partial reaction to the inconsistency of subjective experience and the unreliability of some intuitive-like religious knowledge. The approach has been thought to be very helpful in establishing a more reliable body of knowledge about the natural world based on ordinary sensory experience and logical reasoning. The increased control of the natural world afforded by this approach has strengthened its acceptance throughout modern civilization.

But it also has led to increased conditioning and fixation on the ordinary sensory level of experience that has made it more difficult for modern science and modern civilization to progress beyond the physicalist worldview. Indeed this fixation has been so engrossing that many even now remain firmly committed to the assumption that the material domain of nature is the only domain that exists, even though modern scientific understanding has been progressing beyond this view for the past century. The forefront of modern scientific understanding has advanced to the stage of recognizing the incompleteness of the physicalist worldview associated with this objectified approach to knowledge. At the same time models of the natural world have developed based more on abstract reasoning that are recognizing the necessity of extending beyond the physicalist conception of nature into a less restricted, more mind-like abstract information space or *quantum reality* in the pursuit of more inclusive coherent explanations of nature.

Still, ordinary observations and common subjective reports—as well as the disintegrated status of modern civilization in general—strongly suggest that the vast majority of the scientific and general population function in the ordinary waking state of consciousness dominated by surface sensory experience of the material domain. This has contributed to the ordinary, empirical, consensually validated experience and corresponding logical understanding that mental life—mind and consciousness—exists only in the brain. It also has strengthened the belief that the objective material domain is the only world that *really* exists, and that mind and consciousness are products of brain functioning that are yet to be understood and explained. It supports the interpretation of naive experience that the mental is somehow a product of the physical.

Indeed the contemporary interpretation of the concept of *naturalism* seems to be based significantly on common subjective experiences. At this stage of societal development it unfortunately remains to be the types of sensory experience and abstract reasoning associated with the ordinary waking state of consciousness. Ordinary sensory experience is identified as natural, based on it being common in today's world. Contrary empirical experience is discounted because it has not been adequately validated experimentally or experientially within the limited framework of the objective means of gaining knowledge as generally defined, accepted, and practiced.

But also, as noted in the Prologue, there appears to be persistent intuitive feelings shared by most people of more interconnected nonlocal phenomena in nature, and that mind and consciousness are more than material objects in some vague, foggy, undefined but essential way. They also are frequently associated with the intuitive sense that the universe is fundamentally orderly rather than random—and further that the order is due to some causal agency. How do such intuitive feelings relate to theories extending into a more inclusive nonlocal picture of nature as an underlying information field or quantum mind?

Nonlocal fields, brains, and minds. While mainstream modern science remains fixated on the local physicalist view, quantum and unified field theories have progressed beyond it. The implications of these expansive theories are becoming acknowledged, but are not yet generally recognized or appreciated for their important logical implications about levels of *reality*. Especially, the most significant implication reflected in some of the research on quantum *reality*—that mind is a nonlocal field of existence underlying matter—still is a stretch even for many prominent contributors to these cutting edge theories. But if elementary particles and fundamental fields, and thus brains, stretch from local to nonlocal, what about minds also?

A major challenge is to reconcile the common experience that the individual mind feels like it is inside the head with the conceptual understanding of an expanded nonlocal quantum *reality*—and even quantum or sub-quantum mind—not confined to the head. Already nonlocality is established logically and empirically. How come this is not an ordinary experience of *objective reality*?

There is a correspondence between levels of *reality* and levels of subjective experience of what *reality* is—what the primary locus of experience is. The sensory system that is part of the human brain and nervous system is built to sense the level of macroscopic objects. The range of normal experience in the auditory modality, for example, is between 20-20,000 Hz. But we know that there are finer-grained forms of sound, such as the wider range of sounds that our pet dogs and cats can hear. Also amplifiers permit us to hear sounds outside the normal range, and microscopes allow us to see things the unaided eye cannot see.

Sensory experience is not of atoms or elementary particles, even though the macroscopic sensory system is sensitive enough to sense their effects. For example, the visual sense responds even to a single photon of light. However, it is not built to see a photon—we don't directly *see* light, at least in our ordinary world of experience, but rather the results of electromagnetic processes that illuminate objects via light reflecting off of them.

The macroscopic sensing system may be said to be designed to interact with the gross material domain. But there are subtler levels of creation that we are not commonly familiar with if our capacity for experience is undeveloped and restricted to the gross level only. When such gross experiences predominate in daily life and there are no or only rare and fleeting subtler experiences, it supports the belief that the material domain is the only domain of existence. When subtler sensory experiences are rare and elusive, and when the belief predominates that the physical world is the only domain, the common conclusion is that mental experience occurs in the brain. This is the characteristic understanding, and paradigm blindness, that conditions experience in the ordinary waking state of consciousness.

The material domain is a quasi-closed cosmological system characterized by Einstein locality and material causality associated with the four fundamental force fields within a quantum gravity-like field or aether. The corresponding range of sensory experience is generally of macroscopic objects that are relatively independent in time and space; and the corresponding inner experience is that the mind exists only in the brain. When there is no deeper understanding and predominant subtler experience of nature, the belief that this domain comprises the entire universe becomes firmly established. Most everyday experiences in our modern civilization including modern science concern only this domain. There are many unanswered and unanswerable questions within the constraints of this fragmented view of the nature of matter, mind, and consciousness.

As has been summarized in the past few chapters, recent advances in modern science have moved toward more holistic understanding of a subtler level of *reality* that underlies the material domain. Given its emphasis, for the most part modern science has made these advances primarily via abstract intellectual analyses to develop more logically consistent understanding of natural phenomena—on occasion reluctantly pushed by experimental findings such as nonlocality. This is expected, because even the level of ordinary thinking is a much subtler level of *reality* than the level that is perceived via ordinary sensory experience. Because there has not been equivalent development of subtler levels of experience, the subjective mind is still widely believed and experienced to be localized in the brain, and the primary locus of experience remains the physical.

Expanding The Means of Gaining Knowledge

Individuals who don't have clear experience of the mind as a background or field underlying matter need to do their empirical research of transcending sensory and mental experience, using systematic subjective technologies in the *inner laboratory* of their own minds. This results in expanded understanding and experience of the mind as more fundamental than matter, and not entirely localized in the brain. Examples of such subtle experiential development will be discussed at length in the chapter on higher states of consciousness, drawn from research on practitioners of the Transcendental Meditation technique as well as historical figures, who have provided first-person accounts.

In sharp contrast, in much of modern education based on modern science students are not even told about the underlying unified field. Most unfortunate, however, is that students are not taught systematic subjective technologies to integrate and develop further the direct experience of consciousness itself—the basis of all means of gaining knowledge. The system of object-based education strongly emphasizes intellectual development through critical analysis, problem solving, and memorization of facts. It also is associated with secularist and physicalist worldviews. Throughout secondary and college education, students are inculcated in cultural and moral relativism, which have underpinnings in fundamental randomness and the meaninglessness of life.

Especially at the college level this training sometimes contributes to styles of student life that are fraught with self-destructive behaviors and attitudes—exhibited in alcohol and drug abuse, sleep deprivation, poor diet and eating habits, and high-risk social behavior accepted as the normal 'college experience.' In its current state, modern education as exemplified in modern college life can be quite damaging to students, and in the opposite direction of what reasonably might be hoped for in order to fulfill the aspirations of 'higher education.' As originally conceived, higher education wasn't intended just to produce better intellectual understanding but rather to result in the spontaneous ability to live a more integrated, balanced, responsible, and fulfilled life.

> "Higher education should be for higher consciousness. Higher education should not be dedicated only to some fragmented knowledge of the laws of nature, such as that offered by traditional scientific disciplines, but to enlightenment, the permanently established experience of that universal state of consciousness from which all the laws of nature arise. The truly educated man...ceases to create problems and suffering for himself and his surroundings. Gaining the support of natural law for his every endeavor, he is a joy to himself and a joy to his environment."—*Bevan Morris, Vedic educator (p. 7)* [18]

In pursuit of more comprehensive—and inevitably more abstract—theories of nature, modern science has been relying less on ordinary sensory processes and more on abstract reasoning processes. It is not surprising that at some point the highly mathematical, intellectualized research has advanced to the stage where it is quite difficult to maintain connection to the sensory-based view of the world associated with ordinary experience. This has been a fundamental challenge in modern science in the past century, exemplified by the difficulties surrounding the concept of the process of measurement in quantum theory—directly comparable to the historical mind-body problem—and a major contributor to the fundamental tear in the psychosocial fabric of modern civilization referred to in the Prologue. It relates to the apparent differences between nature—especially our own nature—and our beliefs, assumptions, and conceptions about it. Modern scientists now intently focus their intellectual abilities to resolve the differences.

However, the intellect or reasoning aspect of the process of knowing relaxes and disembeds from its ordinary activity through the process of transcending. When the intellect disembeds from its fixation on a particular limiting paradigm or set of beliefs in the process of transcending thought, then the inaccurate aspects of the assumptions such as knots of misunderstanding, stress, cognitive distortions, and inaccurate biases can be unraveled and eliminated more efficiently. Inconsistencies on deeper levels of feeling and intuition result in belief systems that are not in tune with the most abstract and fundamental principles and laws of order. As the mind and body of the investigators become more stress-free and function in a more refined manner—and when the brain becomes more orderly and coherent—the deeper levels of reason and intuitive sense become increasingly orderly and also flexible enough to encompass or become attuned to a more integrated, all-encompassing understanding. Experience of natural law is then recognized to be not based just on logical reasoning and ordinary sensory experience.

The objective of this chapter has been to begin relating levels of *reality* to developmental levels of subjective experience associated with the means of gaining knowledge—sensory experience, reasoning, intuition, and also transcending. Hopefully it provides somewhat more substance to the concept that *reality* is in the 'mind's eye,' according to the state of consciousness of the *knower*. As Maharishi recently put it:

"The power of light is definitely more than the power of darkness." [19]

"In the darkness, problems are very real. But for one who knows how to light the lamp, there is no problem." [20]

To 'light the lamp' refers to cultivating the regular direct experience of consciousness itself, the 'lamp at the door' (refer to Prologue) that illumines the black box of the mind. It means to develop the highest state of consciousness, in which both the diverse relative field and the unified field of all the laws of nature become integrated in the knowledge and experience of the totality of nature. In that state, the natural primary locus of experience is unity, the ultimate totality.

A major issue threading through this discussion is whether *objective reality*—indeed nature itself—exists independent of the observer. The discussion began to place this issue into a developmental context associated with the state of the *knower* and corresponding levels of subjective experience. Other key planks will be laid down that are relevant to this issue after a more inclusive range of levels of *reality* is enumerated in Chapter 11.

At this stage we can summarize the discussion by pointing to the emerging understanding of three levels of *reality*: 1) *material reality* associated with object independence and locality, 2) *quantum reality, quantum mind* associated more with conceptual, informational, or *mental reality*, object and process interdependence, and nonlocality; and 3) the transcendental *reality* of the unified field. These generally can be related to the historical philosophical approaches of material realism, idealism, and also somewhat to an expanded meaning of transcendentalism. From the reductive perspective, embedded in material realism is idealism, and embedded in idealism is the transcendent. From the holistic perspective, it is the opposite sequence.

As will be increasingly evident in upcoming chapters, the local/nonlocal duality—reductionism and holism together—can be found to characterize all three of these levels of nature. This duality can be identified in the quantum/field and wave/field dynamics of gross and subtle levels of nature, on the even more abstract level in terms of individual/cosmic mind, the distinction of attention/consciousness and point/infinity, as well as in the self-referral singularity/infinity on the level of consciousness itself and the unified field. The distinction of individual and cosmic mind on the level of the nonlocal highly-interactive field of mental *reality* will help clarify the issue of the phenomenal independence of observed and observed and the contrast of objectivity and subjectivity in the ordinary gross relative domain. The relationship between objective, subjective, and transcendental levels—and progress toward their integration—is clearly expressed in the following 1963 quote from Maharishi about the pursuit of unified field theory in modern science:

"Certainly, in his attempts to scientifically establish the unified field theory, Einstein seems to have been clearly aware of the possibility of one ultimate basis of all diversity, one common denominator of all creation. At least he was trying to establish one element at the basis of all relative existence. If and when physical science arrives at what Einstein was trying to pinpoint by his unified field theory, one element will be established as the basis of all relative creation. With the rapid pace of development of nuclear physics, the day does not seem to be far off when some theoretical physicist will succeed in establishing a unified field theory. It may be given a different name but the content will establish the principle of unity in the midst of diversity, the basic unity of material existence.

The discovery of the field of this one basis of material existence will mark the ultimate achievement in the history of development of physical science. This will serve to turn the world of physical science to the science of mental phenomena. Theories of mind, intellect, and ego will supersede the findings of physical science. At the ultimate or the extreme limit of investigation into the nature of reality in the field of the mind will eventually be located the state of pure consciousness, the field of the transcendental nature lying beyond all relative existence of material and mental values of life. The ultimate field of Being lies beyond the field of mental phenomena and is the truth of life in all its phases, relative and absolute. The science of Being is the transcendental science of mind. The Science of Being transcends the science of mind which in its turn transcends the science of matter which, again, in turn, transcends the diversity of material existence.

Being is the ultimate reality of all that exists; It is absolute in nature. Everything in the universe is of a relative order, but the truth is that eternal Being, the ultimate life principle of unmanifested nature, is expressing itself in different forms and maintaining the status quo of all that exists. The absolute and relative existence are the two aspects of eternal Being; It is both absolute and relative." (pp. 32-33) [21]

In this quote can be seen a delineation of the three fundamental fields of phenomenal existence extensively described in ancient Vedic science and Maharishi Vedic Science and Technology: gross relative that can be associated generally with the level of the body or material *reality*, subtle relative that can be associated generally with the level of mind or mental *reality*, and their transcendental basis, the *unified field of consciousness* or pure Being. Historically modern science, including classical physics, focused only on the gross relative field of material *reality*.

Over the last half of the past century theories have been emerging that posit an underlying unified field, and in the past few years theories of a nonlocal field of information space, quantum *reality*, implicate order, or quantum mind have also been emerging in conjunction with quantum and quantum gravity theories. But in modern science the ontological delineation of and relationships between the three fundamental fields are only just beginning to be unfolded—as discussed in the past few chapters. There remains an explanatory gap between matter and the level of information space from which matter is theorized to be generated in some quantum gravity theories. There also is an explanatory gap between matter (as well as information space) and the unified field. In Maharishi Vedic Science and Technology, these explanatory gaps can be related to experiential gaps between body, mind, and consciousness, which are bridged in higher states of consciousness. These profound developments toward a holistic unified understanding of nature are an important focus of the next part of this book.

Given the epistemological emphasis in modern science on the intellectual aspect of the means of gaining knowledge, recent attempts to formulate a unified field theory have focused in the mathematical applications of intuitive-based assumptions such as symmetry and super-symmetry that have little direct or indirect experimental validation. The most comprehensive attempts to build a unified field theory extend completely into abstract conceptual, mathematical space or infinite possibility space. These models—working toward unified field theory—are highly consistent logically and mathematically. But at this stage they are too complicated to use for calculating specific predictions, and cannot yet be linked empirically to our ordinary four-dimensional world. Descriptions of the unified field from intuitive-based perspectives as well as logical reasoning—and how they begin to link up with Maharishi Vedic Science and Technology—are key issues in this next chapter, which focuses on conceptual understanding of the 'nature' of the unified field.

Chapter 9 Notes

[1] Bernard, T. (1947). *Hindu philosophy.* Delhi: Motilal Banarsidass Publishers.

[2] Wilber, K. (1998). *The marriage of sense and soul: Integrating science and religion.* New York: Random House.

[3] Penrose, R. (2005). *The road to reality: A complete guide to the laws of the universe.* New York: Alfred A. Knopf.

[4] Zee. A. (1986). *Fearful symmetry.* New York: Macmillan Publishing Company.

[5] Bohm, D. & Hiley, B. J. (1993). *The undivided universe.* London: Routledge.

[6] Smolin, L. (2001). *Three roads to quantum gravity.* New York: Basic Books.

[7] Chalmers, D. J. (1996). *The conscious mind: In search of a fundamental theory.* New York: Oxford University Press.

[8] Penrose, R. (1994). *Shadows of the mind: In search of the missing science of consciousness.* New York: Oxford University Press.

[9] Wilber, K. (1984). (Ed.). *Quantum questions.* Boulder, CO: Shambhala.

[10] Hagelin, J. S. The Physics of Flying. *The video magazine,* Vol. 7, Tape 1, N- 38, Maharishi University of Management.

[11] Davies, P. (1992). *The mind of God: The scientific basis for a rational world.* New York: Simon & Schuster.

[12] Einstein, A. As quoted in Thinkexist.com <http://en.thinkexist.com/quotation/15496.html>

[13] Hawking, S. (2001). *The universe in a nutshell.* New York: Bantam Books.

[14] Hodgson, D. (2000). The Easy Problems Ain't So Easy. In Shear, J. (Ed.) *Explaining consciousness—The hard problem.* Cambridge, MA: The MIT Press, pp. 125-131.

[15] Greene, B. (2004). *The fabric of the cosmos: Space, time, and the texture of reality.* New York: Alfred A. Knopf.

[16] Hill, C. (2000). *Teaching symmetry in the introductory physics curriculum.* FERMI-PUB-00/27-T;EFI-2000-3; arXiv.physics/0001061 v12 7 Feb.

[17] *Maharishi International University Catalogue, 1974/75.* Los Angeles: MIU Press.

[18] Morris, B. (1981). Message from the President. In *Education for Enlightenment: An Introduction to Maharishi International University.* USA: MIU Press.

[19] Maharishi Mahesh Yogi (2003). Maharishi's Global News Conference, October 15.

[20] Maharishi Mahesh Yogi (2004). Maharishi's Global News Conference, April 29.

[21] Maharishi Mahesh Yogi (1963). *Science of Being and art of living.* Washington, DC: Age of Enlightenment Press.

Chapter 10

The Whole Point: Ultimate Unity of Silence and Dynamism

In this chapter, we will apply reason and intuitive-based knowledge discussed in the past chapter to describe the incomprehensible, indescribable 'nature' of the unified field. We will consider the unified field as the field of all possibilities, the source of creation, and the infinitely correlated, self-interacting, self-referral, simultaneous coexistence of part and whole, point and infinity, infinite silence and infinite dynamism. At the end of the chapter, direct empirical means to validate the unified field will be discussed. The overall main point of this chapter is that the unified field is the source of everything, directly experienceable as the essential 'nature' of consciousness itself.

This chapter focuses on describing the ultimate wholeness of the unified field. It may be useful at the onset to rephrase a point from the Prologue that language, limiting by its very nature, cannot capture the indivisible unified field. Apparent inconsistencies inevitably arise, perhaps most evident in integrating conceptions of an ultimate infinitely dense *contraction* or singularity of *nothing* with an infinite *expansion* of the unified field as *everything*. To be all-inclusive, the unified field must be simultaneously *beyond* infinitesimally small and infinitely large. In ancient Vedic literature, it is referred to as *smaller than* the smallest and *bigger than* the biggest:

> *"Anoraniyan Mahato-mahiyan (Katha Upanishad. 1.2.20)—(The Self is) smaller than the smallest and bigger than the biggest."* (p. 18) [1]

According to the reductive perspective as reflected in some quantum gravity theories progressing toward unified field theory, observing, probing, and measuring reach an ultimate impasse at the Planck scale where super-unification occurs. In these theories the Planck scale is the point at which space and time end, or at least no longer have conventional meaning. If the unified field somehow underlies and is beyond this level, and there is no possibility of either direct or indirect measurement of it, how can it be distinguished from *nothing*?

Does Nothing or Everything *Matter*?

Historically all objects in physical existence have been thought to be various forms of matter and energy. From the unified field perspective that the essential substrate of the physical universe is beyond matter and energy, this could be regarded as implying that this ultimate field might be literally *nothing*. [2] [3] The question might arise whether an underlying undifferentiated field *really* exists. Does *everything* come from *nothing*?

Direct empirical means to validate the unified field still may not be apparent. If this is the case, then the best that can be done at this point in our discussion is to establish its existence indirectly, through aspects of the process of knowing associated with logical reasoning and intuitive sensibilities—inasmuch as it certainly cannot be validated via the ordinary senses. This chapter will examine the 'nature; of the unified field primarily based on these means of gaining knowledge, which were discussed in the previous chapter. Then we will revisit what still might seem to be an intractable paradox of its direct empirical verification within our own individual conscious experience.

A completely unified field would be beyond any form or conception of space and time—infinite and eternal. However, these descriptors also might be thought of as somehow applying to *nothing*. Conceiving of the unified field as *nothing* is from a reductive perspective, whereas the unified field as *everything* is from a holistic perspective. In this chapter these perspectives will be associated with the 'dual nature' of the ultimate singularity, unity, wholeness, or Oneness—a unity beyond even consistency and inconsistency according to the logical mind and discriminative intellect. This can be likened to the mathematical concepts of the empty set, or zero, or one (in the sense of Oneness or Totality). It also is reflected in the contrasting philosophical terms of the eternal *Void* and eternal *Being*, to be discussed later in the chapter.

The unified field can be understood to be nothing in the sense that it is *no thing*. But does *no thing* mean *nothing*? The term *nothingness* may give a little better sense of it, in that the unified field can be thought of as sort of *something*, at least a 'nothing-*ness*' rather than literally *nothing*. The vacuum state, the least excited state of a quantum field, was first described as an empty field of *nothing*. Subsequent research indicated that the vacuum state had an effect on physical processes and thus is *something*. Even the notion of a field implies *something*, and so does the notion of something that can be *unified*. Other descriptors may help distinguish more clearly the unified field from just *nothing*.

All possibilities. Consistent with a quantum theoretic perspective, the unified field has been described as a field of *all possibilities*, from which is derived any particular possibility and any actuality. It seems intuitively sensible that anything would come from *something*, rather than that anything can come from *nothing*. It further seems reasonable that *any* possibility could come only from *all* possibilities, rather than from *nothing* with *no* possibilities.

This relates to the issue of the initial conditions of the universe. If the deterministic sequence of the evolution of the post-big bang (manifest) universe became known, and also any 'pre-big bang dynamics' within the Planck scale were known, still the shape and nature of the universe would depend on what the initial conditions were 'when all this began.' If the initial conditions were also somehow known, it would be possible in principle to identify everything that has happened, as well as will happen. Of course there are serious concerns in modern science whether it will ever be possible to know what the ultimate initial conditions were, and thus things may never be completely predicted in practice, even if completely predictable in theory.

However, even if the initial conditions cannot be known—due to inherent randomness, the uncertainty principle, or unfathomable complexity—it seems intuitively sensible that they at least would need to contain the possibility for existence as we know it now, discussed earlier in connection with the anthropic principle and the principle of decoherence. If there is an arrow of time as implied by the 2nd 'law' of thermodynamics, and if orderly laws or principles shape phenomenal *reality*, then it suggests an extreme context dependence—even infinite context if that makes any sense—in which the initial conditions included possibilities for everything in the entire past, present, and future. [4] The unified field thus would need to contain immense, even infinite diversity of contextual possibilities. It would be inclusive of the total range of everything possible.

The strong version of the anthropic principle is sometimes interpreted as suggesting that the universe anticipated human life. This notion is reflected in the following quote from physicist and mathematician Freeman Dyson:

> "The more I examine the universe and the details of its architecture, the more evidence I find that the universe in some sense must have known we were coming." (p. 4) [5]

At least the initial conditions of the universe would seem to have included the possibility of intelligent human life. It is in this line of reasoning that the unified field can be described as a field of *all possibilities.*

Source of creation. The unified field as the field all possibilities also relates to it as the *source of creation,* associated with unified field theory as a 'theory of everything.' All things that exist are thought to come from some source. All things are thought to be forms of energy, shaped by the laws of nature, and both energy and the laws of nature ultimately also would have their source. But it might be asked whether it is necessary to posit a source for everything. What if things just exist in themselves and act according to their own natures, without having a source? It seems logical that anything that has no source would be independent of everything else. All relative things interact with other relative things and are changed by the interactions—whether objects, processes, event histories, information flow, or whatever. This is supported empirically by observations that at least all the things we experience appear to have a source and to interact and change relative to each other. Relative things, as relative, don't seem to be able to be their own source, and don't have their own nature independent of everything else.

Paradoxically, the ultimate source of all relative things would contain the duality of being its own independent source and also the source of everything else in relative degrees of dependence. It must be both absolutely non-interacting, and also relatively interacting, with everything emerging within it. This suggests that the unified field fundamentally is a coexistence of opposites.

> "[T]he laws of physics have a twofold job. They must provide the simple patterns that underlie all physical phenomena, and they must also be of the form that enables depth—organized complexity—to emerge. That the laws of our universe possess this crucial dual property is a fact of literally cosmic significance."—*Paul Davies, theoretical physicist and cosmologist (p. 138)* [6]

"Swarupe avasthanam (Yog-Sutra,1.3)—Self established in itself; and ...Vritti sarupyam itah atra (Yog-Sutras, 1.4)—Reverberations of the Self emerge from here (the self-referral state) *and remain here* (within the self-referral state.)" (p. 18) [1]

Further, as the source of everything, all manifestations in nature would be fluctuations of the unified field itself, at various frequencies and patterns of vibration. In order for there to be fluctuations or vibrations, there needs to be something that fluctuates, and somewhere it fluctuates. This also distinguishes the unified field from literally nothing—in that it can be said to fluctuate within itself. An example is zero point motion or quantum vacuum fluctuations.

In ancient Vedic science the unified field is the infinite source of all finite phenomenal manifestations in nature. Everything that exists is created from and remains within the unified field. Whatever energy and intelligence appear to be drawn from it in the process of manifestation of creation does not in any way diminish its eternally infinite status. It can appear to express itself infinitely, and also to remain completely unchanged and undisturbed. This quality of the unified field is described in the following verse from the Vedic text of the Upanishads:

"Purnam adah purnam idam purnat
purnam udachate,
purnasya purnam adaya purnam
evavashishyate.
(Brihadaranyak Upanishad, 5.1.1)

That is full; this is full. From fullness, fullness comes out. Taking fullness from fullness, what remains is fullness." (p. 354) [7]

Infinite dynamism. Throughout nature, higher energy is associated with smaller time and distance scales. For example, shorter wavelength light in the blue spectrum has more energy than longer wavelength red light, the microscopic nuclear level is much more powerful than the molecular level, and the energy of a hypothesized string loop or brane near the ultramicroscopic Planck scale is thought to be quite immense. If it were possible for matter to be compressed smaller and smaller toward a dimensionless point, the pressure, temperature, and energy would approach infinite magnitude—the infinitely dense singularity.[8] This suggests that the unified field or infinitely dense singularity is an unlimited source of energy and power. It seems reasonable that if it were *nothing*, it would have no energy and no potential, including no potential to explode in a big bang—or implode in a big condensation. The quantity—absence of quantity—of *nothing* is mathematically represented by *zero*. It is perhaps easier to distinguish *nothing* from *everything* when *nothing* is represented by zero. An unbounded infinite unified field is the opposite of *nothing*, even though at the same time it is *no thing,* or nothing-*ness.* But in the sense of 'no thing' or beyond all things, it might be represented by zero in the sense of no manifest thing.

However, apparently something can be derived from *nothing* in the objectified axiomatic perspective of mathematics, such as the empty set in set theory (although empty and set could also be thought of as opposing or contrasting concepts). The opposing concepts of zero or nothing and infinity are used in similar ways. Mathematically, infinity can take on values such as plus or minus infinity, which don't seem logical as physical quantities in nature. For purposes of calculation it is sometimes easier to use infinity as a number in place of a very large and cumbersome finite number. It also sometimes is used as the most general case, in that if the formulas work for infinity then in some cases they will work for any smaller finite number. But again it is assumed in using infinity in these ways that it doesn't represent a *physical quantity* but rather an abstract conceptual tool for purposes of calculation.

There is great reluctance to the notion of an infinite quantity of something in nature other than as a mathematical concept, such as that the unified field has an infinite amount of energy. As alluded to in earlier chapters, when infinity is the outcome of mathematical calculations intending to represent physical quantities, it is frequently concluded that there must have been a calculation error or that the model itself is incorrect. Obtaining an infinite amount of some physical quantity is thought to be a logical impossibility. It is an unacceptable result for mathematical models attempting to identify measurable quantities, and inconsistent with the concept of a measurable quantity for which some form of objective evidence can be obtained. Mathematically the range of probabilities of the occurrence of some event is between 0 and 1, no probability and theoretical 100% certainty. A probability of infinity doesn't make sense; any measurement or outcome of calculations representing a physical quantity would need to be finite.

Thus, for example, the law of conservation of energy is undefined and cannot be applied with respect to the universe as a whole. It is not consistent to say that the whole universe has a particular energy amount—including an infinite amount. This might by thought of as implying that the amount is a definite quantity that is not changing, which is contradictory to the understanding of the inherent dynamism of nature. It also would mean that the universe is in a definite discrete state associated with particles and forces, contradictory to the quantum model of an unlimited number of possible states. In addition, the idea of a measurable infinite quantity of physical energy would contradict the limitations of measurement according to the uncertainty principle. In the more abstract conceptions of mathematics there are countable and uncountable infinite values; but in the more concrete conceptions of physics an infinite value of a physical quantity is considered to be inconsistent.

It might be pointed out that modern scientists tend not to talk of anything that implies an ultimate absolute—such as nature as infinity, or God as ultimate—at least in part because these concepts seem to stop further intellectual analysis, as well as to stop measurement and thereby scientific progress. It also may be due in part to the belief that modern scientists have the most advanced knowledge and if they don't know something, then no one does—an unfortunate tyranny of brilliant but restricted intellects. With the advent of unified field theory, hopefully the habit of just intellectualizing about nature will be superseded by systematic investigation of the source of all the laws of nature in the unified field through direct experience.

The dilemma of infinite quantities in physical space and time is similar to the issue in quantum theory about how to collapse the abstract mathematical wave function into a concrete classically observed physical object. As discussed in Chapter 6, one approach to resolve this issue is the contemporary theory that the smallest possible size in creation is not a dimensionless point but rather the Planck length, such as in string and quantum gravity theories. This keeps everything finite—even if immensely high finite values of temperature, mass, and energy—and avoids the apparent contradiction of *infinite quantities*. Greene comments on this issue:

> "We believe that the outrageous initial state of *infinite* energy, density, and temperature that arises in the standard and inflationary cosmological models is a signal that these theories have broken down rather than a correct description of the physical conditions that actually existed. String theory offers an improvement by showing how such infinite extremes might be avoided; nevertheless, no one has any insight on the question of how things actually did begin... In the context of string/M-theory, our cosmological understanding is, at present, just too primitive to determine whether our candidate 'theory of everything' truly lives up to its name and determines its own initial conditions, thereby elevating them to the status of physical law. This is a prime question for future research." (pp. 355-356) [9]

From the reductive perspective of the ultimate discreteness of nature there still could be a theoretically measurable infinite extent of space and time, such as traveling around a circle forever. Circular motion could continue eternally and add up toward infinite distance, but also would be limited and bound to the circle—not infinite in other dimensions. Distance and time could go on infinitely long even under conditions in which space has discrete finite boundaries.

It might be thought that ultimate discreteness resolves the concern of infinite regress. The issue of infinite regress is thought to arise only when considering a continuous variable, not with an ultimately discrete variable as the smallest possible unit that nature gives us. This is in the context of mathematical abstraction; but when considering the physical world, as mentioned in Chapter 8 any *finite* model of the world as ultimately discrete would seem to lead to infinite regress. Discreteness of any type of object implies some kind of distinguishing boundary—typically in a spatial dimension, as in the Planck size. The discrete finite object with a boundary—circle, sphere, loop, string, brane, or whatever—at least can be conceptualized as becoming smaller and smaller until infinitesimally small. At some stage motion of an object such as a vibrating string or brane would be just a rotation, or possibly a pulsation within itself. This reductionism also could go on forever. A conceptual way out is to go beyond any infinitesimal motion into a continuous field with the *potential* for motion. Rather than to say that the unified field has an infinite quantity of energy, it is sometimes described more abstractly as infinitely dynamic in the sense of *infinite potential* for energy, without specifically assigning it an infinite quantity of energy. Another term for the source of creation as infinite potential is *omnipotent*.

Non-changing, infinite silence. Nature is described as inherently and constantly dynamic. As infinitely dynamic, the unified field also would need to be non-moving or static, in the sense that there would be nowhere it could move outside of itself—because there isn't anything outside of it. It also would be non-changing if it were to maintain *forever* its constant infinitely dynamic status. The attribute of *inherent dynamism*—the constantly changing nature of creation—would necessarily require an underlying *non-changing* attribute in order for the change to be *constantly* or *ever*-changing. It seems logical that the basis of the ever-changing nature of creation also would be ultimately a field of non-change. Implicit in the concept of perpetual change is non-change. Motion implies rest—the basis of motion is stillness, the basis of speech is silence. 'Ever-changing' implies a transcendental base that never changes.

In order for the unified field to be a field of absolute non-change that underlies the ever-changing nature of creation, it also would be beyond the possibility of being changed, destroyed, or turned into something else—beyond everything that is subject to change. It would be non-changing, impenetrable, indestructible, indivisible, without parts, and in this sense *undifferentiated*. Paradoxically its 'nature' thus would be that it is simultaneously a field *of infinite dynamism* or *infinite potential* and also a field of absolute *undifferentiated non-change*. It would be the absolute never-changing basis of relative ever-changing nature. A related term for infinite eternal non-change is *infinite silence*. As Maharishi recently stated:

"Only absolute silence can be all-pervading." [10]

It may be helpful to comment here on the relationship of the unified field as an absolute non-changing background of relative change to the concept of background independence of relativistic space-time, which relativity and loop quantum gravity theories posit. The level that underlies ordinary conventional space-time would be a nonlocal, highly interactive but still *relativistic* field. This subtler information space or mental *reality* would not be absolute. The only absolute non-changing background would be the all-encompassing unified field underlying all relative levels. *That* underlying background, upon which all manifest forms and phenomena would be dependent, is sometimes referred to in Maharishi Vedic Science and Technology as unmanifest transcendent pure Being. It would underlie both subtle relative mental space and gross relative physical space.

The attribute of being a non-changing background for relative change is implied in models of cyclic big bangs and collapses, in that some kind of non-changing substrate would need to continue across explosions and collapses—or implosive creations and dissolutions—if it is an *eternal cyclic* process. There would be no basis for such an underlying continuity if the ultimate unified field were literally *nothing*—as well as if the physical universe were fundamentally random. As the non-changing basis for all change, the unified field then indeed necessarily would be an *absolute frame of reference* and ultimate *preferred perspective* from which to evaluate all relative change. In Maharishi Vedic Science and Technology the absolute background, frame of reference, or observer perspective is transcendental consciousness—the unified field of consciousness—the inherently self-aware background of all individual observers.

The attribute of an observer dependent cosmological system attempting to be formulated in loop quantum gravity theory, and the attribute of the unified field as an absolute frame of reference, can be related to the discussion in Chapter 2 about transcendental consciousness. When the mind comes to a state of complete rest and transcends all mental activity, the underlying, silent, non-changing foundation of mental activity is 'directly experienced.' Correspondingly the knower needs to have some persistent non-changing basis in order to build continuity of knowledge. Each individual is aware of his or her own existence, and this underlying *I* at least has some aspects of a non-changing frame of reference from which to experience change. When the mind is completely silent and still, the individual can directly experience the underlying non-changing universal basis of knowledge and the non-changing basis of the *I*—the *knower,* the universal Self, the unified field of consciousness. This will be discussed in more detail in Chapters 11-17.

Self-sufficient, ultimate cause. If the unified field were the ultimate source of everything, it would not depend on anything other than itself for its existence. It seems intuitively sensible that it then would be *self-sufficient.* The phenomenal process of manifestation from the unified field to relative levels of existence would not involve anything outside of the unified field, because there is no thing outside of it. If it contains every thing, it would have to contain and determine its own initial conditions. [9]

This means that the infinite unified field would have inherent creative potential to manifest— infinite organizing power. It would be the ultimate cause of all effects in nature, but in itself it would not have a cause other than itself. If the created universe functions within a system of cause and effect, then the unified field would seem to be the ultimate or first cause of all effects. Every relative thing that exists in nature would be an effect that is caused by the unified field, but it itself would be uncaused. In other words it would be its own *self-sufficient, ultimate cause.* Alternatively it could be described as both its own cause *and* effect, or *beyond* cause and effect.

Perfect order. The principle of sequential symmetry-breaking from the unified field into the fundamental forces of nature implies that the unified field would be a state of *perfect order*—the most integrated, holistic, and perfectly symmetrical state, the source of all order in the universe, the state of lowest entropy. Spontaneous symmetry breaking results in the phenomenal appearance of breaking perfect order into relative degrees of order, associated with different perspectives on undivided wholeness. None of the progressively more concrete, more restricted relative levels would express the total degree of orderliness of the unified field itself.

Order at least appears to be ubiquitous in nature. If it can be found everywhere, then it makes sense that there would be a universal source of order that coordinates the orderly functioning at all levels of diverse manifestation throughout the universe. That source of order would be the highest level of order, or perfect orderliness. This is suggestive that order is built into the very nature of creation itself, and that the source of creation would be an underlying field of perfect order. The field of perfect order would be unmanifest with respect to the field of phenomenal manifest creation. It could be described as perfectly symmetrical in it unmanifest state, but broken into asymmetries in its relative manifest state—consistent with spontaneous symmetry breaking, the 2nd law of thermodynamics, the principle of decoherence, and the arrow of time.

The thermodynamic concept of entropy might suggest that the undifferentiated state of the unified field is a state of complete disorder. The concept of entropy is a classical perspective associated with the statistical motion of large collections of particles, perhaps a function of the limitations of calculation due to the immense complexity of large systems. It doesn't apply on the level of individual atoms—in this sense, there is no entropy in an atom or electron. There are four 'laws' or principles of thermodynamics: 1) energy is conserved in all physical interactions; 2) everything tends toward greater entropy or disorder; 3) entropy is reduced with reduced motion, or temperature, and 4) the *zeroth* law of thermodynamics that if two systems are in thermodynamic equilibrium with respect to a third system, they are also in equilibrium with each other.

The 1st law of thermodynamics states that all motion or interaction between physical objects results in no loss in the total amount of energy. This might seem to suggest that there is a fixed and definite total amount of energy in the universe that never changes. On the other hand, the total amount of energy is undefined in physics, and cannot be defined due to the constant dynamic change inherent in the universe. This principle does not concern how energy became existent in the universe in the first place, and it is not obvious how to reconcile it with the theory of the unified field as an infinite source of energy.

The 2nd law states that everything tends toward greater disorder, that entropy always increases. This principle also doesn't concern how order became existent in the first place, but it does imply that the initial state of the universe must have been a state of high order and low entropy. Similarly it is not obvious how to reconcile it with the theory of the infinite unified field as a state of complete homogeneity—unless complete homogeneity is considered to be a perfectly ordered state rather than a completely random, incoherent, high entropy state.

The 3rd law' of thermodynamics states that decreased activity—associated with decreased temperature in material systems—results in decreased entropy, related to increased order. This suggests a fundamental negentropic process in nature. But to be consistent with the 2nd law of thermodynamics, the order would have to be limited and localized; and taking into account everything involved the total system would become less orderly and more entropic. How this might fit with any theory of the totality of the universe following orderly patterns, such as a cyclic model of the evolution of the universe, also is not obvious. These are ways of restating some of the challenges to contemporary physical theories introduced in Chapter 7.

From the reductive perspective of an infinity of dimensionless or near dimensionless points, the unified field might be thought of as consistent with a state of disorder, in the sense that an infinite number of possible rearrangements could be made without affecting its overall properties. But it also would seem to satisfy all possible symmetries—perfect symmetry. It then could be considered a state of perfect order in the sense that any change would appear to disturb its perfect symmetry and result in relative asymmetries within it—as in symmetry breaking. As an infinitely dynamic field, however, it could be conceptualized in terms of infinite energy being exchanged constantly. But it could also be thought of as a field of non-change; and according to thermodynamics, there would be no entropy.

These points are again suggestive of the unified field as a coexistence of opposites that is both a perfectly ordered state and also undifferentiated homogeneity. This is reflective of the simultaneous locality/nonlocality inherent in the unified field—the simultaneity of the reductive dimensionless point value of the unified field and the holistic, infinite, undivided wholeness of everything. It suggests that symmetry breaking and thermodynamic principles of entropy apply to relative systems, but the absolute nature of the unified field would transcend these relative notions. It is with respect to some relative utilitarian criteria that the unified field would be describable as either entropic or negentropic. The description of the unified field as a state of perfect order can be viewed as more consistent with such criteria—which may further help distinguish it from literally nothing.

The perspective that the unified field is perfectly orderly also seems consistent with the notion that it is a field of all possibilities. From all possibilities the relative fields sequentially manifest into progressively limited and restricted levels, from the subtlest level of mind to the grossest level of matter. Each more expressed level reflects increasing differentiation or discreteness and, in some relative sense, increasing entropy. In these relative levels entropy increases through time according to the 2nd law of thermodynamics. This is associated with the concept of the arrow of time that time is uni-directional in the forward direction. However, if according to the 2nd law of thermodynamics entropy always moves from lower entropy and higher order to higher entropy and lower order, it would suggest that the beginning of the process that is subject to this law would have to be a state of the highest possible order and lowest entropy—a state of *perfect order*.

But also, in accord with the 3rd law of thermodynamics, the cosmic system would allow for reestablishment of order—negentropic behavior—on the relative levels through reduction of activity, recontacting the underlying absolute field of non-change. These dynamics are reflected in the process of transcending, discussed in more detail in Chapter 19. They also may be expressed in cyclic models of creation and dissolution in which the universe dissolves back into the unified field that reestablishes its perfect orderliness as the unmanifest platform for the next cycle of manifestation. These negentropic processes could occur universe-wide, as well as in terms of an individual via transcending mental activity and settling down to pure consciousness itself. In the overall tendency toward increasing entropy according to the 2nd law of thermodynamics, there is also a tendency toward decreasing entropy in living systems. The unified field would be perfectly orderly, and would be the source of both the tendency toward disorder and the tendency toward order—possibly to reestablish perfect order, suggestive of a self-referral, eternally self-reordering, self-correcting principle in nature.

As a perfectly ordered state, the unified field also could be described as the simplest possible state. On the other hand, as the field of all possibilities it might be described as infinitely complex. The contrast of simplicity and complexity is another angle on the 'nature' of the unified field as a coexistence of opposites—infinite simplicity and infinite complexity at the same time—or in other words, transcending the duality of simplicity and complexity into the holistic unitary state of all possibilities simultaneously.

This contrast is significant to the contrast of parts and whole that is a difficult issue in modern scientific thinking about emergence. From a reductive perspective the whole is an integration of parts. From an emergent perspective the parts integrate into a whole that is more than the sum of the parts, from which properties blossom that are not reducible to its parts. Neither of these perspectives capture the 'nature' of the unified field as a unitary field that is prior to any parts and from which all parts emerge. If the unified field is all-encompassing, nothing more integrated can emerge from it. Only various limited degrees of wholeness can emerge from it, none of which create the wholeness which then controls the parts, as would be necessary in the reductive and non-reductive emergence perspectives—to be discussed in later chapters.

As also will be brought out further in later chapters, at some point modern science needs to take seriously the 'nature' of the unified field as the source of order in nature. There is ultimately no way to get around this point about the 'nature' of the unified field. The universe is not fundamentally random or derived from a highly disordered state, and the unified field is not literally *nothing*. Its 'nature' is perfect order and all possibilities. This has tremendous—some might say *unbelievable*—implications for human life. It opens the possibility for an uplifting, hopeful, integrated view of life, and contrasts with the fragmented thinking that has led to existential meaninglessness associated with the physicalist worldview. It also establishes the basis for a systematic science of ethical and moral behavior, based on natural psychophysical and behavioral laws. It establishes the basis for systematic means to develop life in accord with the totality of natural law through realignment with the perfect orderliness of the unified field—to be unfolded in later chapters.

Infinite correlation.   The test of Bell's theorem demonstrated a remarkable nonlocal correlation in nature that is not accountable for within the limitations of local causality basic to the space-time model of Einstein relativity. It reflects a degree of interconnectivity and interactivity built into the fabric of nature that goes beyond the ordinary material level of object independence, a more holistic level of quantum coherence somewhat akin to a holographic system. The concept of relativity can be maintained, but the limitation of Einstein locality cannot, at the nonlocal level of nature. The unified field would have this holographic interconnectivity to the *infinite* degree. It would integrate completely contrasting and opposing qualities—complete coordination of the extreme opposites of point/infinity, infinitesimally local/infinitely nonlocal, smaller than the smallest and bigger than the biggest.

Even further, from the perspective of the unified field the wholeness of the universe would contain every possible point in the universe, and every possible point would contain the entirety of the wholeness. It would be infinitely differentiated into points, with every point containing the whole. This 'attribute' of the wholeness in every point—the transcendental integration of infinity and point—is sometimes described as *infinite correlation*. It would be both infinite diversity and infinite unity at all times and places, simultaneous and sequential, local and nonlocal—beyond the concepts of time and space, independence and dependence, diversity and unity.

> "[E]ternal possibility and modal differentiation into individual multiplicity are the attributes of the one substance."—*Alfred North Whitehead, mathematician and clergyman (p. 177)*[11]

Infinitely self-interacting. If the unified field were self-sufficient, its own cause and effect, and infinitely correlated, it seems intuitively logical that it would be *infinitely self-interacting*. All change would originate from and take place within the unified field, ultimately caused by the unified field interacting with itself. It would be independent, uncaused, and by virtue of its own self-interacting dynamics, the origin and cause of all dependent effects in relative creation. Anything that exists and interacts with other things would be ultimately nothing other than the unified field interacting with itself.

The concept of a self-interacting system relates to non-Abelian mathematics, non-commutative logic, and non-linear geometry (refer to Chapter 5). In a commutative process, elements of an equation can be exchanged without changing the outcome, such as $2 + 1 = 1 + 2$. In a non-commutative process, this symmetry doesn't hold—such as, for example, combing your hair after putting on your baseball cap yields a different result than the typical opposite sequence. A non-Abelian field such as the unified electroweak field, gluon field, or super-unified gauge field interacts with itself through its non-linear dynamics. New processes are created from the field interacting with itself. A self-interacting field is modeled by non-linear equations and topologies that *reflect back on themselves* and create new structures that become part of the whole picture; and frequently such non-linear equations are impossible to solve except approximately.

A self-interacting field somehow creates distinctions and new relationships within itself. In other terms it has dynamic creative potential and the ability to discriminate and evolve new orderly systems of relationships. The ultimate unified field, as an *infinitely* self-interacting field, would be able to create any actuality, to manifest any possibility from and within itself. It then would be both infinite non-changing silence *and* infinite self-interacting dynamism, completely undifferentiated indivisible oneness *and* the infinitely dynamic basis of all differentiation. It would be the non-changing source of all the laws of nature that govern orderly change *and* it would discriminate distinctions within itself that manifest into the diversity of changing forms and phenomena found throughout nature. It would be the non-changing infinitely dynamic source of the trinity of substance, function, and form—or observer, process of observing, and observed—that comprises objects anywhere and everywhere in the phenomenal universe, *and* remain ultimate oneness—simultaneously three in one unity.

Because the unified field would be the source of all order in the universe, and that order would emerge somehow only within it and due only to its self-interacting dynamics, it could be said to contain attributes of orderliness or negentropy that are typically associated with the concept of intelligence. It thus seems reasonable that the unified field also would be a field of intelligence, at least in the sense of a self-sufficient source of order and the source of all the laws of nature. Quantum physicist and unified field theorist John Hagelin points out:

> "These laws of nature formally express the order and intelligence inherent in natural phenomena. If there were no laws of nature, there would be no consistent patterns of natural behavior, and nature would be unintelligible. If, as particle physicists believe, all the laws of nature have their dynamical origin in the unified field, then the unified field must itself embody the total intelligence of nature's functioning." (p. 9) [12]

Intelligence also connotes *intentionality*—purposive action or agency. Is there a reasonable intuitive sense in which the unified field can be attributed the property of intentionality? Certainly in the sense of the initiation of orderly action the unified field can be thought of as intentional—as the ultimate, singular, uncaused cause or origin of all action. Can it also be thought of as intentional in the sense of having a purpose or design that refers to a specific outcome, an *intended* result, as the anthropic principle suggests? Can the concept of agency be attributed to an orderly—even mechanistic—system such as the system of natural laws, or only to some form of living being, such as a human or God? This issue can be seen more fundamentally to relate to the concept of freedom of choice, or free will. As mentioned in the discussion of the anthropic principle in Chapter 7, the choices for the agency of the universe are typically identified to be either God or Nature. If the entirety of nature functions at the behest of God, then God also would be the all-powerful or omnipotent, omnipresent agent of creation. Whatever other attributes or qualities are attributable to God, at least that aspect of God that is omnipotent and omnipresent would be the ultimate unified field and an intentional field of intelligence. In other words it would have the quality of *omniscience*—an intuitive-based attribute of the concept of God. This has important implications for religious understanding. *I may think* I am just a small part of God's creation, but omniscient, omnipotent, omnipresent God is always *in me* and *all of me*; ultimately *I* cannot be separate from God as the unified totality and source of all that exists, though *I* may appear to be separate at certain stages or states of knowledge and experience.

> "Even if we don't know what the laws of nature are, or where they have come from, we can still list their properties. Curiously, the laws have been invested with many of the qualities that were formally attributed to the God from which they were once supposed to have come... First and foremost, the laws are universal... The laws are taken to apply unfailingly everywhere in the universe and at all epochs of cosmic history. No exceptions are permitted. In this sense they are also perfect... Second, the laws are absolute. They do not depend on anything else...the laws, which provide correlations between states at subsequent moments, do not change with time... [W]e arrive at a third and most important property of the laws of nature: they are eternal... Fourth, the laws are omnipotent... They are also, in a loose sense, omniscient, for, if we go along with the metaphor of the laws "commanding" physical systems, then the systems do not have to "inform" the laws of their states in order for the laws to "legislate the right instructions" for that state... A schism appears, however, when we consider the status of the laws. Are they to be regarded as discoveries about reality, or merely as the clever inventions of scientists?... Physicists talk about planets "obeying" Newton's laws... This gives the impression that the laws are somehow "out there"... Falling into the habit of this description, it is easy to attribute an independent status to the laws. It they are considered to have such status, then the laws are said to be transcendent, because they transcend the actual physical world itself."
> —*Paul Davies, theoretical physicist and cosmologist (p. 82-84)* [6]

On the other hand, if the ultimate agency were just Nature, the laws of nature, then the potential bases for intentionality would seem, according to most modern scientific thought, to be either fundamental determinism or fundamental randomness. It isn't obvious how either of these satisfy the intuitive concept of agency in terms of freedom of choice or free will. With respect to complete determinism there is thought to be no choice involved—only action according to the determined chain of cause and effect. With respect to fundamental randomness there also would be no choice—only capricious uncertainty, no preference associated with the free will of an agent. Thus if nature is run solely by the laws of nature within the unified field and not by some agent such as God, it seems reasonable to attribute the concepts of intentionality and also intelligence to it as well. These concepts would not necessarily include free will, but rather be characterized in terms of the initiating source of action. In either case—Nature or God as the agency—there is some reasonable basis to identify the unified field as a field of intelligence.

The concepts of the unified field as a field of perfect order and of all possibilities also seem consistent with the concept that it is a field of unlimited intelligence or *omniscience*. However, if the unified field can be thought of in terms of the coexistence of opposites, then perhaps it might be thought of as somehow simultaneously having both determinism and free will—to be discussed in Chapter 18.

Self-referral. There is another fundamental connotation of the concept of intelligence: the sense of self-knowingness or self-awareness. Is this connotation of intelligence also attributable to the unified field? This issue can be viewed as related to the distinction between *self-interaction* and *self-referral*. We already discussed the unified field as being infinitely self-interacting, inherently dynamic, and in some sense inherently intelligent—at least perfectly orderly and the source of order in nature. Is it also inherently self-referral, in the sense that it is self-aware?

If the unified field is an aspect of an omnipresent, omnipotent, omniscient God—or vice versa or both—then it seems sensible to attribute self-awareness to it, inasmuch as this is also an intuitive attribution about God. But what if the unified field is only the self-interacting dynamics of the laws of nature? In what sense can the unified field be said to be self-aware or self-referral if it is not God? This cuts to the core of the definition of awareness or consciousness itself, which will be an important topic in the next part of the book.

One approach to this issue is to identify the characteristic marks that something is conscious or self-aware. As described in the Prologue, typical signifiers that an entity is conscious relate at least in part to the signifiers of life such as intelligence, intentionality, attention, and the survival instinct. These can be related to the notion of discriminative action by an entity to maintain its own existence, to survive. Can the unified field be said to maintain its own existence? This might be thought of in part as an empirical question of whether the universe has existed and continues to exist in some form or another. It would seem consistent with ordinary direct sensory experience that the universe does continue to exist in some sensed form—it is likely that most all of us would agree—and it does this by virtue of being a manifestation of nothing other than the infinite, eternal, self-interacting unified field because nothing else exists.

In addition to this means of empirical verification, the means would include logical consistency as well as intuitive sensibility. Intra-subjective and inter-subjective agreement of the continuing existence of creation in some phenomenal form through logical and intuitive means of verification also seems consistent with it as maintaining its own existence. To conclude empirically, logically, or intuitively that the universe does not exist at least phenomenally seems inconsistent. If the signs of its existence—that is, creation and created beings such as us—are maintained through time; and in addition are maintained logically only by its own self-interaction, then the unified field would seem to be self-maintaining—at least one of the signifiers of self-awareness.

Are there other signifiers of the unified field being self-aware or conscious of itself? Some approaches to consciousness assume that an entity has to have a body that supports conscious experience in order to be conscious. In what ways might the unified field satisfy this assumption? If the unified field is the source and basis of everything that exists, it necessarily would be the source of all bodily forms that support conscious experience. The unified field could be thought of as its own body—the basis itself of the abstract principle of form, the essence of any particular body. Alternatively it could be taken that the form or body of the unified field is the entire manifest universe as an integrated unit—an embodied, unitary, completely self-sufficient system or organism. These points suggest that the unified field might be thought of in some sense as having a body that could support consciousness, if consciousness requires a body.

However, as will be discussed in upcoming chapters, most functional approaches to consciousness would add the additional requirement that the body needs to have a high enough degree of complexity and interactivity for conscious experience to emerge. But there would be no degree of complexity and interactivity higher than the total integration of everything in the unified field. Any localized functional component of the universe would be less complex and less interactive than the system in its totality as infinite and eternal. The unified field would necessarily contain all the components—all of the substances, functions, and forms—comprising any of its parts. There would be no greater wholeness, no greater capability—including the capability of self-awareness—than the totality of capability of the unified field. This suggests that the unified field certainly is complex enough to support consciousness.

Thus there are at least some indirect empirical as well as logical and intuitive bases for the unified field as being self-referral, conscious of itself, whether as God or as Nature. From the above discussion, the unified field of natural law—or that aspect of God that is omnipresent, omnipotent, and omniscient—can be thought at least in some fundamental sense to be self-referral, to be conscious of itself. These and related points will be unfolded more completely in subsequent chapters. They cut to the core of our assumptions about the nature of our individual selves, individual consciousness, and our relationship to the unified field and to universal consciousness. As Maharishi recently stated:

"The unified field knows everything. It does everything. It administers everything." [13]

Beyond infinite regress. The notion that space and time don't exist either before or outside of the big bang may seem to have an uncomfortable sense of incompleteness about it. Even if the closed universe has an infinite attribute, such as that it goes on indefinitely in time, we are still able to have some conceptual sense of time as existing before the big bang, as space existing outside of the edge of the expanding universe, or of some subtler field at a smaller scale than the Planck scale—smaller than the smallest. Although 10^{-33} cm is almost inconceivably tiny, we can at least conceptually speculate about there being a 10^{-34} or $10^{-100009}$ and so on. The human mind or intellect seems to have an innate intuitive sense of something outside whatever limitation it can conceive. Indeed any discrimination or delineation tacitly implies something beyond or outside of it. How do we get beyond any challenge of infinite regress that the mind is able to conceive?

Even though a main goal of modern science has been to identify universal invariant laws of nature, some modern thinkers including some scientists seem quite reluctant to accept any form of knowledge as absolute. Emphasis has been on the relative nature of knowledge and the assumption that nothing can be known with absolute certainty. This was alluded to in Chapter 1 with respect to modern scientific methodology increasing the relative truth-value of knowledge but not proving anything with absolute certainty. This relative view depends on what means of gaining knowledge we count as scientific. Certainly knowledge can be understood to be relative in that it is conditioned by culture and historical framework, and especially by the state of the knower.

However, a statement such as 'nothing can be known with absolute certainty' is itself an absolute statement. [14] As discussed in connection with Gödel's incompleteness theorem in Chapter 9, no conceptual system is complete in the sense that it can prove its own consistency. Assertions, such as axioms, that cannot be proven within the system are unavoidable. Indeed the concept that everything is relative implies a single attribute of relativity that is everywhere as a universal constant that permeates and transcends everything. Every cogent system of knowledge contains absolute statements that imply an absolute transcendental foundation to its principles. Thus the 'nature' of the unified field also would seem to be that it could be described as simultaneously relative *and* absolute, manifest *and* unmanifest, tangible *and* transcendental.

In order for the unified field to be the ultimate source of everything, and for there to be absolutely no thing outside or beyond it, it seems intuitively logical that it would necessarily have no boundaries at all—that it transcends all boundaries. The concept of infinity is of being unlimited, unbounded, beyond finite values. In the abstract field of mathematical sets, there are many types of infinities, such as the set of countable numbers and the larger set of real numbers, both of which have an infinite number of elements in the set. *Cantor's alephs* describe a hierarchy of infinities. In each case, however, the concept of infinity is associated with something that is infinite in some way but not in others, always within an overall framework or limitation of some kind. Even the concept of an empty set can be thought of in this way. In this limiting framework infinity is a mathematical concept that can be manipulated as if it were a number—such as infinity plus, times, squared, to the infinite power, and so on.

The expression *unified field* also can be thought of as implying a defined domain, realm, region, or space—a *field,* within which everything is *unified*—and thus some sense of limitation or boundary. It is an expression of the wholeness of nature from the perspective of parts that become unified, more characteristic of a reductive rather than holistic view of unity. It reflects the inherent nature of the intellectual mind or discriminative faculty that attempts to identify things through language and symbol, and that can only point to the ultimate unity beyond—and existing prior to—any and all processes of unification.

The only way out of an infinite regression of infinities seems to be an all-inclusiveness that is not infinite with respect to a limited frame of reference or set of properties or principles. It would transcend *all* properties, principles, frameworks, measurements, definitions, or conceptualizations of any finite value as well of any infinite value. It would be beyond any and all conceivable limitations and non-limitations—beyond any concept of infinity or any set or class of infinities (or whatever the latest intellectualization happens to be that is superimposed on the underlying silence). It would be beyond the conceptualizing mind. It would not be directly accessible within the inherent discriminative function of the conceptualizing intellect and intellectual knowledge.

Words or concepts cannot capture *that* completely unbounded transcendent infinity. Historically from a reductive and negational perspective it sometimes has been labeled the eternal *Void* or *nothing-ness*—to contrast it with the ordinary perspective, as beyond everything knowable through the senses and mind. From a holistic perspective it sometimes has been labeled eternal *Being,* with emphasis on it being the transcendent essence of everything. In an attempt to capture both the negation and the positive essence together, terms such as *Being-Void* have been proposed. [15]

At that ultimate transcendental level, the distinction of Nothing—Void or 0 (Zero) and Everything—Unity, Being, or Singularity (One)—can be thought of as a difference of perspective and language. At *that* level, Zero is One. However, in the sense of no possibilities versus all possibilities, or as completely entropic and disorderly versus perfect order, the holistic perspective of Oneness, pure Being, and all possibilities would seem more descriptive. The term that is most applicable in a particular setting would depend on its intent or purpose. In general, the term *Infinite Eternal*, or just *Infinity*, referring to all infinities, perhaps does a little better job than most terms in incorporating negation and essence together. Toward the end of the book will be introduced the principle of the correspondence of sound and form in Vedic language. We will then discuss how this principle results in more profound language representations, a key to the fundamental developmental significance of Vedic language beyond language as conceived by the human intellect.

The unified field as *Infinity*—inclusive of all infinities and all finite values—resolves the challenge of infinite regress in its various forms. For example, it resolves the problem of smaller and smaller constituents. Progress in modern physics has been toward smaller and smaller fundamental constituents of nature, from what used to be thought of as 'irreducible' or 'uncuttable' atoms to subatomic and elementary particles, force fields, unification of force fields, strings, loops and spin networks, bits of information, and so on. Ultimately there must be an end to this regression, and the unified field as *Infinity* ends it. It also resolves the search for first cause, as well as for initial conditions. Infinity means beyond both infinitesimally small and infinitely large, at the same time beyond ultimate differentiation and ultimate integration, beyond reductionism and holism.

Moreover it provides a resolution of the quandary of the spontaneous emergence of life—as well as consciousness—from non-living matter. It is a common physicalist belief that life begins at about the cellular level, and that molecules, atoms, and all more fundamental objects and processes are non-living. If the cell is more fundamentally a collection of non-living matter and force particles, how does non-living matter create living cells? If the unified field as *Infinity* contains all possibilities and actualities, then it would necessarily contain all of the attributes that comprise life, as well as the ability to organize these attributes into living organisms. At least all the basic ingredients, and the capability to organize them, would be inherent in the unified field—as well as any form or manifestation of life. This is certainly in the direction of suggesting that the unified field itself could be described as a living wholeness of *Being* in some non-trivial sense.

At least since Aristotle, descriptions of the universe as a living pulsating organism have been somewhat common in Western thought, and ubiquitous in Eastern thought. Accordingly life could be inherent in and begin at the level of the unified field, rather than only at the molar level of the cell. This will be discussed in the next chapter, and at length in Part II of the book in connection with the concepts of emergence and upward and downward causation in models of consciousness and the brain. Alternatively of course the unified field could be described as beyond life and death, beyond living and non-living.

Further, *Infinity* resolves the problem of the infinite regression of homunculi. This issue is sometimes characterized in terms of a little person inside the body and the mind that views the screen of the mind, which then requires a viewer of the screen, and so on ad infinitum. Any form of infinite regress is resolved by an all-encompassing, all-inclusive *Infinity*. As suggested by these points, the intuitively rational understanding seems to be that the ultimate unified field is all-inclusive *Infinity*. Further, as will be discussed later, the unified field as a self-referral field of consciousness is the most consistent explanation of consciousness.

The coexistence of opposites of infinite dynamism and infinite silence is identified in Maharishi Vedic Science and Technology as the 'nature' of the unified field. It is also identified as the essential 'nature' of consciousness itself. The unified field described in modern science parallels the source of creation described in great detail in ancient Vedic science—*Atma*—which fully integrates these points into the unified field as the source of both objective and subjective nature. Many of these empirical, logical, and intuitive points about the 'nature' of the unified field are summarized by Maharishi in the following quote:

"The unbounded field of Being ranges from the unmanifested, absolute, eternal state to the gross, relative, ever-changing states of phenomenal life, as the ocean ranges from the eternal silence at its bottom to the great activity of ever-changing nature on the surface of the waves. One extremity is eternally silent, never-changing in its nature, and the other is active and ever-changing... The active, ever-changing phase of the ocean represents the relative phase of Being, and the ever-silent aspect of the bottom of the ocean represents the never-changing, eternal absolute state. This is the relationship of Being with the world of forms and phenomena in which we live. Both these states, the relative and the absolute, are the states of Being. Being is eternally never-changing in Its absolute state, and It is eternally ever-changing in Its relative states... An example will illustrate the nature of Being in a more comprehensive manner. As we have seen, the oxygen and the hydrogen atoms in one state present the qualities of a gas, in another they combine and exhibit the qualities of water, and in yet another they exhibit the qualities of solid ice. The essential content of gas, water, and ice is the same, but it changes in its properties. Even when the properties of gas, water, and ice are quite contradictory to each other, the essential constituents, H and O, are always the same... The unity of Beingness, without undergoing any change in Itself, assumes the role of the multiplicity of creation, the diversity of Being. The absolute, assuming the role of relativity, or unity appearing as multiplicity, is nothing else but the very nature of absolute Being appearing in different manifestations... The reality of duality is unity. Even though different in their characteristics, absolute Being and relative creation together form the one reality."(pp. 31-36) [16]

A Mathematical Glimpse at the Unified Field

In pursuit of a completely unified field theory, a level of abstraction beyond physical *reality* and time and space is necessary. In the abstract language of mathematics, the concepts that can be used to model a unified field theory are available in the *Lagrangian* formulation. [12] It is a mathematical formulation of an ultimate field of existence and the orderly dynamism—or intelligence—in the field. The Lagrangian formulation is fundamental to almost all physical theories, in order to derive the equations of motion or interaction of particles in the theories.

In very compact form, the Lagrangian applied to unified field theory contains two terms. [12] The first term, denoted as *phi*, can be described as a classical conception of a static, space and time translation invariant field—a non-changing field of *existence*. Added to this is a second term representing orderly dynamism or change, denoted as *II*. This term can be thought of as representing the inherent capability of the field to generate orderly change in the field. Hagelin associates this with the most abstract interpretation of the *quantum principle,* in terms of the discriminative property or *intelligence* that delineates and structures dynamic action in the field:

"To some extent, we can trace this property of intelligence to the fact that the unified field, beyond its mere existence, has a very precise and definite mathematical structure. This structure is typically defined in terms of symmetries of the field—invariance with respect to a set of internal and external transformation, such as Lorentz invariance, super-symmetry, modular invariance and gauge invariance. External symmetries, such as Lorentz invariance, describe the behavior of the field under transformations of space and time—translations, rotations and boosts. Internal symmetries, such as gauge invariance, refer to transformations among the various internal degrees of freedom of the unified field—bosonic and/or fermionic. The precise mathematical structure of the unified field serves as an unmanifest blueprint for the entire creation: all the laws of nature governing physics at every scale are just partial reflections or derivatives of this basic mathematical structure. However, this view of intelligence in terms of the classical symmetries of the unified field is a rather passive and inert one. The term 'intelligence' achieves its full significance only at the quantum-mechanical level of description, in which the field acquires a degree of *dynamism, discrimination and creativity* not present at the classical level." (p. 9) [12]

In this Lagrangian formulation, the quantum dynamism of the unified field is a product of applying the *canonical commutation* relation. This incorporates the quantum mechanical concept of zero-point motion—the inherent dynamism of the field that distinguishes the inertia of the field from its own motion at the ground state of the field. Applied to the compact Lagrangian, it results in the two terms—*phi* and *II*—becoming incompatible, asymmetric, and non-commutative. This is the abstract mathematical basis for the field to differentiate itself into specific shapes. Directly derived from this is the quantum mechanical superposition or coexistence of all possible shapes of the field in its ground state, as well as the inherent indefiniteness of the field associated with Planck's constant and the uncertainty principle. [12]

The Lagrangian formulation places the unified quantum field in *Hilbert space*, a complex vector space of infinite dimensions—an infinite collection of points that comprise all states of a quantum mechanical system. It is a mathematical space in which state vectors or points can be added or subtracted linearly, which leads to the principle of *superposition* or simultaneous coexistence of possibilities, related to the dynamism of the field. It also can characterize invariance of the field under unitary transformations of states of the field with no loss of information—the non-changing attribute of the wholeness of the field space. Further the structure of Hilbert space relates to an *inner product* that introduces the concept of length and magnitude that determines physical properties such as energy, momentum, and position of objects in the field. [12]

Hagelin uses the Lagrangian formulation in Hilbert space to present a unified field theory that more explicitly includes principles identifiable with the concepts of the knower and the process of knowing—not merely the known. [12] The *knower* quality of the field is interpreted as the property of the Hilbert *space* of states to be a non-changing or unmanifest background for all possible unitary transformations or states of the field, while itself remaining completely unchanged. It is the uninvolved 'observer' of all transformations that, through its dynamic orderliness associated with the discriminative role of the inner product in evolving the quantum mechanical system, determines the physical manifestations of the system. The *process of knowing* quality of the field is related to quantum mechanical *observables* that serve as quantum mechanical operators in Hilbert space, generating changes of one state into another in unitary transformations. The *known* is interpreted as the stable quantum mechanical *states* themselves. These individual points in Hilbert space represent isolated possibilities in the field of all possibilities. Hagelin further explains:

> "These quantum-mechanical observables represent all the properties of a quantum-mechanical system that can be known—its energy, momentum, angular momentum, etc., depending on the details of the quantum-mechanical system. These quantum-mechanical observables correspond to operators in Hilbert space, and can therefore be viewed as infinite-dimensional matrices. These operators, it must be emphasized, are distinct from the classical quantities they represent. The latter depend intimately upon the state of the system, whereas the quantum-mechanical observables do not. They have a more universal status, and are associated with the process of *gaining knowledge*, i.e., of extracting information about the quantum-mechanical system... Without the quantum-mechanical observables, and the transformations they generate, the Hilbert space would

be completely inert. These transformations map the space onto itself, relating points in Hilbert space to other points. The set of all such transformations that leave the space invariant serves to define the space—its symmetries and its structure. They provide the dynamical means through which the Hilbert space knows itself—through which the quantum-mechanical space of all possibilities becomes aware of its infinite structure... In the context of a unified quantum field theory, these stable vibrational states of the field play an especially fundamental role: they comprise the elementary particles and forces of nature. For example, in the framework provided by the superstring, all the elementary particles and forces governing physics below the Planck scale correspond to massless vibrational states of the string, or string "fundamentals." The higher-energy modes, or string "harmonics", correspond to heavier particles with masses... These stable vibrational states of the string thereby provide the stable foundation on which the entire material universe is constructed—the elementary particles and forces of nature. They ultimately underlie the behavior of macroscopic, bulk matter, which has a tendency to hide the essential, abstract nature of the field from which it arises." (pp. 19-21) [12]

This interpretation of unified field theory attempts to incorporate the process of knowing and the knower into it, at least in the language and perspective of mathematical conceptualization. The knower is related to the Hilbert *space*, the process of knowing to the quantum mechanical *observables*, and the known to the quantum mechanical *states*. This mathematical representation of a completely abstract field of existence, with the addition of dynamic order or intelligence associated with the quantum principle, can be elaborated to derive all energy states, Hamiltonian states, and all particle states, including all particles, fields, and their interactions.

In this elaboration of the Lagrangian any particle can be translated into any other particle—a graviton is not different from a photon, for example. Even fermions and bosons—the most fundamental difference in particle theory—become unified and indistinguishable. The Lagrangian formulation in Hilbert space can model mathematically how from their indistinguishable unified basis all the various particles that comprise the vast diversity of nature can be derived. This Lagrangian formulation provides a unified field theory that in mathematical terms is beginning to match up much more fully with the unified field described in ancient Vedic science. Hagelin also explains that it directly predicts the five fundamental spin states of elementary particles. [12] This is an important topic to be discussed further in Chapter 12 that links this mathematical glimpse at the unified field with the levels of nature outlined in ancient Vedic science. But first, in Chapter 11 we will explore the full range of levels of nature as outlined in ancient Vedic science, which similarly identifies five basic constituents or elements as fundamental levels of existence.

Empirical Validation of the Unified Field

As the ultimate source of everything, the unified field would be the source of existence and matter as well as the source of intelligence and mind. It would be the ultimate source of all order in nature, objective and subjective. All aspects of knowledge and the known, all means of gaining knowledge or processes of knowing, and all knowers also would be contained within it, derived from it, and ultimately nothing other than it. This is nicely articulated from the angle of the mathematical description of the Lagrangian formulation in Hilbert space. [12]

With progress in theories of the unified field as the single underlying constituent of creation beyond matter and energy, modern science has reached the limit of objective investigation—referred to in the quotes by Maharishi in Chapters 2 and 9. It will continue to refine understanding of the effects of the unified field in relative creation. But as an unmanifest, transcendent, self-referral field beyond space and time, the unified field cannot be investigated from an outside objective perspective. It is not possible to probe or measure the unified field employing some tool or method of investigation phenomenally appearing to exist outside it. The unified field cannot be known directly by anything other than *itself*; if it is to be known, it has to be known by itself. Ultimately the *knower* of the unified field must be the unified field itself.

In order to understand the unified field as a self-referral field of consciousness—the *knower*— a means of gaining knowledge is needed that allows the unified field to be explored directly, from the inside, beyond the individual mind but from within it. It is in the equivalence of the unified field as a self-referral field *and* of the essential nature of individual awareness as infinitely self-referral universal consciousness that the possibility of direct empirical investigation of the unified field arises. If there is ultimate equivalence of the knower—whether individual or universal—and the unified field itself, it could only be *known* to itself through *direct experience* of itself. It would not be known by knowing it through the intellectual mind, but by *being* it directly—beyond knowing it through the mind as an object of experience in the mind. Knowing the unified field directly is a completely self-referral process, in which the three aspects of known, process of knowing, and knower fundamental to ordinary experience settle back into their unified source.

Experienceability of the unified field. Because there is no outer objective means to verify the existence of the unified field, evidence for it comes from observation of its indirect effects and from logical and intuitive consistency—as discussed in Chapter 9. The underlying transcendental unified field is ordinarily hidden because of the nature of the senses and the mind. The senses and the mind could be said to be built as the means to experience the relative field of life. The unified field is beyond the range of the senses and the mind, which are in the phenomenal relative field of existence—it cannot be perceived, comprehended, or known by the mind.

However, if the unified field is a self-referral field of consciousness—aware of itself—and also the simplest state of individual consciousness is the same self-referral field of consciousness, then it follows that it is directly 'experienceable' through transcending the mind. Fundamental to Maharishi Vedic Science and Technology is that each one of us as individual sentient beings is essentially that ultimate transcendental unified field—the unified field is consciousness itself. It is possible to *experience directly* the unified field because the individual 'experience' of consciousness itself is nothing other than the point value of the unified field—self-referral consciousness itself. That *direct experience* is the ultimate empirical verification of the unified field. It is 'directly experienced' in the simplest, least excited, most intimate, inner, transcendental state of our own individual consciousness as universal consciousness.

Not only is 'direct experience' of the unified field as transcendental consciousness the means to validate empirically the unified field of consciousness, but experience of all phenomenal manifestations in relative existence as emerging from and nothing other than the unified field of consciousness can be said to be its complete 'objective' empirical validation. Progress through higher states of consciousness that objectively demonstrates consciousness as the source of mind and matter will be discussed in later chapters. This empirical verification is the purpose of *Yoga,* which has been revived in the scientific framework of Maharishi Vedic Science and Technology as the *TM and TM-Sidhi program*—to be discussed in Chapter 21.

Throughout history most every society has held conceptions about an absolute level of life. Highly developed individuals have repeatedly reported validation through direct empirical experience of their essential connection to the totality of Infinity, the eternal Void, or absolute Being— described according to their cultural and experiential development. It sometimes is characterized as absolute bliss consciousness—*Sat-Chit-Ananda*:

"Sat-Chit-Ananda—It is Sat which never changes, it is Chit which is consciousness; it is Ananda which is bliss." (p.22) [16]

That ultimate field has been referred to in detail via the language of modern science in Maharishi Vedic Science and Technology, which includes systematic developmental technologies to validate it personally through one's own transcendental consciousness. Initially the experience of transcending simply may be of deep mental relaxation and inner silence. With regular practice, eventually within the least excited state of individual awareness, the unified field and all its phenomenal manifestations are unfolded—the entire cosmos within and without. The individual finds in his or her own unbounded awareness and essential self the wholeness of life in its full grandeur and significance—fully lived as the individual's natural daily experience in the state of enlightenment. Examples of progressive stages of clarity of that experience in higher states of consciousness will be discussed in Chapters 19-21.

The next chapter, Chapter 11, lays down important planks by describing the phenomenal sequence of manifestation or materialization of the unified field within itself in more detail, enumerating the manifest levels of subtle and gross relative creation. It outlines the full range of levels of nature in the holistic approach of Maharishi Vedic Science and Technology, drawn from that aspect of ancient Vedic literature called *Sankhya*.

Chapter 10 Notes

[1] *Celebrating perfection in education: Dawn of total knowledge* (1997). India: Maharishi Vedic University Press.

[2] Guth, A. H. (1997). *The inflationary universe: The quest for a new theory of cosmic origins*. Cambridge, MA: Perseus Books Group.

[3] Greene, B. (2004). *The fabric of the cosmos: Space, time, and the texture of reality*. New York: Alfred A. Knopf.

[4] Smolin, L. (2001). *Three roads to quantum gravity*. New York: Basic Books.

[5] Dyson, F. (2004). The world on a string. *New York Times*, Vol. 51, No. 8, May 13.

[6] Davies, P. (1992). *The mind of God: The scientific basis for a rational world*. New York: Simon & Schuster.

[7] Maharishi Mahesh Yogi (1996). *Maharishi's Absolute Theory of Defence: Sovereignty in Invincibility*. India: Age of Enlightenment Publications.

[8] Hawking, S. (2001). *The universe in a nutshell*. New York: Bantam Books.

[9] Greene, B. (1999). *The elegant universe: Superstrings, hidden dimensions, and the quest for the ultimate theory*. New York: Vintage Books.

[10] Maharishi Mahesh Yogi (2004). Maharishi's Global News Conference, December 12.

[11] Whitehead, A. N. (1925). *Science and the modern world*. New York: The Free Press.

[12] Hagelin, J. S. Restructuring Physics from Its Foundation in Light of Maharishi's Vedic Science. *Modern Science and Vedic Science*, 3, 1, (1989, 3-72.

[13] Maharishi Mahesh Yogi (2004). Maharishi's Global News Conference, June 30.

[14] Charleston, D. E. (1984, Spring). Personal communication, Norman, OK.

[15] Shear, J. (2002). A quale for pure consciousness? Plenary session, Is There Pure Consciousness?, Thursday, April 11. Toward a Science of Consciousness "Tucson 2002" Conference, April 8-12.

[16] Maharishi Mahesh Yogi (1963). *Science of Being and art of living*. Washington, DC: Age of Enlightenment Press.

Chapter 11

Sequential Enumeration of Undivided Wholeness

In this chapter, we will explore the expanded range of nature in the holistic approach of Maharishi Vedic Science and Technology outlined in the Sankhya system, one aspect of ancient Vedic literature that describes the sequential manifestation of levels of phenomenal creation. It provides a coherent, integrated view of the relationship of gross objective and subtle subjective domains of nature as phenomenal reverberations of the transcendental domain of the unified field. The subtle subjective levels are described as comprising an inner dimension that manifests within the undivided wholeness of the unified field, extending from consciousness to the most expressed, inert level of rocks and earth—which ultimately remain nothing other than fluctuations of the unified field. The overall main point of this chapter is that Vedic science articulates a holistic structure of the full range of nature that includes and extends the ontological levels in modern science.

As modern science has investigated larger systems in the universe, from the solar system to the perhaps trillions of galaxies and immense system of galactic clusters, it also has been drawn to smaller and smaller time and distance scales in attempting to comprehend the full range of nature. In probing subtler levels at smaller scales there is simultaneously an expansion to larger, more abstract fields that permeate the cosmic system. At the current state of modern scientific knowledge this involves a shift from classical physical *reality* to a subtler, mathematical information *reality,* and further to the all-encompassing unified field. But how these relate to mind and consciousness is yet to be worked out in modern science. Chapters 1 and 2 outlined an initial model of subjective levels of the means of gaining knowledge, and Chapter 3 outlined objective levels of time and distance scales. In modern science these domains have not been integrated into an overall structure of levels of nature. Maharishi Vedic Science and Technology provides an integrated description of levels of *reality* that addresses this and many other key questions about the range and structure of the cosmos.

This chapter presents an outline of the full range of the objective and subjective structure of phenomenal nature in the holistic approach of ancient Vedic science and Maharishi Vedic Science and Technology. In this outline the simultaneity of bigger and smaller is a general characteristic of the levels of nature. It is ultimately the simultaneity of infinite nonlocality and locality, eternity and instantaneity, in the self-interacting infinity-point dynamics of the unified field. This understanding is based on direct experience of the self-referral dynamics of the unified field within the inner laboratory of the minds of Vedic scientists who practiced the systematic subjective approach of gaining knowledge in ancient Vedic science.

Beginning with the undivided wholeness of the unified field, the holistic approach describes phenomenal creation as progressive limitations of infinite unity into relative manifest levels of diversity. Some of the levels relate to the fundamental force fields identified in modern science; however, a much more comprehensive range is included. Importantly the subjective and objective domains are completely integrated, and the process of knowing and knower—not just the objective known—are accounted for in the structure. In the structure of the levels of subjectivity and objectivity, the linkage can be explained between outer objects in nature and their corresponding inner qualitative experiences or *qualia*—how experienced objects happen to 'feel like' they do, as they seem to do for most all of us.

An important Vedic reference for the structure and range of levels of nature is *Sankhya*. Sankhya refers to the principle of *enumeration* inherent in the unified field that structures categories, components, or levels and their ordered sequence in phenomenal creation. The Sankhya system describes the phenomenal process of manifestation as a systematic unfolding of potentialities into actualities of levels of nature according to deterministic laws.[1][2] A similar description of the structure and range of natural law is contained in the aspect of Vedic literature called *Ayur Veda*, as referenced for example in *Caraka Samhita*.[3]

The Sankhya system in the Vedic literature is one of the six *Darshanas*—perspectives from which the totality of nature can be viewed. The Darshanas are sometimes associated with different experiential *realities* in various states of consciousness. Each of the six Darshanas contains the totality of knowledge reflected in the full range of experience in higher states of consciousness, and also brings out a particular view. They are also sometimes referred to as the six systems of Indian philosophy. In this book the Sankhya system is discussed because its systematic enumeration or delineation of ontological levels of nature is useful for connecting with, clarifying, and completing the picture that has been unfolding in modern science. As used in this chapter, however, the specialized perspective of the Sankhya system also draws from general perspectives reflected in the other Darshanas.

For example the Sankhya system is sometimes interpreted as describing the fundamental duality of pure existence and pure intelligence (*Purusha* and *Prakriti*) as a contiguous duality, which is an important experience in one of the states of consciousness described in ancient Vedic science. In order not to lose the fundamental unity or non-duality of nature from the perspective of the ultimate unified field, however, in this chapter the duality is also connected to the completely unified perspective. Thus pure existence and pure intelligence are described in terms of the self-referral unity of infinite silence and infinite dynamism that is associated more with the non-dual perspective of totality described in the Darshana of *Vedanta*. This point about the application of Sankhya in this book is included for those who are familiar with the Darshanas, and otherwise can be considered a technical point.

In enumerating and superimposing phenomenal divisions of levels onto the indivisible, non-dual wholeness of the unified field, it is useful to appreciate again that the levels remain within the unified field and are nothing other than the unified field interacting with itself. The relative manifest levels can be thought of as modes of excitation or vibration of the unified field. But it is not that they are expressed outside of the unified field or that they become something other than the unified field. They are limited expressions or derivatives within the unified field—specific levels of phenomenal *reality* appearing as relatively independent processes and objects in manifest creation, according to the state of consciousness and ability of perception of the observer. The degree of *reality* attributed to the levels is related to the perspective, and more fundamentally the state of consciousness, of the observer. The different perspectives or *realities* associated with the full range of states of consciousness will be the focus of Chapters 19-20, and then the different states of consciousness will be described in some detail.

The Inner Dimension

The overall structure of subtle levels of natural phenomena is characterized by what might be called a vertical dimension of inner depth, or the *inner dimension*. It is not vertical in just the sense of up and down or higher and lower, although these descriptors will later be shown to have relevance. It is also a dimension of outer and inner, shallower and deeper, grosser and subtler. The concept of subtlety is meant both in a qualitative sense of refinement, as well as in a quantitative sense of larger and smaller time and distance scales (refer to Chapter 8). The levels in the structure can be thought of as linear, sequential, with subtler levels encompassing grosser levels—hierarchically embedded in a non-linear, simultaneous, indivisible wholeness.

The ordered structure doesn't just mean levels like a multi-level office building, but also like levels of the ocean which express different qualities—from the choppy shallow surface to the silent motionless depths. Another meaning is added by surface/core, or peripheral/central, as in levels or strata of a golf ball or Earth. However, somewhat paradoxically the core of the inner dimension is bigger than the outer surface—indeed infinite, while the surface appears to be finite.

Still another important connotation is reflected in terms such as concrete/abstract, tangible/intangible, coarse/fine, and crude/refined—related to gross/subtle. In a similar way that water is in some sense subtler or less coarse than rock, air than water, and space than air, deeper levels are subtler or finer than surface levels. (Interestingly, these also can be related quite tangibly to the simultaneity of smaller and bigger—in the sense that the biggest mounds of earth, mountains, are smaller than the ocean, which again is smaller than the atmosphere, the air, which again is smaller than space. But these are all at the same ontological level, the gross relative domain of nature).

The inner dimension also relates to degrees of potential and power—similar to the gross sensory level of ordinary objects which contains deeper cellular, molecular, atomic, sub-atomic levels and the immensely powerful underlying quantum fields. In addition diversity is more pronounced toward the surface where differences between things are more apparent, and unity is relatively more prominent at the depths where everything ultimately contracts into oneness. From the deepest to the surface levels, nature appears to become more concrete, specialized, diverse, limited, structured, constrained, and less powerful in its phenomenal expression.

The levels also can be thought of as concentric, one completely engulfing the other—from transcendental Infinity to the subtle nonlocal levels, to the gross local levels. It can be likened to sheaths or levels of density of mediums or aethers, with subtler levels completely permeating the grosser levels, and the grosser levels emerging ontologically from the more abstract subtle levels.

A crucial point is that the phenomenal manifest levels of nature can be said to be limitations, restrictions, or partial realizations of the unified field. The unified field contains all possibilities, all abilities, all powers, all creativity, all intelligence, all order. It is not just a static homogeneous ground state that requires something more than itself to know or express itself, as implied in a reductive perspective. All phenomenal manifest *realities* are partial reflections of the *total reality* of the unified field. As Maharishi states:

"Nothing is impossible for total natural law." [4]

Phenomenally the subtlest manifest levels most fully reflect the total potential of natural law, and progressively grosser levels reflect lesser degrees. The full range is from the total potential of natural law in the self-referral dynamics of the *transcendental* domain of the unified field to increasingly limited expressions in the *subtle relative* domain, to the least expressed potential at the most concrete, most limited *gross relative* domain. The grossest layer of the gross level includes ordinary rocks and earth, in which the total potential of natural law is most hidden. However, the total potential of the unified field necessarily would be present at every level, even the grossest surface level. This level is nothing other than the infinite potential of the unified field, but simultaneously also in its most restricted phenomenal expression—from the perspective of ordinary experience, completely localized, discrete, and inert.

Thus the universe can be said to be closed and open at the same time—phenomenally closed on its crusty material surface and infinitely open at its foundation. The phenomenal structure of relative creation is that, going from gross to subtle, it becomes smaller and smaller—more fine-grained—while at the same time becoming bigger and bigger—more expanded and nonlocal—until it comes to the ultimate degree of smallness of a point and the ultimate degree of infinite extension, at the same time. But also, it is infinite at every point of every level. As a simple analogy, a wave is bounded on the surface and unbounded at its deeper basis, while all the time remaining the entire ocean at every level. In a sense we can say that inside each wave is the entire ocean, and the entire ocean makes up every wave.

Thus the inner depths of the universe are much bigger than the outer surface. Although the universe's gross outer surface with its immense expanse of galactic clusters appears quite vast, its subtle inner layers are much vaster, and its innermost core is absolutely infinite—permeating even its gross outer surface. The fundamental seamlessness of creation is that all phenomena are nothing other than fluctuations, vibrations, or manifestations of the unified field reverberating within itself. When that indivisible unified field appears to manifest into the diversity of phenomenal creation, then it takes on various relative limitations that divide it up into levels and parts. These phenomenal levels are structured into an orderly sequence from the most to the least like the undivided wholeness of the unified field. In Sankhya the undivided wholeness of the unified field is *enumerated* into 25 phenomenal categories or levels that cover the full range from transcendental to subtle to gross domains. Maharishi states:

> "In its analysis of life and creation, Sankhya establishes 25 categories as lying at the basis of the entire creation and of the process of cosmic evolution." (p. 480) [1]

Two of the levels are associated with the unmanifest *transcendental* domain; eighteen are associated with the manifest *subtle relative* domain; and the remaining five are associated with the manifest *gross relative* domain. This is associated with the consciousness-mind-matter ontology. The levels fit into a chain of causation consistent with the logic that something cannot come from literally nothing. The most fundamental level is uncaused but is the cause of all more expressed levels. Each level is classified in terms of whether it is uncaused or caused, and whether it causes or doesn't cause other levels or modes of being. Levels that are caused and that cause other levels are sometimes referred to as *evolvents,* and levels that are caused but don't cause other levels are sometimes called *evolutes.* An evolvent is a level of nature that serves as the basis for additional ontological levels to emerge, and an evolute is a level of nature from which no additional ontology emerges.[2] However, evolutes have innumerable permutations from which different properties and attributes emerge in the same ontological level.

The full range of levels is verifiable through direct empirical experience when the mind of the investigator has developed a high degree of coherence or subjective consistency and is free enough from the overshadowing, distorting influence of stress and strain to experience them clearly. In the ordinary waking state of consciousness, closely associated with the physicalist worldview, this overshadowing influence restricts most empirical experience to the macroscopic level of the gross relative domain only. With increasing clarity of deep inner experience the full range of levels or categories are unpacked, disentangled, and identified—while at the same time in higher states of consciousness their fundamental unity at every level of nature is also open to spontaneous experience. Generally, development through higher states of consciousness also simultaneously involves increasing unity and increasing clarity of diversity. Maharishi points out:

> "The teaching concerning these categories can be verified by direct experience through the practice of Transcendental Meditation, in which the mind travels through all the gross and subtle levels of creation to the state of pure transcendental consciousness." (p. 480) [1]

If the phenomenal structure of levels is not clearly experienced and delineated, it can lead to conceptions such as of an *entangled hierarchy*, or *world knot*.[5] This is the current state of modern science with respect to the phenomenally gross, subtle, and transcendental levels. The physicalist view has been almost entirely focused on attempting to identify the structure of the gross material domain. In this paradigm the subtler subjective domain is thought to be a product of the gross levels. Because of this restricted understanding and experience, it has been difficult to make progress on delineating levels and relationships of objective and subjective *reality*—let alone experiencing their ultimate underlying transcendental basis. However, important progress is being made, as summarized in past chapters of this book, especially with the advent of unified field theory.

Singularity, Duality, and Trinity within the Transcendent

For purposes of explaining the process of manifestation of phenomenal creation, the undivided wholeness of the unified field—the ultimate singularity—is intellectually analyzed into parts. It is the nature of the intellect that there is an inherent implication of duality in the concept of singularity or oneness. In conceptualizing, identifying, or labeling anything, there is a tacit implication of some relative contrasting principle. In identifying singularity, a contrasting non-singularity is implied.

In Sankhya the most fundamental enumeration is into the duality of *Purusha* and *Prakriti*. This duality within the unmanifest, completely unified, transcendental domain of the unified field can be thought of as the self-referral coexistence of opposites of infinite silence and infinite dynamism. But this is an intellectualization into duality of the infinite, eternal, non-dual, indivisible unity. As Maharishi recently stated:

"Silence and dynamism—they are one thing, not two things." [6]

Purusha. The first level, Purusha, is described as the eternal *silent witness*—the absolute non-changing background of all creation. It is identified as the source of consciousness, or consciousness itself, and as Maharishi states, "the basis of the subjective aspect of life." (p. 481) [1] It is uncaused and doesn't directly cause anything else; but it is the basis of the orderly manifestation of phenomenal creation, as well as the eternal persistence of the essence of being.

"Purusha, or Cosmic Spirit, is the transcendental Reality which comes into direct experience during Transcendental Meditation at the point where even the subtlest level of creation is transcended and pure transcendental consciousness alone remains." (p. 481) [1]

From the perspective of duality, Purusha can be associated more with infinite silence than infinite dynamism. As the basis of consciousness, Purusha can be described simultaneously as causal of all phenomenal existence, and epiphenomenal as the 'silent witness.' Alternatively everything else can be described as epiphenomenal.

Prakriti. The second level, *Prakriti*, is described as Nature, or cosmic substance, "the primal substance out of which the entire creation arises." (p. 481) [1] Prakriti comes from the roots *pra*—before or first—and *kr*—to make or produce. It can be identified as the self-interacting basis from which all phenomenal creation is produced and returns. It is described as unmanifest pure potentiality, but also is sometimes described as the first manifestation in creation in the sense that it is the totality of everything in creation in its potential state of abstract self-interacting vibration, prior to manifestation. From the perspective of Prakriti as the basis of the existence of creation, it is described as not conscious other than in its relationship to Purusha; it thus is sometimes described as the basis for the *objective* aspect of life. However, Purusha and Prakriti are prior to the phenomenal distinction of subjective and objective. Prakriti is uncaused, and can be said to be the cause of phenomenal existence. From the perspective of duality, it is associated with infinite dynamism, identified as the Veda itself, the source of the laws of nature. Perhaps a way of getting a deeper sense of the *unity-duality* concept in the self-referral, self-interacting dynamics of the unified field is in the distinction of infinity and point. On the one hand it can be likened to infinity contracting to a point of singularity (neither an exact depiction nor drawn to scale!).

On the other hand, because the unified field is self-interacting, at the same time it also can be conceptualized as infinitely expanding from point to infinity.

The range from infinity to point and point to infinity incorporates all possibilities—everything and nothing—all possible change, including non-change. Conceptually the nature of infinity is that every point contains the entirety of infinity. The point value contracts (or expands) the entirety of infinity into it, and the infinite value expands (or contracts) every point into it, at every point. At that infinitely self-interacting level, every instant is eternity, and eternity is instantaneous.

Because both infinity and point are beyond the relative limitations of time and space, we can conceive of instantaneous reverberation, vibration, or pulsation from infinity to a point and back to infinity. It is sort of like taking attention from the whole ocean to an individual wave and then back to the whole ocean at infinite speed, such that simultaneously both the entire ocean and individual wave together are lively in attention, superposed in each other—when all the time the wave is nothing other than a phenomenal part of the ocean. Simultaneous instantaneous pulsation from infinity to point and back again, eternally self-referring back and forth infinitely fast, is a way to picture infinite dynamism, infinite creative potential, *infinite organizing power*. But infinite dynamism also implies infinite silence at the same time—complete immovability, complete non-change, nothing always happening.

Both point and infinity, contracting and expanding, separating and conjoining, dividing and uniting, diversifying and unifying, can be said to arrive at the same oneness. Unifying more and more results in ultimate Oneness; and dividing more and more, to the infinite degree, transcending any degree to divide further, also results in Oneness. This can be described in terms of Infinity being both infinite expansion to unboundedness and infinite contraction to singularity, as well as of the point contracting beyond infinitesimal magnitude to infinite singularity, or expanding to infinity of points—simultaneously complete reductivism and complete holism. Taking it a step further, imagine the unimaginable event of infinite dynamism and infinite silence, infinite pulsation or liveliness of infinity to a point and back, never going anywhere or doing anything, expanding more. Because it already is infinitely expanded and infinitely contracted simultaneously, for heuristic purposes we could say that the only 'place' it could *appear* to contract or expand into would be into a finite value, into phenomenal manifest existence.

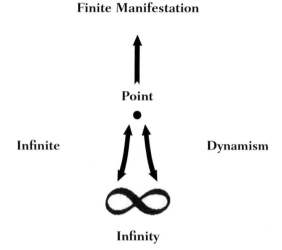

Once again, imagine the unimaginable event of infinite dynamism and infinite silence contracting more. We could say that the only 'direction' it could expand or contract into would be complete stillness, absolute non-change—eternal infinite silence.

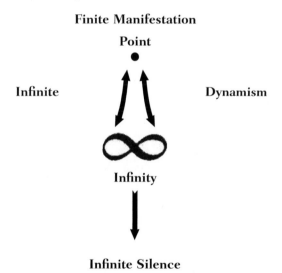

As a final wholeness point, the graphic in this example may be misleading. The *point* depicted refers to an infinite singularity or wholeness of everything; and the depiction of *infinity* refers to infinity of points—infinite self-referral of wholeness and part, point and Infinity. Again for heuristic purposes we can say that the infinite silence of Purusha goes beyond any duality whatsoever, beyond existence and intelligence, beyond even the singularity of unity or ultimate Oneness. It is beyond any conception or experience of infinite and finite, cause and effect, present, future, and past—beyond the *face of the deep,* beyond the *lamp at the door* of consciousness, of infinitely dynamic silence into infinite eternal silence. In the chapter on higher states of consciousness, subtleties of 'experience' associated with these self-referral dynamics within the unified field will be discussed. Maharishi provides a clear summary of the infinity-point duality in the ultimate unity of Infinity:

> "Infinity, fully awake to itself, is fully awake to its infinite value. At the same time, it is awake to its point value. In this we find the dynamism of infinity converging to a point and a point expanding to infinity. This [is the] infinite dynamism of the self-referral nature of pure consciousness..." (pp. 65-66) [7]

The unmanifest trinity. Implicit in the concept of duality in the spontaneous phenomenal unfoldment of creation from unity to diversity is *trinity*. Conceptualizing ultimate Oneness as the unmanifest duality of Purusha and Prakriti naturally implies a third aspect—namely, their relationship—from which arises the fundamental trinity of knower, process of knowing, and known appearing throughout this book, and throughout phenomenal creation.

Sankhya describes the self-referral, self-interacting dynamism within the unmanifest duality of Purusha and Prakriti as three-in-one dynamics. This fundamental three-in-one dynamic is the unmanifest structure of Prakriti, or Nature. The unified or 'one' aspect of the three-in-one relationship—the unity of knower, process of knowing, and known, the nature of consciousness—can be associated with Purusha. The potential of the diversified 'three' aspect of the indivisible three-in-one relationship can be associated with Prakriti. However, at the level of Prakriti these three basic aspects of Nature are said to be in perfect balance with respect to each other. When they are in perfect balance and there is no relative disturbance between them, they remain unmanifest potentiality. When the balance appears to be disturbed and there is relative activation between them, then relative motion and change begin and the unified field can be said to manifest into progressive, relative, limited levels of subtle and gross creation within itself. In other terms the ultimate perfect symmetry is spontaneously broken into phenomenal manifestation as the relative field of diversity.

The three basic aspects of Prakriti or Nature can be associated with knower, process of knowing, and known. In their most fundamental and abstract sense as the 3-in-1 self-referral, self-interacting dynamics of Prakriti or Nature, they can be related to the terms *Rishi*, *Devata*, and *Chhandas*; and their unified state is called *Samhita*. Maharishi describes this in the following quote:

> "...[K]nowledge means VEDA, unity is called SAMHITA, knower is called RISHI, process of knowing is called DEVATA, and the known is called CHHANDAS." (p. 72) [8]

In their capacity to change from unmanifest potentialities into subtle actualities as phenomenal manifestation begins, they are termed the *Gunas*—derived from the concept of a strand, twirl, or wind of thread, cord, or string. The three gunas are *Sattva*, *Rajas*, and *Tamas*. Each has specific potentialities, qualities, forces, or powers that unfold in various permutations and combinations into the vast diversity of creation—discussed in more detail in the next chapter.

The three gunas are fundamentally inseparable, and coexist in every interaction or transformation at all of the levels of phenomenal manifest creation. They can be said to be either manifest in their activated relative states or unmanifest in their infinitely self-interacting state, but they can be said to never lose their power or become non-existent. They correspond to the basic dynamical principles of the creative, maintenance, and destructive or dissolution operators that carry out all phenomenal change throughout nature.

From one perspective these three can be understood as having the respective qualities to illuminate, to activate, and to hide or obscure. From another perspective sattva guna and tamas guna can be understood as principles or forces of expansion and contraction activated by rajas guna. Their essential self-interacting dynamic is the basis for the infinite diversity of patterns of motion and pause, action and inaction, go and stop, dynamism and silence, day and night, life and death, manifestation and dissolution that can be witnessed in all levels of phenomenal nature.

Sattva guna is the force in nature associated with the concept of *essence, purity,* or *illumination*—it *reveals* the essential nature of life. It is *uplifting,* on the subtle mental level in the sense of increasing purity, lightness of being, refinement, and reduced dullness; and on the physical level both in the sense of illumination or light, refinement, reduced heaviness or density, and upward or ascending movement such as of smoke from fire. It is responsible for equanimity and balance that brings fullness or completeness, rather than activity or inertia. It is associated with the maintenance operator, upholding and uplifting the newly developing state in all transformations or change in nature. It also can be associated more with the concept of the subject or knower, as well as the principle of infinite dynamism.

Rajas guna is the force in nature associated with the concept of *energy* or *activation.* It is the force or power that does work, that drives activity. It is responsible for activating the other two gunas to produce all motion throughout manifest creation; and thus is associated with the creative operator, impelling expression, manifestation, or change in nature.

Tamas guna is the force in nature associated with the concept of *inertia* or *restraint*; it restrains the other two gunas from revealing and activating. It binds and resists motion or change, on the material level in the sense of inertia, mass, and related downward or descending movement such as of water. It restricts motion or change in all transformations in nature. As the destructive or dissolution operator, however, it also has the important effect of eliminating incompleteness. Ultimately it also can be associated more with the principle of infinite silence.

In the overall structure of manifest creation, different levels express different relative percentages of the gunas. Generally with activation by rajas guna, sattva guna predominates in the subtle relative levels of creation, and tamas guna predominates in the gross relative levels. Importantly, however, there is a natural optimal relative percentage of the three gunas at every level of manifestation and in every transformation throughout nature. This relative balance supports progressive change—according to the law of least action—in the integrated process of dissolution of a state, maintenance across states, and creation of a new state.

Activated by the force of rajas guna, generally sattva guna can be described more in terms of conjoining, contracting, or unifying, and tamas (and also rajas) guna in terms of separating, expanding, or diversifying—although they also can be viewed in the opposite perspective depending on the particular process under consideration. Together they automatically carry out all the natural cycles of activity and rest, pleasure and pain, and life and death throughout phenomenal creation. Purusha and Prakriti with its three gunas make up the entire creation. Regarding the fundamental role of the gunas, Maharishi explains:

> "The entire creation is the interplay of the three gunas. When the primal equilibrium of sattva, rajas and tamas is disturbed, they begin to interact and creation begins. All three must be present in every aspect of creation because, with creation, the process of evolution begins and this needs two forces opposed to each other and one that is complementary to both... Sattva and tamas are opposed to each other, while rajas is the force complementary to both. Tamas destroys the created state; sattva creates a new state while the first is being destroyed. In this way, through the simultaneous processes of creation and destruction the process of evolution is carried on. The force of rajas plays

a necessary but neutral part in creation and destruction; it maintains a bond between the forces of sattva and tamas. Thus all three gunas are necessary for any state of manifested life... Mathematically, the gunas may combine with each other in six possible ways:

1. Sattva dominates, rajas is secondary
2. Sattva dominates, tamas is secondary
3. Rajas dominates, sattva is secondary
4. Rajas dominates, tamas is secondary
5. Tamas dominates, sattva is secondary
6. Tamas dominates, rajas is secondary.

Combinations 2 and 5 are not possible because of the contrast in the nature of sattva and tamas. Thus the three gunas have only four possible combinations... This is the fourfold order in creation. Every species, whether vegetable, animal, or human, is divided into four categories according to the four divisions of the gunas, which determine the natural mode of activity of each category." (pp. 269-270) [1]

> "Once again, it seems that there is a threshold for complex behavior—that is reached as soon as one has at least four states... [A]dding more complexity to the underlying rules does not yield behavior that is ultimately more complex."—*Stephen Wolfram, computer scientist (p. 79)* [9]

A major point to be made here is that the gunas can be understood to represent the most fundamental forces of nature. Their first phenomenal manifestations are the most abstract forces or potentialities in subtle relative creation, prior to the manifestation of objects in nature. At this level, the gunas are impulses of intelligence that comprise the fundamental psychological and energetic forces operating in mental processes such as love and indifference, harmony and conflict, motivation, affect, and cognition. They further manifest into the fundamental physical attractive and repulsive forces or energies that are the basis of all change in physical objects associated with our ordinary empirical world. They also can be related to the three categories of the manifestation of time—present, future, and past—as well as the three spatial dimensions—the x, y, z axes or up/down, forward/backward, and right/left.

This description of the unmanifest balanced state of Prakriti and its phenomenal manifest expression in sattva, rajas, and tamas guna also can be compared to the four laws of thermodynamics. In this comparison, the zeroth law of thermodynamics can be compared to the balanced state of the three gunas, the 1st law with rajas guna, the 2nd law with tamas guna, and the 3rd law with sattva guna. On the subtle relative or subjective domain, these principles reflect dynamic principles of change, but not *thermo*-dynamic principles of change characteristic of the gross relative domain associated with Einstein locality, the speed of light, and particle interactions.

In Maharishi Vedic science, subjective mind and objective matter are completely integrated and complementary. Mind is more fundamental than matter, but both operate according to the same laws of nature at different degrees of abstraction and subtlety. This provides the basis for understanding the relationship between objects in nature, their corresponding subjective experiences or qualia, and consciousness itself underlying subjectivity and objectivity—to be discussed in the next part of the book.

All phenomena in creation are due to the self-interaction of Prakriti—in its unmanifest state, the infinite, eternal, omnipresent, omnipotent, omniscient unified field of natural law. Purusha is the fundamental animating principle that can be associated more with infinite silence, and Prakriti is the fundamental cosmic substance that can be associated more with infinite dynamism. Both together can be said to create and permeate the entire phenomenal universe. As the basis of the 3-in-1 self-referral state of consciousness itself, Purusha is the eternal, uninvolved, non-changing silent witness—the knower—that never ceases to be and is never completely overshadowed or veiled by the phenomenal expressions of the eternal cosmic substance of Prakriti. Maharishi quotes the Bhagavad-Gita on this point:

"Guna gunesha vartanta iti matwa na sajate. (Bhagavad-Gita, 3.28)

The three Gunas...behave amongst themselves, and the Samhita level remains uninvolved with their behaviour." (p. 540) [10]

As to how Purusha and Prakriti relate to the unified field described in Chapter 10, we can say that the unified field is a term associated with the unified value of Purusha and Prakriti, from the perspective of manifest diversity being unified at its source. It is a general term that includes both the unmanifest unified (samhita) value of Prakriti along with the uninvolved Purusha value (although sometimes the Prakriti value is a bit more emphasized in the context of objectified physical science). Although Purusha/Prakriti is ultimately unity, in order to explain phenomenal manifest diversity the duality of Purusha and Prakriti is introduced. In order to explain the apparent qualities of the unified field as simultaneously unmanifest and containing the potentiality and actuality of manifest creation, Purusha and Prakriti are described as contiguous. But the contiguity is a phenomenal *reality*. Ultimately it is the non-dual universal Self—Totality.

Levels of the Subtle Relative

Mahat. The next level of the 25 levels enumerated in Sankhya is called *mahat*—literally the 'great principle,' sometimes characterized as *cosmic intelligence*, or the *principle* of intellect. This is associated with the first motion toward manifest relative creation from the state of perfect balance of Prakriti—into the relative interaction of the gunas of sattva, rajas, and tamas. It is the subtlest relative expression of the gunas, where the revealing, illuminating force of sattva guna most predominates by virtue of the disturbance of perfect balance due to the activating force of rajas guna. It can be described as almost pure sattva, and as associated with the experience of bliss. It is the beginning of the differentiated attributes of the gunas, when the self-interacting dynamism of Prakriti, or Nature, is first expressed in relative phenomenal manifestation. At this level the perfect balance begins to be hidden by tamas guna, and the eternal infinite appears to take on asymmetries and directionality in terms of relative time and space.

Mahat can be characterized as the universal or cosmic level of intelligence, the manifest seed form of intelligence and orderliness that structures the direction of evolution of the entire universe. It is described as the cosmic will in operation, taking a specific direction toward manifestation.[1] It is the laws of nature in their first stage of phenomenal manifest expression. It is sometimes compared to the swollen state of the ocean's surface prior to the formation of a wave.[2] Maharishi further describes this first expression into the subtle manifest creation in the following quote:

"The first manifestation of creation is the self-illuminant effulgence of life. This is the field of established intellect, or the individual ego in its own established state. This self-illuminant effulgence of life is called the Veda. The second step in the process of manifestation is the rise of what we call vibration, which brings out the attributes of prakriti, or Nature—the three gunas. This point marks the beginning of the *functioning* of the ego. Here experience begins in a very subtle form: the trinity of the experiencer, the experienced, and the process of experience comes into existence. This is the beginning of action in the process of creation. Just before the beginning of action, just before the beginning of the subtlest vibration, in that self-illuminant state of existence, lies the source of creation, the storehouse of limitless energy. This source of creation is the Veda..." (p. 206) [1]

Mahat also is characterized as the level of the cosmic *I* in its own manifest existence as a universal field of integrated wholeness—the universal *I exist*, the universal *I am*, without the distinct sense of any other thing existing. In Maharishi Vedic Science, because of its functional role as the basis of individual experience, this level is sometimes identified as the *cosmic ego*. This will be discussed in the chapter on the levels of subjectivity associated with individual mind. Mahat is classified as an evolvent: it is caused by Prakriti, and it causes more expressed manifestations or ontological modes of being.

Ahamkara. The next level enumerated in Sankhya is called *ahamkara*—which can be interpreted to mean *I do or perform*. It is characterized as the *individuating principle* in nature, "the principle responsible for the individuation of mahat." (p. 482) [1] It is the level at which the universal *I* or cosmic intelligence as *subject* identifies its own cosmic existence as *object* at the subtlest level of manifestation. It is the phenomenal sense of *I* as doer, which begins to discriminate some other object upon which to act; it is an individuation of *amness* into the possibility for experiencing, acting, or doing that influences some other object. Ahamkara also is classified as an evolvent: it is caused, and it causes more expressed modes of being. It is the basis for the distinction of subjective and objective existence. In Maharishi Vedic Science and Technology this level is sometimes identified with the abstract principle of discrimination, and is associated with *cosmic intellect*.

As a technical point, in some references in ancient Vedic literature *mahat* is associated with intellect (*buddhi*) and ahamkara is associated with ego. In Maharishi Vedic Science and Technology, ego is more fundamental than intellect. This relates to the extremely subtle dynamics of intellect, which can be said to involve simultaneous unifying and analyzing in the act of discrimination as the background for the sense of discreteness or *individual* unitary wholeness. The levels of cosmic ego and intellect combined together can be associated with the term *buddhi*, the discriminative faculty that distinguishes phenomenal *reality* from Purusha. This point will be unfolded further in Chapter 17. In terms of physical processes in nature, buddhi also can be associated with the most abstract sense of the *quantum principle*, [II] and ultimately the conceptual duality of point and infinity.

Manas. The next category or level in Sankhya is called *manas*, from the root 'to think,' and characterized as *cosmic mind*. Maharishi explains that manas:

> "...provides the object for the individuating principle, ahamkara. In the state of manas, the urge of prakriti towards manifestation becomes clearly defined." (p. 482) [1]

Manas is the level of nature that can be described as the subtle subjective platform for experiencing other objects and processes that are limited aspects of the universal *I* or experiencer. In terms of individuality it can be related to the concept of the *'screen of the mind'* upon which experiences of other objects and processes become clearly represented, and with which to various degrees the individual *I* of ahamkara identifies in the progressive manifestation of subject-object relations. Manas is classified as an evolute: it is caused, but does not cause other more expressed manifestations to come into being ontologically.

In this sense manas is one of the levels of subjectivity. However, all the levels of subjectivity taken together are commonly identified by the general term 'mind.' To elaborate a bit more on this point of language, with respect to the levels of subjectivity manas is identified specifically as the level of mind. However, the term mind also commonly is used to refer to all the levels of subjectivity of the subtle relative domain—cosmic mind in the sense of cosmic subjectivity (*citta*). This general use of the term mind is to distinguish it from the cosmic body and the transcendent *universal Self* (used to refer to the level of the transcendental domain only, but also sometimes used to refer to all three domains as a unified whole inasmuch as all the levels are contained within it and there is no thing outside of it). Thus the term manas refers to the specific meaning of mind as one of the levels of subjectivity, in contrast to its common general use as all levels together.

The sequential relationship of mahat, ahamkara, and manas—cosmic ego, intellect, and mind—is associated respectively with relative degrees of sattva guna, rajas guna, and tamas guna, within the overall subtle level in which sattva guna predominates. In this sense cosmic ego is more closely related to the knower, cosmic intellect with the processing of knowing, and cosmic mind with the known—in conjunction with the existence of 'other' objects of experience in the phenomenal manifest duality of subject and object. The fundamental patterns of singularity, duality, and trinity reverberate through all levels of phenomenal creation.

Indriyas.   The next ten categories or levels of nature are collectively called the *indriyas,* meaning capacities or abilities. They include the five basic abilities to sense or know and the five basic abilities to do or act. The sensory or knowing capacities, the *gyananendriyas,* are the five sensory processes. These include the power to hear (audition, *srota*), to touch (somatosensory ability, *tvak*), to see (vision, *caksus*), to taste (gustatory sense, *rasana*), and to smell (olfactory sense, *ghrana*). The capacities of action, the *karmendriyas,* include the power to express (*vak,* such as speech), to grasp (for example, with arms and hands, *pani*), to move (motility such as, for example, with legs and feet, *pada*), to execute (such as rejection, excretion, and elimination, *payu*), and to generate (such as procreation, *upastha*). The indriyas are not the physical instruments through which these powers of sense and action are expressed, such as ears or arms. The indriyas are functional capacities associated with manas, the level of mind in which the senses unite. The indriyas are classified as evolutes: they are caused, but don't cause other ontological levels.

At this subtle level of phenomenal manifestation, physical objects have not yet emerged. However, the subjective capacities to sense and to act imply corresponding objects to sense and upon which to act. As these capacities manifest in nature, their corresponding objects manifest almost simultaneously with them. The fundamental division of the three gunas then becomes expressed in terms of manas, the mind that collects together sensory input and coordinates output; indriyas, the organs of sense and action that draw input and express output; and the objects that are sensed and acted upon in nature—the *tanmatras* and *mahabhutas* (the next levels). Both the senses and their corresponding objects are sequential with respect to degree of subtlety. The degree of subtlety from most to least of the five senses, for example, is in the sequence of hearing, touch, sight, taste, and smell. Each correspond to increasing degrees of limitation, restriction, and concreteness associated with their corresponding phenomenal objects.

Tanmatras.   The next five categories or levels are collectively called the *tanmatras,* the subtle essences, constituents, or elements that make up the subtle objects of creation. Directly corresponding to the five sensory capacities (indriyas), they include the essence of sound (*shabda,* hearing), touch (*sparsa*), form (*rupa,* sight), flavor (*rasa,* taste), and odor (*gandha,* smell). The five-fold tanmatras are very abstract energy patterns or processes—also called impulses of intelligence—that combine to construct subtle phenomenal forms of manifest existence. They represent the most expressed levels of the subtle relative domain of nature, associated generally with the subtle subjective level as distinguished from the gross objective level. The tanmatras are classified as evolvents: they are caused and they cause more expressed ontological levels.

The tanmatras are sometimes referred to as the subtle essences of the elements of space, air, fire, water, and earth—which may not give a proper sense of their characteristics or functional meanings. The tanmatras are not material elements and are not made of discrete quantized particles associated with ordinary classical concepts of mass and inertia or of quantum concepts in the gross relative domain. They are very abstract nonlocal impulses or waves of intelligence and energy that comprise subtle objects or forms; they permeate gross matter but generally are not detectable within the domain of objective material creation.

Being in the subtle relative domain, the tanmatras reflect a higher degree of sattva guna and its attribute of illumination or light—but not in the sense of the grosser electromagnetic fields and photon particles of light. They have local properties, not as particles, but more in the sense of individual waves in an unbounded ocean or field. They can be said to be 'objective' in the sense that they build into subtle forms or objects that the senses and mind experience. But they are not 'objective' in the sense of the gross material domain of our familiar objective material world. With respect to defining the levels of objectivity and subjectivity, Maharishi explains that the tanmatras:

> "...mark the dividing line between the subjective and objective creation. In the process
> of evolution, as the influence of tamas increases, the subjective creation comes to an end
> and the objective creation begins. The tanmatras, forming as they do the basis of the five
> elements, lie in the grossest field of the subjective aspect of creation." (p. 483) [1]

In this sense the subjective aspect of creation refers to the subtle relative domain. In this domain, sattva guna—associated with impulses of intelligence—generally is more prominent than in the gross objective level dominated by tamas guna and associated with inert matter with mass and magnitude in ordinary conventional space-time. From the five tanmatras, through the increase of tamas guna emerge the gross elements of nature—the *mahabhutas* (discussed next)—which complete the full range of nature from transcendental to subtle to gross.

To summarize the levels thus far, in Sankhya the unmanifest transcendental level of nature can be understood to be comprised of the non-dual duality of Purusha and Prakriti. The eighteen levels of the manifest subtle relative domain of nature include mahat (cosmic intelligence), ahamkara (individuating principle), manas (cosmic mind), the ten indriyas (sensory and action capacities), and the five tanmatras (subtle essence elements). This subtle relative domain of nature has been completely overlooked in the physicalist worldview common in modern science. The primary focus in modern science has been the gross relative domain, and more recently also—at least conceptually—the unified field itself with respect to a 'theory of everything.' However, important modern scientific theories and research are beginning to glimpse the subtle, nonlocal, interdependent levels of nature—introduced in previous chapters in the context of *quantum reality* and *quantum mind*, and discussed in more detail in later chapters.

Levels of the Gross Relative

Mahabhutas. The remaining five of the 25 levels enumerated in Sankhya are collectively called the *mahabhutas*. The term mahabhuta is from *maha* (great, universal), *bhu* (curving back, giving form, to happen, occur, exist; *bhut* (creation), and *ta* (finished, created). The mahabhutas are frequencies or vibrations of Prakriti or Nature in its grossest, most concrete, objective phenomenal expression. They are the gross elements or constituents that make up the gross objects of phenomenal experience. They are the five basic constituents or elements that manifest from the five subtle essence elements (tanmatras). They combine to form all physical or material objects that directly correspond to the sensory and action capacities (indriyas).

The mahabhutas are the gross constituents of objective material creation. They are associated with the classical concepts of space, air, fire, water, and earth—but again this terminology can be interpreted in a simplistic and misleading manner. They refer to the abstract structures of natural law that organize the gross domain into objects with the properties of vacuity, mobility, luminosity, liquidity, and solidity that manifest as forms of conventional space, air, fire, water, and earth. They comprise the entire manifest gross relative creation that most closely corresponds to the objective level of nature investigated in the physical sciences. All objects in the gross relative domain are combinations of the mahabhutas or gross elements of nature, which compose the ultramicroscopic, microscopic, macroscopic, and ultramacroscopic objects of our ordinary waking state experience studied in the objective approach of modern science. The mahabhutas are classified as evolutes: they are caused, but don't cause other ontological levels or modes of being. They combine in innumerable ways, from which emerge new patterns and effects throughout the material universe; but no new and different ontological levels of existence are *created* from them.

Reflecting the integration of objective and subjective aspects of nature, in Sankhya the important attributes that delineate five mahabhutas—and five tanmatras—are directly tied to their properties as objects of sensory experience. The constituents from which objects are built directly correlate with their subjective perceptibility—as the *qualia* of subjective experience. This is in contrast to the modern physical analysis of material objects in terms of their atomic and subatomic structural features, such as in the standard periodic table. The three subtle aspects of subjectivity—cosmic ego, intellect, and mind—along with the five gross and subtle objects of the five senses—space, air, fire, water, earth—are sometimes collectively called the *eight-fold Prakriti* or Nature, which comprises all of relative manifest existence.

The mahabhutas are functional elements characterized by specific inherent properties. They reflect the same sequence of degrees of subtlety with respect to each other as do the tanmatras and the indriyas, only in the gross relative domain rather than the subtle relative domain. Each mahabhuta precipitates from the immediately preceding one, and manifests an additional attribute or property—an additional limitation or specific quality—along with the general attributes of the others. Each more expressed level is completely encompassed in the previous level, as a further limitation that brings out an another quality due to its additional limitation. Relationships between the mahabhutas and particle-force theories from physics are discussed in Chapter 12.

Akasha (Space). From the root 'to appear,' *akasha* relates to the abstract principle of *vacuity*, and is most akin to the concept of ordinary or conventional space. Every gross object in the gross relative domain is subject to and shaped from akasha. In Sankhya, akasha corresponds to the sense of hearing—it doesn't manifest sensory attributes of the other mahabhutas in that it cannot be touched or felt, seen, tasted, or smelled. This association between hearing and space contrasts with the modern understanding that hearing is mediated by compression and rarefaction of molecules such as in air—to be discussed shortly.

Vayu (Air). From the root 'to blow,' vayu relates to the abstract principle of *mobility* or motion, and the related functions of pressure and impact, compression and rarefaction; it is most akin to the concept of air. In addition to being heard—as in the rushing of wind—it primarily has the somatosensory attribute of touch, or being felt. But it does not manifest form, flavor, or odor and thus cannot be seen, tasted, or smelled.

Tejas (Fire). From the root 'to be sharp,' tejas relates to abstract principles of *luminosity*, *form*, and *transformation*; it is associated with fire. The inclusion of the concept of fire as a fundamental element or constituent of material creation clearly suggests the more abstract functional nature of the mahabhutas, not a superficial, crude, tangible classification sometimes mistakenly attributed to primitive pre-scientific cultures. Tejas manifests the sensory attributes of sound, touch, and form and is able to be heard, touched, and seen; but not flavor or odor and thus cannot be tasted or smelled.

Apas (Water). From the root 'water,' *apas* relates to the abstract principle of *liquidity* or *fluidity.* It can be heard, touched, seen, and tasted but does not have the attribute of odor and thus cannot be smelled.

Prthivi (Earth). From the root 'broad or extended,' *prthivi* relates to the abstract principle of *solidity*; it is associated with earth. It manifests all of the five sensory attributes of sound, touch, form, flavor, and odor. It is the grossest and most limited, concrete, tangible, and inert of the 25 phenomenal levels of nature described in Sankhya. In the following quote Maharishi summarizes the subtle and gross levels of creation manifesting from the level of Prakriti:

> "Creation begins with prakriti, or Nature, which expresses itself in the three gunas, sattva, rajas, and tamas. As the process of creation continues, the three gunas manifest as 'mahat tattva', the principle of intellect. This further manifests as 'aham tattva', the principle of mind, which in its turn manifests as the five 'Tanmatras', from which arise the five senses. Then, as the process of manifestation continues, the five Tanmatras manifest into the five elements, which combine to constitute the entire objective creation." (p. 424-425) [1]

In Sankhya the entire gross relative domain or material creation is composed of the five mahabhutas or gross elements. They are direct physical evolutes of the five tanmatras or subtle elements. In the sequence of manifestation, the grossest tanmatra (gandha)—the subtle essence of earth—is further expressed into the subtlest element of the mahabhutas—akasha, or conventional space—and so on. However, it is important to emphasize that the tanmatras are subtler than any object or field in ordinary material creation, the gross physical domain.

The mahabhutas represent the grossest expression or endpoint of phenomenal creation. No new ontological level of creation emerges from the mahabhutas, although from their combinations and modifications emerge innumerable compound qualities and phenomena. All gross objects contain these five basic elements, expressed in varying degrees. They are embedded in the subtle and transcendental levels of nature, but their nature as energy and intelligence—and ultimately as nothing other than consciousness itself—is hidden on the gross level and phenomenally expressed in them the least, due to the predominance of tamas guna.

As mentioned in Chapter 4, the abstract concept of akasha or space can be understood to have meaning in each of the three fundamental gross, subtle, and transcendental domains. In the transcendental domain, it is associated with pure existence—an undifferentiated, immovable, all-pervading, infinite background continuum. In the subtle and gross relative domains, its meaning is *relative* to the limited nature of the objects in these domains. It is not so much that space has completely different fundamental levels. Rather, objects that exist and move in space can be characterized in terms of degrees of limitation or concreteness, and this can be described as conditioning the textural quality of existence as space at these relative levels. It is relativistic space-time in subtle and gross domains due to the association with the corresponding objects made of fields or aethers of energy and matter. But ultimately space and time are absolute, in the sense of the infinite and eternal continuum of the transcendental domain of the unified field.

Space is first conditioned or limited into a field within which subtle objects exist and appear—a nonlocal, interdependent, self-interacting, and unrestricted but relativistic field. Objects in this subtle field are impulses of intelligence and extremely abstract energy—they have specific forms with phenomenally localized and individual sensory qualities, but are non-material. Movement in this subtle field can be likened to the way in which waves move in the ocean, but much subtler—associated with the speed of *thought*, which is practically instantaneous, rather than the speed of light. The closest concept in modern science to this field seems to be the notion of nonlocal information space and nonlocal mind. The notion of subtle space—the tanmatra of space—is much subtler than conventional gross space, and does not have the same limitations (although it still can be said to have some limitations in that it is relative and thus not absolute and infinite; phenomenally, it is not yet the completely unified field).

At the gross mahabhuta level the space element of akasha is a further limitation of the subtle tanmatra or essence element of earth. At this stage of manifestation the principle of vacuity of akasha is conceived as having an additional, more concrete textural quality of *porosity*—a sort of foamy vacuity which may correspond to concepts in quantum gravity and string theories such as space-time foam. It is at this level that akasha can be related to the limitations of relativistic space-time associated with Einstein locality, the speed of light, and material causality. Although in modern science it is thought that space is only a relational phenomenon, it is nonetheless associated with specific, tangible, texture-like properties or limitations—characterized by Einstein locality and the speed of light. It is in this relative sense that akasha—the mahabhuta of space, the gross level of space—historically has been associated with the concept of *aether,* and also in contemporary cosmological theories with the concept of membranes.

The akasha mahabhuta is not described as having a particulate structure in the sense of quantum gravity and string theories which posit space-time as fundamentally discrete Planck-size quanta. It also is not characterized specifically as mediated by a particle such as the hypothesized graviton. In *Vaishesika,* another one of the six Darshanas in Vedic literature that identifies fundamental constituents of nature from a different but complementary perspective, there is also a delineation of the five mahabhutas. In this perspective the four mahabhutas other than akasha are identified as *paramanus,* meaning the smallest possible divisions of matter. They can be associated more fully with the concept of quantization—in the sense of particulate attributes—than can akasha. As discussed in Chapter 7-8, the difficulty of developing a theory of quantum gravity in physics may be because space most fundamentally is *not* a completely quantized field.

The mahabhuta of space is also described as having qualities such as non-interrupting, frictionless, flexible, expanding, dividing, separating, lightweight, immaterial, and the basis for vibration. The concept of vibration, fluctuation, or frequency—and the related concept of motion—is intimately related to the concept of space or aether. In Maharishi Vedic Science and Technology akasha corresponds to the auditory sense—hearing. In the human body—as well as in many organisms that have bodies built of gross matter—the auditory sense functions through the mechanical operation of the ear. Vibrations are transmitted through a medium, such as air or water, that displaces specialized receptors in the cilia of the semicircular canal built to discriminate sound qualities such as frequency, pitch, and so on. In this model sounds cannot be heard in empty space because there is no medium to transmit sound energy through space.

In Maharishi Vedic Science and Technology, however, vibration and sound are inseparable: any vibration produces a wave pattern or sound, even in so-called empty space. Space is never completely empty; it is just that the medium or aether that transmits vibration is subtler than the capability of detection. All vibration produces sound and potentially can be heard depending on the capability of the system of sensory reception. It is only due to the restrictions of its association with the gross body made of gross elements that subtle sounds or vibrations cannot be heard. In contrast, however, in the subtle operation of the senses associated with sensing objects made of the subtle elements (tanmatras) there is no such restriction, and audition is possible that does not require mechanical operation of the bodily structure of the outer, middle, and inner aspects of the ear in the gross physical body. It is in this framework that hearing corresponds to akasha or space. Obviously this has major implications for the potential range of human perception—to be discussed at length in later chapters.

In Maharishi Vedic Science and Technology the mahabhutas are sometimes described as dimensionless points, in the same sense as the classical *point particle* conception used in calculations of motion in non-relativistic and relativistic classical physics (refer to Chapters 6-8). The phenomenal manifestation of objects with magnitude or form involves the concept of volume of space, which is based on the delineation of space into the three spatial dimensions necessary to establish volume and magnitude. Superimposed on abstract pure existence is the concept of relative or conditioned space as a phenomenally manifest field—a limitation of unmanifest pure existence into fields or levels of manifest existence.

As pure existence is intellectually delineated into the three dimensions or axes, and further delineated into the more concrete concepts of vacuity and porosity, the relative concept of space becomes conditioned into a porous texture-like three-dimensional field or aether that is the mahabhuta level of space. It is the medium within which the grosser elements congeal, condense, quantize, or compactify into discrete material forms made of particles. It can be said to have distinctive qualities or limitations with respect to more concrete objects in it. It is in this sense that gross and subtle space are *relativistic*—but only gross space is associated with Einstein locality, the speed of light, and the related value of the gravitational force.

Emerging within the mahabhuta of space are the four other, more limited, more concrete mahabhutas, the paramanus characterized by having extension or magnitude in gross space. It is these four paramanus that comprise all concrete objects in the gross relative domain that can be experienced through the gross functioning of the senses. Each paramanu is characterized by an additional limitation, an additional quality or attribute that can be perceived through its corresponding sense. For example, air in the form of wind can be touched or felt but does not have form, flavor, or odor; a strawberry has touch, form, flavor, and odor.

These qualities—directly associated with subjective qualia—manifest according to the subtle elements or impulses of intelligence (tanmatras) that are active or expressed in the mahabhuta. In akasha or space, only the quality of sound (shabda) is active, and the others are dormant. The following table lists the mahabhutas in sequence of phenomenal manifestation and their corresponding sensory associations and active subtle essence elements (tanmatras):

Space (akasha) with sound (shabda)—hearing
Air (vayu) with sound and touch (sparsha)—hearing and touch
Fire (tejas) with sound, touch, form (rupa)—hearing, touch, and sight
Water (apas) with sound, touch, form, flavor (rasa)—hearing, touch, sight, and taste
Earth (prithivi) with sound, touch, form, flavor, odor (gandha)—hearing, touch, sight, taste, smell

The abstract concept of the mahabhuta of vayu (air) is described as having additional qualities such as lightness, minuteness, and flexibility. Tejas (fire) is described as hot, sharp, light, bright, and transforming. Apas (water) is attributed qualities such as lubricating, flowing, fluidity, unctuousness, coldness, sliminess, and structuring or cohesiveness. Prithivi (earth) is associated with qualities such as being heavy, rough, substantial, coarse, hard, structuring, and concrete. These qualities are expressed in all the diverse forms in nature that are composed of these gross elements. Their qualities are very important in that aspect of Vedic literature associated with natural health care, Maharishi Ayurveda—to be discussed in the next part of the book.

The senses are capable of operating at the level of gross objects or at the level of subtle objects, depending on the developmental state, level of functioning, or degree of refinement of mind and body of the experiencer. This depends on the relative balance or imbalance of the three gunas, and especially the influence of the hiding, obscuring, or restricting quality of tamas guna. It is the over-abundance of tamas guna in the mental and physical structure of the individual that restricts phenomenal experience to only the gross relative objective domain of nature. In this state, the subtle objects of sense, which naturally contain more richness, beauty, and profundity of experience, are overlooked and rarely experienced, appreciated, or even believed to exist at all.

In the overall structure of levels enumerated in the discipline of Sankhya in Vedic science, the senses from most gross to most subtle are smell, taste, sight, touch, and hearing. This means that, for example, touch is even subtler than sight. Whereas in the gross relative domain the objects of sense are mediated by the gross elements or mahabhutas, subtle sensory experiences have the much subtler and more refined tanmatras as their objects of experience.

The graph below is one way of depicting the 25 levels of the full range of nature enumerated in Sankhya, with Vedic terms for each level or category:

Transcendental   **Purusha** *(Cosmic Spirit)* / **Prakriti** *(Cosmic Substance)*
3-in-1 Self-Referral Dynamics

Rishi
(associated with Sattva guna)

Devata   **Chhandas**
(associated with Rajas guna)   *(associated with Tamas guna)*

Mahat *(Cosmic Ego)*

Subtle   **Ahamkara** *(Cosmic Intellect)*

Manas *(Cosmic Mind)*

Indriyas *(Sense and Action Capacities)*   **Tanmatras** *(Subtle Elements)*

Srota *(Hearing)*	**Vak** *(Speech)*	**Shabda** *(Sound) (Space)*
Tvak *(Touch)*	**Upastha** *(Procreate)*	**Sparsa** *(Touch) (Air)*
Caksus *(Sight)*	**Payu** *(Excrete)*	**Rupa** *(Form/ Sight) (Fire)*
Rasana *(Taste)*	**Pani** *(Grasp/ Arms)*	**Rasa** *(Flavour/ Taste) (Water)*
Ghrana *(Smell)*	**Pada** *(Move/ Legs/ Feet)*	**Gandha** *(Odor) (Earth)*

Gross   **Mahabhutas (gross elements)**

Akasha *(Space)*

Vaya *(Air)*

Tejas *(Fire)*

Apas *(Water)*

Prthivi (Earth)

For purposes of sequential enumeration the levels of creation are depicted as a linear vertical sequence. As discussed at the beginning of this chapter, the levels also can be described in terms of an inner dimension of depth akin to a sequence of concentric sheathes, embedded in and encompassing each other. A simple analogy is of an iceberg, the solidified state of water, compared to a bit subtler level of a wave or current of water, reflecting different states of activity or flow within the ocean of water. With respect to states of consciousness, the spontaneous experience of the concrete, localized perspective of the iceberg as the most tangible or *real* level is analogous to the ordinary waking state of consciousness. The spontaneous experience of the whole ocean as the most *real,* with degrees of subtle and gross states of activity within it as limited conditional levels of *reality*, can serve as an analogy to the state of unity consciousness—the unbounded ocean of consciousness in motion.

In *one* sense, all the 25 levels are nothing other than the unified field of consciousness, nothing other than the self-referral dynamics of consciousness itself. In *another* sense, Purusha (silent witness) is the essence of consciousness, and Prakriti (cosmic substance) and everything created by Prakriti are insentient limited values that reflect consciousness only due to their connection with Purusha.

The infinitely self-interacting dynamism within the unified field can be described as a cyclic vibration, pulsation, or reverberation at infinite frequency from point to infinity and back. It is self-referral because there is nothing else but itself to which it can refer—it is inherently self-aware. Because the reverberation is infinitely fast and infinitely extensive—so to speak—it is unmanifest. Phenomenal manifestation into relative creation within the unified field can be described as limited modes of the infinite self-referral reverberation or cycle into finite degrees of reverberation at more and more expressed levels and larger and larger time and distance scales. When phenomenally there are finite levels and the wholeness of Infinity is overshadowed, it is as if there is now something *other* that exists, and to which can be referred—at which it can appear to be no longer primarily *self-referral* but rather primarily *object-referral* because of the apparent existence of *other* independent objects. Ultimately, however, all manifest creation is self-referral.

The self-referral silence and dynamism can be said to manifest into finite cycles of the flow of energy and intelligence in the relative subtle and gross levels of manifest creation. This is manifest in the multitude of cyclic phenomena throughout nature, expressing silence and dynamism sequentially, including the cycles of rest and activity, night and day, and possibly big bang explosions or implosions, and dissolutions.

The levels of individual mind will be discussed in detail in the upcoming chapters on human psychoarchitecture. A crucial point to be made here is that the structure of levels that describes the entire cosmos from transcendental to subtle to gross is the same structure in an individual human being. The cosmic structure of nature is fully expressed in the structure of each of us as individual human beings. This point is the key for understanding individual mind and consciousness, and the fundamental cosmic nature of the individual—to be described most tangibly in the final part of the book. The point value of cosmic ego is individual ego, the point value of cosmic mind is individual mind, and so on. The point value of universal consciousness is individual consciousness—which is ultimately consciousness itself, the one ultimate witness of all individual as well as universal experience (Purusha). At the transcendental level, individual consciousness and universal consciousness are the same thing—one wholeness of consciousness, the unified field of consciousness—point and infinity at the same time and as the same thing.

This point about the overall design of the universe is reflected in the inspiring religious statement that man is made in the image of God (Genesis 1:26).[12] It also is reflected in the Vedic reference that the microcosm of the individual reflects the macrocosm of the universe. [13] This will be exemplified in a profound way in the relationship of the Veda and human physiology introduced toward the end of the book.

At the most concrete perspective of levels of manifest phenomenal existence, all objects existing in the gross relative domain are composed of combinations of the mahabhutas, the gross elements. This includes all ordinary objects associated with classical or conventional space-time that are experienced via the ordinary senses. Embedded in each of these ordinary objects are all the subtler levels. Therefore gross objects also can be said to have their existence on the subtler levels. However, this is not the case with subtle objects that are not also expressed on the level of the mahabhutas. In ancient Vedic science, many objects in the subtle relative domain are not manifest in the gross relative domain—indeed even most of them aren't. The material domain is the crusty surface of a vastly more expansive universe with many layers of subtle creation, with objects of sense composed of the subtle essence elements (tanmatras). The gross and subtle objects are embedded in the even subtler subjective levels, all embedded in and nothing other than the all-encompassing, transcendental unified field of consciousness itself.

Because the five senses are subtler than both the gross elements (mahabhutas) and the subtle essence elements (tanmatras), they have the potential to sense both of these levels. In this meaning of the senses we are not referring merely to the ears, skin, eyes, tongue, and nose and their physiological mechanics in the brain, but rather a much subtler meaning. In the ordinary waking state of consciousness, however, sensory experience is almost exclusively limited to experience of macroscopic objects in the gross relative domain only. In this ordinary state of perceptual experience, subtle experiences are overshadowed, hidden, and blocked by the overshadowing influence of incoherent, inconsistent, deeper levels of subjectivity. At this stage of development experiences of the subtle relative domain are typically unclear, confusing, elusive, intangible, and rare—if experienced at all. In higher states of consciousness the subtler levels of nature are clearly experienced and there is the functional capability to sense and act in both gross and subtle levels.

The unified field contains all power and all capability—it is not that the unified field cannot intuit, feel, think, sense, and act on its own. As the field of omniscient, omnipresent, omnipotent *Being*, it can do *anything*. From within itself it expresses itself phenomenally in progressively increasing limitations, even to the extreme of appearing lifeless and insentient. Correspondingly, as described in ancient Vedic science, in higher states of consciousness individuals naturally can do things seemingly impossible from a more restricted and less developed level. As referred to in Chapter 1, what was initially experienced and reasoned as incorrect, impossible, or even mystical at one stage eventually becomes obvious in the natural development of higher stages. These overall issues will be discussed at length in the upcoming descriptions of higher states of consciousness and technologies for full human development, including the practice of *Yoga*.

The physicalist worldview is that consciousness and mind emerge from matter. Although this is commonsensical from the ordinary objective perspective, in Maharishi Vedic Science and Technology the opposite sequence is more accurate. The self-referral unified field manifests phenomenally into the more limited subtle relative domain associated with mind, which further manifests into the most limited gross relative domain. All the time and distance scales identified in modern science—inclusive of galactic clusters galaxies, solar systems, planets, nations, states, communities, families, individuals, organ systems, cells, molecules, organic and inorganic compounds, elements, atoms, elementary particles, quanta, strings, branes, and quantum fields—comprise the gross relative domain. Although there are many degrees and layers in this vast range, they are all are identified as the gross relative domain. The subtle relative level—we could say at a higher vibratory frequency—is said to be much, much more vast.

The primary focus on the objective gross relative domain of nature characteristic of the physicalist worldview is certainly reasonable given the belief that it is the only domain of existence. In this belief system the intelligent negentropic activity of living organisms is observed on macroscopic and large microscopic levels, but is not observed at any smaller time and distance scale. At smaller microscopic and ultramicroscopic levels only the behavior of non-living physical energy and matter has been observed. Given this observation, it is also quite reasonable to believe that life and conscious intelligent behavior emerge only at larger time and distance scales. However, accounting for the emergence of life and consciousness from non-living physical matter and energy has been, at best, quite challenging. This is an important topic in the next part of the book: Psychology Unbound.

Perhaps with these planks on the bridge to a holistic unified conception of nature, a crucial additional plank now can be put in place. The reason that material objects in nature that function at smaller microscopic and ultramicroscopic scales do not exhibit behaviors characteristic of life, intelligence, and consciousness is that these objects are composed of limited gross elements in which these capabilities are hidden and unexpressed—that is, they are made of the mahabhutas. Although ultimately nothing other than the total potential of the unified field itself, they are the most limited and most inert levels of phenomenal expression of the unified field. The total potential of natural law exists in them, but phenomenally only in latent form, hidden due to the predominance of the hiding influence of tamas guna. The phenomenal experience of objective physical *reality* is eventually recognized, in the highest state of consciousness, as a partial view of nature. In the highest state of consciousness the total potential of the unified field is directly experienced at every level of nature along with the phenomenal limitations of each level.

Because of the dominance of tamas guna, objects or systems that are built of the gross elements do not 'on their own' have the capability to exhibit conscious behavior—even though well-designed computers, cellular automata, or Turing machines made of gross matter may mimic such behaviors quite effectively on the level of gross behavior. This is because the subtler subjective levels are dormant in them. Macroscopic systems composed of the gross elements exhibit conscious behaviors only when they are part of organisms that also have active subtle subjective levels. This will be discussed in detail in later chapters on the mind-brain relationship and emergent theories of consciousness, especially Chapters 13-18.

Within this expanded framework of gross, subtle, and transcendental domains as the full range of objective, subjective, and pure subjective levels of nature, it might be helpful to consider again some of the fundamental issues and paradoxes associated with different interpretations of quantum theory summarized at the end of Chapter 5. One core issue is the independence of objects—a pretheoretical assumption and fundamental tenet of classical modern science. From the discussions in this chapter, as well as in Chapters 9-10, we now can put this issue into a more expanded context. The independence of objects is a phenomenal *reality* that depends on the predominant level or locus of subjective experience. From the perspective of the gross relative domain, objects appear to be relatively independent of each other. From the perspective of the subtle relative domain, objects appear as separately identifiable, independent objects but also their nonlocal nature and relative interdependence appear more prominently. From the perspective of the transcendental domain, objects are simultaneously individual and universal, point value and infinite value at the same time, and their ultimate underlying unity predominates.

Do the attributes of objects corresponding to their experiential qualia inhere in the objects themselves, or are just static properties inherent in the objects and dynamic properties attributed to them via the process of observation (the distinction of classical versus quantum objects)? In the sense that there are relatively independent gross and subtle objects, both the static and dynamic attributes corresponding to their experienced qualia appear to be inherent in the objects themselves. However, this is conditional to the level of nature that is the underlying frame of reference or observer perspective.

Within the gross and subtle fields where phenomenal objects exist and individual observers experience them, their static and dynamic attributes are not products of the process of observing them. However, the particular qualia that are the product of the observing process do depend on the nature of the observer. As an extreme example, whether this book phenomenally appears to exist as a 'book' is due to the nature of the observer—an amoeba would likely not even observe it as a separate object at all. But then again, ultimately all independent objects and all independent subjects or observers are nothing other than phenomenal fluctuations of the infinite unified field, the universal Self. Individual minds don't create the outer objective world. But what the objective world is understood and experienced to be is a creation of the interaction of phenomenal objects and the state of consciousness and capability of perception of individual observers.

The question of whether the wave function precipitates into a physical collapse due to some kind of instantaneous unmediated interaction with an individual observer also can be placed in this expanded context. The answer to this question depends again on the underlying assumptions, including the level of nature being assumed. It is, however, perhaps most clear to think of the wave function as a mental model in the mind of the observer associated with the observer's state of knowledge before and after an observation—as described in Chapters 4-5. In this sense we can say that there is a collapse of the wave function in that the knowledge state of the observer changes from a probability function to a discrete answer. However, this is not a collapse of the state of the observed object from a quantum wave function of some kind into a discrete gross physical object being observed. The gross object has its physical manifestation in the gross relative domain whether it is being observed by an individual or not. It also has its subtle manifestation whether observed by an individual or not. But again, gross and subtle objects exist by virtue of the unified field of consciousness which is their essence, and don't exist separate from that.

However, from another angle we can sort of say that there is a collapse (condensation) of the unified field of nature into subtle and gross domains in which individual objects exist and in which how they appear to a particular observer depends on the consciousness of the observer. But in this, we are referring to the universal process of manifestation of any object in phenomenal creation, associated with the concept of sequential symmetry breaking. Any phenomenal object is a collapse in the sense of a phenomenal limitation of the unmanifest, all possibilities of the infinite unified field of nature into finite actualities. This universal 'collapse' involves phenomenally subtle and gross ontological levels of *reality*. Universal consciousness appears to manifest or condense (or 'collapse') into the finite, relative, subtle level of nature, within which are individual minds that distinguish themselves from other objects. This subtle, nonlocal (wave-like) level of nature further appears to manifest (or collapse) into the finite relative (particle-like) gross level of creation, in which material objects appear to be highly localized and independent of each other.

These levels of nature, or condensations of all possibilities of the unified field, can be thought of as universal collapses or symmetry breaking that are dependent on the observer in the sense that all things are nothing other than the unified field of consciousness that is the substance and form of all creation—the universal observer or silent witness (Purusha). But these collapses into phenomenal *reality* are not due to the act of observation by an *individual* observer's conscious mind. To put this in other terms—which may or may not be clearer—there is an 'objective' reduction of possibilities into discrete phenomena. This process is independent of the apparent experience of *reality* of any individual observer. Ultimately it is dependent on the universal subjectivity of the unified field of consciousness, and in this sense it is not just an 'objective' reduction.' It is a subjective reduction into the objective level of nature. These abstract points will be developed further in the next part of the book, which focuses on the structure of mind, human psychoarchitecture.

The 'collapse' of infinite possibilities into finite actualities can be delineated into three levels. One level can be identified as the collapse of wholeness to a point that relates to the self-referral mechanics of the entire creation (at every point in creation), associated with fluctuations within universal consciousness. A second level can be identified as the collapse of the state of the knowledge of an individual upon observation or attending to a particular phenomenon, which can be said to create the objective nature of the phenomenon in terms of individual experience. A third level can be identified as the collapse of subtle waves into gross particles, which occurs as an objective reduction whether or not any particular conscious observer is involved in the natural process of subtle creation manifesting or condensing into gross creation.

Another important issue is whether there is a nonlocal field of existence that underlies conventional space-time that is not mediated by the four fundamental forces. This is directly related to the further question whether superluminal communication is possible. In the expanded framework we have been discussing, the answer to these questions is yes, even though this may not be verifiable within the ordinary objective means that modern science has recognized thus far. It is, however, empirically verifiable through direct experience when observers develop more of their innate potential to experience the full range of phenomenal creation—gross, subtle, and transcendental. It also is indirectly verifiable through the reasoning aspect of the process of knowing—which cutting edge theories in modern science are now positing in order to build logical and coherent explanations of even the gross physical world. In addition it is beginning to be examined more rigorously in the experimental laboratory. Objective evidence is accumulating, exemplified by research on the influence of attention as well as by research in so-called paranormal phenomena—referred to briefly in the next part of the book.

Of course a particularly important issue is whether nature is fundamentally random and indeterminate, or deterministic even if only probabilistically determinable. This is the same question as whether—most fundamentally—there are universal deterministic laws of nature. It also deeply relates to the question of whether there is free will. These issues will be addressed in Chapter 18 after establishing more planks on the bridge to unity. We will then consider how probabilism, universal determinism, and free will coexist.

Clearly the Sankhya enumeration of the inner dimension that includes gross, subtle, and transcendental domains of *reality* greatly expands the range and possibilities in nature far beyond the current paradigm in modern science. Is there a way to map the two systems of the overall structure of nature on to each other in order to bring a deeper sense of their consistency and *reality*? In Chapter 12, we will consider specific links between the structure in modern science and in Maharishi Vedic Science and Technology. All these intellectual points and conceptual discriminations, however, are to fill in planks on the bridge to unity in order that unity is appreciated more fully. They are concepts, and crossing the bridge to unity involves going beyond all concepts, experiencing one's core Self prior to any concepts—being oneself completely, one's Self, the totality, *Brahman*.

Chapter 11 Notes

[1] Maharishi Mahesh Yogi (1967. *Maharishi Mahesh Yogi on the Bhagavad-Gita: A new translation and commentary, chapters 1-6.* London: Penguin Books.
[2] Bernard, T. (1947). *Hindu philosophy.* Delhi: Motilal Banarsidass Publishers.

[3] Sharma, P. V. (1981) (Ed., Trans.) *Caraka Samhita,* Vol. 1. Varanasi: Chaukhambha Orientalia.

[4] Maharishi Mahesh Yogi (2003). Global News Conference, June 25.

[5] Edelman, G. M. & Tononi, G. (2000). *A universe of consciousness*: *How matter becomes imagination*. New York: Basic Books.

[6] Maharishi Mahesh Yogi (2004). Global News Conference, March 17, MOU Channel.

[7] Maharishi Mahesh Yogi. (1985). Inaugural address of His Holiness Maharishi Mahesh Yogi. In *Maharishi Vedic University Inauguration.* Washington, DC: Age of Enlightenment Press, pp. 65-66.

[8] *Maharishi Vedic University: Introduction.* (1994). Holland: Maharishi Vedic University Press.

[9] Wolfram, S. (2002). *A new kind of science.* Canada.

[10] Maharishi Mahesh Yogi (1996). *Maharishi's Absolute Theory of Defence: Sovereignty in Invincibility.* India: Age of Enlightenment Publications.

[11] Hagelin, J. S. Restructuring Physics from Its Foundation in Light of Maharishi's Vedic Science. *Modern Science and Vedic Science,* 3, 1, (1989, 3-72.

[12] *Holy Bible*: (1973). Genesis 1:26. Nashville, TN: Dove Bible Publishers, Inc.

[13] Nader, T. (2000). *Human physiology. Expression of the Veda and the Vedic Literature.* Vlodrop, Holland: Maharishi Vedic University.

Chapter 12

The Eight-Fold Nature of the Phenomenal Universe

In this chapter, we will compare the structure of the physical universe as understood in modern science to the sequential expression of the eight-fold Prakriti or Nature in ancient Vedic science and Maharishi Vedic Science and Technology introduced in Chapter 11. We will explore direct correspondences between the gross relative domain and fundamental particle-force fields that emerge in sequential symmetry breaking according to big bang cosmological and unified field theories. This extends the discussion of the nature of space and time in Chapters 8 and 11. The overall main point of this chapter is the remarkable but entirely expected correspondence between phenomenal levels of nature in modern physical theories and in ancient Vedic science.

This final chapter in the Physics Unbound part of the book is followed in the next part—Psychology Unbound—with an exploration of human psychoarchitecture and levels of mind. We will examine functional and structural models of mind and relate them directly to Maharishi Vedic Science and Technology. The holistic framework of transcendental, subtle, and gross levels of phenomenal nature will be applied to address long-standing issues about the relationship of consciousness, mind, and body. Then we will have established sufficient planks on the bridge to unity to explore the most practical topics of higher states of consciousness and their systematic development.

Until recent decades physical theories viewed the material domain as the entire universe. It now is theorized to manifest from *something* like an infinitely dense singularity of *nothing* that underwent spontaneous sequential symmetry breaking in a hot big bang. The conditions 'prior' to the big bang, as well as the initial conditions when it began, are not known. It is believed that somehow the big bang happened, and space-time began expanding. In sequential stages, spontaneous phase transitions occurred at which particle-force fields differentiated from each other, associated with decreasing temperature and increasing particle mass. In the current cold state of the physical universe all interactions take place through four fundamental forces.

With the advent of unified field theory, however, that infinitely dense singularity is perhaps better characterized as *everything* rather than nothing. From the perspective of the unified field as everything, the theorized forces can be said in some sense to have pre-existed in the super-unified state, but did not differentially function until their corresponding stages of symmetry breaking. Initially in 10^{-43} seconds the universe was 10^{-33} cm in size, the Planck scale. Consistent with quantum theory, however, there was at least an instantaneous quantum jump directly from the super-unified state to the Planck size and time, if it started from nothing, but not a smooth expansion. Also, in these theories all subsequent expansion—and all change—involve Planck-size quantum jumps.

But also, the unified field must continue to exist as the source of continuously occurring quantum vacuum fluctuations or jitters, zero point motion, or inherent dynamism. This means that the unified field is something more than the unification of the fundamental forces—if it continues to remain as an underlying basis even after the big bang (or big condensation) occurs and the forces diversify. The underlying unity doesn't go away when diversity begins: all diversity is within the ultimate unity. This is crucial for understanding the source of order in creation, and will be referred to throughout this chapter.

"A common misconception is that the big bang provides a theory of cosmic origins. It doesn't. The big bang is a theory...that delineates cosmic evolution from a split second after whatever happened to bring the universe into existence, but it says nothing at all about time zero itself... [T]he big bang leaves out the bang. It tells us nothing about what banged, why it banged, how it banged, or, frankly whether it ever really banged at all. " *Brian Greene, theoretical physicist (p. 272)*[1]

Sequential Symmetry Breaking

Consistent with unified field theory, before symmetry breaking the force fields were in an undifferentiated highly-ordered state of super-symmetry—even perfect symmetry, perfect order. Sequential phases of symmetry breaking spontaneously occurred as the temperature cooled down and the universe expanded. This can be likened to the phase transitions of H_2O from gaseous steam to liquid water to solid ice as temperature drops. In each case symmetry is reduced; a gas is more symmetric than a liquid or solid. Increased symmetry means that more transformations can be applied—something can be looked at from more angles—without changing its appearance.

According to the theory, there were three phase transitions of symmetry breaking (refer to Chapter 3). The first phase transition broke super-symmetry into the gravitational force and the grand unified force. The gravitational force is described as attractive, relatively weak, and mediated by the hypothesized graviton, a boson particle. According to super-symmetry, hypothesized gravitino sparticles also differentiated at the same time. Gravitinos are fermions that mediate transformations of gravitons into X and Y bosons associated with the grand unified force. Interactions at this phase are sometimes characterized as a hot plasmic soup of incredibly high energy particles that banged into each other as the universe expanded. The kinetic energy or temperature of these fundamental particles was too high for composite particles such as protons and neutrons that compose atomic nuclei to form.

In the theorized second phase (about a hundred-thousandth of a second later and at about 10^{-27} cm), the grand unified force broke into the strong nuclear and electroweak forces. The strong nuclear force is described as relatively very strong, very short-ranged—limited to ultramicroscopic distances as the 'glue' that binds together quarks into composite particles that build into atomic nuclei—and mediated by the bosonic gluon particle. At this phase transition, gravity and the strong and electroweak nuclear forces began to function differentially, with interactions mediated by gravitons, gluons, and electroweak intermediate vector bosons.

In the third phase transition (about a hundredth of a second later and at about 10^{-16} cm), the electroweak force differentiated into the weak force and the electromagnetic force. The weak force is described as relatively weak and short-ranged—less than the nucleus of an atom—and mediated by intermediate gauge vector bosons (X and Y particles). It is involved in several types of interactions, including decay of quarks from massive to less massive states which emits radiation such as sunlight.

The electromagnetic force is described as relatively strong, long-ranged, and mediated by the bosonic photon particle. At this phase transition, attractive and repulsive electric and magnetic charges began having differential effects. Historically electricity and magnetism were identified as two different forces, reflecting the understanding prior to recognizing their symmetry. Like gravity, the electromagnetic force decreases with the square of the distance from the point center of its source. These long-range forces account for most activity in the physical universe. The short-range strong nuclear and weak forces were proposed later, in order to explain the dynamics within atomic nuclei.

Greene provides a description of presumed events following the big bang:

"For the next three minutes, as the simmering universe cooled to about a billion degrees, the predominant nuclei that emerged were those of hydrogen and helium, along with trace amounts of deuterium ('heavy' hydrogen) and lithium. This is known as the period of *primordial nucleosynthesis...* About a billion years later, with the universe having substantially calmed down from its frenetic beginnings, galaxies, stars, and ultimately planets began to emerge as gravitationally bound clumps of the primordial elements. Today, some 15 billion or so years after the big bang, we can marvel at both the magnificence of the cosmos and at our collective ability to have pieced together a reasonable and experimentally testable theory of cosmic origin." (p.347) [2]

"Even though in the particle physics realm gravity is the mysterious weak cousin of the better understood other forces,...gravity's infinite range and solely attractive nature gives it the edge it needs to be the dominant force. The strong and weak force, both much larger than gravity at the size of the proton or smaller, disappear entirely when two particles are separated by as small a range as the size of an atom. Even the electromagnetic force, with its own infinite range, has both attractive and repulsive aspects. Averaged over the large number of subatomic particles that comprise a star, planet, or asteroid, the attractive and repulsive contributions cancel out, yielding no net electromagnetic force at all. So gravity finally gets the attention that our senses suggest that it should."—*Don Lincoln, particle physicist (p. 447)* [3]

Higgs field theory. To explain the spontaneous phase transitions of symmetry breaking and the phenomenon of particle mass, a new type of field—called the *Higgs field*—has been proposed in recent decades. Higgs field theory is a way to address how particles acquire mass, through interactions with this field. The symmetry breaking due to the Higgs field is a phase transition that occurs at a particular energy level or temperature. The Higgs field is theorized to be mediated by elementary non-zero mass Higgs particles—which have not yet been found.

According to the theory, in the third phase of symmetry breaking into the weak and electromagnetic forces, a Higgs field condensed to a nonzero value when the temperature of the universe dropped to about 10^{15} degrees. This created a *Higgs ocean*—analogous to how steam condenses into liquid water when temperature is reduced. The Higgs ocean can be thought of as a viscosity or aethereal field throughout space that resists change in motion, giving the property of *mass* to particles. Fundamental particles acquire resistance to change—whether acceleration or deceleration—by interacting with Higgs fields. This adds to the total mass of the particles, along with contributions from other forces such as the strong nuclear force that builds composite particles; before their corresponding phase transitions, particles apparently were massless.

Higgs theory is considered one of the key concepts proposed in the past century in theoretical physics. The physical forces we apply in our daily life, such as pushing a button on a keyboard or dancing a jig, to change the direction and velocity of physical objects are to counteract the resistance to change that comes from the Higgs ocean. [1]

The theory of the Higgs field further posits that different particles interact with the Higgs ocean to different degrees, accounting for the wide range of particle masses. For example, it doesn't interact with the photon (which means the photon is massless), whereas the heaviest quark interacts strongly with the Higgs ocean (about 350,000 times stronger than the interaction with an electron) and thus is relatively massive. But it is not known how these theorized interactions might take place, and the reasons are not known for how the known particles happen to acquire their observed masses.

Another Higgs field—grand unified Higgs—was proposed to explain the earlier second phase of symmetry breaking when the universe was at the much higher temperature of about 10^{28} degrees. At this phase the grand unified Higgs field began to interact differentially with the strong nuclear and electroweak forces. In addition the Higgs field associated with the gravitational force in the first phase of symmetry breaking concerns an important theory attempting to explain the shape of space-time resulting from the big bang—called *inflationary theory*. [4]

"Its successes notwithstanding, the [big bang] theory...had trouble explaining why space has the overall shape revealed by detailed astronomical observations, and it offered no explanation for why the temperature of the microwave radiation, intently studied ever since its discovery, appears thoroughly uniform across the sky. Moreover... the big bang theory provided no compelling reason why the universe might have been highly ordered near the very beginning, as required by the explanation for time's arrow... These and other issues inspired a major breakthrough...known as *inflationary cosmology*... Inflationary cosmology modifies the big bang theory by inserting an extremely brief burst of astoundingly rapid expansion during the universe's earliest moments..."—*Brian Greene, theoretical physicist (p. 14)* [1]

Inflationary cosmology. From a reductive perspective of an infinitely dense singularity blasting out in a big bang, a key issue is the mechanics of how the universe has expanded in the way that cosmological evidence indicates. In this framework of understanding, the universe as it exists today is an exquisitely balanced interplay of kinetic energy from the big bang that is expanding the universe and the attractive gravitational force pulling it back together. But how might it have first exploded?

Einstein's formulation of general relativity predicted that space, and thus the entire universe, would either shrink or stretch. Because this contrasted with his firm belief that the universe is balanced, static, and unchanging, he added another term to the equation—the *cosmological constant*. This allowed the equation to have a negative value, which means that gravity also could be *repulsive* rather than just its familiar attractiveness. The strength of gravity is not influenced only by mass, but also energy and pressure. Negative pressure contributes a repulsive force that can counteract the attraction. If the value of the cosmological constant were carefully chosen, repulsive and attractive forces could balance out, resulting in a stable static universe. Einstein didn't specify the origin of the force related to the cosmological constant. When cosmological evidence demonstrated that the universe is *not* static but rather expanding—which the original formulation of relativity theory allowed—Einstein withdrew the cosmological constant from his theory of gravity.

"An attractive force makes objects tend to come closer together. Thus, after a long time, one would expect the various bits of matter that comprise the universe (i.e., the galaxies) would have all come together in a single lump. Given that we observe this not to be true, if we know the mindset of the astronomers of the early 1920s (during which time this debate raged), we can come only to one conclusion. While there certainly was discussion on the issue, the prevailing opinion was that the universe was neither expanding nor contracting, rather it was in a "steady state." Accordingly, Einstein modified his equations to include what he called a "cosmological constant'... In a steady state universe, the strength of the repulsive cosmological constant is carefully tuned to counteract the tendency of gravity to collapse the universe... In 1929, Hubble presented initial evidence, followed by an improved result in 1931, which suggested that the universe was not static, but expanding very rapidly... The fact that the universe was not static caused Einstein to remove from his equations the cosmological constant, calling it "the greatest blunder in his life."—*Don Lincoln, particle physicist (pp. 447-449)* [3]

However, the discredited cosmological constant was later revived, associated with the Higgs field and the modification of standard big bang theory called inflationary theory. This theory is thought to help explain important cosmological concerns, including the flatness and horizon problems mentioned in Chapter 7—but there are still significant issues to be resolved about these problems.[5] According to the theory, for an extremely brief time period of 10^{-35} seconds at the outset of the big bang, gravity became a repulsive force that drove the emerging universe into a vast expansion. This inflationary event acted like a Higgs field—in this case called the *inflaton field*—contributing a uniform negative pressure to space that produced a repulsive force. This repulsive force was so strong that the universe underwent a colossal expansion, by a factor estimated to be as much as 10^{90} or more of its initial volume. This inflationary expansion was much faster than the speed of light but not inconsistent with it, because the speed limit of light applies to motion *through* space whereas the expansion of space in inflationary theory refers to the inflation *of* space itself. However, the colossal expansion is *not* theorized to have been instantaneous.

One major concern of inflationary theory is that it postulates a total amount of matter and energy in the universe that is considerably more than the tally from the visible objects—which contribute about 5% of the total needed. Astronomical research also suggested that additional matter is needed to have a strong enough gravitational attraction to hold galaxies together. This research led to the theory of *dark matter*. Although dark matter has not been found yet, by its theorized gravitational effects it has been estimated to account for approximately an additional 25% of the total needed in inflationary theory. But this still leaves about 70%. An accurate measurement of how much the expansion of the universe has been slowing down due to gravitational attraction was thought to provide an alternative independent estimate of the total amount of mass and energy in the universe. This estimate was calculated by measuring how fast different galaxies at different distances are receding from our galaxy. However, the calculations led to the conclusion that the rate of expansion is *accelerating*, not decelerating.

To account for this, the cosmological constant was revived—but not with the same numeric value Einstein had calculated for a static universe. From cosmological measurements of the recession rates of supernova, it was estimated that the rate of expansion of the universe requires a cosmological constant associated with an amount of *dark energy* that contributes about 70% of the critical mass and energy. This fits nicely with the remaining amount needed according to inflationary theory. Thus contemporary theories suggest that the total amount of matter and energy in the universe is made up of about 5% visible matter, 25% dark matter, and 70% dark energy.[4]

" Starting in 1998, observations indicated the remarkable fact that the expansion of the universe appears to be accelerating, not slowing down. The theory of general relativity allows for that; what's needed is a material with a negative pressure. Most cosmologists are now convinced that our universe must be permeated by a material with negative pressure that is causing the acceleration we're now seeing. We don't know what this material is; we refer to it as "dark energy." Even through we don't know what it is, we can use general relativity itself to calculate how much mass has to be out there to cause the observed acceleration, and this number turns out to be…just what was missing from the previous calculations! On the assumption that this dark energy is real, we now have complete agreement between what the astronomers are telling us the mass density of the universe is and what inflation predicts."—*Alan Guth, cosmologist (p. 294)* [6]

According to these theories all these types of energy interact with the gravitational field. However, the particles that make up dark matter have not been found, and the nature of dark energy is not known. Although experiments underway for several years have not found evidence for dark matter and dark energy, new approaches in the next few years are believed to have a better chance (see Chapter 4). It also is not yet known how come the universe appears to have very little antimatter and is almost all matter, inasmuch as contemporary theories call for an equal amount of matter and antimatter; but some promising ideas also have been proposed.

Explosive inflationary theory is sometimes described as the consensus view in contemporary theories of cosmology, in part because it offers an explanation for how stars and galaxies formed. According to the theory, the inflationary explosion stretched the quantum jitters of the gravitational field, resulting in tiny lumps of gravitational wave energy distributed throughout space that made the relatively homogeneous field of space-time slightly lumpy. Through gravitational attraction the lumps collected into clumps of energy and matter, and eventually into stars and galaxies. Remarkable evidence supporting this model of the development of stars and galaxies has been obtained from precise measurements of small variations in microwave radiation temperature in different areas of the universe, using satellite-based telescopes. [1]

Explosive inflationary theory also provides a framework for explaining how the very early universe was in a state of low entropy. A low-entropy, highly-ordered state is necessary to be consistent with the arrow of time, the 2nd law of thermodynamics which states that entropy increases through time, as well as aspects of unified field theory. According to explosive inflationary theory the colossal inflationary expansion smoothed out space and created a homogeneous gravitational field with much lower entropy than what gravity could have produced without the colossal inflation. After the inflationary explosion the entropy due to gravity producing non-symmetric lumps and clumps of matter in space has increased the total entropy in the system—consistent with the 2nd law of thermodynamics. However, there is no time asymmetry in inflationary theory, or in physical laws generally, so the basis for the 2nd law of thermodynamics is yet to be addressed.

Other theories are emerging that propose mathematical relationships suggestive of dark matter and dark energy can be accounted for in less exotic ways. Other forces also have been proposed, such as a 'technicolor' force mediated by particles called 'techniquarks.' [7] One alternative to explosive inflationary theory proposes that the accelerating expansion has occurred in a steadier manner over perhaps billions of years. In this theory there is no cosmological constant of accelerating expansion; the current expansion will end at some time, moving into other phases that ultimately constitute a cyclic pattern. This is in contrast to explosive inflationary theory, which predicts that the universe could eventually result in complete dissipation. This alternative approach is sometimes called *quintessence theory*. [8] The term quintessence is associated with early Greek thought and with the notion of a fifth fundamental element in nature in addition to air, fire, water, and earth. Contemporary quintessence theory suggests that the universe may not dissipate forever. [9 10] According to this theory a new fundamental field serves as the cosmological constant. It is an alternative explanation of dark energy theorized to produce a negative pressure in the universe that drives its current phase of accelerating expansion. [5]

An important issue for explosive inflationary theory is what triggered the inflationary expansion. An elaboration of inflationary theory proposes that it emerged from a *'preinflationary'* period in which the gravitational field and the Higgs inflaton field were bumpy, chaotic, and highly disordered. In this highly disordered randomly fluctuating state, just due to chance a small fluctuation eventually produced the precise values needed to initiate inflationary expansion. When the 'right' values randomly occurred the expansion spontaneously took place; and the subsequent chain of events led to the formation of stars and galaxies. In the preinflationary period the random fluctuations of the gravitational and inflaton fields of chaotic energy eventually triggered the inflationary expansion.

This *almost* brings us back to time zero and space zero 'when' the theorized big bang is thought to have begun. According to quantum uncertainty, everything would have been in an indefinite state of potentiality without any definite characteristics—presumably even without a probability wave function of differential probabilities of being in a particular state or location or velocity or anything else because no specific probabilities or actualities yet existed.

> "The data we've gathered in the last decade have eliminated all of the theories of cosmic evolution of the early 1990s save one—a model you might call today's consensus model... But I have a different view... The standard model, or consensus model, assumes that time has a beginning, which we normally refer to as the Big Bang. According to this model, for reasons we don't quite understand, the universe sprang from nothingness into somethingness, full of matter and energy, and has been expanding and cooling for the past 13.7 billion years. In the alternative model, the universe is endless. Time is endless in the sense that it goes on forever in the past and forever in the future. And in some sense, space is endless... More specifically, this model proposes...the evolution of the universe is cyclic... People have considered this idea as far back as the beginning of recorded history. The ancient Hindus, for example, had a very elaborate and detailed cosmology based on a cyclic universe... It's as if our three-dimensional world were one face of a sandwich with another three-dimensional world on the other face. These faces are referred to as *orbifolds* or *branes*... In the cyclic universe, at regular intervals of trillions of years, these two branes smash together... [W]hen the radiation and matter are thinned out, the energy associated with the force between these branes takes over the universe. From our vantage point on one of the branes, this acts just like the dark energy... Where the two branes come together, it's not a contraction of our dimensions but a contraction of the extra dimension... Only the extra dimension contracts. This process repeats itself, cycle after cycle."—*Paul Steinhardt, cosmologist (pp. 299-308)* [11]

However, it is inconsistent to say that there was *nothing*. In this hypothesized pre-inflationary period we still would have quantum fluctuations of highly disordered gravitational and inflaton fields, and thus also at least some meaning of pre-space and time in which the random fluctuations occur. Also it would seem that there must have been the *possibility* of everything that exists, as well as their apparent *unequal* probabilities—such as not only the existence of the inflaton field but also its specific properties that led to inflation when the 'right' values randomly occurred.

> "What is a big deal...is how you got something out of nothing. Don't let the cosmologists try to kid you on this one. They have not got a clue either—despite the fact that they are doing a pretty good job of convincing themselves and others that this is really not a problem. "In the beginning," they will say, "there was nothing—no time, space, matter, or energy. Then there was a quantum fluctuation from which..." Whoa! Stop right there. You see what I mean? First there was nothing, then there is something. And the cosmologists try to bridge the two with a quantum flutter, a tremor of uncertainty that sparks it all off. Then they are away and before you know it, they have pulled a hundred billion galaxies out of their quantum hats... You cannot fudge this by appealing to quantum mechanics. Either there is nothing to begin with, in which case there is no quantum vacuum, no pre-geometric dust, no time in which anything can happen, no physical laws that can effect a change from nothingness to somethingness; or there is something, in which case that needs explaining"—*David Darling, astronomer (p. 49)* [12]

How a chaotic pre-inflationary period reconciles with the developing theories of information space—described in Chapters 6-8—and with unified field theory also calls for explanation. The theories of information space underlying physical space-time propose levels of nature at or underlying the Planck scale that are not just bumpy chaos and random disorder. This has important implications for the inflationary theory of the hot big bang. For one thing it would seem that an information space that generates conventional space-time wouldn't necessarily be hot. But more importantly, information implies order. This is consistent with the unified field as the source of order in the universe. If the unified field is a highly ordered super-symmetric state, then the theory of the pre-inflationary period that low entropy came out of inflationary expansion would seem to suggest the inconsistency that something existed *prior* to the unified field—namely the highly disordered, randomly fluctuating gravitational and inflaton fields from which the low-entropy state emerged. Penrose makes a similar point in emphasizing the necessity of accounting for the initial state of entropy with respect to its consistency with the 2nd law of thermodynamics:

> "[P]articular problems in the standard model of cosmology tend to be singled out by inflationists, these all being issues that are indeed related to the initial precision in the early universe... the horizon problem, the smoothness problem, and the flatness problem. In the standard model, these issues are handled by 'fine-tuning' of the initial Big Bang state... The claim is that the need for such fine-tuning of the initial state is removed in the inflationary picture, and this is regarded as a more aesthetically pleasing physical picture... There are certainly some elements fundamental to the inflationary picture whose aesthetic status is somewhat questionable, such as the introduction of a scalar field...unrelated to other known fields [referring to the inflaton Higgs field] and with very specific properties designed only for the purpose of making inflation work... Let us consider the horizon problem... There is...something fundamentally misconceived about trying to explain the uniformity of the early universe as resulting from a thermalization process...whether this is a uniformity in the background temperature, the matter density, or in the space-time geometry generally... For, if the thermalization is actually doing anything (such as making the temperatures in different regions more equal than they were before) then it represents a definite increasing of the entropy... There are certainly deep puzzles relating to the peculiarly constrained state of the early universe. But these constraints are fundamental to the very existence of the Second Law of Thermodynamics... All thermalization processes *depend upon* the Second Law... Moreover, all spontaneous symmetry-breaking processes and all phase transitions (these being needed for inflation) take place only by the good grace of the Second Law. These processes do not explain the Second Law; they *use* it... If we want to know why the universe was initially so very very special, in its extraordinary uniformity, we must appeal to completely different arguments from those upon which inflationary cosmology depends." (pp. 754-757) [5]

A more inclusive way of looking at these issues is that the theory of the pre-inflationary period can be viewed as another angle in the attempt to understand the subtle domain of nature possibly underlying the Planck scale. This is consistent with the model of a subtle level of information space underlying gross space-time, as described in prior chapters. This subtle level as a pre-inflationary period would include the order in nature that creates the gravitational field, the inflaton field, and vacuum fluctuations or inherent dynamism. It would be a highly ordered field that is unmanifest with respect to gross or conventional space-time, but the basis of it, again underlain by the unified field. In other words these theories are progressing toward the understanding of three domains: the local field of conventional physical space-time; a pre-inflationary, non-local, subtle field of information space or field of mind; and the perfectly ordered unified field.

It is helpful to keep in mind that the three-in-one self-interacting dynamics described in Maharishi Vedic Science and Technology repeat at each level of manifestation—from the unmanifest unified field into the subtle relative domain, and again from the subtle relative domain into the gross relative domain. Because modern science does not have an understanding of these three levels of nature yet, there is a blurring of the mechanics. Modern cosmology is attempting to characterize how the gross relative level manifests; but this needs to be consistent with the unified field level. In addition the intermediate subtle relative level, which can at this stage be associated with information space, seems to be entirely left out of current cosmological theories.

> "If we need a pre-spatial level of reality to account for the big bang, then it may be this very level that is exploited in the generation of consciousness. That is, assuming that remnants of the pre-big bang universe have persisted, it may be that these features of the universe are somehow involved in engineering the non-spatial phenomenon of consciousness. If so, consciousness turns out to be older than matter in space, at least as to its raw materials."—*Colin McGinn, philosopher (p. 103)*[13]

To summarize, four fundamental quantum forces and a multitude of actual and hypothetical particles that mediate them have been identified in modern physics. How particles attain mass is not known yet, but is theorized to be due to Higgs fields. How the universe initially banged is not known, but is theorized to be a product of the inherent dynamism of random vacuum fluctuations. How orderly forms of energy and mass are created from randomness is not known, but is theorized to be an inevitable eventual product of random vacuum fluctuations, inflationary expansion, and gravity. How this relates to the unified field as the highly ordered low-entropy basis of everything in nature—as well as to an underlying information space posited as generating physical space-time—is not addressed. Additional cause for pause is that empirical observations only account for an estimated 5% of the theorized total amount of matter in the universe, and the other theorized 95 % of dark matter and dark energy has yet to be found.

In these developing theories of the origin and structure of the universe, three fundamental issues are coming into view a little more clearly: from whence the order, from whence the dynamism or activation, and from whence the mass? These three fundamental issues are beginning to match up with the fundamental qualities or forces of sattva guna, rajas guna, and tamas guna in the holistic view of ancient Vedic science described in Chapter 11—to be explored in more depth soon. Such notions as gravitational attraction, repulsion, and the Higgs inflaton field are beginning to look somewhat similar to these three fundamental principles (gunas).

> "[D]oes the notion of *symmetry,* so prevalent in many ideas for probing Nature's secrets, really have the fundamental role that it is often assumed to have? I do not see why this need always be so. It does not necessarily strike me that basing particle physics on some large symmetry group (which is part of the GUT [Grand Unified Theory] philosophy) is really a 'simple' picture, as far as a fundamental physical theory is concerned. To me, large geometrical symmetry groups are complicated rather than simple things. It might well be the case that there are fundamental asymmetries inherent in nature's laws, and that the symmetries that we see are often merely approximate features that do not persist right down to the deepest levels."—*Sir Roger Penrose, mathematician and cosmologist (p. 746)*[5]

Spinning the Web of the Universe

In Maharishi Vedic Science and Technology the three fundamental forces (gunas) are manifestations of the three-in-one self-interacting dynamics of the unified field—also directly associated with the fundamental trinity of the knower, process of knowing, and known. The link between these three fundamental forces and the unified field has been described by unified field theorist and Vedic scientist John Hagelin.[14][15] According to Hagelin, mathematical derivations from applying the quantum principle to the Lagrangian of the super-string (refer to Chapter 10) involve these three components, and also specifically predict only the five fundamental spin states as their dynamic expression in the physical universe. Hagelin also suggests that these spin states match up with principles in the enumeration of the full range of nature in ancient Vedic science.

The hypothetical particles in super-symmetric and big bang theories described earlier—such as gravitons, gluons, quarks, and their partner sparticles—are a product of mathematical equations. At this point they are not directly translatable into known physical phenomena. The evidence for them is their mathematical consistency—experimental evidence for them has not been found. They relate to what are called 'internal' symmetries, and are distinguished by mathematical classifications of quantum numbers, which are quantities used to identify the different particles As described in Chapter 9, quantum numbers are numerical schemes to keep track of symmetry and invariance principles—related to mathematical degrees of freedom. Quantum numbers quantify quantum properties into discrete values—like charge, which can be thought of as lumps or packets of plus or minus a whole number or integer (But inside a proton is theorized to be quarks, for mathematical purposes assigned fractional values of charge; if quarks exist in nature, they haven't been observed yet). Generally agreed upon quantum numbers include lepton number, baryon number, parity, charge conjugation, electric charge, strangeness, flavor, color, and internal angular momentum or spin. The most fundamental quantum property is the distinction between bosons and fermions, which relates to internal spin.

The term *spin*, a key concept in particle physics, is used in several ways (refer to Chapters 3 and 9). It is quantified in terms of angular momentum (momentum = mass x velocity, which means it has direction; angular momentum = momentum about a point). External types of spin or angular momentum relate to the spins that large classical objects in our ordinary world of experience such as golf balls, baseballs or planets might have, including revolving about another object (orbital spin) or rotating about its own axis (rotational spin). External spin is well understood in classical mechanics.

Spin as internal angular momentum in quantum mechanics can be thought of more as an analogy. It is based on the mathematical model of angular momentum, but it doesn't mean that the particles are spinning around in a particular manner. On the other hand it isn't necessarily the case that internal spin has no meaning with respect to physical space-time—it is just that how it is related is not currently known. The term is used because the mathematical relationships it represents are the same relationships as ordinary objects that do have external angular momentum or physical spin.

As quantum objects subject to the measurement limitations in quantum theory, spin can be measured only about one axis—not simultaneously other axes—due to the uncertainty principle. Internal spin is quantified in units of ½ Planck's constant (6.6 x 10 –34 joule seconds): Spin 1 involves two units of ½ Planck's constant, spin ½ involves one unit of ½ Planck's constant, and so on.

As introduced in Chapters 3 and 10, there are five types of internal spin: 0, ½, 1, 3/2, and 2. These are the five spin states derived from the Lagrangian of the superstring (also noted above). These spin states are thought to be qualitatively different concepts that cannot be characterized in terms of a single quantitative scale. Nonetheless one way to help conceptualize internal spin is that it is analogous to discrete units or multiples of speed of rotation around a central point. Another helpful analogy, related to rotational symmetry, is described by Hawking:

> "All particles have a property called spin, having to do with what the particle looks like from different directions. One can illustrate this with a pack of playing cards. Consider first the ace of spades. This looks the same only if you turn it through a complete revolution, or 360 degrees. It is therefore said to have spin 1. On the other hand, the queen of hearts has two heads. It is therefore the same under only half a revolution, 180 degrees. It is said to have spin 2... The higher the spin, the smaller the fraction of

a complete revolution necessary to have the particle look the same. But the remarkable fact is that there are particles that look the same only if you turn them through two complete revolutions. Such particles are said to have spin ½... [A]ll known particles in the universe belong to one of two groups, fermions or bosons. Fermions are particles with half-integer spin (½), and they make up ordinary matter. Their ground state energies are negative. Bosons are particles with integer spin (such as 0, 1, 2), and these give rise to forces between the fermions, such as the gravitational force and light. Their ground state energies are positive. Supergravity theory supposes that every fermion and every boson has a superpartner with a spin that is either ½ greater or ½ less than its own. For example, a photon (which is a boson) has a spin of 1. Its ground state energy is positive. The photon's superpartner, the photino, has a spin of ½, making it a fermion. Hence, its ground state energy is negative... In this supergravity scheme we end up with equal number of bosons and fermions." (p. 48-50) [16]

In the classification of internal spin states, spin 2 has five possible values (+2, -2, +1, -1, 0); spin 3/2 has four (+3/2, -3/2, +1/2, -1/2); spin 1 has three (+1, -1, or 0); Spin ½ has two (+½ up or -½ down); and spin 0 has one. The hypothesized field of the spin 0 Higgs boson is theorized to be responsible for mass, which is resistance to change in motion. Only spin 1, ½, and 0 have been found in nature; the others are mathematical postulates needed for unification in theories of super-symmetry (which may or may not exist in manifest creation).

Half-integral spin fermions create the vast diversity of behavior throughout objective creation. These are discrete matter particles that can be clearly distinguished from each other. Spin ½ is associated with the heavier mass baryons (such as the proton) and the lighter mass leptons (such as the electron). Spin 3/2 relates only to the hypothesized gravitino, thought to be responsible for unification of the spin 2 and 1 particles, which would connect the gravitational force to the other three forces. As mentioned before, fermions repel each other, or diversify, in the sense that they cannot occupy the same energy state or configuration; whereas bosons have the fundamental statistical property of unifying, or collecting together. Whole number or integral spin bosons are the basis for coherent phenomena—such as laser light. They are the force carrier, messenger, or exchange particles that are less discrete and cannot be distinguished. Spin 2 relates to gravity and is associated with the hypothesized graviton particle. Spin 1 relates to the electromagnetic, weak, and strong nuclear forces and is associated correspondingly with particles including the photon, intermediate gauge vector boson, and gluon. Spin 0 particles are not directly related to any one of the fundamental forces; they are associated with hypothesized Higgs particles and mesons.

Although a specific physical interpretation of internal spin that associates these spin states with well-defined properties in conventional space-time has not been developed, the different spin states are thought to have physical consequences. However, the concept of internal spin is fundamental in super-symmetric theories for classifying quantum particles and force fields. The differentiation into bosons and fermions—associated with internal spin—is theorized to emerge at the first stage of spontaneous symmetry breaking, the most fundamental distinction according to super-symmetry.

Do spin states correspond to fundamental constituents in ancient Vedic science? In the early history of Western thought, four fundamental elements were identified—air, fire, water, and earth. For example, records of this delineation appear in the writings of Empedocles, and also in Aristotle's work. Particle physicist Don Lincoln explains further:

"Empedocles...believed that the things we observe could be made from a suitable mix of four elements: *air, fire, water* and *earth*. His elements were pure; what we see is a mix, for instance, the fire that we observe is a mixture of *fire* and *water*. Steam is a mix of *fire, water,* and *air*. This theory, while elegant, is wrong, although it did influence scientific thinking for thousands of years. Empedocles also realized that force was needed to mix these various elements. After some thought, he suggested that the universe could be explained by his four elements and the opposing forces of *harmony* and *conflict* (or *love* and *strife*)." (p. 4) [3]

"According to the Vedic Literature, there is a universal field of consciousness or intelligence that lies at the basis of creation. From this field of consciousness the first stage in the creation of the material universe is the emergence of five primal entities... [S]imilarly, in physics the unified field of all the laws of nature is considered to be the foundation of the entire physical universe. From the framework provided by unified field theory such as superstring theory, all elementary particles and forces can be clearly categorized into five fundamental categories distinguished by their spin."—*Nirmal Pugh and Derek Pugh, Vedic Researchers (pp. 210-211)* [17]

By interpreting too literally these ancient attempts to understand the essence of matter, some modern scientists have discounted them as primitive, long out-dated modes of thought. Some even have expressed concern about the revived interest in such ideas within the 'New Age' spiritual movement, exemplified by what Lincoln calls their dangerous "canonization." (p. 7) [3] However, a deeper interpretation of these elements and forces—such as that harmony and conflict relate to the behavioral tendencies in nature that define bosons and fermions—might suggest a high level of intuitive *prescience* rather than undeveloped *pre-science* in ancient science.

Importantly ancient Vedic references in Sankhya included the concept of *space* to the four elements to total five fundamental elements. This may be related to the ancient term quintessence. Hagelin has noted that the five fundamental elements in ancient Vedic science appear to have a direct correspondence to the theorized five fundamental spin states in modern particle physics. [13][14] In this correspondence the gravitational field, associated with spin 2 gravitons, corresponds to the fundamental constituent of *space*. The spin 3/2 gravitino field corresponds to *air*, which connects space and the other constituents—especially fire, the next in the

".. [H]uman behavior resembles nature because it is part of nature and ruled by the same laws as everything else... The parallels between organization of a life and organization of electrons are not an accident or a delusion, but physics."—*Robert B. Laughlin, theoretical physicist (p. 201)* [18]

sequence. The gravitino mediates between gravity and super-symmetry, uniting fermions and bosons. The spin 1 force fields, mediated by the photon, correspond to *fire* and are associated with chemical transformations, light, and with the sense of sight. The spin ½ matter fields, such as protons and neutrinos, correspond to *water*. The spin 0 matter field, associated with inert Higgs bosons that give mass or inertia to the matter fields, corresponds to the fundamental constituent of *earth*. Hagelin draws additional parallels that strengthen these correspondences:

"This correspondence is even more striking in the context of a super-symmetric theory, where there is a natural pairing of the five quantum-mechanical spins into three types of N=1 superfields (see Figure 10). The spin-2 graviton and the spin-3/2 gravitino become unified in the context of the gravity superfield, the spin-1 force fields and spin-1/2 "gauginos" combine to form gauge superfields, and the spin-1/2 matter fields and their spin-0 supersymmetric partners give rise to matter superfields." (p. 75) [14]

Hagelin identifies a comparable system of pairings in ancient Vedic science of space and air, fire and water, water and earth—associated respectively with the terms vata, pitta, and kapha introduced in earlier chapters.[14] These three groups have fundamental application to human physiology in the approach to natural medicine of *Maharishi Ayurveda*—to be discussed in Chapter 21. This grouping of fundamental constituents is described in great detail in ancient Vedic literature, especially in the ancient Vedic text that focuses on health, *Caraka Samhita*.

We will return to this important correspondence of five fundamental categories later in the chapter. But first we will lay down a few more planks concerning the structure of relative creation and the place of the five fundamental constituents or elements as enumerated in the holistic approach of ancient Vedic science.

The Eight-fold Nature in Ancient Vedic Science

The holistic view of ancient Vedic science begins from the perspective of infinite eternal unity, and describes phenomenal creation as an unfolding sequence of progressive condensations, limitations, or localizations into tangible forms. The unified field is inherently conscious, inherently orderly, and inherently dynamic. It is *everything* in unmanifest potentiality, and also within it are all phenomenal actualities. It is eternally silent non-changing pure Being; and within that non-changing eternal silence is the ever-changing diversity of phenomenal creation.

Non-duality, duality, and trinity. Again summarizing key points from Chapters 10-11, to explain the process of phenomenal manifestation, the 'nature' of the unified field is described as the simultaneity of infinite silence and infinite dynamism, wholeness and part, infinity and point. The infinite silence is *infinitely* dynamic—which means it is infinitely silent. The Infinity can be thought of as an infinite number of points, each point being the singularity that is the totality of Infinity. In each point is the infinite totality, and the infinite totality contains infinity of parts. It is non-changing silence that also simultaneously is ever-changing dynamism; it is infinite wholeness that also simultaneously is infinity of points. Thus it is described as the coexistence of opposites. But it is not two things—it is non-dualistic. Remaining non-dual or beyond all duality, to explain phenomenal manifestation it is attributed two coexisting opposite qualities as its inherent 'nature'—infinitely dynamic silence.

To help explain how the opposites *coexist*, the unified field is described as infinitely self-referral—infinitely referring back to itself, within itself. Maharishi has explained that Self-referral means non-changing *Self* and infinitely dynamic *referral*. The liveliness of infinite self-referral can be described as infinite reverberation of infinity to a point and point to infinity. In the process of *finite* manifestation these apparently opposite values become expressed more tangibly.

The range of phenomenal manifestation is from universal consciousness or pure Being to apparently no consciousness, no intelligence, and no life in inert matter—from totality to the least expression of totality. At each level there is an expression of wholeness and point values: universal and individual, eternity and instantaneity, infinite and finite, nonlocality and locality, silence and dynamism, non-change and change, unity and diversity, whole and part, consciousness and attention, intelligence and energy, freedom and boundary, simplicity and complexity, as well as light and dark, day and night, life and death, substance and form, function and structure, wave and particle, holism and reductionism, equality and hierarchy—and so on and on. The unmanifest dynamics of infinity and point, and their relationship, are the basis for all contrasts in finite nature—with the infinite degree of contrast in the dual 'nature' of the unified field prior to any finite manifest form. The first phenomenal expression of *finite* values from Infinity is the closest to Infinity—most reflective of the full infinite value of intelligence and energy. There is a phenomenal sequence of expression from Infinity to the least infinite point value, while simultaneously at every level of this sequence there is the dynamic interplay of infinity and point. The opposite values of infinite silence and infinite dynamism, infinity and point, coexist at every level.

The process of infinity and point, eternity and instantaneity, beginning to be expressed in finite values establishes the phenomena of finite space and time. It can be said that the basis of finite space is infinity and point, and the basis of finite time is eternity and instantaneity. The first step toward manifestation into finite time and space can be thought of as the coexistence or duality of *both* eternity and instantaneity, *both* infinity and point, the full range of possibilities.

If we think of phenomenal manifestation as beginning at some point in time, then we can say it is at the delineation of eternity into instantaneity—the eternal *now*. The *phenomenon* of time has a direction, from the eternal instant toward the next instant, and away from the prior instant—past to present, to future—the arrow of time. These three phenomenal aspects of time also can be related to the three fundamental forces of tamas, sattva, and rajas.

Likewise finite space can be thought of as beginning at that point or instant in time. The phenomenon of finite space is a limitation of infinite space—infinity of dimensionless points. The totality of space is already present in Infinity. In this sense space doesn't begin at a point and expand out in all directions but is already everywhere, taking on phenomenal limitations down to a dimensionless point—so to speak. The subtlest manifestation of finite space is the closest possible to the infinite self-referral of infinity and point simultaneously. It is the maximum integrated, interdependent expression of nonlocality and locality on a finite level. It then sequentially manifests into more predominance of the phenomenal properties of locality and independence.

In the three-in-one self-interacting dynamics of the unified field, the three fundamental forces—sattva, rajas, and tamas—also can be thought of as expressing the three different dimensions of volume—the three axes of up/down, forward/backward, and right/left. But this volume includes all finite space in the simultaneity of infinity and point. From the perspective or reference frame of the dimensionless point, three points or axes are needed to generate the volume of three-dimensional space. From the perspective of Infinity, all finite space is a limitation superimposed on Infinity into the three-in-one dynamics of the basis of relative creation.

From the unmanifest, self-referral, self-interacting level, infinitely self-interacting time and space manifest into the finite relational dynamics of finite space-time. Progressive phenomenal expressions take on more limitations, down to the point of complete limitation. Each progressive level of phenomenal manifestation embodies the infinity-point, whole-part dynamics in lesser degrees, from universality, nonlocality, and interdependence to the most limited expressions of individuality, locality, and independence. At the same time, however, each progressive limitation ultimately is nothing other than the unified field—eternal and infinite through and through, even while appearing to be finite, localized, and independent. This is directly experienced and appreciated in the highest state of unity consciousness.

In the holistic view of the unified field as the simultaneous coexistence of opposites the universe would not have to begin from an infinitely dense singularity or literally nothing that blasted out in a big bang, but rather many 'places' or points simultaneously. This eliminates paradoxical issues in a reductive conception of nature as emerging from an infinitely dense singularity, conceptualized as a point—or Planck-size quantum—from which space and time expand outward. Space and time don't have to be thought of as emerging from a singular dimensionless point or Planck-size quantum—which impels questions such as what existed before it, what does it expand into, or what remains when it contracts. Time is eternal and space is infinite. Finite values are a phenomenal limitation of the infinite eternal unified field. This also is relevant to the contemporary model of space as 'flat' in the sense of extending in all three directions without being curved. This overall point about the most fundamental nature of space is discussed in the following quote of theoretical physicist Brian Greene:

"Normally, we imagine the universe began as a dot...in which there is no exterior space or time. Then, from some kind of eruption, space and time unfurled from their compressed form and the expanding universe took flight. But if the universe is spatially infinite, *there was already an infinite spatial expanse at the moment of the big bang.* At this initial moment, the energy density soared and an incomparably large temperature was reached, but these extreme conditions existed everywhere, not just at one single point. In this setting, the big bang did not take place at one point; instead, the big bang eruption took place *everywhere* on the infinite expanse. Comparing this to the conventional single-dot beginning, it is as though there were many big bangs, one at each point on the infinite spatial expanse. After the big bang, space swelled, but its overall size didn't increase since something already infinite can't get any bigger. What did increase are the separations between objects like galaxies (once they formed)... An observer like you or me, looking out from one galaxy or another, would see surrounding galaxies all rushing away, just as Hubble discovered... Bear in mind that this example of infinite flat space is far more than academic. We will see that there is mounting evidence that the overall shape of space is not curved... [T]he flat, infinitely large spatial shape is the front-running contender for the large-scale structure of space-time. (pp. 249-250) [1]

Infinite space can be thought of as flat and infinitely extensive. With respect to finite space in the sense of phenomenal manifestation, however, space can be thought of as curved also. The notion of the curvature of space—such as into a torus or sphere, or both if a sphere can be conceived in terms of curving back on itself—relates to finite limitation of infinite self-referral. In order to explain finite creation it can be said that Infinity curves back onto itself in infinite self-referral.

"*Prakritim swam avashtabhya visrijami punah punah* (Bhagavad-Gita, 9.8)

Curving back upon My own Nature, I create again and again." (p. 37) [19]

The self-referral, self-interacting dynamics of the unified field curving back onto itself is fundamental to the entire process of creation at all levels of manifestation. This curving back onto itself—while on the unmanifest level is infinite self-referral—Self-referral—on the finite manifest level can be described as creating a mandala form. This also is associated in Vedic literature with the concept of *Hiranya garbha,* [20] sometimes called the cosmic egg. This is described as the manifest form of the cosmic expanse in the infinity of space and time—curving back onto itself in the creation of finite space and time. On the *gross level* of space and time, this dynamic of curving back on itself can be associated with the concept of a point particle, infinity of points, Planck-size quanta, and atomic structures.

This is another expression of the coexistence of opposites of infinity and point values as the 'nature' of the unified field. From the absolute perspective of Infinity every point is the center of everything; every point *is* everything. Every point is *infinitely special* because it is the totality that contains everything. From the relative perspective of finite creation a point can be relatively special in the sense of being in a particular relation to other things, such as the gravitational center of the Earth, or of our galaxy, or of all galaxies (called the *Brahmasthan*, to be discussed in Chapter 22). But the center of the Earth moves through different points in time, which doesn't seem to make a difference with respect to the laws of nature. Thus every point can be thought of as not at all special, relatively special with respect to other things, while also at the same time *infinitely* special as being everything (or from a reductive negational perspective, nothing special).

In the finite relative field there are asymmetries which can be thought of as yielding relatively 'special' positions or directions in space. For example, relative to the human body, the head and the heart are in some ways more significant than the feet. The front is more significant than the back, in terms of the general direction for motion and interaction with the environment, as indicated by sensory orientation. Also the right side is frequently stronger than the left—chirality or handedness in the human body. Similarly, relative to the Earth, east is 'special' in the sense that it is the direction from which the Sun rises—significant to biological life on Earth in numerous ways. There are many implications of these relative directional differences built into the finite relative structure of nature expressed throughout the phenomenal universe. The three axes associated with direction in space—up/down, front/back, right/left—also can be associated with the three fundamental principles or forces of sattva, rajas, and tamas (the gunas), each having particular significance throughout finite space-time.

Infinity appears to manifest, localize, concretize, condense, and materialize itself into finite discrete values of intelligence and energy. Time and space as measurable quantities reflect an intellectualization and objectification into discreteness or localization imposed on Infinity—like a wave or an even more condensed iceberg in the ocean, both of which remain nothing but ocean water. We can think of drops of water building up to make an ocean, or we can think of the ocean as dividing up into drops of water, but there are not drops or waves of water in the ocean without the ocean.

The 'nature' of Infinity is the simultaneity of infinite value and point value. In the understanding that phenomenal manifestation happens, then the manifestation would happen everywhere as a phenomenal limitation of Infinity into finite values. There needs to be a phenomenal limitation of Infinity in order to have forms and objects distinguishable from each other. If there is phenomenal manifestation, it has to be finite and involve some kind of limitation. But at the same time it never loses its unbounded unlimited infinite status. The universe appears as finite because it appears to have limitations and thereby forms, structures, and functions that are distinguishable. The finite appears as *relatively real,* or *relatively unreal* (or both simultaneously) depending on the state of consciousness of the observer—whether the ever-changing, phenomenal, manifest, finite appearances or the never-changing, unmanifest Infinity is prominent.

In the sense of space we are discussing, space is perhaps better interpreted more abstractly as *existence*—mentioned in Chapter 4. Levels of space, levels of nature, levels of existence, or levels of *reality* are relative to their limitations—defined relative to the limited nature of the objects and processes in them—within the eternal Infinity. In this holistic view space and time are most fundamentally infinity and eternity. Infinite space means the infinite self-interacting simultaneity of infinity and point; within that eternal Infinity are finite levels of space. The *subtlest finite* levels are characterized by the least restriction—almost simultaneity of point and infinite values—and they are nonlocal. At the *grosser* levels, infinite values are hidden and point values are more obvious and distinct—they are localized. Objects in subtle space-time are more interdependent, and at the gross level of space-time they appear to be more discrete and independent.

From another angle, space-time can be thought of as an aether-like substance, medium, or field with degrees of simultaneity of the values of infinity and point, eternity and instant. The subtle substance or field has degrees of refinement, density, viscosity, or limitation. In this way of looking at it, space and time are relative to each other, but more fundamentally concern the degree or level of subtlety of the field or medium, which relates to the degree of interaction of space and time, the degree of simultaneity of infinity and point, eternity and instant. This is another way of characterizing the levels of the inner dimension described in Chapter 11.

The gross relative domain is a field or medium characterized by the limitations of Einstein locality and the speed of light—within which the point value rather than the infinite value of space-time is the most prominent, and within which objects appear to have local independent existence. This is the crudest, least refined, most dense phenomenal field of infinite eternal space-time. It has been the focus of the physical sciences. Contemporary scientific theories are now going beyond this conventional understanding and experience of space-time and are glimpsing the subtle nonlocal level of space-time. Much further, they also are glimpsing mathematically the transcendent level of space and time in unified field theory. The holistic view of space and time in ancient Vedic science is able to link the level of experience of locality with the level of Infinity—and also the ultramicroscopic Planck scale with the unbounded wave field level—through an understanding of the gross, subtle, and transcendental domains. As will be discussed further in the next part of this book, this also can be related directly to body, mind, and consciousness.

There are other important implications of the holistic view of time and space. For example no additional or higher-order spatial dimensions beyond the ordinary three are necessary to account for phenomena that appear to interact nonlocally (mentioned in Chapter 8). They are subtle phenomena that don't have the particular limitations that define what we are conventionally accustomed to calling space-time—identified as gross space or conventional space, and associated with the material domain of nature and Einstein locality. The subtle nonlocal field of space or existence does not have to be conceived as some *extra* dimension of space-time. Rather it refers to a level or medium of existence within infinite eternal space-time that is subtler than conventional space-time. The four-dimensional nature of space-time is sufficient to provide the experiential framework for the five senses of perception. That is, the nature of the phenomenal world and the nature of subjective experience match or correspond with each other.

Both finite time and space emerge as limitations of the three-in-one self-interacting dynamics of the unified field, as the expressions of the three fundamental qualities or forces of Prakriti or Nature—sattva, rajas, and tamas. This structures a direction to time, and the volumetric dimensionality of space. They also can be associated with the relative concepts of inward, outward, and center point of reference—related to the inner dimension. This importantly relates to what is attributed to be the primary locus of experience. Phenomena associated with the gross relative domain—characterized by Einstein locality, the speed of light, relativistic space-time in terms of the limitations of the gravitational field, and Planck-size quantum units of space-time—relate generally to a particular experiential frame of reference or locus of experience. They can be thought of as the limitations of an observer's frame of reference to the ordinary gross relative domain, corresponding to the limitations of the level of development or state of consciousness of the observer. These particular limitations generally define experience in the ordinary waking state of consciousness. In higher states of consciousness the primary locus of experience is not the gross relative domain, and not independent objects made of quantized particles. Correspondingly, capabilities of experience are not nearly as limited—to be discussed in the chapters on higher states of consciousness and systematic technologies to develop full human potential.

From the view of conventional, material, or gross space and time the basic *metric* for any measurement is distance such as in centimeters and duration such as in seconds. Einstein's relativistic theories importantly revealed the interrelationship and complementarity of space and time as *space-time*. In these theories the speed of light is an absolute value associated with conventional space-time that defines all interactions in the material domain. This relational nature of space and time implies that space-time is some kind of aether-like medium or field; but space was thought to be empty space with no substance to it. However, quantum theory has revealed more clearly that space is not 'empty,' in that it at least is said to contain vacuum fluctuations or zero point motion. Unified field theory further reveals that the basis of space-time is a field of all possibilities and the source of all order in the universe. Rather than being empty and nothing, it contains everything. From the perspective of levels of space-time or existence, the distinguishing feature can be thought of as the degree of simultaneous interrelationship—or expressed degree of self-referral—of infinity and point, whole and part, unity and diversity, interdependence and independence, universality and individuality, or nonlocality and locality. This can be thought of as the *metric* for the inner dimension of gross, subtle, and transcendental domains.

In the transcendental domain there is absolute *infinite* self-referral—Self-referral—and in the subtle and gross relative domains there are finite relative degrees of the expression of self-referral. In the subtle relative domain it can be thought of as a more limited referral to individual self, and in the gross relative domain it can be thought of as the most limited into a point particle. To keep referring back to the holistic view, these levels of self-referral are ultimately nothing other than phenomenal expressions of Self-referral.

To summarize, general relativity, the phenomenal independence of objects from each other and from the subject, Einstein locality, the speed of light and the Planck scale all characterize the texture of the encompassing aether or medium at the level of the gross physical space-time domain. A more interdependent non-local medium characterizes the subtle space-time domain. In this domain, the texture of the space-time field would be less localized, less independent, not particle-like or particle interaction causality; and more wave/field-like. Subtle phenomena would be relatively more spread out in time and space with respect to the perspective of ordinary gross space-time, and narrow down into more discreteness as the information/energy is transduced from subtle to gross domains. The entire field or medium with all objects in it would have these characteristics, comparable to but different from gross relativistic space-time. The whole conception of measurement would require methods other than the phenomenally objective gross relative domain. Further, this suggests that if individual observers were able to be sensitive enough to experience the subtle level, they could potentially pick up information/energy processes as they condense toward gross expression in a manner that historically has been categorized as impossible with respect to the four known forces in the physical world. In this understanding, nature is much more interconnected, and processes thought to be impossible from the limited perspective of classical and to some degree quantum worldviews become rational and logical, including subtle causal relations associated with ancient Vedic technologies described as natural experiences in higher states of consciousness.

Three-in-one into five and eight. As outlined in Chapter 11, the *inner dimension* is phenomenally linear and sequential in terms of concentric levels, aethers, mediums, or layers, but embedded in a non-linear, simultaneous, and ultimately indivisible, infinitely self-interacting wholeness. The indivisible unity can be thought of as having an inherent duality—but each part refers back to the other infinitely fast, so they are the same thing. Again, the inherent duality also can be thought of as having an inherent trinity, involving the two parts and their relationship—the three-in-one infinitely self-referral, self-interacting dynamics of the unified field. The unified wholeness (Samhita) of these infinitely self-referral dynamics can be delineated into the three-in-one self-interacting dynamics of the unified field—associated with knower, known, and process of knowing; subject, object, and their relationship; and directly related to sattva, tamas, and rajas. The terms *Vishnu, Shiva,* and *Brahma* are also used in ancient Vedic science to identify these three universal principles of natural law—to be discussed more in the final part of the book.

The inherent trinity, along with their state of unity or singularity, can be counted as *four*—three parts in one wholeness. This is related to the four fundamental aspects of the Veda: *Rk Veda, Sama Veda, Yajur Veda* and *Atharva Veda.* The first Veda is the wholeness or *samhita,* the other three being the *samhita* of each of the three parts. As will be discussed in Chapter 23, the phenomenal elaboration of wholeness into parts is the structure of the Veda itself, the mechanics of the expression of all the laws of nature—as well as the fine fabric of our own consciousness.

The phenomenal process of the diversification of the one into three—knower, process of knowing, and known along with their unified state—can also be thought of in the opposite unifying direction of self-referral—the unification of three-in-one. These four, in both the diversifying and unifying directions, identify eight fundamental aspects. This is another elaboration of the self-referral dynamics of infinity to a point and point to infinity described earlier that unfolds into the eight-fold Prakriti or Nature. It is the self-referral dynamic at each stage of delineation or manifestation that structures the next sequential stage, level, or textural medium—sequential creation through self-referral dynamics. As Maharishi states:

"In terms of dynamism, nature is divided into eight values" [21]

This delineation of the eight-fold Prakriti is expressed in the structure of the samhita of the Rk Veda, the first Veda. The Rk Veda has 10 divisions, chapters, or *mandalas*, eight of which directly relate to the eight-fold Prakriti. The eight-fold Prakriti—also called *Apara Prakriti*—relates to the phenomenal, changing values in nature. Added to this is the non-changing value of Prakriti—called *Para Prakriti*—resulting in nine values of the dynamic changing and silent non-changing Prakriti. These are witnessed by a tenth quality—*Purusha*—infinite silence, the silent witness of all the changing and non-changing qualities. These 10 values or qualities comprise the 10 mandalas of the Rk Veda which describes the entire structure of nature.

"The ten *Mandals* of Rk Veda contain all knowledge of the <u>sequential</u> evolution of Veda... The tenth *Mandala* displays *Purusha*—the supreme state of evolution. This presents the complete display of Natural Law, from point to infinity." (p. 170) [19]

As described in the past chapter, in the subtle relative domain the trinity is manifest as the subjective structure of cosmic (and individual) ego, intellect, and mind. This trinity is further expressed into the five modalities of sense and of action, corresponding to the five fundamental constituents that make up the objects of sense. Again, the three levels that are the basis of subjective experience, in combination with the five objects of experience corresponding to the five senses, make up the *eight-fold Prakriti*. In Maharishi Vedic Science and Technology this eight-fold structure refers to ego, intellect, mind, and their objects of sense—space, air, fire, water, and earth—which comprise the entire phenomenal universe (refer to Chapter 11). This eight-fold structure also is reflected in the first verse of the Rk Veda. The first eight syllables of the verse lay out the structure of creation—*Agnim ile purohitam*—containing the eight aspects of ego (ahamkara), intellect (buddhi) , mind (manas), space (akasha), air (vayu), fire (agni), water (jal), and earth (prthivi) (refer to Chapter 11). [19] There is a repeating pattern of the eight-fold structure throughout the 192 syllables of the first sukta (set of verses) of the Rk-Veda (3 x 8 x 8=192).

Hagelin brings out another quite interesting correspondence between this structure and the expanded Lagrangian formulation of the super-symmetric unified field (refer to Chapter 10). In the four-dimensional expansion of the Lagrangian, there is also a structure of 192 terms. In Chapters 23-24, these patterns will be shown to repeat themselves in the structure of the Veda and Vedic literature, in the phenomenal structure of the universe, and also in the structure of human physiology. [22]

Sattva, Rajas, and Tamas as Fundamental Forces in Nature.

As described in Maharishi Vedic Science and Technology, when the three-in-one samhita of Prakriti begins the phenomenal stir or limitation within its unmanifest state of equilibrium, the three fundamental forces or gunas become differentially active in relation to each other. These three principles, qualities, or forces can be thought of as the creation, maintenance, and destruction or dissolution operators that structure all change. Their interactions shape all phenomenal levels of finite manifest creation. They are never separated from each other, but their relative prominence is generally distributed throughout creation from the subtle level where sattva is most prominent to increasing expression in the gross level where tamas is most prominent.

Again, the three-in-one dynamics of the unified field relate to different terminology at the transcendental, subtle, and gross domains of nature. With respect to the unmanifest level of the transcendental domain, the three principles or qualities of nature relate to the terms *rishi*, *devata*, and *chhanda*s. At that level, infinite and point values are the same—we can say that infinite self-referral predominates. As the three fundamental qualities appear to take on finite manifestation in the subtle relative domain, they relate to the terms *sattva*, *rajas,* and *tamas*. At this level infinity and point values relate to nonlocal and local processes, wave-field dynamics, individuality and universality, in which nonlocal or wave properties can be said to predominate. As they appear to take on additional limitations in the gross relative domain, they relate to the terms *vata*, *pitta*, and *kapha*. At this gross level individual units and collections of individual units, and point-particle, quantum, and atomic properties predominate. The three qualities are the same—a repeating pattern across the three domains—but specific features emerge in their progressive manifestation, limitation, and condensation into more tangible discreteness.

Each of the three fundamental forces can be understood in terms of unifying or diversifying, expanding or contracting, attracting or repelling, depending on the object, direction, or perspective taken. With respect to the unmanifest transcendental domain, the unifying principle can be thought of in terms of the dynamics of infinity to a point. But in that infinitely self-referral level, the point contains infinity, so it simultaneously unifies point to infinity. Likewise the diversifying principle divides a point into infinity, but simultaneously places infinity into each point. In that level, it may be easier to conceptualize the point as a singularity containing everything rather than a point as a dimensionless point containing nothing. As the three-in-one self-interacting dynamics are activated in relative levels of creation, the unifying and diversifying principles take on more concrete meanings. Generally sattva is associated more with unifying, and rajas and tamas more with diversifying—even though all three are never completely separated from each other and function together as an indivisible whole.

Sattva guna, as the unifying principle in the subtle relative domain, can be said to unify oneness toward Oneness, individuality toward universality; individual point value, individual self (individual ego, to be discussed in the next part of the book) toward universal wholeness, universal Self. It is associated with the maintenance operator in that it maintains the direction of progress toward infinite unity and wholeness, from finite to infinite value of lively dynamic silence.

As that which guides all action in nature in an evolutionary direction toward increasing unity, sattva can be associated with the principle of *dharma*—upholding or maintaining action toward unity and the ultimate fullness, wholeness, oneness, and completeness of existence. Sattva is expanding in the sense of refining toward ultimate unity—toward Oneness, Infinity, Unity, the universal Self. But it also can be thought of as contracting in the sense of attracting or pulling back or unifying to ultimate unity or Oneness—centering onto the universal Self. With the increasingly limiting influence of tamas, however, it can be understood to unify into oneness in the sense of individuality. This relates the unifying value to the unitary sense of the individual self—centering onto the individual self, individual ego. At the much more tangible gross level of nature, it can be conceptualized as unifying into a dimensionless point value. In the gross relative domain where locality or the point value is more prominent, sattva can be associated with the attractive force of gravity—centering onto the center point of an object or local field.

Thus the gravitational force that is ubiquitous in the physical universe is a gross expression of a much more abstract principle of the attractive or unifying force of sattva, encompassing both gross and subtle, objective and subjective, levels of nature. This principle relates to the attractive force commonly associated with the concept of love, demonstrating again the limitations of Einstein's notion of the physical universe as reflected in his statement: "Gravity cannot be held responsible for people falling in love." [23] Einstein's point is certainly valid within the restricted gross meaning of the concept of gravity. But in the completely integrated understanding in ancient Vedic science, it is the same universal principle or force of attraction manifesting on different levels of phenomenal nature. On the gross level it is the familiar concept of gravity as precisely characterized in Einstein's formulation; but this is not the whole picture. On the subtle level it is the same underlying principle that is associated with the attraction of love—individual love, and ultimately Universal Love. This understanding of 'gravity' as a manifestation of the universal principle of sattva obviously is not verifiable within the boundaries of ordinary sensory experience and Einstein locality that views the physical as the entirety of nature. It extends beyond Einstein locality, into the subtle relative field of nonlocality of which Einstein apparently had little inkling. It is in this expanded and completely integrated understanding of nature that Maharishi has stated that the unifying principle underlying the concept of gravity unifies all diversity in nature:

> "The force of gravity will be ultimately found to be the unified field administering the whole universe." [21]

The universal principle of sattva also can be related to the 3rd 'law' of thermodynamics—decreased activity associated with decreased temperature in material systems. This typically results in decreased entropy or increased order, a fundamental negentropic process in nature. This is consistent with sattva as the principle that maintains the evolutionary direction of nature toward higher-order levels of unification—gross to subtle to transcendental. It is related to the fundamental principle that maintains the orderly evolutionary direction of change throughout the entire phenomenal creation—the principle of *dharma*.

Rajas guna activates the other two forces; it can be associated with the creative operator. It expands in the sense of moving toward manifestation and diversity; but it also can be understood in terms of contraction or diversifying toward finite limited expressions of infinity and point. It can be related to the expression of the inherent dynamism in the unified field, as the force that activates in the direction of finite manifestation and diversity. As the quality or force that activates the other two gunas, rajas can be associated with the concept of *karma*—the universal law of action and reaction throughout nature—corresponding to the law of cause and effect in modern science. Rajas also can be related to the 1st 'law' of thermodynamics—all motion or interaction between physical objects results in no loss in the total amount of energy. At its grossest level it can be associated with the kinetic energy that expresses inherent dynamic activation in the gross physical universe—which follows the law of the conservation of energy.

Tamas guna contracts in the sense of inertia, restraint, or resistance to change—but also expands or unfolds in the sense of moving into diverse expressions in more concrete manifestation, toward increasing inertness. It can be associated with the destructive or dissolution operator, holding back progress. It also can be taken to be toward changelessness, toward hiding or unmanifesting—and ultimately eternal non-change, or infinite silence. Tamas in some sense can be related to the 2nd 'law' of thermodynamics—everything tends toward greater disorder, entropy always increases. In this sense it leads to the least amount of change, the most inert states of nature in which the most localized point value predominates and in which the infinite value is most hidden. At its grossest level tamas can be associated with the accretion of mass or inertia that counteracts change—resisting change in momentum whether from an inertial state or changes of speed or direction, either speeding up or slowing down. The theory in modern physics of the Higgs field to account for particle mass is in the direction of trying to account for the principle or force of tamas on the gross level of nature.

In modern physics the most fundamental distinction or division in nature is between bosons and fermions. In super-symmetric string theory, fermions are left-moving or counterclockwise spin and bosons are right-moving or clockwise spin. This distinction is also associated with statistical properties of being able to occupy the same state or not. However, these properties are not thought of as forces related to the distinction between bosons and fermions, but rather only what might be called empirical statistical patterns. Bosons can be said to reflect the inclusive, unifying or attraction principle, and fermions can be associated with the exclusive, diversifying, or repulsive principle.

Thinking of points as making up an abstract field (perhaps a little different than the concept of field typically used in physics), if each point has a certain property then the overall field also has the property. In this notion of a field, the field can be thought of as a type of substance or medium that exhibits a particular property or properties at each point in the field, which gives a texture or dynamic conditional quality to the field.

According to ancient Vedic science all the types of finite relative fields in manifest creation are composed of the interplay or dynamics of the three fundamental forces of sattva, rajas, and tamas interacting with each other to various relative degrees of predominance. As described in Chapter 11, all objects in the gross relative domain are made of these three forces, with tamas generally predominating. But the subtlest layer of the gross relative would be most characterized by sattva with respect to the progressively grosser layers in the gross relative field. The subtlest layer of the gross relative field is the gross mahabhuta of space. Space contains the potential for all the grosser layers that emerge from it—air, fire, water, earth—but exhibits primarily just the subtlest quality. It is the least restricted, most symmetric of the five layers of the gross relative domain, the field of conventional space-time.

One way to look at the structure of the gross relative domain as quantized is that when each point in a field or medium has a quality of attraction or gravity, pulling toward itself from all directions—so to speak—and points in the field are differentiated or separate from each other in some sense, then the points would pull on each other. If there were only two points, they would gravitate back together to become one point. But when the pull of each point on the points adjacent to it is from all directions—for all practical purposes of infinite extent—then they would pull against each other in the sense that a point on one side would pull in the opposite direction of the point on the other side, in all directions. There would be opposing pulls that would establish each point as a specific point in a local field. The point could be thought of as becoming quantized into discrete differentiated points. These quanta would depend on the strength of their attraction to each other, combined with other influences or forces that might work to counteract the strength of the attraction. In the gross relative domain the strength of attraction can be associated with the gravitational constant. This would be a textural quality of the field, and it would influence the size of the quantum, along with any counteracting influences.

The gravitational constant can be related to the force of attraction or sattva on the gross material domain. The abstract unifying principle of sattva thus can be characterized according to the level of nature it applies. In the gross relative domain it can be associated with the concept of gravitational attraction to a point—the center point. In the subtle relative domain, however, it can be associated with the unifying value associated with the concept of attraction experienced as love, directed either to an individual, family, kin, humanity, or God, all moving toward the totality of all existence—the center or heart of everything, toward the ultimate unity or Oneness.

Correspondingly the influences counteracting the force of gravity would seem to be directly related to the speed of light and Planck's constant. The point particle field is said to be inherently dynamic. This is quantified in terms of the Planck energy, the amount of energy that is inherent to each quantum, directly related to the speed of light. It seems quite reasonable that this can be related to rajas or activation. Correspondingly the property of viscosity or resistance to motion associated with Higgs theory seems directly related to the principle or force of tamas, and possibly to Planck's constant. Thus the three components from which the quantized *Planck length* is derived—gravitational constant, light-speed, Planck's constant—may have direct correspondence on the level of the gross material domain with the natural dynamics of the abstract forces of sattva, rajas, and tamas—the gunas in ancient Vedic science.

The Five Mahabhutas

The three inseparably interacting qualities of the gunas function on the subtle relative domain to guide impulses of intelligence and energy through the subjective levels to objective action or behavior. In the gross relative domain they also function as fundamental physical forces that shape the five fundamental constituents or mahabhutas of gross space, air, fire, water, and earth. Again, it is important to avoid interpreting the mahabhutas too concretely or literally. Even though making up the gross relative domain, they still are quite abstract principles. For example, the mahabhuta of air not only refers to what we ordinarily think of as air, but more fundamentally to the abstract principles or qualities that manifest as gaseous processes, and also agglomerations into matter. As noted in Chapter 11, the abstract nature of the mahabhutas may be more obvious with respect to the 'element' of *fire*—which doesn't carry the tangible connotations ordinarily associated with air, water, or earth. The mahabhuta of fire refers to the underlying laws of nature involved in *transformations* of objects through processes such as radiation, combustion, oxidation, and illumination. Likewise the mahabhuta of *space* refers to the abstract principle of vacuity, *air* to mobility, *water* to liquidity, and *earth* to solidity.

In the transition from the subtle relative domain to the gross relative domain, the subtle essences—tanmatras—are further conditioned or limited into the gross mahabhutas. The mahabhutas are ontological precipitants or evolutes of the tanmatras. The general feature of the progressive ontological emergence of the gross relative domain from the subtle relative domain is that tamas—the force of inertia or restraint, or mass—becomes more prominent.

Gross space or akasha is the limitation of the level of the tanmatras or subtle essences into the gross relative domain—due to a relative increase of tamas that transforms the tanmatra of earth into the mahabhuta of gross space. The five mahabhutas come out of the underlying subtle essence of the earth tanmatra. The earth tanmatra is the unified basis or medium of all objects, particles, and physical forces in the gross relative domain. It thus can be thought of as kind of a unified field, but only with respect to gross relative domain of nature. It can be distinguished from the unified field of *everything* as the unified basis of the material domain only. Deeper than the gross domain is the subtle level of nature made up of the subtle essence tanmatras, senses, mind, intellect, ego—and ultimately consciousness itself. The unified field of everything is the transcendental unified field of consciousness—the universal Self, pure Being. This identifies the gross, subtle, and transcendental levels of nature, which we have been working on clearer appreciation of in this book. So far the discussion has emphasized the gross objective domain. In the next part of the book, we will focus more on the subtle subjective domain. The change from gross to subtle also can be associated generally with the change from objective to subjective levels of nature, depending on definitions used.

In the increasing limitation or condensation of consciousness into matter, phenomena in gross or conventional space are the most limited. Again, the *grossest* of the tannmatras—the essence of the earth constituent—is further restricted by the predominance of tamas into the *subtlest* level of the gross relative domain—the *mahabhuta* of space. The mahabhuta of space contains all the five qualities of the subtle essence tanmatra of earth. The change from the subtle essence elements of the tanmatras to the gross constituents or elements of the mahabhutas involves predominantly nonlocal wave dynamics *sort of* condensing or compactifying into predominantly local point-particle dynamics. The nonlocal properties of the tanmatra of earth—containing all the five essences on the subtle level corresponding to the five senses—concretize into the local aether of space.

It again may be useful here to discuss briefly wave mechanics with respect to subtle and gross levels. An ordinary gross wave is a progressive vibration through a medium, such as air or water, without corresponding progress of the parts or particles themselves. The medium undulates in opposite directions, curving back onto itself through sequential points in time as impulses of propagating energy. This can be thought of as a pulsation that gets extended as it moves through space and time into a wave pattern—potentially expanding and contracting in all three dimensions simultaneously. An ordinary *physical* wave made of matter, as a pattern of the movement of particles such as an ocean wave, is a collection of undulating localized water molecules. A wave on the subtle level of nature is nonlocal, and not a collection of particles or molecules of matter.

The five mahabhutas or gross constituents of nature can be thought of as point-particle fields with progressive limitations, each more expressed one embedded in the previous one. They also can be thought of as progressive layers of gross space-time, each one taking on an additional specific quality from which is expressed a variety of different material phenomena. The fields are mediums which embody the specific qualities at each point in the field, as if each point is an individual point but all points exhibit the same inherent quality. The five mahabhutas can be thought of as point-particle fields in degrees of specificity from the least limited, most symmetric of gross space to the most limited, least symmetric of gross earth. The manifestation of these processes in terms of the gross senses is the phenomena of ordinary space, air, fire, water, and earth. From the mahabhuta of gross space the four other mahabhutas emerge—the *paramanus or* smallest indivisible units described in Chapter 11. The progressive unfoldment of the five mahabhutas involves increasing restriction of motion, localization, and independence—denser packing and rigidity, so to speak, associated with increasing tamas akin to the concept of mass.

Within the framework of relativistic space-time associated with Einstein locality, one way to think about the paramanus is that they are structured by the space-time gravitational field being further limited, drawing into its point value—sharply collecting into or curving back onto itself. In the sequential manifestation from transcendental to subtle to gross, the localized gravitational field can be thought of as curving back onto itself or as compactified into discrete forms that function as if independent and self-contained. The mahabhutas are sometimes described as dimensionless points. As paramanus, they exhibit the properties of discrete material particles, gaining mass and extension in space, which also characterize Planck-size quanta in quantum theory.

These mechanics of manifestation in the gross relative domain may relate directly to the contemporary theoretical research exemplified in M-theory, string theory, and loop quantum gravity theory. These research approaches appear to be attempting to characterize how nature structures Planck-size fundamental particles from dimensionless points. In the context of the discussion here, the theories seem to be attempts from a reductive perspective to explain how spatial extension in conventional local space-time is manifested in the transition from subtle to gross creation—from predominantly nonlocal wave-field mechanics to point particles and Planck-size quanta. These mechanics are reflected in wave-particle duality, and the challenge of trying to model the simultaneous local-nonlocal fabric of nature.

The subtle nonlocal field or level would not have the limitation of the speed of light. If non-material, nonlocal, superluminal, subtle phenomena exist, then they would have neither physical mass, nor quantum-sized particle nature, nor be subject to the gravitational force associated with gross or conventional space-time. This doesn't mean that they are not relative, not finite, not deterministic, and not subject to an attractive force; but rather that they aren't as limited as objects built of gross matter. They would be made of an even subtler aether-like 'substance' more akin to *mindstuff* than *matterstuff*.

Again, we can think of three levels of curving back or referral processes. At the transcendental level of the unified field the referral process is *infinite* self-referral, or *Self*-referral. At the subtle relative or subjective level the referral process tends to be characterized more in the sense of referral to the individual self—finite individual self-referral moving toward infinite self-referral or Self-referral as individuality is established in universality. At the gross relative or objective level the referral process can be characterized as point-particle referral—referral into a finite local point value, or object-referral. In this progression to increasing limitation, the 'curving back' eventually appears to curve back onto the local point value at the gross level of existence, and further to the Planck-size quantum and the atom. This is a highly localized form that appears to limit or cut itself off from the underlying nonlocal and infinite levels of nature in which it is embedded. The less limited subtle levels—as well as the transcendental level—become *unexpressed* or *hidden* in this gross, discrete, highly localized form—due to the hiding influence of tamas.

The gross level expresses *least* the inherent qualities of life and intelligence. It can be characterized as being objectified and inert—not exhibiting subjectivity and not exhibiting life. Gross objects take on the appearance of being independent of each other—exhibiting localized interactions characteristic of classical physics. They are characterized as having a localized particle aspect, but also more fundamentally a nonlocal wave aspect—that is, a quantum wavelet. The particle aspect characterizes its gross appearance, and the nonlocal wave its less expressed subtle appearance that guides the more concrete particle aspect, such as in Bohmian mechanics.

Further, gross objects made of gross particles appear as if they don't have the ability to initiate action or movement—they are described as inert matter, with no power of agency. They can be said to be inherently dynamic, but don't appear to exhibit agency, intentionality, or intelligence 'on their own.' With respect to living organisms made of these inert particles, it is the influence of the energy and intelligence on the subtle relative domain that guides their orderly, intelligent, negentropic gross behavior. This was discussed in prior chapters, and will be discussed in more detail in the next part of the book.

Each of the five gross elements (mahabhutas) exhibits a specific quality that corresponds directly to one of the five sensory capacities. As the limitations of the five elements increase, their gross objective forms express more specificity, which correspond to tangible sensory qualities. In this enumeration the mahabhuta of space is associated with and is necessary for the sense of hearing, air with touch, fire with sight, water with taste, and earth with smell. Whatever additional limitation characterizes the next grossest level of nature within the gross relative domain, it brings out its corresponding sensory quality. Space relates to hearing, air adds touch; fire adds sight; water adds taste, and earth adds smell. The grossest mahabhuta of earth expresses all five qualities, and thus can be sensed via all five sense modalities. All objects in the gross relative domain of nature are composed of these five fundamental principles or constituents in various degrees and percentages. They constitute the level of nature in which the localized particle or point value predominates. Thus the mahabhutas are progressively associated with increasing tamas, increasing limitation, increasing entropy, increasing sensory specificity, and with decreasing degrees of freedom and symmetry.

As physical *realities* of the phenomenal world of ordinary experience, the five mahabhutas must in some way correspond to the emergence of quantum particle fields and particle states in sequential symmetry breaking into levels of physical matter as conceptualized in modern physics. The current state of knowledge in modern physics may not be quite developed enough to establish the precise correspondences and relationships between the known particle and force fields and these levels of nature as described in ancient Vedic science. However, if both are describing the same natural world, it seems reasonable to suggest that they will match up. It is interesting to speculate further on how these two different descriptions of the gross physical level of existence in nature correspond.

In ancient Vedic science, the qualities of the five mahabhutas are expressed in sequential enumeration. This means that the mahabhuta of gross space contains the properties of the other mahabhutas, but expresses the qualities associated with space. To link this system to the fundamental physical forces as identified in modern science, it seems reasonable that the mahabhuta of space would be most closely associated with the gravitational force, but also would encompass the other fundamental physical forces—strong nuclear, weak, and electromagnetic. Likewise the mahabhuta of air would express the gravitational and strong nuclear forces, but also would incorporate the weak and electromagnetic forces. The mahabhuta of fire would express the gravitational, strong, and weak forces, but also include the electromagnetic force. In this comparison, the mahabhutas of water and earth would express all the four forces but be most associated with the electromagnetic force.

Space (Akasha). In the gross relative domain all gross objects are expressions of the mahabhuta of space—the subtlest mahabhuta. The dynamic properties of objects existing in this level of manifestation have the limitation of the speed of light, or Einstein locality. All gross movement of energy and mass in conventional space-time appears to reflect this limit, and is subject to the local relativistic nature of space and time in the gross relative domain. This limit also can be related to the Planck scale, zero point energy, and the uncertainty principle, the defining features that make up the texture of the medium or aether of gross space-time.

Conventional or gross space in terms of the field of Einstein locality can be thought of as the field of existence in which objects are limited to a certain range of interactions—namely, local interactions limited by the speed of light. The limitation from the subtle relative domain to the gross relative domain can be associated with the localization of action in space-time to the gravitational field characterized by the speed of light—the speed limit of material objects with mass and the speed of massless particles. A massless particle expresses the point value of nature in its most limited sense. But again, it also is underlain by a nonlocal wave in the subtle relative domain. Gross or conventional space can be characterized as an extremely refined, rubbery kind of aether that can expand, contract, and be curved and shaped by the more limited objects that it comprises and that appear to occupy it—especially massive objects such as material forms.

Air (Vayu). The next limitation within gross space in ancient Vedic science precipitates from the mahabhuta of space into the mahabhuta of air. In the increasing limitation of the underlying gravitational field it is the nature of the gravitational unifying force to attract points of space-time together into clumps or regions of more and less compression of the space-time field. This further precipitates into a gaseous state which expresses the principle of air—and eventually into more tangible material forms and objects. Air has the quality of expansion to fill the *available* space. Unlike space, however, fundamental to the nature of air is that it cannot permeate other objects *in* space. This means that it expresses the additional boundary or limitation of *impermeability.* In the increasing localization and independence of phenomena, space condenses into a more limited field or medium—a more concrete substance or gaseous field. This more limited, more concrete mahabhuta of air is also characterized by the property of transferring energy through it via the mechanics of compression and rarefaction in particle wave motion, within the limiting property of impermeability. This also can be related to the principle of *prana* in ancient Vedic science, sometimes associated with breath.

As discussed in Chapter 11, it is interesting that in ancient Vedic science the sense of hearing is associated with the mahabhuta of gross space. But we know from sensory physiology that gross hearing is mediated by sound waves in a medium of some kind that involves propagation through compression and rarefaction of impulses of energy into the sensory mechanisms that make up the middle part of the ear. Because the gross sense of hearing hasn't been found to be able to function in so-called empty space, a more limited medium seems to be required. Hearing involves the three-dimensional nature of space in some more concrete medium—associated also with the principle of air. In the gross relative domain, phenomena are characterized in terms of particle and wave properties. But the waves associated with sound energy are characterized as local phenomena in an ordinary physical field of particles that mechanically interact with each other. They are not the same as the more abstract wave aspect of the quantum field characterized by nonlocal interactions. In ancient Vedic science the mahabhuta of air corresponds to the gross sense of touch. As described above, the nature of air is that it fills the available three-dimensional space—within the constraints of gravity—but it has the additional limitation of not being able to permeate objects. These are properties of a gas, also associated with the physical mechanism of pressure, compression and rarefaction.

What might limit gross space into the property of impermeability and the physical mechanics of compression and rarefaction involved in the propagation of sound energy through space associated with gross hearing and touch? With respect to particle physics, the fundamental force that binds or glues particles into atomic nuclei and compounds is the strong nuclear force. It seems reasonable that this force relates to the principle of air in ancient Vedic science. Thus the mahabhuta of air would express the gravitational force along with the strong nuclear force (but again also including the weak and electromagnetic forces not yet expressed or sensed).

Fire (Tejas). In ancient Vedic science the next level of the gross relative domain is the mahabhuta of fire, corresponding to the sense of sight. The mahabhuta of fire is associated with luminosity, form, and transformation, related to heat and temperature as well as radiation, combustion, and oxidation—including related processes of digestion in the physical body. Fundamental to *fire* is oxygen, a core element associated with the principle of air, involved in combustion.

When there are aggregates of points as volumes in space-time that cannot penetrate each other, as in air, their agitation increases when the space they occupy is further limited. When motion of air or a gas is further restricted, pressure and activity rise, measured as increased temperature or heat. At certain degrees of temperature—as well as other factors related to energy levels—particles are radiated in the form of kinetic energy. This process of transformation involves interaction with 'air' in 'space.' As fire, it results in a gaseous 'upward' motion such as of smoke, and release or radiation—heat and luminance—but also 'downward' motion of ashen remains—expressed in radiation, and in ashes. The energy locked up in a substance is released as kinetic energy in radiation and light, the remains of which is a denser, inert, less dynamic form of matter. The release of energy in radiation results in a lower energy state in what remains. As Greene points out:

"Gravity is a universally attractive force; hence, if you have a large enough mass of gas, every region of gas will pull on every other and this will cause the gas to fragment into clumps... Even though the clumps appear to be more ordered than the initially diffuse gas—in calculating entropy you need to tally up the contributions from *all* sources... For the initially diffuse gas cloud, you find that the entropy decrease through the formation of orderly clumps is more than compensated by the heat generated as the gas compresses, and, ultimately, by the enormous amount of heat and light released when nuclear processes begin to take place." (p172) [1]

Continuing the comparison with fundamental physical forces, the mahabhuta of fire thus seems to be associated with the gravitational, strong nuclear, and especially weak forces—which involve quark decay and radiation, such as sunlight. Also, in ancient Vedic science the mahabhuta of fire would include electromagnetic forces and be related to chemical processes as well, but not yet expressed. The radiation of photon energy has an obvious relationship to the sense of sight.

Water (Apas). The next level of manifestation associated with an increase of tamas is the mahabhuta of water, directly related to the concept of liquidity or fluidity, and corresponding to the sense of taste. It has the freedom of flow or movement to fill the available space within the limitations of its permeability, but because of its lower kinetic energy and higher mass, only sort of 'downward' motion, due to the attractive force of gravity on its increased mass. For example a liquid conforms to the shape of the bottom of a container, but not its entire volume in the manner of a gas. The liquid state, such as water, has additional limitations over fire, air, and space. There is less internal motion, less heat, and additional restriction into a liquid flow rather than a gaseous expansion. Again Greene discusses relevant points with respect to symmetry principles:

"On a molecular scale, for instance, ice has a crystalline form of H_2O molecules arranged in an ordered, hexagonal lattice... The overall pattern of the ice molecules is left unchanged only by certain special manipulations, such as rotations in units of 60 degrees about particular axes of the hexagonal arrangement. By contrast, when we heat ice, the crystalline arrangement melts into a jumbled, uniform clump of molecules—liquid water—that remains unchanged under rotations by any angle, about any axis. So, by heating ice and causing it to go through a solid-to-liquid phase transition, we have made it more symmetric... Similarly, if we heat liquid water and it turns into gaseous steam, the phase transition also results in an increase in symmetry. In a clump of water, the individual H_2O molecules are, on average, packed together with the hydrogen side of one molecule next to the oxygen side of its neighbor. If you were to rotate one or another molecule in a clump it would noticeably disrupt the molecular pattern. But when the water boils and turns into steam, the molecules flit here and there freely; there is no longer any pattern to the orientations of the H20 molecule and hence, were you to rotate a molecule or group of molecules, the gas would look the same. Thus, just as the ice-to-water transition results in an increase in symmetry, the water-to-steam transition does so as well." (p. 253) [1]

The abstract principle of liquidity embodied in the mahabhuta of water expresses the concept of *flow*—the movement of some form of energy through or along a specific path, such as a current of water in a river. With respect to fundamental physical forces, this seems to be most closely associated with the electromagnetic force—the motion or flow of electrons. The outer shell of charged atoms allows electrons to flow. This current or flow, such as through a medium of copper wire, is from negative to positive and positive to negative electrical charge. Electric current flows easily when the electrons are loosely held. Mediums that hold the electrons more tightly are insulators, in which the flow of electrons is restricted. In this comparison the mahabhuta of water would express all the properties of all the four fundamental physical forces—and specifically the electromagnetic force, with a bit more emphasis on electricity.

Historically electromagnetism was thought to be two separate forces—electricity and magnetism—before their underlying symmetry was recognized. This symmetry so intimately connects electricity and magnetism that they are not characterized as differentiating through symmetry breaking, as are the other fundamental forces. It isn't that at a particular stage of the cooling of the universe the electromagnetic force differentiated into electricity and magnetism. Electric and magnetic properties are intimately related, and are basically one fundamental force. However, materials can appear to exhibit electricity or magnetism, as well as both or neither.

Prthivi (Earth). The grossest level of the gross relative domain is the mahabhuta of earth, associated with solidity, the most limited, most inert state, and related to the sense of smell. The mahabhuta of earth has all the five sensory qualities active: it can be heard, felt, seen, tasted, and smelled. Matter associated with the principle of earth has no directional freedom, in the sense that it doesn't flow—except to the degree that it is mixed with the other mahabhutas such as water or air. It involves various degrees of crystalline structures, with relatively rigid and fixed alignment of parts. The solidity of the mahabhuta of earth represents increasing limitation over a liquid form—such as water into ice when the temperature and motion associated with heat or fire is reduced into a colder, more rigid, less dynamic state.

The mahabhuta of earth is the point that is the most fixed, solid state of nature—the phenomenal level in which matter is expressed most and consciousness is expressed least. It is the endpoint of the process of manifestation. With respect to correspondence with the sequence of fundamental physical forces, it would seem that the mahabhuta of earth is most associated with the magnetic force—although tangibly expressing all of the fundamental physical forces and all of the other mahabhutas. Is there some reasonable basis to associate the mahabhuta of earth more with magnetism and the mahabhuta of water more with electricity?

An electric current flows across objects between charge sources. Attraction and repulsion between two charges occurs in a *straight line* between the two point sources of the charges. Electric currents generate magnetic fields. In contrast to the electric force, the magnetic force is a dipole system in which the opposites of attraction and repulsion—north and south poles—are contained in one source and travel in a more defined *circular* path—a path that curves back onto itself. The endpoint of the magnetic field is itself; in magnetism there is both north and south poles in one point source. The magnetic force most tangibly turns back on itself in a closed circular loop around an electric current—in a perpendicular direction to the current flow. This more contained quality of the magnetic force can be thought of as a further limitation compared to the more free flow of the electric force, even though they are intimately involved with each other.

All matter exhibits magnetic properties in the presence of a magnetic field, and can be classified in terms of degrees to which it is attracted or repulsed by it, depending on the alignment of atoms in the material objects. In some cases the attractive and repulsive forces cancel each other, resulting in net neutral magnetic properties. The association of the earth mahabhuta with magnetism doesn't mean that all materials made of earth are magnets—although they all interact to some degree with magnetic fields. It is that the abstract principle associated with mahabhuta of earth can be related to the underlying laws of nature that are expressed as magnetism, a bit more closely than with the mahabhuta of water. Again, the level of the mahabhuta of earth is the expression of the combination of all the five mahabhutas. All the fundamental forces are differentially functioning at this level of nature. In this delineation the magnetic force is based in the electric charge—which is consistent with the theory of electromagnetism. The electromagnetic force is expressed within the framework of the other three forces—weak, strong, and gravitational.

However, there is an important difference with respect to the underlying basis of the four fundamental physical forces in modern physics and the underlying basis of the mahabhutas in ancient Vedic science. The level of nature that is the immediate basis for the five mahabhutas is the subtle level of nature, and specifically the earth tanmatra. However, in modern particle theory, the underlying basis of the four fundamental forces and related particles now presumably would be the unified quantum field. In ancient Vedic science there is an enumeration of three fundamental domains—gross, subtle, and transcendental. Recognizing these three levels of nature or levels of *reality*—and especially the subtle relative domain—provides the needed bridge to account for many paradoxical phenomena that have plagued the objective approach of modern science—such as the mind-body problem, wave-particle duality, the measurement problem, locality and nonlocality, and similar unresolved issues. These will be discussed at length in the next part of the book, in the context cognitive science and neuroscience.

The five spin states. As described earlier in this chapter, the five states of internal spin also seem to have a direct correspondence to the five fundamental constituents or principles of space, air, fire, water, and earth. Only elementary particles associated with the spin states of 1 and ½ have been experimentally confirmed; elementary particles associated with the spin states of 2, 3/2, and 0 have not yet been found in nature. (Evidence for spin 0 compound particles such as the pi meson has been found, but not the elementary spin 0 particles such as the spin 0 Higgs particle). Another way of saying this is that antiparticles and super-symmetric sparticles have not been found yet in nature. However, the model of five fundamental spin states is strongly supported by the mathematics of the theories that predict their presence in nature. Given that the five spin-states model does reflect the structure of nature, let's consider other possible relationships with the fundamental principles in ancient Vedic science associated with the five constituents of space, air, fire, water, and earth—which of course we do know exist in phenomenal nature. To do this, we need more detail on the five spin states.

In the model of five spin states, spin 2 has the highest degree of freedom. Spin 2 values can be + or- 2, + or – 1, and 0, totaling five different possibilities. Spin 3/2 has four (+ or – 3/2, + or – ½); spin 1 has three (+ or – 1, 0); spin ½ has two (+ or – ½); and spin 0 has only one possibility.

There a nine possibilities in the five spin states (+2, -2, + 3/2, -3/2, +1, -1, +1/2, -1/2, 0) made up of three distinctions of fundamental properties that mathematical point particles in conceptual space seem to exhibit (although again, physical interpretations of them in conventional space-time are not known yet). The most fundamental distinction is between the particle families of bosons and fermions—integer versus half-integer spin (Integer spins 2, 1, and 0 particles are bosons; and half-integer spins 3/2 and ½ particles are fermions). Also there is the distinction of the value of the spin (2 or 1, and 3/2 or ½), and the distinction of the opposite signs of plus and minus which relate to opposite directions of spin in mathematical space. Adding to these three binary classifications the no-spin state (spin 0) makes for the nine possible spin states. Of the particles that have been found in nature, however, there are five possible states rather than nine. As mentioned in Chapter 3, there is at least presently little if any evidence for super-symmetry as a basic feature of the natural world, which directly relates to the higher-order spin types, dark energy, and string theory. [24] [25]

The five types of internal spin also relate to three different types of rotational symmetry or invariance, as well as different types of mathematical fields. Rotational invariance in mathematical space concerns the transformations necessary to reestablish the same appearance (refer to the analogy of rotating face cards by Hawking earlier in this chapter). Different types of mathematically defined fields have to do with a directional component of the point particle field. The reference used here is four-dimensional space—three spatial and one time dimension.

Spin 2, associated with gravitation, relates to a *tensor* field (rank 2) in which there is magnitude and direction in all three spatial axes (plus time) associated with every point in the field. For this spin state, rotational invariance means that a 180-degree spin results in a return to the same appearance or original state. Spin 3/2, associated with connecting gravitation to the other forces, relates to a *pseudo-tensor field*. It is a tensor field (rank 2) but with a change in orientation about the axis of rotation to the opposite sign. In this case a 180-degree rotation results in change in orientation or the opposite sign of the axis of rotation. Thus rotational invariance involves a 360-degree spin in order to return to the original state. These tensor fields seem to correspond to properties of the mahabhutas of space and air. They also may correspond directly to the three-dimensional dynamics of the senses of hearing and touch.

Spin 1, associated with electromagnetism, relates to a *vector* field (rank 1 tensor) in which there is magnitude and a particular direction in one axis—a directional force field. For this spin state, a 360-degree spin also results in a return to rotational invariance of the original state. Spin ½, associated with matter fields, relates to a *pseudo-vector* field involving a vector with opposite sign. For this spin state, when the directionality of the field is rotated 360 degrees, there is a change to opposite sign, so a 720-degree rotation is needed to return to the original state for rotational invariance. Spin 0, associated with the Higgs field and particle mass, relates to a *scalar* field (rank 0 tensor) which has only magnitude and no directional meaning—no internal spin.

The mahabhutas of space and air may correspond to tensor fields. The mahabhuta of fire, and to some degree the mahabhuta of water, may correspond to vector fields. The mahabhuta of earth may correspond to a scalar field—the most inert, least dynamic level of nature characterized by magnitude but no inherent directional component.

These speculations about the relationship between particle physics and mahabhutas represent an important direction in developing a precise and unequivocal matching of modern physical theories and ancient Vedic science. Although the precise matching of fundamental forces and spin states in modern physics with the five fundamental constituents in ancient Vedic science has yet to be established definitively, even at this point the possible correspondences encourage additional research. This might require that physical theories advance further the understanding of the physical meaning of quantum numbers, especially internal spin. Nonetheless, hopefully this chapter brings out sufficient correspondences between accepted principles in modern physical theories and in ancient Vedic science, such as the Sankhya enumeration of levels of nature, that will lead to their integration and will yield major scientific advances toward a more holistic appreciation of the natural world and our place in it.

The next chapter begins the Psychology Unbound part of the book, in which the focus shifts to the deeper, subtler, subjective domain. We will begin this second part of the book with a discussion of consciousness, applying criteria from contemporary work on the so-called *hard problem* of consciousness. The issues considered are fundamental for resolving long-standing dilemmas about objectivity and subjectivity and the relationship of individual body, mind, and consciousness. A crucial plank is a definition of consciousness that bridges *individual* conscious experience and *universal* consciousness.

Chapter 12 Notes

[1] Greene, B. (2004). *The fabric of the cosmos: Space, time, and the texture of reality*. New York: Alfred A. Knopf.

[2] Greene, B. (1999). *The elegant universe: Superstrings, hidden dimensions, and the quest for the ultimate theory*. New York: Vintage Books.

[3] Lincoln, D. (2004). *Understanding the universe: From quarks to the cosmos*. Singapore: World Scientific Publishing Co., Pte. Ltd.

[4] Guth, A. H. (1997). *The inflationary universe: The quest for a new theory of cosmic origins*. Cambridge, MA: Perseus Books Group.

[5] Penrose, R. (2005). *The road to reality: A complete guide to the laws of the universe*. New York: Alfred A. Knopf.

[6] Guth, A. (2003). A golden age of cosmology (pp. 285-296). . In Brockman, J. (Ed.). *The new humanists: Science at the edge*. New York: Barnes & Noble Books, p. 294.

[7] Smith, C. L. (2003). The Large Hadron Collider: The edge of physics. *Scientific American*, Special Edition, May.

[8] Quintessence. (2000, November) *Physics World*.

[9] Caldwell, R. R., Dave, R., & Steinhardt, P. J. (1998). *Phys Rev Lett., 80*, 1582.

[10] Zlatev, I., Wang, L., & Steinhardt, P. J. (1999). *Phys Rev Lett, 82,* 896.

[11] Steinhardt, P. (2003). The Cyclic Universe. In Brockman, J. (Ed.). *The new humanists: Science at the edge*. New York: Barnes & Noble Books, pp. 297-311.

[12] Darling, D. (1996). On creating something from nothing? *New Scientist,* Vol. 151, No. 2047, 14, September.

[13] McGinn, C. (2000). Consciousness and Space. In Shear, J. (Ed.) *Explaining consciousness—The hard problem*. Cambridge, MA: The MIT Press.

[14] Hagelin, J. S. Is Consciousness the Unified Field? A Field Theorist's Perspective. *Modern Science and Vedic Science, 1, 1,* (1987), 29-87.

[15] Hagelin, J. S. Restructuring Physics from Its Foundation in Light of Maharishi's Vedic Science. *Modern Science and Vedic Science, 3, 1,* (1989), 3-72.

[16] Hawking, S. (2001). *The universe in a nutshell*. New York: Bantam Books.

[17] Pugh, N. D., and Pugh, D.C.,(1999). Unveiling Creation: Eight is the key. Fairfield, IA: Sunstar Publishing Ltd.

[18] Laughlin, R. B. (2005). *A different universe: Reinventing physics from the bottom down*. New York: Perseus Books Group.

[19] *Celebrating Perfection in Education: Dawn of Total Knowledge*. (1997). India: Age of Enlightenment Publications (Printers).

[20] Radhakrishnan, S. (1978). *The principle Upanishads*. USA: Humanities Press International (First published 1953 by George Allen and Unwin, Ltd.).

[21] Maharishi Mahesh Yogi (2004). Maharishi's Global News Conference, June 23.

[22] Nader, T. (2000). *Human physiology: Expression of Veda and Vedic Literature,* 4th Edition. Vlodrop, The Netherlands: Maharishi Vedic University.

[23] Einstein, A. (2006) as quoted in Albert Einstein home page, URL = <www.humboldt1.com/~gralsto/einstein/quotes.html>

[24] Woit, P. (2006). *Not even wrong: The failure of string theory and the search for unity in physical law*. New York: Basic Books.

[25] Smolin, .(2006). *The trouble with physics: The rise of string theory, the fall of a science, and what comes next*. New York: Houghton Mifflin Co.

Part II

PSYCHOLOGY UNBOUND:
FROM INFINITY TO HERE

This part of the book establishes planks that link progress in modern scientific theories of body, mind, and consciousness to the unified field. Progress toward a general model of the mind in psychological and cognitive science and neuroscience is summarized. The challenge of finding the place and role of consciousness in functional and structural models of the mind is described, and is resolved in the expanded framework of gross relative, subtle relative, and transcendental levels of nature.

Further, how individual psychology can be systematically unbound and naturally developed to higher states of human consciousness are described. Examples are included of phenomenal reports of experiences in higher states of consciousness, drawn from research on practitioners of Maharishi Vedic Technologies and from accounts in historical literature. Systematic technologies for the full development of individual and societal life in higher states of consciousness are introduced.

Chapter 13

Fundamental Issues in a Scientific Theory of Consciousness

In this first chapter of Part II, fundamental issues concerning a scientific theory of consciousness are introduced. Examples are given of how the issues are addressed through application of the full ontological range of levels of nature in the inner dimension described in Chapter 11, based on the holistic approach of Maharishi Vedic Science and Technology. The chapter places the third-person, reductive, fragmented perspective in modern science into a more developmentally significant first-person, holistic perspective based in the unified field. We will consider how the so-called 'hard problem' of consciousness is resolved in the natural course of developing higher states of consciousness. The overall main point of this chapter is that the 'nature' of consciousness is both individual and universal.

Consciousness is perhaps the most challenging concept in the history of human thought. This is not surprising—even expected—if consciousness is beyond the conceptualizing mind. Based on the discussion so far in this book, it also might not be surprising that there are three major perspectives that attempt to address the nature of consciousness: *structural*, *functional*, and *experiential*. Embedded in this can be found the fundamental trinity of known (structure), process of knowing (function), and knower (experiencer) within the three-in-one self-referral dynamics of the unified field. A general correspondence with its phenomenal enumeration in the gross relative, subtle relative, and transcendental domains of nature also becomes apparent.

Progress in modern science applying a reductive approach to attempt an explanation of consciousness basically is following the path of outer to inner, grosser to subtler—exemplified by the search for the essential constituent of matter and energy in physics, described in Part I: Physics Unbound. Chapters in this section of the book, Part II: Psychology Unbound, will show the direction of this progress toward the consciousness-mind-matter ontology in ancient Vedic science and Maharishi Vedic Science and Technology that emphasizes consciousness as the unified field underlying the laws of nature and their phenomenal expressions.

Focusing on structural analyses, neuroscience and related fields have been attempting to identify the locus of consciousness in the physical brain. Encountering great difficulty locating and explaining consciousness within the confines of the brain, attempts in recent decades inspired by the computer as a model of the mind have additionally emphasized a more abstract, disembodied functional analysis, such as in the psychological, cognitive, and information sciences.

However, in the physicalist paradigm or worldview, any functional model contains the same difficulties as the structural model, because function must be realized or instantiated in a physical structure.[1] A fundamental challenge is that the concept of *function* is abstracted from its physical realization, but everything must be fundamentally physical. There can be no place or existence of function outside of the physical, because nothing is outside of the physical. Therefore abstract functional organization must be just a theoretical concept. But in the structural approach the mind is thought to be in the brain, and brains as well as all material objects have been shown most fundamentally *not* to be physical matter. The notion that *function* is an abstract principle beyond its realization in material structure is a hint that mental concepts might actually exist *somewhere* outside or beyond the physical. Developments in quantum theories already summarized support the direction of understanding that this *somewhere* is a subtler mental or information space *level of reality* deeper than physical matter. This will be unfolded more fully in this part of the book.

Encountering great difficulty trying to identify either the location or function of consciousness using either structural or functional reductive objective approaches, there is growing appreciation of the importance of an *experiential* investigation of consciousness. [2] It seems quite reasonable that an experiential first-person investigation would be useful in order to explore inner mental or information space—the *quantum mind* associated directly with first-person subjective experience. This could be viewed as even more direct empirical investigation than assumptions of an independently existing world to be investigated objectively from a third-person perspective. [3]

Further, if the assumption that the universe is orderly is to be taken seriously, it would seem that the orderliness applies to both first-person subjective as well as third-person objective domains of nature. A systematic subjective means of gaining knowledge is consistent with this view. If there is an underlying quantum *reality* or quantum mind, then systematic first-person means would be more simple and direct for gaining reliable and accurate knowledge of it than fundamentally fragmented third-person objective means. This subjective methodology would need to isolate the substrate of conscious experience from the inside—through actual first-person experience in the inner laboratory of the mind of the scientist (refer to Chapter 2).

Indeed to assume a single, unified, holistic, self-aware basis of existence verifiable by direct empirical experience—discussed in Chapters 10-11—could be viewed as the most simple, parsimonious, and methodologically efficacious *scientific* view. To assume that it is more parsimonious to view nature as fundamentally fragmented, to assume obvious differences between ordinary everyday conscious experience and physical matter are artifactual or non-existent, and to consider third-person reductive experimentation as the only viable methodological approach could be seen to be quite *unnatural*. [1]

In Maharishi Vedic Science and Technology it is only through a direct, systematic, first-person, experiential approach that understanding and satisfying appreciation of the nature of consciousness develops. This is a logically consistent and clearly articulated alternative to the major structural, functional, and introspective approaches in modern science as it is currently practiced. [4] In this consciousness-based Vedic approach, both structural and functional attributions about consciousness are identified as associated mainly with experiences in the ordinary waking state of consciousness. An introspective approach of thinking about conscious experiences is also typically associated with the ordinary waking state of consciousness. Introspective, phenomenological approaches involve thinking about different levels of mind, with frequently only a foggy picture of their hierarchically embedded structure—accompanied by no systematic means to experience levels of mind and typically no reports of direct experience of consciousness itself.

Recognition that consciousness itself is distinct from its structural and functional aspects is fostered through direct experience of the fourth state of consciousness—transcendental consciousness. This was introduced in Chapter 2 as a primary means of gaining knowledge in Maharishi Vedic Science and Technology. The experiential bridge between individual consciousness in the ordinary waking state of consciousness and universal consciousness is crossed through direct first-person experiences of transcending the mind. This systematic subjective means is not, however, an introspective or self-reflective approach.

In this chapter core issues in modern science toward a science of consciousness are examined with the helpful addition of Maharishi Vedic Science and Technology. Definitions of consciousness and the relationship between body, mind, and consciousness are main topics.

Definition of Consciousness

As referred to in the Prologue and in Chapter 2, it is most often the case that meanings of the term *consciousness* are associated with the functional notion of *awareness of* some object of experience. It is frequently linked to the ability to be aware of and interact with one's surroundings, emphasizing the general concept of *wakefulness* or *alertness*, such as in contrast to deep sleep and coma. Another common association is with *awareness of* the external as well as internal environment—including sensations, perceptions, thoughts, feelings, memories, and awareness of individual self. These meanings are closely related to the concept of attention—frequently voluntary control of attention, but also involuntary control involving non-conscious or unconscious processing that direct conscious attention. Another more specific meaning emphasizes reportability (the ability to give a verbal report of the experience), as well as introspection or self-reflective thought about one's own inner subjectivity—closely related to language processes.

In an effort to highlight the distinction between functional and experiential aspects of consciousness, some go so far as to associate the term *awareness* just with the functional aspects, reserving the term *consciousness* for the experiential aspect.[5] This could be a quite useful distinction if awareness meant *conscious of*, and consciousness meant *consciousness itself*.

A related distinction has been drawn between psychological processes, referring to information processing functions whether consciously experienced or not, *and* phenomenal experience, referring specifically to the conscious experiential aspect. This distinction reflects recognition that consciousness is not fully accounted for by its functional aspects. Present in it can be found at least the slightest of glimpses that consciousness is beyond the structural body and the functional mind.

Applying this psychological-phenomenal distinction, the psychological or functional aspects have been labeled as *easy problems* of consciousness. Though they may not be particularly easy to address, they are said to be easy in the sense that they are the types of problems amenable to research and explanation within objective methodology. These issues are now typically examined using a combination of structural and functional approaches—applying the objective third-person perspective. They are distinguished from the ominous *hard problem* of consciousness. This refers to challenging issues that, due to the relationship of thinking to consciousness, inevitably arise in attempting to reason or think through an explanation of consciousness based on some form of introspective phenomenological analysis.[5]

The classification into the easy and hard problems of consciousness is considered by some to be artifactual and unproductive. It is sometimes suggested that modern science hasn't solved even the easy problems so there is little basis for distinguishing easy and hard ones,[6] or that solving the easy problems will solve the hard problem.[7] It is further suggested that we should be methodologically parsimonious and assume that conscious experience is 'identical to neurally instantiated cybernetic functions'—the *functional identity hypothesis*—until there is enough evidence to distinguish them.[8] This attempts to close the explanatory gap by squashing out any conceptual or experiential differences between consciousness and (grey) matter. Typically these positions are based on a pretheoretical commitment to physicalism, paradigm blindness with respect to counter-evidence that doesn't fit the paradigm, and especially—again not surprisingly—first-person reports of the inability to experience consciousness itself introspectively.[5,7,9]

The conceptualizing mind cannot fully grasp what consciousness is, because it is a limited sub-set of the wholeness that is consciousness. Inasmuch as consciousness is ultimately beyond conceptualization and explanation, it tends to remain a *hard problem* until there is clear *direct experience* of consciousness itself in its transcendental state beyond the conceptualizing mind.

Fully respecting the ineffable nature of consciousness from the perspective of the conceptualizing mind, nonetheless the *direct experience* of consciousness itself is not hard to have—indeed it is in some sense the simplest possible experience. It is hard and difficult only if a systematic natural methodology to transcend mental activity is not available, or has not been applied. The Transcendental Meditation technique is a simple, effortless, reliable procedure based on psychophysical principles of natural law inherent in the mind and body that foster regular transcending.[10] The principles and mechanics of this methodology will be described in more detail in the chapters on technologies to develop higher states of consciousness.

To again briefly describe the transcending process, consciousness itself is experienced through settling down the mind to lesser and lesser states of mental activity. In this natural and effortless settling process to the self-referral state, the limiting focus of ordinary thinking and perceiving relaxes, and the mind settles and expands into less bounded modes of subjective experience. Settling down into unbounded wakefulness void of mental activity, the simplest and least excited ground state of the mind is directly experienced—transcendental consciousness. In other words the overlooked knowledge and experience of consciousness itself is within our easy grasp through systematic, effortless, first-person, subjective means available in Maharishi Vedic Science and Technology (refer to Prologue).

Initially that experience might not seem at all like something that would be the ultimate source of creation—the unified field of nature. Ancient Vedic science and Maharishi's profound articulation of it in Maharishi Vedic Science and Technology record in detail how repeated transcendence eventually demonstrates empirically that it is. Direct empirical investigation of the transcendental level of nature eventually clarifies the relationship between the conceptualizing mind (buddhi), which attempts to explain consciousness, and its source in consciousness itself (Purusha). Even a little bit of *that* experience removes doubt and greatly facilitates understanding and appreciation of the essential nature of consciousness.[10] This developmental technology has fundamental implications for understanding consciousness, which need to be incorporated into an adequate definition of consciousness.

Given the 'nature' of consciousness itself as beyond thought, attempts to define it inevitably fall short. Because understanding and experience of consciousness change in different states of consciousness, a definition might ideally relate to different states of individual observers—a most daunting issue in developing a general definition. On the other hand a definition of consciousness as the unified field of nature is likely to seem counterintuitive for people not accustomed to clear experience of transcendental consciousness, and may not even make sense compared to what is familiar. A conceptual understanding of consciousness as the unified field of nature would reasonably be a hard problem for those who rely on ordinary object-referral sensory experiences and reasoning as the only bases for identifying *reality*. There can be such a strong fixation on intellectual analyses of experience—and belief in a closed physical universe—that systematic development beyond the ordinary waking state of consciousness is given little if any consideration as means to gain knowledge of what consciousness is and where and how it can be located.

A reasonable starting point for a definition of consciousness might seem to be one that reflects a broad consensus of experience in the general population. A simple definition of this type is consciousness as *awareness of* some object of experience, typical of dictionary definitions. However, for the most part this meaning is associated with limitations of the ordinary waking state of consciousness. This frequently carries with it the belief that consciousness is only in the brain, quite misleading with respect to the essential nature of consciousness as the unified field. It is challenging to comprehend the universal aspect of consciousness as pure self-referral awareness when experiences are in the context only of the ordinary object-referral waking state in which individual consciousness predominates. Both aspects need to be incorporated into a definition of consciousness, in order for the definition not to be so misleading. Perhaps a good starting point is to build a definition that recognizes the developmental distinction between individual consciousness and the universal level of consciousness in its pure self-referral state—identified as the fourth state of consciousness. This would reflect the simultaneous point-infinity, individual-universal nature of consciousness itself. A definition of this type might associate consciousness with either meaning, then add the distinction of universal and individual. Here is an example of a definition that applies this strategy:

Consciousness is wakefulness, alertness, or awareness itself; in its simplest self-referral state it is the unbounded, universal, transcendental essence of nature, and in the ordinary waking state it is the bounded individual awareness of some object of experience in nature.

Maharishi provides a fuller exposition of consciousness in the following quote:

"Consciousness is that which is conscious of itself. Being conscious of itself, consciousness is the knower of itself. Being the knower of itself, consciousness is both the knower and known. Being both knower and known, consciousness is also the process of knowing. Thus consciousness has three qualities within its self-referral singularity—the qualities of knower, knowing, and known—the three qualities of 'subject' (knower), 'object' (known), and the relationship between subject and object (process of knowing).

Wherever there is subject-object relatedness; wherever subject is related to object; wherever subject is experiencing object; wherever subject (knower) is knowing object, these three together are indications of the existence of consciousness.

The universe with its observer expresses the three values of observer, process of observation, and object of observation; therefore it is the indicator of the existence of consciousness. The universe, with its observer, is the expression of consciousness in its self-referral state. The observer, being conscious of the universe, is conscious of his own self-referral state.

The reality that the universe is the observer himself is the reality of the total disclosure of consciousness; it is the total potential of consciousness; it is the total reality of consciousness.

When we say total reality of consciousness, we mean consciousness in its self-referral state, where consciousness knows itself and nothing else. This state of consciousness is pure consciousness. Another state of consciousness is when it knows other things; then it is known to be object-referral consciousness, because all objects can only be perceived by virtue of the intelligence quality of consciousness, which creates the observer and process of observation within the singularity of the self-referral state of consciousness.

This establishes that the object-referral state of consciousness is also within the self-referral state of consciousness.

This reality of consciousness, that by nature consciousness is self-referral and object-referral at the same time, makes it obvious that the nature of consciousness is both singularity (self-referral), and diversity (object-referral). All knowledge about this field of consciousness is the basic knowledge of the Ultimate Reality at the basis of all manifest creation, which is complete knowledge of the transformation of singularity into diversity.

Complete knowledge of consciousness is the complete knowledge of the basic reality of life, which is available to everyone in the field of one's own Transcendental Consciousness through Transcendental Meditation." (pp. 53-56) [11]

Criteria for a scientific theory of consciousness identify crucial issues necessary to address in order for modern science to make substantive progress toward a coherent, completely unified understanding of nature that includes both objective and subjective phenomena. These include the issues of what is conscious and what is unconscious, the relationship between consciousness and a nervous system that embodies it, and the dynamic correspondence between matter and body, mind, and consciousness. Another crucial issue is the development of higher states of consciousness. Unfortunately many efforts to explain consciousness don't even include this issue, though it is critical in resolving the 'hard problem' of consciousness, and the 'easy problems' as well.[25] Consciousness is indeed a hard problem when experience is constrained only to the ordinary waking state, starkly reflected in characteristic statements such as the following:

> "Everyone knows what consciousness is: It is what abandons you every evening when you fall asleep and reappears the next morning when you wake up." (p. 3) [12]

The Conscious/Unconscious Distinction

From the holistic view of Maharishi Vedic Science and Technology that consciousness is the unified field of nature, the task of explaining consciousness requires accounting for phenomena that do not appear to be conscious. This is opposite of the impossibly hard task of accounting for consciousness on the basis of non-living physical matter—which structural and functional approaches must do. This is sometimes called the *generation problem*—related to "...how experience is generated by certain particular configurations of physical stuff." (p. 269) [13] It also is directly related to the *combination problem* of how to combine basic units of either physical or mental stuff into consciousness. [13] Both of these important problems are resolved by the holistic approach that consciousness is the unified field or universal Being. As will be discussed in upcoming chapters, it is also related to the *binding problem* in neuroscience of how neural activity and associated brain functions bind into unitary, coherent, goal-directed conscious behaviors in a living biological organism.

> "The whole of science revolves around this question of enduring organisms." —*Alfred North Whitehead, mathematician and clergyman (p. 194)* [14]

Attempts to explain consciousness in modern science, for the most part, still apply a reductive strategy within the physicalist worldview. In this view higher-order processes are thought to *supervene* or depend on non-living physical mechanisms—in the sense that phenomena including life, intelligent behavior, and consciousness can be accounted for from more basic lower-order physical processes. Chapters 4-10 summarized developments in contemporary quantum field theories extending past this view in pursuit of a more coherent understanding of mind and matter. However, it is largely within this view, and the challenges that consciousness brings to it, that criteria for a scientific theory of consciousness have been developing.

At the current stage of modern scientific knowledge of the physical world there are a few basic constituents in nature that are thought to be primitive, irreducible, or non-reductive, such as mass, charge, spin, and space-time—which basically means that modern science can't yet explain them. In Maharishi Vedic Science and Technology consciousness is not just non-reductive—but also ultimately beyond explanation. Charge, spin, and other primitives—as well as all other physical as well as mental processes—depend on consciousness and are ultimately reducible to it. There are no elements or aspects of nature more fundamental than consciousness itself. Consciousness is the simplest possible state of everything in nature, the only ultimate non-reductive constituent. Everything can be said to *supervene* in the sense of depend on the unified field of consciousness. In another sense, nothing exists that is more than, additional to, or unexpected with respect to consciousness itself as the unified field of nature, the source of everything.

If consciousness is the unified field, then how does something appear to be not conscious? If everything is consciousness, how can anything *not* be conscious? One approach to address this core issue is the Sankhya system of levels of nature in ancient Vedic science introduced in Chapter 11. Sankhya describes progressive levels of phenomenal limitation from the unified field of consciousness to the grossest domain of inert matter. Although everything is made of the unified field and is nothing other than the unified field interacting within itself, as a field of *all possibilities* it is possible even to hide its essential nature as consciousness and *appear* as not conscious, in the same way that it hides infinity in the finite.

One simple illustration of this is the analogy of various states of H_2O. Remaining H_2O, it can be found in its three basic states of solid, liquid, or gas, each of which exhibit quite different properties. Also, the degree to which H_2O is translucent or opaque is affected by its degree of purity—the amount of particulate matter or pollutants it contains. Remaining H_2O, its state or condition influences its degree of translucence. The first part of this analogy can be associated more with state changes related to the three ordinary states of consciousness. The second part can be likened to development of higher states of consciousness. As consciousness appears to take on different qualities of mind and matter, the phenomenal degree to which its pure nature is reflected is changed. Its appearance changes in the ordinary states of consciousness, and also changes with degrees of refinement or purity in higher states beyond the three ordinary states of sleep, dreaming, and waking.

A limitation of this analogy is that much of what reduces the translucence of H_2O comes from something other than H_2O—impurities such as minerals. In the case of consciousness itself as the unified field, ultimately there is nothing else that could make it appear less or not conscious; whatever covers over the essence of consciousness and results in it appearing not conscious somehow must come from—and must be—consciousness itself. In addition the essential nature of consciousness must not be affected or lost by the process of appearing less or not conscious. This is unique to consciousness, applying to no other phenomenon in nature. The resolution of these issues is nothing less than a comprehensive accounting of the manifestation of phenomenal creation from the unified field of consciousness, and the simultaneous coexistence and maintenance of these apparently opposite states and conditions. This is described in detail in ancient Vedic literature—a vast treasury of knowledge that opens all possibilities to human experience through its developmental technologies. In this book, at best only a few pieces of the whole picture can be presented.

Maharishi uses the analogy of colorless sap as the basis of all parts of a tree in order to clarify further these fundamental issues. Remaining colorless sap, it appears to transform into the different parts of the tree with their qualities of color, texture, flavor, and fragrance quite different from the colorless sap. But in probing deeply into a root, branch or flower petal, the underlying colorless sap is again found. Remaining colorless sap, it also at the same time appears in diversified states such as colorful branches, leaves, blossoms, fruits, and seeds. Remaining colorless sap, it appears to go through progressive expression in different phenomenal states—from seed to tree to seed, and from a single seed to a multiplicity of seeds.

Even in everyday usage consciousness is conceptualized as common to the three ordinary states—sleep, dreaming, and waking *states of consciousness*. This implies an underlying consciousness across these psychophysiological states, relevant to the distinction of conscious and unconscious processes. The typical description of sleep or coma, for example, is that consciousness is not present. There is also considerable research evidence in cognitive science and neuroscience that suggests even most psychological information processing occurs in a manner that is described as outside of ordinary conscious awareness. This evidence is consistent with one aspect of the overall view being presented here—to be discussed in later chapters. Even if the bulk of psychological processes are not experienced in consciousness, however, this doesn't mean that consciousness is less fundamental than or somehow emerges from unconscious psychological and physiological processes.

Unconscious processing can be described as a type of conscious processing, rather than consciousness being either present or entirely absent. The distinction need not be categorical, but more like a continuous scale from consciousness to minimal or no obvious conscious experiential aspect. In the sleep state there could be some small degree of conscious experience, for example—and even in deep sleep or coma. Additionally, lower forms of life could have some minimal amount of conscious experience. This can be taken further to include everything in creation—although counting a rock as a conscious entity may seem quite a stretch.

Analogously all ordinary classical objects are composed of elementary particles and even nonlocal fields. Although these underlying components and processes are not directly observed, they are believed to exist at more fundamental levels of the objects. The ordinary senses are capable only of experiencing the macroscopic classical level of objects, not their sub-phenomenal components with respect to the ordinary sensory level. But as has been discussed in prior chapters, their underlying structural and functional relationships are thought in some sense to be more *real*—at least in the important sense that including them leads to more coherent, accurate, and reliable scientific theories, explanations, and predictions.

This relates to an important general principle about subjective experiences corresponding to structures and functions throughout nature. At least in most instances the system of attention in information processing—in humans and likely all beings—does not involve direct experience of its mechanical and functional operations. For example the process of digestion generally presents only its final product to experience.[8] If things are going well with the bodily processes involved, they function smoothly and, for the most part, feelings of satiation, energy, and clarity of mind are the experienced results. Many bodily functions don't even include sensory receptors and neural links that allow them to be available to gross attention and phenomenal experience. As another example, ordinary visual experiences include practically none of the complex electrochemical as well as feature detection and gestalt organizational processes involved in the corresponding psychophysiological functioning; the same is the case with memory processes. For the most part only the final picture is presented to phenomenal experience.[7 15] This will be discussed further in a later chapter, including in connection with *theatre* models of mind and consciousness.

> "Much of our cognitive life may be the product of highly automated routines. When it comes to talking, listening, reading, writing, or remembering, we all are like accomplished pianists... How our brain performs these demanding tasks remains largely unknown to us... In action as well as in perception, it appears as if only the last levels of control or of analysis are available to consciousness, while everything else proceeds automatically."—*Gerald M. Edelman, neurobiologist, and Giulio Tononi, neuroscientist (pp. 57-58)*[12]

This important design feature of phenomenal experience—that much of its mechanics are not consciously experienced—was introduced in Chapter 11. As mentioned there, the distinction between what is conscious and what is not conscious can be related to the phenomenal duality of Purusha and Prakriti. The indivisible singularity of the unified field can be analyzed in terms of this fundamental duality, the origin of phenomenal differences within the unified field. In this duality, Purusha is the silent witness that is pure consciousness; and *all* the diverse manifestations of Prakriti or Nature can be considered *not* conscious. Prakriti or Nature as manifested in creation is phenomenally conscious only when presented to Purusha, to consciousness itself, the silent witness of all dynamism and change in nature. All the levels of subjectivity can be said to conduct their dynamic processing *unconsciously*. It is the outcomes, results, or fruits of their processing that are presented to consciousness, not their functional or structural mechanics. From this perspective neither body nor mind *as such* are conscious, but both are involved in presenting information to consciousness. What information is presented is determined by the myriad of functional dynamics of the mind and body that localize conscious experience into states of consciousness, and their interaction with the environment—to be discussed with respect to the concept of attention in Chapter 18.

The subjective experience that information is *presented to* consciousness indicates that consciousness is something other than the processes involved. Stages of input, computation, and output processes themselves do not have conscious experience as a concomitant property—nor does consciousness emerge from them. Conscious experience *does not* accompany the processing that leads to phenomenal experience, either on the input or output side of processing stages. In this sense consciousness is not identified with functional analysis and synthesis of objects of experience in any of the various stages of processing. Consciousness is the experiencer of the results of these mechanisms and processes, not an experiential property of the stages. It does not result from, emerge from, or involve itself in any of them. Consciousness is the undivided wholeness that is the basis of the parts; the parts don't combine together to create the wholeness of consciousness—to be discussed in detail in the chapters on models of mind.

In this context it may be more helpful to turn the picture around and view consciousness as shining out into the mind and illuminating objects of experience in the mind. As will be discussed in Chapter 18, all the levels of subjectivity and all of their many stages of processing information comprise a limiting channel that localizes and individualizes conscious experience depending on the state of consciousness of the sentient entity. All the levels of subjectivity and objectivity structure the individualization of consciousness—like the wave is structured from levels of the ocean. They don't generate or create consciousness, in an analogous way that waves don't generate the ocean.

In terms of a developmental framework, when the physical and mental levels of the individual function incoherently and are inconsistent, a rougher or coarser-grained presentation is made of the results of processing—sometimes so much so that it is as if nothing is presented to consciousness. Alternatively it could be said that less of the light of consciousness shines through the levels of mind when they are distorted, unrefined, and underdeveloped. In the ordinary sleep state, for example, even the experience of continuity of self is described as absent or minimal (refer to the quote earlier in this chapter). But it is not completely lost, and it quickly is identified as part of subjective experience when ordinary thinking and behavior is reengaged. When body and mind are functioning at their finest and most orderly or optimal degree, however, continuous inner wakefulness—even more fundamental than the sense of self—is spontaneously maintained during sleep. This more developed experience during sleep, not based on self-reflective thinking, is called *witnessing*—to be discussed in the context of higher states of consciousness in Chapter 20. In the process of developing higher states of consciousness, increasing clarity of underlying unbounded inner wakefulness during sleep, dreaming, and ordinary waking naturally unfolds.

The conscious-unconscious distinction can be viewed in general terms as a continuous variable with pure consciousness at one end and virtually no conscious experience at the other end. The differences are related to the degree to which consciousness is hidden or overshadowed—while never losing its essential nature as the unified field of consciousness. The three ordinary states of consciousness are classified phenomenologically in terms of the degree of alertness or awareness: none or virtually none in deep sleep, some awareness of illusory objects of experience in dreaming, and more awareness of the external and internal environment in the waking state. This suggests at least some degree of conscious experience, whether immediately reportable and clearly recognized or not, in all of these states. A growing body of phenomenal reports of a variety of conscious experiences even in deep sleep and coma supports this understanding. [16] [17] [18] [19]

Obvious in the waking state, there are voluntary and involuntary attentional processes that influence what is presented to consciousness. These attributions of what is conscious and unconscious concern attentional mechanisms, and it is perhaps clearer to discuss such differences in terms of processes of attention rather than consciousness. Attentional processes in part have to do with what is presented to consciousness within the limitations of the state of consciousness of the individual. Many of these attentional processes don't change in the development of higher states of consciousness, although the types and range of empirical experiences do—including what objects and activities draw attention. These points will be elaborated in the upcoming chapters on human psychoarchitecture and on higher states of consciousness.

However, in the highest state of consciousness—unity consciousness—all unconscious processes that present information to consciousness turn out to be nothing other than consciousness itself. Their essential nature as the unified field of consciousness becomes directly experienced. In that highest state both simultaneously are experienced: at the same time, objects appear to be non-conscious and also are experienced as nothing other than consciousness itself. Prior to that unified state of experience, objects are experienced only as separate from consciousness—or as not consciousness—and are *presented to* consciousness.

Panpsychism? The holistic perspective that consciousness is the essence of everything might be mistakenly classified as *panpsychism*—the view that conscious mind is in everything. But this would overlook the nature of the unified field of consciousness to remain unity and express diversity simultaneously—to be undivided wholeness of consciousness and also appear as the phenomenal diversity of creation including inert matter. Objects in creation can appear conscious or not, and ultimately be the unified field of consciousness itself. Panpsychism does not seem to be understood in this way. Some interpret panpsychism to mean that a type of mind-like quality is a universal property of matter, or that there is a quantized proto-mentality in all levels of matter—to be discussed in Chapter 14.[20] This interpretation of panpsychism has a similar problem as structuralist and functionalist perspectives of how consciousness could be generated from more basic parts—even if the parts are some form of mental phenomena. This relates to the *generation* and *combination problems* referenced earlier. [13] This panpsychic view is quite distinct from the view described in ancient Vedic science and Maharishi Vedic Science and Technology, in which consciousness is one indivisible wholeness and there is no separate 'consciousnesses.' Rather, the task in Maharishi Vedic Science and Technology is to explain the opposite issue of how the parts come out of the whole—how the whole appears phenomenally in terms of parts in manifest creation, even appearing to be non-conscious parts such as inert matter. One angle for explaining this was introduced in Chapter 11 in the discussion of Sankhya (and also in a quote by Bohm in Chapter 8), but other angles also will be discussed to make the issue clearer.

Consciousness and its Embodiment

There is a subtle issue in this discussion, alluded to in Chapter 2 as arising from the notion of a non-conceptual experience of consciousness itself influenced by the state of the body. This might seem to suggest that something other than consciousness itself exists that makes up the body, and that consciousness in some way depends on it. The relationship of consciousness to the brain and body is a core issue in a scientific theory of consciousness. As discussed here the issue appears to contrast the phenomenal experience of consciousness in an individual with consciousness itself. Isn't a core meaning of consciousness that it involves *individual experience*?

Universal consciousness is the essential characteristic of individual consciousness. But in the ordinary waking state of consciousness its universal nature is overshadowed. It is always there as the essential nature of individual consciousness, and nothing needs to be done to 'experience' it except settle down the mental activity in ordinary individual conscious experience to its least excited ground state. Here are three quotes from Maharishi relevant to this issue. In the quotes, consciousness itself relates to the term *Being*:

"Being is not appreciated by the mind, although It is its very basis and essential constituent. Because It is at the very root of everything, It is, as it were, supporting the existence of life and the creation and not exposing Itself. The great dignity, the great splendor and grandeur of Its...omnipresent nature is present in man as the basis of ego, intellect, mind, senses, body, and surrounding. But it is not obvious; It underlies all creation... It is the omnipresence of Being that is responsible for hiding Being behind the scenes... It lies out of the realm of time, space, and causation, and out of the boundaries of the ever-changing phenomenal field of creation." (p. 25) [21]

"Since Being is of transcendental nature, It does not belong to the range of any of the senses of perception. Only when sensory perception has come to an end can the transcendental field of Being be reached. As long as we are experiencing through the senses, we are in the relative field. Therefore, Being certainly cannot be experienced by means of any of the senses. This shows that through whatever sense of experience we proceed, we must come to the ultimate limit of experience through that sense. Transcending that, we will reach a state of consciousness where the experiencer no longer experiences... The word "experiencer" implies a relative state; it is a relative word. For the experiencer to be, there has to be an object of experience. The experiencer and the object of experience are both relative. When we have transcended the field of the experience of the subtlest object, the experiencer is left by himself without an experience, without an object of experience, and without the process of experiencing. When the subject is left without an object of experience, having transcended the subtlest state of the object, the experiencer steps out of the process of experiencing and arrives at the state of Being. The mind is then found in the state of Being which is out of the relative... The state of Being is neither a state of objective nor subjective existence, because both of these states belong to the relative field of life. When the subtlest state of objective experience has been transcended, the subtlest state of subjective experience also has been transcended. This state of consciousness is then said to be pure consciousness, the state of absolute Being... The transcendental state of Being lies beyond all seeing, hearing, touching, smelling, and tasting—beyond all thinking and beyond all feeling." (pp. 45-46) [21]

"Those whose hearts and minds are not cultured, whose vision concentrates on the gross, only see the surface value of life. They only find qualities of matter and energy; they do not find innocent, ever-present, omnipresent Being. The softness of Its presence is beyond any relative degree of softness. They do not enjoy almighty Being in Its innocent, never-changing status of fullness and abundance of everything that lies beyond the obvious phase of forms and phenomena of matter and energy, and of mind and individual... Pure

Being is of transcendental nature because of Its status as the essential constituent of the universe. It is finer than the finest in creation; because of Its nature, It is not exposed to the senses which primarily are formed to give only the experience of the perception of the mind, because the mind is connected for the most part with the senses." (pp. 24-25) [21]

Here is an additional quote in which Maharishi explains how pure consciousness, which always underlies ordinary conscious experience, is missed:

"When a flower is seen, then only the flower remains in the mind, as if the mind had been completely annihilated, void of its own glory, and the glory of the flower had overtaken it— as if the flower had overshadowed the glory of the mind itself. The experiencer is missing, only the sight remains and the object... This is called objective life, material life. Matter remains dominant... Anything that we experience, that alone remains in the mind. The value of the experiencer is missing. Pure bliss consciousness, that bliss consciousness of absolute nature, has been overshadowed by the impression of the object." (p. 294) [22]

As discussed in Chapter 2, a core aspect of Maharishi Vedic Science and Technology is the systematic subjective means of gaining knowledge to go beyond the ordinary waking state of consciousness. The Transcendental Meditation technique is a systematic technology that reliably produces direct experience of consciousness itself, in which consciousness is disembedded or isolated from its identification with ordinary individual mental activity. This direct 'experience' can be said to be beyond space and time, and beyond individuality. Another way of putting it is that the direct experience of consciousness itself involves simultaneously individual and universal conscious experience. The universal value of consciousness is not overshadowed by the individual values of consciousness.

Discussions about the nature of consciousness frequently involve the mind-body problem. From a developmental perspective, however, the mind-body problem confounds mind and consciousness. As will be discussed later, the mind-body problem is best associated with how gross relative and subtle relative levels of nature link up with each other. In this sense the mind-body problem has little to do with what is and what is not conscious—neither of them being conscious *as such*.

The issue here, however, is whether consciousness depends on anything other than itself— especially any kind of a substrate or body—related to the question of how consciousness can appear not to be conscious. Individualized consciousness appears to depend on the body in the ordinary waking state. But ultimately consciousness doesn't depend on any substrate—or we can say that it is its own substrate. A series of analogies that Maharishi uses can help illustrate this quite subtle point. One analogy is of a movie screen and the movie projected onto it. When we are absorbed in the plot and action of a movie, there may be no thought or attention to the underlying screen upon which the movie is projected. If the movie were to fade out slowly, we would then see the screen. In some sense the screen was in our awareness all the time, in order for the movie to be projected onto it. But it was hidden and not prominent enough in our attention to notice; and in this sense, we were not aware of it. It might be said that we were conscious, and aware and not aware of the screen. This analogy also can be used to describe the process of settling down and transcending mental activity to experience the underlying ground state of consciousness itself— like fading out the action of the movie to reveal the underlying screen.

Another relevant analogy Maharishi uses is of sunlight reflected in a mirror. When a mirror is covered by layers of dust the reflected light is of a different quality than the direct sunlight. As the dust is cleaned off the reflected light from the mirror and the direct sunlight merge together, and the difference between them is no longer identifiable. However, there remains an individual reflection of sunlight off the mirror, in the sense that the mirror still reflects sunlight. This analogy also is used in describing the process of developing permanent higher states of consciousness.

A third and similar analogy Maharishi uses, mentioned earlier, is of a wave settling back into the ocean. In this analogy, however, as the wave settles down—likened to fading out the movie on the screen or cleaning the dust from the mirror—the wave is found to be essentially nothing other than the ocean. In the screen analogy the movie is quite different than the screen. In the mirror analogy the clean reflection can be said still to be an individual localized reflection of the direct sunlight, even if no longer identifiable as separate from it. The wave analogy is more illustrative of the issue under consideration, in that when the wave settles down there can be said to be no difference between it and the ocean.

In the ordinary waking state of consciousness there is identification as an individual being with individual experience. This identification can be so pronounced that it seems to overshadow completely the most fundamental nature of the individual conscious self as consciousness itself. When mental activity settles down and is transcended, the essential nature of the individual self as consciousness itself is directly experienced—and recognized more clearly with repeated direct self-referral experience.

In that experience individual subjective levels of the senses, mind, intellect, and ego or self are not active in processing information—they do not form that direct experience. Even the experiencing self, associated with the mechanisms involved in localizing awareness to objects of experience, settles back into consciousness itself, and consciousness stands in its own self-referral light—it is not experienced with or through object-referral mental activity. In other terms the first-person *direct acquaintance* with consciousness [2] settles into consciousness itself—complete self-referral consciousness.

This also was discussed in Chapter 2, in which the ordinary waking state of consciousness was characterized as the reflective object-referral experience of a separate object of experience, process of experience, and experiencer. In the process of transcending, these three separate aspects of experience systematically settle down to subtler and subtler levels, until they merge together into the *self-referral* state—or Self-referral. It can be likened to points on a loop of string that gets smaller and smaller, becoming closer and closer, until the loop curves back onto itself in complete self-referral singularity beyond extension in space and time. Again, the unified field of consciousness as the field of all possibilities can do anything on its own, including appear to be individual consciousness or even non-conscious. But it can also 'experience directly' itself.

Recalling discussions of infinity and a point from prior chapters, the individual or point value is inherent in the unified field, coexisting with the infinite value. The unified field of consciousness is the simultaneous coexistence of the opposite values of point and infinity, singularity and infinity, individuality and universality. When the individual can directly experience both at the same time, that experience is direct empirical evidence that the essence of the individual is the unified field of consciousness itself. In terms of the analogy of the loop of string, the loop both remains in its infinitely self-referral state of singularity, while also at the same time *appearing* to extend or unfold itself self-reflectively into the string loop with separate points on it, a simultaneous superposition of the two states.

It might be pointed out, however, that the screen, mirror, and wave analogies all have limited application to the issue here. The screen is different from the observer of the screen, the mirror is different from the sunlight, and the wave is caused by something other than the ocean, such as the wind or the gravitational pull of the Moon. In no case is it obvious how the phenomenal appearance—movie, screen, or wave—completely manifested from *and* is completely due to what underlies it. In the unique case of consciousness itself, its manifestation as individual consciousness is the same as universal consciousness—both refer to the same consciousness. Although the above analogies may give a better sense of the fundamental issues, aren't we basically still in the same dilemma of how consciousness can *not* know its universal nature in its individual state? What else, seemingly something other than consciousness itself, individualizes consciousness and makes it appear not to be conscious?

The ultimate resolution to this issue is in higher states of consciousness. In the ordinary waking state of consciousness the experience and understanding is typically that consciousness depends on the body. In the early stages of development toward permanent higher states of consciousness the experience is either awareness of some object of experience, or transcendental consciousness—but not both at the same time due to the inflexibility of the body and mind to support experience of both simultaneously. In the next stage or state of progress, pure consciousness is maintained along with individual awareness of objects of experience. In the highest state of development every level that comprises individual experience is directly experienced as emerging from and nothing other than universal consciousness. In *that* state all experienced objects—including the body—are experienced by consciousness itself in terms of nothing other than consciousness itself; and it continues whether there is an individual body or not. That experience is the completely unified, infinite self-referral dynamics of consciousness. The evolutionary process to the highest state of consciousness can be viewed as progress from individuality to infinite singularity—oneness to Oneness. That is the ultimate verification and resolution to the relationship of individual body as an apparent substrate of individual consciousness to universal consciousness, and how consciousness itself can appear to be not conscious. The planks required are in the developmental sequence of the full range of states of consciousness, described in some detail in later chapters. It might be pointed out here that each state of consciousness has its own legitimate empirical *reality*. But higher states can be said to be more *real* in the sense that they encompass the lesser developed states and much more.

This also is the basis for resolving what has been identified as the '*other minds*' problem. This is the issue of how it is that there appears to be other minds besides *my own* in the universe. Also, it relates to how it is that these other minds seem to be able to agree with *me* concerning many things about the universe in *my* experience in *my* mind—but *I* don't seem to have any direct access to the experiences had by these other minds. In other terms it is the issue of how there are *third-person* and *first-person* perspectives in the first place.

In the highest state of development referred to above, all aspects of *my* experience are experienced to be nothing other than universal consciousness itself. All levels of *my* individuality are traced back to consciousness itself. Likewise all levels of other individuals are traced back to and experienced in terms of nothing other than consciousness itself. In the infinitely self-referral dynamics of consciousness itself, in the ultimate unity of the unified field, the underlying bases for the third-person and first-person perspectives bridge into that unity. In later chapters we will discuss in more detail the mechanics of how individuality comes from universality, and how they coexist in consciousness itself.

The function of consciousness. In the context of this general discussion it may be helpful to point out that the phenomenal discrimination in Sankhya of Purusha and Prakriti also relates to the important issue of the functional or causal *role* that consciousness plays in mental and biophysical processes. This is crucial to the meaning of consciousness as *awareness of* some object of experience, because this meaning carries with it the connotation of *intentionality* in terms of a functional role.

The direction in the structural perspective is toward more basic underlying biophysical processes, even recently to the quantum level. The difficulty in this is that at some point these biophysical processes go beyond any familiar notion of the brain into nonlocal fields, no longer keeping mind and consciousness on the macroscopic or even microscopic levels. As described in Chapter 7, there is a theoretical leap into a much more abstract disembodied functional perspective not embedded in any specific form of physical substrate. It then is impossible in the functionalist perspective to fit back into the hypothesized closed chain of physical causes where the function of consciousness would have to be implemented. There is no gap or opening for a causal effect of consciousness to be inserted into a closed chain of cause and effect that began long before conscious beings evolved, at least according to prevailing beliefs in modern science.

> "From a third-person perspective consciousness seems to do nothing, but from a first-person perspective there seems to be little of importance in human life that we can do without it."— *Max Velmans, research psychologist (pp. 325-326)* [23]

In the physicalist view this closed universe seems to be able to run on its own, without the need for anyone or anything to know, observe, or be *aware* of it—without the need for any role of consciousness. The functionalist perspective needs to identify the role of consciousness in an apparently closed causal nexus that doesn't require it. The direction of functional models is that consciousness is thought to have a general role as a workspace for integration of information processing functions [24]—to be discussed in an upcoming chapter. But there seems to be *no actual causal role* for it in the chain of causation, and there seems to be *no actual place* for it in the closed physical universe. Given these challenges it seems necessary to conclude that consciousness is an inexplicable epiphenomenon, with no causal effect or *real* existence at all. If so, then consciousness seems highly counterintuitive. What is the sense of having it in the design of the universe if it has no role? But on the other hand, many feel that it is the most fundamental, direct, undeniable fact or directly felt *reality* of our existence and the basis of our ability to experience and know anything—including that we exist. [25]

In the Sankhya system described in Chapter 11, from the Purusha level consciousness is the uninvolved silent witness of everything in nature; and from the Prakriti level it is the essence of and pervasive cause of all change. It thus can be said to be both the cause of all change and the uninvolved non-changing witness or *experiencer* of all change. It can be said to be epiphenomenal with respect to everything in creation 'other than itself'—so to speak—but also everything 'other than itself' can be said to be epiphenomenal with respect to it. These two diametrically opposite positions are completely reconciled in higher states of consciousness, and otherwise may be quite unfathomable.

This same issue also can be generalized to the role or purpose of the entire creation, as well as to the fundamental meaning of causality. At this point in the discussion, consciousness can be said to have no specific role in the various stages of information processing in individual experience. But it can be said to have a universal role as the source of everything, the essence of all experience, and the cause of all phenomena. In the chapter on the structure of Veda toward the end of the book we will discuss the fundamental point Maharishi has made about the functional role of consciousness that, 'knowledge is structured in consciousness.'[25]

It is possible that the concept of the uninvolved witness of Purusha might be misconstrued as having the *homunculus problem*. On the contrary it resolves the major concern of the homunculus concept of an infinite regress of individual observers—by a single universal observer or experiencer of the entire universe, consciousness itself, the source of all sentience, capable on its own of experiencing itself as well as 'anything else.'

Perhaps here is a good point to discuss briefly several other problems that a scientific theory of consciousness needs to address—including what are sometimes called the *not mental problem*, the *no sign problem*, and the *completeness problem*.[13] The *not mental problem* refers to what possible grounds there might be for associating basic units of the physical with the concept of mental. The *no sign problem* refers to lack of evidence of anything other than a physical dimension to the fundamental elements of nature. The *completeness problem* refers to the apparent completeness of the chain of physical causes that leaves no room for a causal role of consciousness. These problems are reflected in the following statement by philosopher William Seager:

> "I often have to worry about whether my car will start, but I thankfully don't have the additional worry about its failing to start even when there is absolutely nothing mechanically wrong with it but just because it 'feels like' staying in the garage today!" (p. 279)[13]

These problems are due to limiting presumptions in physicalism, reductionism, and the objective third-person means of gaining knowledge common in modern science. Even in this overall physicalist view, however, the need to go beyond the physical in explaining physical phenomena is becoming recognized, as is evident in quantum gravity and unified field theories. These theories are starting to address the *not mental* problem by way of the concept of fundamental information bits underlying physical matter, and the *psi* wave in Bohmian mechanics, which have somewhat of a mental quality about them. However, they reflect only the faintest of glimpses into this subtler level of quantum mind or mental *reality* deeper within the physical.

In terms of the *no sign* and *completeness* problems, the basic issue is the degree of subtlety of our methods of validation. These problems also relate directly to the measurement problem in quantum physics, as well as related concerns of both classical and quantum ignorance. With respect to classical ignorance it is the inability to measure potentially subtle causal influences due to their complexity and signal-to-noise ratios at incredibly tiny scales of measurement. With respect to quantum ignorance it is the apparent inherent randomness of nature at the fundamental quantum level. This is quite relevant to the 'cars don't have feelings' example above. It is reasonable not to worry about the car having the causal power to choose whether or not to start. However, it does seem reasonable to be concerned that in a causally closed physical universe there seems to be no room for the power to choose on the part of the owner of the car either.[26]

However, there is plenty of logical room within modern scientific theories to extend beyond a causally closed physical universe. For example, just because occasions when the car won't start seem later to be due to a gross mechanical problem with the car, this does not at all exhaust possible explanations of the cause of the car not starting. According to the correspondence of gross and subtle behavior, the cause of the gross effect actually could be an underlying subtler cause. Belief that there is no underlying subtler field that could have a causal effect, inability to measure subtler effects, and the high correlation or coherence between gross and subtle, all give the impression that the gross level is all that is available and needs to be posited and explained.

Numerous subtle mental phenomena are now being investigated that seem inexplicable within the physicalist worldview of a causally closed universe. But such phenomena need to be experienced reliably enough to be consensually validated by modern scientists. Through Maharishi Vedic Science and Technology, expanded means for systematic empirical validation of the subtler and transcendental levels are available. Although it is a major challenge due to paradigm blindness, this book at least makes an attempt to describe these systematic means with sufficient consistency and plausibility for researchers to take action in order to test them empirically. In so doing, more clarity of the reasonableness and necessity of positing a subtler level of mental *reality* that has subtle causal connections to gross physical *reality* will develop. Modern scientists need to be open-minded enough to examine these deeper experiences.

To be clear, however, this is not to assert that cars have a mental life of their own. As combinations of many physical parts, they have no mental life other than the degree to which collections of pieces of iron, silicon, or plastic can be said to have a mental life, which is practically speaking non-existent with respect to the gross material domain. Rather it is to assert that innumerable influences can contribute to gross physical behavior, including subtler influences that may not be measurable in the gross objective domain of nature using objective probes and measurement instruments. Modern science is increasingly recognizing that quite small influences can have large effects on gross physical behavior. This now extends even to the empirical finding that local phenomena have nonlocal influences, which of course are quite challenging to measure locally.

What about Occam's razor? On the surface, the theory of subtler causes underlying gross effects may seem to be inconsistent with the scientific principle of parsimony associated with *Occam's razor*. This principle has to do with a value judgment that theories are better if they depend on fewer assumptions to explain a phenomenon. As the incompleteness of our current

> "[T]he mind should not multiply entities beyond necessity. What can be done with fewer...is done in vain with more."—*William of Occam, 14th Century philosopher (p. 269)* [27]

physical explanations are becoming more apparent with cutting edge theories extending beyond matter, and as anomalies inexplicable without additional assumptions accumulate, the next step toward understanding the subtle level of nature becomes more evident. This important plank, extending into the underlying mental *reality*, is clearly developing in modern science.

It may be worth pointing out, however, how a first-person holistic perspective completely changes the picture with respect to interpreting Occam's razor when compared to the third-person objective perspective currently typical of modern science. A first-person holistic perspective directly supports the essential parsimony of ancient Vedic science. From this view any third-person explanation would be considered not as good as a first-person explanation because objective third-person perspectives require assuming the existence of others. Also the reductive perspective is based at least partially on the assumption that higher-order processes are reducible to lower-order processes. This is not a necessary assumption in the holistic perspective—and even is inaccurate from that perspective.

Phenomenologically the direct experience that *I am* is the only empirical fact that involves no assumptions. The only assumption then needed in a 'theory of everything' is that *I create all appearances*. Indeed there is *no independent evidence* to prove or disprove any other assumption. In this view the first-person holistic perspective is consistent with the principle of Occam's razor. It also is not solipsistic, at

> "...[O]ur conscious experience is the only ontology of which we have direct evidence."—*Gerald M. Edelman, neurobiologist, and Giulio Tononi, neuroscientist (p. 35)* [12]

least in its usual sense, though it might at first seem to be. Solipsism is usually understood to be the egoistic philosophy that the self is all that exists, and nothing else can be proven to exist. In this form it has long been rejected. But in the first-person holistic perspective discussed here, the *I* or self being referred to is not the *I* associated with individualized awareness in the ordinary waking state. It is prior to individuality, in the sense that it does not assume the distinction of universality and individuality. From that perspective there is no independent evidence to prove *my* consciousness is individual or distinct from any other consciousness—in contrast to the usual notion of solipsism that tacitly assumes it. Consistent with Occam's razor, the most parsimonious theory of the fundamental *reality* then is the holistic approach in ancient Vedic science, succinctly captured in the statements drawn from the Vedic texts called the Upanishads:

"I am That, Thou art That, all this is That, That alone is, and there is nothing else but That." (p. 34) [21]

These phrases and their sequence can be viewed as integrating the first-person (*I am That*), second-person (*Thou art That*), and third-person (*All this is That*) perspectives into the completely unified three-in-one self-referral perspective (*That alone is, and there is nothing else but That*). From that ultimate holistic perspective—the unified field of consciousness—there is no independent evidence of anything, and no other assumptions are independently provable, the most consistency with the principle of Occam's razor. Additional assumptions may have their value to explain more comfortably the relative *reality* of the phenomenal diversity of nature in certain relative states of consciousness—which is what much of this book does. But they would not have the same simple, direct, unassuming status as the one immediate fact that *I am,* and the one assumption that, as universal consciousness, *I create all appearances*. That assumption is said to be empirically verifiable in the completely unified state of consciousness. This is beautifully summarized in a single Vedic expression:

"*Aham Brahmasmi*
(Brihad-Aranyak Upanishad, 1.4.10)
I am totality." (p. 181) [28]

These points are elaborated in the chapters on higher states of consciousness. They are included here to emphasize the plausibility—as well as simplicity, profundity, and directness—of systematic subjective means in Maharishi Vedic Science and Technology for understanding and experience of consciousness.

The Correspondence of Body, Mind, and Consciousness

In considering the consciousness-mind-body relationship it may be useful first to clarify some key terms—especially the terms *subjective* and *objective*, and *mind* and *body*. As anticipated in Chapter 1, throughout the book the

> "Confronted with the hard problem of the relation between conscious experience and the physical world, our first move is to postulate the presence of a third aspect of reality, besides the dimensions of space and time that underlie our descriptions of the world in terms of physics."—*Piet Hut, astrophysicist, and Roger N. Shepard, cognitive and evolutionary psychologist (p. 318)* [3]

distinction between subjective and objective has been applied in somewhat different ways. We have been moving toward being able to apply them to the distinction between gross relative and subtle relative domains of nature. This reflects how Maharishi uses the terms in the quote in Chapter 11. The quote identified a dividing line between the gross objective domain and the subtle subjective domain in distinguishing the gross elements (mahabhutas) and subtle essence elements (tanmatras). It helps clarify the material and mental domains.

In this broad sense *objective* refers to the material domain associated with the fundamental quantum fields as particle fields in the model of a closed, or quasi-closed, universe limited by Einstein locality—everything that is made of physical matter (mahabhutas). To contrast it most simply, *subjective* then refers to everything else. This includes all the levels of nature that are subtler than the material domain: the subtle essence elements (tanmatras) and all the levels associated with subjective mind including senses, mind, intellect, ego, and consciousness itself—indriyas, manas, ahamkara, mahat, Prakriti, and Purusha as enumerated in Chapter 11.

In Maharishi Vedic Science and Technology the essence of all the gross and subtle levels is the unmanifest transcendental unified field (Prakriti and Purusha together), beyond all the relative levels of objectivity and subjectivity. Although that unmanifest transcendental level can be said to be beyond objective and subjective, it is sometimes placed with subjective in comparison to the objective material domain. It also then is distinguished from the subtle subjective domain by identifying it as *pure subjectivity*. The distinction of gross, subtle, and transcendental then can be directly related to the terms objective, subjective, and pure subjective.

There is, however, another more specific sense of the objective-subjective distinction that concerns further delineation of the subtle relative domain. This relates to the *object-subject* distinction, in the sense of what objects can be observed through the senses in the mind. Frequently this distinction is associated with the body-mind contrast, but it can be made more specific. In this distinction the *object* refers to any object of sense—everything made of the gross elements (mahabhutas) *and/or* the subtle essence elements (tanmatras). The *subject* refers to everything else—senses, mind, intellect, ego, and consciousness. The object-subject distinction can be said to originate in the unmanifest duality of Prakriti as pure intelligence and Purusha as pure existence, coming from the completely unified coexistence of opposites of infinite dynamism and infinite silence. On the subtle level this object-subject duality manifests in the discrimination by cosmic intellect (ahamkara) of cosmic ego (mahat). From this discrimination comes the cosmic mind (manas) and senses (indriyas) for observing the objects of sense, and their corresponding objects inclusive of both the subtle and more expressed gross objects in phenomenal nature (tanmatras and mahabhutas).

The difference in these two meanings of the objective-subjective dichotomy is in where the subtle essence elements (tanmatras) are placed. In the first usage they are counted as *subjective*, part of the subtle relative domain distinct from the gross relative domain (as in the quote by Maharishi from Chapter 11 mentioned above, and perhaps easier to understand within the context of physicalism unaccustomed to the subtle level). In the second usage they are counted as *objective* in that they are subtle *objects of sense*. In the first usage the trinity of observed, process of observing, and observer relates directly to gross, subtle, and transcendental. In the second usage the *observed* is every observable gross *or* subtle object, the *process of observing* is the levels of mind, and the *observer* is consciousness itself.

The terms *body* and *mind* similarly have two basic referents. In one usage *body* refers to the biophysical structure in the gross relative domain (made of mahabhutas); and *mind* refers to the more abstract functional processes associated with the subtle relative domain (made of all the subtle subjective levels combined). In a second usage body refers to both gross *and* subtle elements (mahabhutas and tanmatras), and mind refers to all the subtler subjective levels. This means that the body can be understood to have both its familiar classical aspect made of gross elements (mahabhutas), and a subtler aspect made of the subtle essence elements (tanmatras). In the Darshana of Yoga the gross aspect of the physiology is called *sthula sharir*, and the subtle aspect is called *linga sharir*. These points are important for understanding the correspondence of body, mind, and consciousness and resolving the perennial mind-body problem that will be discussed from several angles in this part of the book.

Maharishi describes a *one-to-one correspondence* between mind and body, indicating that there is a precise and systematic relationship between them. This is associated with the point made in Chapter 11 that the gross relative or physical domain is built from the underpinning subtle relative domain. All activity in the gross body, including neurophysiological activity of the brain and nervous system, is underlain by corresponding activity in the subtle aspects of body and mind. The gross material aspects of the body express the subtle aspects in the gross material domain. Subtle mental activity gets transduced into gross neural activity that is then expressed in gross behavior. However, subtle activity need not be expressed in the gross body and behavior.

A popular analogy to illustrate this point is of a radio that transduces high frequency electromagnetic signals into acoustic signals received by auditory mechanisms in the body, which again transduce the signals into sound information that can be heard somewhere in the mind of the listener. When the radio can no longer receive radio-wave signals or send acoustic-wave signals, the connection between the broadcast and the listener is broken. The signals are still being broadcast, but can't be heard via the radio.

Similarly subtle processes in body and mind can be operative, but the gross aspects of the body may not function properly to express them in gross behavior. When gross aspects of the body are functioning properly, gross behavior is expressed. When parts of the gross body are not functioning properly, and can no longer maintain their communicative link with the subtler aspects, there is no expression of the relationship in gross behavior.

An observer who is limited to observations of only gross behavior could not sense that the subtler aspects are still operative when the gross body can no longer express them in gross behavior. When gross behavior is absent or dysfunctional due to gross sensory or brain malfunction, and when it is believed that the physical is all that exists, and when there is no objective evidence of anything else, then it is reasoned that there is nothing deeper to it. This is tantamount to making the error of assuming that the broadcast system is in the radio receiver.

> "Matter is an admirably calculated machinery for regulating, limiting, and restraining the consciousness which it encases... If the material encasement be coarse and simple, as in the lower organisms, it permits only a little intelligence to permeate through it; if it is delicate and complex, it leaves more pores and exits, as it were, for the manifestation of consciousness... Matter is not that which produces consciousness, but that which limits it and confines its intensity within certain limits: material organization does not construct consciousness out of arrangements of atoms, but contracts its manifestation within the sphere which it permits. This explanation...admits the connection of Matter and consciousness, but contends that the course of interpretation must proceed in the contrary direction... If, e.g., a man loses consciousness as soon as his brain is injured, it is clearly as good an explanation to say the injury to the brain destroyed the mechanism by which the manifestation of consciousness was rendered possible, as to say that it destroyed the seat of consciousness... If the body is a mechanism for inhibiting consciousness, for preventing the full powers of the Ego from being prematurely actualized, it will be necessary to invert...our ordinary... ideas."—*Ferdinand Canning Scott Schiller, philosopher (p. 132)* [29]

The one-to-one correspondence of body and mind is reliable in intact living and functioning bodily systems, and is reliably broken when the body doesn't work. It also has fairly high specificity, related to functional localization in parts of the brain and corresponding gross behaviors. This is analogous to the tweeter of a radio malfunctioning such that sounds reproduced by the radio leave out high frequency components. But this by no means completely rules out the possibility of the subtle aspects of body and mind functioning even when their gross body counterparts don't function. In living organisms there cannot be gross behavior that does not include both gross and subtle functioning of body and mind, but there can be subtle functioning that is not expressed through the gross body in gross behavior. This will be discussed in more detail in a later chapter. It is open to direct empirical verification in the natural development of higher states of consciousness.

An additional point needs to be made about the simultaneous universal and individual aspects of body, mind, and consciousness. The Sankhya enumeration (Chapter 11) of the full range of gross, subtle, and transcendental levels identified these levels as the phenomenal structure of the entire manifest creation and its unmanifest transcendental basis. The key point was made that each of the 25 levels has its cosmic or universal status—they can be thought of as universe-wide at their own levels. For example the level of mahat was described as cosmic ego, the level of ahamkara as cosmic intellect, and the level of manas as cosmic mind. But at each of the levels there are *both local and nonlocal* components, both individual and universal—a manifestation of the infinity-point dynamics of the unified field. In other words the structure of levels of the entire cosmos is mirrored in the structure of consciousness, mind, and body of the individual human being. This profound point is captured in the Vedic verse as quoted by Maharishi:

"Yatah pende, yatah bramande.
The individual is cosmic."* [30]

With respect to the structure of the individual, the transcendental and subtle aspects of universal nature correspond to *jiva* or individual soul—associated with *linga sharir,* the subtle individual physiology. Added to this to form an individual human being is *sthula sharir,* the individual gross physiology. Although the unified field of consciousness and all of its phenomenal manifestations are nothing other than consciousness itself, all relative manifestations of it can be said to be conscious only in the sense that they are presented to consciousness—presented to the one ultimate universal witness (Purusha). All the levels of body and mind are not conscious 'on their own'—so to speak—but become conscious in the sense that the results of their activity can be presented to consciousness. This applies to both gross and subtle aspects of body and mind.

> "When one studies the evidence for reincarnation, again its is surprising how much high quality evidence there is. Thousands of cases have been painstakingly researched by highly qualified academic scientists... The findings are persuasive, even overwhelming. Once more it reinforces the notion that the soul, the identity, continues after death. It makes us realize that we humans are much more than we have been taught... The new physics will have to account for a "field of consciousness,"... It is but a short step from this point to realize that the new science begins to resemble, or at least support, some of the tenets of religion. Not any specific dogmatic religion, but the general principles underlying all religions. As we stumble our way towards the ultimate "unified field theory," we may find to our surprise that it bridges the gulf between science and religion. As our understanding becomes more profound, the differences between these two belief systems may become smaller. There is, after all, only one universe. There is only one true description of it at the deepest level... It would offer an integrated view of the universe and of man. This would be the ultimate "unified field theory."—*Claude Swanson, applied physicist (pp. 13-14)* [31]

Most fundamentally consciousness can be said inherently to present itself to itself—its infinite self-referral nature. It also can express and experience itself as individual or universal consciousness, both simultaneously, and as individual in terms of universal in higher states of consciousness. In addition the results of psychophysiological processes, and their subtle and gross objects of sense, can be presented to consciousness or not, depending on attentional functioning in the various individual states of consciousness.

But the mechanics of the structural and functional operations of body and mind are not typically presented to individual consciousness. From this perspective consciousness seems to have no specific function in the sequence of information processing events as ordinarily conceived in functionalist accounts of individual experience. The processing carried out in body and mind can be said to be all that is needed to fulfill their functional roles. This is consistent with the functional identity hypothesis in the sense that all functional aspects of experience are in the functions themselves. In this sense consciousness itself is not involved in the processing functions. There is, however more to it with respect to the relationship of individual and universal, which will be discussed further in connection with causality, determinism, free will, and related topics in

Chapter 18. The levels of mind and body—and all their functions and structures—are limitations of the unified field of consciousness itself, the one ultimate observer or witness. As the field of all possibilities, the unified field of consciousness is able to carry out any function and have any experience possible, on its own. Ultimately everything that happens in the phenomenal universe takes place within the self interacting, self-referral dynamics of the unified field of nature.

By way of summarizing key points in this chapter, let's apply them to common experiences in the ordinary waking state of consciousness. In this state consciousness is associated with awareness of objects of experience. In order to be aware of external objects the objects have to be within the typical range of the gross senses—for example, with respect to distance, speed, size or scale, spectrum of light, sound frequency, and so on. The range of the gross senses defines the macroscopic scale. There is a fundamental correspondence between the gross senses and the macroscopic properties or qualities of gross objects, such that the individual and the objects can interact reasonably well with each other at the gross level of behavior.

Also, for the most part internal bodily processes can be sensed in a manner that provides information for general maintenance and control of the gross bodily system and its behavior, without extraneous and inefficient input. The finer-grained levels of gross objects of sense—at smaller time and distance scales than the macroscopic scale—are not experienced (although a somewhat extended range can made available by different forms of amplification such as microscopes).

With respect to inner subjectivity, virtually none of the mechanical operations of the senses are presented for conscious experience, rather only their products that support behavior—the phenomenal *objects of sense*. The same is the case with the mechanics and dynamic processes involved in attending to emotion, thinking, memory, discrimination, as well as more integrative feelings and intuitions. Only the final products show on what might be called the Cartesian theatre or screen of the mind (manas) to consciousness. The mechanics of how attention is directed is not experienced, for example.

However, for most individuals in the waking state of consciousness, subtle objects of sense (objects made of tanmatras, but not mahabhutas) are rarely experienced, because the senses usually function too coarsely to register them. In individual cases where the senses are refined enough to register subtle objects, this still does not mean that the microscopic or ultramicroscopic levels of gross objects are sensed. We can say for practical purposes related to ordinary understanding and experience that no corresponding sensory mechanisms are present to sense them (nor instruments to amplify them enough); and they are not necessary to sense in order to function generally in the control of gross behavior in our everyday, macroscopic, objective life.

In the ordinary waking state of consciousness there is thus a close correspondence between objects in the outer environment, their neuroanatomical representations, their sensory and cognitive representations, and phenomenal experience. When the gross body stops functioning, gross behavior stops and the senses no longer receive information about the outer physical world through the gross body. The correspondence between the gross environment, the gross aspect of the body, *and* the subtle aspects of the body and the senses is broken.

To be clear about the overall direction discussed in this chapter, Maharishi Vedic Science and Technology generally can be said to be *monistic*. However, it is not materialistic monism, nor monistic idealism, but rather an ultimate monism of pure *Being*—called *Atma* or *Brahman* in Vedic literature, the Universal Self, the unified field of consciousness. Perhaps a bit more explanatory than monism, it can be described as *non-dualistic*. As ultimate, undivided, infinite, eternal wholeness, it can be intellectually delineated as sequential levels of *reality* in terms of the *dualism* of mind and matter, and more completely as the *trinity* associated with the manifestation of mind and matter within consciousness itself. But these levels of *reality* are relative and conditional. They are due to various perspectives in different developmental states of consciousness, superimposed on the ultimate indivisible wholeness of the unified field. The trinity can be associated generally with structural, functional, and transcendental experience perspectives; gross, subtle, and transcendental levels of nature; and also with known, process of knowing, and knower—in the three-in-one self-interacting dynamics of the unified field of consciousness.

These points provide an overall framework for understanding when activity in body and mind will be accompanied by conscious experience. We can now illustrate with key examples how theoretical approaches to consciousness and levels of mind under the umbrella of modern scientific thinking can be viewed in this holistic framework—the focus of Chapters 14-17.

Chapter 13 Notes

[1] Velmans, M. (2000). Intersubjective Science. In Varela. F. & Shear, J., (Eds.) *The view from within: First-person approaches to the study of consciousness*. Bowling Green, OH: Imprint Academic Philosophy Center, Bowling Green University, pp. 299-306.

[2] Chalmers, D. J. (1996). *The conscious mind: In search of a fundamental theory*. New York: Oxford University Press.

[3] Hut, P. & Shepard, R. N. (2000). Turning 'the Hard Problem' Upside Down and Sideways. In Shear, J. (Ed.). *Explaining consciousness: The hard problem*. Cambridge, MA: The MIT Press, pp. 305-322.

[4] Hardcastle, V. G. (2000). The Why of Consciousness: A Non-issue for Materialists. In Shear, J. (Ed.). *Explaining consciousness: The hard problem*. Cambridge, MA: The MIT Press, pp. 61-68.

[5] Chalmers, D. J. (2000). Facing Up to the Problem of Consciousness. In Shear, J. (Ed.). *Explaining consciousness—The hard problem*. Cambridge, MA: The MIT Press, pp. 9-30.

[6] Churchland, P. S. (2000). The Hornswaggle Problem. In Shear, J. (Ed.). *Explaining consciousness: The hard problem*. Cambridge, MA: The MIT Press, pp. 37-44.

[7] Dennett, D. C. (2000). Facing Backwards on the Problem of Consciousness. In Shear, J. (Ed.). *Explaining consciousness—The hard problem*. Cambridge, MA: The MIT Press, pp. 33-36.

[8] Clark, T. W. (2000). Functionalism and Phenomenology: Closing the Explanatory Gap. In Shear, J. (Ed.). *Explaining consciousness—The hard problem*. Cambridge, MA: The MIT Press, pp. 45-59.

[9] James, W. (1890). *The principles of psychology*. New York: Holt.

[10] Maharishi Mahesh Yogi (1967). *Maharishi Mahesh Yogi on the Bhagavad-Gita: A new translation and commentary, chapters 1-6*. Baltimore, MD: Penguin.

[11] *Maharishi Vedic University: Introduction*. (1994). Holland: Maharishi Vedic University Press.

[12] Edelman, G. E. & Tononi, G. (2000). *A universe of consciousness: How matter becomes imagination*. New York: Basic Books.

[13] Seager, W. (2000). Consciousness, Information, and Panpsychism. In Shear, J. (Ed.). *Explaining consciousness—The hard problem*. Cambridge, MA: The MIT Press, pp. 269-286.

[14] Whitehead, A. N. (1925). *Science and the modern world*. New York: The Free Press.

[15] Velmans, M., quoted in <http://www. goldsmiths.ac.uk/departments/psychology/staff/velpub.html>

[16] van Lommel, P., van Wees, R., Meyers, V., & Elfferich, I. (2001). Near-Death Experience in Survivors of Cardiac Arrest: A Prospective Study in the Netherlands. *Lancet, 358*, 2039-2045.

[17] Cardena, E., Lynn, S. J., & Krippner, S. (Eds.) 2000. *Varieties of anomalous experience: Examining the scientific evidence*. Washington, DC: American Psychological Association.

[18] Bailey, L. W. & Yates, J (Eds.). (1996). *The near-death experience: A reader*. New York: Routledge.

[19] Ring, K. & Valarino, E. E. (1998). *Lessons from the light: What we can learn from the near-death experience*. Portsmouth, NH: Moment Point Press.

[20] Hameroff, S. R. & Penrose, R. (2000). Conscious Events as Orchestrated Space-Time Selections. In Shear, J. (Ed.). (2000). *Explaining consciousness—The hard problem*. Cambridge, MA: The MIT Press, pp. 177-195.

[21] Maharishi Mahesh Yogi, (1963). *Science of being and art of living*. Washington, D.C.: Age of Enlightenment Publications.

[22] Maharishi Mahesh Yogi (1986). *Thirty years around the world—Dawn of the Age of Enlightenment*. I. The Netherlands: MVU Press.

[23] Velmans, M. (2000). The Relation of Consciousness to the Material World. In Shear, J. (Ed.). *Explaining consciousness—The hard problem*. Cambridge, MA: The MIT Press, pp. 325-336.

[24] Baars, B. J. (1997). *In the theatre of consciousness.* New York: Oxford University Press.

[25] *Maharishi International University Catalogue* (1972). Los Angeles: MIU Press.

[26] Stapp, H. P. (2000). The Hard Problem: A Quantum Perspective. In Shear, J. (Ed.). *Explaining consciousness—The hard problem*. Cambridge, MA: The MIT Press, pp. 197-215.

[27] Pigliucci, M. (2002). Neuro-Theology, a Rather Skeptical Perspective. In Joseph, R. *NeuroTheology: Brain science, spirituality, religious experience*. San Jose, CA: University Press, pp. 269-271.

[28] *Inaugurating Maharishi Vedic University* (1996). India: Age of Enlightenment Publications.

[29] Braud, W. (2002). Brains, Science, Nonordinary & Transcendent Experiences: Can Conventional Concepts and Theories Adequately Address Mystical and Paranormal Experiences? In Joseph, R. (Ed.). *NeuroTheology: Brain, science, spirituality, religious experience*. San Jose, CA: University Press, pp. 123-134.

[30] Maharishi Mahesh Yogi (2003). Maharishi's Global News Conference, December 12.

[31] Swanson, C. *The synchronized universe: New science of the paranormal*. Tucson, AZ: Poseidia Press.

Chapter 14

Quantum Information-Based Theories of Consciousness

In this chapter, examples of progressive approaches toward a scientific theory of consciousness that are beginning to take modest steps beyond structural and functional perspectives will be reviewed. Included are the Orch OR model of consciousness as a non-computable quantum space-time geometry from Hameroff and Penrose,[1] a theory of consciousness as the specifier of classical reality from quantum reality in Hilbert space from Stapp,[2] and nonreductive functionalism and dual-aspect monism from Chalmers.[3] We will explore the connection between these quantum information-based approaches and the holistic approach of Maharishi Vedic Science and Technology. The overall main point of this chapter is that theories now emerging represent attempts to describe consciousness beyond the closed reductive physicalist paradigm.

Given the planks laid out so far in this book, hopefully it will be recognized as consistent in the context of the present discussion not to examine in great detail reductive structural and functional theories attempting a scientific explanation of consciousness. Although relevant to neurophysiology, cognitive processes, attention, and psychoarchitecture—to be discussed in more detail in the next chapters—they are prevented by the limitations of their accompanying physicalist views from getting to the essential nature of consciousness.[4] This chapter discusses prominent examples of theoretical attempts to progress beyond the functionalist-physicalist perspective in explaining the relationship between matter and consciousness.

Also hopefully it won't be long before the many brilliant thinkers in these areas of investigation recognize the limitations of these structural and functional analyses as approaches to consciousness. Some of the planks summarized in this book to foster this recognition are popping up in many fields on the cutting edge of contemporary knowledge development. They have been described extensively in the ancient Vedic literature, and have been clarified in scientific language by Maharishi over the past 50 years.

Consciousness as Non-Computable Space-time Geometry: The Orch OR Model

The *Orch OR Model* is an attempt to develop a scientific theory of consciousness that combines quantum physics and neurobiology, proposed by Sir Roger Penrose and anesthesiologist Stuart Hameroff.[1] The physics aspect of the model has to do with the collapse of the quantum wave function, theorized in this model to be a *real* objective phenomenon in nature—called objective reduction (OR)—introduced in Chapter 5. Objective reduction is the hypothesized process through which superposed quantum wave functions are reduced to discrete classical macroscopic objects available to ordinary phenomenal experience without necessarily involving the conscious attention of an observer. The 'Orch' part of the name—short for *orchestrated*—refers to the hypothesis that the OR is orchestrated by cytoskeletal microtubule proteins in the brain's neurons that 'tune' the quantum oscillations leading to OR.[1]

As discussed in Chapters 4 and 5, the collapse of the wave function of a quantum system—the OR in this model—is described in orthodox quantum theory as instantaneously occurring in the measurement process. In other interpretations of quantum theory the collapse occurs spontaneously under conditions in which random effects predominate in the environment—associated with the principle of decoherence. According to the Orch OR model, under special conditions in which a quantum system is well isolated from random environmental effects and remains coherent, it is theorized to *self*-reduce or collapse spontaneously in an orchestrated non-random fashion. This is held to be an objective reduction, not directly occurring by virtue of the process of observation by a subjective observer as in orthodox quantum theory. It is theorized that somehow this is capable of producing a conscious moment in at least some living organisms.

The orchestrated OR is said to involve fundamental space-time geometries. It is non-random but also *non-computable*, meaning that the orchestrated OR cannot be derived entirely from preceding states through any computation method. The concept of non-computability, associated in this model with Gödel's incompleteness theorem, is theorized to be a special property of some conscious states. It is thought to be unique to humans and other animals, in contrast to machines. This model extends the assumption of determinism into levels of nature beyond the probabilism of orthodox quantum theory, as the result of "...some presently unknown 'non-computational' mathematical/physical (i.e., 'Platonic realm') theory..." (p. 180),[1] which is thought to be essential to consciousness.

> "In the OR description, consciousness takes place if an organized quantum system is able to isolate and sustain coherent superposition until its quantum gravity threshold for space-time separation is met; it then *self*-reduces (non-computably). For consciousness to occur, *self*-reduction is essential, as opposed to the reduction being triggered by the system's random environment... We take the *self*-reduction to be an instantaneous event—the climax of a *self*-organizing process fundamental to the structuring of space-time—and apparently consistent with a Whitehead 'occasion of experience.'... Only large collections of particles acting coherently in a single macroscopic quantum state could possibly sustain isolation and support coherent superposition in a timeframe brief enough to be relevant to our consciousness." (p. 178)[1]

The theory favors cytoskeletal microtubules in the interiors of living cells such as neurons as the likely site for the orchestrated ORs associated with fundamental events or occasions of conscious experience. Microtubules are microscopic hollow cylinders comprised of 13 longitudinal protofilaments, arranged in a slightly twisted hexagonal lattice that repeats every 3, 5, 8 or other sets of rows. They provide the skeleton of the cell, and also are thought to have information processing roles in cognition. Microtubules are favored because of:

> "1) high prevalence, 2) functional importance (for example regulating neural connectivity and synaptic function, 3) periodic, crystal-like lattice dipole structure with long-range order, 4) ability to be transiently isolated from external interaction/observation, 5) functionally coupled to quantum-level events, 6) hollow, cylindrical (possible wave-guide), and 7) suitable for information processing." (pp. 184-185)[1]

According to the model quantum coherence is isolated in brain microtubules until a threshold is reached, when there is a *self*-collapse OR that "creates an instantaneous 'now' event..." (p. 187)[1] Microtubule-associated proteins that attach to microtubules 'tune' or orchestrate the self-collapsing OR events. This process involves both quantum and classical computations at time and distance scales consistent with quantum gravity interactions and with time frames typically associated with conscious attentional processes—in the range of about 500 msec.

In the model, cascades of ORs give rise to the experience of a stream of consciousness. Discrete ORs, each reflecting a momentary 'now' conscious event, build up into a cascade that produces the forward flow or stream of consciousness in time. This is likened to descriptions in Buddhist texts of consciousness as collections of mental events, each appearing and disappearing somehow with no duration in time, that produce the subjective experience of continuous time.

Consistent with this model, lower life forms such as worms could have brief conscious experiences—very primitive glimpses of their next action in space-time.

This model can be viewed as another attempt to account for conscious experience by proposing some unknown and undefined form of experiential medium. Because the experiential medium is built up from fundamental quantum mechanical events in physical space-time, the theory is physicalistic to the degree that quantum mechanics can be said to be physical. However, consciousness is elemental in the theory in the sense of non-computable. But it is not associated with a subtler mental *reality* underlying physical *reality*. As an account of consciousness emerging from the physical it doesn't address the ontological possibility of mind as a substrate in information space underlying physical space-time, or as existing and functioning apart from the physical brain. The aspect of the theory that is closest to an underlying mental level of *reality* seems to be the concept of the isolation of quantum coherence across microtubules. Each microtubule spontaneous self-reduction produces a 'now' conscious experience that cascades in an extended field effect of groups of microtubules into a stream of consciousness. What it is that maintains the quantum coherent state resulting in a spontaneous non-computable but deterministic OR accompanied by a moment of conscious experience doesn't seem to be articulated. What experiences the moment and how it becomes subjective conscious experience also aren't articulated.

However, the model can be said to progress a little further than the strict physicalist view of a closed causal physical universe in that it is based on a fundamental, deterministic non-computability in nature thought to be associated with consciousness. In this sense it also can be said to have a non-reductive feature. It theorizes where consciousness might come into play in the physical brain, and it suggests it accompanies self-collapsing ORs that provide the organism with a glimpse into its next action. But it doesn't seem to explain how the organism becomes an integrated functioning whole that is able to use such a glimpse to perform unitary goal-directed behavior—where the glimpse actually resides in the physical, who or what experiences it as a conscious stream, and how the conscious moments combine with cognitive functions in unitary behavior.

Consciousness as the Specifier of Classical Reality from Hilbert Space

Another approach to a theory of consciousness interprets quantum theory as not only about the quantum wave function (the Schrödinger equation) but also fundamentally about subjective experience. Physicist Henry Stapp, developer of this theory, points to the importance of the experiential component of the collapse of the wave function in the following statement:

"The core idea of Bohr was to recognize that physics is basically about our experiences... But von Neumann, and more unambiguously Wigner, went the next step and brought consciousness into the theory as a causal agent that actively did what needed to be done to make the theory work at the ontological level in the same way that it worked at the practical level. The ontology then included conscious experience as the co-equal partner of a more shadowy world of 'possibilities,' or, in Heisenberg's terminology, of 'objective tendencies' for transitions from the possible to the actual... Each actualization event has its physical side, which is just the collapse of the wave function itself, and also its experiential side." (p. 212) [2]

According to Stapp the cause of the wave function collapse from shadowy potentia to concrete actuality is:

"...not to be found in that physical part of the quantum ontology... [F]rom the purely physical standpoint the collapse seems to come from nowhere, as an unpredictable and undetermined 'bolt from the blue.' Something is needed to...bring 'classicality' into the dynamics, and it needs a 'cause' for the collapse, and it needs a reality to complement the 'potentia'... It must be something that exists, and the only thing that we know exists, besides the physical part of reality...is the experiential part..." (p. 213) [2]

As to the function of consciousness, Stapp asserts that consciousness is needed because:

"...the local-reductionistic laws of physics, regarded as a causal description of nature, are incomplete... The physical part of reality represents merely the possibilities for an actual experience, not the actually experienced reality itself." (p. 213) [2]

Consciousness has a role in actualizing 'classicality'—consistent with orthodox interpretations of quantum theory. But new points are brought up in an attempt to address how the actualization event involves subjective intentionality, where consciousness fits into the event, and the ontological substrate of consciousness. According to the model, the actualization event is a totality that:

"...contains the slowly changing fringe of the experience that constitutes the 'I', or 'psyche', which is felt as the experiencing subject and actualizer. The experiencing subject is *part of the thought*, not an outside observer of the thought: if the 'I' were not *part of the thought* then there could be in the thought no awareness of 'I' as the background relative to which the focus of the thought is the foreground. Thus it is not that the thought belongs to an 'I', but rather that an 'I' belongs to the thought." (p. 213) [2]

This importantly reflects a more explicit attempt to consider the actual role of subjective experience in physical processes, beyond a general recognition of consciousness as involved in the collapse of the quantum wave function. It also emphasizes that the subjective *I*, the experiencer, is embedded in thought processes, and it further begins to consider the causal elements of subjectivity associated with the individual *I* involved in the actualization of a physical event from quantum possibilities. But due to its reductive perspective, it doesn't capture the point in Maharishi Vedic Science and Technology—to be discussed in more detail in the next two chapters—that the *I* is a more fundamental holistic basis that underpins thoughts, of which thoughts are more specific expressions. It reflects the ordinary subjective experience that consciousness is embedded in mental activity, rather than the opposite holistic view. In the holistic view thoughts (mind) are embedded in the *I* (ego), which is embedded in consciousness itself.

Also it seems not to identify a place where these thoughts exist outside of the quantum level of physical possibilities. It seems to place them in an undefined experiential *reality* that is derived from the quantum level, but that is not the physically actualized level of classical *reality* either. It identifies subjective experience as *real*, but can't quite seem to find its place or relation to physical existence. As to the general whereabouts of subjectivity and of consciousness, Stapp posits that:

"...the whole process is represented in the Hilbert space in which the quantum analogue of matter is represented. But rising out of the matter-like aspects of nature lies another dynamics governed by the experiential aspects of nature... What is important is the presence in the physical substrate of potentialities for quantum actualizations of experiential structures... *Is experience a fundamental element of nature, or is it derivative, or emergent?* It is fundamental because the fundamental realities are experiential." (p. 214) [2]

Thus the theory identifies one fundamental aspect of *reality* as experiential, embedding it in an underlying quantum substrate of Hilbert space. Experiential *reality* is identified as a co-equal partner of physical *reality*, both deriving from the more fundamental substrate of possibilities of mathematical Hilbert space. These quantum potentia in Hilbert space represent the quantum analogue of matter, and presumably the quantum analogue of experience. Three ontological 'realities' are recognized: quantum possibilities in Hilbert space, and the physical and experiential *'realities.'* This is in the general direction of three levels of nature in Vedic science.

In the model, consciousness is placed in the experiential side, somehow differentiated from the physical material side, in its own unarticulated ontological *reality*. This might seem like psychophysical parallelism, with both physical and experiential underlain by quantum potentia in Hilbert space. However, there is at least a one-way causal influence of experiential *reality* on actualized physical *reality*. The causal power to collapse the wave function from the quantum possibilities in Hilbert space into physical 'classicality' is placed on the experiential side, not on the physical side where the closed chain of causes is typically located. This recognizes the incompleteness of the picture of physical causation in the physical world as experienced. It states that experiential and actualized physical *reality* are co-equal partners, but also it states that experiential *reality* in some way actualizes physical *reality*. Because this causally ties physical matter to experiential *reality*, it is not psychophysical parallelism in the sense that they are correlated but not causally related; it rather can be viewed as a type of psychophysical *interactionism*.

It further identifies consciousness in experiential *reality* as if it were an ontologically general element in nature. But as human consciousness, Stapp describes it as an emergent property:

> "...[T]he particular sort of consciousness that we human beings experience is emergent, because it represents a highly evolved form of the general ontological type. The complexity of a human experience is a consequence of the complexity of the body/brain that supports the physical activity. The complexity of the physical carrier has undoubtedly co-evolved with the complexity of the associated experiential reality." (p. 214) [2]

This suggests that consciousness exists as a general property derived from Hilbert space potentia that has its own ontological status contiguous with physical actuality. But it requires a complex physical carrier corresponding to equally complex experiential *reality* to appear in its highly evolved human form. It seems unclear in the theory whether consciousness exists anywhere in its general form, but rather only according to the level of evolution of its physical carrier. This seems to be a functionalist-physicalist emergent view of consciousness—discussed in Chapter 13. It reflects the challenge of trying to understand the mind-body problem within the overall perspective of reductionism and localization of mind generally in some level of neural function.

However, this theory is a bit more holistic in the sense that it explicitly identifies an underlying substrate to experience—as well as to matter—of quantum potentia in the infinite possibilities of Hilbert space. The infinite possibilities of Hilbert space as shadowy quantum potentia underlie physical and experiential *reality*, thoughts, the sense of *I*, consciousness, and also classical material *reality*. The subjective derives from Hilbert space, as does physical matter, but also causes the collapse of the quantum wave function into classical physical *reality*. This represents some progress toward at least identifying that there are these three levels of nature, and that consciousness is an essential fundamental element, but not yet recognized to be the most fundamental. It at least gives the status of *reality* to the subjective or experiential, in addition to the physical. But it is far from developed enough to match up with the sequence in Maharishi Vedic Science and Technology (described in Chapter 11): consciousness itself (Purusha), the self-interacting dynamics of the unified field (Prakriti), individual ego and thoughts as levels in the mind, the subtle physical levels, and then gross physical matter of which the classical macroscopic level is a part.

The theory has some aspects that are reminiscent of the Hilbert space formulation of the Lagrangian by John Hagelin, which was introduced in Chapter 10.[5] If the concept of an infinite dimensional Hilbert space is likened to the level of Prakriti, then the theory is starting to approximate the sequence in Maharishi Vedic Science and Technology of the field of all possibilities (somewhat similar to infinite dimensional Hilbert space), then next the subjective domain of mind, and then the classical objective domain of gross matter. However, the theory does not place consciousness as underlying these levels, which is the core of ancient Vedic science and Maharishi Vedic Science and Technology. Rather it associates consciousness with the fringe of thoughts in an experiential *reality* that seems to float somewhere between quantum potentia and its physical carrier of the body/brain in physical *reality*. This seems to be more accurately depicted as a model of where the mind or mental *reality* is than of consciousness.

Perhaps the most interesting contribution of this theory is that it explicitly recognizes the importance of some form of conscious free will in addressing even physical explanations of the universe. The theory attempts to counter the two factors of local determinism and quantum chance that threaten to eliminate free will entirely. One strategy in the theory toward doing this is to point out that choices are made at the level of the organism as a whole, not at the level of neurons in the brain. As Stapp states:

> "One's fate is not controlled exclusively by mechanical local deterministic laws, or by an avalanche of microscopically entering chance elements that would make a mockery of the idea of personal choice... If the quantum events in the brain occurred at the level of the neurons then the choices would be blind, for the consequences of each individual choice would be screened from view by the inscrutable outcomes of billions of similar independent random choices... The conditioning for this event is an expression of the values and goals of the whole organism, and the choice is implemented by a unified action of the whole organism that is normally meaningful in the life of the organism." (p. 210) [2]

The interesting question here is where the 'organic whole' of the organism exists (similarly unspecified in the Orch OR model). The idea of meaningful action of the whole organism is suggestive of a classical behavior level of functioning of the individual. But this level is usually understood to be firmly ensconced in the closed causal nexus of the actual physical universe. So it would seem difficult for this level to be the site where free choice is made. If free will and free choice exist in human life, then the universe must be quite different than the closed chain of physical causation as theorized in modern science, which this model recognizes.

Perhaps what is meant is that there is a level of the organism as a whole that is at a smaller time and distance scale than macroscopic behavior, as well as microscopic neurons. For example it could be somewhere in the "quantum soup." (p. 210)[2] But one difficulty with this is that the quantum soup is thought to be random, so it doesn't seem to help much as the site of free choice—unless the fundamental core of free will is randomness, which Stapp argues against. The other possibilities would seem to be an even smaller time and distance scale than the quantum soup—impossible in most modern scientific theories because there is thought to be nothing more fundamental—or somewhere in between the neuronal level and the quantum soup.

What we are looking for is the site of conscious free will—its ontological level. In this theory it is based on an organic wholeness of functioning that chooses classically meaningful behavior based on its own past history and memory, its own perception of needs in the classical environment, and its ability to reason and calculate optimal choices according to its own perspective.[2] Where do all these abilities—memory, perception, reason ordinarily identified with the mind—come together in an organic wholeness of feeling when the classical behavioral level, the neuronal level, and the quantum soup don't seem to be able to accommodate them?

The tacit implication is that there is a level of mental *reality* where conscious choice is made that is somewhere deeper than classical *reality* and somewhere less deep than the infinite random possibilities of the quantum soup—but that still operates as a functional wholeness. This might seem to suggest that the element of chance in the quantum soup would still be a fundamental aspect in the selection of a choice by the organism as a whole, and thus might not really be a freely made choice. However, there is a suggestion in the theory that the quantum soup may not be completely random. Stapp notes:

> "In my own opinion this occurrence of pure chance is a reflection of our state of ignorance regarding the true cause, which must in any case be nonlocal, and hence both difficult to study and quite unlike the local issues that science has dealt with up until now..." (p. 211) [2]

Stapp further poses the possibility of:

> "...replacing the element of pure chance by a nonlocal causal process that makes the felt psychological subjective 'I', as it is represented within the quantum-theoretic description, rather than pure chance, the source of the decisions between one's alternative possible courses of action." (p. 211) [2]

This could be interpreted as implying that decision making, feelings, and the psychological *I* exist in a deterministic and causally efficacious but nonlocal field that is somehow in the quantum soup. This would be placing mental processes at a much deeper level than even experiential *reality* as derived from quantum potentia, even possibly associating some properties of quantum potentia as a field of mind—at least the non-random, nonlocal parts of the quantum soup. These are potential implications of speculative aspects of the theory, of course, but represent further progress toward the holistic view of matter, mind, and consciousness.

Importantly Stapp's theory can be viewed as a positive step in the direction of the structure of levels described in the Sankhya system in Chapter 11. It at least acknowledges and begins to consider key issues not even mentioned in most all the other models and approaches. As will be summarized in Chapter 14, similar correspondences between theories of mind in cognitive science and neuroscience and the levels of mind in Maharishi Vedic Science and Technology will add additional clarity to the overall direction in which modern science is progressing with respect to the place and role of both mind and consciousness.

Nonreductive Functionalism

Another prominent approach toward a scientific theory of consciousness attempting to extend beyond structural and functional approaches is the *nonreductive functionalism* of philosopher David Chalmers.[36] It emphasizes the uniqueness of conscious experience and the inability of a strict reductive physicalist perspective to account for it. Key concepts in this approach include *structural coherence*, *organizational invariance*, and *property dualism*.

Structural coherence. The concept of structural coherence refers to the correspondence between cognitive representations of perceptions and their phenomenal experience. It identifies awareness, in the sense of being *aware of*, as the psychological correlate of phenomenal conscious experience. In this functional sense, awareness refers to being aware of an object of experience such that it is accessible to behavioral control. Each detail of cognitive representation has a corresponding detail in awareness and in phenomenal experience, demonstrated by the ability to react to it in behaviorally appropriate ways (except of course unconscious perceptual or cognitive processes). For example visual experience of a book is accompanied by functional perception of the book, which allows controlled behavior such as turning the pages, reading, and so on.

In the phrase structural coherence, *structure* refers primarily to functional relations of features in an abstract conceptual space; and *coherence* refers to the high consistency across cognitive representations in awareness that aid purposeful action and phenomenal experience of them. The principle of structural coherence states that when there is an experience, there is an awareness of the content of the experience, demonstrated by the ability to act on it. The relationship also goes the other way, in that when there is awareness of some object of experience, there is usually a corresponding conscious experience of it. For example the cognitive representation in awareness of a friend smiling is typically accompanied by the *experience* of a smiling friend.[3]

These coherent relationships are related to correspondences between abstract functional spaces. Cognitive representations of qualitative features of objects can be thought of as functionally distributed in a conceptual space, such as tri-partite functional color space. Other qualities such as intensity also have a functional analogue in conceptual space. These cognitive representations have been found in some cases also to have spatial counterparts in the neuroanatomy—as in brain maps, such as in the somatosensory cortex. They also correspond with spatial structure in the outer physical environment. Chalmers states:

> "So alongside the general principle that where there is consciousness, there is awareness, and vice versa, we have a more specific principle: the structure of consciousness is mirrored by the structure of awareness, and the structure of awareness is mirrored by the structure of consciousness. I will call this the *principle of structural coherence.* This is a central and systematic relation between phenomenology and psychology, and ultimately can be cashed out into a relation between phenomenology and underlying physical processes." (p. 225)"[3]

The overall point here is that the outer objective environment, the objective physical environment in the brain, the inner subjective environment of cognition, functional awareness, control of behavior, and also phenomenal experience all empirically cohere and are not arbitrary with respect to each other. It is empirical evidence of orderly relations across body and mind, structurally and functionally, that can serve as the basis for psychophysical laws of nature that can build into a scientific explanation of consciousness. This insightful and important principle of structural coherence in nature is consistent with Maharishi's general principle of the one-to-one correspondence of body and mind referred in Chapter 13.

Organizational invariance. The concept of organizational invariance holds that consciousness depends on functional organization only, not on what forms of physical substrate make up or realize the functions.[3][6] The assumption is that qualitatively equivalent conscious experiences will occur across different but functionally equivalent physical systems. In other words consciousness is an inevitable product of the right type of functional organization—apart from its physical instantiation. However, the organization must have sufficient detail or be *fine-grained* enough to support the full range of behaviors.

Although there is some empirical basis for this principle, it relies more on a conceptual and logical analysis, and is illustrated by the use of thought experiments.[3][6] In one thought experiment neurons in the brain of a conscious human being are imagined to be progressively replaced by other processing units such as silicon chips, toward creating a functional isomorph that no longer has a biological substrate—that is, a robot. According to the principle of organizational invariance, the functional isomorph would have conscious experiences that are the same as the human.[3]

Another thought experiment to illustrate the principle may be a little more challenging to imagine. In this second thought experiment a huge collection of many billions of individual humans is used to model the functional organization of the neural network in the brain of an individual human (no matter that it may be impossible to implement in practical terms); and each neuron becomes represented by an individual human in a manner that fully corresponds to the functional relationships of neurons in the brain. [3] The issue is whether this collective network or group brain would somehow produce its own consciousness (in addition to the individual consciousness of each participant). The principle of organizational invariance again would predict that consciousness automatically will emerge from such a group organization. [3] Where in the universe the new consciousness might have ontological existence is quite a mystery.

Let's compare the conclusion in these thought experiments with the holistic understanding of consciousness described in this book. In sharp contrast to the prediction based on organizational invariance, the prediction from the holistic Vedic perspective in the context of the first thought experiment would be that the silicon-brained functional isomorph—even if capable of replicating much of human behavior—would not have consciousness other than the degree of consciousness expressed in silicon (for all practical purposes, none). Likewise the group brain isomorph—in the second thought experiment—also would not produce a new and separate group-based consciousness. In Maharishi Vedic Science and Technology consciousness is fundamental, it is uncaused. There is only one consciousness, the unified field of consciousness. It does not emerge from functional organization or anything else; it is undivided, whole, and uncreated. As described in Chapter 13, everything supervenes in the sense of depends on consciousness.

In analyzing the undivided wholeness of the unified field of consciousness for purposes of explaining phenomenal creation, it can be understood in terms of the coexistence of opposites of infinite dynamism and infinite silence, which can be related to Prakriti or pure intelligence and Purusha or pure existence. Pure existence can be said to be the most abstract conception of structure, and pure intelligence can be said to be the most abstract conception of function. The more concrete structural and more abstract functional processes in nature both come from consciousness itself. They localize the unified field of consciousness into levels of phenomenal manifestation. They 'create' consciousness only in the sense that they express the phenomenal individuation of consciousness into forms that appear to be conscious in different conditions or states of consciousness. This neither involves creating more consciousness, nor new 'consciousnesses.'

In the group brain isomorph example, undoubtedly there would be a field effect created by a group organization of individual conscious beings—but not a new and separate consciousness, and not even a new group mind. This thought experiment example is relevant to an understanding of the important principle in Maharishi Vedic Science and Technology called *collective consciousness*—to be discussed in Chapter 21.

Considering functional organization in either of these thought experiments, eventually the issue of the physical substrate in which the functional design is implemented or instantiated has to be considered—and especially the level of subtlety of the physical substrate. This issue may be a bit more challenging to recognize in the second thought experiment, in which humans play the role of neural information processing units; but then behavioral equivalence is also harder to imagine in this example.

The crux of the matter with respect to organizational invariance is the requirement that the functional isomorph be fine-grained enough to support the full range of behavior of conscious humans. The issue of fine-graining is basically the same issue of subtlety or levels of nature discussed throughout this book. In these thought experiment examples it needs to apply not only to the degree of fine-graining of the functional organization—its detailed interconnectedness and specificity—but also importantly to the degree of fine-graining of behavioral equivalence and of the physical substrate that instantiates the functional organization. At some point the analysis must get to smaller and smaller time and distance scales of the physical medium in which the functional organization is realized and the behavioral equivalence is demonstrated. In the physicalist view of nature there needs to be a physical substrate for any functional organization. A functional approach cannot be fully divorced from a structural approach, because functions don't exist outside of structure and must be implemented in a *physical* structure. Eventually the degree of fine-graining has to do with the time and distance scale at which functional organization is implemented and behavioral equivalence is assessed *in the physical*.

Again let's consider the first thought experiment example in which silicon chips replace neurons—first in terms of the approximate level of fine-graining currently possible. It would be expected that the electrochemical activity of the silicon would not match up particularly well with neural activity, and might disrupt or even prevent it—similar to a lesion or tumor. This would degrade expressions of cognitive processing and eventually any expression of subjective experience in gross behavior, the details of which would depend on the brain areas involved. Perhaps the nervous system could compensate via redundancy to minimize the expression of cognitive and behavioral deficits. But at some point in the replacement process the system would no longer be able to maintain its functioning and the human body would expire.

In order to make silicon chip-neuronal connections more seamless, the silicon replacement chips would need to replicate DNA functions, even to quantum levels. To get finer-grained structure, function, and behavioral equivalence as well as comparable cognitive representation, the entire history of the individual including for example the entire memory content and history of intentions—whether accessible and acted upon or not—also would need replication. Because this individual history is causally connected to everything else in the past light cone that included this individual, the record of everything in this light cone would need to be replicated as well.

> "The enormous variety of discriminable states available to a conscious human being is clearly many orders of magnitude larger than those available to anything we have built. Whether we can verbally describe these states satisfactorily or not, billions of such states are easily discriminable by the same person, and each of them is capable of bringing about different consequences."—*Gerald M. Edelman, neurobiologist, and Giulio Tononi, neuroscientist* (p. 32)[7]

One could take it so far as to say that in order to have a complete fine-graining there might need to be a replication of the entire history of the physical universe, including the initial conditions making that particular individual entity possible. Even then, however, a new consciousness would not be created. Rather, what would be created is a new gross individual structure that could possibly serve as the localized embodiment of a particular state of consciousness, and that some other conscious entity might chose to inhabit.

It is possible that the structure necessary for a functional replication of the conscious human entity at the quantum field level of fine-graining would in fact require something that is at least quite similar to human physiology. But let's assume there still might be a range of possible structural configurations that could do it, and the equally conscious entity would no longer have a human body. Assuming this isomorphic body was fine-grained enough that the mind-body connection could be maintained, still any organizational or structural difference would show itself in some behavioral difference if behavioral equivalence also were assessed at a fine-grained enough level. This might be revealed, for example, in reports of global feelings, memories, or sense of self. It likely would feel different subjectively for the conscious entity, similar to how a prosthetic arm feels different—only perhaps on a much finer level of feeling. However, even this level of fine-graining would not replicate the underlying subtle subjective, psychological functions as described in the past few chapters. Even replicating everything down to the quantum level, no level of mind would have been replicated yet. A vastly higher degree of fine-graining would be needed in order to create a functional isomorph on the subtle levels of non-local information space, mental *reality,* or quantum mind that has fine-grained behavioral equivalence. Drawing upon discussions in prior chapters about the levels of objective and subjective *reality*—and what is the subtlest substrate—the fine-graining could go down through the microscopic levels, the DNA level, and ultramicroscopic levels and underlying quantum fields to non-material information space or nonlocal quantum mind. However, what is not recognized in a physicalist view is the subtle mental domain. In order to build a conscious entity in the gross material world which incorporates all the underlying levels of creation, every level of fine-graining would need to be replicated, gross and subtle.

The notion that different functional organizations will produce the same consciousness if fine-grained enough—organizational invariance—suggests that consciousness is some general property that can be instantiated in different ways. But if there is an invariant *functional identity* between conscious experience and the physical, different ways of instantiating the function would produce different experiences. They would not produce some generic property of phenomenal experience. Again, fine-graining of behavioral equivalence and of physical structure need to be at the same level as the fine-graining of organizational invariance.

An exact replication of organization at every level of fine-graining of the function in different structures would seem impossible in a physicalist view. [8] At some point the structures and the functions interact with each other to such a fine level that small differences in structure would influence small differences in function. According at least to the functional identity hypothesis, this would necessitate small differences in conscious experience. [9] From the perspective of extreme fine-graining, a human and a robot isomorph would necessarily not be functionally equivalent and not have identical experiences. The only way to get the same conscious experience would be if structure and function were identical at *every* level, which might be thought of as having the same identical human at two macroscopically different physical locations—which doesn't make sense in ordinary physical space-time. (Maybe we are unique individuals after all.)

Robot empowerment? As an additional point of consideration, the notion of fine-graining relates to the increasing abstraction and subtlety we have been discussing for several chapters. As mentioned earlier in this chapter, the most abstract notions of structure and function can be related to pure existence and pure intelligence. We could further relate pure existence to the point value—as the infinite singularity of existence—and pure intelligence to the infinite value—as the ultimate discrimination into infinity of points. We can conceive of a sort of tension or reverberation between the infinite value and the point value in order to maintain their superposition and not collapse into one or the other, keeping both lively at the same time—even though most fundamentally they are completely unified. That liveliness is a way of describing the alertness or wakefulness of consciousness itself—as a coexistence of opposite values. It also can be described as the source of the inherent power or evolutionary drive throughout nature toward higher levels of the expression of alertness or consciousness.

In this quite abstract way of understanding structure and function, a third factor can be delineated: that which powers, drives, or activates structure to function. It is neither just structure nor just functional organization, but also a dynamic factor or energy that is needed to activate the system into dynamic functioning. The three aspects of structure, functional organization, and dynamic activation are all required. This brings us back again to the three-in-one self-interacting dynamics of the unified field—associated with the three forces of tamas, sattva, and rajas guna.

Practically applying this trinity to the function of a computer robot, for example, the structural component of hardware—its existence—and the functional component of software—its intelligence—can both be present, but the computer still doesn't work until it is plugged in to a power supply. In the case of a human, we can say that it is the energy such as from digestion and respiration that fuels the system to function. But this isn't the whole story, in that the fuels can be present, the functional organization can be present, and the bodily structure can be present, without the system functioning such as in a recently expired body. This will be discussed in later chapters, and is initially mentioned here only to bring up the issue of what drives systems to function, relevant to this general discussion of whether artificially intelligent machines could become conscious.

How many machines initiate their functioning entirely on their own, without being powered to do so from outside the machine (even though underlain by the inherent dynamism of the quantum field)? How many machines have created themselves on their own, or have reproduced themselves without the design, construction, and programming intelligence coming from outside the machine? These questions might give us pause in considering organizational invariance, especially because the universe has given birth to much more complex living organisms than the machines now extant—namely, such as us. Apart from whether robotic machines would have phenomenal experiences, research in artificial intelligence has made little progress on fundamental issues such as these. Also these issues are exponentially more challenging when more relevant degrees of finer-graining are attempted.

Considering any degree of fine-graining as possible, of course conscious organisms can be created. After all, God—or Nature if you prefer what might seem to be the impersonal—can be said to have done it with each of us. However, not just gross matter at all its time and distance scales down to the Planck scale, but also all the subtler levels associated with the subjective levels of information space and nonlocal mind would have to be replicated. Even then, however, no new consciousness or consciousnesses would be created.

" [T]he technology of electronic computer-controlled robots will *not* provide a way to the artificial construction of an *actually* intelligent machine—in the sense of a machine that understands what it is doing and can act upon that understanding. Electronic computers have their undoubted importance in clarifying many of the issues that relate to mental phenomena (perhaps, to a large extent, by teaching us what genuine mental phenomena are *not*), in addition to their being extremely powerful and valuable aids to scientific, technological, and social progress. Computers, we conclude, do something very different from what *we* are doing when we bring our awareness to bear on some problem."—*Sir Roger Penrose, mathematician and cosmologist (p. 393)* [10]

Fortunately long before we as humans are able to create anything at the subtle subjective levels from the gross, outside, objective, third-person perspective, we will be able to create virtually anything that is possible from the inside first-person perspective by *mere intention*—mind over matter. This will be discussed in upcoming chapters. With that requisite level of integrated knowledge and experience, and its corresponding degree of fulfillment, whether a robot can be conscious will long have become moot. Hopefully we will soon spend less time, energy, and intelligence tinkering with robots, as well as the human gene pool, in the sandbox of matter and apply them more pragmatically for systematic natural development of our unlimited inner potential, utilizing the total potential of natural law in our own consciousness and in our own daily life.

Property dualism. The *nonreductive* part of nonreductive functionalism relates to the concept of property dualism.[3] This importantly recognizes that there are some properties of phenomenal experience that are not completely reducible to the physical. It describes consciousness and phenomenal experience as not entirely accountable through the physical. But it doesn't account for them in terms of a functional explanation either. Basically property dualism can be viewed as an attempt to deal with the underlying mental *reality* as nonreductive with respect to its physical properties, without giving it full status as an ontologically *real* level of existence—apparently because this would be too risky in terms of the mainstream belief in physicalism. Although it acknowledges something that is explanatorily nonreductive with respect to the physical, it doesn't accept it fully as a non-physical mental *reality*. In respect of first-person experiential evidence, it tries to carve out some very tiny middle ground between body and consciousness by virtue of a constrained explanatory dualism that is nonreductive, but not a separately identifiable ontological level of *reality*. Thus it is described as *dual-aspect monism*, for the most part the bottom line of which is considerable faith in physicalism, or materialistic monism. Consciousness is thought to be nonreductive with respect to physical explanations of nature, but not quite yet acknowledged as an underlying field of existence or level more fundamental than the physical. It strives to be explanatorily dualistic, but not to embrace an ontological dualism.

This approach emphasizes that consciousness is an unmediated fact of experience; but with respect to the physical, it seems explanatorily and causally irrelevant. The challenge is that in being causally irrelevant with respect to the physical, consciousness is also irrelevant in terms of its possible functional role. The implication is that consciousness is epiphenomenal. [3 11]

> Epiphenomenalism acknowledges that mind is a real phenomenon but holds that it cannot have any effect on the physical world. Epiphenomenalism views the brain as the cause of all aspects of the mind, but because it holds that the physical world is *causally closed*—that is, that physical events can have only physical causes—it holds that the mind itself doesn't actually cause anything to happen that the brain hasn't already taken care of. It thus leaves us with a rather withered sort of mind, one in which consciousness is, at least in scientific terms, reduced to an impotent shadow of its former self... Because it maintains that the causal arrow point in only one direction, from material to mental, this school denies the causal efficacy of mental states... To deny the causal efficacy of mental states altogether is to dismiss the experience of willed action as nothing but an illusion... And that raises an obvious question: What possible selective advantage could consciousness offer if it is only a functionless phantasm?"—*Jeffrey Schwartz, research neuropsychiatrist, and Sharon Begley, science writer/editor (p. 39-40)* [12]

In broad outline, consciousness is being placed deeper and deeper into the individual experiencer, but with still no clear identification of its location or role anywhere in objective or subjective processes in nature. This does, however, represent further movement toward recognition of the inability to explain consciousness in terms of either its concrete physical structure or its abstract function. Appropriately, consciousness continues to shine through the explanatory gap.

There is an additional double-aspect feature of this proto-dualistic approach that takes it a step further. The double-aspect feature is that information space can be realized both physically and phenomenally. This conception is similar to the models described in Chapters 6 and 7 about quantum gravity and quantum bits of space-time underlain by qubits of information space. In this, it goes a bit further in identifying a correspondence between functional information space and the phenomenology of the observer. In typical reductive fashion, however, it suggests that information space may have more fundamental *micro-phenomenal* or *proto-phenomenal* properties—sort of intrinsic bits of *phenomenological-stuff* that combine somehow into experiences. [3 13]

These *quantum information-based* models conceive of physical matter as having a substrate of pure information. Hopefully it is becoming clearer that in these models mental *reality* is being disembedded from matter—at least in conceptual understanding. There is slow movement toward recognition that the mind actually exists as a non-local level of *reality* underlying physical matter. So far, however, this recognition is just in terms of abstract information space-bits.

Eventually consciousness is disembedded from both matter and mind—structure and function—isolated into itself. This is understandably a hard problem for the reductive-oriented thinking mind to grasp. This thought-full (full of mental activity) objectified means to get to the essential nature of consciousness can be viewed as much less parsimonious and efficacious—as well as much less integrating and fulfilling—than the systematic technology of transcending thought that is the core methodology in Maharishi Vedic Science and Technology.

This chapter briefly reviewed three prominent theories in order to exemplify contemporary work toward a scientific theory of consciousness that attempts to go beyond structural and functional approaches. With all due respect, however, this work has not even established a foggy outline of a theory of mind; and there is virtually no hint of even a theoretical understanding of consciousness itself with respect to Vedic science. As will be exemplified in the next chapter, in psychological and cognitive science, and in neuroscience, at least an initial functional outline of the levels of mind has been emerging—even if there is little progress in understanding its relationship to consciousness itself.

In this next chapter we will discuss examples of functional models of the mind in psychological and cognitive science. Fortunately we are getting a bit closer to exploring the practical and personally relevant topic of first-person experiences of the development of the levels of mind in higher states of human consciousness.

Chapter 14 Notes

[1] Hameroff, S. R. & Penrose, R. (2000). Conscious Events as Orchestrated Space-Time Selections. In Shear, J. (Ed.). *Explaining consciousness—The hard problem.* Cambridge, MA: The MIT Press, pp. 177-195.

[2] Stapp, H. P. (2000). The Hard Problem: A Quantum Perspective. In Shear, J. (Ed.). *Explaining consciousness—The hard problem.* Cambridge, MA: The MIT Press, pp. 197-215.

[3] Chalmers, D. J. (1996). *The conscious mind: In search of a fundamental theory.* New York: Oxford University Press.

[4] Hut, P. & Shepard, R. N. (2000). Turning 'the Hard Problem' Upside Down and Sideways. In Shear, J, (Ed.) *Explaining consciousness— The hard problem.* Cambridge, MA: The MIT Press, pp. 305-322.

[5] Hagelin, J. S. (1989). Restructuring Physics From its Foundation in Light of Maharishi's Vedic Science. *Modern Science and Vedic Science,* 3, 1, 3-72.

[6] Chalmers, D. J, (2000) Facing Up to the Problem of Consciousness. In Shear, J. (Ed.). *Explaining consciousness—The hard problem.* Cambridge, MA: The MIT Press, pp. 9-30.

[7] Edelman, G. E. & Tononi, G. (2000). *A universe of consciousness: How matter becomes imagination.* New York: Basic Books.

[8] McGinn, C. (2000) Consciousness and Space. In Shear, J. (Ed.). *Explaining consciousness: The hard problem.* Cambridge, MA: The MIT Press, pp. 97-108.

[9] Hardcastle, V. G. (2000). The Why of Consciousness: A Non-Issue for Materialists. In Shear, J. (Ed.) *Explaining consciousness—The hard problem.* Cambridge, MA: The MIT Press, pp. 61-68.

[10] Penrose, R. (1994). *Shadows of the mind: A search for the missing science of consciousness.* New York: Oxford University Press.

[11] Seager, W. (2000) Consciousness, Information, and Panpsychism. In Shear, J. (Ed.). *Explaining consciousness—The hard problem.* Cambridge, MA: The MIT Press, pp. 269-286.

[12] Schwartz, J. M. & Begley, S. (2002). *The mind & the brain: Neuroplasticity and the power of mental force.* New York: ReganBooks.

[13] MacLennan, B. (2000). The Elements of Consciousness and Their Neurodynamical Correlates. In Shear, J. (Ed.). *Explaining consciousness—The hard problem.* Cambridge, MA: The MIT Press, pp. 249-266.

Chapter 15

Psychological and Cognitive Models of Mind and Consciousness

In this chapter, functional models of mind in psychological and cognitive science are overviewed with respect to their relationship to ancient Vedic science and Maharishi Vedic Science and Technology. We will focus on where and how consciousness fits into the models. Key issues include sequential stages of processing, levels of depth of processing, and the relationship between function and structure. The overall main point of this chapter is that the direction of progress in models of the mind is clearly toward the psychoarchitecture of the inner dimension in the approach of Sankhya, an aspect of ancient Vedic science outlined in Chapter 11.

For the most part modern science has functioned within the experiential limitations of the ordinary waking state of consciousness. Relying primarily on highly intellectualized reductive analyses within the ordinary waking state, it has been quite difficult to establish an integrated theory of mind. This especially has been due to the objectifying habit of directing attention outward toward the outer material domain of nature, rather than inward to experience directly the levels of mind and their underlying basis.

In the attempt to remain objective, frequently Western conceptions of the mind have used mechanistic analogues. Cartesian hydraulic descriptions of behavior as being driven by fluids in tubes, associative networks paralleling principles of chemistry and resembling electrical switchboards in behavioral psychology, cognitive models of cybernetic control systems, and computers and holograms provided metaphors by which theorists conceptualized the mind. [1] However, three thousand years of Western thought has not resulted in the discovery of a general model of mind. A general model would provide the basis for a mental taxonomy, and also an epistemological framework for knowledge development—neither of which have been achieved through the objective approach in modern science. But during this relatively long time certain conceptions have repeatedly arisen that now can be viewed as converging on a general model—reflected in progress over the past century in psychological and cognitive science, and also neuroscience. As will be demonstrated in this and the following two chapters, importantly this developing model of mind is coming closer to the levels of mind in Sankhya and Maharishi Vedic Science and Technology, introduced in Chapter 11 in the context of an 'inner dimension' as part of the structure of phenomenal nature.

To get a glimpse at the levels of mind let's first take a minute to reflect on our own inner experience. The outer visual environment is displayed before our eyes, and stimuli enter and are somehow processed in our mind as perceptions. We may or may not have quick emotional reactions to the input, but likely think about the perceptions and qualia, evaluate them, and perhaps consider how to respond. We may notice deeper feelings, such as subtle impressions of appreciation, comfort, uncertainty, or enthusiasm for new understanding. Further we may notice that these feelings relate to an even deeper sense of what is important in our lives, and who we are. We may even notice that on the deep interior fringe of the perceptions, thoughts, and feelings is our inner sense of existing right now as an individual self—and *possibly* even universal Self. These experiential levels of mind relate to the *inner dimension* introduced in connection with the full range of levels of nature enumerated in the Sankhya system of ancient Vedic science.

This chapter focuses on how theories of mind in psychological and cognitive science are beginning to be consistent with the inner dimension in Vedic science, in terms of functional and structural notions of *depth of processing*. As a starting point for reviewing overall progress in the development of these modern scientific models of mind, we will begin with the extreme outer perspective of *behaviorism*. This approach, the most prominent paradigm in scientific psychology at the onset of the 20th Century, attempted to eliminate mind and consciousness entirely from scientific investigation. For heuristic purposes the major alternative model concurrent with the behaviorist paradigm—here referred to as the Freudian model—also will be discussed briefly with respect to the place and role of consciousness.

The Behavioristic Approach

Although consciousness was the constitutive issue at the start of scientific psychology—as well as now—the behavioristic paradigm attempted to redefine the field in positivistic terms as the study of only observable behavior. References to mind and consciousness were eliminated in favor of a focus only on observable behavior—the *black box* approach discussed in Chapter 2. In attempting to be as rigorous as possible, behaviorists sought to assume a strict third-person perspective and to make the fewest number of assumptions in attempting to explain the behavior of living organisms. The focus initially was on comparatively simple animal behavior under highly controlled laboratory conditions. This paradigm took the position that the concept of mind may not be needed to explain behavior, and indeed that mind and consciousness may not even exist.

This seemingly simple approach started with observable stimuli or inputs into an organism and observable responses or outputs from the organism. Implicit in this was that an organism can receive stimuli, process them in the nervous system, and respond to them in consistent ways for its own purposes as an integrated functional unit—thus tacitly assuming much of what scientific psychology needed to explain. From the perspective presented in this book, starting from the level of outer gross macroscopic behavior would be the least simple in that all the subtler levels are embedded in and reflected in this grossest level of nature.

In between observable stimuli and observable responses was the black box of the whole organism, the contents of which were not observable from the ordinary objective third-person perspective. If the concept of a psychological inside or content in the black box was necessary at all, it was basically assumed to be a blank slate upon which associations built up between observable stimuli and responses. Consistent with the classical model of cause and effect in physical science, the task was to build scientific theories of lawful relationships between macroscopically observable stimuli (S) and macroscopically observable responses (R), with as few assumptions and inferences as possible:

$$\text{S (observable)} \quad — \quad \text{Black Box} \quad — \quad \text{R (observable)}$$
$$\text{(unobservable)}$$

Methodological differences in behavioral research led to two basic models of associative learning or conditioning. Classical, respondent, or Pavlovian conditioning emphasized associative learning based on pairing unconditioned stimuli (US) that evoked unconditioned responses with conditioned stimuli (CS), called US-CS learning. In this model, learning was a process of associating a new stimulus with an old stimulus that already produced a particular response, conditioning the response to the new stimulus also. A typical example is conditioning salivation to a bell by pairing the bell with food. Operant, instrumental, or Skinnerian conditioning emphasized pairing selected behaviors with rewarding stimuli such as food in order to reinforce the behaviors, called R-S learning. In this model, behaviors chosen by the experimenter were learned by the subject through reinforcing them with stimuli the subject inherently valued, evidenced by its behavior. A typical example is training lever pressing behavior in rats with a food reward.

These two basic theories were more a product of methodological differences than inherently different learning processes. Fundamental similarities in these two models suggest that they are different perspectives of a general cybernetic model involving feedback principles. [2,3] In such a cybernetic model, a system changes its output in order to control input—including effects of its own behavior—based on *internal* criteria or reference points. As a mechanistic example, a thermostat can 'measure' environmental temperature, compare it to an internal reference or set point, and activate heating or cooling units that shut off when the set point is reached. In the case of non-living control systems the reference or set point is selected by the designer or user of the system and built into the system. The system is a servomechanism, serving the designer or user.

In the case of living organisms the reference points are not set by some external user; rather they are inherent in the system itself. The living organism thus could be described as a *self-servomechanism*, inasmuch as it serves to regulate input for its own benefit. [4] This can be understood in terms of an *inner purpose*. It can be related to homeostatic balance in the internal and external environment of the organism, associated with individual and species survival. It also can be interpreted in a developmentally higher framework—to be discussed in the upcoming chapter on theories of human development and higher states of consciousness.

As an aside, perhaps the only theory in the history of scientific psychology that was proposed as the basis for creating an ideal society was Skinner's operant conditioning model. [5] This model focused on experimental control of stimulus input to shape response output. In attempting to apply the principle to the practical goal of improving human society, however, it generated major controversy. A primary issue was who would have the power to control the input into human organisms, and especially to choose the behaviors to be shaped. There was great concern whether the power to control behavior would be used for the good of the whole society or abused by certain groups or individuals—a major concern given human history, especially in the 20th Century (and continuing to be quite relevant with respect to technologies such as genetic engineering, which attempts to change natural set points inherent in living organisms).

Highly controlled research in these models of learning or conditioning led to several inferences—*hypothetical constructs* or *intervening variables*—about what is inside the black box:

S — **hypothetical constructs** — R
intervening variables

An important step from associative behavioral models of learning to cognitive models came with the experimental verification that stimulus-response relationships didn't just depend on principles of conditioning such as contiguity and repetition. Behavioral responses also were found to be influenced by the *information* in the stimuli—related to conditional probabilities of events. These research findings could not be accounted for in strict associative models. This meant that inside the black box of living organisms are complex information processors with channel capacities requiring decision making skills, rather than just S-R associative connections.

Another significant factor driving this change was the need for applications of psychological research relevant to war technology. There was a strong demand—and associated funding—for research to build better interface strategies between humans and machines. This fostered research on human information processing of sensory-perceptual, memory, and decision making functions in *real* world conditions—such as how to minimize mistakes in reading radar screens, or how to train complex new skills efficiently. Another major factor was development of the computer as a mechanistic physical model of computational processes in the black box of the human mind.

Without reviewing the huge amount of data, the research demonstrated that concepts such as innate drives, selective attention, memory, and eventually cognitive representations and reasoning processes seemed to be necessary inferences in order to account for lawful patterns of relationships between stimuli and responses in the black box:

$$\text{S} — \quad \begin{array}{c} \text{memory} \\ \text{selective attention} \\ \text{cognitive processes} \\ \text{innate drives, etc.} \end{array} \quad — \text{R}$$

With the inclusion of such constructs and inferences—suspiciously like *mental* processes—mind and consciousness crept back into psychological science. In recent years the nature of consciousness again has become a central issue on the cutting edge of this field and other related fields.

The Functional, Cognitive Approach

Using the computer as a mechanistic physical analogy of the mind, models of what happens inside the black box took center stage. This was based on indirect evidence from the third-person perspective outside the black box (but of course inside the black box of the investigators).

For the most part psychological and cognitive research has focused on functional analyses of mental processes, whereas neuroscientific research has emphasized structural analyses. These different research strategies reflect the mind/body problem. The functional approach attempted to investigate mental processes by abstracting them out of—and not primarily trying to understand them with respect to—the brain. In other terms the inside of the black box was examined from both the psychological functions of the mind—analogous to computer software—and the physiological structures of the brain and nervous system—analogous to hardware. Within the generally accepted physicalist worldview, however, this was a distinction of research strategy, inasmuch as psychology was thought to be nothing other than physiology, mind was nothing other than body, function had to be instantiated in physical structure.

The functional cognitive models attempted to identify stages of information processing, and where consciousness might fit in and play a role. The following is a simplified summary of progress toward a functional model of the mind, focusing on the location and role of consciousness.

"Functionalism, or mentalistic materialism...denies that the mind is anything more than brain states; it is a mere by-product of the brain's physical activity... The materialist position goes so far as to deny the ultimate reality of mental "events"... If we hold tenaciously to such quaint notions as experiential reality, consciousness, and the ontological validity of qualia, it is only out of ignorance; once science parses the actions of the brain in sufficient detail, qualia and consciousness will evaporate just as the "vital spark" did before biologists nailed down the nature of living things. Materialism certainly has one thing going for it. By denying the existence of consciousness and other mental phenomena, it neatly makes the mind-matter problem disappear." —*Jeffrey Schwartz, research neuropsychiatrist, and Sharon Begley, science writer/ editor (pp. 38-39)* [6]

Stage analytic models. A basic observation in cognitive research was that the human information processing system is limited in the amount of information it can process at any given time. Initial models attempted to identify stages through which multiple inputs were narrowed down into the selection of a single discrete behavioral output. These 'bottleneck' models were based, for example, on studies of sensory channels of attention. This research is exemplified by dual task and dichotic listening experiments such as the 'cocktail party effect.' Processes that are involved in different tasks and that could be done simultaneously were generally theorized to imply separate functions or stages of processing. Here is a simple depiction of one prominent capacity-limited model, psychologist Donald Broadbent's 'filter theory' of attention: [7] [8]

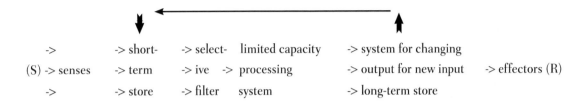

->	-> short-	-> select-	limited capacity	-> system for changing	
(S) -> senses	-> term	-> ive ->	processing	-> output for new input	-> effectors (R)
->	-> store	-> filter	system	-> long-term store	

Analogous to computers with parallel inputs, some of the stages were thought to involve relatively unlimited parallel processing (stages indicated by multiple arrows in the above model). But the stages associated with conscious attention were thought to be limited to serial processing, like the serial central processor of a computer (in the above model the stages with single arrows, the limited capacity perceptual system and the effectors). Information comes into the system in parallel, and goes through selective filtering that narrows processing down to a specific response.

Consciousness was hypothesized to be associated with serial central processing stages, involved with the formation of semantic associations in active short-term memory (limited capacity processing system) and selection of output requiring attentional resources (effectors). This reflects the general association of consciousness in the cognitive models with attentional *awareness of* objects of experience, which is characteristic of the ordinary waking state. There was virtually no distinction in these initial cognitive models between attention and consciousness.

Extensive cognitive research attempted to establish the sequence, functions, and limitations of processing stages inside the black box from input to output, stimulus to response. Generally the stages included: sensory input channels, sensory information memory (very brief, a couple hundred milliseconds or less), short-term memory (up to about 30 seconds) and a highly limited central processing function associated with conscious attention, various types of long-term information stores such as episodic and semantic memory (potentially permanent and unlimited), response selection, and response output channels.

The filter theory depicted above implied that conscious attention is associated with the capacity-limited stages of perceptual input processing and response output selection. However, research on activation of long-term memory demonstrated that perceptual analysis even to the level of semantic meaning is not as capacity-limited as was first thought, doesn't necessarily involve attentional resources, and takes place outside of conscious experience. Incorporating these findings, conscious attention was then placed later in the sequence of stages, theorized to be associated more with just the late stage of deciding upon and selecting response output than with earlier stages of perceptual input. Importantly cognitive models then placed access to long-term memory as an earlier peripheral stage of processing, *prior* to the central processing stage associated with short-term memory and conscious experience. In other words the research suggested that access to long-term memory did not involve conscious attention and experience.

As an example of supportive research findings, in one study city names were paired with mild electric shock to condition a galvanic skin response (GSR) to the names alone. In a subsequent dichotic listening task subjects showed GSR to city names even though they were unable to report hearing the city names at all. [9] These and similar findings led to models in which activation of meaning in semantic memory was placed earlier in the sequence and not conscious. Consciousness was associated only with later stages in the sequence. [10] [11]

Consistent with applying the ordinary discriminative intellect as a primary tool for knowledge acquisition, the growing body of cognitive research was interpreted as indicating a fundamental discrimination of modes of processing into a *two-process* theory. The first mode of processing was described as automatized, capacity-unlimited, non-attentional, and unconscious—called *automatic* or preattentive processing. This mode of processing was said to develop with stimulus consistency and extensive practice. The other mode was described as effortful, highly capacity-limited, controlled by focal attention, and conscious—called *controlled* or effortful processing. This mode was said to appear in variable stimulus conditions or in learning a new task. [12] [13] [14] [15]

The automatic mode was associated with unconscious or non-conscious processing, and the controlled mode was associated with conscious attention. Conscious attention was related specifically to a limited-capacity central processor, and use of this central processor was a core identifying feature of a conscious experience. Consciousness was conceptualized to be in a specific discrete functional stage of processing. The view of some theorists, exemplified in the following quote from Posner and Warren, was that a common difficulty in psychological literature was the failure to understand that, "...conscious awareness is a discrete event that plays a specific role within the stream of information processing." (p. 180) [16] This is an example of the common reductive mental set that associated consciousness with a specific stage of functioning—presumably also associated with specific areas of neuroanatomy. It has been a misleading feature of many cognitive and neurocognitive models of mind.

An important implication of the automatic mode of processing in two-process theories was that the entire sequence of stages from stimulus to response could become automatized. This means that conscious attention was not necessarily involved in *any* of the stages between input and output in these linear models of automatic processing. Models of information processing that included consciousness were no longer viewed as simple linear sequences. The horizontal conception of early and late stages in linear stage models was combined with what can be depicted as a vertical notion of shallow and deep processing related to the degree of conscious attentional resources applied to a task—called *depth of processing*. Consciousness was associated with the deep stages of attentional processing in the central processor, directly related to conscious decision making. Unconscious processing was associated with shallower automatic processing, including the stages of semantic memory access and some types of learned action sequences and reflexive action conducted even without any deeper conscious processing.

In addition to instinctual and reflexive fixed action sequences not involving conscious attention, cognitive tasks initially requiring conscious attention could become automatized through learning and conditioning. These processes then no longer require conscious attention and intentional control. Automatic processing accounted for a considerable portion of information processing inside the black box, but *outside* of conscious experience. In other words the inside of the black box was partitioned into shallower non-conscious levels and deeper conscious levels; and under certain conditions of learning, the processing mode shifted from conscious effortful to automatic unconscious processing. Thus in addition to a linear, 'horizontal' model of stages from stimulus to response, there was theorized to be a 'vertical' mode' of depth of processing associated with conscious attention. Consciousness was placed in the deeper controlled processing mode.

A similar concept of depth or *levels of processing* developed in studies of perceptual-cognitive analyses of the activation of long-term memory storage. [17] Stimulus inputs that received conscious attentional processing in so-called primary, active, working, or short-term memory after automatic activation resulted in stronger and typically longer lasting memory traces or representations coded in secondary, long-term memory. In these theoretical developments can be seen inklings of the *inner dimension* in models of the mind, reflected in adding a vertical dimension of levels or depth of processing to the linear model of sequential processing stages.

Operational criteria for the two modes of automatic and controlled processing involved capacity or resource limitations, intentionality effects such as whether a subject could avoid or improve upon performance of a specific task, and reportability or availability of the contents of processing for verbal report. With respect to capacity effects, both task parameters and subject parameters were recognized to be significant factors. Task parameters included such things as sensory modality, number of stimulus items, duration of presentations, signal-to-noise ratios of stimuli and background, and consistency in the roles of stimuli across practice trials. Subject parameters such as motivation and previous experience also impacted processing capacity. The complex concept of intentionality associated with conscious control was difficult to operationalize experimentally. Likewise attempts to define availability for verbal report were somewhat difficult to operationalize, due to questions of the accuracy of the reports influenced by motivational, introspective, and other factors. As with the capacity criterion, research using the criteria of intentionality and reportability were influenced by many task and subject parameters. For the most part the research applying these criteria appeared to fit more a gradation or continuum model—or resource pool model—than a model of discrete modes. It suggested that human information processing doesn't fit models conceptualized in terms of such rigid dichotomies. [18] [19] [20] A prominent resource-allocation model by psychologist Daniel Kahnemann reflected this point. [12] [21]

Levels of depth of processing. Nonetheless two-process theories of consciousness have a long history in psychological theories of mind. One example is from the 'founder of psychology,' Wilhem Wundt. [22] His theory delineated an *apperceptive* focal attentive process in conscious awareness and an *apprehensive*, inattentive, vague periphery. In this model, development of an automatized mode of processing involved 'retrogradation' in which an initial conscious task faded out with practice until the act became habitual and performed as a physiological reflex not involving consciousness. [23]

In the model of the 'law of two levels' proposed by Wundt's student, E. B. Titchener, [24] conscious tasks were described similarly as 'degenerating' with habitual practice into an acquired automatic reflex with no affective components such as feelings of effort. In this theory the conscious sensation associated with the task was thought to continue in the *margin* of conscious experience—no longer in the center of conscious attention but not completely outside of conscious attention. The learned habit did not become unconscious, but rather was relegated to the *periphery* of awareness, no longer in focal awareness and no longer associated with a sense of effort or control. A similar model of *conscious fringes* or a region of *conscious inattention* also was put forward by the 'founder of American psychology' William James. [25]

Generally these early psychological models did not make a strong distinction between the two modes of processing, but rather viewed them in terms of a graded region of conscious attention with a focal center and a peripheral fringe. In the initial cognitive models that came later, however, the peripheral fringe was associated only with more discrete shallow levels of processing and the focal center was associated only with deeper levels of conscious attention. But as noted above, subsequent cognitive research also supported more of a gradation or continuum model.

Of particular importance in this discussion of levels and depth of processing is that in the earlier psychological models the fringe region had connotations in addition to the shallow levels associated with automatic processing. In these earlier models the notion of a conscious fringe region carried with it a fringe associated not only with shallow preattentive input processing but also with mental processes *even deeper* than the focal conscious region—that is, not only an outer fringe but also an inner fringe. They contained an extended notion of depth of processing in the sense of an *inner* experiential fringe even deeper than the focal central region of conscious attention. Thus in these early models there was a shallow outer margin or fringe to conscious attention and a deep inner margin or fringe, with conscious focal attention in the center. Although also hinted at in some of the more recent cognitive theories, the deeper inner fringe region was less explicit in these later mechanistic computer-like models. This is an example of increasing objectification from models of mind to models of discrete information processing stages in cognitive theories.

The deeper inner fringe region in the earlier models can be interpreted as the extended context of experience, inclusive of subtler affective feelings underlying focal conscious attention. In this view the graded field of attentional experience extends beyond the few clearest and most prominent features in the focus of attention into subtler undertones of feeling that contribute to personal judgments, attributions, and intentions. This is in the opposite direction of depth than processes that are retrograded, degenerated, or automatized into the shallow outer conscious fringe on the stimulus input side of the conscious central processing function. Thus in the early models there was automatic, non-conscious processing, then a deeper focal conscious region, and then an even deeper inner fringe not in focal attention but still conscious.'

As an example of the deeper inner fringe region underlying focal conscious attention, when sitting in the library there is a tacit monitoring of the noise level of the room, the temperature and lighting, one's general bodily state and posture, and many other factors in the external as well as internal environment, including feelings of security and degrees of inner comfort. This provides a baseline for recognizing changes in the ambient context. These inner feelings, for example, involve a tacit sense that the room is stable and not shaking from an earthquake, which contributes to an inner sense of security. This inner sense of security allows cognitive resources to be applied more fully to focal mental tasks, such as reading and studying. As another example, consider a child playing in the backyard when the mother is home versus when the mother is away. The feeling of safety and security in the knowledge that mother is at home supports attention to more creative play. When the mother is away, it is likely that the child would have more attentional resources applied to heightened environmental vigilance, and thus less mental resources available for creative play.

In addition the conscious inner fringe involves even deeper patterns of feeling and self-attribution. These deeper inner fringe experiences relate to concepts such as inner locus of control, self-esteem, and self-concept. They provide an undertone and general context that influence focal attentive decision making and resulting behaviors. The deeper inner fringe of feelings can be associated with self-worth influenced, for example, by patterns of parental care.

From this perspective human information processing is influenced by numerous deeper interior as well as shallower exterior psychological factors that make up the entire conscious experience, whether in focal or peripheral attention. A more integrated, interactive information processing system is suggested that cannot easily be modeled in terms of discrete modes of processing with well-defined capacity limitations and functional roles—even if there is some degree of specialization and specificity at various stages. Both a linear, horizontal dimension of early and late stages, and a vertical inner dimension of shallow and deep processing, are involved in a richer, more integrated model of mental functioning inside the black box. This is somewhat reflected in contemporary '*theatre*' models such as the *global workspace* theory of consciousness by psycholinguist Bernard Baars.[26]

In Baar's model, conscious experience is described as a global workspace for information processing—analogous to the center stage of a theatre. This model characterizes conscious central processing as the center stage, with focal and fringe regions. In this particular model there is a distinction between attention and conscious experience. Unconscious components of attention off-stage and outside of conscious experience, called context operators, direct competing information from sensory processes, thinking, affective feelings, and intuitions onto or off of center stage. Compared to stage analytic models, this model incorporates a broad range of cognitive and affective processes that contribute to a more functionally interactive processing system, with less linearity of processing stages.

In Baar's model, consciousness is thought to play a generalized role as a global workspace that facilitates information processing and decision making. It is said to support the dissemination of integrated consciously experienced information to a wide array of unconscious memory, motivation, decision making, and action systems. Exteroceptive and interoceptive sensory processes, ideas, interpretations of conscious contents, and motivational and other functions are all described as *outside* of conscious experience. This aspect of the model is consistent with one of the basic points made in Chapters 11 and 13 that much psychological processing is carried out unconsciously, and that only the results of the processing are presented to conscious experience—that is, to center stage. Also the categories of processing in this model are less discrete than in stage analytic models. The model does not clearly depict either levels of depth of processing or stages that reflect the flow of information through the system. While retaining some sense of depth and some sense of stages of processing, the processing is not organized into sequential stages or levels. The focus of the model is on consciousness as center stage. It reflects somewhat more than stage analytic models the phenomenal experience of focal conscious attention.

However, in the context of the overall approach to consciousness and mind being presented in this book, this global workspace theory might be viewed as a model of mind rather than of consciousness. The concept of center stage would be associated with the screen of the mind rather than with consciousness (more related to the level of *manas* in Sankhya). The closest analogue to consciousness itself—the knower—in the theatre analogy would be not center stage but rather the *audience*—the silent uninvolved witness of the stage (Purusha). In this particular application of the theatre analogy, however, the audience is identified as unconscious memory, reflex action codes, and other aspects of information processing. There seems to be no direct analogue at all for consciousness itself in the model, but rather only of the focal and fringe region of attentional experience—which is what the mind presents to consciousness.

Objective emotions and subjective feelings. The concept of a deeper *inner conscious fringe* can be related to an interesting debate in the psychological literature about the primacy of either *cognition* or *affect*. Affect generally includes both emotions and feelings. The debate can be described in terms of whether cognitions structure affect or vice versa, and which one is the deeper level. On one side of the debate were theorists who thought that emotions depend on how we interpret a situation, the meanings attributed to it—usually cognitive psychologists. Emotions depend on the cognitive appraisal or evaluation of an event, based on past learning and life experience.[27] On the other side of the debate were theorists who felt that emotions, feelings, or affect are more fundamental than cognitions—usually humanistic and social psychologists.[28] They pointed out that cognitive evaluations significantly depend on deeper underlying feelings. They suggested that feelings are more fundamental and more powerful contributors to our sense of who we are as human beings than cognitive evaluations. Situations are interpreted cognitively in a manner to support or explain the underlying feelings. In this view, it is the underlying feelings that shape the cognitions. Cognitive psychologists were criticized for largely overlooking these deep inner feelings as key contributors to meaning and significance.

Both sides of the debate made strong points—and of course there were others who discounted the debate altogether, viewing it as an artificial or over-simplistic contrast. The model of levels of subjectivity, and the notion of a deeper inner conscious fringe, can help reconcile the two sides of

> "We sometimes delude ourselves that we proceed in a rational manner and weight all of the pros and cons of various alternatives. But this is seldom the actual case. Quite often "I decide in favor of X" is no more than "I liked X"... We buy the cars we "like," choose the jobs and houses we find "attractive," and then justify these choices by various reasons."—*Robert B. Zajonc, cognitive social psychologist* [28]

the debate and clarify the various interior levels of the mind. In order to do this it is instructive to make a distinction between *emotions* and *feelings*. To emote means to show, display, or express feelings. *Emotions* can refer to the observable behavioral expression in the body of unobservable inner feelings and thoughts in the mind and subjective experience. Theories of emotion typically relate emotions to hormonal and other physiological functions in the body, expressed in patterns of body language, gestures, postures, facial expressions, and verbal utterances such as crying, sighing, laughing, and so on (such as the James-Lange, Cannon-Bard, Schacter-Singer, Soloman Opponent Process, Plutchnik's Emotions Wheel theories). [29] These patterns of emotional behavior are highly consistent across most cultures.

Feelings can refer to subtler, deeper, inner processes that cannot be observed directly by others from the outside. In inner subjective experience, deep inner feelings can penetrate to the foundations of our individuality. Sometimes they may be strongly expressed in emotional reactions in the body observable by others, and at other times they may be only minimally expressed. But in terms of levels of depth of our interior life, they are frequently experienced as deeper and more powerful in shaping behavior than thoughts, opinions, evaluations, and cognitive interpretations. Both the inner level of thinking and the more expressed level of emotional behavior seem too shallow to capture the richness of the deeper inner fringe level of feelings.

In this understanding of affect, cognitive processes can be said to be deeper and more primary when the debate is between *cognition* and *emotion*—that is, emotion in the sense of *emotional behavior*. Emotions are physiological and behavioral reactions of the body that are significantly determined by the cognitive lens through which we interpret life events. Emotional behavior is significantly a product, expression, or manifestation of the inner mental interpretation or cognitive appraisal we give situations in our lives. On the other hand feelings can be said to be deeper and more primary when the debate is between *cognition* and *feeling*. Feelings relate to the deepest levels of subjectivity; the cognitive level of mind and thinking is a more expressed level of subjectivity. To be discussed shortly, some recent models attempting to account for the entire individual self—both physiologically and psychologically—make this distinction between emotions and feelings. For example, With respect to the model described in a previous chapter, Damasio states:

> "It is through feelings, which are inwardly directed and private, that emotions, which are outwardly directed and public, begin their impact on the mind." (p. 36) [30]

Ordinary logic and reason on the cognitive thinking level of the mind are frequently subject to the influence of underlying feelings. As referred in the Prologue, we use our cognitive abilities to interpret, argue for, defend, and sometimes rationalize what we feel about something. For example, on occasion we may use our intellectual thinking and logic in a biased manner to support a person we love from getting into trouble, sometimes even ignoring or misinterpreting the facts in service to our underlying feelings. This is relevant to the modern scientific view of the unreliability of subjectivity and the need to be rigorously objective. To be discussed in a later chapter, for the most part these inconsistent patterns of thinking and feeling emerge due to the overshadowing influence of stress and tension that distort these levels and interfere with their optimal, most coherent, and consistent functioning.

The identification of feelings as underlying thoughts adds another plank to the concept of levels and depth of processing. It further emphasizes subjectively experienced levels of depth in addition to the meaning of depth associated with the amount or capacity of attentional resources. The concept of an experiential fringe in some sense *interior* to and deeper than focal attention can be associated with these inner levels of feeling. The inner feelings may be only on the vague interior fringes of experience, in comparison to clear thoughts in the focus of attention—on center stage. However, generally they are not experienced as shallower levels but rather deeper levels that significantly influence the tone and context of focal attention. Though they can be difficult to describe or articulate clearly by verbal report—feeling more like flowing waves than discrete well-defined thoughts—they are frequently experienced as core aspects of our inner subjective experience.

> "In most states of consciousness, there is an awareness of being situated or located in time and space and an awareness of our bodies, types of awareness that are clearly based on many different sources of information. There is also a conscious fringe, which has to do with feeling of familiarity, of being right or wrong, of being satisfied or not. There can be, as well, all those refined discriminations that are the essence of culture and art."—*Gerald M. Edelman, neurobiologist, and Giulio Tononi, neuroscientist (p. 22)* [31]

At this stage of the discussion we can summarize generally the developing functional model of mind in terms of several broad stages and levels of depth. The earlier, shallower, peripheral levels extend from sensory input and perceptual analyses even through activation of semantic meaning in long-term memory. Added to these processing stages are deeper, more central processing functions related to active short-term memory and conscious attentional processes associated with decision making and verbal report. Central conscious processing has a conscious fringe of peripheral awareness on the input side, associated with more automatized processes including well-learned associations between sensory input and behavioral output. This is the more shallow or outer level of subjectivity, more closely aligned with the senses. However, central conscious processing has an even deeper, inner conscious fringe associated with deeper, subtler feelings—including the sense of self—which underlie and influence focal conscious attention as well as behavioral output. These deep feelings shape and reflect longer-term personality traits.

The human information processing system is highly interactive, incorporating both sequential bottom-up sensory input processing stages and also deeper top-down stages that influence the input stages. The overall system is highly reactive to environmental input; but more fundamentally, it is proactive—controlled in a top-down manner from deeper thoughts and feelings—in implementing external behaviors to achieve its own internal, self-serving goals.

In horizontal linear stage analytic depictions where early stages associated with stimulus input are at one end and behavioral output is at the other end (such as the filter theory illustrated earlier), it is somewhat awkward to capture levels of *depth*. Because stimulus input and behavioral output both refer to interaction with the outer environment—outside of the black box—a model that recognizes this can be used which makes the depiction of levels of depth inside the black box clearer. These notions of a sequence of levels and depth of processing are incorporated into the following example which depicts the levels of depth of processing from a functional cognitive perspective. This depiction summarizes results of relevant psychological research over the past 150 years toward a model of mind, briefly overviewed in this chapter. It shows that psychological theories of mind are beginning to approximate the sequence of the inner dimension of levels of subjectivity or psychoarchitecture outlined in the Sankhya system in Chapter 11. The sequence from outermost to innermost in that system included: the organs of sense and action (indriyas), thinking mind (manas), intellect and deeper feelings (ahamkara), and sense of self or ego (mahat). Whereas in the Vedic system consciousness is the deepest level, in the above depiction consciousness is not yet disembedded from deeper levels and thus is not recognized as underlying them. This is typical of experience in the ordinary waking state of consciousness.

Sensory World

Sensory Input (S) —   — Behavioral Output (R)

Sensory input and storage processes

Long-term memory

Peripheral conscious attentional fringe
Short-term Memory (activated long-term memory)

Focal conscious attention

Conscious fringe of deep inner feelings

Conscious sense of self

Relationship of the functional model to a structural model. The above depiction primarily draws upon functional and to some degree phenomenological evidence of depth of processing, where consciousness is associated with the deep central level. However, the functional model of levels of depth takes on a different meaning when placed in the structure of the brain. In such biologically-based structural models of mental processing, structural and functional notions of depth become mixed together. This situation can be seen in models from so-called *depth psychology*, associated with the general conception of mind reflected in the work of neurologist and psychoanalyst Sigmund Freud.[32] This different concept of depth, with a stronger emphasis on structure, is related to the Freudian model of mind as a product of the body. It reflects more strongly the influence of Darwinian evolutionary theory.

The Freudian model of depth. The approach reflected in the Freudian model did not emphasize functional information processing stages that result from stimulus input, as do the cognitive models. Rather it emphasized analysis of the inner psychological dynamics and how they relate to bodily functions and needs. It was motivated by concern for understanding mental dysfunction, neurotic behavior, emotional problems of everyday life, and especially inner mental conflicts leading to behaviors that individuals weren't generally able to explain or understand about themselves. These mental processes were thought to be for the most part outside of conscious experience.

Emphasizing levels of *unconscious* psychological processes, the Freudian model added the preconscious and the unconscious as levels underlying conscious experience. Deeper levels were associated with inner preconscious processes that can become conscious, as well as even deeper unconscious psychological processes that influence conscious perception and behavior. Shallower, more surface levels of processing were associated with conscious processing related to interaction with the outer environment through sensory perception. In Freud's theoretical scheme these three levels—conscious, preconscious, and unconscious—had a correspondence to the concepts of ego, superego, and id, but not a precise match. For example, superego functions were hypothesized to span all three levels. [32]

In contrast to historical, behavioral, and later cognitive models, the Freudian model was concerned with deeper interior subjectivity and much less with shallow peripheral levels involved in perception of the environment. In the Freudian model, the *shallower* level was the conscious level, the next deeper level was the preconscious—unconscious but accessible to consciousness—and the *deepest* level was the unconscious. This reflects a different functional meaning of depth. In both the historical models and the cognitive models, functional depth was associated for the most part with *conscious* processes. But in the Freudian model, greater depth was associated with *unconscious* processes. Here is a simplified depiction of these models in terms of the location of consciousness:

Historical two-process models	Cognitive models	Freudian model
Senses	Senses	Senses
Unconscious preattentive/automatic processes		
Conscious peripheral fringe	Conscious fringe or region of inattention	
Focal conscious attention (central processor)	Focal conscious attention	Conscious
Deeper conscious fringe	Deeper conscious fringe	

In Freudian theory, levels underlying the conscious level include the preconscious, the unconscious, as well as innate psychobiological drives.

The Preconscious
(unconscious but available to consciousness)

The Unconscious
(psychological)

Biological instincts and drives

Still another deeper level of the unconscious could be added to the Freudian model. Freud's later theories incorporated a contribution of species-wide cultural experience, in the direction of Jung's archetypal theory of the *collective unconscious*. [33] It contained a genetically inherited psychological unconscious level that reflected the heritage of human culture—sort of a genetic psychological memory of the species—called "primaeval phylogenetic experience." (p. 64) [32] This suggested that cultural acquisition of superego experience becomes genetically inherited in the unconscious, which could be interpreted as reflecting an element of LaMarkian evolutionary theory of acquired characteristics. The concept of collective memory will be referred to again in a later chapter.

The concept of a deeper inner fringe relates to feelings that are the undertones of ordinary focal attention, with many degrees of sometimes vague subjective experiences. Their clarity and subtlety relate to development within ordinary states of consciousness, and degrees to which these underlying feelings are foggy due to overshadowing stress and incoherent functioning.

However, all the models above generally reflect the ordinary waking state of consciousness, in which consciousness is not disembedded from the levels of mental activity and is not experienced as underlying them. In this state, consciousness is experienced as mixed with thoughts and feelings. It is through settling down the mind and transcending thoughts and feelings that consciousness itself can be isolated from them. Clarity of the underlying level of consciousness itself grows with progress toward higher states of consciousness. In the Freudian model—as well as the other models—there is no recognition of an extended range of development to higher states of consciousness. The location of consciousness in all of these psychological models reflects the range of perspectives and experiences typical of the waking state of consciousness—for the most part ordinary, but sometimes abnormal and even dysfunctional.

All three types of models described above are fundamentally embedded in a physicalist view in which the brain and body are theorized to underlie psychological conscious and unconscious levels. The earlier two-process models and the more recent cognitive models emphasized a more abstract disembodied conception of function. The biological embodiment of the psychological mind was emphasized in Freudian theory, and is similarly emphasized in contemporary neurocognitive models to be discussed next. The neurocognitive models further emphasize a functional role of consciousness related to a Darwinian evolutionary perspective of biological adaptation. These attempts to place mind and consciousness into the brain and body try to combine the *functional* notion of depth into a *structural* notion of depth, all within a physicalist worldview—largely based on experience in the ordinary waking state of consciousness only.

It might be pointed out that the models discussed in this and following chapters, whether focusing primarily on function or attempting to combine function and structure, are basically attempts to account for mind—they are not primarily models about consciousness. They reflect important attempts to understand human psychoarchitecture, which essentially can be understood to be the study of the nature of individual attention, how consciousness is individualized into an individual sentient entity that is in a *state of consciousness*—to be discussed in Chapter 18.

To summarize the discussion in this chapter, functional and structural models reflect different meanings or dimensions of depth. The general notion of functional depth in cognitive models has to do with degree of conscious attention and conscious experience at various levels or stages of processing. Deeper processing refers to more extensive involvement of focal attention and consciousness. This notion of depth corresponds to the notion of peripheral versus central processing, with deeper referring to more central or focal conscious attention. It also can be associated with a phenomenological or experiential meaning of depth—with peripheral input processing associated with shallow processing and conscious attention with deeper processing. In addition it relates to the functional meaning of higher-order and lower-order levels of processing. Higher-order levels refer to higher levels of cognitive functioning as well as more complex global neural patterns associated with consciousness. Shallow levels refer to lower-order, automatic, instinctual, and reflex-like processes. In this model the deeper higher-order levels bring coherent integrated functioning to the shallower levels in top-down super-ordinate control by a conscious self.

However, in this *functional* meaning or dimension of depth also can be identified an inner fringe associated with even deeper feelings that are conscious but subtler than focal attention. In the Freudian model this inner fringe would be on the side of the preconscious, underlain by the even deeper unconscious. Further, in the Freudian model even deeper than the unconscious are somatic processes of the physical body. This reflects a second meaning of the notion of depth—a *structural* dimension of depth, in which psychological functions are underlain by deeper neurobiological structures in the brain. This structural dimension of depth, also associated with biologically-based neurocognitive models, has to do with higher-order and lower-order levels of physical structure. In this meaning, deeper levels refer to underlying *lower-order* biological and physical levels. Deeper means more fundamental physical levels—from the organismic level to the neural, cellular and molecular levels of the body; and eventually elemental, atomic, subatomic, and even quantum levels.

This can be extended deeper to the qubit or quantum information level, whereupon we encounter the curious situation that this is where information space—and quantum or nonlocal mind—is theorized to exist in cutting edge theories of the foundations of *reality*. Here again we encounter vividly the fundamental mind-body problem, and how contemporary developments in quantum physics are beginning to glimpse its resolution. This progress is based on a more expanded understanding beyond the gross objective levels to subtle subjective levels of nature and nonlocal mind deeper than neural functions in the brain and body (refer to Part I of this book). In Chapter 17 we will refer back to this issue and reconcile the functional and structural notions of depth of processing in the model of the inner dimension in Maharishi Vedic Science and Technology. However, first we will discuss prominent examples of neurocognitive models that attempt to embed functions of mind and consciousness into the structure of the physical brain.

When functional conceptions about the mind are placed into brain structure, the crucial issue emerges of how neurobiological processes believed to supervene in the sense of depend on lower-order physical processes cohere into a functional unit in the performance of unitary behaviors. What is it that draws together fundamentally random physical processes—which don't make decisions and take actions on their own, just passively reacting to change according to the laws of nature without any free will or purposive intention—into a coherent purposeful, self-motivated, intelligent, negentropic behavioral organism? How does top-down functional control of unitary experience and behavior develop in a system that is fundamentally composed of non-integrated lower-order, random processes? To paraphrase, where is the controlling authority?

This crucial issue underlies the question of how to reconcile functional and structural perspectives in a general model of mind. Cognitive and neurocognitive models attempt to account for this by some form of *emergence theory*. This theory proposes that the conscious mind and self as a super-ordinate controlling causal agent *somehow* emerges in biological evolution with sufficient neural complexity in the living brain. How the sense of an integrated self emerges from neural processes as a coherent binding principle for purposeful adaptive behavior is a core concern of the contemporary neurocognitive models discussed in the next chapter—Chapter 16.

Chapter 15 Notes

[1] Kling, J. W. & Riggs, L. A. (1971). *Experimental psychology, Vol. II: Learning, motivation, and memory.* New York: Holt, Rinehart and Winston.

[2] Miller, G. A., Gallanter, E., & Pribram, K. H. (1960). *Plans and the structure of behavior.* New York: Holt, Rinehart & Winston, Inc.

[3] Powers, W. T. (1978). Quantitative Analysis of Purposive Systems: Some Spadework at the Foundation of Scientific Psychology. *Psychological Review*, Vol. 85, No. 5, 417-435.

[4] Boyer, R. W. (1981) *Goal-Directed Behavior*. Unpublished manuscript, University of Oklahoma.

[5] Skinner, B. F. (1948). *Walden Two*. Englewood Cliffs, N.J.: Prentice Hall.

[6] Schwartz, J. M. & Begley, S. (2002). *The mind & the brain: Neuroplasticity and the power of mental force*. New York: ReganBooks.

[7] Broadbent, D. E. (1958). *Perception and communication*. London: Pergamon Press.

[8] Lachman, R., Lachman, J. L., & Butterfield, E. C (1979). *Cognitive psychology and information processing: An introduction*. Hillsdale, N. J.: Lawrence Erlbaum Associates.

[9] Corteen, R. S. & Wood, B. (1972). Autonomic Responses to Shock-Associated Words in an Unattended Channel. *Journal of Experimental Psychology*, 94, 308-313.

[10] Leahey, T. H. (1979). Something Old, Something New: Attention in Wundt and Modern Cognitive Psychology. *Journal of the History of the Behavioral Sciences*, 15, 242-252.

[11] Deutsch, J. A. & Deutsch, D. (1963). Attention: Some Theoretical Considerations. *Psychological Review*, Vol. 70, No. 1, 80-90.

[12] Kahnemann, D. (1973). *Attention and effort*. Englewood Cliffs, N.J. : Prentice-Hall.

[13] Neisser, U. (1976). *Cognition and reality*. San Francisco: Freeman and Co.

[14] Posner, M. I. (1978). *Chronometric explorations of mind*. Hillsdale, N. J.: Erlbaum.

[15] Shiffrin, R. M. & Schneider, W. (1977). Controlled and Automatic Human Information Processing: II. Perceptual Learning, Automatic Attending, and a General Theory. *Psychological Review*, Vol. 84, No. 2, 127-189.

[16] Posner, M. I. & Warren, R. E. (1978). Traces, Concepts, and Conscious Constructions. In Melton, A. W. & Martin, E. (Eds.). *Coding processes in human memory*. New York: Wiley.

[17] Craik, F. I. M. & Lockhart, R. S. (1972). Levels of Processing: A Framework for Memory Search. *Journal of Verbal Learning and Verbal Behavior, II*, 671-684.

[18] Townsend, J. T. (1971). A Note on the Identifiability of Parallel and Serial Processes. *Perception and Psychophysics, 10*, 161-163.

[19] Boyer, R. W. & Charleston, D. E. (1985). Auditory Memory Search. *Perceptual & Motor Skills,* 60, 927-939.

[20] Boyer, R. W. (1984). Conscious Processes in Automatic Visual and Auditory Search. *Dissertation Abstracts International,* Vol. 45, 5B, A AD84-18575, 1605.

[21] Lachman, R., Lachman, J. L., & Butterfield, E. C (1979). *Cognitive psychology and information processing: An introduction*. Hillsdale, N. J.: Lawrence Erlbaum Associates.

[22] Wundt, W. (1912). *Introduction to psychology*. London: Allen.

[23] Wundt, W. (1907). *Outlines of psychology*. Leipzig: Wilhelm Engelman.

[24] Titchener, E. B. (1908). *Lectures on the elementary psychology of feeling and attention*. New York: The MacMillan Company.

[25] James, W. (1890). *The principles of psychology*. New York: Holt.

[26] Baars, B. J. (1997). *In the theatre of consciousness*. New York: Oxford University Press.

[27] Lazarus, R. S. (1984). On the Primacy of Cognition. *American Psychologist, 39*, 124-129.

[28] Zajonc, R. B. (1980). Feeling and Thinking: Preferences Need No Inferences. *American Psychologist, 35*, 151-175.

[29] Bear, M. F., Connors, B. W., & Paradiso, M. A. (2001). *Neuroscience: Exploring the Brain,* 2nd Edition. Baltimore, MD: Lippincott Williams and Wilkins

[30] Damasio, A. (1999). *The feeling of what happens: Body and emotion in the making of consciousness*. New York: Harcourt, Inc.

[31] Edelman, G. M. & Tononi, G. (2000). *A universe of consciousness: How matter becomes imagination*. New York: Basic Books.

[32] Freud, S. (1949). *An outline of psychoanalysis*. New York: W. W. Norton.

[33] Jung, C. G. (1953-1979). *The collected works* (Bollingen Series XX). 20 Vols. Read, H., Fordham, M. Adler, G., & McGuire, W. (Eds.). (R. F. C. Hull, Trans.). Princeton: Princeton University Press.

Chapter 16

Neurocognitive Models of Mind and Consciousness

In this chapter, examples of prominent biologically-based neurocognitive models attempting to identify structures in the brain that carry out the functions of mind and consciousness will be considered. The models include consciousness as a feeling of knowing from Damasio,[1] consciousness as reentrant neural connections from Edelman and Tononi,[2] and the 'I' as a command and control function from Llinas.[3] Major issues in this chapter include functional and structural levels of depth, the theory of emergence of mind and consciousness from neural activity, and Darwinian principles of natural selection. Also the binding problem of how biophysical and more fundamental physical processes cohere into a unitary individual self that expresses intelligent goal-directed behavior—a key issue in contemporary neuroscience—is a focus of this chapter. The overall main point of the chapter is that physicalist models cannot account for top-down causal control of behavior in a goal-directed biological organism, and also cannot account for mind and consciousness.

Neuroscience research focuses on correlating psychological functions with brain structure. Similar to *indirect* observations in particle physics from which inferences are made about underlying unobservable processes, this research on unobservable mental processes is aided by empirical observation of the physical brain using technologies such as electroencephalographic recording (EEG), functional magnetic resonance imaging (*f*MRI), and positron emission scanning (PET). These brain-probing tools provide images of neural activity that indirectly may reflect subjective, psychological, mental processes. The images also are correlated with verbal reports of phenomenal experiences as well as other objective behavioral measures, used to make inferences about functions inside the black box of the mind.

As examples of this type of research, brain-imaging techniques have been used to examine neural correlates of the development from effortful controlled processing to automatic processing—discussed in the prior chapter. EEG findings have shown that initial desynchronization to a new task in widespread regions of the brain narrows down with practice to specific sensory and motor regions of the cortex as automatized performance develops. Cerebral glucose metabolic rate as measured by PET scans similarly show significant decreases accompanying improved performance with extensive practice of video games. Recent *f*MRI studies also show that the breadth of motor cortex involved in a task requiring rapid finger movement decreases with extensive practice. These types of research provide evidence of the correspondence between conscious experience of controlled and automatic processing and neural structures in the brain associated with gross behavioral performance. [2] Neurobiologist Gerald M. Edelman and neuroscientist Giulio Tononi state:

> "These neurophysiological data...provide a first indication that the cortical activity in the thalamocortical system, which is associated with conscious performance, is more widespread and distributed than brain activity associated with automatic performance." (p. 61) [2]

Neurocognitive models reflect recognition that current knowledge is not sufficient for precise identification of mental functions, such as conscious experience, in specific brain structures. But it is sufficient for delimiting and shaping important aspects of the functional models, and it demonstrates the considerable correspondence between brain structures and mental functions—the general one-to-one correspondence between body and mind in Maharishi Vedic Science and Technology discussed in Chapters 11, 13, and 14. The neurocognitive approach can tell us much about the neural architecture of the brain and, indirectly, its correspondence with the mind, such as with respect to gross behavioral deficits consequent to brain damage. As a functional analysis, it can also help in developing a model of psychoarchitecture and the levels of mind—a primary theme of this part of the book.

As might be expected, however, these neurocognitive models are firmly invested in the physicalist view of mind and consciousness. They have not yet begun to incorporate the more advanced theories of the underlying information space or quantum *reality* suggestive of quantum mind that goes beyond the physical—as Chalmer's model discussed in Chapter 14, for example, begins to do. Neurocognitive models will not eventually lead to a complete explanation of mind and consciousness in terms of the physical brain. As discussed in previous chapters on the foundations of physical *reality*, the more fundamental reason for the inability to correlate precisely the mental with the physical is that the mind is not just in the brain but rather is a subtler nonlocal level of *reality*. The brain is the mechanism through which the subtle impulses of information in the subtler ontological level of mind are transduced into gross behavior in the body. This expanded perspective of levels of objective and subjective *reality* appears to be quite difficult to obtain given reductive methodologies and belief in a causally closed physical universe. This is evident in the theories to be discussed below, which draw extensively from the effects of brain damage and deterioration to support physicalist assumptions.

> "We suppose that a physical process starts from a visible object, travels to the eye, there changes into another physical process, causes yet another physical process in the optic nerve, and finally produces some effects in the brain, simultaneously with which we see the object from which the process started, the seeing being something "mental,' totally different in character from the physical process, which precede and accompany it. This is so queer that metaphysicians have invented all sorts of theories designed to substitute something less incredible."—*Bertrand Russell, philosopher (p. 1-2)* [2]

> "If...all functions of the brain/mind can be explained by specifying mechanisms operating in accordance with impersonal universal laws, there is no possibility of a functional explanation of consciousness: the development over time of any system will occur in accordance with the impersonal laws (with or without randomness), so that choice between available alternatives is excluded, and with it any efficacious subjectivity."—*David Hodgson, philosopher (p. 127)* [4]

In other words the neurology of mind and consciousness is the story of the mechanics of how mind and consciousness are expressed in the gross material domain and corresponding gross or conventional space-time. This is interesting research, but neither gets to an explanation of mind and consciousness nor to a unified understanding and experience of nature. For that, there is great advantage by combining the objective third-person methodology of modern science with the direct systematic, subjective first-person methodology in ancient Vedic science and Maharishi Vedic Science and Technology.

Consciousness as a Feeling of Knowing

One prominent neurocognitive theory that attempts to provide a model of the *self* as a whole, unitary, functional organism has been proposed by Antonio Damasio. According to Damasio:

> "In the very least...the neurobiology of consciousness faces two problems: the problem of how the movie-in-the-brain is generated, and the problem of how the brain also generates the sense that there is an owner and observer for that movie. The two problems are so intimately related that the latter is nested within the former. In effect, the second problem is that of generating the *appearance* of an owner and observer for the movie *within the movie*; and the physiological mechanisms behind the second problem have an influence on the mechanisms behind the first." (p. 11) [1]

This theory outlines a functional model of depth in which the sense of self is a deeper aspect of experience, nested in a cognitive representational system akin to the central processor, similar to the concept of the 'screen of the mind.' It brings up directly the challenging issue for physicalist models of how higher-order processes believed to emerge from lower-order processes appear to have a causal influence on the lower-order processes—the issue underlying the theory of *emergence* introduced in the past chapter. This issue cannot be addressed adequately within physicalist conceptions, and is ultimately related to the self-referral dynamics of consciousness itself. These points will be unfolded throughout this chapter.

If this particular neurocognitive theory can be said to address this issue at all, it appears to take all the levels of depth of processing in the model to be causally interactive with each other within the range of gross biology. This includes molecular, cellular, and neuroanatomical levels; but it doesn't address lower-order inorganic physical levels on which the biological processes depend. Quite reasonably, just how and where in the causal physical chain mind and consciousness enter the picture remains a great mystery. At some point quantum and unified field theories will need to be incorporated that go beyond electrophysiological, cellular, molecular, elemental, and even quantum levels of brain processes; and when they do, the issue of what drives the physical processes and binds them together to structure intelligent organismic functioning will become even more evident.

As to the place and role of consciousness, in the neurocognitive model by Damasio:

> "...[S]ome aspects of the processes of consciousness can be related to the operation of specific brain regions and systems, thus opening the door to discovering the neural architecture which supports consciousness. The regions and systems in question cluster in a limited set of brain territories and no less so than with functions such as memory or language there will be an anatomy of consciousness." (p. 15) [1]

To be consistent with planks laid down in the book so far, however, it might be helpful to point out again that the anatomical operation of the brain *expresses* mind and consciousness in the gross objective material domain. The neural architecture does not *generate* consciousness itself, and is not an anatomy *of* consciousness. The brain is the substrate of gross behavior, but not the underlying substrate of mind and consciousness. In Maharishi Vedic Science and Technology, the substrate of the brain is the mind, and the substrate of brain and mind is consciousness itself, the unified field.

Functional levels of depth. In terms of levels and depth of processing, Damasio's model places functional stage analytic distinctions along a continuum between:

> "...a *state of emotion*, which can be triggered and executed nonconsciously; and a *state of feeling*, which can be represented nonconsciously; and *a state of feeling made conscious*, i.e., known to the organism having both emotion and feeling."(p. 37) [1]

The model also contains functional levels of a sense of self, which include a proto-self, core self, and autobiographical self. These levels comprise an integrated life regulation system to support survival of the individual organism. This regulation system contains components such as for the regulation of metabolism, reflexive action, motivation associated with pleasure and pain, as well as "high reason." (p. 54) [1] The model considers consciousness to be a late addition in evolutionary development of the human species that gives it an enormous advantage, allowing the organism to be aware of its biological emotions in the form of inner feelings that help guide survival behaviors. The function of consciousness, associated with high reason and the discrimination of good and bad, apparently is in the service of biological survival as the greatest good.

Damasio states that, in this model, the neuroanatomical structure of self is called the unconscious *proto-self*; it is said to be a:

> "...coherent collection of neural patterns which map, moment by moment, the state of the physical structure of the organism in its many dimensions." (p. 154) [1]

The proto-self is implemented through the interaction of several interconnected neural pathways from the brain stem to the cerebral cortex, especially the brain-stem nuclei, the hypothalamus, and the insular cortex (S2 and medial parietal) in the somatosensory cortices. It is a *first-order unconscious* mapping or representation of the internal state of the organism.

In the model, consciousness is generated from a variety of neural structures and levels in an integrated but transient neural pattern that fuses and blends the participating brain areas together into a *second-order conscious* mapping, which then somehow can feed back to have a causal influence on the first-order mapping of the proto-self. The second-order conscious mapping, associated with *core consciousness* and the *core self*, is hypothesized to include the superior colliculi or tectum, cingulate cortex, thalamus, and prefrontal cortices working together. Prefrontal cortices come into play in the formation of higher-order *extended consciousness*, related to the continuity in long-term memory associated with an autobiographical self.[1]

The overall model in terms of levels of life regulation functions of the subjective sense of *self* in a human organism can be represented in a vertical depiction, *functionally* from lower-order peripheral processes at the top to higher-order conscious processes at the bottom:

Sensory register	(Initial or first-order map of object or inducing input)
Proto-self	(Non-conscious, temporary, coherent neural pattern representing the state of the organism, moment by moment, at multiple levels of the brain, that is then modified and enhanced into a second-order neural map by the inducing input that drives higher-level focused attention; involving non-conscious emotions related to basic life regulation)
Core consciousness	(A fraction of a second pulse of consciousness or sense of self-knowing of the here and now)
Core self	(Conscious second-order basic feeling of knowing that occurs whenever an object modifies the proto-self, constructed anew in each pulse of consciousness but undergoing minimal change in a lifetime.)
Autobiographical self	(Activation of autobiographical memory of individual experience of the past and anticipated future, also generating a pulse of core consciousness similar to inducing input to the proto-self)
Working memory	(Global workspace for extended consciousness)

Extended consciousness	(The sense of self across seconds and minutes in working memory based on the personal past that may encompass memories from the entire life of an individual, required for intelligent planning of complex behavior)
High reason	(Conscious high-order intelligence; moral judgment of good and bad)
Conscience	(Developed conscious sense of good and bad, based on survival)

As can be seen, this model is in the general direction of the levels of mind in the Sankhya system in ancient Vedic science and in Maharishi Vedic Science and Technology—senses, mind, intellect, ego or self, and consciousness. But the underlying nature of consciousness is not recognized and the place and role of consciousness is not clearly identified (it is embedded in higher-order processes generally). The deepest level is the active higher-order cognitive level of conscience, not the level of consciousness itself.

The model proposes basically a type of functional workspace role for consciousness. It is a limited role in that consciousness is essentially a momentary feeling of knowing. But it is expanded somewhat in the concept of extended consciousness connected with experience of the autobiographical self. This is similar to the role of a centralized, short-term, working, or active portion of long-term memory in cognitive models discussed in the prior chapter and common to many cognitive models of mind.

Two other key perspectives in Damasio's neurocognitive model are noteworthy in the current discussion. The first is the hypothesis that, "...consciousness and emotion are *not* separable." (p. 16)[1] This is said to be based on typical findings that emotions are impaired when consciousness is impaired. These reliable correlational findings support the principle of structural coherence described by Chalmers[5] (refer to Chapter 14) related to the one-to-one correspondence of mind and body described by Maharishi[6] (refer to Chapter 13). But they are not an adequate basis for Chalmer's[5] principle of organizational invariance or the more stringent functional identity hypothesis—albeit quite compelling within the physicalist view. Contrary to the assertion quoted above, Maharishi Vedic Science and Technology provides systematic empirical means to isolate and disembed consciousness itself from all mental processes, including emotion and feeling.

Another key perspective in this neurocognitive model about the functional role of consciousness is that it is intimately associated with biological survival. In this model, consciousness is an emergent evolutionary development of brain function that initially was a primitive momentary subjective feeling, serving the purpose of biological survival. Damasio explains this viewpoint:

"If actions are at the root of survival and if their power is tied to the availability of guiding images, it follows that a device capable of maximizing the effective manipulation of images in the service of the interests of a particular organism would have given enormous advantages to the organisms that possessed the device and would probably have prevailed in evolution. Consciousness is precisely such a device... Because survival in a complex environment, that is, efficient management of life regulation, depends on taking the right action, and that, in turn, can be greatly improved by purposeful preview and manipulation of images in mind and optimal planning..." (p. 24)[1]

In this biologically-based reductive approach to the place and role of consciousness, conscious experience is described as a momentary event or pulse of feeling:

"We become conscious when the organism's representation devices exhibit a specific kind of wordless knowledge—the knowledge that the organism's own state has been changed by an object—and when such knowledge occurs along with the salient representation of an object. The sense of self in the act of knowing an object is an infusion of *new* knowledge, continuously created within the brain as long as "objects," actually present or recalled, interact with the organism and cause it to change... In a curious way, consciousness begins as the feeling of what happens when we see or hear or touch..." (pp. 25-26) [1]

In Damasio's model, consciousness is:

"...created in pulses, each pulse triggered by each object that we interact with or that we recall... The continuity of consciousness is based on the steady generation of consciousness pulses which correspond to the endless processing of myriad objects..." (p. 176) [1]

In this model, consciousness is a brief discrete event in a biological device of some kind, associated with feeling a change in the representation of the internal or external environment based on input that induces the change. This is somewhat similar to the notion in cognitive models described in the past chapter that associate consciousness with specific processing stages. Further, in the model stimulus input drives the functioning of the sense of self, emotion, and conscious experience. Emotion brings together the object of experience or inducing input and the existing pattern of the self at the time of input in a representation of their relationship that can become conscious. This is theorized to help maximize the selection of behaviors in the service of biological survival. The continuous flow of inducing input produces the sense of continuity of conscious experience. But the continuity is a contribution from the processing of several objects, adding bits of consciousness to build a subjective stream of conscious experience. In this description of a conscious event can be seen the aspects of object of experience, experiencer, and process of experiencing, but quite circumscribed and limited meanings of these fundamental aspects that come out of an intellectual delineation of the 'nature' of consciousness.

Importantly a clash of levels of depth becomes apparent when this functional neurocognitive model is placed structurally into the neuroanatomy of the brain. In the model, core consciousness is deeper in the sense that it is more elementary than and not as high-ordered as extended consciousness. But in the sense of more central to ordinary subjective experience in terms of a continuous sense of self, extended consciousness is deeper—similar to cognitive models of central conscious attention. The sense of self is identified with consciousness, and both are placed at the deepest *functional* levels of processing in the model.

From a *structural* perspective, however, the top-down functions of conscience, high reason, and extended consciousness are higher-order neural patterns underlain by deeper lower-order biological processes. In a general way this structural notion of depth also can be viewed as consistent with neuroanatomically higher-order cognitive processes in the outer cortical areas of the brain and lower-order processes in the older sub-cortical areas. From this structural perspective of depth, the above functional depiction of the model would need to be turned *upset down* in order to reflect lower-order processes being deeper than higher-order processes. That is, *functional depth* refers to higher-order psychological processes, whereas *structural depth* refers to more fundamental lower-order biophysical processes. How higher-order consciousness, which derives from lower-order neural structure, has a causal influence on neural structure is fundamental to reconciliation of the clash of the structural and functional perspectives of depth of processing. We will reconcile these perspectives about depth in the next chapter. But first we will explore other attempts to account for the unitary sense of self, the emergence of consciousness, and the causal influence of mind and consciousness on matter within a physicalist worldview.

Consciousness as Reentrant Neural Connections

Another important neurocognitive model, proposed by Edelman and Tononi,[2] is of interest because it tries to connect structure and function more closely. This model contains a more abstract notion of structure and a more concrete notion of function, in attempting to identify the neuroanatomy of mind and consciousness. The model emphasizes *process* in conceptualizing how the gross macroscopic brain generates consciousness. Consistent with the previous model, it also emphasizes the necessity of placing consciousness in a biological evolutionary context as an emergent property of neural complexity.

This model also has similarities to the previous model in terms of functional levels of depth of processing. It delineates a lower-order level of *primary* consciousness and a higher-order level of *secondary* consciousness. In addition it reflects a workspace notion of consciousness similar to the theatre analogy. However, it contains a somewhat more detailed hypothesis of how neural structures are thought actually to generate consciousness—which Edelman and Tononi call the *"dynamic core hypothesis."* (p. 111) [2]

The dynamic core hypothesis. The dynamic core hypothesis emphasizes two basic principles: *reentry* and *selectivity*. Reentry concerns the reciprocal interaction of different functional groups of neurons. Selectivity concerns evolutionary development of the brain via interaction with the outer environment, through Darwinian natural selection. These principles are thought to provide conditions under which consciousness emerges when the underlying neural activity is sufficiently complex.

These principles are relevant to the *binding problem*, which on one level deals with how distributed cortical areas with different functional roles cohere globally. This important issue concerns what holds together the neural activity of distributed populations of neurons to create a sense of self and unitary conscious experience. It is akin to the combination and generation problems mentioned in Chapter 13 with respect to lower-order physical processes.

Recognizing that consciousness is not due to a particular neuron or neural group, this model proposes a more specific account of the unitary nature of conscious behavior as generated by a dynamic core of neural connections with reentrant activation. Reentrant activation basically refers to functional interconnections between groups of neurons. Complex, strong, and rapid reentrant neural interconnections across multiple groups in parallel are thought to allow for the stable integration or coherence of function necessary to create consciousness. This is theorized to be achieved by a dynamically changing group of interconnected neurons called a *functional cluster* or *dynamic core*—sort of a mass or swarm of neural activity. This is interpreted as explaining how functional integration takes place *without* a homunculus in the brain associated with a particular neural architecture. The dynamic core is a changing pool of spontaneously active neurons that produce a stable cluster of neural activity, which then guides functional processes. It is theorized that, in a living brain, this dynamic core can serve as the substrate of consciousness when it has sufficient complexity. Complexity is said to be the result of the coexistence of the opposite qualities of functional integration or unification and differentiation or specialization. (Note the similarity to the nature of the unified field of consciousness as the coexistence of opposites discussed in Chapters 10 and 11, but on the concrete, highly limited material level of nature).

According to the model, high reentrant interactions allow complex nonlinear dynamics to build into a system, which can communicate changes in its parts that are relevant or informative to other parts. This can be likened to a self-monitoring system, where one part of the system can simultaneously 'observe' other parts. This is thought to place the 'observer' inside the neurocognitive processing system of the physical brain.

It is hypothesized that the degree of brain complexity in lower animals is sufficient for producing primary consciousness that doesn't have an enduring self-reflective ability or sense of self. But only in the human species is brain complexity sufficient to generate higher-order secondary consciousness associated with self-reflection, extensive long-term memory integration, higher-order language, higher reason, and conscience. [2]

The model is supported empirically by research demonstrations that utilize computerized modeling of large collections of simulated individual neural units. Using the example of a visual discrimination task, separate simulated neural groups for color, for shape, and for motion cohere or coordinate together in correct performance of visual discrimination based on reentrant interconnections. In the computer model no specific area is given the role of coordinating the different functional groups, but the functional core is said to *emerge* when the system has sufficient differentiation of functional groups and also has a high degree of integration by virtue of reentrant activation. Reentrant neural dynamics is said to be in humans the neural basis for the unitary experience of *self* and of consciousness. [2]

Let's consider the principle of reentry a bit more reductively. It is proposed that the dynamic core of interconnected, electrochemical, neural processes generates consciousness at some evolutionary level of complexity. But do cells, molecules, atoms, and quantum fields functioning in a closed causal chain somehow produce conscious mind as an additional level of ontology concomitant with these physical structures? How does this emergent consciousness then gain the causal power to have a super-ordinate top-down controlling influence on the neural processes and the brain as a unitary *conscious self* that directs unitary behavior?

Outside of neural cells in the brain are other tissues such as arteries and veins, skeletal encasement, and sometimes hair. The most expressed level of the physical human organism may be said to be such things as nails, hair, and dried skin—reflecting the least degree of life and the most similarity with inert matter such as rocks and earth. This might be described as the most expressed or grossest level of gross physical existence. Perhaps we might conceive additional outer levels to be the organism as a macroscopic behavioral unit or whole, as well as levels emerging from its larger social context. From interactions with the outer physical and social environment, humans and other species form various layers or strata of social networks—family, community, nation, culture, world system. But these represent ways in which organisms existing in gross physical *reality* interact with each other. There is no *new* ontological level of physical *reality* as a product of these environmental and social interactions that emerges *outside* of and as a product of complex neural activity in the brain.

On the other hand there is also no *new* level of physical *reality* that emerges as a product of neural complexity *inside* neural processes. Reentrant neural activity does not change the internal mechanical functioning of neural cells or their synaptic connections in a manner that generates a new ontological level. There is no functional space or stage in the physical causal chain that somehow unlinks itself, separates, and opens up for the new subjective, causally efficacious *reality* of mind and consciousness to emerge. Neither inside nor outside of neurons is there a super-ordinate causal *reality* created in physical space due to neural complexity that is able to control the physical brain as a global, integrated self—and in the physicalist view there are no other choices. No functional role for such an emergent *reality* seems possible in the closed causal chain.

Further, although reentrant interconnections facilitate a dynamic core of neural activity, this doesn't mean that this abstract structural core automatically has the capability on its own to establish a functional cluster that *directs itself* globally to perform unitary tasks. The binding problem still exists with respect to integrated functioning of the whole system, even though the structural capability may be present for integrated activity. This is a fundamental challenge to neurocognitive models of mind and consciousness based on emergence theory.

The whole is greater than the sum of its parts in the sense that new patterns and capabilities may emerge when the parts unify into a whole. But with respect to gross physical processes, the whole is not greater than the parts in the sense that a complex collection of interconnected parts creates a *new, transcendent level of physical reality that has top-down causal control over the parts and directs unitary behavior of the parts*. New and different physical effects emerge from new combinations of physical objects and processes, but not with causal power over the closed physical chain of events. With respect to brain function, complexity results in new physical effects due to new types of interactions; but no ontologically new level of physical existence with super-ordinate causal power emerges inside or outside of neural structures as complexity increases. New physical effects emerge within physical *reality*—but not new ontological levels of existence.

> "...[T]he most widely accepted models of scientific explanation...are...largely inspired by part-whole reductionism... Indeed, the very idea that the stuff at the bottom (whether it be fundamental laws or fundamental entities) provides the ultimate explanation for all phenomena is simply an expression of this kind of reductionism... A property of an object or system is epistemologically emergent if the property is reducible to or determined by the intrinsic properties of the ultimate constituents of the object or system, while at the same time it is very difficult for us to explain, predict or derive the property on the basis of the ultimate constituents... Ontologically emergent features are neither reducible to nor determined by more basic features. Ontologically emergent features are features of systems or wholes that possess causal capacities not reducible to any of the intrinsic causal capacities of the parts nor to any of the (reducible) relations between the parts. Ontological emergence entails the failure of part-whole reductionism..."—Michael Silberstein and John McGeever, philosophers [7]

To illustrate the point more concretely, a group of mountains conceptualized as a unified mountain range doesn't empower the mountain range by itself to have a controlling influence on each mountain. A forest of trees similarly doesn't control the trees in it—even though the collection of trees has ecological effects in addition to each tree. A network of computers can increase processing power and capability for problem solving, but no super-ordinate conscious mind—either physical or non-physical—is created by pooling processing capacity. Three hands may be better than one in that they may allow greater breadth and efficiency of task performance, including the capability of entirely new tasks. But they don't in themselves produce a new mind.

The whole as a collection of parts does generate new capabilities not present in each part, which can influence the parts. In the physical world the collected whole influences the parts via the same physical mechanisms through which the parts influence each other and the whole. Reentrant interaction—or any other similar feedback, reflective, non-linear, or referential type of interaction between physical parts—by itself doesn't *create* a new level of existence with super-ordinate control over itself or its parts.

For a global influence of the whole directly and simultaneously on each part, or for each part to influence directly the whole, physical connections between them are necessary in a causally closed, local physical *reality*. Complexity increases dramatically with more parts, with more complicated tasks, and with more processing speed. At some point smaller time and distance scales or finer-graining associated with lower-order physical processes at which there is more interconnectedness in nature need to be utilized in order to implement interaction between all parts of a grosser physical system. This is the current direction, for example, in building physical structures that allow faster and more powerful computational functions in information science.

In order to locate where consciousness and its causal influence might come into play in the physical brain, we could similarly analyze each point in the structure of a neuron, or in each stage of processing of one neuron signaling to another neuron or neural group. The only direction to go in this analysis would be to smaller and smaller distance scales of these structures and processes. We could inspect the chemical, electrical, atomic, quantum and quantum field interactions, strings, branes, quantum loops, qubits and the like. In none of this would we find the influence of consciousness in the physical causal chain emerging due to neural complexity. This would bring us down to the self-interacting fields of nature that underlie biological and microphysical processes, and even further to the nonlocal level underlying local physical *reality*—the abstract nonlocal field of information space or quantum mind. Until we go beyond local material *reality*, there is no additional ontological existence, space, or *reality* where mind and consciousness could emerge and influence the causal chain. The only place for such a causally efficacious level of existence is underneath—more abstract and subtler than—the physical level of existence.

To be clear about this point, we are not here describing epistemological or ontological emergence from a bottom-line quantum physical level such as quantized particle-fields or strings or branes at the Planck scale. We are discussing an additional ontological level of existence that is subtler than the physical, that is causally efficacious with respect to the physical, that is nonlocal, and that becomes condensed or compactified into the physical via increasing levels of limitation in the natural course of manifestation from transcendental to subtle to gross creation (refer to Part I of this book). If consciousness is associated with degrees of complexity of neural activity, it is that higher degrees of neural complexity allow higher expressions of consciousness to manifest in the gross physical world. But they do not by themselves *generate* minds or consciousnesses.

As another tangible example, having all the building materials present in one location doesn't spontaneously generate a house. The house is created by coherent action in accord with the super-ordinate guidance of an architectural plan—an intelligently designed guiding image. New qualities or effects emerge with completion of the house, but do not have causal influence on the bricks or other materials of which the house is composed. The house and its emergent properties are effects of implementing the plan. The plan relates to the structuring power of intelligent thinking of the architect or designer. The subtle inner feelings of desire and intent of the potential owners and builders are required for the house to be generated, and can be understood to be the underlying intelligence that makes the parts cohere. Unfortunately building materials collected together don't spontaneously cohere into houses, and houses don't build other houses.

Likewise groups of neurons don't bind together on their own to create a conscious mind. Neural systems function coherently through a guiding influence that directs them to perform planned behavior. When the physical causal chain is analyzed, there is no physical gap in it for this guiding causal influence. It comes from a subtler, subjective level of nature. This is again reminiscent of the guiding or psi wave in Bohmian mechanics described in Chapters 5 and 7. What is relevant about a dynamic core of spontaneously active reentrant neural interconnections is that it may be sensitive enough at a quantum mechanical level to respond to extremely subtle influences from an underlying nonlocal field of information space, quantum mind, or mental *reality* It may allow the expression of subtle mental processes, but doesn't create mind or consciousness.

In computer simulations of the neurocognitive model being discussed here, the requirement of coherent guidance or intelligence was modeled by Edelman and Tononi through adding a simulated *"value system."* (p. 46). [2] In the example of the simulation of a visual discrimination task referred to above, in order to perform a coherent function, the functional cluster is facilitated by programming into it a value system that activates when the system moves in the right direction—according to the value system or plan of the investigators. This is implemented by a diffuse signaling program that favors or strengthens a particular direction toward the 'correct' discriminative response, guiding the system into a globally coherent functional response. It is analogous to a super-ordinate wave or influence of intelligence that facilitates the formation of a coherent discriminative process—a guiding purpose or goal. In the computer model it is programmed into the system from the outside.

If we investigate the source of the intelligence in the software program in the simulated model of a functional cluster of neurons, it won't be located in the software disk itself, but can be said to have been programmed into it as a set of organized instructions from the outside by the programmer. Likewise if we investigate the source of the intelligence in the physical programmer, we can reduce the programmer to cells, atoms, self-interacting fields, all in a closed causal nexus, and won't yet uncover the source of the intelligence. We will get closer to the source of it in the underlying functional information space of the quantum mind, beyond the neural structures, beyond the brain, underneath the physical, and eventually locate it in the transcendental unified field.

The pursuit of non-linear and self-interacting dynamics—including so-called self-organizing systems—as an explanation for the guiding causal influence will lead to smaller and smaller time and distance scales. It could take a long time for those intent on finding the explanation to coherent intelligent behavior by such dynamics to examine the most fundamental levels of physical and microphysical processes. But at some point it will be recognized that deeper, subtler levels of nature underneath the physical will need to be incorporated into the explanations. Such investigations would have to lead through all the gross and subtle levels to their self-referral basis in the unified field itself. That level is the only completely unified self-referral state and the source of all coherence, binding, and intelligent behavior throughout all of nature.

Perhaps at this point some clarification about the concept of reductionism would be helpful. As the term has been used in this book—as well as it is used generally in modern science—it is associated with reducing phenomena to their underlying physical bases. It implies both a process of getting to the bottom line of phenomena, and commonly also the assumption that the bottom line is physical. However, the term could be used more broadly, as reduction to the bottom line—whatever that turns out to be. It doesn't have to carry with it the limitation of the physicalist assumption. The epistemological strategy of getting to the bottom of things is a valuable characteristic of the scientific endeavor. (It even can be applied to describe the process of transcending in which mental activity is reduced to its underlying ground state). But the accompanying assumption that modern science knows that the bottom line is physical already has been shown to be untenable—when microphysical processes attempt to be modeled and explained.

Some may feel comfortable with a distinction between material and physical. For example, associating material with anything that is quantized (as discussed in Chapter 8), and associating physical with anything that exists in relative creation whether gross or subtle, whether matter or mind—akin to Bohm's distinction between explicate and implicate orders. But this certainly is not what is typically understood by the term physical, which generally equates with material.

Similarly the term emergence can be applied in different ways. Although emergence in the ontological sense refers to phenomena that are not reducible to the microphysical, theorists who posit emergence models typically adhere to the belief that the physical is all there is. Although ontological emergence is interpreted to mean new properties and processes emerge that are not reducible to microphysics, it tends to mean emergent levels of organizational function in which causal efficacy occurs in some undefined way. In a foggy manner, emergence theory seems to reflect attempts to come to grips with phenomena that are irreducible with respect to the macrophysical and microphysical, without fully recognizing that there is an underlying subtle ontologically *real* nonlocal level of nature. It thus could be identified more as explanatory emergence, not rising to the much deeper notion of ontological emergence.

In ancient Vedic science new phenomena emerge due to the progressive phenomenal limitations of the transcendental level to the subtle level to the gross level. Emergence in this sense refers to new phenomena that become partial manifest expressions of the totality of the unified field of nature due to new combinations and permutations of constituents at the respective levels of manifestation and limitation. There is an underlying seamless unity from which the vast diversity of nature emerges. The expressed levels of diversity don't create consciousness, mind, or life by virtue of organizational complexity, although more complexity allows more complete expressions of the qualities of the unified field of consciousness to emerge—formed by increasing specificity and limitation. Diversity emerges from unity; parts emerge from the whole; relative emerges from its absolute basis. New ontological levels are limitations that unfold within the ultimate totality.

Neural Darwinism. According to the neurocognitive model being discussed here, the inner purpose, goal, or value system—the programming intelligence—is said to be built into neural activity in living organisms through evolutionary processes of natural selection. This is explained in a theory of neuronal group selection (TNGS), or what has been termed *"Neural Darwinism."* (p. 83) [2] This brings us to the second fundamental principle in this model, the principle of *selectivity.* Let's now consider this key principle of the model, and then later put all this together into the theme of the current set of chapters on structural and functional notions of levels of depth. The primary issue at this point is the deep functional level of conscious self that provides a superordinate top-down binding together of the other levels into a coherent, intelligently functioning, negentropic, goal-directed organism.

In the model under discussion selectivity is a fundamental organizing principle in morphological and phylogenetic development. It is applied in attempting to explain individual choices of behavior, as well as how memory processes are implemented in the brain. According to the model, human information processing does not consist of coded programs based on principles of logic and representational memory characteristic of artificial intelligence in computers—which are programmed into the system from outside. Edelman and Tononi state that memory, as well as functional behavior, are embodied in the organism by selective action patterns that build up through interactions with the physical environment:

> "Logic is not necessary for the emergence of animal bodies and brains, as it obviously is to the construction and operation of a computer. The emergence of higher brain functions depended instead on natural selection and other evolutionary mechanisms... [I]t is essential to grasp that selectionist principles apply to brains and that logical ones are learned later by individuals with brains. Only with such notions in mind can one avoid the paradoxes that result from attempts to explain consciousness solely in terms of computation." (p. 16) [2]

However, many scientists might say that logic is a fundamental principle reflected in orderly natural laws inherent throughout the natural world—including natural selection—which the human mind is able to recognize in comprehending the order. But human logic doesn't create that order. This was a major issue for science discussed in Chapter 9 with respect to whether mathematical principles are inherent in nature or the product of human thinking.

According to Edelman and Tononi, in living systems the 'value system' is embodied in the organism, built at every stage of phylogenetic and morphological development. It is not a code of instructions as if from the outside, but rather is inherent in the system in the sense of being bred into it as the result of natural selection. The purposeful functioning of the system develops in the structure of the system, even psychologically and experientially. The logic or intelligence of the human information processing system is a late evolutionary development that is a product of a long history of natural selective processes. The brain follows the principles of natural selection by amplifying repertoires that match environmental signals better than other repertoires. The orderly structure of the organism, including its eventual ability to think logically, is due to random mutation and a value system based on natural selectivity developed through its evolutionary history.

In the theory many selective processes contribute to the implementation of this value system in the brain. In *developmental* selection, neurons initially branch in many directions, forming extensive pathways of electrical activity that become strengthened or weakened in subsequent development. In *experiential* selection, synaptic selection occurs with behavioral experience, continually modified throughout the life span by successful performance. Reentry results in synchronous activity in neuroanatomically diverse functional groups, binding them together for coherent global output. Through these means the value system is bred into the organism, slowly evolving into an integrated behavioral system with diverse functional capabilities in the service of biological survival. These selective processes are thought to produce the simultaneous differentiation and integration that generate and characterize consciousness.

A key issue related to the binding problem needs to be considered further in this model—in addition to the continuing concern in all functional models that none of the functions seem to require conscious experience. The fundamental issue here is how biological survival becomes an organismic purpose for behavior *in the first place*. How is it that survival becomes *valued* and *favored* in nature? Natural selection is driven by survival of the fittest, but lower-order inert physical objects don't have motivation to survive, choose to do anything, or have any values on their own. How do we get from entropic physics to negentropic biological survival?

The instinct to survive seems like a value or purpose in living organisms. Where does it come from if the underlying materials and processes are all inorganic and fundamentally random? It is difficult to imagine how inorganic compounds would exhibit anything like an instinct for survival. Lower-order physical processes do not exhibit the requirements for negentropic functions to develop on their own. If there is no new ontological, causally-efficacious level of existence emerging either outside of or inside of neurons as they collect into more complex networks, then causal efficacy must somehow be inherent in or equivalent to the existence of a physical neuron—which is how the next model to be discussed addresses this concern—or a non-existent misperception. Although living organisms certainly appear to act as if they have some control over their survival potential, and even further we humans also attribute causal control over our own actions, these empirical and phenomenological observations apparently must be delusional.

The issue we are now considering is directly related to how the sense of individual self is created. The sense of self as a coherent living organismic wholeness seems fundamental to the concept of biological survival that guides unitary behavior. But the sense of self is thought to be a much later and higher-order evolutionary development than the phenomenon of biological life. What the notion of biological survival would mean absent a coherent sense of self needs to be addressed—survival of what, of molecules and atoms? The binding problem is still present with respect to this fundamental issue, in that lower-order physical processes don't on their own bind together functionally in the manner of living organisms that act to maintain their own survival. From whence comes the coherence or intelligence that holds together physical matter into a holistic biological organism or self, and how does the self develop a survival instinct?

Neo-Darwinian evolutionary theory holds that processes of natural selection do not involve an intention or instinct to survive. With random mutation, some organisms happen to be stronger and more capable, and thus more likely to survive. The empirical fact is that some organisms that are products of random mutations happen to be more fit to adapt to environmental demands and survive to contribute to the gene pool, and it is in this way that evolution of species proceeds. In a straightforward understanding of natural selection, there is no need to posit an intention to survive. But many if not all human beings seem to consider it an undeniable empirical fact of their own experience that they do have an intention to survive which influences their own selection of behaviors. Most also make a similar attribution about other living organisms, based on careful observation. Further, many humans hold that their selection of behaviors is not based solely on personal survival or species survival, but on ethical and moral principles that transcend their instinct for biological survival. To reject out of hand these strongly held attributions amounts to denying humans personal agency and free will. In addition, many of what might be considered the fittest, most capable and intelligent humans choose not to be the biggest contributors to the gene pool in terms of progeny. These points suggest a more expanded evolutionary theory is needed, consistent with neo-Darwinism while also recognizing its limitations. This will be discussed further in upcoming chapters in the context of higher human development. Several planks of understanding need to be in place for the limitations of neo-Darwinism in the physicalist worldview to become apparent. Hopefully at least modern scientific progress in the past century beyond physicalism is clear enough to be open to a more inclusive evolutionary theory that respects our own inner empirical experience of efficacious life choices that naturally extend into moral and ethical principles beyond just biological survival.

The notion of an overriding super-ordinate value system that guides or provides intelligent direction to behavior in living organisms has close affinity to the notion of a sense of a *real* self. It even sounds quite a bit like an inner *homunculus*. However, Edelman and Tononi claim that their neurocognitive model eliminates the concept of the homunculus:

> "In our theory of brain complexity, we have removed the paradoxes that arise by assuming the God's-eye view of the external observer and, by adhering to selectionism, we have removed the homunculus." (p. 220) [2]

In contrast to Edelman and Tononi's statement, a more logically consistent alternative is that what has been accomplished in this model is in the direction of helping to explain how the conscious mind, and the inner subjective homunculus or self, can interact with the gross physical level of nature as a biological self embodied in complex reentrant neural processes in the brain for the purpose of gross behavioral functioning. This will become clearer in upcoming discussions in the next two chapters on the nature of the self. Also it will become clearer how this discussion of the sense of self fits into an integrated model of levels of mind. Especially it will be discussed how the sense of self is the deepest level of individual subjectivity that provides the fundamental binding of mental and physical processes based on the super-ordinate top-down guidance or intelligence of a coherent individual experiencer. The 'binding into a biological self' can be understood to be the effect of the guiding influence of an underlying information space, quantum mind, mental *reality*, or subtle field of thought or intelligence that binds parts of physical matter together and guides them in coherent unitary behavior. Neural activity depends on the physical, but consciousness, the self, and the mind do not. Rather the gross physical depends on the subtle mental—both of which depend on the unified field of consciousness itself.

The challenge within a physicalist view to create causally efficacious top-down control of a coherent unitary functional system or self from lower-order neuroanatomical structures is also a major concern of this next neurocognitive model. This model embeds function completely into structure at each level of biological evolution. Attributing the binding of the sense of self and related conscious mental functions to primitive organisms, it is an alternative to emergence theories that posit functional complexity of neural activity as the origin and generator of mind and consciousness. It is a more reductive version of emergence theory that does not place the conscious self as emerging from neural complexity such as reentrant connectivity.

The *I* as a Command and Control Function

A third example of prominent neurocognitive models is from the work of clinical neuroscientist Rodolfo Llinas. This model emphasizes the importance of understanding the mind as the:

> "...product of evolutionary processes that have occurred in the brain as actively moving creatures developed from the primitive to the highly evolved." (p. IX) [3]

This model emphasizes conscious mind as the evolutionary internalization of movement. It attributes a conscious self and coherent goal-oriented behavior to even primitive organisms. Whereas in the previous model the functions of mind, consciousness, and the sense of self are thought to be an evolutionary product of neural complexity, this model identifies rudimentary versions of what is usually thought to be higher-order mental functions as intrinsic to simple neural structures.

This model focuses on the ability of organisms to predict the results of actions and modify evolutionarily 'fixed action patterns (FAPs)' to produce integrated unitary behavioral strategies that fit the environmental context. [3] Llinas describes FAPs as relatively hard-wired, but they can be modified by volitional control according to context. He explains:

> "This "breaking out" or overriding of a given motor event that is constrained by the FAP being executed is accomplished by the thalamacortical system, the self, making volitional choices that arise from weighing the information and predicting consequences of the unfolding context of the given situation. It is the necessary advent of consciousness to an otherwise responsively fixed repertoire of movement." (p. 151) [3]

In this model FAPs have evolved from motor tissue into premotor internal *emotions* that either drive or deter motor actions.

> "In this regard, we may look at emotions as the *global sensation* aspect of FAPs, if not as FAPs themselves." (p. 161) [3]

Further,

> "...given the complexity of the decisions and the speed at which the nervous system must implement a given global strategy, the only solution that will make this work is one in which the animal is *conscious* of the particular emotional state. Why? Because consciousness has the great ability to *focus*–this is why consciousness is necessary. It is necessary because it underlies our ability to *choose*." (p. 169) [3]

Thus consciousness provides a functional space for the self to make efficient volitional choices of behavioral strategies. We will refer back to this super-ordinate notion of conscious choice and the causal power of intentions in unitary goal-oriented behavior shortly.

Functional levels of depth of processing. This model is similar to the previous two with respect to *functional* levels of processing; however, in this model the functional levels are described in terms of memory. The fixed action patterns (FAPs) develop through natural selective processes and allow efficiency of increasingly complex behavior via an integrated set of different levels of memory.

Phylogenetic memory relates to the basic connectivity that is built into the neuroanatomy of the organism. *Dynamic* memory relates to the action of this phylogenetic neural circuitry, the electrochemical structures associated with the basic intrinsic activity of the brain prior to experience. Llinas describes these two levels as *genetic* memory:

> "Genetic memory (long term in species terms) is present at birth, *as memory that occurs in the absence of sensory experience.* Such memories were written directly into our genetic code by the myriad of small mutations occurring in our genome over time and brought to light by natural selection." (pp. 190-191) [3]

According to Llinas, a third level, *referential* memory, involves processes that change due to life experience:

> "Where the first two types represent the memory accrued and pruned over many lifetimes, as an organism's qualities and characteristics that have been naturally selected for, the referential memory represents that which has accrued during development and throughout a single lifetime... Referential memory of the long term variety can be subdivided further into implicit and explicit types... Explicit memory, also known as declarative or conscious memory, generally refers to the memory underlying the conscious recollection of things, such as faces, names of objects, past experiences. It has also been further divided into two possibly distinct aspects of the retrieval process (Schachter 1987): the voluntary, intentional retrieval of a memory, and the subjective, conscious awareness of having remembered it (Tulving 1983). Implicit memory, nondeclarative or nonconscious memory, is the unconscious, unintentional retrieval of memory for performance of a learned activity or skill." (pp. 183-184) [3]

In this delineation of memory processes can be seen the similarity to the other models of functional levels of depth of processing, including non-conscious reflexive, automatic, and higher-order conscious intentional processes. Together these levels of memory comprise a centralized ability to predict the appropriate action patterns and to shut off or modify them by conscious choice in response to environmental change, called the self or *I*. According to Llinas, this self or *I* is the centralized command and control function of the organism as a behavioral unit. In this model, "...the command system is the self (i.e., the centralization of prediction)." (p. 33) [3]

Intrinsic neural dynamism. In the overall current discussion the most interesting aspects of this model are the assertions that living organisms are intrinsically *active* and intrinsically *purposeful* or goal-directed. Llinas' model attempts to account for the spontaneous activation of movement in intelligent goal-directed behavior based on an "internal reckoning—a transient sensorimotor image... " (p. 18) [3] In other words the organism acts based on a plan, embedded in which is an inner purpose or value system in the service of survival. This is centralized in the self, what Llinas describes as a "pre-existing functional disposition of the brain..." p. 8) [3] This pre-existing function anticipates and predicts the results of environmental change. Let's consider further these two assumptions of intrinsic activity and intrinsic purpose in biological organisms.

Spontaneous or intrinsic neural activation is associated in the model with *feedforward* control of movement related to *prediction*—and also related to the *command* function of the self. The *feedback* control of movement, involving fine-tuning of the feedforward movement, is associated with *reflection*—and can be related more to the *control* function of the self. Feedforward command and feedback controls must be centered around a guiding purpose that both energizes the system to behave and that directs it with varying degrees of intelligence toward certain outcome goals—especially the purpose of survival. The model attempts to explain how the global, unitary, behavioral perspective of the organism as a functional whole—including processes of abstraction, sensorimotor images or internal representations, emotion, intention, will, the sense of self, and consciousness—exists at *all levels of neural complexity*.

With respect to intrinsic spontaneous movement, the model emphasizes intrinsic oscillations and rhythmicities in neural processes. Various types of neurons have spontaneous electrical activity in the form of small voltage fluctuations across their membranes—ranging in frequency from less than one to 40 or more cycles per second. These voltage oscillations modulate the responsiveness of the neurons to the much larger synaptic action potentials involved in information transfer across neurons. Llinas hypothesizes that they are the:

> "electrical glue that allows the brain to organize itself functionally and architecturally during development... Neurons that display rhythmic oscillatory behavior may entrain to each other via action potentials. The resulting, far-reaching consequence of this is neuronal groups that oscillate in phase—that is, coherently, which supports simultaneity of activity." (p. 10) [3]

This notion of intrinsic motion could be taken much more reductively. It might even be brought down to the notion in quantum physics of the inherent dynamism of quantum fields—vacuum fluctuations or zero point motion discussed in the previous physics part of this book. But in this model the reductionism cannot go that far down into lower-order physical processes, because the inherent motion is also thought to contain inherent purpose, which is thought not to be attributable to the more fundamental levels of inert matter.

With respect to how this intrinsic movement has an intrinsic *purpose* in biological cellular structures in this model, an analogy is given of circada chirping in rhythm. The first circada may be calling for others, which rhythmically chime in also. According to Llinas:

> "...[T]his unison of many circadas chirping rhythmically becomes a bonding, literally a conglomerated functional state. In the subtle fluctuations of this rhythmicity comes the transfer of information, at the whole community level, to a vast number of remotely located individuals... Simultaneity of neuronal activity, brought into existence not by chance but by intrinsic oscillatory electrical activity, resonance, and coherence are...at the root of cognition. Indeed, such intrinsic activity forms the very foundation of the notion that there is such a thing called our "selves." (p. 10-12) [3]

In this example, however, it is worth recognizing that the information value of the resonance doesn't come spontaneously from the resonance alone. It is due to the ability of the participating organisms to be informed by and perform functions based on the resonance. Each circada has the ability to make sense of and make use of the resonance on some level. Resonance alone doesn't bind anything into a *functionally meaningful* or *purposive* state with respect to the organism as a whole. No functional 'mindness state'[3] is created as a product of either resonant circada chirping or spontaneous intrinsic neural oscillations that resonate with each other. Resonance due to spontaneous entrainment *in itself* and without minds capable at some level of attributing meaning to it does not have functional information value.

Nonetheless according to the model spontaneous oscillatory firing in muscle tissue eventually evolved into internalized thinking. Llinas explains:

> "...[A] physiological tremor is a reflection, at the muscoloskeletal level, of a descending control signal... Yet, during early stages of development, the tremor is not just a reflection. In fact, *the tremor is a property inherent to and exclusive of muscle tissue* (Harris and Whiting 1954). This is known as the myogenic moment of motricity, which occurs during development before motoneurons have even made contact with the muscles they

will later drive... [T]his tremor is "handed" from the muscles to the motoneurons that innervate them, and then to the upper motoneurons that drive them, and further and further "inward" to become the controller system and, ultimately the command system... *[T]hat which we call thinking is the evolutionary internalization of movement.*" (p. 34) [3]

According to this outside-in functional model of evolutionary neural development: "The internally generated initial feedforward component does not require sensory feedback during execution." (p. 34) [3] Rather, the drive to generate organized movement is provided by:"...intrinsic neural oscillations and the specific ionic currents necessary for their generation." (p. 34) [3]

In this model spontaneous action in the form of physiological tremors provides the energy for movement. Its synchrony produces resonance in motor control signals that produce organized movement, evolving through natural selection into the functional control of behavior by a centralized system of prediction or self. Llinas concludes:

"So now we have a wondrous biological "machine" that is intrinsically capable of the global oscillatory patterns that literally *are* our thoughts, perceptions, dreams—the self and self-awareness." (p. 133) [3]

In the model, action is intrinsic—by virtue of spontaneous neural firing—and it is also intrinsically unitary, intentional, and goal-oriented—engaging the whole organism—and somehow controlling and guiding its own behavior:

"In order to place movement within the context of the whole animal, the animal must first be capable of generating some type of internal "image" or description of itself as a whole, and this image must support the strategy around which to organize the tactics of what the animal will do ." (p. 225) [3]

This suggests that the abstract sense of self that is said to be necessary for unitary intention and thereby holistic, coherent, biologically adaptive behavior accompanies and is intrinsic to each evolutionary level of development. Complex behaviors associated with higher-order thought and language are the evolutionary elaboration of simple neural systems that have primitive levels of abstractions, intentionality, sense of self, and consciousness in them. Llinas further asserts that the beginnings of abstraction are built into the morphological structure:

"Elongated animals are basically made of a horizontal stack of "coins," where the neural wherewithal subserving each coin is organized to know about its respective segment and relatively little else. In order to make a complete working animal out of these segments, there *must* be a portion of the nervous system that is *not* exclusively segmental in its organization. This portion of the nervous system can put the many segments together into something that beforehand did not exist: a unified whole... [W]e may consider this the beginning of abstract function as this portion of the nervous system does not relate *directly* to the connectivity of the nervous system at any particular, segmental level. The central nervous system abstracts the fact that the animal is composed of a series of unit segments; ipso facto, the process of intersegmental integration is an abstraction, and represents the beginning of abstraction as a naturally selected biological process. That this is the evolutionary direction is supported by the observation that the central nervous system mushrooms out in front of the spinal cord, polarizing encephalization. We see something important happening: from the animal's very neurological becoming is the fact that the animal can have an internal *representation* of itself not only as a set of parts but as a whole entity. It is here, from this germinal metaevent, that abstraction begins and the self emerges... [T]he intrinsic circuits of the nervous system are capable of generating a premotor representation of what is going on outside. From this, self-referentially in the presence of that motor image, the animal is capable of deciding what to do. The animal is capable of prediction. Run, fight, find food or whatever, functionally the animal *is* the circuit that represents its sensory motor attributes, and this central event is an abstract entity...

And so, from the need to sensorially monitor the world within which an animal may move, the head pole becomes richer in sense organs because of the forward movement of the animal. Further, not only do these individual sense organs become richer in their capability of monitoring the external world, the nerve centers associated with and supporting this rostral pole in turn specialize to perform the rapid, predictive decision making that underlies and maintains the holistic behaviors crucial for survival. But more fundamentally, the experience serves to contextualize and to arouse the unity of sensory activation into one global functional state—something akin to "I feel" that acts to mediate decision making. It is clear from this that the head becomes the seat of qualia, having moved from more caudal regions to be supported by and to drive a richer neural connectivity." (pp. 226-227) [3]

The ability of multicellular organisms to move actively on their own—motricity—involves efficient control of movement; and this requires the ability of prediction. According to the model, the value system of movement for self-survival and ability to predict the results of actions are intrinsic to the organism; they are *not* based on interaction with the environment. As Llinas explains:

"The neural system is not built on the basis of a reflexological view that movement is driven by sensing the external environment, but rather...by central neuronal circuits whose functional properties can and would generate the appropriate patterns of activity to "will" the body into organized movement. This view is referred to as self-referential because although information arising from the external world may be a sufficient reason for organized movement (behavior), such information is not necessary for its actual physiological genesis. " (p. 42) [3]

Again we come up to a challenging circularity, in that movement which involves intention to survive is prior to the intentional control of movement, which then controls the movement. It is reminiscent of the 'chicken and egg' paradox brought up in Chapter 6 in connection with relativistic space, in which space exists only in relation to objects but objects exist in space. In Maharishi Vedic Science and Technology, movement is the *externalization* of thinking—just the opposite of what is proposed in this neurocognitive model.

However, the neurocognitive model tries to address the circularity by placing movement and intention as one thing—neural activity—and as intrinsic to certain neurons. In this model each level of physiological genesis is intrinsically organized self-referentially and holistically in building coherent unitary behavior of the whole organism. This means that, according to the model, the living organism is self-willed and to some degree self-conscious at each evolutionary stage of development. This seems to imply that the survival instinct is inherent in the organism and is not a product of Darwinian evolutionary interaction with the environment.

Further the model suggests that subjective processes including abstraction, emotion, will, and consciousness do not emerge only in *sufficiently complex* neural processes. This goes beyond the version of the functional identity hypothesis that subjective functions, the phenomenological *inside* of neural processes, are nothing other than the neural processes themselves *at a sufficiently complex level*. It suggests that they are *equivalent*—actually the same thing—whether the neural activity is complex or not. Llinas clearly states this contention:

> "Parsimony and serious science clearly indicate that "the bridge," "the mysterious transformation" of electrochemical events into sensations is an empty set. It does not exist: neuronal activity and sensation are one and the same event." (p. 218) [3]

This is a very strong version of functional identity, in which there is no difference between subjectivity and neural activity. Wherever there is neural activity—*even primitively*—subjective mind and consciousness inhere in it. It places subjective processes even into the simplest of neural systems. If each neural action is accompanied by a sensation, and if sensations mean that they are conscious experiences or qualia in some sense, then it would imply that all neural activity involves conscious sensation.

Qualia as fundamental to volitional control. The fundamental issue of the super-ordinate top-down control of the self and the related causal role of consciousness in a physicalist framework comes to a head so to speak in this model in Llinas' attempt to explain the nature and role of qualia—experienced sensations or feelings:

> "Ultimately, motricity is always the product of muscle contraction; we can make no movement through any other means. From this we immediately come to the conclusion that the nervous system works within the context of a final motor effector that is capable of transforming the electrical activity of motor neurons into actual muscular contraction. We may ask by analogy what is the effector, the apparatus for the endpoint expression of *sensory* experience. This is the central question of neuroscience at the present time... [W]e may come to the conclusion that the effectors of qualia are very similar in their neuronal bases to the neuronal bases of motor FAPs—except that they appear to be internalized FAPs... [T]his expression is what we call subjective experience." (p. 208) [3]

Again expressing a very strong functional identity—an equivalence—of subjectivity and neural functions, Llinas asserts:

> "Qualia represent the ultimate bottom line, because sensations themselves are geometric, electrically triggered events. At this time we cannot reduce it any further. If this geometric, functional state is sensation itself, a serious philosophical problem immediately arises. By this definition are not qualia just another example of "that which we have yet to understand'? Or are they perhaps something of a qualitatively different character altogether, something transcending the neurological substrate of neurons and their electrical activity we attempt to hide them behind? I don't think so, for I believe that patterned activity in neurons and their molecular counterparts are sensations." (p. 209) [3]

Also again, however, there is reluctance to consider further reduction into lower-order biophysical and inorganic physical processes. As Llinas explains:

> "At this point there are a few possible scenarios that we can consider. One, as many people believe, is that qualia represent a very profound event in neuronal function dealing with quantum mechanical structures of neurons that include the detailed organization of microtubules and microfilaments... The reason for its dismissal is that the neuronal elements that support sensory activation seem to be quite similar to those that support motor activity. Qualia seem to be related not only to particular neurons per se, but also more to the geometrical, electrical patterns of activity neurons are capable of supporting." (p. 209) [3]

One primary basis for this view is that electrical stimulation of certain brain structures can produce sensation qualia, and disruptions can make them disappear. Because sensory qualia can be eliminated, for example, through barbiturate application or anesthesia without changing neuroanatomy or neural connections at all, Llinas concludes:

> "...[W]e must accept that qualia are triggered by electrical activity in the brain and are made up of events very close in time to the electrical structures that skate over the surfaces of neuronal membranes." (p. 207) [3]

This fleeting electrical field is said to be not the substrate of, or correlated with, qualia but rather what the actual and complete qualia are. However, if subjective qualia are equivalent in some way to physical neurons, then it would seem that the fact that a physical neuron exists would necessitate that there be a subjective quale also. There seems to be no way from the outside to determine and validate this theorized equivalence. It also would suggest that there is no way to stop subjective qualia in functioning neurons. It remarkably seems to imply the challenging conclusion that *all* neural functioning is conscious, quite inconsistent with most cognitive and neuropscyhological theories which hold that even most psychological functioning in the brain is *not* conscious. This seems to be a fundamental challenge to any strong version of the functional identity hypothesis.

The model further states that qualia—in the form of electrical activity in the brain—are the evolutionary precursor of higher-order cognitive processes. They involve diverse collections of neurons firing in resonance, built up from a primitive, qualia-like property of single cells:

> "We have known for a long time that single cells are capable of irritability...hunting for food or fleeing from deleterious or threatening conditions. These observations should remind us that there are in single cells certain abilities related in a primitive way to intentionality, and thus to what may be considered a primitive sensory function. If we are allowed to consider that qualia *represent a specialization of such primitive sensorium,* then it is a reasonable conceptual journey from there to the multicellular phenomenon of "corporate feelings" manifested by higher organisms. If this is something we can live with, then we will understand that *qualia must arise from, fundamentally, properties of single cells...* amplified by the organization of circuits specialized in sensory functions." (p. 212) [3]

Thus qualia, intentionality and indeed conscious subjectivity, and neural activity are equivalent. Qualia do not emerge only as higher-order functions that require neural complexity, but are amplified and elaborated equivalently with neural complexity. According to this model, the processes of sensation, emotion, prediction, intention, volition, and consciousness typically thought to be higher-order functions exist in simple form even in individual neurons.

As to the spontaneous functional coherence of neural processes, Llinas states:

> "...[T]here is a concept that has been lurking in the halls of neuroscience about as long as that discipline has been around. It is the concept of "labeled lines," and it may further help us theoretically remove the ghost in the machine once and for all. The concept of labeled lines states that sensory pathways of all sense modalities encode the specific properties of the world they convey by very specific firings patterns, and that each line or pathway only carries information of that specific modality. In literal terms, these specific patterns *are* the specific sense modality messages from the outside world." (p. 219) [3]

This suggestion reflects a further attempt to equate certain neural activity with qualia by pointing to the close functional correspondence of physical vibratory frequencies with qualities such as subjective magnitude that characterize sensory input processes. It emphasizes the correspondence of objectively measured physical features of stimuli, such as their physical energy, with verbal reports of features of sensations, such as subjective magnitude. This was referred to in Chapter 1 associated with the seminal research by Fechner, Weber, and Helmholtz at the beginning of scientific psychology—the correspondence of the geometry of the outer objective domain with the geometry of functional space in the subjective domain. It was also discussed in Chapter 14 in connection with the principle of structural coherence, and with the one-to-one correspondence of body and mind in Maharishi Vedic Science and Technology in Chapter 13.

But how the top-down super-ordinate influence of the self and of conscious qualia comes out of this is still unclear. The general answer to the binding problem and the self-referential influence of conscious qualia and the self seems to be that they also are somehow intrinsic at every level of neural activity. Llinas asserts that these qualia are said to arise:

> "...from the very properties of physical mechanisms present in the living organism, and more ancient than the cognitive processing of a complex brain. Cause and effect driven from the other end, at least conceptually (p. 211)." [3]

According to Llinas, the binding of biophysical processes and a conscious intentional self or mind inherently resides in the nature of qualia as neural processes:

> "One cannot operate without qualia; they are properties of mind of monumental importance. Qualia facilitate the operation of the nervous system by providing well-defined frameworks, the simplifying patterns that implement and increase the speed of decision and allow such decisions to re-enter (the system) and become part of the landscape of perception... [Q]ualia become exceedingly important tools in perceptual integration; it is the repository of the binding event." (p. 207) [3]

Llinas asserts that the functioning of a neuron produces qualia that somehow make predictive judgments about sensory information that causally influence internal processes of the organismic self and modulate behavior that connects intentional output with sensory input:

> "Along with the eye, the heart, and so on, FAPs and even language may be considered as organs, local modules of function with very specialized capabilities and duties. We may, and I think we *must* come to understand qualia as a sort of master organ, one that allows for the individual senses to operate or co-mingle in an ensemble fashion. Qualia make simplifying, momentary judgments about this ensemble activity, allowing these judgments to be re-entered into the system for the predictive needs of the organism (self). Qualia represent judgments or assessments at the circuit level of the information carried by sensory pathways, or sensations. And these sensations, the integration product of the activation of internal sensory FAPs, represent the ultimate predictive vectors that recycle/re-enter into the internal landscape of the self. They *are* the "ghost" in the machine and represent the critically important space between input and output, for they are neither, yet are a product of one and the drive for the other. And all the while they are simplified constructs on the part of the intrinsic properties of the neuronal circuits of our brains." (pp. 221-222) [3]

But as to the bottom-line understanding of qualia as self-referential neural function, Llinas acknowledges that how neural processes actually refer back to themselves—the self-referential binding—to influence themselves causally and direct the organismic self is still a mysterious trick of nature: "Qualia are that part of self that relates (back) to us! *It is a fantastic trick!*" (p. 207) [3]

This model emphasizes the critical role of qualia, associated with conscious experience, as a functional space right in the center of neural functioning—the critical essence or core space between input and output. But in the model as a physical model of the brain there is no such space: qualia are nothing other than neural impulses. This abstract functional space does not exist in the physical world or in the chain of physical causation. But also at the same time it is attributed mysteriously the power to do all the crucial work of integration, decision making, sensation, and prediction that guides the organism and that is conscious experience. Here again is the black box of the mind and of consciousness, now existing even in primitive organisms with the simplest of neuronal activity. The homunculus remains after elaborate explanations attempting to eliminate it once and for all. The ghost is still in the machine. The ghost has been reduced from neural complexity to single neurons, but it is still clearly present.

This model is useful in the sense that it can be said to be in the direction of placing the intelligent binding and control of the biological machine into simpler, more basic organismic processes—arguing against the neural complexity version of emergence theory. But there is reluctance to take it any further in a reductive direction. This is expected, because further into more basic levels means toward smaller time and distance scales. Underlying simple neurons are physical rather than biophysical processes, in which modern science has long thought to have demonstrated that negentropic processes such as the survival instinct, a purposive organismic self, intentionality, consciousness, even life don't exist.

> "One restricts one's questions to the domain where materialism is unchallenged."—*Richard Lewontin, evolutionary biologist and geneticist* (p. 28) [8]

The framework discussed so far in this book may be said to be *consistent* with the view that the mysterious transformation of electrochemical events into sensations is an 'empty set' and 'does not exist'—when considering *only* the gross physical level and its apparent closed causal nexus. In the gross relative material domain of nature—as distinct from the subtle levels with respect to mind—there is no space or specific functional role for mind and consciousness in neural activity of the brain. This is also consistent with the inability to explain the emergence of super-ordinate top-down functional control by the self within the physical. Mind and consciousness are neither a phenomenological functional space nor an epiphenomenal space *in the physical*. This model adds to the clarity of recognizing that there is no 'space' between neural activity and qualia, no 'ghost' *in the neural machine*. But neither is there causal efficacy or conscious intention in the neurons, even though it inexplicably places it there as a trick of nature. It resides in the ontologically subtler levels of existence that underlie all gross physical phenomena perhaps beyond the level of the Planck scale that causally, intentionally influences brain function.

Again, if the reductionism is taken further into the underpinnings of a single neuron, we would get down to orderly, lawful processes in nature that do provide binding together of inert physical matter. These processes are the fundamental forces—gravity, strong and weak nuclear forces, and attraction in the electromagnetic force. They function in an orderly way according to natural laws, and they also can be said to be inherently dynamic with respect to the inherent dynamism of the vacuum state of quantum fields. At this point, however, an important distinction needs to be made between coherent binding into a goal-directed living organism and consistent patterns of lawful relationships generally in the physical world.

At the sub-biological level, physical matter does bind together—such as mineral compounds, crystalline structures, atoms, as well as planets and galaxies—through the fundamental physical forces of nature. This physical binding is not due to some volitional aspect or purpose associated with individual elements or atoms or collections of them, and it doesn't include such intentional binding processes. However, the binding into living organisms is not caused by these fundamental forces alone. With respect to living organisms, there is an *individualized* global binding together of inert matter into an intention of an individual self to survive. This individualized intention is not just due to the universal 'intention' of nature for binding associated with the fundamental forces and their intrinsic dynamically lawful behavior. In the gross material domain the cause of this individualized binding into a self with survival intention leading to negentropic behavior cannot be found. The binding into unitary psychological experience is some kind of subtler *mind power*, not just ordinary gravity or electromagnetism which binds together constituents of matter throughout the physical universe but doesn't necessarily result in living organisms with the unitary will to survive.

The synthesis, collection, or build-up of individual lower-order processes into higher-order processes is fundamental to gross physical creation. That is, all global physical processes are collections of individual elements. However, no new ontological level of *reality* is created by these collections of elementary particles—only combinations of small gross parts into larger gross wholes. This is what was meant by the statement in Chapter 11 related to Sankhya that the gross elements in the material domain of nature—mahabhutas—are classified as evolutes: they are caused, but don't cause other ontological levels or modes of being.

Implications for the principle of natural selection. It may be further pointed out that placing mind, consciousness, and the sense of self into primitive biological levels—even single neurons—has implications for the principle of natural selection. The key components of natural selection are random mutation and selection of biologically adaptive mutations through survival of the fittest in interaction with the environment. But in the neurocognitive model just discussed, even primitive living organisms move spontaneously on their own in goal-directed behavior, not requiring interaction with the environment for their physiological genesis. This would seem to suggest that the ability to act intelligently and purposely for survival actually *did not* evolve from natural selective environmental interactions, but only became more complex via them. Indeed Llinas has asserted that organized movement is not a product of interaction with the environment.[3]

Also the model seems to suggest that the crucial point in evolution when subjectivity begins is when living organisms develop specialized segmented bodily structures that can communicate and coordinate with each other—through which neurons evolved. However, the rudiments of subjectivity seem to go even deeper in the model—in the sense of biologically more fundamental—if a sense of self is necessary for survival behavior. As noted, the survival instinct, thought to be exhibited in escape-like movement, can be observed even in single cell animals with no spinal structures, segmental bodies, or specialized neural cells. Subjective processes of mind, consciousness, and the sense of self are being placed deeper into the structure of neurobiology. The sense of individual self is held to be necessary to bind or centralize the coordination of actions for integrated survival behavior. Llinas states that there necessarily is "...only one seat of predictions: survival is not random." (p. 38)[3]

Is it that being alive and being able to sense and interact with the environment is equivalent to having subjectivity and a sense of self, and equivalent even in an individual neuron? Is the value system of survival—and globally coherent, unitary, purposeful, intelligent, intentional behavior—somehow inherent in any living organism, simple or complex?

If subjective processes such as the sense of self that bring coherence to movement are inherent in simple living organisms, and somehow develop even without interaction with the external environment, then they would seem to exist without having evolved through the process of natural selection. Apparently spontaneous random mutations occurred that unified physical matter into a functionally coherent self with a value system of survival and at least rudimentary ability to predict the results of its unitary behavior—even before biological mechanisms associated with survival through interaction with the environment came into play.

The question then arises as to how this impacts the distinction between living, biophysical processes and non-living physical processes. Does classifying some physical object as alive inherently mean attributing subjectivity to it also? Again what makes inert parts of matter—that do not cohere on their own—unify into a living organism with the *top-down global intention to survive?*

The model tries to embed the unitary wholeness of the sense of self, as well as survival intention, into each part of the organism—even each neuron. This is said to involve mysterious self-referential neural dynamics in which the whole comes from the parts and then feeds back or bootstraps somehow into a controlling influence on them. The model tries to describe both the parts and the whole with super-ordinate power over the parts as simultaneously emerging at each stage of evolutionary development—primitive as well as complex.

This model can be thought of as a bit closer to ancient Vedic science in the sense that the mind and consciousness are intrinsic to even simple biophysical systems. In Vedic science, however, they become expressed in the gross material domain only when activated by association with the subtle mental levels of nature, whereas this model describes them as active in biophysical organisms generally.[3] The mechanics of how the parts of the organism could possibly have super-ordinate control of it when they are functionally identical to the biophysical and underlying physical processes seems impossible to explain. The coherent unitary control of biological organisms comes from an ontological level of nature that is subtler than the gross material domain; there is no place for it in the physical.

Gross Body and Subtle Mind

We have gone into some detail with these models in order to exemplify the level of detail they reflect and also to illustrate the fundamental inadequacies of these attempts to embody holistic concepts from reductive strategies within the physicalist worldview. In general the game in modern science can be seen as starting from the parts—the ordinary macroscopic world—and trying logically to put them back together into wholeness—ultimately the cosmic egg and the unified field as the source of everything. But when the parts are the bottom line, explaining how they generate super-ordinate control of the parts is doomed to be inadequate.

In the perspective of relative gross and subtle levels of nature, mind and consciousness are latent in even gross physical matter. They begin to be expressed in the gross physical world at a certain level of complexity of physical structure. It is in this developmental sense, but not in the ontological sense, that emergence theory has merit. The requisite complexity is generated by subtle impulses of energy and intelligence binding together the physical parts. The super-ordinate binding influence comes from the more fundamental, subtle, nonlocal, underlying mental reality that guides physical objects to cohere into a living, purposeful, organismic self.

This relates to a fundamental design feature in nature. In the gross relative domain, wholes are synthesized from more fundamental parts. Parts in the form of quantized matter (paramanus and mahabhutas) comprise all gross objects in nature. The unifying or binding of the parts into phenomenal wholes as living organisms comes from the deeper subtle impulses of energy and intelligence in the subtle relative domain of nature. Unless the parts are connected to them, the parts don't cohere on their own into negentropic living organisms. As soon as the connection to the subtle is lost, the binding influence goes away, and the collection of parts fall apart, back into their basically inert, quantized, particulate, non-biological, physical constituents. More complex bodily forms allow more of the subtle energy and intelligence to be expressed in the gross relative domain. But evolution to more complex organismic behavior is a process of expressing higher intelligence, rather than creating it.

Again, in Maharishi Vedic Science and Technology phenomenal creation is described in three levels: 1) the gross relative domain in which localized, quantized parts combine or synthesize into wholes; 2) the subtle relative domain in which individual waves in non-localized fields exhibit high interconnectedness; and 3) the unified field, which is infinitely self-interacting, self-referral wholeness. The subtle and gross levels can be said to be the sequential expression of the three-in-one self-interacting dynamics of the unified field. This also can be associated with the fundamental duality of simultaneity and sequence. In the unified field there is simultaneity—wholeness in every part; and most fundamentally, there are no parts. Parts are only phenomenal expressions that appear to divide up the wholeness into parts. In relative phenomenal creation where there are parts and levels of nature, there is a sequence of levels, which provides the basis for reconciling structural and functional models of the mind and brain.

Another concrete example may help tie together many of the issues in this discussion of biologically-based neurocognitive models of the mind—comparing a group of rocks and a group of rock stars. A group of rocks, in themselves and on their own, don't work together to create or generate a functional unit. They may pile up into a mountain, but the mountain is not a functional unit operating for itself. They may be put together to form a group in the sense of an artistic configuration—from which can emerge gestalt properties of wholeness. They also may be put together to form a functional unit from the perspective of an intelligent biological organism, such as a protective wall or fireplace. These two examples are functional formations imposed on the rocks from some outside intelligent agent.

On the other hand a group of rock stars may decide together, on their own, to unite together into a musical group. Theoretically they can create a functional unit that generates an influence of harmony and makes a positive aesthetic contribution to the social milieu. Both the rocks and the rock stars are objects in the physical world, underlain by lower-order processes including atoms, sub-atomic particles, quantum fields, and perhaps strings, branes, and qubits thought to comprise a closed causal chain. Nowhere in nature have we found a complete chain of evidence of such physical processes on their own combining to create negentropic, self-acting, purposeful behaviors—except in living organisms. As we try to identify precisely the combinations of materials and events that generate a living organism, inevitably we examine smaller and smaller time and distance scales. We have yet to discover the binding source of coherence—the intelligence—in these lower-order natural physical processes and events from which life emerges in the physical world. Indeed the prevailing modern scientific view has been that the ultimate source of physical processes is not coherent intelligence but rather random disorder.

Living organisms beget living organisms; DNA begets DNA. We can break it down more and more, but always the coherence or intelligence that brings the factors and conditions together to reproduce life comes from other living organisms. When this intelligent binding influence is lost, the bits of matter resort to their basic qualities and actions that are entropic and non-active on their own. Rock stars' brains and bodies become rocks again—ashes to ashes and dust to dust. *Something* holds them together in living organisms; and when that something leaves, they dissipate. The processes that underpin selectivity to produce living functional beings do not come from the physical structures alone. The intelligent binding influence that naturally coheres neural events into a purposive organismic self doesn't come from the material particles alone.

Perhaps this is the place in the overall discussion to consider explicitly the distinction between living and non-living. This especially may be the case in light of the above point that the energy and intelligence that activates and binds together gross material particles into organismic selves comes from the subtler level of nature and does not inhere in the gross material objects *on their own*—so to speak. The distinction between living and non-living can be seen as a perspective about nature that is largely an artifact of the ordinary waking state of consciousness and the common view that the gross material domain is the only domain of existence.

> Life on Earth could not have emerged from an organic soup, or an undersea thermal event—at least on Earth. The necessary ingredients for the manufacture of life did not exist on the young planet. Nor was there sufficient free oxygen, and there may have been no free oxygen at all, which is an essential ingredient that makes up the structure of DNA. In fact, almost all the essential ingredients for the construction and manufacture of DNA, were nowhere to be found—at least on Earth—thus refuting any and all notions that Earthly life originated from non-life... In order to account for the missing necessary ingredients for the creation of life, many mainstream evolutionary scientists and astrobiologists assume that complex organic molecules, but not living things—fell to Earth, encased, perhaps, in the debris that bombarded this planet for the first 700 million years after it was formed... However, the first evidence of life on earth appeared immediately following the cessation of that cosmic bombardment, 3.8 billion years ago... These living creatures were already quite complex and fully formed—they were not the product of an organic soup—at least on Earth... There simply was not enough time for life to form from an organic soup, given that complex life appeared immediately following the heating and sterilization of the Earth."—*Rhawn Joseph, neuroscientist (p. 28)* [9]

From the holistic perspective of nature, the unified field as the source of everything can be said to be the field of Life itself, as implied in Chapter 10. All the signifiers of life—such as those mentioned in the Prologue including intelligence, intentionality, selective attention, and the survival instinct—ultimately are located in the infinitely self-interacting dynamics of the unified field. These qualities manifest to various degrees of limitation into individual subjective selves or into objective things made of waves and particles that do not exhibit subjective functions—as described in Chapter 11. In the process of evolution the individual subjective selves may take on gross bodily forms made of inert matter particles. It is in this gross bodily form that we have been accustomed to attributing life to them, and attributing the status of non-life to the matter particles not bound together into gross organismic selves as biological organisms.

The common concept of death is associated with individual organismic selves detaching from the gross bodily form made of matter particles, such that the bodily form no longer expresses the signifiers of life and then dissipates according to the 2nd law of thermodynamics. From the holistic perspective it is a matter of either the expression of life at the subtle and gross levels, or just at the subtle level. All objects have deep within them the qualities of energy and intelligence that comprise life, but may not express them. Everything can be said to be *alive*—as well as to be *conscious* in the ultimate sense of consciousness itself—but express these qualities at various degrees even down to the point of apparent non-living, non-conscious, inert matter such as rocks.

The holistic perspective is quite a departure from the common physicalist worldview and its accompanying notion of life and death. Versions of the holistic perspective have appeared throughout ancient and modern intellectual history, and again are becoming more prominent—exemplified in models that imply that the universe as a whole functions as if it were a biological system, mentioned in Chapter 7. Again, in the holistic perspective, eternity is the basis of time, infinity is the basis of space, and immortality is the basis of mortality—the objective and subjective basis of individual life is universal Life.

In the next chapter we refer back to the issue of reconciling functional and structural models of depth of processing that we had left off in Chapter 15. We then will discuss their relationship to the model of human psychoarchitecture in Maharishi Vedic Science and Technology described in the inner dimension in Chapter 11, and also will describe each of the levels of subjectivity in more detail.

Chapter 16 Notes

[1] Damasio, A. (1999). *The feeling of what happens: Body and emotion in the making of consciousness.* New York: Harcourt, Inc.

[2] Edelman, G. M. & Tononi, G. (2000). *A universe of consciousness: How matter becomes imagination.* New York: Basic Books.

[3] Llinas, R. R. (2001). *I of the vortex: From neurons to self.* Cambridge, MA: MIT Press.

[4] Hodgson, D. (2000). The Easy Problems Ain't So Easy. In Shear, J. (Ed.). *Explaining consciousness—The hard problem.* Cambridge, MA: The MIT Press, pp. 125-131.

[5] Chalmers, D. J. (1996). *The conscious mind: In search of a fundamental theory.* New York: Oxford University Press.

[6] Maharishi Mahesh Yogi (1972). *Science of Creative Intelligence: Knowledge and Experience.* [Videotaped course]. Los Angeles: MIU Press.

[7] Silberstein, M. & McGeever, J. (1999). The Search for Ontological Emergence. *The Philosophical Quarterly, Vol. 49, No. 195, 183-200.*

[8] Schwartz, J. M. & Begley, S, (2002). *The mind & the brain: Neuroplasticity and the power of mental force.* New York: HarperCollins Publishers, Inc.

[9] Joseph, R. (2002). The Myth of the Big Bang: Cosmic Organic Clouds & Creation Science. In Joseph, R. (Ed.). *NeuroTheology: Brain, science, religious experience.* San Jose, CA: University Press.

Chapter 17

Human Psychoarchitecture and Levels of Mind

In this chapter, functional and structural models of the mind and consciousness are reconciled using the approach of Maharishi Vedic Science and Technology introduced in connection with the inner dimension in Chapter 11. Functional levels of mind are related to the structure of gross, subtle, and transcendental levels that encompasses the full range of phenomenal nature. The dynamics of the phenomenal expression of impulses of intelligence and energy through the levels of mind into behavior in the individual self are also discussed. The overall main point of this chapter is that psychological and neurocognitive models of mind are converging on the inner dimension in ancient Vedic science.

In modern science, investigation into nature begins with ordinary experiences of outer objective phenomena and reasoning about the experiences—basically opening the eyes and looking, then thinking about what is seen, then looking again and reconciling the thoughts with the experiences. Based on the relatively high inter-subjective consistency among the majority of scientific observers, this has come to be thought of as a *naturalistic* approach, emphasizing physical matter *naturally* presented to sensory experience—that is, naturally in the ordinary waking state of consciousness using the gross senses, generally within the physicalist worldview. This approach might well be expected to result in object-based views of *reality*, as well as stimulus-based models of psychological processes discussed in the past two chapters. It also is associated with a bottom-up model of psychological functions in which mind and consciousness emerge in some mysterious way from lower-order physical processes.

But in reflecting even a little bit on the subject of the knower—and on our own inner experience—we can appreciate that the knower or experiencer at least seems to have top-down control of the machinery of experience. This top-down control is not accounted for given the fundamentally random activity of the known particles and forces in the theorized causally closed objective physical universe. In the past few chapters we have couched this overall issue in terms of trying to identify what binds inorganic matter into conscious living organisms with an instinct for survival, and what binds neural activity into an organismic self that emits coherent, unitary, goal-directed behavior—associated with the generation, combination, and binding problems.

The solution to these problems is in the holistic understanding of the relationship between gross, subtle, and transcendental domains of nature. The relationship between these domains also is directly related to the mind-body problem and models of levels of depth in the mind—the inner dimension. We have been exploring these issues with respect to structural and functional levels of depth. We left off this topic in Chapter 15 at the point of identifying two meanings to the dimension of depth. One meaning of depth associates deeper with higher-order psychological processes, higher reason, intelligence, the sense of self, and conscious experience. This meaning is reflected in cognitive theories emphasizing abstract disembodied notions of function and top-down control of psychological processes by a conscious self. The other meaning of depth is an embodied structural notion associated with lower-order unconscious and non-conscious levels and underlying biophysical and physical processes. This meaning of depth is more prominent in neurobiological and neurocognitive models of mind in the brain, generally consistent with a

Freudian notion of levels of depth in the broader Darwinian and neo-Darwinian physicalist view. When this structural notion of depth is taken deep enough, however, the theorized abstract level of information space or quantum, nonlocal mind is encountered at about the smallest time and distance scale identified in modern science. This again is the historical mind-body problem—now to be addressed on the basis of a more inclusive range of levels of *reality*. In this structural context the problem relates to the divergent views of mind as a macroscopic phenomenon of ordinary sensory experience embodied in the gross neurophysiology of the brain, and of mind as a nonlocal information space or field underlying physical *reality* and its fundamental force and matter fields.

Hopefully the inadequacy of reconciling the two notions of depth within the physicalist view is becoming evident. In the physicalist view the experiential, functional notion of depth is a phenomenological convention only. Depth *really* means toward lower-order biological processes and eventually sub-phenomenal time and distance scales underlying ordinary macroscopic sensory experience. In this view conscious experiences—including thoughts, feelings, and memories—have no *reality* other than being the epiphenomenal *inside* of neurophysiological processes that more fundamentally are physical processes apparently with no such inside. At some point of complexity of neural systems, conscious experience is said to emerge from and then control the physical. But also some theories hold it to be identical to neural activity in some difficult to imagine way—the functional identity hypothesis. In analyzing what biophysical and physical processes comprise consciousness and mind, nothing accounts for them in the world of physical matter. The phenomenological experience of consciousness appears to exist only at the macroscopic scale of neural processes, in part because it appears to go away when neural processes lose their functional integrity, or in states such as sleep and coma. No logically consistent causal effect of mind and consciousness can be found in this view of a causally closed physical universe.

Historically the strongest challenge to this view was that, experientially and intuitively for many, it just *doesn't ring true*. Many of us are convinced that our mind and consciousness do exist, and that we do make choices that have causal effects in the world. To deny this is to deny what is *naturally* evident in our experience, as well as to eliminate about everything meaningful and significant in our life.

But importantly, now there is another fundamental challenge to this view. It is that modern scientific understanding has progressed beyond the physical in pursuit of a coherent understanding of the foundations of even objective physical *reality*. Our rational theoretical and experimental evidence has led to the conclusion that physical matter is not the most fundamental *reality*. The implication of this conclusion is another angle of the important plank on the bridge to a unified understanding and experience of nature—it is that, relatively speaking, phenomenal experiences including thoughts, feelings, and memories *really* exist somewhere, but not just in the material world and physical brain.

This requires reevaluation of the functional levels of mind described in the past chapters, logically placing them in an underlying mental *reality* more fundamental than the macroscopic, microscopic, and ultramicroscopic levels of physical material *reality*. The neurology of psychological experience then becomes the study of how these subtler mental levels correspond to and are expressed in the gross physical brain and our ordinary macroscopic experience—not the study of mind and consciousness as an ontologically emergent property of the brain. This revolutionary change in view has immense—and wholly positive—implications for human life and its relationship to cosmic life, which will be developed much further toward the end of the book.

Placing mind only in matter leads to an *'entangled hierarchy'* or *'world knot'* that is quite difficult to untie. Mixing functional levels of depth with gross macroscopic and microscopic structural levels ends up with attempts to explain conscious mind as a property of the brain where there is no actual place or causal role for it. This results in complicated explanations of emergence, and hypotheses such as that parts combine into a greater epiphenomenal wholeness that transcends the parts and then inexplicably has causal control over them—in an effort to keep closed what is not a closed causal nexus of the physical universe.

It might be pointed out that the more expanded perspective of gross and subtle levels of *reality* is *consistent* with the conclusion that mind doesn't exist as a function separate from structural processes. But this doesn't mean that mind is epiphenomenal or a non-existent delusional misperception. Models of mind need to be disentangled from gross physical structures, then placed into an expanded structural model of depth that incorporates gross *and* subtle levels. An even deeper perspective of nature—the transcendental level of the unified field of consciousness—is then needed in order to unify matter, mind, and consciousness. We need to understand how to isolate, disembed, and differentiate mind and consciousness from matter, then reintegrate them into the ultimate unity. This will be discussed in terms of higher states of consciousness in Chapter 20.

It also might be pointed out here that throughout human history there have been anecdotal reports of a wide variety of empirical experiences that significantly challenge the physicalist paradigm. Based on pretheoretical beliefs in modern scientific thought, these reports long have been disregarded as irrelevant to scientific progress. They include, for example, such historically controversial topics as *out-of-body*, *near-death*, and *paranormal* experiences. Empirical research on these topics has been progressing in recent years. The high degree of inter-subjective agreement in many of the descriptive reports, and objective validation of specific content in the reports, at a minimum now support these topics as viable and relevant scientific investigations.[1] The growing body of rigorous evidence in these areas of investigation is not explicable within the physicalist view that mind and consciousness are products of localized brain functioning only.[2]

"Through the years, the investigations of experimental parapsychology have become increasingly sophisticated, criticisms of earlier work have been effectively met, and present studies match or exceed those of conventional behavioral and biomedical research in the tightness of their designs and in their safeguard against artifacts and confounding variables. The experimental designs effectively rule out the possibility of ordinary sensory cues, rational inference, and chance coincidence, so that if consistent relationships are found between subjective experiences and distant and shielded target events, such evidence cannot be explained in terms of conventional informational or energetic transfers or mediation, nor can these correspondences…be accounted for in terms of the brain activities or conditions of the research participants… A reasonable and balanced approach to judging claims about these processes would be to examine carefully and dispassionately the published primary reports, read the critics' arguments, read the counterarguments to these, recall your own—and others'—live experiences of similar incidents, and then draw conclusions based upon the fullest possible amounts of evidence and argument."—*William Braud, research psychologist (pp. 127-128)* [3]

However, now a more profound systematic means of investigation is available for empirical validation of mental *reality*, even within our *easy grasp*: systematic means to experience directly the subtle and transcendental levels of nature in higher states of consciousness. Applying the subjective means of gaining knowledge in Maharishi Vedic Science and Technology, the boundaries of the ordinary waking state that frame the physicalist view of nature are transcended. From that, a much more comprehensive understanding and experience of nature *naturally* unfolds—the focus of discussion after a few more planks of understanding on the bridge to unity are in place.

Phenomenal experience of macroscopic objects is the common experience in the ordinary waking state of consciousness, and it is the basis for the assumption in physicalism that the mind is in the gross physical brain. How can this be reconciled with the concept of nonlocal mind or mental *reality* as underlying physical *reality*? This reconciliation requires a more inclusive understanding of levels of depth of processing, ontological levels of *reality*, and levels of experience in a model of the mind that integrates function and structure—the focus of this chapter.

Integrating Functional and Structural Models of the Mind

In order to reconcile the functional and structural models of levels of depth in the mind and body, it might be helpful first to restate and clarify some of the key concepts. Especially significant are the contrasts of shallow/deep levels of processing, peripheral/central processing, lower-order/higher-order functions, bottom-up/top down processing, and the related concepts of downward/upward causation.

Functionally, shallow, peripheral, bottom-up processing refers to reflexive and automatic processing not involving serial central processing, frequently associated with sensory stimulus input but also motor output stages that result in observable behavior. Again *functionally*, deep or top-down processing refers to such phenomena as conscious attention, conscious sense of self, intentionality, and free will. Top-down processing is associated with the concept of *downward* causation or super-ordinate control by a central executive or unitary self that makes decisions and controls unitary behavioral output.

> "In order for there to be downward causation, it must be possible for a whole to determine the behavior of its parts, rather than the other way around." —*Teed Rockwell, philosopher* [4]

Structurally, on the other hand, the shallow/deep dimension generally relates to degrees of complexity of biophysical processes from more complex higher-order neural interactions to their deeper lower-order biophysical and physical bases. Again *structurally*, top-down refers to these higher-order neural interactions influencing the underlying lower-order biophysical processes, associated with downward causation. *Upward* causation relates to lower-order physical and biophysical processes influencing higher-order neural interactions that comprise a conscious self.

An important issue is how to depict these concepts in a two-dimensional graph. Concepts of top and bottom, higher and lower, and levels of depth could easily become confusing in a vertical depiction. The convention used in this chapter to reconcile functional and structural models of depth is a structural model depicting top-down, higher-order, functional levels of processing on the *bottom*, to emphasize that these are the deeper levels of the structure of nature. As summarized in the past chapters, the huge amount of research in the psychological and cognitive sciences and in neuroscience has been converging on a general *functional* model of the mind. In broad outline, the model can be depicted as follows:

Sensory Input (S) ----　　　　---- **Behavioral Output (R)**

Brief sensory specific information stores

Long-term information stores (*including instinctual and reflex action patterns*)

Peripheral fringe of conscious attention (*including learned automatic action patterns*)

Conscious Short-term Memory (*activated long-term memory*)

Focal conscious attention (*center stage in the theatre analogy*)

Conscious inner thoughts and feelings including high reason, conscience, and so on
(*sometimes experienced as a deep inner fringe of attention*)

Conscious sense of self

The following depiction shows the correspondence of this general functional model with the levels of mind, the inner dimension, in Maharishi Vedic Science and Technology described in the Sankhya system in Chapter 11.

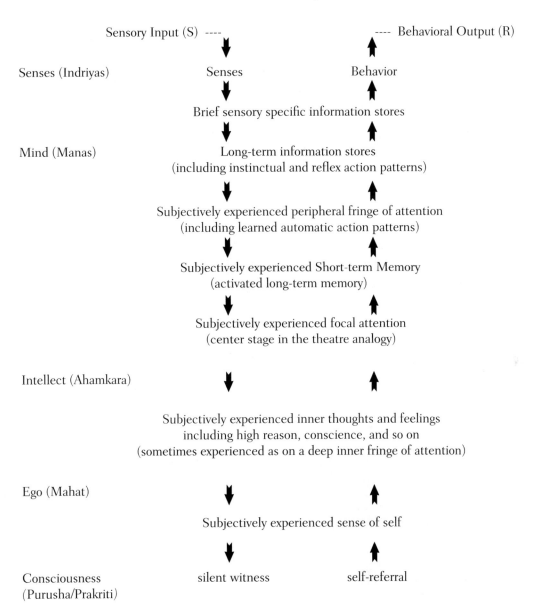

Gross Sensory World

Sensory Input (S) ----   ---- Behavioral Output (R)

Senses (Indriyas)   Senses   Behavior

Brief sensory specific information stores

Mind (Manas)   Long-term information stores
(including instinctual and reflex action patterns)

Subjectively experienced peripheral fringe of attention
(including learned automatic action patterns)

Subjectively experienced Short-term Memory
(activated long-term memory)

Subjectively experienced focal attention
(center stage in the theatre analogy)

Intellect (Ahamkara)

Subjectively experienced inner thoughts and feelings
including high reason, conscience, and so on
(sometimes experienced as on a deep inner fringe of attention)

Ego (Mahat)

Subjectively experienced sense of self

Consciousness   silent witness   self-referral
(Purusha/Prakriti)

When we try to fit this functional model into the physical structure of levels and time and distance scales in macroscopic physical *reality*, however, the best that can be done is to insert conscious mind into a non-existent or epiphenomenal functional space in the brain (the upper levels in bold italics). This reflects contemporary models in neurocognitive science. Notwithstanding that there is no logically consistent place for a causally efficacious conscious mind in a theorized closed causal chain of physical *reality*, it might be depicted like this:

Ultramacroscopic Levels $\quad\quad\sim$10 ? to Infinity?

Macroscopic Levels of Sensory World $\quad\quad\sim$10 –3 to $\sim$10 ?

Sensory Input (S) $\quad\quad$ Behavioral Output (R)

Brief sensory specific information stores

Long-term information stores

(including instinctual and reflex action patterns)

Peripheral fringe of conscious attention

(including learned automatic action patterns)

Conscious sense of self

Conscious inner thoughts and feelings

including high reason, conscience, and so on

Conscious Short-term Memory

(activated long-term memory)

Focal conscious attention

(center stage in the theatre analogy)

Deeper preconscious processes

(unconscious but sometimes available to consciousness

as a deep inner fringe of experience)

Unconscious psychological processes (psychobiological drives)

Electrochemical cellular processes

Microscopic Levels $\quad\quad\sim$10 – 4 to $\sim$10 –8

Molecular processes

Ultramicroscopic Levels

Atomic processes $\quad\quad\sim$10 –9 to

Sub-atomic processes

Quantum field processes

Quantum gravity processes

Quantum information processes $\quad\quad\sim$10–33 (Planck scale)

Quantum mind, nonlocal information space

Hilbert Space

Infinity?

Unified field

In this functional/structural model of the mind as embodied in gross physiology, there is an obvious inconsistency between the levels of subjective experience in what might be called *epiphenomenal functional space* (upper area in bold italics) in the macroscopic physical brain and the nonlocal level of *quantum mind or information space* (lower area in bold italics) posited in models on the cutting edge of quantum physics. This inconsistency is irreconcilable in a causally closed physicalist view in which the mind is only in the material body.

In Maharishi Vedic Science and Technology, the mind is in the subtle non-local domain of nature that underlies the gross material domain, which is further underlain by the unified field of consciousness—the consciousness-mind-matter ontology or consciousness-based understanding of nature. The brain localizes the expression of mind and consciousness in the gross material domain by transducing subtle mental impulses into gross physical behavior. In this holistic approach levels of mind can be placed into levels of nature as generally depicted in the following *structural/ functional* psychoarchitecture of the inner dimension:

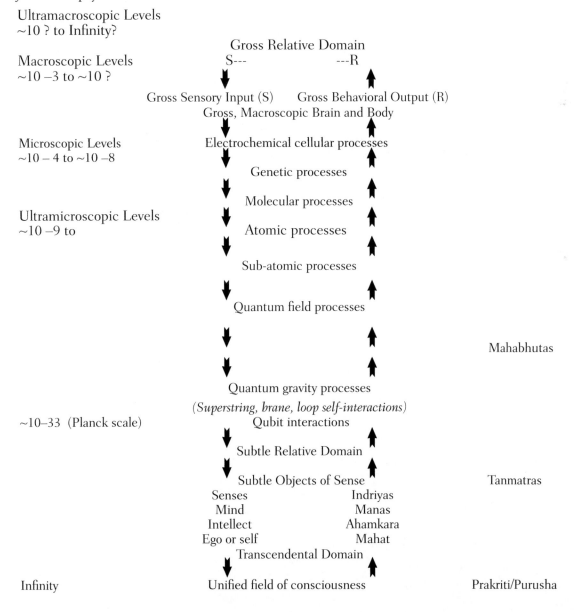

Ultramacroscopic Levels
~10 ? to Infinity?

Gross Relative Domain

Macroscopic Levels
~10 –3 to ~10 ?

S---   ---R

Gross Sensory Input (S)   Gross Behavioral Output (R)
Gross, Macroscopic Brain and Body

Microscopic Levels
~10 – 4 to ~10 –8

Electrochemical cellular processes

Genetic processes

Molecular processes

Ultramicroscopic Levels
~10 –9 to

Atomic processes

Sub-atomic processes

Quantum field processes

Mahabhutas

Quantum gravity processes
(*Superstring, brane, loop self-interactions*)

~10–33 (Planck scale)

Qubit interactions

Subtle Relative Domain

Subtle Objects of Sense   Tanmatras

Senses   Indriyas
Mind   Manas
Intellect   Ahamkara
Ego or self   Mahat
Transcendental Domain

Infinity

Unified field of consciousness   Prakriti/Purusha

To be clear about this overall structure, the material domain includes everything in the *gross* relative levels of nature, where energy and intelligence are quantized into discrete units (mahabhutas and paramanus). This includes the ultramicroscopic, microscopic, macroscopic, terrestrial, extra-terrestrial, and cosmological objects in the objective material universe. It covers all the various time and distance scales from the Plank scale to the galactic expanse of the observable cosmos—everything in modern science that comprises the theorized causally closed objective physical universe.

The entire gross relative domain of nature, existence, or *reality*—including the familiar objective world—contains no processes from which emerge new causally efficacious levels of existence. It is composed of the gross elements of nature (mahabhutas) interacting with each other, but no new ontological level of existence such as a causally efficacious conscious mind is created, generated, or emerges from these interactions. Everything in the objective material domain—including material bodies, brains, and neurons—is made of the gross elements in a vast multiplicity of combinations and permutations. Although ultimately nothing other than the unified field, the gross elements don't have subjective qualities active in them. The gross elements neither exhibit intelligent negentropic behaviors 'on their own,' nor do these subjective functions emerge from them.

The expression of intelligent, negentropic, intentional actions in living organisms with gross physiologies (made of the mahabhutas) is due to the subtle guiding impulses of energy and intelligence coming from the subtle relative level of nature, which includes the levels of mind. These subtle impulses—associated with thoughts and feelings that are transduced into neural processes in the material brain to be expressed in gross behavior—are so subtle as to be undetectable within the measurement capabilities of objective methodology. Their empirical validation is through logical reason and indirect third-person experimental methodologies, and more importantly through first-person direct empirical experience of deeper levels of mind via experiential methodologies. More will be said about both third-person and first-person validation in later chapters, especially related to the ancient Vedic science and technology of *Yoga*.

The underlying and permeating subtle relative domain is similar to the gross relative domain in that it has many layers. But it is *much vaster*, including permeating the gross domain. Objects in this subtle relative domain are made of more abstract non-quantized waves or impulses of energy and intelligence—the subtle essence elements (tanmatras). Like activity on the gross level of creation, the subtle essence elements also interact according to laws of nature that define their qualities and the mechanics of their interactions to maintain the dynamically orderly, quasi-closed, causally deterministic system of the natural world. But they are much *finer-grained* than anything in the gross objective world—more accurately, they are not grainy at all in the sense of being quantized into the Planck size. They are subtler nonlocal wave-field processes not made of particles of matter, and not limited by Einstein locality, light-speed, and quantum gravity.

Both gross (mahabhuta) and subtle (tanmatra) objects are sensed through the five senses (gyanindriyas), which are even subtler than the gross and subtle elements and the objects they compose. Functioning through the gross physiology, the senses interact with the gross objects; and through the subtle physiology, with subtle or gross objects. The senses bring gross and subtle sensory information to the subtler, permeating screen of the mind (manas), within which are the even subtler, permeating levels of intellect (ahamkara) and ego (mahat). The results of these mental processes are presented to and witnessed by consciousness itself (Purusha).

"The surgery had gone smoothly until the late stages... Sarah's heart stopped beating... But the emergency was all over in a minute, for it took no more time than that for the anesthesiologist to defibrillate her... Yet Sarah had something else to show for her surgery experience besides the ache in her side...and the concentric, reddish rings on her chest left by the sting of the defibrillator's paddles...a clear, detailed memory of the frantic conversation of the surgeons and nurses during her cardiac arrest; the OR lay-out; the scribbles on the surgery schedule board in the hall outside; the color of the sheets covering the operating table; the hairstyle of the head scrub nurse; the names of the surgeons in the doctor's lounge down the corridor who were waiting for her case to be concluded; and even the trivial fact that her anesthesiologist that day was wearing unmatched socks. All this she knew even though she had been fully anesthetized and unconscious during the surgery and the cardiac arrest... But what made Sarah's vision even more momentous was the fact that, since birth, she had been blind."—Larry Dossey, physician and science writer (p. 18) [5]

As described in prior chapters, in ancient Vedic science the inner levels of subjectivity—senses, mind, intellect, ego, consciousness—are both universal cosmic levels of nature as well as individual aspects that comprise individual selves. They can be said to be the super-ordinate, intentional cause for individual choice and action in living organisms, ultimately underlain by, having phenomenal existence within and nothing other than, consciousness itself—the unified field of consciousness.

The subtlest relative aspect of these levels of individual subjectivity—the ego—can be said to be the individual agent or doer that directs all individual feeling and thought of the subtle and gross aspects of the individual organism including its behavior. Individual feelings and thoughts are impulses of energy and intelligence shaped by the subtle levels of mind that guide the subtle and gross elements to express purposive behavior through top-down super-ordinate control.

When the subtle levels of individual subjectivity (tanmatras, indriyas, manas, ahamkara, mahat) no longer connect to the gross physiology made of the gross elements (mahabhutas), the gross elements no longer cohere or bind into an embodied individual organism on the gross level. The gross elements then become guided only by their nature as phenomenally inert physical matter—entropically dissipating according to the 2nd law of thermodynamics. The super-ordinate, controlling, activating individual intelligence or self no longer coheres or binds them together into an individual living organism in the gross relative domain.

However, the cosmic structure of nature is that the unified field of consciousness is ultimately the cosmic or universal Self, which binds all of its phenomenal manifestations together in manifest creation. That transcendental domain, sometimes described as pure Being, the unified field of consciousness, is beyond all appearances of opposites such as living and non-living objects in phenomenal creation. Beyond all causal control, function, and structure, it also at the same time can be said to be the ultimate universal agent or cause of change in nature. Omniscient, omnipresent, and omnipotent, it contains all potential to do anything, all action, and all actualized processes and objects throughout nature. As infinite dynamism, it can be said to be the ultimate super-ordinate agent, while at the same time it is the unmanifest, uninvolved silent witness to all phenomenal activity—as infinite silence.

Thus nature can be said to be fundamentally a *top-down* system with respect to its phenomenal manifestation into hierarchical levels of existence in nature. Manifest levels of creation reflect progressively more limited and specialized capabilities, all the way to the least capability—the most restrictions—at the grossest material level of rocks and earth. All of these phenomenal levels can be said to interact with each other, but all these interactions ultimately are caused by the uncaused, uninvolved universal Self.

Modern science, with its objectifying focus only on the known, has not yet developed an understanding of the underlying subtle and transcendental levels. In this gross physicalist framework, which limits *reality* to the objective physical world only, how living organisms effect top-down causal control in the physical when there is nothing other than the physical is quite appropriately irresolvable.

For additional clarity it may be worth contrasting this functional structure or psychoarchitecture of the inner dimension with dualistic theories of mind and body. In *dualistic interactionism*, for example, brain is a *real* material realm of existence, mind is a *real* non-material realm of existence, and they can causally interact with each other; but there is no explanation of how these two realms interact. In recent years, suggestions have been made about the possibility that their interaction relates to quantum physics.[6] But these suggestions quickly run into the fundamental belief in quantum theory of quantum indeterminism and fundamental randomness. The direction toward understanding the *reality* of mind is certainly constructive, but progress has been slow, and continues to be met with considerable disdain among many 'pragmatic physicalists' who have not developed more integrated and ultimately practical empirical experience and understanding.

"Dualistic interactionism holds that consciousness and other aspects of mind can occur independently of brain. In this view, mental states have the power to shape brain or cerebral states—and going further, the mind cannot in any sense be reduced to the brain. Although mind depends on the brain for its expression, brain is by its very material nature not sufficient to explain mind completely... Scientists and philosophers in this camp reject materialism to the point of actually positing a non-material basis for the mind. Even worse, they seem to have a penchant for speaking about the possibility of life after death, something no self-respecting scientist is supposed to do in public... Even scientists and philosophers who question whether simply mapping neural correlates can truly provide the ultimate answer have doubts about dualistic interactionism: neuroscientists may have worlds to go before they understand *how* brain gives rise to mind, but even in a field not generally marked by certainty they are as sure as can be that it does, somehow, manage the trick."—*Jeffrey Schwartz, research neuropsychiatrist, and Sharon Begley, science writer/editor (pp. 45-46)* [6]

By now it hopefully is becoming clearer that ancient Vedic science is a comprehensive, integrated view of nature. It is holistic in the most profound sense of holism, that everything ultimately is nothing other than the unified field of consciousness, the universal Self. It incorporates body and mind into a seamless structure of phenomenal levels of existence from transcendental to subtle to gross domains, and it describes processes that connect each level in the sequential manifestation of consciousness, mind, and body. Apparent gaps between the levels which delineate them ultimately can be thought of as experiential gaps that are a product of incomplete refinement of mind and body associated with undeveloped states of consciousness. Ancient Vedic science and Maharishi Vedic Science and Technology include systematic technologies to develop the ability to experience directly the full range of levels of nature.

In Chapters 19-20 we will discuss the process of evolution described in Maharishi Vedic Science and Technology, covering the full range from the most limited levels of nature to their highest reflection of the unified field of consciousness in the highest state of consciousness. The phenomenal *process of manifestation* from transcendental to gross can be viewed as part of the *process of evolution* from transcendental to gross and gross to transcendental, through levels of nature and states of consciousness. This process encompasses manifestation of the total dynamic potential of the unified field of consciousness even in its grossest expression in material creation—in the form of the human embodiment of the unified field in the highest state of human consciousness. This is implied in the concepts of *point to infinity* and *infinity to a point*, as well as *universality* and *individuality*, discussed in Chapters 11 and made more tangible in upcoming chapters. Maharishi Vedic Science and Technology as a revival of ancient Vedic science is a remarkably subtle, profound, and all-inclusive scientific view of the totality of nature. It is a rational, logically consistent understanding that provides reliable systematic means for its empirical verification. These means are available and need to be incorporated into contemporary educational systems world-wide.

The Flow of Energy and Intelligence through Levels of Mind

In the *top-down* structure of relative creation, ultimately the *top* is the unified field of consciousness, identified as *Atma*—consciousness itself, the universal Self that is the source of all individual selves (which also ultimately is the bottom, as well as all levels in between). That unmanifest field contains all possible phenomenal manifestations of apparent selves and nonselves. In its essential nature as non-changing infinite silence, it also can be said to be ever-changing infinite dynamism. It is the infinitely silent, infinitely dynamic universal Self, vibrating, fluctuating, pulsating within itself at infinite frequency—the infinite basis of eternal time and infinite space that appears to manifest into finite creation and relative time and space. It can be said that the point value of the universal Self is the individual self expressed at the deepest individual level of subjectivity—phenomenally, individual consciousness and individual ego.

All phenomenal manifestations within the unified field are partial, finite manifestations with finite degrees of its infinite possibilities. Its first phenomenal manifestation reflects the highest finite degree of infinite potential, and then sequentially lesser degrees. To illustrate this, Maharishi sometimes uses the analogy of a seed that contains the whole tree; and further, a banyan tree seed whose center is hollow but in that hollowness is everything about the tree.[7] All in all, this hierarchical sequence of finite levels is nothing other than the infinite totality of the unified field of consciousness itself at every level.

In phenomenal manifest form the unified field of consciousness has a structure of levels of depth—described in terms of the inner dimension in Chapter 11, and depicted in the graph earlier in this chapter. The process of manifestation of relative creation within the unified field can be described as a cyclic pulsation of infinite frequency, vibrating or reverberating at more and more expressed, limited levels. In terms of levels of subjectivity it can be described as an outward and inward flow of energy and intelligence through the relative subtle and gross levels of manifest creation.

The remainder of this chapter sketches the dynamic flow of energy and intelligence through this phenomenal structure of levels in terms of human psychoarchitecture, and the predominant functions at each of the levels. The descriptions will focus on these dynamics with respect to the individual self and individual subjective experience. This section describes in more detail the levels of mind or inner dimension. It is important to keep in mind, however, that these are relative descriptions of structure and function generally within the broad framework of the waking state of consciousness—experienced differently in higher states of consciousness.

The flow of energy and intelligence through the levels of individual subjectivity can be described in terms of *outward* and *inward* flows in self-referral cycles or loops of impulses of energy and intelligence. This description of a referral loop can be related to the description in Chapter 12 of emerging and submerging directions of the three-in-one self-interacting dynamics of nature associated with the eight-fold Prakriti. It is another manifestation of the self-referral dynamism of infinity and point, infinite silence and infinite dynamism. We will first discuss the outward flow, and then the process of referring back to its source in the inward flow.

The outward flow. The outward flow is from consciousness itself to ego, intellect, mind, organs of action, and behavior. As the impulses of energy and intelligence manifest, condense—or rise up, so to speak—through the levels, they take on the limitations that predominantly characterize each level. The subtle flow of impulses through these levels is subjectively experienced as feelings, thoughts, and behavior. In simple terms these processes also can be described as associated with the top-down direction of goals, plans, and actions.

At the deepest individual level is the inner sense of self, the level of *ego*. This can be associated with the sense or feeling of existence, of *isness*, and of *amness*—that *I am*—and in more expressed form, *who I am*. This represents the subtlest individual level of nonlocal vibrations or pulsations in relative manifest creation. Manifesting as a more expressed or specific limitation of this holistic feeling is a further sense of discrimination of the self by the level of *intellect*—which can be associated with subjective experiences of *what I am, what I want,* and *what I want to do.* The finest impulses in the intellect can be related to the *psychological heart*—not the physiological heart—the inner sense of emotion or feeling involving desires, wants, wishes, hopes, dreams, intentions, and aims. These can be represented in the concept of a *goal*, a psychological impulse containing the motivation (energy) and purpose (intelligence) for behavior. A goal can be said to contain in abstract form the beginning as well as ending of each impulse, the entire cycle or self-referral loop—so to speak—in seed form from initial desire to fulfillment.

These subjective processes also reflect the interplay or dynamics of the three fundamental qualities or forces of sattva, rajas, and tamas—but now discussed with respect to the inner subjective domain rather than gross outer objective domain. In the subtle domain the three forces can be related to experiential qualities of love, desire, and indifference (refer to Chapters 11-12).

As a unifying force on the subtle level of mind, sattva can be associated generally with the unifying force of love—and the gravitational force on the gross level. As the activating force, rajas can be associated with desire, and with energy/activity on the gross level. As the restraining force, tamas can be associated with indifference, and with mass on the gross level—but also ultimately silence. These three, always functioning together, shape the experiential reality of the unitary wholeness of the sense of individual ego—the sense of individual I. The unitary point or singularity, the locus of experience, at this level is frequently the sense of individual self. When the individuality of I predominates experience, rather than I in its essential nature as the universal Self or *Atma*, this is called *Pragya aparadha,* the mistake of the intellect, introduced in prior chapters and discussed further soon. It is the narrowed down sense of individual self rather than universal Self.

Closely associated with the sense that *I am* are feelings that *I want*—that *I want* to be or have more of something. Initially this is an abstract sense of wanting to be fulfilled, happy, free, in love, peaceful, wealthy, of service, a creator of beauty, and other desires of the heart. Usually these feelings then become expressed in a more concrete, detailed goal or sense of want related to specific objects of sense. As the impulse becomes even more concrete and discrete, it can be said to rise up farther in the mind to a clear, focal level of *thinking*, where it takes shape in terms of not just what to do but also *how* to accomplish it in action. This is associated with the level of the mind (manas). An impulse of thought at the level of the mind can be said to involve some form of *plan* about how to act to fulfill the goal of the (psychological) heart through action.

When the inner impulse of energy and intelligence is deeper within the subjective self, it is commonly described as an inner *feeling*. As this inner impulse becomes more expressed, it is experienced more concretely, associated with the subjective activity and experience of *thinking*. It is the same impulse rising up through the levels of individual subjectivity, at one point expressing more the qualities of feeling and then expressing more the qualities of thinking. We will discuss the relationship of intellect to both feelings and thoughts shortly. Feelings can be related more to the energy or motivation in the impulse, and thoughts can be related more to the expressed intelligence or specific direction in the impulse. Because feelings are embedded in thoughts as a deeper level of subjectivity, both of these qualities of the impulses of energy and intelligence are accessible to experience at the same time. Even at its inception or seed form, a psychological impulse rising up through the inner levels of subjectivity contains all the qualities that become experienced at the various levels—first more in terms of feelings and then thinking. In like manner a seed contains the whole tree in all its detail, manifesting in increasing specificity and concreteness.

The mind then directs the organs of action to interact with the outer environment. Behavior refers to the body's outer action to carry out the inner goal and plan of the heart and mind. Thus we exist as individual beings, we have a feeling about something we want, we think how to act on it, and then we may take action. In this top-down causal direction, the basis of action is thinking, the basis of thinking is feeling, the basis of feeling is being—and the basis of being is consciousness itself, ultimately pure Being.

The inward flow. The body's actions or behavior produce changes in the outer environment, which loop or refer back to become input into the inner subjective levels by way of the organs of sense—the *inward flow* of psychological impulses. Information about changes in the outer environment is received through the organs of sense—hearing, touch, sight, taste, and smell. The inward flow is from the outer environment in the inward direction through the organs of sense to mind, intellect, ego, and consciousness.

It can be said that the incoming information is compared to the original goal and plan that motivated and directed the behavior. The result of this mental evaluation is a feeling according to the degree of success in fulfilling the plan and goal. If the results match the plan and goal, inner feelings of happiness and satisfaction develop. This also influences how we feel about ourselves on the level of individual ego or self.

As a simple example, a feeling rises up in us of wanting to see a particular movie that was recommended by a friend. We think which dvd store may have it, and then take action to go get it. We find out it is already rented, still want to watch the movie, think of another store to check out, go there and find it has been rented there too. We then decide it is too late to watch a movie, we may be a little disappointed, but then end up having a good time just 'laughing, talking, and joking' with each other.[8] We have a goal, we plan how to achieve it, and then act. The results of the action are compared to the original goal and plan, and we experience some degree of fulfillment accordingly. If the results don't match, we may try a new action, a new plan, or possibly a new goal. The experience of some level of fulfillment motivates another cycle to gain even higher levels of fulfillment, repeating the cycle again and again.

This is a simple model of the cycle of the outward and inward flow of the referral loop through levels of individual subjectivity. This cycle of feelings, thoughts, and behavior—goals, plans, actions—naturally develops bigger goals, and hopefully better plans and more effective actions toward higher levels of fulfillment. Higher levels of fulfillment encompass not only one's own individual experience, but also the fulfillment of the entire social environment, even the entire world family. Usually there are several goals and plans at the same time, all in the direction of striving toward more lasting fulfillment—to be discussed in Chapter 20.

These outward and inward psychological processes function according to the state of consciousness of the individual. The state of consciousness has a top-down influence on all relative degrees of happiness and satisfaction in the ups and downs of ordinary life. In higher states of consciousness there is full recognition that happiness and fulfillment are not due to ordinary actions and reactions in daily life, but rather to direct contact with one's own universal Self.

Each of the levels of individual subjectivity involve a certain function or type of mental activity that contributes to unitary subjective experience. A simple way to look at each of the levels is to consider the issue or question primarily dealt with at each level. The outward flow of levels and simple correspondence to questions of who, what, how, where and when can be added to the simple chart of levels—which integrates a vast amount of psychological research:

	Environment		Change in Environment		
OUTER					
	Behavior				
OBJECTIVE		behaving	Action	Results of Action	Where When
	Body		⬆	⬇	
------Senses------		--sensing--	⬆	⬇	--------
INNER			⬆	⬇	
SUBJECTIVE	Mind	thinking	Plan	Evaluation	How
	Heart	feeling	Goal	Fulfillment	What
			⬆	⬇	
	Ego	being	I have	Sense of self	Who
	Consciousness	(Being)	Silent witness	Self-referral	

This individual cycle or referral loop of the outward and inward flow of energy and intelligence mirrors universal cycles, vibrations, and pulsations throughout all levels of nature. It encompasses the minutest fluctuations of nonlocal fields to the more expressed cycles such as of day and night, rest and activity, manifestations and dissolutions—all embedded in the self-interacting dynamics of infinity to point and point to infinity. All of these cycles are essentially nothing other than the self-referral unified field of consciousness interacting with itself, self-referral loops ultimately pulsating infinitely fast and eternally, and infinitely small and infinitely big—smaller than the smallest and bigger than the biggest.

Functions of the Levels of Individual Mind

This simplified model of the inward and outward flow of impulses of energy and intelligence—feelings and goals, thoughts and plans, and actions—serves as a basic framework for considering the functional levels of individual psychoarchitecture in more detail. We will begin with the subtlest level of individual subjectivity, the ego, and then describe the levels of (psychological) heart and intellect, mind, and senses.

The level of ego and being. In Maharishi Vedic Science and Technology the subtlest relative level of individual subjectivity is the level of ego, the experiencer or individual *I* (associated with *mahat,* but also with *ahamkara*). This level integrates the senses, mind, and intellect into a unitary subjective experience of the sense of individual self. In ordinary experience it is the active individual experiencing agent of the changing values of actions, sensations, perceptions, thoughts, feelings, and memories.

In this meaning of the term *ego*, it is the individual sense of *I* that underlies 'I remember', 'I feel', 'I think', 'I perceive', 'I act'—experienced in ordinary waking as the agent, actor, or performer of actions. We could say that it is the level that feels it is acting out the screenplay—the *film*—on the screen of the mind. From the level of individual being or ego, individual goals, plans, and actions are expressed.

This meaning of the term ego is much more fundamental than how the term is commonly used in psychological literature. In more common parlance the ego refers to a collection of memories or attributions that cohere into a sense of individual self. The ego described here is the unitary sense of self that is prior to and underlies all memories, perceptions, thoughts, feelings or any individual subjective experience. It does not emerge as a composite or collection of subjective experiences that congeal into a sense of continuity that gives the appearance of an individual self. Its continuity is more fundamental and enduring than any aspect of the physical body and nervous

368

system, or any history of individual experiences or memories. Its continuity and holism is not a product of the functioning of either the gross or subtle aspects of the individual body. It is a phenomenal level of nature that has its inherent basis ultimately in the simultaneity of individual and universal consciousness, part and whole, point value and infinite value—the coexistence of opposites of the unified field of natural law. This is another example of the holistic top-down structure of nature. The level of ego is at least as much a substantial part of nature as is any individual experience of manifest phenomenal *reality*. In higher states of consciousness beyond the ordinary waking state, the sense of self is identified with its universal value—the universal Self—not just with its individual value as individual self or ego. One way to describe this is that in higher states the ego *universalizes*, rather than is destroyed or eliminated—to be discussed in Chapters 19-20.

The phenomenal level of *individual* ego is the inner sense of being a whole, separate, unitary individual—which also can mean defining oneself as separate from everything else in existence, not fully appreciating one's universal Self. This is also the level at which the dual qualities of unifying and differentiating of consciousness have their subtlest individual manifestation. The individual ego is fundamentally composed of the two tendencies or forces of expanding and contracting, of unifying and differentiating, which is the nature of discrimination in the intellect. This is a way of understanding how the individual ego is structured, how the sense that *'I am'*—that *'I' exist separate from everything else*—develops. It originates in the self-referral dynamics of infinite silence and infinite dynamism as the 'nature' of the unified field of consciousness, and manifests on the level of individual life in terms of individual ego or sense of self—structured by the three gunas of sattva, rajas, and tamas.

Individual ego is the fundamental sense of *self,* as opposed to *other*—a sense of self as different from or separate from other things, emerging from the dual tendencies of unifying and differentiating, of universalizing and individualizing—the individual expression of consciousness itself as the coexistence of opposites. Whenever the coexistence is overshadowed, and there is the sense of self as opposed to other, whenever the self identifies itself primarily as a separate individual, then there is the possibility of pulling into the self or pushing away from the self. There is the possibility of inward and outward, of approach and avoidance, of accepting and rejecting, of receiving and sending, of giving and taking, of having and not having, of attachment and non-attachment, of love to unite and fear of separation, of incompleteness and desire for more, of love and rejection, of attraction and repulsion, and of suffering and fulfillment in terms of individuality.

All the processes of existing or being, wanting, feeling, thinking, sensing, and acting can be said to emerge from the dynamic interaction of the two apparently opposite forces of unifying and diversifying, driven by the activating principle in nature. The duality of *'me'* and *'not me'* is the basis for all individual activity. This duality also can be said to be the basis of all suffering in life (and can be associated in a religious sense to original sin, the mistake of separating oneness from Oneness, separating oneself from the universal God or Godhead, as well as the 'mistake of the intellect', *Pragya aparadha*). It is built into the very structure of individuality, the structure of *individual* ego—most prominently identified as the essence of oneself until it is refined enough to reflect together simultaneously the opposite values of infinite dynamism and infinite silence so that they perfectly coexist in infinite correlation or unity—simultaneously the individual self and the universal Self, point value and infinite value. Development of higher states of consciousness can be described as a process of growth from individual ego, self, or being—also sometimes referred to as *purusha,* the totality of the individual—to universal Ego, Self, or Being—*Purusha,* the totality of everything, pure consciousness, prior to individuality.

Going from point to infinity can be thought of as a process of differentiating or individualizing, and going from infinity to a point can be thought of as a process of unifying or universalizing (refer to Chapter 11). As these basic tendencies become individuated, individual ego becomes the seed for the more expressed levels of individuality in phenomenal nature. This holistic level of individual subjectivity is most closely associated with the point value of consciousness or individual consciousness, embedded in the infinite value or universal consciousness. It can also be related further to the notion of singularity—the singularity or oneness of individual being embedded in the singularity or Oneness of universal Being or *Atma*. The subtlest phenomenal discrimination on the level of individuality is the uniting and dividing that defines individual ego. When this sense of self as a unitary wholeness of experience gains universal status, then the entire universe is experienced as unified and included within it, and there is no longer the primary sense of something 'other' than oneself.

The level of intellect: Heart and feeling, mind and thinking. The faculty of discrimination implied in the sense of individuality of ego or self is called the level of *intellect* (associated with ahamkara, but also to some degree manas). This level is a primary component in decision making. Although in discrimination the principle of dividing is frequently emphasized, the principle of unifying can also be found. The dividing dynamic expresses more diversity, and the uniting dynamic implies unifying.

Discrimination involves separating things from dissimilar things and unifying things with similar things; this is how categories, concepts, theories, and more all-inclusive models or paradigms are built. As a simple example, a yellow apple and a yellow banana can be divided into different categories, one into the category of apples (uniting it with other apples, even red or green ones) and the other into the category of tropical fruits. We could also unify them into the category of yellow things, or into the category of fruits (dividing them from vegetables), and so on.

These subtle dynamics of the intellect are intimately related to the process of manifestation into phenomenal levels of creation, in the outward and inward flow through levels of subjectivity, as well as in the principles of parts emerging from wholeness and wholeness being embedded in and permeating the parts. In the act of discrimination the diversity becomes more expressed and the unity can become more underlying or hidden—but both are present in each act. This is reflected in the process of manifestation and evolution throughout nature, where increasing expressions of diversity become more apparent and the underlying unity becomes less apparent. When the underlying unity becomes so hidden that it appears to be lost to experience, then fragmenting diversity and disintegration predominate—again, *Pragya aparadha*. This also is associated with the experience of the separation of objectivity and subjectivity in the ordinary waking state of consciousness.

These mechanics further can be said to underpin the objectification of knowledge associated with modern science, which relies predominantly on the fragmenting aspect of intellectual analysis not balanced by the holistic experience of the underlying unity of nature. When the underlying unity is secondary or even completely hidden, then fragmented views and experiences of nature predominate—briefly described in the Prologue. Emphasizing only the objectifying approach to knowledge, there is a disintegration of the sense of wholeness of being, which contributes to a loss of meaning and eventually an existential malaise characteristic of our modern civilization.

Phenomenally the level of intellect can be described as having two aspects. The deeper aspect, related to the unifying dynamic, is sometimes described as the level of *heart* (psychological heart). The more expressed aspect is sometimes described as the level of intellect—in the direction of the next level of expression of subjectivity, the level of mind (manas). The contrast also is referred to in terms of the contrast of *heart and mind*.[9] In the Sankhya system of levels of nature in ancient Vedic science, however, there isn't a separate psychological level of heart—the intellect can be said to contain both of these layers, both heart and mind, unifying and diversifying.

This also relates to the common contrast of feelings in the heart and thoughts in the mind, introduced in Chapter 15 with respect to the distinction between feeling and cognition. In Maharishi Vedic Science and Technology heart and feeling are associated with the *finest* levels of activity of the intellect within the ego. Feelings are impulses of energy and intelligence in which the unifying aspect can be said to be a more prominent, and thoughts are the more expressed impulses of feeling in which the diversifying aspect is more prominent, commonly experienced as more concrete, specific, discrete, and well-defined. Thoughts can be said to be further concretization or manifestation of the seed impulses coming from feelings in the ego—and ultimately from consciousness itself. There can be relatively clear and unclear thoughts, as well as feelings; but the general experience is that thoughts are more expressed impulses of feeling, and that deeper than the level of thinking is the level of feeling in the heart.

As discussed in Chapter 9, when we carefully consider what tells us that a point of logic in our intellect is correct, it settles down to an underlying feeling we have deep in the interior of our individual subjectivity. What is it that tells us we are right in believing that a proposition is logically correct? Fundamentally it is that it makes sense, it *feels* true on a very deep level. *I* have a deep inner sense of confidence that it is correct, a deep feeling within *me* of confidence in its self-evident correctness. This is where the level of thinking merges into deeper levels of feeling and intuition. Even in the case of a non-emotional decision such as a point of mathematical logic, at its base is a deep feeling about it—and even deeper, an intuitive sense or feeling of correctness.

Heart and feeling can be said to be associated more with the general direction and motivation for action, related more to the energy or will to act. Mind and thinking can be said to be associated more with the intelligent direction for action. However, impulses of energy and intelligence are integrated impulses—like a seed—that emphasize different qualities or roles as they are expressed, manifested, or made more concrete through the levels of subjectivity. The energetic impulse of desire is intimately intertwined with its directional component or intelligence. Impulses of desire (goals) that have more clarity also spontaneously include the direction to do the right thing (plan) in fulfillment of the desire.

Experientially the (psychological or feeling) heart can be said to be concerned primarily with unifying—or *enfolding*—to unite into wholeness, completeness, fullness, pulling into itself. It can be said to pull or contract deeper into the nature of the self, to remove boundaries that separate things from the self, constantly moving toward wholeness, toward oneness—but also more fundamentally unifying into higher, more all-encompassing values of union.

The dividing dynamic of intellect, more in the direction of the mind, can be said to be primarily involved in expanding—or *unfolding*—into more diversity. It is more involved in differentiating, specifying, analyzing, and structuring boundaries and divisions—but also diversifying into more concrete values. It is oriented more toward going outward, toward separating things, toward becoming more objective, concrete, and detailed, toward more precision in fulfilling the desire in harmony with the environment. The intellect/heart system functions as a unit to express intelligent behavior that evolves the individual self to discriminate more refined, integrated, unified values. These subtle inner dynamics of thought and feeling have their correspondence with repelling and attractive dynamics such as the fundamental concepts of fermionic and bosonic tendencies—as well as gravitational attraction and repelling forces described in Chapters 11-12—on the gross relative level of particle physics, and also with infinite diversity and infinite singularity, or infinite value and point value on the transcendental level

Though both aspects of intellect are involved in the outward and inward flow of mental energy, in a general way the intellect/mind is a little more closely involved with the outward flow—with action and the organs of action. The intellect/heart is a little more involved in the inward flow—with experience and the organs of sense. The intellect/heart is more analogue, qualitative, and holistic; the intellect/mind is more digital, quantitative, and delineating.

From another standpoint, we can say that the mind is more interested in knowledge, in the sense of understanding, or *standing under*, or *disembedding from objects of experience*, eventually leading to separateness from every *object* or *thing*—ultimately going beyond everything. The feeling heart is more interested in experience, taking things in, in the sense of connecting with *or uniting everything into itself*, into wholeness or oneness, and ultimately Oneness.

The levels of ego, intellect/heart and intellect/mind are not completely distinct levels, but rather ranges where the primary qualities of mind, heart, or ego are most prominent. More fundamentally, the level of mind is embedded in the intellect and heart, which is embedded in the ego, which is embedded in consciousness itself. The intellect can be said to extend from the level of mind through the level of heart into the subtlest level of ego, in the sense that differentiating and unifying extend through all these levels (*buddhi*). In terms of the phenomenology of individual experience, the level of intellect where differentiating is more prominent is in the outward direction of the level of mind, and the level where unifying is more prominent is associated more with the inward direction of feeling in the heart. The level from where they emerge to take on these more distinctive qualities is ego. Ultimately the contrast of heart and mind can be associated with individual and universal consciousness, infinity and point, the coexistence of opposites, the 'nature' of the unified field.

Thus the concept of individuality, which defines individual ego, is intimately associated with intellect, in terms of the discrimination of *self* from *other*. The intellect can be understood to encompass the deeper functions of mind, heart, and individual ego. This is depicted in our simple map of subjective levels below by adding the brackets showing intellect to extend from the level of mind through the level of ego.

			Change in Environment		
	Environment				
			↑	↓	
OUTER					
	Behavior		↑	↓	When and Where
OBJECTIVE		behaving	Action	Results of Action	
	Body		↑	↓	
-----------------------	Senses	-----------------------------	sensing	---	----------------
INNER			↑	↓	
	Mind	thinking	Plan	Evaluation	How
SUBJECTIVE (Range of Intellect)	Heart	feeling	Goal	Fulfillment	What
	Ego	being	I have	Sense of self	Who
	Consciousness	(Being)	silent witness	self-referral	

To summarize, as an impulse of psychological energy initially emerges from consciousness, it contains both the motivation to act and the purpose or direction to act (energy and intelligence) within it, like a seed contains all the information for the tree. As it manifests or rises up through the levels of subjectivity, different qualities predominate, whether in terms of feeling or thinking. This also can be likened to the DNA molecule that contains all the codes for the gross physiology. The actual subjective experience of the process associated with expressing or rising up the through the inner dimension of levels of subjectivity depends on the state of inner development, and the clarity of experience of these processes. We will discuss this more in connection with the relationship of intellect and intuition in a later chapter.

To tie up a loose end, Chapters 1 and 2 outlined a simplified model of levels of the means of gaining knowledge that placed consciousness at its basis:

Sensory World
Sensory Experience
Reason
Intuition
Consciousness

This simplified version of levels now can be placed in the more comprehensive description outlined in this chapter and in Chapter 11. Generally speaking, the deepest level of intellect associated with the finest feelings in the heart can be related more to intuition (to be discussed again later); the more expressed level of intellect associated more with thoughts in the mind can be related more to reason; and of course sensory experience with the senses. However, this doesn't do justice to the embedded, hierarchical nature of levels of subjectivity, in which thoughts are fully contained in the feelings, feelings are fully contained in the ego, and the ego is the manifest impulse or fluctuation that is fully contained within consciousness itself.

The level of mind and thinking. Although in the above discussion we have emphasized the relationship of mind as individual subjectivity to the level of intellect specifically, the mind (manas) is also identified as a separate level within individual subjectivity. It might be somewhat confusing to call the mind just one of the levels of subjectivity, especially because the word is commonly used to refer to all the levels of subjectivity taken together. This broad meaning of the word is used to contrast it with, for example, the body—as in body and mind. But when we are identifying the mind as distinct from senses, intellect, and ego, and all of these are defined as subjective levels, then the word also refers to a particular subjective level.

The mind (manas), as a specific level, processes information received from the senses and carries out ordinary cognitive processes such as attending, thinking, analyzing, and reasoning. It guides the organs of action to express goals and plans in behavior. It also guides the organs of sense to obtain input to evaluate the effects of the goals and plans that have been implemented in the outer environment. By managing action, and by controlling the direction of the senses, the level of mind is able to implement the goals coming from deeper in the feeling level of intellect/ heart in effective action in the environment.

Referring back to the theatre analogy in Chapter 16, the level of the mind thus can be said to be the stage, as well as the stage manager, of both sensory input from the environment and behavioral output to the environment. The intellectualizing aspect of the mind analyzes, dissects, categorizes, and conceptualizes. It unfolds, opens up, takes apart, analyzes, and unveils in order to understand with clarity and precision, so it can direct more effective action to implement healthy goals. It incessantly builds mental representations, concepts, ideas, and plans for action.

Sometimes the phrase *screen of the mind* is used to describe the aspect of the mind that is the working space or stage to put together perceptions, memories, thoughts, and feelings into concepts about life and how to function in the environment. At times it may seem like there is kind of an inner screen—sort of like a movie screen—that displays perceptual and conceptual activity in the mind to conscious awareness (this appears to be the experiential basis of analogies such as the theatre model described in Chapter 15).

For example when we have a clear memory of something, it is as if we can view the object or event again in our minds. This is an inner experience on the screen of the mind. It is basically the same screen of the mind that dreams seem to be projected onto during sleep. These dream images are virtually all constructed from our own thoughts, and usually only a portion of the immediate sensory input from the outer environment is incorporated into the dream. They are almost entirely inner creations of the mind, and we experience them on the screen of the mind almost like we were at a holographic movie theatre watching a film.

Using the concepts developed in the mind based on perceptions, memories, reasoning, and other thoughts and feelings, the mind, intellect, heart, and ego build a mental model, picture, paradigm or worldview about the inner and outer world. The mind looks for consistency or lawful relationships between things, with which it builds mental models to guide actions. If these inner mental models are fairly accurate representations of the world, they can then guide more efficient and effective decision making in order to function better in fulfilling goals through behavior.

As parts of the inner mental model seem to represent the world pretty well so that actions are more successful, then less and less decision making about those parts of the model are necessary. Those parts of the model become more automatic, and choices about how to respond to consistent input become simpler. Deeper levels of thinking, feeling, and decision making no longer need to be as engaged or excited in planning those actions. Other aspects of these psychological dynamics will be discussed in the next chapter on attention and memory, as well as in the chapters on higher states.

When input from the environment does not match the mental model, evaluative processes are more fully active, engaging deeper cognitive levels more fully. The mental models, paradigms, theories, and plans are continually tested and updated. Hopefully over time, through careful observation and evaluation in comparing the inner model with the outer environment, the model becomes more accurate. This is essentially what is meant by the *scientific method*, which emphasizes the levels of the rational intellect, mind, senses, behavior, and the outer environment. Modern science tries to create models or theories of the world using very careful logic and observation of nature, repeatedly testing the models to make them more accurate representations of the observed world in the *self-correcting* process described in the Introduction.

However, more than just the intellect and the senses are involved in creating the mental models. To integrate and expand on these models, theories, and paradigms, the levels of the senses, mind, and intellect draw from deeper levels of inner subjectivity. Impulses of thought coming from these deeper levels involve general impressions, hunches, intelligent guesses, and assumptions that may be very difficult to operationalize and test in the localized outer objective world. They extend into levels of deeper feelings involving beliefs, insights, and intuitions. The models or paradigms are made up not just of concepts or ideas we have based on our past experiences and our thinking about those experiences. Our actions include emotional behavior based on both inner thoughts and inner feelings. The mental models or paradigms include patterns of feelings such as moods or longer-term styles of affect involving beliefs, schemas, mental sets, worldviews and more stable patterns of feelings and thoughts that comprise our individual personalities.

As information comes into the mind in the inward flow, it interacts with memory and with impulses going toward expression in the outward flow coming from deeper in the mind, intellect, and heart. From the ordinary waking state perspective, what is experienced on the screen of the mind is a combination of bottom-up input from the senses and the top-down view coming from the deeper levels. What we hear, touch, see, smell, and taste depends on what we feel about and how we understand the outer environment—what our inner mental model, template, or paradigm of it is. Our view of the outer environment changes with our changing ability to experience, understand, interpret, and relate to it. As discussed in Chapter 9, what we observe '*objective reality*' to be is significantly a product of who we are, our level of development.

Models or paradigms involve underlying assumptions, beliefs, and inner feelings that have not been fully tested using modern scientific objective methodology. They are projections of the mind, based on bottom-up input from objective sensory experience, and more fundamentally based on the underlying subjective top-down impulses according to the state of consciousness of the individual that also shape the bottom-up input.

Eventually in higher states of consciousness these processes of feeling, thought, and action are spontaneously conducted in accord with the totality of natural law. In a sense there is then no individual distortions due to stress of the conceptual model through which impulses of energy and intelligence are filtered—there is the direct, unfiltered, undistorted experience of action in full alignment and attunement with the totality of natural law, expressed via the individual mind and body. There is still an individual subjectivity and individual personality, but unfettered and not overshadowed by the accumulated limitations and distortions, resulting in *spontaneous* action in accord with total natural law through the channel of the individual being.

Unity has its phenomenal expression in integrated, orderly, coordinated diversity. The diversity is perfectly coordinated by maintaining connection to the underlying unity at every point—self-referral. The degree to which the unity is lively in the diversity determines whether the diverse, apparently independent parts coexist harmoniously in finite creation. When connection to the underlying unity appears to be lost, then disorder appears more prominent. Partial values of wholeness rather than the total value, partial knowledge rather than total knowledge, partial harmony rather than total harmony, then seem to predominate.

Phenomenal creation can be more—or less—in accord with the evolutionary patterns of the totality of natural law. This depends on, for example, the degree to which human beings with free will can maintain their lively connection and alignment with the totality of natural law spontaneously. When they are not fully connected and aligned with the totality, their actions inevitably are not fully harmonious. When they are in full contact with the totality of natural law—when individual consciousness is fully lively in universal consciousness—then natural law computes and coordinates all the levels of feelings, thoughts, and senses for spontaneous action in accord with the totality. Then the diversity is in full harmony with the unity of nature. The mechanics of these processes will be discussed more in the next chapter.

The level of the senses: Perceiving and acting. The level of the senses primarily refers to the five subjective psychological functions of hearing, touch, sight, taste, and smell—but also to the functions of action (together making up the 10 *Indriyas*). The senses are the most concrete, outermost level of inner subjectivity, connecting the inner level of mind to the objects of the senses in the subtle and gross environment. Through a series of stages of processing, the senses and their organs perform the task of transferring or transducing energy from the environment into psychological information the mind can use in the inward flow, and vice versa with the organs of action in the outward flow.

Interacting with memory, the information coming from the different senses coheres into subjective perceptions on the screen of the mind, resulting in meaningful concepts and categories associated with thinking or cognition and feeling or affect. The relationship of the five senses to the level of mind can be likened to the relationship between the five fingers to the palm—they take in information about objects in the environment which are grasped on the screen of the mind and evaluated by the intellect.

The level of the senses involves a fundamental division related to the distinction described earlier in the chapter of the inward and outward flow of information. The outward flow of energy and intelligence is due to the connection of the mind with the organs of action (karmenindriyas—karma refers to action), and the inward flow is associated with the organs of sense (gyanindriyas—gyan refers to knowledge). In the gross body the organs of action are the organ of speech (tongue, etc.), the organs for receiving and holding (hands), the organs for movement of the body (feet), and the organs for discharge (genitals and anus). [10] The gross bodily organs of sense are the ears, skin, eyes, tongue, and nose and their neural processes—the seats of the auditory, tactile, visual, gustatory, and olfactory processes. The gross bodily organs of sense and action directly correspond with the gross properties, qualities, or elements (mahabhutas—space, air, fire, water, earth). On the subtle level, the senses also are associated with what are sometimes called the sense essences (tanmatras—sound, touch, sight, taste, smell—refer to Chapters 11-12).

In the summary of the 25 levels of nature in the Sankhya enumeration described in Chapter 11, the cosmic level of the mind is identified as an evolute—caused but not causing other levels or modes of being. This means that the level of the mind emerges or evolves from the subtler subjective level of the cosmic intellect; but the level of the mind does not cause the next more manifest level, the level of the senses. The level of the senses also emerges from the level of the intellect, not the level of the mind. In turn the level of the senses is also identified as an evolute. This means that the five senses do not cause the next levels of manifestation, associated with the five subtle essence elements (tanmatras) that compose the subtle objects of sense. The level of the subtle elements also is caused or emerges from the cosmic level of the intellect (buddhi). Thus the level of the mind, the level of the senses, and the level of the subtle elements all emerge from the cosmic level of intellect—the discriminative function in nature. This is the subtle phenomenal expression of the objects of sense, the senses, and the mind that receives input from the senses. It is the psychological basis of the individual experience of objectivity and subjectivity.

The level of mind is experienced as subjective, and the level of the subtle essence elements and gross elements emerging from them can be said to be objective—in the sense that they are objects of sense. The subjective/objective distinction has its deeper basis in the level of ego—in the sense of self and other—and ultimately in the three-in-one self-interacting dynamics of the unified field of consciousness. It is based in the distinctions of unmanifest Prakriti and Purusha, infinity and point, the coexistence of opposites infinite dynamism and infinite silence as the 'nature' of the unified field, and ultimately non-dual Being.

The distinction of subjective and objective experience can be said to be expressed from the dynamics of the cosmic level of intellect within the cosmic level of ego. Also emerging from this level are the five senses and their directly correlated five elements in nature that can be sensed. The objects and their subjective sensory qualia directly match. The objects are built of the qualities of space and sound, air and touch, fire and sight, water and taste, and earth and smell. These five elements, in subtle and gross forms, combine to produce the immense variety of objects and corresponding qualia of phenomenal experience.

As described throughout this book, and especially in Chapters 11-12, the fundamental trinity of knower, process of known, and known is reflected in the sequence of three phenomenal domains of nature: transcendental, subtle relative, and gross relative domains. This enumerates the fundamental three-in-one self-interacting dynamics of consciousness itself. Again, in the transcendental domain the terms used in ancient Vedic science to refer to this fundamental trinity (in corresponding sequence) are *rishi*, *devata*, and *chhandas*—key terms in the Veda (the level of Prakriti or Nature) about which more will be said in the final chapters of this book. In Chapters 11-12, this fundamental trinity was identified with respect to its expressions in the subtle relative domain by the corresponding terms *sattva*, *rajas*, and *tamas*. In the gross relative domain—the domain of ordinary sensory experience in the gross physical body—the corresponding terms are *vata*, *pitta*, and *kapha*.

The terms vata, pitta, and kapha relate to fundamental qualities that comprise all the gross objects of sense composed of the five gross elements (mahabhutas). All objects contain all three qualities, as well as all the five gross elements. *Vata* is most prominent in the qualities of the gross elements of space and air, *pitta* with the qualities of the gross elements of fire and water, and *kapha* with the qualities of the gross elements of water and earth. In addition each gross bodily structure—human, animal, insect, plant—is made up of the predominance of certain permutations of *vata*, *pitta*, and *kapha*. Thus in humans there are basic body types depending on which are primary, secondary, and tertiary. Mental and physical health is related to the proper balance of these fundamental qualities in the basic body types.

In ancient Vedic science and Maharishi Vedic Science and Technology the properties or qualities of objects that make up the natural world—both gross and subtle—directly match the functional abilities of the senses resulting in sensory qualia. Sensory experience is directly related to the effects in body and mind of the objects of sense in the form of smells, tastes, sights, touches, and sounds.

There are basic types or categories that are made up of combinations of gross elements and the predominant vata, pitta, or kapha qualities. With respect to the gustatory sense, for example, there as six basic types of taste: sweet, sour, pungent, astringent, bitter, and salty. There are seven basic color qualia (rainbow, processed through three basic types of sensory receptors), as well as seven basic notes in the auditory tonal scale. These categories are extensively described in ancient Vedic literature, identifying a completely integrated structure of objects and their corresponding sensory qualia. These relationships also reflect fundamental mathematical patterns throughout nature, as well as fundamental artistic and esthetic values in accord with deep laws of nature.

The basic qualities in the objects of sense directly affect body and mind, bringing them into balance for health and vitality, or throwing them out of balance. For example the consumption of foods with certain qualities, or of environmental toxins through respiration, or of other sensory inputs such as musical or visual entertainment all have direct effects on body and mind. In this context the phrase 'we are what we eat' means that what we ingest through any of the senses in our interactions with the environment influences what we become—the overall condition of body and mind. Increased balance and purity in body and mind contributes to the degree of subtlety or fine-graining of sensory, cognitive, and affective functions—eventually becoming refined enough to experience naturally the subtler levels of nature that are ordinarily foggy at best.

This deep correspondence between objective phenomena and subjective experience relates to the fundamental design of the universe that connects objectivity and subjectivity. It ties together actions, the environment, and the effects of actions on the environment and on oneself. It provides an integrated framework for action in accord with the fundamental patterns and relationships throughout nature—that is, action in accord with natural law. There is an immense amount of detail in ancient Vedic science about these fundamental patterns in nature—far beyond the scope of this book. These fundamental correspondences are systematically applied in the integrated consciousness-mind-body approach of Maharishi Vedic Healthcare introduced later in the book.

In this integrated understanding, subjective experiences and their results are determined by the coming together of the top-down design of natural law emerging from the unified field of consciousness, top-down control by the individual being that can act intentionally within this structure, and bottom-up input from the outer environment that at certain stages of development is phenomenally separate from the individual. In Chapter 18 many of the issues covered in the past several chapters will come together in a consideration of individual top-down intentional control, in terms of the concepts of attention and free will. How the structural levels of individual subjectivity and the mechanics of attention and memory allow freedom of choice and action in the deterministic universe will be the focus of discussion. Describing in what ways these experiences change with natural development of higher states of consciousness will be main topics in Chapters 19-21.

Chapter 17 Notes

[1] van Lommel, P., van Wees, R., Meyers, V., & Elfferich, I. (2001). Near-Death Experience in Survivors of Cardiac Arrest: A Prospective Study in the Netherlands. *Lancet, 358,* 2039-2045.

[2] Churchland, P. S. (2000). The Hornswaggle Problem. In Shear, J, (Ed.). *Explaining consciousness: The hard problem.* Cambridge, MA: The MIT Press, pp. 37-44.

[3] Braud, W. (2002). Brains, Science, Nonordinary & Transcendent Experiences: Can Conventional Concepts and Theories Adequately Address Mystical and Paranormal Experiences? In Joseph, R. (Ed.). *NeuroTheology: Brain, science, spirituality, religious experience.* San Jose, California: University Press.

[4] Rockwell, T. (2005). *A defense of emergent downward causation.* 5/16/2005. URL = <http://www.california.com/~mcmf/causeweb.html>

[5] Dossey, L. (1989). *Recovering the soul: A scientific and spiritual search.* New York: Bantam Books.

[6] Schwartz, J. M. & Begley, S. (2002). *The mind & the brain: Neuroplasticity and the power of mental force.* New York: ReganBooks.

[7] Maharishi Mahesh Yogi. (2003). Maharishi's Global News Conference, December 31.

[8] Boyer, L. M. (1988). Personal communication. Fairfield, IA.

[9] Maharishi Mahesh Yogi (1967). *Maharishi Mahesh Yogi on the Bhagavad-Gita: A new translation and commentary, chapters 1-6.* London: Penguin Books.

[10] Sharma, P. V. (1981), (Ed., Trans.). *Caraka Samhita,* Vol. 1. Varanasi: Chaukhambha Orientalia.

Chapter 18

Attention, Memory, and Free Will

In this chapter, we will explore the mechanics of attention and memory in the context of individual psychoarchitecture and its basis in universal consciousness. Key issues are the relationship of the channel or spotlight of attention and of integrated memory processes to consciousness itself. After laying down planks related to these topics on the bridge to unity, we then will consider the perennial issue of free will and determinism. The overall main point of this chapter is that the mechanics of individual experience are embedded in the universal mechanics of nature in the unified field.

At a conference in northern California in 1970 Maharishi was asked about the nature of attention in relation to the levels of the senses, mind, intellect, and ego. His brief reply was, "Attention is *stationary individual awareness*." [1] In this answer three fundamental aspects can be identified: awareness or consciousness, its individuation, and the holding or maintaining of individual awareness. We have already extensively discussed the nature of awareness and consciousness. Let's now consider further the individuation of consciousness in attention, and then the act of focusing attention on a particular object of experience—the act of attending. Later in the chapter, how memory relates to attention and consciousness will be discussed. At the end of the chapter we will deal explicitly with the contrast between universal determinism in nature and free will in individual action, referred to in earlier chapters but not addressed.

The Spotlight of Attention

In order to explain the structure of phenomenal creation, the indivisible wholeness of the unified field of consciousness is intellectually divided into parts. One way this is described in Maharishi Vedic Science and Technology involves starting with the simultaneous coexistence of opposites of infinite dynamism and infinite silence—reverberating between infinity and point at infinite frequency. In that infinitely dynamic silence can be located the basis of time and space—time in terms of simultaneity and sequence, and space in terms of wholeness and parts. It can be delineated further into the three-in-one self-interacting dynamics of the unified field, which manifests into the fundamental forces that shape all levels of phenomenal creation (refer to Chapters 11-12).

The process of phenomenal manifestation can be thought of in terms of increasing excitation or reverberation superposed on the underlying non-changing ground state within the unified field. It also can be thought of in terms of decreasing excitation or reverberation in the sense of finite frequencies that are partial values of the infinite frequency. These represent ways to conceptualize the holistic view of the phenomenal parts of nature emerging from the wholeness, either from the perspective of infinite silence or the perspective of infinite dynamism.

Maintaining unity of the simultaneous coexistence of opposites, the opposites appear to manifest into levels of nature as wholenesses at each sequential level. The concept of sequence is superimposed on the concept of simultaneity, while not losing the simultaneity. It is the same with finite superimposed on infinite, and relative superimposed on absolute. These apparent dualities also can be described as the unified field creating within itself partial wholenesses while continually

379

referring back to the infinite wholeness. Wholeness is maintained at each level of the parts, but also can appear to be hidden in the parts. The parts can appear more prominent and obvious, and wholeness can appear less so, at each level of subtle to gross phenomenal expression. However, at increasingly subtle levels of nature the qualities associated with wholeness become increasingly prominent, and eventually all levels and cycles of manifestation and dissolution are seen in terms of self-referral loops of infinity and point, nothing other than indivisible wholeness.

On the level of individual experience the impression that the parts of nature are more *real* than the indivisible wholeness is associated with the concept of *Maya*. In the course of individual development, the state of experience in which the parts are primary and the wholeness is secondary is related to the ordinary waking state of consciousness. The state in which the wholeness is primary and the parts are secondary is associated with *full enlightenment*. In that state Oneness or indivisible wholeness is the reality.

At each sequential level of phenomenal manifestation, the duality of individual and universal is expressed. Emphasizing universality, the point value can be related to the ultimate singularity of consciousness itself. Emphasizing individuality, the point value can be related to individual consciousness, and further to individual self or ego. From these levels emerge all the more expressed levels of subjective mind and objective body. At the level of mind there is the universal cosmic mind and individual minds. This contrast manifests into the universal environment with its infinite diversity of individual objects and individual subjects that attend to and experience it—the object-subject duality. With respect to individual life, infinity and point can be directly associated with individual conscious attention and universal consciousness.

In the context of individual attention the phenomenal manifestation of levels of nature can be thought of as progressively limiting the indivisible wholeness of the unified field of consciousness into subtle nonlocal wave-like phenomena, and then further into gross particle-like phenomena that form processes and objects in relative creation—including mind and body. All the various levels of mind and body can be said to individualize consciousness into an individual being— the consciousness-mind-matter (body) ontology. This individualized structure of mind and body forms a localized channel of attention that can sense what appear to be separate, independent, discrete objects of experience—analogous to an individual wave emerging from the ocean, or the structure of a flashlight that channels light into an individual beam.

The nature of attention is perhaps second to consciousness as the most challenging of theoretical issues. Consistent with a reductive conceptual approach, cognitive models typically attempt to place attention in particular stages in the sequence of information processing. As discussed in Chapters 14-16, however, the principle of selectivity fundamental to the concept of attention is recognized to be the contribution of many different processing functions distributed throughout psychological and neurophysiological processes.

The analogy of a beam or spotlight is sometimes used to characterize attention in historical, cognitive, and neuroscientific models.[2][3][4][5] The spotlight of attention can be conceptualized as shining on objects and bringing them to consciousness through individual attention in the process of knowing. The spotlight, channel, or beam of attention is the product of all the subtle and gross limitations that individualize experience. To be discussed shortly, the screen of the mind described in earlier chapters refers to a particular level of individual psychoarchitecture, whereas all the levels together structure the individual spotlight of attention. The levels that comprise individual psychoarchitecture progressively shape the output of impulses of energy and intelligence into this limited channel or spotlight through which individual feelings, thoughts, and actions flow. This individual channel also contains the limited input processes involved in sensing objects of experience. Both the input and output aspects of the channel of attention localize the expression of consciousness into specific functions and concrete structures—each phenomenal level a further specification of infinite potential into finite actualities. Each more concrete level appears to hide further the underlying infinite wholeness of the unified field. The grossest parts of nature—phenomenally discrete, individual objects in localized conventional time and space—can appear to dominate and the wholeness becomes secondary, sometimes completely hidden to individual experience.

In theories of attention and consciousness in psychological and cognitive science, attention is rarely distinguished from consciousness. Consistent with ordinary waking state experience, consciousness is confounded with and not disembedded from the ordinary mental activity of feelings, thoughts, perceptions, and sensations. Thus the phrase *conscious attention* is commonly used—apparently due to the infrequency of clear *direct experience* of consciousness itself that is the underlying basis and source of the individual spotlight of attention. In this ordinary state of subjective experience there is a lack of direct experience that isolates consciousness itself from the mental activity of attending and remembering.

In the ordinary waking state the objects of sense that appear to be in the outer objective world independent of the experiencing subject can be said to be brought to consciousness through the process of attention. The unified field of consciousness manifests itself into diverse individuals that eventually evolve to experience universality in individuality. Recontacting, referring back, or 'remembering' by direct experience the holistic universal basis of the individual through transcending all the parts of individual experience eventually reestablishes the completely unified level of consciousness itself as primary in experience. The TM technique is a systematic developmental technology to facilitate this evolutionary process of reconnecting the individual self to the universal Self, for the self to 'remember' itself fully as the Self. It is only through transcending all the parts—all sensations, thoughts, feelings, and intuitions—that the wholeness of nature can be directly experienced as self-referral consciousness itself.

Attention can be described as a process of bringing the phenomenal parts of nature to the screen of the mind to be observed by consciousness—of presenting objects to consciousness. The process of evolution can be viewed as a pulsation of universal consciousness appearing to diversify into individual selves that evolve to experience universality in individuality—infinity to point to infinity, and point to infinity to point. Wholeness appears to divide into parts that lose the wholeness. Eventually through repeated experience of taking attention from the parts to the underlying wholeness, the individual experiencer develops the ability to 'remember' the totality of the wholeness in each part of individual experience—Self-referral consciousness fully enlivened in the individual. In ordinary waking experience attentional processes present objects of experience to individual consciousness. In the process of transcending, attention naturally settles down to its basis in pure self-referral consciousness, which greatly facilitates the process of enlivening universal consciousness in individual consciousness. Bringing the spotlight or channel of individual attention full circle, back to its source in self-referral consciousness—through effortless transcending—is the foundation for spontaneous success in goals, plans and actions. As Maharishi explains:

> "Consciousness is the light by which man looks out onto the world to experience objects. When that same light is turned in upon itself the resulting self-knowledge enlightens all thought and action. Consciousness then knows itself as unbounded, free, the source of all creation. The experiencer is no longer overshadowed by what he experiences. Consciousness ceases to be the servant of thoughts and percepts and becomes their

governor. Any desire projected from this state of self-knowledge achieves its goal without resistance." (p. 123) [6]

In ancient Vedic science the limiting of the simultaneous wholeness into the phenomenal sequence of levels is sometimes described in terms of five coverings, aethers, sheathes, or filters—called *khoshas*.[7] They can be described as functional structures at different levels from subtle to gross that construct individual psychoarchitecture and the phenomenal reflection of individual consciousness. These aethers, coverings, forms, membranes, or structures—which correspond to levels of subjectivity in the Sankhya system outlined in Chapter 11—can be said to localize consciousness into an individual channel of experience or spotlight of attention. Each sheath reflects increasing limitation, concretization, and differentiation from the subtlest to the grossest levels of nature. In the other direction, from gross to subtle, each sheath reflects increasing expansion, permeability, dynamic liveliness, and integration of energy and intelligence.

The most expressed sheath is built of matter particles that exhibit solidity and the least amount of inherent intelligence—where energy and intelligence appear the most independent of each other. The increasingly subtle sheathes reflect degrees of integration of energy and intelligence toward the underlying infinite integration of energy and intelligence in the infinite self-referral of the unified field. The most expressed sheath is called *ananamaya khosha*, associated with the physical form or 'body made of food' built of the five fundamental elements of nature (mahabhutas)—a structure made of *matterstuff*, quantized particle interactions that make up the gross level of nature. It is commonly associated with restricting sensory experience almost entirely to the macroscopic gross relative domain.

Subtler than this most expressed, most tangible level is *pranamaya khosha*, associated with the vital breath, the inward and outward pulsation or reverberation of nature from universality (inward) to individuality (outward). This level can be associated on the gross level with respiration. It also can be understood on the subtle level in terms of nonlocal wave mechanics, pulsations or vibrations of energy and intelligence at the level of the subtle essence elements (tanmatras), which form subtle individual objects of sense—sort of a structure made of *energystuff*.

The next subtler level is *manamaya khosha*, associated with the level of mind, through which very subtle energy and intelligence manifest in thoughts and mental impressions. This level can be described in terms of pure mind—sort of a subtle structure or very abstract substance made of *mindstuff*.

Next is *gyanamaya khosha*, associated with pure knowledge, the subtlest differentiation between the universal value of intelligence and individuality. It can be described as a sheath, structure, or very abstract form made of impulses of intelligence, and as associated with the level of intellect—sort of *intelligencestuff*.

Intimately related to this is the subtlest level of structure or form, called *anandamaya khosha*. This is related to the almost perfectly balanced state of the three fundamental forces (gunas) of nature in the three-in-one self-interacting dynamics of consciousness. It is associated with bliss (*ananda*) overlaying (through *Maya*) consciousness itself—the almost universal value in individuality. It can be associated with the concept of pure individual being or individual ego, the individual reflection directly contacting absolute Being. It is the subtlest, most abstract level of individuality, sometimes described as the body of bliss or bliss-filled body—sort of pure amness or *egostuff*, which is of the nature of bliss. This encouragingly suggests that the most fundamental aspect of individuality is the experience of pure bliss. It can be described as the enlivening of the infinite silence of the universal ocean of consciousness into individual waves of dynamic intelligence. Underlying these delineated levels of objective and subjective existence is consciousness itself, Atma, or pure Being. The general point here is that the discrete channel or spotlight of attention is constructed from all the phenomenal levels of individual ego to individual body. Each level contributes to the individualized channel of attention.

For example there is a strong influence on sensory attention when we are alert and experience the finer qualities of a friend's loving compassion, or are dull, choppy, and a bit disconnected after watching TV or a computer screen all day. The state of our bodies—such as when we are feeling digestive discomfort, sore muscles, or increased vitality after moderate exercise—also influences our attentional focus. The social environment influences whether we feel safe, protected, or enthusiastic about a common goal and open to take in information from the environment. Likewise the gross physical environment—such as very hot or cold temperature, pollution, earthquakes, beautiful vistas, the pristine silence of redwood forests—impacts our ability to focus attention. Any of these levels may appear to be the predominant factor influencing our ability to attend, as well as the selective direction of the spotlight of attention, at a particular time.

The levels function as an interdependent feedback system where each level impacts the other levels. For the most part, however, the deeper the subjective level, the more overall influence it has on determining what we attend to and how we evaluate it. For example it is due to the greater power of our mind over physical matter that we are able to devise means to regulate the temperature of our immediate physical environment (although on occasion the environment may seem to predominate). The deep feeling level in our psychological heart has even more of an overall impact than the other more expressed levels of subjectivity on the selective direction (goals) we take. Who we believe we are on the level of individual ego or self has even more impact.

The object of experience is attended to and evaluated most fundamentally on the basis of the state of consciousness of the individual experiencer—which has the most comprehensive and global influence. This is a core aspect of the *top-down hierarchical* nature of the inner dimension, expressed in an individual channel or spotlight of attention. Top-down and bottom-up influences interact to structure the integrated composition of attentional experience, but top-down influences are the most fundamental and establish the overall selective direction of input and output.

The central attentional mechanism: The level of mind. Attention also concerns the processes through which this individual channel is directed to focus on, and maintains focus on, a particular object or array. This is related to the issue in cognitive models whether there is a specific selective attentional mechanism at a particular stage of processing that does the directing and selecting.

Stage analytic models—described in Chapter 15—conceptualized a specific stage or mechanism that primarily carries out these functions. However, the experimental evidence indicates that the notion of limited capacity—associated with the individuation of attention, which then requires selectivity—is perhaps applicable to all the functional stages of processing. Furthermore, neuroscientific models—such as the ones summarized in Chapters 16-17—have brought out more clearly that the function of selectivity is distributed to some degree *throughout* biophysical and neurological processes and structures, built through the evolutionary history of the species as well as individual personal experience.[89] All these functions and structures contribute to individuation and selectivity that characterize the individual spotlight or beam of attention.

As further examples of factors contributing to the selective channel of attention, visual input receptors in the nervous system cannot take in all the information that impinges upon them. They are sensitive only to a narrow spectrum of wavelengths, as well as having spatiotemporal and qualitative specificity that further define the macroscopic range of gross vision. These limitations constrain what information can be received as visual input through the gross body, significantly contributing to the selectivity of the functional channel of attention.

Attentional processes include efferent neural connections that innervate muscle complexes at virtually every level of sensory function, even before sensory input begins to activate sensory receptors, such as for receptor orientation. Arousal, salience, pre-perception and mental sets, as well as functions occurring after sensory reception such as ideational and decision processes, shape what is experienced in attention. The notion of capacity limitation—which then requires selectivity—is relevant to each of them, influenced by a myriad of processes throughout the integrated functioning of the organism. For instance it is not uncommon to stop walking temporarily while thinking about a particularly engaging issue, which suggests that even walking in some way involves shared attentional resources requiring selectivity.

In cognitive models of attention the most significant capacity limitation is attributed to the serial limitations of a central processing mechanism or executive processor—analogous to the central processing unit of a computer. This is somewhat analogous to the level of the mind (manas) in Maharishi Vedic Science and Technology—which includes the screen of the mind or center stage upon which the results of acts of attending are observed. It is at this level that objects of experience seem to come together to be presented to consciousness. This at least gives the subjective impression of a centralized executive attentional mechanism. As described in Chapter 17, the level of the mind manages the output of energy and intelligence to control the direction of the senses, as well as presenting or displaying the impulses of feelings and thoughts coming from deeper inner levels.

But the *mechanics* of attention generally are not part of subjective experience. Even in the case of what feels like intentional choice, it is the results of selective attentional processes—not the mechanisms or processes themselves—that are experienced on the screen of the mind. When we introspect about how attentional control and selectivity is implemented inside us, there is little sense of it. It feels like our *individual self* or ego makes a choice somehow and somewhere in our *heart and intellect* that is implemented via some automatic unknown means, resulting in experience of the objects to which we choose to attend. Although for the most part we feel that we have overall super-ordinate control of our experience, the essential mechanics of these processes seem to be automatic and beyond our individual experience. We will discuss this issue later in this chapter, in connection with memory function and also free will.

Maharishi's Absolute Theory of Attention. The key points discussed thus far in this chapter about the nature of attention can be found in the 10 principles of *Maharishi's Absolute Theory of Attention* in Maharishi Vedic Science and Technology:

1. "Attention is the flow of consciousness. It flows both within its own unmanifest nature and outward towards greater levels of excitation.

2. In the process of perceiving objects, attention flows from unmanifest to manifest. It enlivens both subject and object, perceiver and perceived.

3. Attention is the link between subject and object; when it goes towards objects, consciousness takes on the form of the object. Attention identifies an object by transplanting it onto the consciousness of the subject.

4. Both subject and object have many levels of existence, ranging from states of greater excitation to the state of least excitation. The object has its surface structure (as perceived by the senses), its molecular, atomic, and subatomic structures, and its basis, the unmanifest vacuum state where all potentialities are

as yet unexpressed. The subject, too, has its levels of existence: perception and thinking, finer and less excited levels of understanding and feeling, and the basis level of the subject, the state of least excitation of consciousness, which is the source of all mental potential.

5.   The world is as we are. How an object is evaluated depends upon the degree to which the subject can appreciate; since knowledge is different in different states of consciousness, it depends on the level of consciousness of the subject.

6.   Only an individual established in unbounded awareness can appreciate the total value of the object at all levels of excitation, from its surface level to its unmanifest level.

7.   Through the process of Transcendental Meditation, consciousness itself becomes the object of attention. Attention is turned back onto itself, enlivening the full creative potential of consciousness, the field of all possibilities. The subject thus established in its supreme value is able to appreciate the supreme value of the object.

8.   This field of all possibilities in consciousness, when fully enlivened by attention, is able to do more than just reflect the qualities of the environment. It can create from within itself and project its creation outside, and then, through the process of perception, take it back into itself, all possibilities of creation and perception are open to it. This characteristic of consciousness to express itself in terms of objects and take the objects back into itself, shows the field of consciousness to be circular—pure intelligence in motion within its own nature.

9.   The most profound value of research is research into one's own nature, into the deepest level of consciousness, in order to locate the field of all possibilities in the unmanifest Self within every human being.

10.   The secret of mastery over nature is mastery over one's Self—the ability to operate from the field of all possibilities within, in order to fulfil any desire or achieve any goal." (p. 152)[6]

One important implication of the above principles of attention is that both the object and the subject are influenced by the process that connects them together. Not only is an impression of the object brought to consciousness through the channel of attention, but also there is an influence on the object when the spotlight of attention shines on it. The explanation of this phenomenon involves the important distinction of gross and subtle levels of nature discussed at length in prior chapters, and it relates to the top-down causal influence of consciousness and mind on the objects of sensory experience.

In order to understand these points, differences between the gross and subtle domains of nature need to be clear. In the physicalist view of a closed causal universe, all interactions in nature are mediated by the known fundamental forces—electromagnetism, weak and strong forces, and gravitation. There is believed to be no other level of existence that involves any additional, subtler form of energy flow; and there is thought to be no evidence that mental attention on an object influences the object through these known fundamental forces.

There is a growing body of interesting cognitive research, however, investigating whether individuals can discriminate above expected chance levels that another individual is placing attention on them.[10] This research suggests that some individuals fairly reliably seem to sense or 'feel' when the attention of another person is on them. The results have been obtained even under highly controlled experimental conditions that would appear to rule out physical means for information transmission between the subjects mediated by the four known fundamental forces. If these findings are accurate and reliable, they will require a more abstract notion of attention and mental energy consistent with the model of mind as an underlying nonlocal field. In such an expanded understanding of levels of nature, nonlocal influences subtler than the four known fundamental forces are posited as mediating the effects of attention between individuals.

Some of this cognitive research apparently indicates that it is not just that sensory attention needs to be placed on another subject. Even mental attention, or thinking about the person with no physical means to attend to or observe another person directly, may be sufficient to enliven or communicate in some subtle way the feeling of receiving attention by another person. This involves a more abstract notion of attention than is associated with directing the five sense modalities, and is even further suggestive of an underlying field of mind involving nonlocal correlations and interactions. It suggests that nonlocal pathways of mental energy mediate subtle effects of attention, analogous to but subtler than a beam of light reflecting off of an object that then activates visual receptors. The mechanics are similar, but the medium is a subtler level of energy flow and causal interaction. A rational explanation for this subtler level of nature was put forth in the Physics Unbound part of this book based on quantum and unified field theories.

With respect to the causal influence of top-down processes of mental attention on the brain, neurocognitive research has demonstrated in recent years that the brain is not nearly as hard-wired as first assumed. Research on the incredible neuroplasticity of the brain further has documented the powerful top-down influence of mental attention on neural processes. As an example of this type of research, two human faces and two houses were presented at the same time, and subjects were cued to take note either of the faces or the houses. Even though the four stimuli were exposed to sensory receptors and the visual system, directed attention on a face increased activation of the specialized cortical area for face recognition. The function of the face recognition cortical area was not fully determined by the stimulus input, but depended on the allocation of voluntary attentional resources. Additional studies show that this also occurs when invoking a memory of a face, without external stimulus input.[10] Studies such as these suggest that mental attention can control sensory processing in the brain, and in some circumstances may contribute in a top-down manner at least as much to the final experienced perception than the bottom up sensory input. Schwartz and Begley point out:

> "Since attention is generally considered an internally generated state, it seems that neuroscience has tiptoed up to a conclusion that would be right at home in the canon of some of the Eastern philosophies: introspection, willed attention, subjective state—pick your favorite description of an internal mental state—can redraw the contours of the mind, and in so doing can rewire the circuits of the brain, for it is attention that makes neuroplasticity possible. The role of attention throws into stark relief the power of mind over brain, for it is a mental state (attention) that has the ability to direct neuroplasticity." (p. 339) [10]

> "The mind creates the brain. We have the ability to bring will and thus attention to bear on a single nascent possibility struggling to be born in the brain, and thus to turn that possibility into actuality and action. The causal efficacy of attention and will offers hope of healing the rift, opened by Cartesian dualism, between science and moral philosophy. It is time to explore the closing of this divide, and the meaning that it has for the way we live our lives." (p. 364) [10]

The Act of Attending

The selective attentional control system in the human organism involves a complex hierarchy of biological reflexes, instinctive tendencies, acquired sequences of behavior such as fixed action patterns modified by past learning (refer to Chapter 15), and well-established patterns of feeling, thinking, and decision making. They include bottom-up control processes that are non-intentional, whether phenomenally experienced or not. In addition they include top-down control processes that are phenomenally experienced as intentional and that seem to be due to volitional choice. Together these control processes establish an integrated set of fundamental programs or functions, regulating output to control input in service of the inner purposes of a unitary individual organismic self. Function and structure are intimately related throughout the organismic system. However, in this integrated purposive system, function may be said to guide structure—intelligence guides energy—in the same way that a lawn mower, for example, is engineered and operated to serve a function intended by intelligent designers and operators.

What guides attention? The concept of attention underscores the overall purposive nature of the human information processing system that flexibly interacts with its environment to produce its own desired experiential states. The human organism can be said to be designed with both high structural specificity and at the same time high flexibility in order to respond and adapt to changing environmental conditions. Of course the gross bodily organism is adaptable within a relatively narrow range of environmental conditions in the gross relative domain of nature—more specifically, to conditions characteristic of Earth.

According to many neuroscientific models the design correspondence between living organisms and their environment theoretically develops via natural selection, underlain by fundamental randomness. Certainly one basic goal of a living organism is survival. However, at least advanced organisms—especially humans—are capable of action that doesn't seem to be characterized adequately by the goal, purpose, or empirical result of biological survival. This suggests that there are more inclusive overriding goals that guide behavior beyond participation in individual and even species biological survival.

> "I believe that the very purpose of our life is to seek happiness. That is clear. Whether one believes in religion or not, whether one believes in this religion or that religion, we all are seeking something better in life. So, I think, the very motion of our life is towards happiness..."—*The Dalai Lama (p. 13)* [11]

Attempts have been made to organize individual purpose into a hierarchical sequence of needs or goals to describe this more expanded understanding of the functioning of the individual human organism. For example in *need hierarchy* theory, psychologist Abraham Maslow proposed a sequence of higher-order needs emerging as lower-order needs are met—from physiological needs, safety needs, social needs, esteem needs, to self-actualization.[12] This theory was expanded to incorporate a *synergistic* perspective in which optimal combinations of needs are developed that support the more inclusive need or goal of self-actualization. This synergistic goal is sometimes described as the insatiable reach toward higher potential as an inherent feature of the human organism.[13] This is clearly an expanded conception of natural selectivity beyond the biological imperative of individual and species survival or the empirical facts of survival of the fittest.

This expanded direction for understanding the relationship of natural selection and a hierarchy of needs associated with self-actualization could be taken still further—placed in the context of the ultimate development of the individual. The theory of the instinct for biological survival developed mainly from observations of lower animal species. However, with respect to the human species it also could be understood in religious and spiritual contexts as extending into the pursuit of the permanent maintenance of life, permanent survival—that is, to *eternal life*. Similarly the need for safety and security can be understood in terms of the ultimate safety and security of fully establishing the individual self in the universal Self—the unified field of consciousness, beyond all the pairs of opposites including life and death. In this context it can be related to going beyond the sense of duality of *self and other* into a completely unified experience of existence—the ultimate sense of safety, security, and survival.

In the upcoming chapter on higher states of consciousness the goal of permanent fulfillment in enlightenment is identified as the overriding direction of evolution and of selective choice and action in nature. In the more limited realm of animal species that have lower capability than the human species, however, individual and species biological survival characterizes the restricted scope of behavior. But biological survival can be seen as a sub-set or special case of the more universal natural intention or goal of permanent fulfillment in the eternal unity of life.

As an additional angle on this expanded understanding of survival, much of human behavior can be understood as attempts to establish some sense of lasting contribution in the general direction of at least a taste of immortality—such as by making one's mark on history. This may be understood in terms of seeking fame and fortune, or raising a family that carries on the family name. It might include, for example, aspirations to establish a record of athletic performance difficult to top, or to create a classic novel or painting that attains a status far beyond its historical period, or to discover a new scientific law or phenomenon, or to build a lasting monument such as a memorial statue, mansion, or grand architectural complex that symbolizes power and permanence. These *outer* objective notions of permanence, if considered the highest goals, would be typical of the ordinary waking state of consciousness, associated with partial limited paradigms of understanding of the full range of natural law and the full range of human potential.

All enter God's Kingdom."—*R. H. Boyer, Biblical researcher*
14

Maharishi has pointed out that the level of evolutionary development of humans is such that all of us are capable of evolving to the highest state of consciousness in which the entirety of cosmic life is fully reflected in individual experience—that the *birthright* of human life is the fully enlightened cosmic individual in permanent fulfillment. This will be brought out profoundly in the chapters on experiences in higher states of consciousness. While generally consistent with the more restricted physicalist worldview that underlies the theory of natural selection in neo-Darwinian biological evolution, it involves a dramatically expanded evolutionary scope with respect to human potential. The survival instinct, hierarchy of needs, and self-actualization can be understood as aspects of a super-ordinate goal of increasing degrees of happiness toward permanent fulfillment. In Maharishi Vedic Science and Technology, this is a universal principle of nature that defines the fundamental functional role of attention and of all feeling, thought, and action: *the purpose of life is the expansion of happiness.*[15] All of the subordinate goals—even those appearing to be in contrast to individual survival or the immediate satisfaction of individual desires—can be understood to be incorporated into the super-ordinate purpose of increasing happiness toward permanent fulfillment.

This universal principle is also described in Maharishi Vedic Science and Technology as the basis for the inherent functional principle of the mind and of attention: *the natural tendency of the mind is toward increasing enjoyment or happiness.*[15] This principle is identified as the super-ordinate guide of all the complex, integrated processes of attention in the individual self—as well as the evolutionary course of the entire cosmos. Consistent with this overarching principle, the act of attending can be described in terms of attention being drawn to a specific object of experience or sensory array, and then further the selected object being held or maintained in attention through a period of time. Attention is drawn by the entire integrated complex of needs, desires, wants, or goals of the individual in a particular environmental context, in the natural pursuit of increasing happiness and fulfillment according to the state of consciousness of the individual organism. This goal of course would include the goal of biological survival, but only as one facet of it, even if most prominent when under threat. What is being described here is that attention is the momentary state of a globally integrated top-down information processing system that is programmed by the universal laws of nature to progress toward a permanent state of survival, a permanent state of oneness with nature—to reflect the totality of natural law. This will be discussed further in the upcoming chapters on higher states of consciousness.

What maintains attention? Now let's consider what holds attention on a particular object or array that the information processing system has been guided or drawn to experience—the stationary aspect of attention as *stationary individual awareness* introduced at the beginning of this chapter. The maintaining of individual attention on a particular object of experience is influenced by the capability to experience deeper, more refined, subtler levels of the object that naturally bring more appreciation, enjoyment, and happiness. When the nervous system is restricted and stressed, the natural flow of attention to more refined enjoyable levels of the objects of sense is interrupted. When stress-free, attention naturally flows to appreciate the universal level of the objects of sense, providing the maximum value of experience. The issue of mental effort, strained or forced attention, and concentration will be discussed in Chapters 21-22. It is a key issue in that aspect of the Vedic literature which deals with the practice of yoga, the Yoga Sutras, one of the six Darshanas introduced in Chapter 11.

In Maharishi Vedic Science and Technology attentional mechanisms are described as naturally and automatically working together to present objects of experience to consciousness that are salient to the immediate sub-ordinate and overall super-ordinate purposes of the organism. This involves the sensory reception of objects of experience and their application to the functions of the mind. The level of mind is involved in controlling the senses, thinking about the sense objects, and analyzing their qualities. The levels of intellect/heart and ego involve critical reasoning about their significance, reflecting upon and determining their possible effects and deciding on their meaning and value to the ego, which leads to intentional action. All of these processes—described in Chapter 17—serve the super-ordinate goal of the organism in the overall pursuit of increasing happiness toward permanent fulfillment.

In Maharishi Vedic Science and Technology the natural tendency of the mind is to be drawn to experiences that are more enjoyable—that increase happiness. Coupled with this natural tendency is the structure of the levels of nature in the inner dimension and also outer objective environment, and specifically that deeper levels of nature are inherently more powerful, holistic, fulfilling, enjoyable, attractive, and charming to experience.

Nature is structured phenomenally in terms of levels, with the deeper levels being more powerful. This is an established scientific principle, exemplified by the atomic level being more powerful than the chemical level and the Planck scale being massively more powerful than the atomic and sub-atomic levels. However, these examples are within the boundaries of the gross relative domain. It is with respect to even subtler levels of the objects of experience that the descriptors of being more fulfilling and charming have more significance. In the growth to higher states of consciousness the deeper, more integrative levels of individuality spontaneously become more lively and fulfilling. The unique aspects of each individual become fully expressed, while at the same time universal love for all the gross and subtle aspects of nature progressively develops.

It can be somewhat difficult to describe the experience of subtler levels of the objects of experience. In the ordinary waking state of consciousness, most all experiences relate only to the gross levels of objects—which covers all layers from ordinary macroscopic to sub-phenomenal microscopic and ultramicroscopic scales. With respect to examples of extending the subtlety of ordinary waking experience, the subtler levels of experience can be depicted in terms of aesthetic or artistic qualities of the gross objects. These are commonly associated with a more powerful subjective sense of refinement, appreciation, expansion, and deepened attributions of profundity, meaningfulness, and beauty. It involves the focus of attention settling down and being drawn to deeper levels of the objects of experience. This process naturally occurs more frequently and more clearly as the mind and body are refined enough—functioning at a fine-grained enough level—to settle into these finer aesthetic values of the objects.

A simple example is the expanded fine feeling level of a mother and father observing their baby's smiles and giggles, related to expansion of the (psychological) heart, the deep fine feeling level. This at times may expand into a holistic feeling of love and delight in all babies—perhaps feeling a wave of attraction and connection of love even for giggling babies in TV commercials. As other examples, at times we are drawn to deeper aesthetic appreciation of works of art or music, or the graceful coordinated movements of a professional athlete in action. These reflect times when we are settled enough to experience finer values of the objects of experience. Such experiences are powerful, moving, enlivening, and naturally attract and maintain our attention—unless some other salient stimuli take precedence, associated with a hierarchy of needs and goals. We may be listening intently to a news story when a favorite song from the neighbor's car radio naturally draws our attention; but then our attention is drawn back as the neighbor's car heads down the street and out of range, or if the newscaster begins to cover an important and personally relevant issue such as an impending severe storm.

These two principles, then, are fundamental to the act of attending: the principle that the natural tendency of the mind is to go to a field of greater enjoyment, and the principle that greater enjoyment is naturally in deeper, more refined or subtler levels of experience of nature. These principles will be discussed in Chapter 21 in connection with systematic natural technologies to develop higher states of consciousness and the crucial principle of effortlessness.

To reiterate the brief discussion in the first part of the Prologue—now in the context of this current discussion of attention—the objectifying approach of modern science has contributed to the conditioning of attention to the outer objective surface level of experience. This is one aspect of the holding of objects of experience in individual awareness in the act of attending. As a result of stress and incoherent functioning, the individual machinery of experience is constrained to the gross surface level of the objects of experience. It is as if frozen on the surface levels of the objects and restricted from appreciating their finer qualities. The natural flow of attention is in the direction of increasing happiness, but the overshadowing influence of stress and strain restrict the natural process of the mind from settling into deeper experience of the objects of attention.

Under such conditioning the deeper, underlying, more holistic qualities become overlooked, illusive, and rare—to the point that even their *reality* becomes doubted. The result is incessant habitual exploration of only the objective surface level of nature. Coupled with beliefs about the normal range of nature being limited to only the gross relative domain, as well as educational training in what and how to focus attention in the objectified means of gaining knowledge in modern object-based education, the objective surface levels predominate in experience. Subtler levels are experienced less frequently. Attention tends to remain on the surface levels, and eventually begins to shift more quickly and impatiently in search of more enjoyable experiences. Attention then becomes associated with patterns that have been characterized in terms of attention deficits and difficulty maintaining attention except for shorter periods or only under more intense stimulus conditions—in the more intense, frantic, transient, and unsuccessful pursuit of increased happiness and fulfillment.

Because the outer surface level of nature is the least powerful—and least fulfilling—dissatisfaction, restlessness, and frustration tend to increase. The focus of attention shifts from one object to another on the surface. As described in the Prologue, this frequently results in shorter attention spans and emphasis on immediate sensory gratification in the search for fulfillment largely restricted to experience in the more mundane and shallow gross surface of life. The holding of attention then involves trying to get more out life through increased intensity and extreme contrasts on the surface level—because deeper levels are inaccessible or don't even seem to exist. Hopefully it is not too big of a leap to recognize how these habitual attentional patterns contribute to existential meaninglessness, social discontent, purposeless, and an accompanying decline of moral and ethical behavior as the gross levels of experience are pushed to the extreme in misapplied attempts to increase happiness in the flatland of material life.

In terms of the concept of the *mistake of the intellect* (*Pragya aparadha*) referred to in earlier chapters, the disruption of the attentional processing of objects such that they are not experienced in their full significance is associated with the ordinary waking state of consciousness. In this state there is a fragmentation of experience into separate objects or the known, the processes of knowing, and the knower. There is *incomplete* processing of the objects of experience; the parts dominate. The subtler, more integrated levels and their universal basis in the unified field are overlooked, and the unity of nature remains unavailable to direct experience, resulting in fragmented, incomplete, and unfulfilling experiences. When this is the habitual pattern, guidance in reestablishing the effortless process of deeper levels of experience is quite helpful. This is fundamental to the simple but subtle mechanics of the Transcendental Meditation technology.

By way of summarizing and applying points in this and previous chapters so far, we can say that a phenomenal object of attention only appears to be a discrete, static, concrete, independent object. Even the gross human body as an 'object' is a changing whirl of molecules, atoms, and flows of abstract energy—all of which are being replaced regularly through time. It is held together by the thought or intelligence of the object, which is based on memory of the object as a consistent history, present, and future through time. Attention to the object by the subject enlivens the object in the subject and enlivens the subject in the object. Attention is drawn or charmed to deeper, subtle levels of the objects, until the object eventually is experienced at its universal level—which is also the universal level of the subject, the universal Self, the source of everything.

In the natural growth to higher states of consciousness the attentional or referral loop from consciousness to independent object and back to consciousness becomes subtler and subtler, more abstract and refined. Ultimately the object is experienced as nothing other than the universal Self—Self-referral consciousness—while simultaneously being observed at all its gross and subtle individual levels. This is the course of manifestation and evolution from subject to object and back to subject, and back to all objects throughout the wholeness of existence. It is a phenomenal process involving different levels of experience of *reality* in different states of consciousness. When the flow of attention is not interrupted by the limitations of individual functioning of the spotlight of attention due to stress, strain, and imbalances in the organism, attention brings the object completely back to its origin in one's universal Self. The object is then spontaneously appreciated in its infinite value of totality and in its individualized phenomenal expression of relative finite values at the same time.

"There are a number of conundrums about the mind which a physical model cannot explain, such as for example the apparently unlimited information processing capacity of long-term memory, the rapidity with which negative occurrences can be known (implying lack of an exhaustive search strategy), and the extreme lability of organization of information in memory. A spatial model of human memory implies serial exhaustive search for information not in memory. Yet often we can know that something is not in memory faster than locating something that is—such as knowing we have not been to Singapore versus knowing we have been to Malibu Beach. Whereas computer memories function on the basis of serial search strategies, the human memory is much more flexible and holistic in its functioning. A fundamental shift in how we conceptualize memory in our cognitive models is needed."—*D. E. Charleston, research psychologist* [16]

Attention and Memory

In Maharishi's Absolute Theory of Attention the process of experiencing a sensory object involves shining the spotlight of attention on it—so to speak—and then experiencing deeper levels of the object through corresponding activation of deeper levels of subjectivity. When the deeper levels of subjectivity are foggy or incoherent, then the object is appreciated at less depth and less value or meaning. The value of each object in terms of the totality or wholeness of knowledge and experience is lost, and only its fragmented, partial, gross values are experienced. In other words the memory of the total value of the object of experience—recollecting the totality of nature or wholeness of the unified field in each part of nature—is overlooked and not fully enlivened.

Regardless of the level of subjective experience and degree of the total value of the object that is appreciated by an individual, the processing of information involves the same general mechanics. The mechanics, especially associated with the processes of memory, reflect an extremely subtle and remarkable understanding that is quite different than computational and structural models in cognitive and neuroscientific theories.

In Maharishi Vedic Science and Technology memory is located in the self-interacting dynamics of the unified field of nature, rather than at a particular level or stage of subjective function or structure. This is not to deny the structural conception of selective attention and memory processes as being built phylogenetically into the entire organism—as described for example in models in Chapter 16 and mentioned again earlier in this chapter. We can say that organisms remember how to respond to environmental changes in part by virtue of their phylogenetic structure—analogous to a hammer sort of 'remembering' its function by its very structure. Organisms' actions are according to their own natures that are built into each level of individual psychoarchitecture and bodily structure. In addition to this neurocognitive model of memory, however, a more comprehensive holistic understanding is described in Maharishi Vedic Science and Technology.

This holistic understanding of memory is such an incredibly expansive and integrative view of nature that it may be quite challenging for many to grasp intellectually. In this holistic understanding the concept of memory and the mechanics of memory processing are directly tied to the nature of the unified field as the field of all possibilities, the field of infinite correlation, that guides all feeling, thought, and action, functioning holistically at the underlying basis of nature itself. These underlying mechanics of memory relate to universal as well as all individual expressions of natural law.

Recall the discussion in the past chapter about the inward and outward flow of energy and intelligence through the levels of individual psychoarchitecture. In the outward flow impulses of energy and intelligence are expressed from the self-interacting dynamics of the unified field of consciousness into the levels of individual ego, heart and intellect, mind, senses, body, behavior, and then into the environment. Correspondingly in the inward flow of information from the environment via the senses, the information goes through these levels of subjectivity and is presented to consciousness. It is at the level of consciousness that the information input interacts directly with memory—not at the levels of the mind or screen of the mind.

In this most fundamental sense of memory, the storehouse of memory is embedded in the total potential of nature at the transcendental level of all possibilities—the level of Prakriti—the home of all the laws of nature. At this level of nature all individual memories including individual episodic memories of specific past events are recorded—as well as the universal memory of the structure of nature, the laws of nature. In this sense memory refers to the abstract underlying laws of nature that compute and bring consistency, coherence, and continuity to all their expressions throughout manifest creation. This ultimate meaning of memory is called in Vedic science the level of *Smriti*— the unmanifest basis for structuring everything that manifests. Maharishi states:

> "The tree is there in the unmanifest seed, but it is not the tree, it is the memory of the tree... [The] whole tree is nothing other than the nothingness in the unmanifest. There is a sequential evolution of nothingness into everything concrete." [17]

Because the universal laws of nature structure all change, they also structure change on the level of individual action, including each individual stimulus-response connection. This means that each and every stimulus event that triggers a subsequent response event is ultimately linked to the continuity of memory at the infinitely correlated level of functioning of the totality of nature. It is only at the universal level of nature that all action and reaction—all karma—can be computed holistically in order to take into account fully the entire history of individual and universal nature throughout all levels, processes, and events. It is only at this ultimate holistic level that nature can compute the consistent history of everything occurring at the same time throughout the entire universe. Maharishi sometimes calls that functional level of nature the *'cosmic computer.'*

In this most comprehensive meaning, activation of memory is not in the individual mind and body as such. Rather it is in the self-interacting dynamics of the unified field of consciousness— the basis of all individual minds and individual actions. It is only when the individual is not fully developed to appreciate these ultimate dynamics of memory that there appears to be an individualized storehouse of memory and individual memorial processes in the mind and body, although the individual seems to have no experiential component or introspective access to these mechanics. Even in cognitive and neuroscientific models of memory, the mechanics of memory processing are outside of introspective experience. One reason for this is that these mechanics operate at a level of nature that is more fundamental than the level of individual mind and ordinary subjective experience—that is, they are beyond the individual mind, in the field of the coexistence of individuality and universality.

Let's now put these points about the mechanics of memory into the context of the model of individual psychoarchitecture that is a major theme in this part of the book, which might make this highly abstract holistic understanding of memory a little more concrete. It also will add to the framework for the next topic: universal determinism and individual free will.

Recall the initial discussion of stimulus and response in the behavioristic approach introduced in Chapter 15. Inside the black box—the gap between stimulus and response—we now have located and unfolded *all the individual and universal levels of nature*. From a general phenomenal perspective we can say that stimulus information flowing inward through the senses is processed at deeper and deeper levels of the mind, intellect, heart, and ego to be presented to consciousness. It is in consciousness that the connection to memory as Smriti is made. From within consciousness, knowledge in the form of thought energy and intelligence emerges to be expressed in the outward flow as a response. The mechanics of memory activation and processing go on automatically and are outside of ordinary individual experience. This is one important way to understand the key principle in Maharishi Vedic Science and Technology that, "Knowledge is structured in consciousness." [18] We can also say that knowledge as contained in the concept of memory is structured in consciousness.

This has some similarities to workspace models of consciousness in cognitive models of the mind (See Chapters 15-16). In the workspace models, activation of memory is outside of individual conscious experience, but consciousness is thought to play a facilitative role in the process of computing response output. In Maharishi Vedic Science and Technology this facilitative computational function is associated with the level of the cosmic computer, the level identified in the Darshana of Sankhya as Prakriti, Nature, the Veda itself—not the level associated with the screen or stage of the mind (manas).

To analyze the nature of consciousness in this description of the mechanics of memory access, it is helpful to refer back to the distinction between Purusha and Prakriti discussed in Chapter 11. Consciousness means not only individual consciousness but also the level of the infinite self-referral, self-interacting dynamics of unmanifest Purusha and Prakriti, the ultimate totality of natural law, the unified field of nature.

Although Prakriti and Purusha together can be said to be the essential 'nature' of consciousness itself, until everything is experienced as nothing other than consciousness itself a delineation can be made in which Purusha is the silent witness of consciousness and Prakriti is not consciously experienced in terms of its self-interacting dynamics. It is when the wholeness value of the phenomenal distinction of Purusha-Prakriti is primary that these dynamics can be directly experienced as the unity of consciousness itself.

The four stages of the gap. Understanding the mechanics of individual memory thus also means understanding the mechanics of transformation that brings continuity through any and all change in nature. Ultimately the mechanics with respect to an individual stimulus-response connection reflect the same universal mechanics of transformation that conduct all phenomenal change throughout any level of creation—any change of any type on any level.

Maharishi has beautifully described these mechanics in the *four stages of the gap.* [19] This meaning of the gap can be described as a very abstract conception of the junction point or gap between any two sequential states of existence—associated with the universal mechanics of transformation throughout nature. One example of a gap is between stimulus and response, from which the concept of the black box of the mind emerged. This would be a gigantic gap, basically including in it all the levels of body, mind, and consciousness—all the levels of nature underpinning the ordinary gross level of macroscopic behavior. We will later consider a profound understanding of the gap within which the universal mechanics of transformation are reflected in the inherent sequential structure of the Veda expressed in Vedic language and the gaps of silence between Vedic words.

Any change throughout nature can be understood most fundamentally in terms of the mechanics of change from an existing state into the next state through a sequence of steps of transformation. There is a systematic dissolution of the old state and creation of the new state. This can be described as a four-stage sequence in which 1) the initial state is reduced or fades into nothingness (*Pradhwamsa-Abhava*); 2) *that* nothingness is the absolute fullness of infinite silence (*Anyonya-Abhava*); 3) infinite silence is infinite dynamism, which through its infinite self-interacting dynamics or infinite creative potential computes and structures the new state (*Atyanta-Abhava*); and 4) the new state manifests (*Prag-Abhava*). In infinite silence (Anyonya-Abhava) is infinite dynamism (Atyanta-Abhava), from which emerge all phenomenal change in nature.

In this delineation, the second stage can be associated with Purusha—infinite silence, the non-changing, uninvolved witness, pure consciousness itself (also sometimes called *Puran)*—and the third stage can be associated with Prakriti, infinite dynamism, Nature, directly associated with memory as used in this discussion.

The meaning of memory used here—*Smriti*—refers to the ultimate capability of nature to 'remember' the totality of all laws of nature and simultaneously apply them to structure consistent histories of change through all levels of time and space. It can be likened to the database and computational ability of nature itself that embeds all possibilities of past, present, and future into every moment of time and space. It is the infinitely self-interacting, infinitely correlated field of all possibilities that is the unified field of consciousness, the total potential of natural law. It is the cosmic computer—computing all *karma*, the lawful relationships of action and reaction—incorporating into every change in nature the total potential of natural law. Maharishi explains:

"*Smriti* means memory. Everything in nature is run by memory... The one law that governs the universe is memory, Smriti...as in 'what you sow you will reap'... Self-referral...always memory of the Self, beyond boundaries... The whole projection of thought, desire, acting is from that level that is always sunk in silence... Memory is Prakriti; consciousness is Purusha..." [20]

Every instant of time is an ending, a transition, and a beginning—the three-in-one self-interacting dynamics of natural law. The entirety of existence is contained in every instant, infinitely complex and holistic, with all of history spontaneously calculated to be consistent in its progressively unfolding change by the cosmic computer—Smriti. Emphasizing the relationship of memory to natural law, Maharishi recently stated:

"On the basis of memory evolution is carried out. It is the memory and memory is not made new. No new memory can be created—mango tree from a mango seed. No new memory is required to make the apple from the apple seed, the banyan tree from the banyan seed." [21]

Maharishi refers to these eternal patterns of nature by referencing the following Vedic verse:

"*Yatha purvam akalpayat.*
(Rk Veda. 10. 190. 3)

The emergence of creation is always set—it is as it was before.

It is on the basis of memory—*Smriti*—that the *Shruti*, the expression of Natural Law, advances in a set sequence—*Anupurvi* of the Veda—the sequence of the Laws of Nature that is always eternal and invincible in Nature... This is how time and again creation is as it was before—all possibilities keep on expressing themselves within the unbounded range of point...and infinity...in the same perfect sequence, again and again." (p. 527) [22]

Every change in nature, including all individual actions and their consequences, is ultimately computed in Smriti, the total memory of natural law that continually expresses anew the eternal sequential patterns of nature. For our purposes here, we can liken the Smriti value of Prakriti to the ultimate storehouse or database of the total interactive memory of the universe—including the universal laws of nature that structure all change in nature. When a specific Smriti or memory is activated, it can be said to be associated with specific vibrations. These vibrations can be understood to be sound vibrations—called *Shruti,* which means 'that which is heard.' Drawing from the total potential of natural law or *Smriti* at the level of Prakriti and Purusha, the new state or form manifests into phenomenal nature as fluctuations or vibrations from *Shruti*, the vibratory component of sound. [23]

Perhaps bringing out additional aspects of the computer-human analogy will help make more tangible these subtle points about the universal functioning of memory. Each personal computer typically has its own memory system, limited by its storage and processing capacities. When the computer wears out, it is discarded. In the physicalist view that human consciousness and memory are products of complex neural processes, it may seem reasonable to compare and even equate computers and humans in this regard. But there is a fundamental difference that relates directly to the crucial issue of the nature of memory in living systems—such as us.

The Internet allows access to a seemingly unbounded database of knowledge through personal computers. This has been accomplished by networking a multitude of computers and their individual memories into an extensive system that dramatically enhances the capabilities of each part of the system. This system can seem to take on a life of its own—so to speak—in that it can continue whether any individual computer remains connected to the system or not. But there is a lot more lively inside the human processing system than in the personal computer and the Internet (refer to discussions in Chapter 13-17). Embedded in and active in each of us as living beings are the subtle levels of mind and their basis in the unified field of consciousness. Our individual memories are directly connected to and are activated in the cosmic computer, the home of all the laws of nature, the unified field of consciousness, inside each of us. We can directly experience this level as the least excited state of our own mind.

The personal computer is not actively connected to the cosmic computer in the same way that we as human beings are (even though the physical structure of the personal computer, like all things, ultimately also is nothing other than the unified field). Our memories are directly linked to the entire history of the universe in the memory of the cosmic computer, Smriti, the totality of natural law. Individual human memory is embedded in Smriti; the cosmic computer is within each of us at the core of our individual being. This is quite different from linking the memory of a personal computer with the Internet. The Internet is a constructed collection of parts on the gross physical level of nature, outside of each separate personal computer with its own memory that is linked to it—not an inherent holistic underpinning of the parts that can be accessed within each part on its own.

More will be said about the subtle dynamics of Smriti or memory in the final chapters of the book, associated with Vedic language. In those chapters we will discuss *Vedic cognition*—cognition of the Veda as Shruti or the vibration of the self-interacting dynamics of the totality of natural law. The key point here is that memory is what brings continuity to the sequence of change in nature, on both universal and individual levels. This holistic understanding of the mechanics of memory ties even individual memories directly to the total memory of nature associated with the universal laws of nature in the field of all possibilities, the field of infinite correlation, the cosmic computer, the unified field. *Everything* is automatically computed and recorded in the cosmic computer, and is applied at every instant to structure the integrated wholeness and consistency of all the laws of nature. Every individual change in nature has to be consistent and interactive with the simultaneous totality of all change at each moment in the universe.

Using this general framework for understanding individual psychoarchitecture and the mechanics of attention and memory unfolded in the past few chapters, let's now consider the challenging concept of free will. This discussion of free will also serve as a summary of many of the key planks that have been laid down throughout the book. This discussion establishes a basis for understanding how the notions of universal determinism in nature and individual free will can be reconciled.

Free Will and Determinism

As mentioned in the Prologue, modern science has strengthened our collective belief in universal order in nature, as well as in universal laws of nature that can be known and applied to improve our life. However, some contemporary theories posit that there is a limit to the orderliness and predictability of nature—specifically that complete determinism does not extend into lower-order physical processes at the smallest time and distance scales of the quantum levels of nature. Prominent interpretations of quantum theory such as the orthodox interpretation describe nature as *fundamentally* random and indeterminate—quantum uncertainty or ignorance (refer to Chapters 4-7). These interpretations are supported by the uncertainty principle, the inability to explain and predict processes at these fundamental scales, as well as the success of the probabilistic model of quantum theory in accounting for a tremendous range of natural phenomena.

Considerable thought has been given in the past century toward comprehending how nature seems to reflect precise order on the classical level of nature but only probabilistic order on the quantum mechanical level. In Chapter 5 we pointed out the great reluctance of Einstein, for example, to accept interpretations of quantum theory based on probabilism as a complete theory of nature. One reason is that it would appear to require that nature is fundamentally random and not completely deterministic, in contrast to his belief that God does not play dice. In Chapter 7 we also noted the challenge to the concept of free will if the only two choices for understanding fundamental change in nature are either classical determinism or quantum chance. A completely deterministic physical chain of cause and effect suggests no possibility of inserting into the unbroken chain a freely made choice. On the other hand chance also suggests no ability to influence the causal chain by a free choice, inasmuch as it would not be by choice but rather random, indeterminate, and capricious.

However, recent theories of the foundations of time and space propose a nonlocal information field underlying or within localized conventional space-time. As discussed in Chapters 7-8, this abstract underlying information space is not necessarily beyond all cause-effect relations, even if beyond cause-effect relations within the limitations of Einstein locality and particle interaction causality. Extremely subtle, nonlocal but deterministic influences could mediate cause-effect relations in this more abstract field of existence that are not measurable within the framework of Einstein locality and the known fundamental forces. This may involve nonlocal causal relations that are of the nature of an underlying field of mind rather than of physical matter. Again, the notion of the implicate order in Bohmian mechanics is an example of this direction.

As exemplified in the Sankhya system of levels of nature introduced in Chapter 11, Maharishi Vedic Science and Technology extensively describes this subtler-than-matter level of nature. In this approach causal relations extend into levels of nature underlying the physical, and directly relate to the power of mental intention to initiate and direct action on the gross level of material *reality*. Such causal influences are at least virtually undetectable within the limitations of objective methodology according to the current mainstream understanding in modern science. Incorporating this subtler, nonlocal, mental level of nature, quantum processes may be deterministic rather than random—even if still probabilistic due to classical ignorance and their unfathomable complexity. If nature is most fundamentally not random but rather is deterministic even if only probabilistically determinable, the issue of how a free choice is possible in a deterministic physical and mental universe certainly calls for explanation.

The experience of free will. On the subtle inner level of mental experience, what are key components of a free choice? People experience themselves as choosing freely when the decision is not thrust upon them from some outside source—including outcomes of random processes. The external context certainly constrains the options available in a free choice. Likewise the internal context constrains the options. Both the external and internal contexts theoretically can be traced back in a continuous chain of causes. Where *I* am now is determined to a large degree by where *I* was just a moment ago, earlier today, yesterday, to my birth, my parents, their parents, and so on. What *I* experience in myself as options in making a free choice is influenced by my intelligence, education, past general and specific experiences, state of mind, to some degree what *I* have eaten, how rested *I* am, and so on. What *I* choose is conditioned by my entire history, apparently also linked in an unbroken causal chain to the entire history of the universe. This unbroken causal chain can be thought of in terms of local physical processes and causes, but also can be thought of in terms of nonlocal mental processes and causes.

In the current models in modern science the mental is a product of the physical, and it is not easy to comprehend how the mental could add to or change the chain of physical causation. However, the chain of physical causation is interpreted only in terms of gross physical processes. This is a core feature of the reductive physicalist paradigm. However, subtler non-material processes are beginning to be recognized, such as in theories of quantum mind and Bohmian mechanics. In this way of understanding, change in the physical can be influenced and guided subtly but powerfully by the underlying mental. Because the underlying mental level is theorized to be nonlocal, highly interactive, and more holistic in its functioning than the physical, the factors determining outcomes might well be only probabilistically calculable due to their unfathomable and, practically speaking, infinite degree of complexity. In this framework universal determinism and causal relationships would apply, and probabilism would also apply but due to unfathomable complexity rather than a fundamental random element (refer to discussions in Chapters 7-8).

If all the external objective and internal subjective precedents were taken into account, the room or degrees of freedom left within which a 'free' choice could be made would seem to be narrowed down quite a bit. These precedents would include everything *I* could think of considering in my choice at the time, as well as everything else that has had some causal contribution to who, what, and where *I* am in the present moment—incorporating the entire history of the universe—even though *I* may not be able to incorporate them into my thinking and planning of action.

Even if *I* were able to incorporate in my choice everything from the past by my volition and intelligence, on what basis would the choice *I* make be consistent or inconsistent with my history? If *I* feel it is inevitably consistent, then where is the freedom, if *I* end up having to choose something that was entirely destined by my past? If *I* feel it is inconsistent, then am *I* in a dysfunctional state of some kind and thus unable to make a free choice, in which case the choice was not freely made by *me*? Would it all come down to an almost inevitable choice consistent with my history, but the last bit or degree of freedom was somehow random—a quantum jitter? It might then be free in the sense of not predetermined, but would it be a choice that *I* attribute to' *myself* as freely making?

Any choice that *I* would attribute to *myself* as making would seem to be a choice that *I* experience as a conscious choice *I* am aware of intending to make. *I* can discriminate actions *I* take that *I* don't attribute to *myself* as choices *I* am making. In the case, for example, of a reflexive reaction to brush away a fly, *I* can think *I* am doing it or can think it is happening as an automatic reaction not accompanied by a sense of intentional or volitional choice. In either case *I* attribute that *I* am doing it, but only in the first case do *I* attribute to it that *I* *freely choose* to do it. But even in this case *I* still don't seem to have any experience of the mechanics of the choice. It just is a choice *I* feel *I* am consciously making and that aspects of my mind and body unconsciously implement in some unknown, non-experienced automatic way.

Also *I* have the inner sense that *I* can choose to submit to random chance, but the chance resolution is not accompanied by the sense *I am choosing* the specific outcome of chance—the chance result. There is a difference in my inner experience between random chance and free will.

In free will *I* have the inner sense of choosing or determining the action to take. A core aspect of free will is that it *feels like* a choice *I* consciously make and take. In addition to the choice involving *my* conscious awareness somehow, it also involves *my* inner sense of *doership* or agency— *I* identify with the choice as something that *I* have the power to make. The choice comes from something deep in *me* that is associated with *my* own conscious experience. *I* attribute agency to *myself*, as opposed to any other agency or determining cause of the choice. Free choice thus seems to involve *my* sense of being conscious, of being an individual, of being an agent, and of being free to exercise *my* conscious individual agency.

This brings us to the fundamental issue of who *I* am, or what *I* is. Considering for now that *I* refers to *me* as an individual conscious being, then *I* am a product of everything that has made *me* who *I* am. Am *I* also something more than this?

I can do only the things that *I* am capable of doing given who and what *I* am. If *I* somehow do something that *I* am not capable of doing given who or what *I* am, then the doing would not be a choice *I* would identify with making, but rather something controlling *me* from the outside that is determining the action *I* take. If *I* attribute agency to *myself*, then that which makes the choice has to be within *me*, not outside of *me*. That is, the subjective past and present factors that influence *my* choice will have come somehow from within *me*.

If *I* act according to universal deterministic laws of cause and effect that influence *me*, those laws somehow have to be within *me* rather than just outside of *me* and imposing their determination from outside, if *I* attribute a choice to *me*. Embedded somehow in *my* individual consciousness as at least part of who *I* am must be the universal laws of nature and the universal consistent history of all existence—the cosmic computer—if *my* choices and actions are consistent with them and are accompanied by *my* attribution of will or choice to *myself*.

But if *I* have to act according to the universal laws of nature in an unbroken chain of cause and effect that are even within *me*, then what is it that allows *me* a sense of freedom of choice? Are *my* individual choices completely determined by those laws as they manifest from within *me*, and *my* sense of choice is an artifact or epiphenomenon that somehow has been built into *me* in the universal determinism? Is it that as an individual the universal laws of nature have determined that *I* would falsely attribute agency to *myself*? Or perhaps it is that the laws of nature determine that *I* actually do have free will in some way. If so, how is this consistent with universal determinism and an unbroken chain of cause and effect? If there is some actual freedom in *my* choice, and the freedom is not due to chance if *I* attribute correctly the choice to *me*, then from whence does it come within *me*, in the framework of universal determinism?

If the laws of nature are absolute laws that don't change—Smriti—and they apply to *me* as well as the choices *I* make, then *I* cannot violate the laws—they are fundamental to who and what *I* am. They are *my* own nature, inherent in the core of *my* individual conscious self. But different laws can come into play depending on the choices *I* make. The consequences of those choices, and the laws which implement the consequences, would be inevitable. But *I* also must be in some way *more* than the laws and their manifestation—if my sense of free will is veridical.

According to the principle of *karma*—the universal cause-effect relations of action and reaction—there is free choice of action, but the consequences of the actions are set by the laws of nature. Regarding this point, Maharishi paraphrases a verse from the Bhagavad-Gita:

"You have control over action alone, never over its fruits..." (p. 133) [24]

This means that different laws of nature automatically and inevitably apply to produce the results, depending on the choices made. *I* can choose the action to take; and according to *my* choice, the laws are set and will have their inevitable results. Those results, computed by the totality of natural law, might not be exactly what *I* thought *I* was choosing to have happen, because *I* didn't—and couldn't possibly as an individual—take into account all the infinity of contributing past and present factors that narrow down my choices.

The principle of karma relates directly to universal determinism and the causal nexus of nature. Maharishi has pointed out that determinism relates to the past and the future, and free will relates to the present—the eternal *now*. [25] All past events follow an unbroken chain of causation based on universal laws of nature. Computation of the unbroken chain of causal events that brings continuity to past and future—the functioning of the cosmic computer, Smriti—occurs in the instantaneous present. The cosmic computer—the ultimate switchboard of nature—is infinitely correlated, infinitely self-interacting. But it is more than all finite probabilities and actualities. If it is *infinitely* self-interacting, then it is beyond all boundaries, completely unbound. The eternal present or *now* is the field of all possibilities, and it is forever free to create whatever actualities phenomenally exist. In some unfathomable way it contains both instantaneous freedom and sequential determinism at the same time.

In addition to the principle of karma, there is the universal principle of *Dharma*. Dharma refers to action in accord with the evolutionary direction of nature. This principle can be associated with the *arrow of time*—the principle that time is unidirectional in the forward direction from past to present to future. It is also closely associated with the principles that the purpose of life is the expansion of happiness, and that the natural tendency of the mind is toward increasing happiness. These principles were discussed earlier in terms of the fundamental mechanics of attention.

Nature is structured to foster the expansion of happiness through progressive stages toward realization of the total potential of natural law to be reflected fully in individual life. Choices of actions that are in accord with the progressive direction of nature toward higher evolutionary stages and states, in accord with the force of evolution or Dharma, result in the expansion of happiness. Choices resulting in actions that are in a less progressive, less evolutionary direction inevitably and eventually result in less happiness. It is determined that action will lead eventually to complete realization of the universal infinite value of happiness reflected in the individual. As Maharishi has stated: *"Karma* according to *Dharma."* [26]

To what degree this is occurring in *my* choices, actions, and consequences, and when its full realization will occur, depends on the choices freely taken and the subsequent choices in reaction to their inevitable consequences. Choices are made according to the state of consciousness of the individual. In higher stages, spontaneously there is more capability to evaluate what actions will most likely lead to desired outcomes. But because the precedents and determinants of outcomes incorporate an unlimited number of factors, *I* as an individual cannot take all the factors into consideration, however carefully *I* may try. The karmic influences are unfathomable with respect to the capabilities of the *individual* mind and intellect.

The computation of cause-effect relations is conducted at the universal basis of nature beyond individual mind and intellect—deeper inside them. The universal level of Smriti, the cosmic computer, computes the continuity of all expressions in nature. The totality of the laws of nature that compute cause and effect is inside *me*, not just outside and separate from *me*. It is both in my subjectivity and in the objective world, at the same time. But ultimately, *I* am even more than the cosmic computer.

Because individual mind and intellect are expressions of individual consciousness, and individual consciousness is nothing other than the unified field of consciousness at its transcendental basis, the individual can transcend individual mind and intellect to experience his or her universal basis in consciousness itself. *I* can experience *myself* as being more than all the laws of nature and all its manifest values, including its fundamental orderly laws of cause and effect. In *that* experience *I* can access the total potential of natural law that spontaneously computes action in accord with all the laws of nature. Individual memory is not sufficient to compute the continuity of past, present, and future. But the infinite organizing power of the unified field of consciousness—which ultimately *I am*—does it automatically.

The unified field of consciousness is the field of all possibilities. Repeated direct experience of it expands the possibilities for spontaneous action in accord with the totality of natural law. This results in increasing freedom, increasing free will, increasing spontaneous ability to make choices that are in accord with the evolutionary direction of natural law and that result in increased success, happiness, and fulfillment. Increasing degrees of the sense of free will is associated with higher states of consciousness. In the highest state of consciousness there is total freedom of choice, all possibilities or potential. The total potential of natural law automatically computes the results, which is in accord with individual and universal wishes at the same time through the self-interacting process of infinite correlation. It is simultaneously a completely free choice beyond all limiting values, ever new in its phenomenal expression, and automatically ever consistent with the eternal past, present, and future of the universe—ancient and never new.

From this discussion we can say that free will involves 'will' in the sense of the agency or power to utilize the laws of nature in the evolutionary direction, and 'freedom' in the sense of increasing possibilities toward all possibilities—which is the nature of consciousness itself, including *my* consciousness. It is the essence of my individual consciousness, which is universal consciousness. The field of all possibilities is beyond the relative field of cause and effect. Established in the field of all possibilities, the unified field of consciousness, the individual experiences complete freedom from all boundaries and limitations, and simultaneously experiences spontaneous action in accord with the total potential of the universal laws of nature, computed automatically by the cosmic computer—Smriti—in the unified field of consciousness, which is *my* own consciousness.

In *that* state the *I* of the individual self is completely established and awake in the *I* of the universal Self, the unified field of consciousness, the total potential of natural law—*I am that*, *I* am the universal Self. In that state *I* am one and nothing other than the absolute, eternally free, uninvolved silent witness of nature (associated with Purusha) and at the same time the agent for all cause-effect relationships throughout nature (associated with Prakriti). *I* am the master of natural law; *I* can do anything. Then *I* no longer identify *my* individual self as primary but rather *my* universal Self as primary, of which *my* self is phenomenally an individual manifestation—the superposition of individuality and universality. *I* then experience universal freedom and the freedom to act in accord with the totality of natural law *simultaneously*. *I* have *both* universal determinism and free will—ever bound to be ever free. *My* increasing alignment with total natural law brings increasing freedom in all possibilities, until *I* fully integrate complete freedom and completely determined action in accord with total natural law simultaneously.

Ultimately we can associate free will with the unified coexistence of opposites of infinite silence—complete freedom— and infinite dynamism—infinite self-referral action in accord with total natural law. The 'free' part of free will is *my* unbounded universal Self, the field of all possibilities, and the 'will' part is the totality of natural law within *me*. The completely unified coexistence of these apparently opposing values is the 'nature' of *my* consciousness. *I* have freedom of choice, and *I* am determined that *I* will choose increasing happiness to establish ultimate fulfillment and freedom. *My* choices determine when. The direct and ultimate confirmation that *I* have free will naturally comes in *my* highest state of full enlightenment, unity consciousness. Maharishi simply and beautifully summarizes the *whole point* about free will and determinism in his description that they are two sides of the same coin. [25]

In the next two chapters systematic development of higher states of consciousness to the pinnacle of human development and complete freedom in unity consciousness is discussed in some detail. Reports of experiences in the various states of consciousness are included to provide a more tangible sense of the developmental progression.

Chapter 18 Notes

[1] Maharishi Mahesh Yogi (1970, August), Personal communication, Teacher Training Course. Humboldt State College, Arcata, CA.

[2] Hernandez-Peon, R. (1966). Attention, Sleep, Motivation and Behavior. In Bakan, P. (Ed.). *Attention.* New Jersey: D. Van Nostrand.

[3] Broadbent, D. E. (1981). Association Lecture: From the Percept to the Cognitive Structure. In Long, J. & Baddely, A. (Eds.). *Attention and perception IX.* Hillsdale, N. J.: Erlbaum.

[4] Laberge, D. (1981). Automatic Information Processing: A Review. In Long, J. & Baddely, A., (Eds.). *Attention and perception IX.* Hillsdale, N. J.: Erlbaum.

[5] James, W. (1890). *The principles of psychology.* New York: Holt.

[6] *Creating an Ideal Society: A Global Undertaking.* (1977). W. Germany: MERU Press.

[7] Sharma, P. V. (1981). (Ed., Trans.). *Caraka Samhita,* Vol. 1. Varanasi: Chaukhambha Orientalia.

[8] Edelman, G. M. & Tononi, G. (2000). *A universe of consciousness: How matter becomes imagination.* New York: Basic Books.

[9] Llinas, R. R. (2001). *I of the vortex: From neurons to self.* Cambridge, MA: MIT Press.

[10] Schwartz, J. M. & Begley, S. (2002). *The mind & the brain: Neuroplasticity and the power of mental force.* New York: ReganBooks.

[11] Dalai Lama & Cutler, H. C. (1998). *The art of happiness: A handbook for living.* New York: Riverhead Books.

[12] Maslow, A. (1962). *Toward a psychology of being.* Princeton, N. J.: Van Nostrand.

[13] Hampden-Turner, C. (1981). *Maps of the mind: Charts and concepts of the mind and its labyrinths.* New York: Collier Books.

[14] Boyer, R. H. (2003). Personal communication, Bellingham, WA

[15] Maharishi Mahesh Yogi (1963). *Science of being and art of living.* Washington, D.C.: Age of Enlightenment Publications.

[16] Charleston, D. E. (2005, Spring). Personal communication, Fairfield, IA.

[17] Maharishi Mahesh Yogi (2003). Maharishi's Global News Conference, October 15.

[18] *Maharishi International University Catalogue, 1974/75.* Los Angeles: MIU Press.

[19] *Celebrating Perfection in Education: Dawn of total knowledge* (1997). India: Age of Enlightenment Publications.

[20] Maharishi Mahesh Yogi (2003). Maharishi's Global News Conference, November 10.

[21] Maharishi Mahesh Yogi (2004). Maharishi's Global News Conference, March.

[22] Maharishi Mahesh Yogi (1996). *Maharishi's Absolute Theory of Defence: Sovereignty in Invincibility.* India: Age of Enlightenment Publications (Printers).

[23] Maharishi Mahesh Yogi (2003). Global News Conference, October 15.

[24] Maharishi Mahesh Yogi (1967). *Maharishi Mahesh Yogi on the Bhagavad-Gita: A new translation and commentary, chapters 1-6.* London: Penguin Books.

[25] Maharishi Mahesh Yogi (1971). *Free will and determinism.* Videotaped talk from Symposium on the Science of Creative Intelligence, July 24, Amherst, MA.

[26] Maharishi Mahesh Yogi (2005). Maharishi's Global News Conference, April 13.

Chapter 19

Transcendental Consciousness: The Bridge to Unity

In this chapter, we will begin a consideration of the seven states of consciousness in ancient Vedic science and Maharishi Vedic Science and Technology. The chapter focuses on the ordinary three states of sleeping, dreaming, and waking and their underlying basis in the fourth state, transcendental consciousness, the first higher state of consciousness The other higher states will be discussed in the next chapter. Descriptions of higher states of consciousness are supplemented by first-person reports of experiences, taken from research on practitioners of TM and its advanced programs as well as historical and contemporary accounts from other sources. The overall main point of this chapter is that the basis for the natural sequence of development through the seven states of consciousness is the direct experience of transcendental consciousness.

In ancient Vedic science and Maharishi Vedic Science and Technology, the active ingredient in facilitating development of higher states of human consciousness is transcendental consciousness, the direct experience of the unified field of natural law, the field of all possibilities and source of all actualities in phenomenal creation. It is the simplest ground state of individual mind, the state of complete inner silence. Initially direct experience of transcendental consciousness may seem like nothing—a blank, flat, inner state void of sensory and conceptual content. Eventually, however, that inner state reveals itself to be the essence of oneself, the source of all objects in phenomenal creation, and the mechanics of their manifestation—the unified totality of everything in nature, the universal Self. The entire course of evolution of higher states of consciousness is unfolded within the ground state of the mind, self-referral transcendental consciousness. Regular practice of the TM technique efficiently fosters transcendental consciousness.

In Maharishi Vedic Science and Technology, the full range of levels of nature and the evolutionary sequence associated with them can be summarized in five concepts:

Atma, transcendental consciousness, the universal Self on the individual level;

Veda, the unmanifest basis of phenomenal creation, the home of all the laws of nature;

Sharir, the individual body reflecting the structure of the entire cosmos, the cosmic individual;

Vishwa, the entire structure of the phenomenal universe, the cosmic body;

Brahm, the Self on the cosmic level, the universal Self, permanently living the totality in fully awake consciousness.

The source, course, and goal of evolution can be described as unfolding from Atma to Veda, to Sharir, to Vishwa, to Brahm—from point to infinity and infinity within each point. From the perspective of the ordinary waking state of consciousness, individual evolution is systematically facilitated by settling back the individual self to its basis in self-referral transcendental consciousness—the universal Self—like a wave effortlessly settles back into the ocean; then progressively experiencing all levels of the ocean of the cosmos within its self-referral dynamics.

From the active mental state of ordinary thinking and perception, this development involves individual attention settling down within itself, curving back upon itself, and transcending mental activity to experience directly its self-referral basis in *Atma*, transcendental consciousness. Embedded in that direct experience is the totality of natural law, the home of all the laws of nature, *Veda*. That direct experience spontaneously structures attunement with the total potential of natural law through total mind and brain functioning in the individual body, *Sharīr*. Individual consciousness progressively expands to encompass the totality of the phenomenal universe, *Vishwa*—enlivening the cosmic status of nature in the individual. This culminates in fully awake consciousness of the ultimate unity of individual self and all existence as the universal Self in daily life in the highest state of full enlightenment, *Brahm*. One again becomes fully awake to what one always is, individual consciousness 'remembers' that it is universal consciousness: *Atma* is *Brahm*.

This overall developmental progression can be understood simply in terms of actualizing the inherent natural tendency and desire to know and be oneself fully. It relates to progressive stages of purity of individual mind and body. This was referred to in Chapter 2 in terms of eliminating distortions and limitations due to accumulated stress and tension that interfere with living in accord with the total potential of natural law. All in all, it is the story of relatively increasing happiness, success, and fulfillment to the ultimate permanent state that is *infinitely fulfilling*.

Maharishi Vedic Science and Technology describes this phenomenal evolutionary progression through seven states of consciousness. The delineation of states of consciousness provides a framework for intellectual understanding of the progress of subjective experience to full enlightenment. The descriptions offer helpful milestones and guideposts that point to and support the direction of progress. However, they are general descriptions, and are not necessarily experienced in precisely the same way as individuals who report the experiences might describe or explain them. Also, it is possible to have glimpses and periods of the experience of higher states even before one's current state is established and stabilized. There is generally, however, a phenomenal sequence of growth in which one state or stage is the platform for the next stage, with characteristic experiences associated with each state.

The delineation into states of consciousness is a way to understand the levels and range of development, and to help clarify related experiences. Of course intellectual understanding of higher states is not the same as actually being in and spontaneously living them. The descriptions relate to experiences that are far beyond what has been envisioned in theories of human development in modern science. Hopefully the content in this book presented so far has fostered the understanding and clarity of vision to look across the bridge to unity so that these higher states of consciousness can be understood to be the natural inevitable development of full human potential that all of us will eventually achieve and validate directly in our own personal experience.

Theories of the Endstate of Human Development

Revered individuals throughout history have described exalted inner experiences as a great source of meaning in their lives and inspiration for their contributions to the sciences, arts, humanities, and religion.[1] Recognizing the significance of these typically fleeting and rare experiences, theorists have sought to identify their role in understanding the range of human potential.[2][3][4][5][6]

Major contemporary developmental theories have attempted to characterize the most advanced permanent stage or *endstate* of ontogenetic development. Applying objective research methodology, the theories have focused initially on stages of growth in adolescence and early adulthood within the range of ordinary experience in the general population. Subsequently they have extended into more advanced stages in adulthood. Prominent contemporary theories will be overviewed briefly in order to provide a context for understanding their natural extension into the much more expansive development of higher states of consciousness.

Formal operations or abstract reasoning. One of the most influential theories of human development has been the perceptual-cognitive theory of developmental psychologist Jean Piaget.[7][8] This theory, introduced in Chapter 1, proposes qualitatively distinct stages of cognitive development that emerge through a child's active perceptual experience of the world. The stages are hierarchical in that each successive, more complex stage integrates the earlier stages while reorganizing and transforming them. The stages are theorized to form an invariant irreversible sequence, each stage serving as the necessary condition for the next stage. According to the theory, however, the stages are neither inherent nor inevitable. They are said to develop based on appropriate experiences involving the interaction of universal, genetically determined physiological structures and universal features of the objective environment.

In this theory the initial stage is the *sensorimotor* stage (typically ages 0-2), associated with cognitive processing of experiences that are closely linked to relatively simple sensory functions and corresponding overt behavior. The next stage is the *preoperational* stage (ages 2-7), in which cognitive processes related to overt behavior are internalized but are still primarily linked to concrete sensory experiences. In the third stage, called *concrete operations* (ages 7-15), abstract cognitive operations can be performed on representations of concrete sensory experiences, an early form of *abstract representational thought*; but these abstract cognitive operations or mental representations still cannot be integrated simultaneously into higher-order problem solving skills. At the next stage, *formal operations* (typically 15-20 years of age, given appropriate experience), higher-order cognition develops. This is described as involving abstract, symbolic, representational thought rather than only mental representations of concrete sensory experiences and overt behavior. In this stage, problem solving can incorporate exploration of abstract logical possibilities rather than just concrete experiences. As referred to in Chapter 1, formal operations is considered to be the necessary stage of development for the hypothetico-deductive problem solving that typifies scientific thinking. Piaget's theory proposed the stage of formal operations—cognitive development associated with abstract rational thinking or representational thought—as the highest stage or endstate of human development.

One important reason formal operations or representational thought was proposed as the endstate of human development is that modern education is heavily committed to intense intellectual development of abstract reasoning as primary means of gaining knowledge. In modern education the focus of training is almost entirely on building abstract reasoning skills.[9] As mentioned in the Prologue and Chapters 2 and 18, this focus trains mental attention to remain in the object-referral or representational mode of thought. Much of the time students are engaged in active mentation that keeps attention in an outward objectifying direction. By force of training and habit, deeper, subtler subjective experiences are de-emphasized and often completely masked. In Maharishi Vedic Science and Technology, by transcending all mental activity including intellectualized reasoning, there is an unfreezing of development and higher levels naturally unfold.[10]

Beyond formal operations and abstract reasoning. Several contemporary theories propose adult growth beyond formal operations, called post-formal or post-representational development.[11][12] [13][14] Related development of advanced moral reasoning, post-conventional development, is associated with proposed higher levels of growth similar to the concept of self-actualization.[11][15][16] This higher level is characterized by high self-esteem, ego integration, creativity, moral vision, and respect for others and nature. Developmental research has indicated that perhaps less than 1% of the general population achieves a mature level of self-actualization.[17]

One prominent theory of post-formal development by developmental psychologist Jane Loevinger focuses on advanced levels of integration of the individual self.[18][19] This hierarchical theory of levels or stages not only encompasses stages of cognitive development paralleling the stages in Piaget's theory, but also emotional, moral, ego or self, and social development.

In the initial *pre-social/symbiotic* level, sensory processes predominate and there is little or no differentiation between sensory experience and the inner sense of self. In the next *impulsive* and *self-protective* levels, individuals lack self-control, are primarily concerned with short-term self-gratification, and have difficulty taking the viewpoint of others. In the next level, the *conformist* level, there is a shift from primary concern for physical rewards and punishments to social approval and belongingness. Further growth to the *conscientious* level involves the ability to take oneself and the world as objects of thought—associated with representational or formal operational thought. At the higher *individualistic* level, concern for individual self development and psychological causality emerges, associated with increasing tolerance for different contexts and points of view. The most advanced *autonomous* and *integrated* levels are characterized by deep appreciation for the autonomy and identity of self and of others. Individuals at these highest levels are inner-directed, self-actualizing, and exhibit greater adaptation and resilience. These levels are thought to involve advanced representational thought and reasoning ability, and also post-representational, self-actualizing and integrative experiences.

Another prominent contemporary theory, focusing on moral judgment, also extends into post-representational and post-conventional development. Proposed by developmental psychologists Lawrence Kohlberg and R. A. Ryncarz, this theory extends the Piagetian stages of perceptual-cognitive development to a higher stage, described as a 'natural law' orientation toward ethical and moral questions. [20] This stage is characterized by belief that:

> "...human responsibilities, duties, and rights are not arbitrary or dependent upon social convention but are objectively grounded as laws of nature." [20]

This highest stage identified in the theory is thought not to be achievable only through formal operational thought. Rather it is theorized to require some form of transcendent experience involving a sense of connection or unity between the individual self and the cosmos.

In Maharishi Vedic Science and Technology, post-formal, post-representational, and post-conventional development describe processes of growth within the ordinary waking state of consciousness in the direction toward higher states. Development beyond the ordinary waking state to the next permanent higher state of consciousness covers a huge range of experience. This transition is facilitated by the direct experience of transcendental consciousness, identified in ancient Vedic literature as the fourth state of consciousness.

The experience of transcending thought has been proposed to be as fundamental for promoting development beyond representational thought or abstract reasoning as language and symbol use are for facilitating growth beyond the sensorimotor level to the cognitive level of formal operations and abstract representational thought. In this sense, however, transcendence does not mean integrative experiences such as communion with nature or a greater appreciation of humanity and the world family. It means transcending the individual mind completely, and directly experiencing its basis in the unified field of consciousness. [21]

Reports of transcendent experiences appear in the literature of many cultural traditions, [1] [22] but ancient Vedic records contain the most extensive and detailed accounts. [14] [15] [25] Until recent years, however, it had been quite difficult to investigate transcendent experiences systematically using formal objective methods. This was due significantly to lack of a comprehensive theoretical framework to interpret the reported experiences, lack of experimental paradigms to examine the reports formally, and especially lack of systematic means to replicate them reliably under testing conditions. [1] It has been estimated that as little as one-tenth of one percent of the college population, for example, may have such experiences. [18]

However, theorists focusing on human development have not delineated higher states beyond post-formal stages, tending to lump together higher state experiences into a general transpersonal framework. The TM technique has been an important catalyst to extend research into higher states of consciousness by providing a reliable, repeatable methodology through which large numbers of regular practitioners have reported frequent experiences of transcendental consciousness. [1][25][26][27][28] An extensive body of research has accumulated over the past 40 years based on the investigation of reports of transcendental consciousness, and in some cases experiences of higher states of consciousness as well. This research indicates that TM practice produces beneficial physiological, psychological, and behavioral outcomes expected to be associated with higher stages of human development. It strongly supports the TM technique as an unprecedented technological advance in facilitating growth beyond representational thought. [1][29][30]

Brief overview of the full range of human potential: Seven states of consciousness.

As noted in Chapter 1, in the course of development of experience and reason, our experience of the world curves back to connect with our reasoning about what is *real*. Sometimes an especially significant experience leads to deeper levels of understanding, but also increased clarity of reasoning and understanding leads to more openness to pursue deeper experiences. It is with this intent that descriptions of higher states of consciousness and examples of corresponding experiences are presented. However, fundamentally it is deeper direct experience that promotes clarity and growth of knowledge and experience toward higher states.

The natural inherent desire to know oneself fully is the impetus for all thought, action, achievement, and fulfillment in life, impelling each of us through progressive stages of self-knowledge and experience—described in terms of a sequence of states of consciousness. Each state can be said to have its own *reality*. Our view of the world, or what *reality* is to us, is most fundamentally a reflection of our state of consciousness. Vedic educator Bevan Morris states:

> "There are seven distinct states of human consciousness, each with its own physiology and each with its own world of experience." [31]

With consistent reports and corroborative findings of advanced experiences since beginning to teach the TM technique in the 1950s, Maharishi has brought out ancient Vedic descriptions of the full range of human development in the sequence of seven states of consciousness. The states include four higher levels distinct from the three psychophysiologically defined states of sleep, dreaming, and waking. These first three states largely define the common range of experience in ordinary daily life. Progress to higher states is systematically facilitated through the fourth state, transcendental consciousness. Repeated experience of this fourth state expands individual awareness, purifies, refines, and balances the nervous system and body, and establishes the permanent basis for higher states. It is the bridge through higher states of consciousness to complete unity in full enlightenment.

The fifth state, termed *cosmic consciousness*, is described as a permanent state of enlightenment in which transcendental consciousness is an all-time underlying background of daily life. It is characterized by inner freedom, bliss, fulfillment, and self-realization. Further development spontaneously leads to the sixth state, termed *refined cosmic consciousness* or *God consciousness*. This state is characterized by the ability to perceive and appreciate the most refined value of all objects of experience throughout nature, the most sublime appreciation of relative phenomenal creation—one might say, of the highest relative qualities of God's creation. Development culminates in the seventh state of consciousness, termed *unity consciousness,* in which the individual self is fully awake to its universal status as the infinite totality of natural law—the absolute reality of infinite, eternal Being, the universal Self, Brahm.

Below is a listing of the sequence of these seven states of consciousness. [32] The seven states can be distinguished by the experience of self and environment (subject and object) reflected in each state. They can be placed on a continuum from virtually no wakefulness to infinitely full wakefulness:

Waking *(Jagrat Chetana)*—individual self and relative environment

Dreaming *(Swapn Chetana)*—illusory individual self and illusory environment

Sleep—*(Sushupti Chetana)*—virtually no experience of self or environment

Transcendental consciousness *(Turiya Chetana)*—unbounded wakefulness, universal Self only

Cosmic consciousness *(Turiyatit Chetana)*—universal Self and a separate relative environment

Refined cosmic consciousness or God consciousness *(Bhagavad Chetana)*—universal Self and maximum relative value of environment

Unity consciousness *(Brahmi Chetana)*—object and subject are one; the environment is nothing other than the universal Self

The natural course of evolution through these states of consciousness to permanent full enlightenment is the process of crossing the bridge to unity that is described in this book. Some might react to the apparent inconsistency of describing this process in terms of a bridge or a *path* to somewhere other than where one already is if the goal is one's own universal Self which is always *everywhere*. This is similar to the issue discussed in Chapter 13 that if everything is nothing other than the unified field of consciousness, how can something appear not to be conscious? What does it mean to travel on a path to somewhere that is omnipresent? This apparent inconsistency is due to the limited experience of the ordinary state of waking consciousness in which one is as if cut off and separate from one's inner universal level of experience—when one is ignorant of one's universal Self as the core of one's individual self. It is in this sense that it can be said to be traveling from local individuality to nonlocal universality. Maharishi explains:

"For anyone to be, it is only necessary to be. No path to one's own Being could be thought to exist...because the very conception of a "path" takes one's self out of one's Being. The very idea of a path introduces the conception of something far away, where Being is the essential *oneself*. A path means a link between two points, but, in omnipresent cosmic Being, there cannot exist two different points or states. Omnipresent means "present everywhere"; It pervades everything, and, therefore, there is absolutely no question of a path." (p. 270) [33]

"Certainly, to talk in terms of "path" of realization of one's own Being seems to be unjustified, but because all the time in our life the attention is left outside in the gross relative field of experience, we are as if debarred from the direct experience of the essential nature of our own Self, or transcendental Being. That is why it is necessary to bring the attention to the transcendental level of our Being. This bringing the attention is said to be a way to realize. Thus, although we find the idea of a path to realization absurd metaphysically, it is highly significant on a practical level." (p. 278) [33]

An important point to recognize in considering higher states of consciousness is that a wide variety of understandings and teachings have emerged throughout history that reflect views from different states of consciousness. The overarching systematic framework of seven states of consciousness in ancient Vedic science and Maharishi Vedic Science and Technology provides a means to reconcile these various views and teachings and show their underlying compatibility within the full range of human development. This provides some protection from making the mistake of attempting to apply the *reality* of one state to the *reality* of a different state, which has muddled much understanding about enlightenment in higher states of consciousness and how to development it. This was introduced in the Prologue in Maharishi's statement: "Knowledge is different in different states of consciousness" (p. 278) [34] In the following quote, Maharishi elaborates on this important point:

> "The statement that is true in one level of consciousness is not true in another level of consciousness. The truth of one state of consciousness has nothing to do with the reality of another state of consciousness... This has been the reason that the whole teaching of evolution is found in a muddle and a mess. The teaching starts from the waking state to transcendental consciousness, from transcendental consciousness to cosmic consciousness, from cosmic consciousness to god consciousness, from god consciousness to unity consciousness, and so on. These partitions were not made. And that is why some teaching which was thought to be good, even though it applied to one state of consciousness, was adopted for another state of consciousness, and the whole thing became a muddle... But now, what we have done is, we have put in milestones." [35]

The milestones associated with the delineation of seven states of consciousness described in Maharishi Vedic Science and Technology can be associated generally with the six Darshanas introduced in Chapter 11. Sometimes also described as the six systems of Indian philosophy, the Darshanas are *Nyaya, Vaisheshika, Sankhya, Yoga, Karma Mimansa,* and *Vedanta*. While it is important to appreciate that each Darshana embodies the full range of development through the seven states of consciousness, it also focuses on a specific range of experience associated with a particular state. The first three Darshanas focus on intellectual descriptions of the full range of nature and the means of gaining accurate knowledge, and the last three Darshanas focus more on experiential technologies to develop higher states directly.

Thus Nyaya, Vaisheshika, and Sankhya can be said to focus more on perspectives associated with the waking state of consciousness, in which the individual value of consciousness predominates to various degrees. Yoga focuses more on direct experience of pure universal consciousness in the fourth state of consciousness and union of individual consciousness with universal consciousness in the fifth state of consciousness, in which both universal consciousness and individual consciousness are contiguous experiences, toward the ultimate unity. Karma Mimansa focuses more on refining individual experiences toward their more universal value, the sixth state. Vedanta, which refers to the end of the Veda, focuses on the predominant experience of unity, in which all individual experience is spontaneously appreciated in terms of universal consciousness—nothing other than the non-dual universal Self, eternal Being, Brahm.

The Ordinary Three States: Sleep, Dreaming, and Waking

Experience in the ordinary three states of consciousness is so familiar that little needs to be said about them in the context of this discussion of higher human development. They can be placed in the progression of seven states of human consciousness as the least developed states. They are the focus of a quite large literature of psychological, cognitive, and neuroscientific research. Historically they were the only psychophysiologically defined states of consciousness. In recent decades interest has grown in altered states of consciousness, such as trance or drug-induced states, which are often mistaken for higher states of consciousness. In the majority of instances, they refer to unusual experiences within the waking and dreaming states. To some degree this interest has at least fostered increased acceptance of the possibility of natural higher states. Psychophysiological and phenomenological research has now defined a unique fourth state of consciousness, and also is working to identify empirically verifiable characteristics of higher states. Aspects of this research will be referred to in this chapter.

The sleep state reflects fundamental cyclic patterns in nature of activity and rest—day and night, summer and winter, life and death, manifestations and dissolutions, and perhaps big bangs and crunches—of the entire universe through eons of time. Ordinary deep sleep can be considered to be the least developed state in the sense that it is characterized as involving the least degree of awareness of self and the environment. As noted in Chapter 13 and to be discussed again shortly, however, the experience of sleeping changes in higher states.

In the ordinary dream state, for the most part there is only an illusory awareness of self and of an imaginary environment that dances on the screen of the mind. However, the dream state can be said to be phenomenally higher than deep sleep in that it involves more alertness or wakefulness. These two states are known to rejuvenate mind and body for more effective thought and action in the waking state, in which there are higher degrees of awareness of self and the environment. The waking state provides much more opportunity for growth of knowledge, action, achievement, and fulfillment through interaction with the environment.

The common experience of the ordinary waking state, however, is that the individual self is experienced as separate from the objects experienced in the outer objective environment—independent and highly localized in space and time. In this state nature is fragmented into a limited, localized individual self and the outer localized natural world it observes. The individual self is experientially ungrounded and not connected to its source—the total potential of natural law that governs both the subjective life of the individual and the shared outer objective environment.

Typically the individual in the waking state looks 'out there' into the external objective world, and attention tends to be absorbed by the tangible objects and activities in the immediate local environment. Sometimes it is directed past the immediate surroundings, such as to a panoramic horizon, the limitless sky, the starry galactic expanse, or becomes absorbed internally into imagination, thought, and feeling. Still, the typical experience is of separateness from the environment. There may be inspiring intuitive glimpses that everything is all somehow connected on a deeper level—perhaps in the unified field, or Nature, or God's omnipresent grace. But generally daily life goes on with its ups and downs in the search for more permanent happiness in our fragmented, chaotic, perplexing but also poignant and fascinating world as it is ordinarily known.

As the process of physiological maturation develops into adulthood, life frequently settles into a pattern of daily activity and rest that is bound by the cycle of the three ordinary states of consciousness. Ordinary waking experience is fragmented into the objects or the observed, the process of observing, and an isolated individual observer—with no direct experience of or use of the infinite power of their unified basis. Through ordinary sensory experience and reason, our knowledge of life can progress toward emotional and psychological maturity, within the pattern of activity and rest, wakefulness and sleep. As our body ages and loses flexibility, our minds also tend to go through comparable changes.

However, when the inner experience of transcending is added that goes beyond all limitations of

"The only way out is in."—Jerry Jarvis, Vedic research [36]

the mind, and when the body correspondingly experiences much deeper rest than deep sleep, higher levels of maturation and integration begin to unfreeze, opening up new possibilities for natural growth to continue. The three ordinary states of consciousness, although constantly changing, are underlain by non-changing pure consciousness that is transcendental with respect to them. Maharishi explains:

> "...[W]aking, dreaming, and sleeping states of consciousness are like curtains which hide the essential nature of that continuum of consciousness which knows no change and is always the same, constantly existing in its unmanifest value throughout time and space. It can be located but it is not located in the character of the waking, dreaming, or sleeping states." [34]

The Fourth State: Transcendental Consciousness

Transcendental consciousness is described as the simplest state of consciousness. It is the ground state underlying all mental activity of intuition, feeling, thinking, perception, and sensation—the transcendental basis of individual mind, the most settled state of the mind. It is the unified state of observed, processing observing, and observer. Adding regular transcending to the ordinary daily cycle of sleeping, dreaming, and waking is a profound and most direct means to facilitate natural growth to higher states.

For the most part, in the ordinary waking state when the mind starts to settle down toward less activity, alertness or wakefulness recedes and there is a natural drifting off into sleep. At times there may be a degree of alertness in this drifting process, sometimes described in terms of a hypnogogic state that may involve experiential qualities of both ordinary waking and sleep. The process of transcending also can involve similar experiences, sometimes including drifting into the sleep state if there is tiredness such that alertness is not maintained. But when mind and body are less tired, more settled and refined levels of mental activity naturally can be experienced while alertness expands. When the incessant activity of thinking recedes but alertness is maintained and expanded, then the mind experiences finer and finer levels of activity. Sometimes it settles into the self-referral state of consciousness itself, with no object-referral, reflective, or self-reflective mentation. Maharishi describes this experience during TM practice:

> "The Transcendental Meditation technique is an effortless procedure for allowing the excitations of the mind gradually to settle down until the least excited state of mind is reached. This is a state of inner wakefulness with no object of thought or perception, just pure consciousness aware of its own unbounded nature. It is wholeness, aware of itself, devoid of differences, beyond the division of subject and object—transcendental consciousness. It is a field of all possibilities, where all creative potentialities exist together, infinitely correlated but as yet unexpressed. It is a state of perfect order, the matrix from which all the laws of nature emerge, the source of creative intelligence." [37]

There are many levels of clarity of the refined process of transcending. Occasionally such experiences can occur spontaneously. This is typical of many of the historical reports of transcendence, as fleeting but sometimes life-transforming experiences that afterward are longed for and sometimes mistakenly attempted to be replicated through intensity of feeling and effort. They can be fostered with various degrees of efficiency, reliability, and ease through mental practices, depending on the degree to which the practices apply the natural tendencies and processes of the mind (discussed in the past chapter, and to be discussed further in an upcoming chapter). Whatever methods are prescribed in the various mental practices, the direct experience of transcendental consciousness—the fourth state of consciousness— involves transcending mental activity.

The physiological correlates of transcendental consciousness have been investigated over the past 40 years. Key indicators associated with subjective reports of transcendental consciousness include quiescence of breath, with episodes of virtual suspension of breath rate, high altitude alpha in the frontal cortex, and increased global coherence in EEG patterns. With respect to the brain correlates of transcendental consciousness, Maharishi states:

"In Transcendental Meditation, there is no one point of experience, so no *one* area of the brain functions. It is functioning of the total brain that experiences the unbounded field in transcending boundaries. It utilizes the whole brain... There is no other experience, but experience of unbounded consciousness itself, that has all aspects of the brain... Total natural law is lively when the total brain functions." [38]

Chapter 2 included a brief description of various degrees of clarity of transcending experiences. The clarity of these experiences varies with the state of mind and body at the particular time—especially whether rested or fatigued—as well as other developmental and health factors. Here are several reports suggestive of various degrees of clarity of transcending. The first one is a report from a subject in a laboratory study of the physiological effects of TM practice [39] who displayed marked declines in breath rate indicative of a deeply refined and restful state:

"I experience a state of complete silence devoid of any motion, a state of unbounded and total ease in deep relaxation. There are no thoughts, no feelings or any other sensations like weight or temperature. I just know "I am." There is no notion of time and space but my mind is fully awake and perfectly clear. It is a very natural and simple state." [40]

The following report is from a long-term regular TM practitioner:

"My experiences during TM practice vary somewhat each time, as well as over the months and years, influenced by many factors including whether I've been staying up too late, eating properly, or overspent my energy by too much physical work or exercise. Sometimes I have lots of mental activity, ideas, and fleeting impressions that flow through my mind, while generally feeling settled and relaxed. Sometimes things settle into a deeper state for nearly the entire session, during which I have virtually no noticeable mental activity; it is somewhat like being asleep, but I know I'm not asleep inside, and there is no experience of waking up afterward. The state seems to be beyond time, seemingly beyond almost any localized experience of any kind. Also sometimes when I am most clear, relaxed, and alert, I can experience directly a wonderful spark of bliss and inner fulfillment as I transcend all mental activity into clear, silent, unbounded awareness. This can be momentary, or last for minutes. Sometimes for extended periods of time I am bathing in complete, inner bliss and unbounded silence." [41]

Here are several reports suggestive of transcending from historical references. The first is from William Wordsworth, the noted American poet, who describes the following experience suggestive of transcending, in his "Tintern Abbey:"

" ...that serene and blessed mood,
In which the affections gently lead us on,
Until, the breath of this corporeal frame
And even the motion of our human blood
Almost suspended, we are laid asleep
In body, and become a living soul." (p. 156) [42]

Another report of an experience suggestive of transcending was provided by the French playwright, Eugene Ionesco.

"Once long ago, I was sometimes overcome by a sort of grace, a euphoria. It was as if, first of all, every motion, every reality was emptied of its content. After this, it was as if I found myself suddenly at the center of pure ineffable existence. I became one with the one essential reality when along with an immense serene joy, I was overcome by what I might call the stupefaction of being, the certainty of being." (pp. 150-151) [43]

Plotinus, the 3rd Century Alexandrian philosopher, described the state of a person who:

"...had attained unity and contained no difference... [T]here was within him no movement, no anger, desire, reason, nor thought... [He] was tranquil, solitary, and unmoved... He was indeed in a state of perfect stability, having, so to say, become stability itself." (p. 157) [44]

Georg W. F. Hegel (1770-1831), the eminent German philosopher, referred to a:

"...wholly abstract universality...transcendent, self-consciousness, which is identical with itself and infinite in itself..." (p. 39, p. 89) [45]

Henri F. Amiel, the 19th Century Swiss philosopher, offers the following account of being in a state of 'consciousness of self:'

"Feeling of repose, even of quietude... Desire and fear, sorrow and care do not exist any longer. One feels oneself existing in a pure form, in the most ethereal mode of being, namely the consciousness of self. One feels happy, in accord, without any agitation or tension whatsoever... [T]his divine quietude, this state as of the ocean in repose which reflects the sky and is self-possessed in its profundity... [T]he soul is no longer anything but soul and ceases to feel itself in its individuality, in its state of separation. It is something that feels the universal life." (p. 581) [46]

Franklin Merrel-Wolff, the 20th Century American philosopher, also describes an experience that seems to indicate transcendental consciousness:

"I found myself at once identical with the...Silence... I felt and knew myself to have arrived, at last, at the Real. I was not dissipated in a sort of spatial emptiness, but on the contrary was spread out in a Fullness beyond measure... It is as though the 'I' became the whole of space. The Self is no longer a pole or focal point, but sweeps outward, everywhere, in a sort of unpolarized consciousness, which is at once self-identity and the objective content of consciousness. It is an unequivocal transcendence of the subject-object relationship." (pp. 37-48) [47]

J. A. Symonds, the 19th Century British essayist, describes what appears to be a clear experience of transcending to the Self:

"Suddenly at church, or in company, or when I was reading, and always, I think, when my muscles were at rest, I felt the approach of the mood. It consisted in a gradual but swiftly progressing obliteration of space, time, sensation and the multifarious factors of experience which seem to qualify what we are pleased to call our self. In proportion as these conditions of ordinary consciousness were subtracted, the sense of an underlying or essential consciousness acquired intensity. At last nothing remained but a pure, absolute, abstract Self. The universe became without form and void of content. But Self persisted, formidable in its vivid keenness." (pp. 29-30) [48]

This next report is from the 19th Century British poet Alfred Lord Tennyson:

"A kind of waking trance I have frequently had, quite up from boyhood, when I have been all alone...as it were out of the intensity of the consciousness of individuality, the individuality itself seemed to dissolve and fade away into boundless being, and this not a confused state, but the clearest of the clearest, the surest of the surest...utterly beyond words, where death was an almost laughable impossibility, the loss of personality (if it were so) seeming no extinction, but the only true life... I am ashamed of my feeble description. Have I not said the state is utterly beyond words?" (p. 268) [49]

In the following quote, attributed to the 3rd Century Taoist philosopher Ko Yuan, a state of pure stillness is described:

"Now the spirit of man loves Purity, but his mind disturbs it. The mind of man loves stillness, but his desires draw it away. If he could always send his desires away, his mind would of itself become still. Let his mind be made clean, and his spirit will of itself become pure...

In that condition of rest independently of place how can any desire arise? And when no desire any longer arises, there is the True stillness and rest." (p. 696) [50]

The next quote is an excerpt from *The rose-garden of mystery* by Sufi poet Mahmud Shabistari:

"When you and your real self become pure from all defilement, there remains no distinction among things, the known and the knower are all one... There is no duality... 'I' and 'you' and 'we' and 'he' become one. Since in the Unity there is no distinction, the Quest and the Way and the Seeker become one." (p. 111) [51]

Here are excerpts of a description from an 18th Century Zen Buddhist saint, Hakuin Zenji:

"Before long you will find that the mind-nature has become settled in you—like a great rock, immovable and peaceful... All your usual, everyday consciousness will cease... Now your eyes will be opened to see the truth that this universe and the absolute are identical..." (pp. 98-99) [52]

Finally here are four accounts from TM practitioners, the first one describing spontaneous experiences outside of meditation, and the last one describing development over several years of TM practice:

"I started meditating and started studying art at the same time. I started out with very tight small figures and then there was more freedom, and then there really started to be a dramatic difference that my art teachers started to notice. I was planning to go to an advanced meditation course so I had to finish all my semester's work at school a month and a half early. So I did a tremendous amount of work writing papers, drawings, paintings, etc. I would do it once and it would come out. I had never experienced painting, drawing, or writing like that—they always seemed a belabored effort, and to take a lot of time too. My professors noticed and I did too, the art work wasn't tight. It had more freedom or unity to it and it looked better artistically and had better composition according to the professors. The speed with which I produced everything was phenomenal. It was like I was on this flow" (p. 152) [53]

"In meditation it happened that the mental activity was revealed as insubstantial, like a gauze of mist. They then cleared, as clouds would clear, and then vanished—I had transcended. Everything I felt myself to be had vanished, the relative "me" was gone and

I remained alone. I then realized , I *knew* myself to be infinite, unbounded, eternal. I was bliss. I was consciousness, pure and silent. *I WAS THAT...* This experience has changed my life immeasurably. I no longer see myself as this small body behaving, thinking, speaking, as others see me—I am alone, myself in fulfillment." (p. 50) [54]

"There has been a very beautiful transformation in and out of meditation over the past month. During meditation the experiences of being the whole universe started to occur more and more often. It reached its climax in one meditation when I had the overpowering realization that I was so unbounded and so unlimited that anything I wanted could easily be obtained. I kept feeling more and more expanded, and the feeling of bliss kept growing and becoming more powerful within me. I then had (in a more concrete way than ever before) the realization that I was everything and knew everything there was to know." (p. 5) [54]

"In the early days of meditation there was an experience of bliss, a subtle perception of a kind of glow, and then I would slip into a state which was just silent, without any thoughts. Often, at that time, I would notice it only after I came out of that state and there was a feeling of bliss. As time went on, I began to notice with greater clarity that point from which I slipped from the finest level of thinking to that inner ocean of silence. Also, there seemed to be greater liveliness during the experience itself. I was not just noticing it as some gap in time, in which I knew I had not been sleeping, but rather there was an inner wakefulness and a great feeling of unboundedness... What has changed is that, at first the experience was just silence, but over the years it has become fuller, with some dynamism within it. This inner dynamism is not mental activity, not thoughts. I guess I could describe it as an awareness of a great potential activity. This interaction within that unbounded field of silence I experience as giving rise to and structuring all activity in the universe." [40]

With respect to practice of the TM technique, the process of transcending involves many levels of subjective experience, from just closing the eyes and thinking to the entire range of subtler, refined thoughts and feelings, and the transcendental state beyond thought. No particular pattern of experiences can be expected, except generally the natural tendency for mental activity to settle down if it is allowed to do so. Typically the boundaries of discrete thoughts and feelings ease or melt into deeper, more expanded levels of relaxation, inner peace and unbounded silence.

Although the mechanics of the technique don't change through time, regular practice typically results in increasing frequency, clarity, and length of transcendental consciousness. Initially, however, as well as at other times depending on the conditions of mind and body, transcendental consciousness may be just a momentary or brief experience. It is generally not that the mind immediately settles down to transcendental consciousness and remains there during an entire session of the practice (although it could be that). Transcending is typically a dynamic process involving a range of experiences, with periods of more and less active mentation in the direction of increasing subtlety and refinement, with shorter periods of clear transcendental consciousness. It commonly results in a deep sense of ease, clarity, and freshness of mind after TM practice

It is frequently the case that when an individual begins regular practice of transcending, the mind and body are not at the stage of development where the mental activity of thinking can occur simultaneously along with transcendental consciousness. This is because of lack of flexibility of mind and body due to the overall developmental state, related to the restrictions of accumulated deep-rooted stress and strain. In this state the individual identifies with individual ego and individual consciousness, and is as if experientially cut off from the underlying universal level of consciousness and the universal Self. Maharishi contrasts these levels in the following statement:

"Self has two connotations: lower self and higher Self. The lower self is that aspect of the personality which deals only with the relative aspect of existence. It comprises the mind that thinks, the intellect that decides, the ego that experiences. This lower self functions only in the relative states of existence—waking, dreaming and deep sleep... The higher Self is that aspect of the personality which never changes, absolute Being, which is the very basis of the entire field of relativity, including the lower self... A man who wants to master himself has to master the lower self first and then the higher Self. Mastering the lower self means taking the mind from the gross fields of existence to the subtler fields, until the subtlest field of relative existence is transcended." (p. 339) [55]

In the ordinary waking state of consciousness—associated with the lower self—the mind and body don't maintain direct experience of transcendental consciousness, the higher or universal Self. Under such circumstances the individual is said to go beyond or transcend the limitations of ordinary mental activity in order to experience consciousness itself. Transcendental consciousness is a unique fourth state because the mind and body are not accustomed to experiencing it and are not able to experience it within the boundaries of the ordinary three states—it thus can be said to be experientially transcendental with respect to them. Over time, with repeated experience of transcendental consciousness, the mind and body develop the spontaneous ability to maintain various degrees of mental activity along with underlying transcendental consciousness. Eventually this development includes the spontaneous experience of maintaining pure consciousness, the universal Self, during waking, dreaming, and also deep sleep.

The predominant locus of awareness. As alluded to in the first part of the Prologue and in Chapter 2, human development can be understood in terms of the predominant locus of awareness, related to different lenses, mind-sets, schemas, or worldviews through which the world is experienced and understood. In this way of looking at it, the early Piagetian stages of perceptual-cognitive development can be associated with the sensory level of functioning, and the later stages with deeper cognitive levels to formal operational thought, the predominant focus of which is mind, and intellect to some degree.

From birth to early adulthood, cognitive development is conceptualized in terms of growth from more concrete to increasingly abstract functional stages. These stages also generally correspond to structural maturation of the human nervous system. Growth spurts in the brain and proportional increases in EEG alpha production, for example, tend to correspond with shifts in levels of cognitive-structural development. [56] [57] [58] The overall direction is toward increasing differentiation and hierarchical integration. [59] The maturation of the nervous system appears to provide the 'deep structure' for higher-order cognitive processing. [1]

A model by developmental psychologist and Vedic researcher Charles N. Alexander [1] [10] [60] shows a correspondence between the levels of mind in Maharishi Vedic Science and Technology and the sequence of developmental stages in theories proposed by Piaget [7] [8] and Loevinger, [18] introduced earlier in this chapter. This model importantly proposes growth to enlightenment as the natural extension of the developmental stream toward actualization of full human potential. It shows how normal human development as described in contemporary theories naturally extends into the higher states of consciousness identified in Maharishi Vedic Science and Technology—with higher integrative experiences and transcendence. As described in Chapter 1, ordinary growth of knowledge in the waking state unfolds from the interaction of sensory experience and reason. As life experience progresses, the ability slowly develops to be able to observe and appreciate deeper aesthetic values of sensory objects. Accompanying this is more comprehensive reasoning and understanding, creating a knowledge system of increasing depth and appreciation. This is significantly guided by intuitively-based paradigms and worldviews, within the ordinary waking state of consciousness of the individual.

This natural pattern of development of experience and reason generally involves growth from outer to deeper inner levels. The pattern is evident in the overall pursuit of knowledge in modern science. As noted in Chapter 9, modern scientific progress has been proceeding from outer to inner—from the objective surface level of nature to smaller time and distance scales, and toward the deeper underlying levels of matter, non-local quantum fields, even more abstract information fields, and now the nonlocal field of quantum mind, toward a comprehensive understanding of the all-encompassing unified field. But this involves active sensory and reasoning processes.

This outer-to-inner development is greatly facilitated when regular experience of the fourth state of consciousness is added. For example, with experiences of transcendental consciousness after starting TM practice, it is common for artists to report improved color schemes, sensory acuity, and increased creativity. Businesspeople report greater mental and physical energy along with less tension, greater ability to be task-oriented, increased creativity, and improved relations with co-workers. Athletes report improved mind-body coordination and balanced alertness during performance. Scientists report more coherent thinking, broader comprehension, and deeper insights. Such anecdotal reports of development associated with the different levels of mind have been substantiated by a large body of well-controlled studies on the beneficial psychological, behavioral, and social effects of regular TM practice. This research documents a wide range of developmental effects at each level of mental functioning, including sensorimotor, mind-body coordination, cognitive, intellectual, affective, and ego and self development—all progressing at the same time as a result of the systematic technology of effortless transcending.

These findings are consistent with the understanding that the transcending process involves progressive enlivenment of the full range of deeper and more holistic levels of mind. This is associated with the natural process of refinement of mind and body as mental activity settles down to deeper levels, corresponding with deep levels of physiological rest. The holistic nature of these results across a wide subject population suggests that quite fundamental processes of development are activated through this systematic technology. Perhaps it should be pointed out, however, that this doesn't mean that formal operational thought or logical reasoning—the highest stage in Piaget's theory[78]—is a prerequisite for experiences and development of higher states of consciousness. The transcendental state always underlies all sensory and mental activity and is available for direct experience any time. It is more fundamental than representational and self-reflective thought, and does not depend on them. It thus is possible for the individual to be fully awake to the underlying inner level of transcendental consciousness at any ontogenetic or physiologic developmental stage. Although perhaps quite rare in modern civilization, theoretically a child yet unable to demonstrate formal operational reasoning could experience and even maintain transcendental consciousness (Note the report by Tennyson quoted above that suggests such experiences).

Regular experience of the fourth state of consciousness—going beyond the ordinary cycle of waking, dreaming, and sleep—not only activates outer-to-inner development of the mind. It also activates development from the inside out, *inner-to-outer* development—from the holistic basis of the unified field of consciousness progressively into the subtle and gross levels of individual mind and body. Settling back the individual self to the universal Self begins a process of development in which the infinite silence (Atma) and infinite dynamism (Veda) of the universal Self become *infused*—so to speak—into individual subjectivity and the levels of individual mind and body

(Sharir). This brings about increasing connection and oneness with the entire cosmos (Vishwa), and eventually establishes in the highest state of consciousness the ultimate, eternal, fully awake status of the universal Self in individual life (Brahm).

This natural inner-to-outer development occurs along with outer-to-inner development, but underneath, deep inside—in the background of the ordinary states of waking, dreaming, and sleeping. It grows along with the increasingly refined activity of individual feelings, thoughts, and actions—at the basis of the individual ego or self. It is an expansion of individual consciousness that accompanies individual ego, heart and intellect, mind, senses, body, and behavior, while all of these levels move toward functioning in a more coherent and integrated style. This natural inner-to-outer development eventually establishes the fifth state of consciousness, in which the individual self or ego is fully lively in the universal Self, the silent background or foundation of all individual experience. This infusion of the influence of the perfect orderliness of the universal Self or unified field eventually resolves all psychological doubts and dissolves all physiological stress.

As referred to in Chapter 9, there can be said to be two gaps between the three fields of life, which can be identified as body, mind, and consciousness, or gross relative, subtle relative, and transcendental. In the natural course of full human development, these gaps—as experiential gaps—are completely bridged when the seamless unity of life becomes primary. In the progress of inner-to-outer development, the experiential gap between individuality and universality is bridged as the individual self or ego establishes itself permanently as the universal Self in the first stage of enlightenment. Upon this non-changing absolute background or platform of experience of transcendental consciousness, continued progress toward full enlightenment involves progressive refinement of experience that eventually results in bridging the experiential gap between the universal Self and all objects of sensory experience, such that everything is experienced in terms of one's universal Self. This progression will be described in more detail in the next chapter.

"From my point of view as a physiologist, what Maharishi has accomplished is the single most important scientific discovery of our age, or for that matter any age. Perhaps there is no more appropriate time in history for such a discovery, a time when the laws of physics are on the verge of a long sought unification, and yet a time when the entire world lives in fear of total annihilation. For the first time in thousands of years, this ancient [Vedic] tradition of knowledge has been revived in completeness, a tradition that in fact includes the profound science of physiology... It had for centuries been totally misunderstood by scholars and laymen alike in both the East and the West. The revival of this science has resulted in an unprecedented advance in our understanding of human consciousness and in the availability of a set of procedures for the development of an extraordinary state of neurophysiological functioning—a state in which the awareness is established in the unified field of natural law, and in which activity and behavior are thus spontaneously in accordance with all the laws of nature... It represents the ultimate development of what we ordinarily consider the most valuable qualities of human life. It is something real and natural and develops systematically in a continuous and progressive manner on the basis of neurophysiological refinement, utilizing the existing mechanics of human physiology."—*R. Keith Wallace, physiologist and Vedic researcher (pp. i-iii)* [61]

In the ordinary three states of consciousness, the underlying fourth state is hidden from experience; ordinary experience is not open to its universal basis—unmanifest is hidden from manifest—until the manifest levels are transcended. The processes that conduct all activity in manifest creation—the three gunas or fundamental qualities of nature—predominate in ordinary experience. Higher states of consciousness involve going beyond the activity of the three gunas, settling down to the universal ground state of nature, the universal Self.

Maharishi emphasizes two particular verses in the Vedic text the *Bhagavad-Gita* that clearly identify the importance of going beyond the relative values of the three gunas:

"Be without the three gunas...freed from duality, ever firm in purity, independent of possessions, possessed of the Self" (p. 126) [55]

"Established in [the Self]...perform actions having abandoned attachment and having become balanced in success and failure..." (p. 135) [55]

Permanent establishment of the inner balance and non-attachment due to bliss consciousness that comes from direct contact with the cosmic value of the self—the universal Self—characterizes the next stage of evolution. This inner balance and non-attachment due to the infusion of bliss into the deepest level of individual self establishes the platform of permanent unbounded awareness in the next higher state of consciousness.

"The ever changing is the form: the *changeless is the reality*. It is the awareness *that remains continuous and unbroken throughout waking, dreaming and sleep*...just as space remains continuous and unbroken within a house, through the wall and outside the house (and in the sky)." [62]

Chapter 19 Notes

[1] Alexander, C. N., Boyer, R., & Alexander, V. (1987). Higher States of Consciousness in the Vedic Psychology of Maharishi Mahesh Yogi: A Theoretical Introduction and Research Review. *Modern Science and Vedic Science*, Vol. 1, No. 1, January, 89-126.

[2] Plato (1901). *The republic of Plato*. New York: Colonial Press.

[3] Hegel, G. W. F. (1931). *The phenomenology of mind*. New York: Humanities Press.

[4] James, W. (1929). *The varieties of religious experience*. New York: Modern Library.

[5] Maslow, A. (1962). *Toward a psychology of being*. Princeton, N. J.: Van Nostrand.

[6] Walsh, R. & Vaughan, R. (1993). *Paths beyond ego*. New York: Putnam.

[7] Piaget, J. & Inhelder, B. (1969). *The psychology of the child*. New York: Basic Books.

[8] Piaget, J. (1972). Intellectual Evolution from Adolescence to Adulthood. *Human Development*, 15, 1-12.

[9] Alexander, C. N., Heaton, D. P., & Chandler, H. M. (1994). Advanced Human Development in the Vedic Psychology of Maharishi Mahesh Yogi: Theory and Research. In M. E. Miller, M. E. & Cook-Greuter, S. R. (Eds.), *Transcendence and mature thought in adulthood*. Lenham, Maryland: Rowland and Littlefield Publishers, Inc., pp. 39-70.

[10] Alexander, C. N. (1982). Ego Development, Personality and Behavioral Change in Inmates Practicing the Transcendental Meditation Technique or Participating in Other Programs: A Cross-Sectional and Longitudinal Study. Doctoral dissertation, Harvard University. *Dissertation Abstracts International*, 43 (2), 539B.

[11] Alexander, C. N. & Langer, E. J. (Eds.) (1990). *Higher stages of human development: Perspectives on adult growth*. New York: Oxford University Press.

[12] Commons, M. L., Richards, F. A., & Armon, C. (1984). *Beyond formal operations: Late adolescent and adult cognitive development*. New York: Praeger.

[13] Sternberg, R. J. (1990). *Wisdom: Its nature, origins, and development*. New York: Cambridge University Press.

[14] Arlin, P. K. (1989). Problems Solving and Problem Finding in Young Artists and Young Scientists. *Adult Development*, 1, 197-216.

[15] Maslow, A. H. (1976). *The farther reaches of human nature*. New York: Penguin.

[16] Pascual-Leone, J. (1990). Reflections on Life-Span Intelligence, Consciousness, and Ego Development. In Alexander, C. N. & Langer, E. J. (Eds.). *Higher stages of human development. Perspectives on adult growth*. New York: Oxford University Press.

[17] Cook-Greuter, S. R. (1990). Maps for Living: Ego Development Stages from Symbiosis to Consciousness Universal Embeddedness. In Commons, M. L., Armon, L., Kohlberg, F. A., Grotzer, T. A., & Sinnot, J.D. (Eds.) *Adult development: Models and methods in the study of adolescent and adult thought*, Vol. 1 New York: Praeger, pp. 119-146.

[18] Loevinger, J. (1976). *Ego development: Conceptions and theories.* San Francisco: Jossey-Bass.

[19] Snarey, J., Kohlberg, L., & Noam, G. (1983). Ego Development in Perspective: Structural Stage, Functional Phase, and Cultural Age-Period Models. *Developmental Review, 3,* 303-338.

[20] Kohlberg, L & Ryncarz, R. A. (1990). Beyond Justice Reasoning: Moral Development and Consideration of a Seventh Stage. In Alexander, C. N. & Langer, E. J. (Eds.). *Higher stages of human development,* New York: Oxford University Press, pp. 191-207.

[21] Chandler, H. M., Alexander, C. N., & Heaton, D. P. (2005). Transcendental Meditation and Post-Conventional Self Development: A 10 year Longitudinal Study. Applications of Maharishi Vedic Science. *Journal of Social Behavior and Personality, 17,* [Special issue], 93-121.

[22] Pearson, C. A. (2002). *The supreme awakening: Maharishi's model of higher states of consciousness applied to the experiences of individuals through history.* Doctoral dissertation, Maharishi University of Management, Fairfield, IA. Printed by UMI Dissertation Services, Ann Arbor, MI.

[23] Huxley, A. (1945). *The perennial philosophy.* New York: Harpers.

[24] Jung, C. G. (1969). *The structure and dynamics of the psyche: Collected works,* Vol 8. Hull, R. F. C. (Trans.). New Jersey: Princeton University Press.

[25] Orme-Johnson, D. W. (1987). Medical Care Utilization and the Transcendental Meditation Program. *Psychosomatic Medicine, 49,* 493-507.

[26] Orme-Johnson, D. W. (1988). The Cosmic Psyche: An Introduction to Maharishi's Vedic psychology— The Fulfillment of Modern Psychology. *Modern Science and Vedic Science, 2* (2), 113-163.

[27] Orme-Johnson, D. W. (1995). Summary of Scientific Research on Maharishi's Transcendental Meditation and TM-Sidhi Program. *Modern Science and Vedic Science, 6* (1), 60-155.

[28] Wilber, K. (1998). *The eye of spirit: An integral vision for a world gone slightly mad.* Boston: Shambhala.

[29] Alexander, C. N. & Boyer, R. W. (1989). Seven States of Consciousness: Unfolding the Full Potential of the Cosmic Psyche in Individual life through Maharishi's Vedic Psychology. *Modern Science and Vedic Science, 2* (4), 325-371.

[30] Orme-Johnson, D. W. (2000). An Overview of Charles Alexander's Contribution to Psychology: Developing Higher States of Consciousness in the Individual and the Society. *Journal of Adult Development, 7* (4), 199-215.

[31] Morris, B. (2004). Maharishi's Global News Conference, May 5.

[32] *Maharishi Vedic University: Introduction.* (1994). Holland: Maharishi Vedic University Press.

[33] Maharishi Mahesh Yogi, (1963). *Science of being and art of living.* Washington, D.C.: Age of Enlightenment Publications.

[34] Maharishi Mahesh Yogi. (1972). *The Science of Creative Intelligence: Knowledge and Experience* (Syllabus of videotaped course). Los Angeles: Maharishi International University Press.

[35] Maharishi Mahesh Yogi (1970, August 7). *Expanded awareness: The basis of ideal relationships.* (Videotaped lecture). Humboldt State University: Arcata, CA.

[36] Jarvis, J. (2003, Summer). Personal communication, Malibu, CA.

[37] *Creating an Ideal Society: A global undertaking.* (1977). Rheinweiler, W. Germany: MERU Press.

[38] Maharishi Mahesh Yogi (2005). Maharishi's Global News Conference, April 20.

[39] Kesterson, J. B. (1986). Changes in Respiratory Control Pattern During the Practice of the Transcendental Meditation Technique. Doctoral dissertation. Maharishi International University, Fairfield, IA. *Dissertation Abstracts International,* 47, 4337B.

[40] Alexander, C. N. (1997). *The seven states of consciousness in Maharishi's Vedic Psychology.* Unpublished manuscript. Maharishi University of Management.

[41] Anonymous (2003). Personal communication, Fairfield, IA.

[42] Wordsworth, W. (1798/1979). Tintern Abbey. In Abrams, M. H. (Ed.). *The Norton anthology of English Literature,* Vol. 2 (4th Edition). New York: W. W. Norton.

[43] Ionesco, E. (1971). *Present past past present: A personal memoir.* New York: Groove Press.

44 Katz, J. (1950). (Trans.). *The philosophy of Plotinus: Representative books from the Enneads.* New York: Appleton-Century-Crofts.

45 Hegel, G. W. F. (1892). *The logic of Hegel.* Wallace, W. (Trans.). (2nd Ed. Rev.). London: Oxford University Press. [Trans. from the *Encyclopedia of the Philosophical Sciences.* 1st ed. publ. 1873].

46 Amiel, H. F. (1935). *The private journal of Henri Frederic Amiel.* V. W. Brooks & C. V. W. Brooks (Trans.) New York: Macmillan.

47 Merrel-Wolff, F. (1973). *Pathways through to space.* New York: Julian Press.

48 Symonds, J. A. (1895). *A biography.* London: H. F. Brown.

49 Tennyson, H. (1899). *Alfred, Lord Tennyson: A memoir by his son* (New Ed.). London: MacMillan.

50 Ko Yuan. The Classic of Purity and Rest. *The texts of Taoism.* Legge, J. (Trans). New York: Julian Press.

51 Smith, M. (1994). *Readings from the mystics of Islam.* Pir Publications.

52 Hakuin Zenji (1963). *The embossed tea kettle: Orate Gama and other works of Hakuin Zenji.* Shaw, R. D. M., (Trans.). London: George Allen and Unwin.

53 Mason, L.I. (1995). Electrophysiological Correlates of Higher States of Consciousness During Sleep. (Doctoral dissertation, Maharishi International University, Fairfield, IA). Printed 1998 by UMI Dissertation Services, Ann Arbor, MI. *Dissertation Abstracts International,* Vol. 56-10B.

54 *Enlightenment and the Siddhis: A new breakthrough in human potential.* (1977). West Germany: MERU Press.

55 Maharishi Mahesh Yogi. (1967). *Maharishi Mahesh Yogi on the Bhagavad-Gita: A new translation and commentary, chapters 1-6.* London: Penguin Books.

56 Epstein, H. T. (1974). Phrenoblysis: Special Brain and Mind Growth Periods. Human Mental Development II. *Developmental Psychobiology, 7,* 217-224.

57 Epstein, H. T. (1980). EEG Developmental Stages. *Developmental Psychobiology, 13,* 29-63.

58 Matousek, M. & Peterson, I. (1973). Frequency Analysis of the EEG in Normal Children and Adolescents. In Kellaway, P. & Peterson, I. (Eds.). *Automation of clinical electroencephalography.* New York: Raven Press.

59 Werner, H. (1957). The Concept of Development From a Comparative and Organismic Point of View. In Harris, D. B. (Ed.). *The concept of development.* Minneapolis: University of Minnesota Press.

60 Goodman, R. S., Walton, K. G., Orme-Johnson, D. W., & Boyer, R. (2003). The *Transcendental Meditation* Program: A Consciousness-Based Developmental Technology for Rehabilitation and Crime Prevention. *Journal of Offender Rehabilitation,* Vol. 36, Numbers 1/2/3/4, 1-33.

61 Wallace, R. K. (1991). *The neurophysiology of enlightenment.* Fairfield, IA: Maharishi International University Press.

62 *The Concise Srimad Bhagavatam,* 9.3 (1989). S. Vendatesananda (Trans.) Albany, NY: State University of New York.

Chapter 20

Full Enlightenment:
The Fifth, Sixth, and Seventh States of Consciousness

In this chapter, we will complete the discussion of the seven states of consciousness in ancient Vedic science and Maharishi Vedic Science and Technology. The chapter focuses on the three higher states that are based on permanent establishment of the fourth state, transcendental consciousness. This natural progression through the seven states of consciousness culminates in the supreme state of unity consciousness. The overall main point of this chapter is that full enlightenment in unity is the natural course of evolution in which each individual becomes permanently established in the infinitely fulfilling totality of the universal Self, Brahm.

Transcendental consciousness, the fourth state of consciousness, has been shown to be accompanied by a profound state of psychophysiological rest. As with ordinary rest, this deeper rest activates natural healing mechanisms in mind and body, dissolving accumulated deep-rooted stress and strain that block optimal functioning. As deep-rooted stress and strain are eliminated, mind and body become stronger and healthier, and the individual is able to experience more refined states of mental activity as well as increased frequency of transcendental consciousness. In addition mind and body develop increased stability and flexibility to be able to sustain the inner silence of transcendental consciousness as a permanent background even in active states of mind.

In the one-to-one correspondence of mind and body discussed in earlier chapters, each state of consciousness has its own specific style of neurophysiological functioning. In order to maintain transcendental consciousness along with the other three states of consciousness, a new and more flexible, integrated style of neurophysiological functioning develops in which the three states coexist with the fourth state. This integration is cultured gradually and systematically through alternating regular experience of transcendental consciousness with the ordinary three states.

> "...[Steadiness in yoga, or union] is firmly rooted, being well-attended to for a long time without interruption and with devotion." [Yoga Sutras, 1:14] (p. 28) [1]

This more integrated style of functioning develops based on the enormous adaptive capability of the human nervous system. This adaptability is seen even in the ordinary waking state, when the nervous system supports several levels of mental activity at the same time. For example, during the process of perception, thoughts may accompany perceptions, decision making may underlie and direct thinking, and these processes may in turn be guided by delicate feelings—all occurring at the same time.

The integration of transcendental consciousness with mental activity represents a further extension of this natural adaptive capacity. In this process it is natural for individuals to have momentary experiences of transcendence, as well as of higher states, before the new stage of development becomes permanent.[2] Psychophysiological indicators of this development include theta and alpha EEG during sleep along with delta EEG, and frontal EEG coherence continuing even in the performance of specific mental tasks during the waking state.

The Fifth State: Cosmic Consciousness

Typically experiences of transcendental consciousness along with mental activity begin to occur under circumstances in which mind and body are deeply settled and relaxed, such as during TM practice. Next they may begin to occur during settled moments in ordinary waking activity, and possibly in phases of light physical activity or even dreaming. Because sleep is the least alert state of consciousness, the experience of transcendental consciousness during sleep develops more slowly. Experiences of unbounded awareness along with mental activity are natural experiences that typically develop over time. Increasingly the deepest inner sense of who one is becomes permeated by nonlocality and fewer restrictions, and eventually complete unchanging unboundedness. The individual ego or sense of self as if merges into, or 'remembers' itself as, the universal Self, as the unbounded unchanging background for daily living. Maharishi has described this state as the state of 'steady intellect' in permanent transcendental consciousness:

> "How then is the 'steady intellect' maintained...when the mind, established in pure consciousness, is yet engaged in activity? Because in this state the mind has become transformed into bliss-consciousness, Being is permanently lived as separate from activity. Then a man realizes that his Self is different from the mind which is engaged with thoughts and desires. It is now his experience that the mind, which had been identified with desires, is mainly identified with the Self. He experiences the desires of the mind as lying outside himself, whereas he used to experience himself as completely involved with desires. On the surface of the mind desires certainly continue, but deep within the mind they no longer exist, for the depths of desires which were present in the mind have been thrown upward, as it were—they have gone to the surface, and within the mind the finest intellect gains an unshakable, immovable status... This is the 'steady intellect' in the state of...cosmic consciousness." (pp. 150-151) [3]

When individual mind and body are free from the accumulation of deep-rooted stress and strain, it is capable of simultaneously experiencing transcendental consciousness along with the other three states throughout the entire 24-hour daily cycle. This style of neurophysiological functioning spontaneously supports and sustains the fifth major state of consciousness—*cosmic consciousness*. It is the first stage of enlightenment, characterized by the spontaneous maintenance of the blissful state of transcendental consciousness at all times, under all circumstances. Maharishi identifies this state as permanent freedom from the boundaries of individual life, by virtue of establishing the individual self as awake in the permanent bliss of transcendental consciousness, the universal Self:

> "The state of cosmic consciousness is inclusive of transcendental consciousness as well as of consciousness of the relative order; it brings cosmic status to the individual life. When the individual consciousness achieves the status of cosmic existence...a man is ever free, unbounded by any aspect of time, space or causation, ever out of bondage." (p. 145) [3]

> "The intensity of happiness is beyond the superlative. The bliss of this state eliminates the possibility of any sorrow, great or small... No sorrow can enter bliss-consciousness, nor can bliss-consciousness know any gain greater than itself. This state of self-sufficiency leaves one steadfast in oneself, fulfilled in eternal contentment." (p. 424) [3]

Here are several reports suggestive of experiences in the direction of cosmic consciousness, the first from an advanced TM practitioner of the gradual growth of pure consciousness in daily activity:

"Gradually over the years, as the experience of pure consciousness became increasingly familiar in meditation, I began to experience it not just as a state with no thought but rather as having no boundaries; then as unbounded, beyond the limitations of my individuality; then as the unbounded, unchanging essence of my existence... Also, there is less of a contrast between activity and meditation. Sometimes during the day with varying degrees of clarity, my awareness is this unbounded wholeness of my Self, quietly accompanying the thoughts and feelings in my daily life. It is not a mood or conception about myself, it is a natural, spontaneous state in which I am fully my unbounded self."[4]

This next report describes the growth of blissful inner awareness for more effective daily activity:

"...[T]he experience of bliss-consciousness has become more clear, intense, and stable not only during meditation but also during activity. Now I find that a soft but strong feeling of blissful evenness is present most of the time in both mind and body... This evenness is so deep and stable that it is able to maintain its status even in the face of great activity... The evenness cushions one against all possible disruptions and makes all activity easy and enjoyable... Also, I have found that I have become more efficient in activity...and that somehow my responsibilities seem to arrange themselves so that they can be accomplished with very little doing. In this way activity has become more and more effortless while leading to greater accomplishment." (p. 79)[5]

The following are additional historical and contemporary accounts of experiences suggesting growth toward this advanced state. The first report is from Ray Reinhardt, the contemporary stage actor, who describes a clear but infrequent similar experience during activity:

"There are two stages to having the audience in your hand. The first one is the one in which you bring them along, you make them laugh through sheer skill—they laughed at that, now watch me top it with this one. But, there's a step beyond that which I experience, but only two or three times. It is the most—how can you use words like satisfying? It's more ultimate than ultimate: I seemed to be part of a presence that stood behind myself and was able to observe, not with my eyes, but with my total being, myself and the audience. It was a wonderful thing of leaving not only the character, but also this person who calls himself Ray Reinhardt. In a way, I was no longer acting actively, although things were happening: my arms moved independently, there was no effort required; my body was loose and very light. It was the closest I've ever come in a waking state to a mystical experience." (p. 43)[6]

Edmund Husserl, the German phenomenologist, refers to becoming a transcendental observer to ordinary limitations of the individual self:

"I reach the ultimate experiential and cognitive perspective thinkable. In it I become the disinterested spectator of my natural and worldly ego and its life... The transcendental spectator...watches himself, and sees himself also as the previously world-immersed ego." (p. 15)[7]

Romain Rolland, a 20th Century Nobel laureate in literature, describes experience of the Self as separate from body and thinking:

"I have an immediate and direct certainty of this eternity... I feel very clearly that the Self is independent, not only from my body, but from my thoughts." (p. 209)[8]

Chuan Tzu, one of the 4th Century founders of the Taoist school, describes the process of stabilizing inner silence and its natural enrichment of daily life:

"He who practices the Way does less every day, does less and goes on doing less, until he reaches the point where he does nothing, does nothing and yet there is nothing that is not done." (p. 235) [9]

"Emptiness, stillness, limpidity, silence, inaction are the root of the ten thousand things... Come forward with them to succor the age and your success will be great, your name renowned, and the world will be united. In stillness you will be a sage, in action a king." (p. 143) [9]

Finally, here is a description of inner experience from 13th Century Franciscan nun Angela of Foligno:

"And although I can receive sorrow and joy externally in some little way, yet inwardly my soul is a chamber in which no joy or sorrow entereth, nor delight of any virtue whatsoever, or of anything that can be named; but here entereth into it all that is good. And in this... is the whole Truth, and in it I understand and possess the whole truth, that is in heaven and in earth...together with so great a certainty that in nowise, were the whole world to say the opposite, could I believe otherwise" (p. 96) [10]

One important criterion of growth toward stabilization of cosmic consciousness is the spontaneous experience of witnessing, and especially witnessing sleep—inner wakefulness as an underlying continuum during the inertia of deep sleep. The experience of inner wakefulness during sleep is a natural development that underlies the ordinary experience of sleep. It is not that there is no longer the experience of sleeping, but rather an additional, natural, spontaneous, underlying inner wakeful witnessing of ordinary sleep—as well as of dreaming and of daily waking activity. This experience is clearly distinguishable from such experiences as light sleep, or of lucid dreaming. [11] [12] The experience of witnessing sleep is of a deep level of continuity of unbounded awareness and inner blissfulness along with and underlying the usual experiences of the sleep state. Here are several reports of this type of experience, the first reports from TM practitioners who report clear unbounded awareness while sleeping and dreaming:

"I have often clearly experienced pure consciousness, or inner wakefulness continuing all throughout a night's sleep, so that even though I was sleeping, inside I felt awake." (p. 687) [13]

"Often during dreaming I am awake inside, in a very peaceful, blissful state. Dreams come and go, thoughts about the dreams come and go, but I remain in a deeply peaceful state, completely separate from the dreams and the thoughts." (p. 295) [14]

"I have a feeling of separateness and a sense of watching myself dreaming. Realizing that I am dreaming, but then there is something larger than me that isn't dreaming and is motionless and silent and is just observing that experience. There is a part of me that is completely separate from all the dream activity and is quiet and is watching over everything." (p. 28) [15]

The American poet Henry David Thoreau also wrote of a similar experience of wakefulness during sleep:

"I am conscious of having in my sleep transcended the limits of the individual... As if in sleep our individual mind fell into the infinite mind." (p. 157) [16]

The following is an historical anecdote from Sarah Edwards, the wife of Jonathan Edwards, the 17th Century American theologian and philosopher:

"Last night was the sweetest night I ever had in my life...without the least agitation of body during the whole time. Part of the night I lay awake, sometimes asleep, and sometimes between sleeping and waking. But all night I continued in a constant, clear, and lively sense of...an inexpressibly sweet calmness of soul... I think that what I felt each minute was worth more than all the outward comfort and pleasure which I had enjoyed in my whole life put together. It was a pleasure, without the least sting, or any interruption... There was but little difference, whether I was asleep or awake, but if there was any difference, the sweetness was greatest while I was asleep." (pp. 276-277) [17]

This next report from a long-term TM practitioner describes witnessing the dream state:

"When everything is clear and silent and there's just a lot of happiness and light, that's when I'm really alert, and then when a dream comes it almost feels like a wave—like when you are out on a very calm ocean—like when you can see the wave coming from a distance. Sometimes when I am in that space of myself and I'm like the ocean, I can feel the dream is coming at me from a distance, and then it comes on me, and I have the dream, and while I am having the dream I can observe the dream... I have a feeling of separateness and a sense of watching myself dreaming, realizing that I am dreaming, but then there is something larger than me that isn't dreaming and is motionless and silent and is just observing that experience. There is a part of me that is completely separate from all the dream activity and is quiet and is watching over everything." (p. 28) [15]

Here is another report from a long-term TM meditator regarding the growth of witnessing, in this case the natural experience of pure consciousness during activity:

"Over the years, there has been steady growth in the degree of witnessing. In the background of my daily experience, behind the feelings and thoughts of who I am as an individual and what I am doing, there is an increased unboundedness. The limitations of my individuality are being permeated by a sense of complete, unchanging silence and wholeness of wakefulness—a direct experience that I am not just my body and not just my mind, but rather essentially pure wakefulness without any boundaries. At times it feels sort of like I can't get wet anymore because I am beyond the body that gets wet or dry, or that wherever I go I am in a sense already there, that change is only on the surface of my life. There is an increasing sense of inner peace, inner freedom, and more and more an underlying continuity of my unbounded Self through waking, dreaming, and sleeping." [18]

A growing body of physiological and psychological research shows that the predicted characteristics of cosmic consciousness increase with regular practice of TM and its advanced programs. These findings suggest development toward stress-free physiological functioning, increased inner freedom, happiness and fulfillment, increased spontaneous skill in action, and improved social behavior. These findings are also positively correlated with reports of stabilization of transcendental consciousness in daily activity and reports of growth of witnessing sleep.

Survey research has indicated as high as 80% of long-term TM practitioners report some degree of experience of witnessing sleep. [14] [19] Studies also have shown positive correlations between reported frequency of experiences of witnessing and length of time practicing the TM program. [20] These reports have been found to be positively correlated with indicators of psychological development, such as creativity [21] and efficiency of cognitive processing. [20]

In the natural expansion of the state of consciousness and growth in the deep inner experience of witnessing, the individual self with all its levels of activity from individual inner ego to outer behavior is increasingly enriched. It is not so much that the level of individual ego is eliminated or goes away (although it can be thought of in this way), but rather at its deepest level it expands to the state where it naturally is established in the underlying unbounded, fully wakeful universal level—the universal Self—rather than the ordinary waking state sense of a localized individual self. It can be described as a process of unifying with the infinite value that is always in each point value, no longer being overshadowed by identification with the point value of individual ego and individual consciousness.

All the levels of subjectivity continue their activity and their developmental refinement, but also the background of pure consciousness becomes the permanent predominant inner locus of experience. In the analogy of the film projected onto the screen in Chapter 13, awareness of the underlying screen as the unbounded infinite silent witness of the movie develops—along with enrichment of the movie on the screen, the activity of individual daily life in the waking, dreaming, and sleep states of consciousness. It isn't that the sense of individual ego is erased or eliminated. Rather the universal value of ego or self, the universal Self, is enlivened underneath individual ego and eventually encompasses it fully. The primary locus of experience is universal Self, and this unbounded eternal Self engulfs the localized ego, like the radiance of the sun engulfs starlight.

To further clarify the inner experience of witnessing, it may be helpful to distinguish it from other similar sounding concepts that have been given increasing attention in psychological literature, such as experiences of being in the 'zone' or of 'flow' in the waking state, as well as experiences of 'lucidity' in the dream state. These experiences are frequently described as involving a sense of ego separation or watching of ordinary activity. The experience of being in the 'zone' is described as a state of quiet observation of the smooth flow of one's own dynamic behavior. It is as if one's behavior takes place automatically, and is frequently associated with peak performance.

These experiences represent progress in development of mind-body coordination, associated with high levels of skilled performance through extensive practice—frequently associated with high levels of physical conditioning. Although quite positive experiences in the direction of higher development, they rarely are accompanied by reports of clear *unbounded* awareness or transcendental consciousness. In the natural experience of witnessing activity indicative of cosmic consciousness, the individual self becomes established and awake in the never-changing unbounded universal Self. It is a clear experience of unbounded wakefulness in addition to and beyond individual self observation, memories, feelings, thoughts, perceptions, and bodily sensations.

Development of the fifth state of consciousness spontaneously establishes complete inner freedom from being overshadowed by the inevitable ups and downs of individual life characteristic of various degrees of suffering. Individual experiences of suffering or happiness are associated with what we identify ourselves to be. The relative field of existence is ever-changing. Success and failure come and go, like sunshine and rain, according to the results of our current and past actions. If we identify with these changing processes, we ride the waves of change, sometimes the peak and sometimes the trough.

Identifying with the body we experience ourselves as going through the cycle of growth, youthful vitality, tiredness, aging, and death. If we identify with our jobs our house, our car, our family, our life partners, or our inner thoughts, feelings, or memories, then we see ourselves as going through the temporary ups and downs associated with them—and eventually they all fade away. But behind all of these changes is a deep inner sense that we are the same self that experiences or witnesses all these changes—as who *I am*. At its deepest essence, that sense of self is eternal and never-changing—it is pure Being, and it will never go away, never be lost, never end, never be annihilated. It is associated with pure bliss; it is unbounded awareness, the universal value that underlies all individual experiences in life. As we grow toward cosmic consciousness the experience of happiness is based on inner fulfillment and bliss through direct contact with unbounded awareness—not outer success in relative life. There is increasingly less reliance on outer performance and successful outcomes as the basis for attributions of happiness and success, but rather inner wholeness and spontaneous experience of bliss consciousness in the midst of any outcome on the relative level of behavior. At the same time increased success in daily life—generally along with improved health and enriching relationships with others—is also a product of inner development toward higher states of consciousness.

As inner bliss permeates the mind, successes and losses due to action— although fully experienced and enjoyed without restriction—lose their power to overshadow the inner sense of universal Self. Happiness comes from within, as inner bliss, and is less and less dependent on fulfillment of desires on the level of action. At the same time individual action continues to be vigorous, engaging, goal-oriented, and gratifying. The incessant changes associated with the ups and downs of relative life come and go, but deep inside the self is the continuity of *I* that goes on and on—increasingly established in its universal value, the universal Self.

When the nervous system is functioning at more crude levels due to accumulated stress, strain, and imbalance, successes and losses from actions typically overshadow the underlying inner sense of bliss. As the nervous system functions more coherently, fuller values of inner bliss naturally accompany all experiences. As we begin to experience more inner bliss and less the relative mixed results of our actions, success and suffering don't leave as deep of impressions in memory that ordinarily become the basis for more desires and future action.

In the cycle of outward and inward flow through our individual psychoarchitecture, there is an outward flow of feelings and thoughts, goals and plans, into action; and an inward flow of evaluation of the results of the actions and some degree of fulfillment (refer to Chapters 17-18). This is also described as the *cycle of action, impression, and desire*. [3] As the underlying experience of inner bliss and fulfillment permeates this cycle, the results of actions leave less binding impressions in memory. This is sometimes described as '*roasting the seeds of desire.*' When individual consciousness permanently expands to universal consciousness, when we become established permanently and spontaneously in bliss consciousness, then the cycle of action, impression, and desire loses its power to overshadow the unbounded inner contentment of pure consciousness. We naturally attain permanent freedom and liberation from the binding influence of action—due to permanent fulfillment, permanent bliss consciousness amidst the ups and downs of active relative life.

It is not, however, that the ordinary cycles of the outward and inward flow of energy and intelligence stop, or that gross behavior completely changes from the perspective of outside observers, or that all desires cease. There is said to be no particular behavioral indicators of an individual who is permanently established in enlightenment. [3] With that inner development we continue the activity of daily life. There is, however, a spontaneous direction for such highly developed individuals to be motivated to engage in activity for the benefit of all humankind rather than exclusively for personal gain—because of the underlying inner fulfillment, in which one's own life is already established in the benefit that would come from achievement of additional individual goals. [3]

Maharishi has described the inner life of individuals in cosmic consciousness as deep silence, unshakable strength, eternal contentment and freedom, and unbounded bliss. [3] It is complete freedom from limitations, and spontaneous ability to act in attunement with the total potential of natural law. These inner qualities are accompanied by spontaneous skill in action, in which the individual effortlessly fulfills his or her own desires while simultaneously supporting the interests of others and of the society as a whole. As will be discussed in a later chapter, Maharishi emphasizes that even a small number of individuals established in cosmic consciousness, the state of enlightenment, spontaneously living in harmony with the total potential of natural law, harmonizes the diverse tendencies in civilization and creates the basis for a peaceful society.

In cosmic consciousness the universal Self provides an absolute internal frame of reference or background from which the changing phases of sleep, dreaming, and waking life as well as the entire cosmic display of phenomenal nature are spontaneously and silently witnessed. [3] Maharishi emphasizes that this natural experience is not based on any desire, mood, self-reflection, being mindful of oneself, or any other form of mental intention. It is not an intellectual reflection about oneself. More fundamental and intimate than the intellect, it is not based on intellectual understanding or any inner attitude about life. It is a spontaneous experience associated with a new integrated style of stress-free mental and physiological functioning. The individual unfolds the natural inherent ability to maintain unbounded inner wakefulness at the deepest level of the mind while also engaged in the mental activity of feeling, thought, and perception at more expressed levels. A powerful and full range of emotions can be experienced and expressed in daily life, but they are not accompanied by any sense of feeling overshadowed or suffering deep inside.

The natural tendency of the mind to go toward a field of greater happiness that underlies the process of effortless transcending also is said to be the principle that stabilizes the fifth state of consciousness. The individual naturally wants to sustain the blissful experience of transcendental consciousness as much as possible. Through neutralization of stress and the development of increasingly refined and integrated neurophysiological functioning, both inner unbounded bliss and mental activity are maintained simultaneously. This state is called cosmic *consciousness* because it simultaneously encompasses the full range of cosmic existence—the absolute, non-changing universal Self along with all the changing relative values of the individual self.

However, in this state there remains an experiential gap between these two levels. The individual self is awake to the universal Self, but is a silent observer or witness to *everything else*. The experience is cosmic in that the entire range of ever-changing and never-changing nature is included, but is experienced in terms of the dual nature of the universal Self, the ultimate non-changing pure subjectivity (Purusha or Silent Witness) and everything else (Prakriti or Nature) as the constantly changing outer objective creation. [3]

This natural state may be clarified further by contrasting it with the characteristic waking state experience of the independence of object and subject fundamental to the modern scientific concept of objectivity. Recall the point in Chapter 1 that in ordinary waking experience it is as if the natural world were enveloped in a glass bubble the scientist peers into from outside. Although Einstein, for example, believed that the independence of subject and object was a crucial foundation for science, in quantum theory this particular form of the assumption of independence of observer and observed is no longer viable. Now, however, we can consider a deeper form of separation or gap that is a defining feature of experience in the fifth state of consciousness. In cosmic consciousness the gap is between the non-changing universal Self, which the individual self now completely connects with, and the relative ever-changing world that is witnessed (Purusha-Prakriti). This is in contrast to the separation of objectivity and individual subjectivity or inner mental activity typical of the ordinary waking states of consciousness.

From the historical perspective of modern science, the independence of subject and object is associated with the subjective sense that the objective world is somehow more *real*—in the sense of more substantial, reliable, and consistent—than subjectivity. This has to do with the objectification of nature, associated with the experience that inner subjectivity varies dramatically—such as between sleep and waking—and is more changeable than the apparent outer natural world.

Relatively speaking the objective world seems to remain as it is whether we are in one mood or another, asleep or awake, experiencing it or not. In the constantly changing experiences of the three ordinary states of consciousness, subjectivity varies significantly and thus seems in this sense to be less stable and reliable than the objective world. This is due, however, to inability to experience the infinitely stable underpinning of pure subjectivity, the universal Self. The objective world is ephemeral—always variable and constantly changing—thus not the basis for permanent stability.

Even within the range of experience of ordinary waking there is a fundamental sense that conscious experience is somehow separate from the objective world. The full significance of this inner sense of the independence of the subject from the objects of experience becomes clear in the fifth state of consciousness. In this state the individual self no longer identifies with the constantly changing values of feelings, thinking, and perception associated with ordinary experience of the objective relative levels of the individual self. Rather the individual self *realizes* and is spontaneously established in full wakefulness of the universal Self, the eternally non-changing background and source of all relative fields of phenomenal change—without even thinking about it.

The separateness in ordinary waking between the inner individual self and the outer natural world is not the same separateness as in the fifth state, cosmic consciousness. In this higher state the separation is between the unbounded universal Self that each individual fully is established in and everything else in relative existence—including the mental activity of individual subjectivity as well as everything in the outer natural world. In cosmic consciousness the individual realizes himself or herself to be the underlying, unbounded, non-changing field of pure subjectivity, the universal Self or Atma, the unified field of consciousness itself. In the absolute non-changing inner silence of the universal Self all the changing relative values of life are witnessed. Thus in cosmic consciousness there can be said to be a separation of subject and object in experience. But this gap in experience is a completely different and much more profound state than in the ordinary waking state experience of subject-object independence.

In this first stage of enlightenment, the fifth state of consciousness, the infinite, eternal, non-changing universal Self is *real*—always existing and never changing. It is who *I really am*. In this state the ephemeral relative world of change can be said to be comparatively *unreal*—always changing and never the same. This is basically the opposite of the experience of consistency of objectivity and inconsistency of subjectivity in the ordinary waking state of consciousness.

The aspiration of scientists to be *objective*, in the sense of consistent and unbiased, is fully realized in the non-changing inner awareness of the individual self as the universal Self in the state of enlightenment—permanent subjective consistency. As mentioned in Chapter 2, cosmic consciousness can be understood to be the eternally stable and consistent platform of pure subjectivity from which undistorted, unbiased observations of nature can be made. Systematic unbiased observation of nature can be said to begin in its full sense in this first stage of enlightenment, cosmic consciousness, which fosters natural development to even higher states of consciousness.

In cosmic consciousness the individual self is no longer primarily identified as an individual self but rather as the universal Self. This is sometimes described as the essential difference between the ordinary waking state of consciousness, which has been characterized as a state of ignorance (of the universal Self; unillumined darkness in the black box of the mind), and the state of enlightenment, which is the union of the individual self with the universal Self. This importantly leads to consideration of the developmental approach in ancient Vedic science and in Maharishi Vedic Science and Technology known as *Yoga*. As Maharishi points out:

> " Knowledge in its entirety comprises both understanding and experience. Therefore in order to gain fulfillment a man must necessarily acquire both understanding and experience of the relative and the Absolute... It follows that the wisdom of Sankhya, which

brings liberation through *understanding* of the relative and the Absolute, and the practice of Yoga, which brings liberation by providing direct *experience* of these two spheres of existence, are both paths to enlightenment..." (p. 249) [3]

Recall that in Chapter 11 the aspect of Vedic literature called the *Darshanas* was introduced. Each Darshana offers a comprehensive description of the full range of nature from a particular perspective or emphasis. In that chapter the focus was on enumerating the full range of levels from the perspective of the Darshana of Sankhya. Again the six Darshanas are *Nyaya, Vaisheshika, Sankhya, Yoga, Karma Mimansa, and Vedanta.* The first three Darshanas emphasize pure knowledge and the intellectual aspect of total knowledge in enlightenment. Nyaya focuses on the science of reasoning and the fundamental means of gaining knowledge. Vaisheshika focuses on the distinguishing qualities of objects of experience. Sankhya focuses on enumerating their fundamental components. The other three Darshanas emphasize the organizing power and the experiential aspect of total knowledge. Yoga is the applied science that focuses on direct experience of the unified field of consciousness—Atma—and perception of all the different gross and subtle relative levels of nature. Karma Mimansa focuses on the close study of action, including an inquiry into dharma—action in accord with the total potential of natural law. Maharishi states that the Darshana of Vedanta refers to the "...end of the Veda, final knowledge of the Veda...life itself as it is lived naturally on the level of Being." (p. 491) [3]

The term yoga means the state of union. It is associated with the root *to yoke,* to unite or establish union. The Darshana of *Yoga* refers to the systematic technologies that develop the state of *union* of individual self and universal Self. One important emphasis in yoga is growth from the ordinary waking state to cosmic consciousness. More will be said about the Darshana of yoga in the next chapter. It is brought up here in order to contrast ordinary waking and cosmic consciousness, and also as a bridge to discuss the mechanics of the development of the remaining two higher states of consciousness.

In cosmic consciousness, identified as the state of union or yoga—the first state of enlightenment—union involves identification of the individual self with the universal Self. As noted earlier, in this state there can be said to be a phenomenal separation or experiential gap between the unbounded universal Self and *everything else* in the relative world of experience. Maharishi has explained that the individual self is united with the non-changing universal Self, but there is still a *gap* between the non-changing universal Self and all the ever-changing levels of relative existence witnessed by the universal Self. The highest state of consciousness involves not only union of the individual self with the universal Self, but also union with *everything else.*

Progress beyond cosmic consciousness involves progressive refinement of experience of the outer relative field of nature toward bridging the gap between the infinite, non-changing, unified, inner bliss of the universal Self and the ever-changing finite field of relative diversified existence—*everything else.* This transition is said to involve the spontaneous ability to observe increasingly refined levels of relative creation on the basis of the non-changing platform of inner fulfillment in cosmic consciousness. This development is associated with the Darshana of Karma Mimansa that focuses on the analysis of action in the eternal silence of cosmic consciousness, which fosters natural development to the sixth state of consciousness. Maharishi explains:

"When this experience of the Absolute has become permanent, Self-awareness is naturally maintained through all the waking, dreaming and deep sleep states of consciousness. One experiences oneself as separate from activity. As one lives this life of non-involvement, of natural non-attachment, one's intellect begins to inquire: 'Is this the truth of life? Has this sense of separateness or non-attachment anything to do with real life, or is it an escape from life? Is the reality of life a duality—the duality of Being and activity?" (pp. 249-250) [3]

As his practice advances, every seeker must necessarily reach this experience; and if he is to proceed smoothly on his path, unhindered by doubts, he must possess this knowledge." (p. 250) [3]

Knowledge and experience of separation of one's Self, the universal Self, and activity in the relative field of life is brought out in the philosophy of the two fullnesses in the Upanishads:

"...[T]his is full and That is full, 'purnamadah purnamidam'—That transcendental un-manifested absolute eternal Being is full, and this manifested relative ever-changing world of phenomenal existence is full. The Absolute is eternal in its never-changing nature, and the relative is eternal in its ever-changing nature.

This living Reality of two fullnesses in cosmic consciousness finds its consummation in the grand Unity of God-consciousness." (pp. 250-251) [3]

The Sixth State: Refined Cosmic Consciousness, God Consciousness

As mentioned in Chapter 2 modern science conceptualizes creation as structured in levels from the gross macroscopic level through molecular, atomic, and subatomic structures at smaller time and distance scales to abstract quantum fields and now ultimately the completely abstract unified field. As discussed at length in this book, in ancient Vedic science the full range of these phenomenal levels and structures span three fundamental categories—gross relative, subtle relative, and transcendental domains or levels of nature.

To its great credit modern physics has established the theoretical understanding upon which it is attempting to characterize in mathematical form the transcendental level, the unified field of nature. It has accomplished this theoretical understanding based on reductive analysis of the gross level of matter to its most fundamental constituents, and then through abstract mathematical reasoning to the underlying unified field.

At the current stage of this theoretical progress, fundamental issues remain about how to connect theories of the unified field with the classical macroscopic and microscopic world. Much of the cutting edge of contemporary theoretical work in modern physics is trying to define more precisely the unified field, and to fill in the gap between these two realms. This work recognizes the necessity of accounting for mind and consciousness in modeling how all levels of nature are ultimately unified. Theories are now developing which propose that matter is underlain by a subtler non-local information space or quantum field of nonlocal mind that has properties suggestive of a relative manifest field, further underlain by the completely abstract unmanifest unified field.

In a somewhat analogous sense this theoretical and intellectual understanding can be compared to the duality that characterizes spontaneous inner experience in cosmic consciousness. In this state the individual self is fully established in the universal Self, pure Being, the unified field of natural law. That pure wakefulness of Being is the non-changing eternal silent witness of the ever-changing relative field of existence. The natural process of growth from the fifth state of consciousness to the sixth state involves progressive refinement in the perceptual experience of levels of nature, such that the gap associated with the duality of universal Self and everything else is narrowed. This natural developmental process also can be described as continued infusion of the nature of pure Being into more expressed levels of subjectivity. In cosmic consciousness the level of individual self or ego is saturated with and fully established in the universal Self. In the natural unfoldment of the next higher state, this inner-to-outer growth extends into more expressed subjective levels including the level of sensory perception. The gap between the unified field and ordinary classical experience can be related to the gap in experience of the means of gaining knowledge in modern science—ordinary sensory experience and intellectual reasoning. This gap is expressed in the mind-body problem, the measurement problem, and their accompanying explanatory gaps. The means to bridge the gap is the direct experience of the unified field of consciousness and deep empirical experience of all the subtle levels of nature underlying ordinary classical sensory experience. These experiences naturally unfold in the regular process of effortless transcending and refinement of mind and body, in which the full range of levels of nature is experienced and the gap is filled in and bridged.

The sixth state of consciousness is sometimes referred to as refined cosmic consciousness because it involves maximum refinement in the experience of the relative ever-changing field of activity witnessed by the infinite unbounded self-referral consciousness of the universal Self. Feeling and sensory perception reach their most sublime level in this state. The highest, most refined values of the phenomenal relative objects of experience become appreciated.

"In the state of cosmic consciousness, the Self is experienced as separate from activity. This state of life in perfect non-attachment is based on bliss-consciousness, by virtue of which the qualities of the heart have gained their most complete development. Universal love then dominates the heart...the silent ocean of bliss, the silent ocean of love, begins to rise in waves of devotion. The heart in its state of eternal contentment begins to move, and this begins to draw everything together and eliminate the gulf of separation between the Self and activity. The Union of all diversity in the Self begins to grow." (p. 307) [3]

"Before gaining that state, this finest value is hidden from view because our vision falls only on the surface of the objects. When only the surface value of perception is open to our awareness, then the boundaries of the object are rigid and well-defined—the only qualities that are perceived are those which distinguish the object from the rest of the environment. However, when the unbounded awareness becomes established on the level of the conscious mind—we have seen that this is the fifth state of consciousness—then the perception naturally begins to appreciate deeper values of the object, until perception is so refined that the finest relative is capable of being spontaneously perceived on the gross, surface level." [22]

This highly developed state of consciousness is also referred to as God consciousness. This phrase recognizes the natural ability of the individual in this state to be able to perceive directly the entire range of phenomenal nature. This includes the subtlest, most all-encompassing expressions of relative creation, the most refined manifest impulses of natural law that permeate all the immense diversity of subtle and gross objects in nature. The ability to appreciate the entire creation unfolds to its maximum degree. There is a complete opening of love for the beauty, harmony, grandeur, and holiness of nature. Unrestricted appreciation develops of the all-encompassing, awe-inspiring perfection of creation that can be only received in reverence and fullness of devotion. In religious terminology it is total acceptance in the unbounded fullness of one's heart and mind of the omnipresent, omniscient, omnipotent glory of the Creator. Maharishi explains:

"In such a state of integrated life where behaviour is in perfect harmony and where all the planes of living are infused with...universal love for everything...every perception, the sound of every word, the touch of every little particle and the smell of whatever may be, brings a tidal wave from the ocean of eternal bliss. Every rising thought, word or action is a rising of the tide of bliss... Unless one is cosmically evolved, unless one lives the eternity in the day-to-day transitory activity of life, it is not possible to overflow on the level of universal love... A man who has not risen to cosmic consciousness, who is shrouded by selfish individuality and who is only awake in the identity of his individual self, cannot have a clear and significant conception of love and devotion...that reaches eternity." (p. 250) [23]

" The world today has a very vague conception of God. There are those who like to believe in God, those who love God, and those who want to realize God. But even they do not have a clear conception of what God is. The word God has remained for the most part a fanciful, pleasant thought and a refuge during the suffering and misery in life. And, for the custodians of many strange religions, the word God is a magic word, used to control the understanding and religious destiny of many an innocent soul. God, the omnipresent essence of life, is presented as something to fear... The kingdom of God is the field of all good for man. God is to be realized, not to be feared." (p. 271) [23]

Maharishi describes the gradual development of the sixth state of consciousness in terms of the natural progression of the most refined feelings of love:

"In its most infant state, love finds an expression on the lap of mother, in the sweetness of the mother's eyes. It grows in toys and playfields, in the sweetness of friends and folks of society, it grows in the sweetness of husband and wife. With age and experience, the tree of love grows, it grows with the growth of life and evolution and finds its fulfillment in the eternal love of the omnipresent God, which fills the heart and overthrows the darkness of ignorance. And then, in the illumination of the universal love, the abstract love of God finds concrete expression in everything." (pp. 18-19) [24]

Maharishi further explains that this natural unfoldment involves higher integration of neurophysiological functioning. Based on the permanent unshakable stability and unbounded awareness of cosmic consciousness, the refined activity of devotion cultures the senses of perception until the subtlest level of any object is automatically appreciated in its fullest relative value:

"This integration of functions on the physiological level is brought about by a mental activity of ultimate refinement. In order to define activity of this quality, we must analyze the whole range of activity. The activity of the organs of action is the most gross, the activity of the senses of perception is more refined, the mental activity of thought is finer still, and the activity of feeling and emotion is the finest of all. One could further classify different levels of quality in emotional activity, such as anger, fear, despair, happiness, reverence, service and love.

The activity of devotion comprises the feelings of service, reverence and love, which are the most refined qualities of feeling. It is through the activity of devotion that cosmic consciousness develops into God consciousness." (p. 315) [3]

This natural refinement of mind and body is associated with specific physiological changes, including a highly refined mode of functioning of the digestive system. When mind and body are stress-free, the most refined product of the human digestive system supports the most refined levels of perceptual experience. In historical Vedic literature this most refined digestive product is sometimes called *soma*, and is the subtlest physiological basis of the refined perception that develops in the growth of refined cosmic consciousness or God consciousness. In the following quote, Maharishi explains that the natural process of growth in experiences of refined perception involves physiological changes:

"...[A[normally functioning nervous system, free from stress and strain and any abnormality, produces a chemical called soma... If there are no restrictions, no inhibitions, then awareness is unbounded, and when this unbounded awareness is maintained spontaneously at all times, then the nervous system is functioning normally... Now, the best product of such a normally functioning digestive system is soma... So soma is that which helps all the fundamentals of individual consciousness rise above boundaries, and have an unbounded status...in that unbounded self-awareness the perception is very rich—the perception is richest!" (p. 153) [25]

Here are examples of experiential reports suggestive of the development of refined perception. The first is again from an advanced TM practitioner.

"...[A]t the same time as I felt my power and inner strength increase, I also noticed a totally new feeling of softness and sweetness develop. There were days when I felt my heart melting as if I could take everything in creation into myself and cherish it with the greatest love. Often I would have long periods of the day when everything I saw seemed to be glowing with divine radiance." (p. 81) [26]

Another similar report by a TM practitioner describes experiences of growth in terms of refinement of feeling:

"The value of love and devotion is growing more every day, accompanied with the heart expanding to encompass the whole universe. Also, there is a great radiance in my heart. The heart is many times like a window to the Self, like a vast ocean which contains all; and all is known through the heart... There is bubbling bliss flowing and enlivening everything which I perceive. Beaming, radiant bliss pours forth from my Being. Sometimes there is a desire on the mental level, which drops into the heart, like a drop into an ocean, turning into waves of bliss and giving rise to the feeling "I can do anything." (pp. 83-84) [26]

In the process of development toward refined cosmic consciousness the natural ability grows to perceive subtler aspects of the environment, and correspondingly greater spontaneous appreciation and devotion for the grandeur of the essence of all objects of experience. Here is another account of increasingly refined perception that has been provided by an advanced TM practitioner:

"Generally, whenever I put my attention on an object, I become aware of the subtler qualities of the objects around me. For instance, when looking at a tree, I first become aware of the object as it is—a concrete form bound in space and time. But then I perceive finer aspects of the object co-existing along with its concrete expression. On this subtler level, objects are perceived as almost transparent structures of soft light (unlike harsher, normal daylight) through which the very essence of life appears to flow. Perceiving these finer aspects of creation completely nourishes the finest aspect of my own being." [4]

A long-term TM practitioner also provided this description of refined perception:

"It is like pulling away a veil of concrete qualities. When that happens, a dimension of life essence becomes more lively... The bioenergy [of a plant] becomes more real...comes to the surface...along with the leaf and the stem. It is not blocked by the objectivity of matter. Even air has a concrete quality to it when experiencing this dimension. Air is heavy when compared to the air perceived in this subtler dimension. Lightness pervades my heart, my physiology...but it has no heaviness in it. The light is sweeter, there is light essence in it itself. It is life itself...its own dynamism is there. Looking at that light nourishes the self and makes me happy. When one sees it, one hears, feels, tastes, and smells it at the same time... As the life force rises within me without obstruction, the perception happens... Subtle perception is not supernatural or unnatural. It is completely natural perception... It is part of who you are. When the physiology is functioning properly, it is there. Toxins in the body block the perception." [27]

English Unitarian minister J. Trevor provided this account of subtle perception:

"...I was in Heaven—an inward state of peace and joy and assurance indescribably intense, accompanied with a sense of being bathed in a warm glow of light, as though the external condition had brought about the internal effect—a feeling of having passed beyond the body, though the scene around me stood out more clearly and as if nearer to me than before, by reason of the illumination in the midst of which I seemed to be placed. This deep emotion lasted, though with decreasing strength, until I reached home, and for some time after, only gradually passing away." (pp. 396-397) [17]

J. G. Fichte, the 18th/19th Century philosopher, also describes refined perception:

"The universe appears before my eyes clothed in a more glorious form. The dead inert mass, which only filled up space, has vanished; and in its place there flows onward, with the rushing music of mighty waves, an endless stream of life and power and action, which issues from the original Source of life... (pp. 150-151) [28]

Contemporary English poet Kathleen Raine gives an account of momentary refined perception that occurred spontaneously during settled awareness. Though apparently an unusual and rare experience for this person, the account describes aspects of refined perception clearly.

"There was a hyacinth growing in an amethyst glass; I was sitting alone... All was stilled. I was looking at the hyacinth, and...abruptly I found that I was no longer looking *at* it, but *was* it; a distinct, indescribable, but in no way vague, still less emotion, shift of consciousness into the plant itself... I dared scarcely to breathe, held in a kind of fine attention in which I could sense the very flow of life in the cells. I was not perceiving the flower but living it. I was aware of the life of the plant as a slow flow or circulation of a vital current of liquid light of the utmost purity. I could apprehend as a simple essence formal structure and dynamic process. This dynamic form was, as it seemed, of a spiritual not material order; or of a finer matter, or of a matter itself perceived as spirit. There was nothing emotional about this experience which was, on the contrary, an almost mathematical apprehension of a complex and organized whole, apprehended as a whole. This whole was living and as such inspired a sense of immaculate holiness. Living form—that is how I can best name the essence of soul of the plant. By "living" I do not mean that which distinguishes animal from plant or plant from mineral, but rather a quality possessed by all these in their different degrees. Either everything is, in this sense, living, or nothing is; this negation being the view to which materialism continually tends, for lack, as I now knew, of the immediate apprehension of life, as life. " (p. 119) [29]

This anecdotal report illustrates important features of the subtle level of nature—referred to in Chapters 11-13. Note qualitative differences of the object as a subtler level of being or essence of energy of utmost purity that contains a greater sense of wholeness. As pointed out in the account, the sense of essence or life of the object was not a contrast of living or inert, but rather a deeper sense of wholeness of being—a sense of holiness and essence of form.

The gross form of the object (hyacinth) was present in attention, but simultaneously there was much deeper appreciation of finer levels of the object. This appears to be a clear description of experience on the subtle level of nature. As finer perception grows and stabilizes in the development toward the sixth state of consciousness, experiences naturally and spontaneously contain deeper appreciation of the wholeness or holiness of nature in ordinary gross objects of experience.

It is interesting to note Kathleen Raine's reference in her hyacinth experience to a different quality of light as a flow of energy—"a vital current of liquid light of the utmost purity." Others also have noted that experience of the subtle level of objects involves a different quality of light. Some have described the subtle level of objects as having an inner luminance or radiance (e.g., Fichte above), more of an inner essence of radiance than reflection of light off of objects of experience typical of ordinary gross experience. The subtle level of objects of experience also is described as having a kind of transparency, as a flow of energy or essence rather than the concreteness or inertness of gross matter. With respect to the quality of light, it also has been described as being purer, less grainy, more flowing, and in some sense more holistic—less tangible compared to ordinary gross matter but more substantive, meaningful, and reflecting the deep inner quality of things. Anecdotal reports of near-death experiences also frequently contain descriptions of a subtler field permeated with pure, radiant, even divine light that is qualitatively different from sunlight or electric light—as if it were the essence of light itself. This is an experiential description of the subtle essence element or *tanmatra* of fire—associated with the principle of luminosity.

It also is interesting to consider the properties of light from the perspective of this subtle essence element level that permeates the gross level. With respect to the gross level of nature, the speed of light is thought to be an absolute value that does not change relative to the speed of motion of any gross object or subject. This is suggestive that light itself is an underlying field existing throughout the subtle level, and thus more fundamentally does not propagate in the sense that it appears to do in the gross particulate or quantized level of nature associated with our ordinary experience and understanding of quantized light.

The overall issue of experiencing the subtle level of nature was introduced in Chapter 10-11. In the Sankhya system of ancient Vedic science, all the gross objects of experience made of matter have the subtle levels of nature embedded in them. This is said to be open to direct experience as a natural by-product of the development and purification of mind and body in higher states. Although rare, such subtle experiences can also occur at earlier stages of development. These inspiring experiences can be understood to be glimpses into experiences in higher states of consciousness. Because they tend to be rare and fleeting experiences, there is frequently doubt about their *reality*, although when actually having the experiences there is frequently an accompanying deeper sense of *reality* than with ordinary tangible experiences in the gross relative domain.

Not only are the gross objects experienced at the subtle level of their deeper essences, but the natural ability also develops to experience the subtle objects of perception that are not manifest on the gross level. There is an experiential distinction of the gross elements of nature (mahabhutas) and their essences or subtle elements (tanmatras). In ancient Vedic science all gross objects of sense are comprised of the gross elements and the subtle essence elements, as well as having in them the even subtler levels of mind and the unified field of consciousness itself. With respect to perception of subtle objects of sense, going from gross relative to subtle relative can be likened to going from water to air—only even a much more striking contrast. From the gross level it is like looking through the water at objects of experience; but out of the water things are clearer. There appears to be no graininess in subtle objects of perception—beyond particle fields characteristic of independent objects in the gross relative domain.

Also the subtle sense of the passage of time is said to be different. At this level of experience, the sense of time is more like being completely in the eternal present, or timelessness. The flow of time is less bounded and restricted compared to the ordinary sense of clock time on the classical level of gross space and time. It is somewhat akin to the sense of time, or lack of it, that accompanies deep absorption in an object of experience (maybe even an engaging book). There is less sense of time passing, of time lost, of limitations and shortage of time. Temporal divisions that mark the sense of ordinary time are less prominent, and the underlying unity of the timelessness of eternity is said to be much more prominent. There also are said to be corresponding differences in the sense of space, as well as distance in space through time, changing qualities related to different aethers or mediums of space-time or levels of existence.

In this subtle level of phenomenal creation there is also increased integration of energy and intelligence. Complicated organismic systems are not needed in order to sustain life in the subtle domain of relative existence. Dynamic orderliness characteristic of life is more expressed; it is not hidden and latent as it is in more inert forms composed of particles of gross matter. The strong distinction of living and non-living is most particularly relevant to the gross relative domain. All subtle functions and forms express a deeper, fuller quality of life inherently and spontaneously, as if they radiate life itself and are self-luminous. This suggests that the subtle level of bodily structure or form does not require systems of circulation, respiration, or digestion in the way that gross bodies do. At the subtle level it is not a matter of movement of inert particles that involve inertia, friction, heat, and so on. The subtle aspects of individuality inherently express dynamic energy and intelligence. They have the character of mental reality and thought forms rather than gross physical *reality* and material forms. The subtle senses function more directly, not via their corresponding gross bodily organs made of matter particles. The subtle essence elements (tanmatras) are inherently more complete in sensory quality and meaning. Movement is flowing, wave-like, with the capability of permeating gross objects without restriction. This dynamic subtle movement doesn't require mechanical and thermodynamic processes that gross physical motion does.

This does not mean, however, that the subtle level of experience is less *real* because it is less tangible than the gross level. It is relatively much, much more *real*—embodying a fuller, deeper, richer sense of *reality*. When such subtle experiences are rare or unfamiliar, they may seem sort of dream-like or ephemeral, and there reasonably may be questions about their *reality*. Experiences sometimes interpreted as of the subtler levels of nature in fact may be just dreams or illusory images. It initially even may difficult to distinguish subtle experiences from strong imagination and dream images. But eventually these differences are sorted out with regular experience of the subtle levels of nature—associated with increased refinement of physiological functioning—within the context of integrated, balanced, natural development toward higher states of consciousness.

In ancient Vedic science, even in ordinary motion in gross space and time it is the subtle mind that initiates and directs the gross body to move via complex classical mechanics of physical motion. When subtle mind and gross body separate, the components of the gross body follow the 2nd law of thermodynamics—reverting back to the entropic patterns typical of phenomenally inert gross matter. The mind also initiates and directs movement of subtle forms, but subtle motion can be understood to be through impulses of thought directly, rather than via transduction mechanisms associated with electrochemical neural activation in the gross body.

Movement in the subtle level of nature is less restricted, more nonlocal and holistic. Action is said to involve propagation directly through the power of thought impulses and intentions that do not require classical physical mechanics. This may be difficult to comprehend without empirical observation of subtle phenomena. For those who have had experiences such as out-of-body experiences, or subtle experiences that naturally occur in the process of development to higher states of consciousness, it is easier to accept that under certain circumstances the subtle aspects of the senses and mind can function fully without the heavy overlay or sheath of the gross physical body and the gross sensory organs. It is said to be the increased refinement of mind and body that spontaneously allows subtle experiences to occur. These natural experiences of finer and finer relative levels of creation fill in the experiential gap between the absolute universal Self and 'everything else' that characterizes the state of cosmic consciousness. Through time the gap is completely filled in with natural development of the seventh state of consciousness, unity consciousness.

Ancient Vedic science records a vast diversity of objects of sense not expressed on the gross level (made of tanmatras, not mahabhutas), sometimes characterized as subtle *celestial* life rather than gross terrestrial life. These phenomena are often mentioned in religious texts associated with various levels of *heaven* and inclusive of various types of angelic beings. It is beyond the scope of this book to discuss these subtle realms of nature outside of the more inclusive scientific framework the book articulates, other than to point out that such experiences are said to be open to direct personal validation in the *natural* unfoldment of the sixth state of consciousness. But they involve higher levels of development far beyond the ordinary waking state of consciousness, and generally have been quite rare in modern civilization. More will be said about this topic, however, in the context of Vedic language and the cosmic nature of human physiology in a later chapter.

The following report describes growing appreciation of the divine embedded in ordinary experience that shines through at times. This report is from Jonathan Edwards, the 18th Century American clergyman and president of the school which developed into Princeton University:

"...I walked abroad alone, in a solitary place in my father's pasture, for contemplation. And as I was walking there, and looked up on the sky and clouds; there came into my mind a sweet sense of the glorious majesty and grace of God... After this my sense of divine things gradually increased, and became more and more lively, and had more of that inward sweetness. The appearance of everything was altered: there seemed to be, as it were, a calm, sweet cast, or appearance of divine glory, in almost everything. God's excellency, His wisdom, His purity and love, seemed to appear in everything: in the sun, moon and stars; in the clouds, and blue sky; in the grass, flowers, trees; in the water, and all nature... I often used to sit and view the moon for a long time, and so in the daytime spent much time in viewing the clouds and sky to behold the sweet glory of God in these things." (pp. 101-102) [30]

20th Century Bengali poet Rabindranath Tagore provided the following account:

"I was watching the sunrise from Free School Lane. A veil was suddenly withdrawn and everything became luminous. The whole scene was one perfect music—one marvelous rhythm. The houses in the street, the men moving below, the little children playing, all seemed parts of one luminous whole—inexpressibly glorious. The vision went on for seven or eight days. Everyone...seemed to lose their outer barrier of personality; and I was full of gladness, full of love, for every person and every tiniest thing... That morning...was one of the first things which gave me the inner vision, and I have tried to explain it in my poems..." (p. 24) [31]

Here are additional reports, in this case of celestial perception:

"In 1973 in Switzerland, there was a soft explosion of light that took on the form of an angel that came through the window. It was milky white...of pure sweet nectar... She had wings, full dress of white with gold trim. She was of soft light. After about 90 seconds, she left." [32]

"As I was practicing the sidhis [advanced TM program to be discussed in the next chapter], I started to see the qualities of water as a field of blue...the qualities of earth was green, and permeated me...and fire as a field of red... Suddenly the field started to swirl into a whirlpool, swirling faster and faster. The whirlpool dynamics shifted from a field of red into a garment of red. Then the garment of red had gold-specked diamonds, and I looked up into the face of the divine mother goddess, exquisitely beautiful...she faded away, leaving me in an ecstatic state." [32]

In the following quote Maharishi summarizes the glorious level of development achieved in the stabilization of God consciousness:

"Perfect mental and physical health is natural in this state of fulfillment of life...where God-consciousness permeates all daily experiences and activities, where universal love flows in and overflows from the heart and where [cosmic] intelligence fills the mind. In such a state of integrated life where behaviour is in perfect harmony and where all the planes of living are infused with [God] consciousness, universal love for everything flows..." (p. 250) [23]

The Seventh State: Unity Consciousness

Although the profound level of spontaneous experience in daily life associated with the sixth state of consciousness is far beyond what modern science attributes to ordinary classical *objective reality*, ancient Vedic science describes even further development to achieve the endstate of human evolution. The highest state of consciousness, the state of full enlightenment, is identified as *unity consciousness*, the seventh state of consciousness. Progressive stabilization of this highest state also will be described briefly.

Maharishi states that the total value of the individual self as the universal Self, first experienced in transcendental consciousness and stabilized in cosmic consciousness, becomes fully awake in its totality in the seventh state of consciousness. [3] The completely holistic value of existence is the natural primary experience. The universal Self is directly experienced as permeating every aspect of subjective and objective existence. Everything is experienced as having the same infinite status as one's own universal Self.

In that highest state the experiential gap between the universal Self and everything else—characteristic of cosmic consciousness and narrowed to its smallest difference in the progressive refinement of perception in refined cosmic consciousness or God consciousness—is totally bridged into unity. That state is characterized by the incomprehensible and indescribable experience of the infinite value of existence permeating every point of existence, of eternity primary in each moment of time—infinite, all-encompassing self-referral consciousness.

In the sixth state experience has reached its most sublime relative level. Every object is experienced in terms of the maximum value of fulfillment of love and appreciation. However, experience still involves a slight gap between the entirety of relative creation and the universal Self. There remains the slightest difference in experience of the universal Self and the grandness and perfection of the totality of Nature—of God's creation. This slightest separateness supports the unbounded devotion and reverence in refined cosmic consciousness or God consciousness. In unity consciousness this experiential separateness or division is completely bridged and there are no gaps in experience of the infinite Oneness and wholeness of Being. As Maharishi explains:

"This seventh state of consciousness could very well be called the unified state of consciousness because in that state, the ultimate value of the object, infinite and unmanifest, is made lively when the conscious mind, being lively in the unbounded value of awareness, falls on the object. The object is cognized in terms of the pure subjective value of un-bounded, unmanifest awareness... In this unified state of consciousness, the experiencer and the object of experience have both been brought to the same level of infinite value and this encompasses the entire phenomenon of perception and action as well. The gulf between the knower and the object of his knowing has been bridged. When the unbounded perceiver is able to cognize the object in its total reality, cognizing the infinite value of the object, which was hitherto unseen, then the perception can be called total or of supreme value. In this state, the full value of knowledge has been gained, and we can finally speak of complete knowledge." [22]

Here are reports in the direction of unity, the first ones from advanced TM practitioners:

"...[A]ll the beautiful indications of rising consciousness which I had been experiencing seemed to crystallize, and a new reality seems to be dawning in my daily life. I feel an underlying continuum of quiet bliss and fullness, of infinite and universal love. Often the deep silence of my Self seems all-pervading, everywhere the same. Objects seem transparent, and I perceive unboundedness, the unmanifest, in everything I see. At such times I feel infinitely full and enveloped in softness. Perception is often very glorified and rich... I feel very self-sufficient and self-contained; activity makes no impression on the growing wholeness which I feel. At the same time, however, I feel a growing intimacy with everything. Nothing seems foreign to me; I feel at home with everything, everyone, and with any situation. I feel truly invincible, for I feel established in my own unbounded Self." (p. 78) [26]

"I began experiencing bliss consciousness intensely during activity at times... I experienced both myself and the objects of perception as being made of bliss consciousness. When objects are perceived in this light, they seem much more wonderful and satisfying than normal, and the whole process of perception becomes very soothing and fulfilling. In this way, everything becomes extremely valuable because it is so delightful. Also everything seems to belong to oneself in a way because everything is made of the same stuff as one's Self—absolute bliss consciousness. Everything is wealth when bliss consciousness is the dominant value of all objects of perception." (pp. 78-79) [26]

"A continuum of awareness has become more predominant. The sense of events and time has given way to an interconnectedness within a whole... Nothing can hide the Absolute within everything. Everything seems to serve as a window to perceive its true nature within... Activity is characterized more and more by the feeling of being sunk in the Absolute... It is truly a precious phenomenon that is occurring, and I feel it not just for me, but for everyone." (p. 80) [26]

"I have experienced being infinite and containing within me the entire universe and all activity. In this state it is possible to experience life from any vantage point I choose. Once I wanted to see the whole manifest universe, and suddenly it was as if I were looking down upon it from some great vantage point outside the universe... My vantage point was outside the universe, but the universe was contained within me as well. There was no separation. Everything was magnificently connected to everything else, and it was, all at the same time, deeply silent pure bliss, and the most intense activity imaginable. The activity was increasing, and yet the silence was not disturbed at all. I felt that I could do anything, know anything, be anything, and yet I felt that I didn't have to do anything, because everything was already done with utmost perfection. My individuality was not separate from cosmic activity, and the perfect order of the activity on that level gave a great security and supreme bliss. I have had many experiences similar to this of the perfect unity of life, but at this point in my development of consciousness this sort of experience does not last all the time. The experiences come and go, but always have a transforming value that beautifully affects all of my perceptions and my daily life." (p. 51) [15]

Similar glimpses of the experience of unity can be found in the writings of Hegel, who emphasizes the ultimate unity of consciousness and the objective world:

"...Existence and self-consciousness are the same being, the same not as a matter of comparison, but really and truly in and for themselves..." (p. 279) [33]

"...divest the objective world that stands before us of its strangeness, and as the phrase is, to find ourselves at home in it; which means no more than to trace the objective world back to...our innermost self." (p.335) [34]

Gustave Flaubert, the 19th-Century French novelist, also describes experiences of the unity of knower and known:

"It is true, often I have felt that something bigger than myself was fusing with my being... It was like an immense harmony engulfing your soul with marvelous palpitations, and you felt in its plenitude an inexpressible comprehension of the unrevealed wholeness of things; the interval between you and the object, like an abyss closing, grew narrower and narrower, until the difference vanished, because you both were bathed in infinity; you penetrated each other equally, and a subtle current passed from you into matter while the life of the elements slowly pervaded you, rising like a sap; one degree more, and you would have become nature, or nature become you... Immortality, boundlessness, infinity, I have all that, I am that!" (p. 31) [35]

Writing in the early 20th Century, Edward Carpenter described his experience of the mechanics of unification of knower, process of knowing, and known:

"The object is suddenly seen, is felt, to be one with the self... The knower, the knowledge, and the thing known are once more one... This form of Consciousness is the only true knowledge—it is the only true existence... There is consciousness in which the subject and object are felt, are known, to be united and one—in which the Self is felt to be the object perceived..." (p. 220) [36]

Again, Franklin Merrel-Wolff summarizes his experience toward unity from a state descriptive of the duality that is experienced in cosmic consciousness:

"At the time of the culminating Recognition I found myself spreading everywhere and identical with a kind of 'Space' that embraced not merely the visible forms and world, but all modes and qualities of consciousness as well... That totality was, and is, not other than myself, so that the study of things and qualities was resolved into simple self-examination. Yet it would be a mistake to regard the state as purely subjective. The preceding Recognition had been definitely a subjective penetration, and during the following month I found myself inwardly polarized to an exceptional degree. In contrast, the final Recognition seemed like a movement in consciousness toward objectivity, but not in the sense of a movement toward the relative world-field. The final state is, at once, as much objective as subjective, and also as much a state of action as of rest... Speaking in the subjective sense, I am all there is, yet at the same time, objectively considered, there is nought but Divinity spreading everywhere..." (pp. 66-73) [37]

Maharishi describes unity consciousness as the state of perfection, the supreme level of development in which every phenomenally separate object of experience is spontaneously known, experienced, and attributed to be nothing other than the infinite value of the universal Self. There are no gaps, no separations, in the eternal oneness of Being:

"In that perfect liberation he leads the life of fullness and abundance. His vision is such that it quite naturally holds alike all things in the likeness of his own Self, because he himself and the vision that he has are the expression of the Self... The pairs of opposites, such as pleasure and pain, which present great contrasts on the lower levels of evolution, fail to divide the evenness of his vision. To make such a vision more comprehensible to the ordinary level of consciousness, it may be compared to a father's even vision toward a variety of toys which, to the vision of his child's undeveloped consciousness, will present great differences... No diversity of life is able to detract from this state of supreme Unity. One who has reached It is the supporter of all and everything, for he is life eternal. He bridges the gulf between the relative and the Absolute. The eternal Absolute is in him at the level of the perishable phenomenal world. He lives to give meaning to the paean of the Upanishads: "purnamadah purnamidam"—That Absolute is full, this relative is full." (pp. 448-449) [3]

In unity consciousness every object of perception is fully understood and experienced to be nothing other than the universal Self. Unity, wholeness, oneness of everything is the primary locus of experience, permeating all phenomenal diversity and change. This highest state of consciousness is the emphasis of the Darshana of Vedanta, living the totality of the Veda in daily life, Brahm. It is not an intellectual understanding of unity, but rather spontaneously living the ultimate oneness of everything. Maharishi explains that on the basis of the profound intimacy between the knower and the known in refined cosmic consciousness or God consciousness, complete unification of subjective and objective existence is simply a matter of time. [3]

Maharishi sometimes describes that ultimate destination across the bridge to unity as experienced in terms of levels of attention. The initial experience of unity is said to experienced in the primary focus of attention. In focal attention the experience of unity merges with the knowledge of unity. The object in focal attention is experienced not just at its subtlest relative value, but further as having the same infinite status as oneself, the universal Self. The phenomenal object is experienced directly in all its gross and subtle levels as being fluctuations of one's eternal non-changing Self, the unified field of consciousness, pure Being. *Atma*, transcendental consciousness, the universal Self, is directly experienced as the primary essence of every level and object of experience. Everything is one's Self, Brahm, naturally and spontaneously living the infinite totality of Life in everyday experience.

"Esha Brahmi sthitih Parth nainam prapya vimuhyati."
(*Bhagavad-Gita. 2.72*)

"Once achieved, it is never lost—life in enlightenment—life of the individual a lively field of all possibilities—the ability to achieve anything through mere desiring.

"This is life in fulfillment, the goal of all life enjoyed in the practicalities of daily living." (p. 184) [38]

The experience of unity in focal attention is validated by the understanding and confirmation from teachers as well as from ancient texts. As this completely unified experience stabilizes in focal attention, it spontaneously extends to secondary and tertiary objects of attention, eventually encompassing all levels of experience. Then the entire phenomenal display of nature in all its infinite diversity is fully known to be the play of the infinite dynamism and infinite silence of non-dual pure Being. In unity consciousness all of phenomenal creation is directly known to be nothing other than the play and display of *oneself*, the universal Self or pure Being. In this ultimate holistic perspective, one's own self—identified as a limited and localized individual in the ordinary waking state of consciousness—is the universal Self at its ultimate core and on its surface. The infinite value of the universal Self permeates all experience of individual self and everything else; ultimately, there is nothing else. What was thought to be the objective world separate from oneself in the ordinary waking state of consciousness is appreciated as an apparent phenomenal *reality* that for a time overshadowed and hid the ultimate Reality that all is *my* Self.

The seventh state of consciousness, the unification of everything in the universal Self, is sometimes also identified as *Brahman consciousness*, the fully awake state of Brahm, the ultimate unified Reality. The term *Brahm* can be translated as 'coming together;' it is the coming together of Atma, Ved, Sharir, and Vishwa in Brahm—one infinite wholeness of life, fully awake to itself in all its phenomenal point and infinite values.

The process of evolution through states of consciousness can be viewed as a process of eliminating the layers of restrictions, or purifying the levels of self, which appear to overshadow the universal Self and structure the phenomenal experience of the individual self. The nature of existence is that it is simultaneously point and infinity, universal and individual. Through *Maya*, there is a manifestation or limitation of the universal Self into individual selves and separate objects they experience. The phenomenal process of evolution is then one of going through all the apparent levels of limitations of universality into individuality by expanding individual experience until it once again encompasses and fully recognizes its own universality. The course of evolution is from the least degree of recognition—the most restricted states of experience—to the infinite completely unrestricted degree. In terms of individual beings and bodily structures as reflectors of states of consciousness, the range encompasses every phenomenal level of gross, subtle, and transcendental domains. Known, process of knowing, and knower are completely unified. The bridge is crossed, and eternal infinite indivisible Unity is permanently all pervading.

It might be useful to contrast the theory of evolution currently prominent in modern science with ancient Vedic science. The fragmented, reductive, materialistic modern scientific perspective is associated with Darwinian evolutionary theory. In this theory individual living beings were created spontaneously through random activity of non-living physical substances and evolved through random mutation and the survival instinct, or more precisely the empirical fact of survival of the fittest. As outlined in prior chapters there is no coherent explanation of the origin of non-living substances, nor how non-living substances somehow developed into living organisms. Also there is no coherent explanation of how primitive living organisms developed top-down selective control of behavior and what is experienced as an intentional desire or value of survival, nor the necessary place and role of mind and consciousness. Conscious individual beings develop in the process of species reproduction, live out their individual life spans, and are completely extinguished at the death of the individual bodily form. It is all based on fundamental randomness, and there is nothing more to it. There is no continuity of individuality across life forms, and no purpose to the evolutionary process. There is evolutionary continuity only in the sense of species survival—as long as the random mutations that have occasioned different species happen to support adaptation to existing environmental conditions. In this view as generally accepted, humans spontaneously evolved from monkeys or other similar beings, which evolved from simpler animals down to single-cell organisms, from organic and inorganic compounds, from atoms and fundamental matter and force fields, from quantum fields, and ultimately from literally nothing.

In Vedic science individuality is inherent in universality, and ultimately they are the same thing—the inherent point-infinity nature of the unified field, the coexistence of the opposites of infinite dynamism and infinite silence. There is no origin to individuality completely separate from universality. In terms of the phenomenal manifestation of infinite wholeness of the unified field into its finite parts, the unified field or universal Self appears to manifest into parts that maintain wholeness and into an infinite variety of parts that appear not to maintain wholeness.[39] From the holistic perspective, universal consciousness can be said to take on the primary experience of individuality for a while, then re-establishes the primacy of universality.

With respect to the finite parts, there is an apparent process of evolution in which the parts go through levels of development to again fully reflect the infinite wholeness. This process of evolution involves phenomenal manifestation of individuality in terms of an individual soul or being that takes on different bodily forms—like wearing different layers of clothes. This phenomenal process involves continuity of evolution through different forms, until the individual develops and recognizes in itself its full potential as the universal Self, at which time the individual self fully remembers its universal Self in full enlightenment, the pinnacle of evolution in the highest state of consciousness. In this pinnacle of evolution the individual's inherent universal status is no longer overshadowed, and is directly and spontaneously lived.

The inherent purpose of this evolutionary process is the expansion of happiness, the full expression of the infinite wholeness of life, ever continuing to reveal its infinite, universal, phenomenal fullness through each of its infinitely diverse, phenomenally individual expressions. It is a process of eternally proving the inconceivable phenomenon of the equivalence of parts and whole, individuality and universality, point and infinity, infinite dynamism and infinite silence.

The entire phenomenal course of development through the seven states of consciousness can be described simply as a process of growth from seed to tree to seed, infinity to point to infinity— universal to individual singularity and then back to universal singularity. There is an apparent relatively *real* but ultimately *unreal* diversification of absolute unity or indivisible wholeness into relative manifest existence, and of the parts appearing to exist as something other than the whole. It thus can be described as a process of appearing to forget and then remember the Self in the self as the Self. But the forgetting is a phenomenal *reality*—we could say, an illusory *reality* compared to the non-changing universal Self. As infinite eternal self-referral consciousness, the Self is never completely unaware of Itself, and divides itself into relative parts in terms of phenomenal experience only in states of consciousness that appear not to reflect the full wakefulness of pure Being, of fully awake Brahm. In the following quote Maharishi discusses the distinction between the never-changing reality and the ever-changing phenomenal world:

"The discrimination between the different phases of life, leading to the conclusion that the whole field of life is a field of perishable nature, is the first lesson on the intellectual path of enlightenment. It must first be known that the world is not real, even though it seems to be. The mind concludes that these things are always changing and that which is always changing has no lasting status of its own... On the sensory level, however, the world seems to be real. Through the intellect we decide that because the world is ever-changing, it cannot be real; the real is described as that which will always be the same. But the world cannot be dismissed as unreal, because we *do* experience it... We experience that the wall is here, that the tree is there. We cannot say that the tree is not there. If we say that the tree is unreal, we will have to say that it does not exist, and we are not in a position to make such a statement. We acknowledge that the tree is there, but we must also say that it is always changing. Because it is always changing it is not real, but, because it is there, for all practical purposes, we have to credit the tree with the status of existence... What is that status between real and unreal? The phenomenon of the tree is there, even though it is not real. So the tree has a "phenomenal" reality. In Sanskrit it is called *mithya*. The word is "mithya," phenomenal, not really existing. The conclusion is, thus, that the world is neither real nor unreal... A strong cultured mind analyzes his life in the world with discrimination and eventually comes to the conclusion that the world is mithya—or, only a phenomenon." (p. 278) [23]

Maharishi further describes the phenomenal world in terms of the concept of Maya:

"It is in this conceptual (intellectual) aspect of intelligence within the nature of pure unity that the wise locate the existence of Maya, and enjoy deriving the creation from the field of Maya... Maya is a concept, which is the lively awareness of two values: dynamism of Rishi, Devata, Chhandas [trinity of knower, process of knowing, and known], and silence of Samhita [oneness]... Maya is understood to be the source of creation. Its seat is in the relationship of Samhita with Rishi, Devata, and Chhandas. Because Samhita of Rishi, Devata, and Chhandas is the eternal unified reality, the seat of Maya (relationship) in it can only be a conceptual reality, and this concept is the rightful status of Maya... Its first display is in the wakefulness of self-referral consciousness and the self-interacting dynamics within its nature, which appears as Rishi, Devata, and Chhandas within the singularity of Samhita. It is derived from the relationship of Rishi, Devata, and Chhandas with Samhita... Maya is the nature of Brahm (totality), inseparable from it. It enjoys all credit for creation...

Mayadhyakshen prakritih suyate sacharacharm.
(Bhagavad-Gita, 9. 10)

Under my presidentship (my) nature creates all creation." (pp. 322-323) [40]

Another way to describe the development from the ignorance of ordinary waking consciousness—in which one believes one is only an independent individual self—to full enlightenment in the unified state of consciousness is in terms of progressive stages of experience of the gap. In the ordinary waking state of consciousness a gap is experienced between the ordinary outer natural world and ordinary inner experience of oneself and one's inner thoughts and feelings (refer to Chapter 1). The phenomenological basis of the mind-body problem associated with the distinction between objective and subjective, it also is embedded in the measurement problem and the difference between potentiality and actuality in quantum physics. More recently it has been clarified somewhat as the explanatory gap between consciousness and matter, also associated with the so-called hard problem of consciousness. In these conceptions of a gap, however, consciousness and mind are attributed only individual status, as a by-product or epiphenomenal appurtenance of the individual nervous system, a product of fundamentally random evolutionary events in nature. This is sometimes associated with an existential view of individual life with no meaning beyond the fact of its phenomenal existence—as if entirely disconnected from the universal value of life—flatland reductionism, the physicalist worldview, devoid of any fundamental meaning.

Modern science is now going beyond this disjointed view of nature in pursuit of unified field theory. But the knowledge of the unified field is divorced from the ordinary experience of the gross classical *reality* of everyday life—due to fixation on the objective known, and overlooking the process of knowing and the knower. There is beginning to be intellectual recognition of the ultimate unity, but not direct experience of it. Because there is little or no experience of even the subtle levels of nature that connect the unified field with the ordinary experience of classical *reality*, the ultimate unity of life is theoretical and not yet generally experienced as a practical empirical fact. This emphasizes the profound value of the developmental technology of the TM and TM-Sidhi program, which systematically allows the mind to experience all the gross and subtle levels and the unified field of natural law in the simplest state of one's own consciousness.

As described earlier, however, even in the first stage of enlightenment there can be said to be an experiential gap—between the universal Self and everything else. This also can be thought of in terms of the gap between the non-changing Absolute and ever-changing Relative. In the sixth state of consciousness, refined cosmic consciousness or God consciousness, this gap is reduced to its slightest value by virtue of the ability to perceive the most profound relative value of all levels of nature, and in unity consciousness it is completely bridged.

This experiential gap also can be associated with the description of the four stages of the gap in past chapters. It is sometimes described as the *junction point* between unmanifest and manifest, potentiality and actuality, infinite value and point value, and consciousness and body. In the ultimate oneness of unity consciousness, the relative level associated with individuality and the experience of individual consciousness as the point value of the infinite value becomes established in infinite universal consciousness. Maharishi sometimes describes this as the 'lamp at the door,' the light of consciousness that simultaneously illuminates and unifies both the finite Relative and the infinite Absolute (refer to Prologue).

Being in the physical presence of a fully enlightened individual. In the fully awake state of enlightenment there is individuality—the person still exists in the phenomenal world as a separate individual human—but this is not the *whole* story or the *real* story. The inner experience and knowledge is of fully awake universal Being, beyond individuality, beyond differences and sameness. The phenomenal individual aspect is said to be *Lashya avidya*, the remains of the predominant identification with individual consciousness in Pragya aparadha, the remains of the unenlightened state—the remains of ignorance.

The experience of being an individual, with an individual body and mind, as well as the experience of phenomenal creation as an objective and subjective *reality,* is a conceptual *reality* associated with the ever-changing field of relative phenomenal existence. Because it is changing, it doesn't have the ultimate status as that which is never-changing, and in this 'ultimate perspective'—that is, beyond all perspectives—it can be described as relatively unreal. As Maharishi has recently said:

"Unity is real. Diversity is conceptual." [41]

Relative phenomenal 'reality' is certainly *real* within the realms of concepts and conditional *realities*. The understanding is not to undermine the sense of *reality* in states of consciousness that are not the ultimate. These conditional perspectives remain conditionally *real,* and they all can be experienced by an individual in the highest state of consciousness. But they can be said to become merely a tiny part of the ultimate infinite wholeness of experience. In the completely unified state in which universality is the primary natural experience, any sense or form of individuality is secondary—except we could say the infinite singularity or universal individuality of the Oneness of eternal Being. The individual self, mind and body, can be said still to be present, but the totality of pure universal Being is fully awake and fully lively, permeating every level of individuality—whether any semblance of individuality is present or not. That level is so far beyond ordinary physical experience that it is impossible to describe or fathom, but it inevitably is what we already are. If it is to be described, it is said to be *infinitely fulfilling.* [3]

Because pure Being is beyond any specific manifestations, however, it still may be hidden from the perspective of other individuals observing someone who is enlightened. One views others from one's own state of consciousness—unenlightened individuals view enlightened individuals from their unenlightened state. From the view of individuality, an enlightened person still has an individual self, performs individual behaviors, experiences the consequences of actions, and generally displays the personal habits and range of personality that characterizes individual life. This includes reacting to information—as well as misinformation—from others, and sometimes even appearing to make what ordinarily might be seen as factual mistakes. This might include such things as dropping a pen, responding to a question in a manner that suggests it might not have been heard clearly, making a comment that may be incorrect grammatically, or suggesting that some action be taken that appears not to bring the intended result. It is not that individuality vanishes, but rather becomes permeated by fully awake and fully lively universality. The enlightened individual is simultaneously uniquely individual and fully universal, infinity and point.

In that state individual life is fully in tune with and conducted by the evolutionary patterns and cycles of nature operating on their own, without any overshadowing influence of individual stress, limitations, and boundaries of attachment. There are individual desires and intentions, but not overshadowing attachment to the outcomes of actions. The finite, ephemeral, mortal, conditional, individualized values are relatively *real* and continue to be expressed on the relative levels of phenomenal existence. Acting through the three gunas—the creative, maintenance, and dissolution operators—Nature carries on the entirety of phenomenal life in its self-interacting dynamics at every level of relative life—individual and universal. [3]

"When you are stabilized in your own Self, then there is no otherness, you are everything. If you abide in your Self you are like space and there is no duality left. You are as expansive and as subtle as space, and that is liberation. You are not conditioned by any name or form. If you are like space, what is the point of going elsewhere? The space which is here is also everywhere else... First of all, you abide in your own Self and transcend it, and in transcending, you will realize your Ultimate. The words emanating here are not borrowed knowledge, which is available in scriptures and other books; this is from direct experience... You must thoroughly understand what you are, or what you could be when nothing is. When nothing is, you still are. What is that you? It is all one, and when everything is, still you are; that is understandable, but when nothing is, how can I be?"—Sri Nisargadatta Maharaj (pp. 98-99) [42]

"After the dissolution of the universe, when no further vestige of creation was apparent, what remained is my perfect state. All through the creation and dissolution of the universe, I remain forever untouched. I have not expounded this part: my state never felt the creation and dissolution of the universe. I am the principle which survives all the creations, all the dissolutions. This is my state, and yours, too, but you don't realize it because you are embracing your beingness."—*Sri Nisargadatta Maharaj (p. 6)* [42]

The degree of pure Being and inner silence that one experiences in the physical presence of a fully enlightened individual depends on one's own stage of development. The enlightened individual may seem ordinary in some ways, and extraordinary in other ways. Frequently, however, subtler qualities of deep inner silence begin to be appreciated after a little while of being in his or her presence. On occasion there also may be glimpses of the incredible inner peace, fulfillment, and completely life-supporting wholeness or holiness that spontaneously emanates from the fully enlightened individual. However the experience is interpreted, it consistently has a transforming, awakening influence.

From the inside, the fully realized and fulfilled individual is benevolently not attached to the localized conditional values that comprise individual life, or the imaginings and representations that others tend to project, but rather is steeped through and through with the infinite totality of life—like a drop of water in the ocean, or a ray of sunlight in the brilliant radiance of the sun. So to speak, the fully awake individual, fully established in the oneness of Unity, has turned away from relative ever-changing creation toward never-changing, infinite, eternal Being. The localized life of individual ups and downs, of individual joys and sorrows, are surface waves that come and go in the unbounded ocean of the radiance of infinite eternal Being.

In that perspective, with respect to the relationship of the TM technique and the nature of individuality, Maharishi recently explained:

> "It doesn't work on the individual level, it is on the cosmic level... TM is not my creation; TM has been throughout the ages immemorial... We don't give importance to the individual. We give importance to the transcendental level... It is the same reality that is eternal... My life is that level where unmanifest prevails through all differences in manifestations." [43]

Maharishi has pointed out that even the entire process of evolution through the seven states of consciousness can be said to be phenomenal in the supreme level of development. States of consciousness are phenomenal delineations of the ultimate non-changing reality of pure Being that is always Oneness. The point value always has been and always is the self-referral wholeness of Infinity—eternal Being beyond all manifest and unmanifest existence, beyond or prior to the notion of absolute and relative—infinite eternal Oneness of Being.

The fully awake state of Brahm is total absorption in Oneness, beyond all the pairs of opposites, including infinite dynamsim and infinite silence, beyond the concept of states of consciousness entirely. That is sometimes referred to as *Parabrahman*, [44] the eternal, infinite, indivisible Absolute beyond the coming together of parts into the ultimate unified field, beyond all notion of duality. It is the ultimate non-dual Infinity, the Total Reality beyond any notion of individuality, universality, and even consciousness itself.

> *"Aham Brahmasmi*
> (Brihad-Aranyak Upanishad, 1.4.10)
>
> I am totality." (p. 181) [45]

In the last two chapters of Part II: Psychology Unbound we focus on systematic natural means to facilitate development through the seven states of consciousness to permanent full enlightenment. The Vedic technologies and their mechanics, especially the TM and TM-Sidhi program including 'yogic flying,' as well as other advanced programs, will be examined in more detail.

Chapter 20 Notes

[1] Patanjali (1982). *Patanjali's yoga sutras: With the commentary of Vyasa and the gloss of Vachapati Misra.* Prasada, R. (Trans.). New Delhi: Oriental Books Reprint Corporation.

[2] Alexander, C. N., Boyer, R., & Alexander, V. (1987). Higher States of Consciousness in the Vedic Psychology of Maharishi Mahesh Yogi: A Theoretical Introduction and Research Review. *Modern Science and Vedic Science,* Vol. 1, No. 1, January, 1987, 89-126.

[3] Maharishi Mahesh Yogi (1967). *Maharishi Mahesh Yogi on the Bhagavad-Gita: A new translation and commentary, chapters 1-6.* London: Penguin Books.

[4] Alexander, C. N., Davies, J. L,, Dixon, C. A., Dillbeck, M. C., Oetzel, R. M., Druker, S. M., Muehlman, J. M., & Orme-Johnson, D. W. (1990). Growth of Higher States of Consciousness: Maharishi's Vedic Psychology of Human Development. In Alexander, C. N. & Langer, E. J. (Eds.). *Higher stages of human development: Perspectives on adult growth.* New York: Oxford University Press, pp. 286-341.

[5] *Creating an Ideal Society: A global undertaking* (1977). Rheinweiler, W. Germany: MERU Press.

[6] Richards, G. (1977). The World a Stage: A Conversation with Ray Reinhardt. *San Francisco Theater*

Magazine, Winter.

[7] Husserl, E. (1950/1970). *The Paris lectures.* Kostenbaum, P. (Trans.). The Hague: Martinus Nijhoff.

[8] Starr, W. T. (1971). *Biography of Romain Rolland.*

[9] Chuan Tzu (1968). *The complete writings of Chuan Tzu.* Watson, B. (Trans.). New York: Columbia University Press.

[10] Angela of Foligno (1888). *The book of visions and instructions.* (Trans. by a secular priest). Leamington Art and Book Co.

[11] Mason, L.I. (1995). Electrophysiological Correlates of Higher States of Consciousness During Sleep. Doctoral dissertation, Maharishi International University, Fairfeld, IA. Printed 1998 by UMI Dissertation Services, Ann Arbor, MI. *Dissertation Abstracts International,* Vol. 56-10B, p. 5797.

[12] Alexander, C. N., Boyer, R. W., & Orme-Johnson, D. W. (1985). Distinguishing Between Transcendental Consciousness and Lucidity. *Lucidity Letter,* 4 (2), 68-85.

[13] Orme-Johnson, D. W. (1977). The Dawn of the Age of Enlightenment: Experimental Evidence That the Transcendental Meditation Technique Produces a Fourth and Fifth State of Consciousness in the Individual and a Profound Influence of Orderliness in Society. In Orme-Johnson, D. W. & Farrow, J. T. (Eds.). *Scientific research on the Transcendental Meditation program: Collected papers,* Vol.1. Rheinweiler, W. Germany: Maharishi European University Press, pp. 671-691.

[14] Alexander, C. N., Cranson, R. W., Boyer, R. W., & Orme-Johnson, D. W. (1987). Transcendental Consciousness: A Fourth State of Consciousness Beyond Sleep, Dreaming, and Waking. In Gackenbach, J. (Ed.). *Sleep and dreams: A sourcebook.* New York: Garland Publishing, pp. 282-315.

[15] Alexander, C. N. (1997). *The seven states of consciousness in Maharishi's Vedic Psychology.* (Unpublished manuscript). Maharishi University of Management, Fairfield, IA.

[16] Thoreau, H. D. (1929). Blake, H. G. O. (Ed.). *Early spring in Massachusetts: From the journal of Henry D. Thoreau.* Boston: Houghton Mifflin.

[17] James, W. (1902/1982). *The varieties of religious experience.* New York: Penguin.

[18] Anonymous (2004, June). Personal communication. Fairfield, IA.

[19] Gackenbach, J., Cranson, R., & Alexander, C. N. (1986). *Lucid dreaming, witnessing dreaming, and the Transcendental Meditation technique: A developmental relationship.* Paper presented at the annual convention of the Association for the Study of Dreams, Ontario, Canada.

[20] Cranson R. W. (1989). Intelligence and the Growth of Intelligence in Maharishi's Vedic Psychology and Twentieth Century Psychology. Doctoral dissertation, Maharishi International University, Fairfield, IA. *Dissertation Abstracts International,* 50, 08 A.

[21] Orme-Johnson, D. W., Clements, G., Haynes, C. T., & Badawi, K. (1977). Higher States of Consciousness: EEG Coherence, Creativity and Experiences of the Sidhis. In Orme-Johnson, D. W., & Farrow, J. T. (Eds.). *Scientific research on the Transcendental Meditation program: Collected papers,* Vol.1. Rheinweiler, W. Germany: Maharishi European University Press, pp. 705-712.

[22] Maharishi Mahesh Yogi (1972). *The Science of Creative Intelligence: Knowledge and Experience,* Lesson 23-6 (Syllabus of videotaped course). Los Angeles: Maharishi International University Press.

[23] Maharishi Mahesh Yogi, (1963). *Science of being and art of living.* Washington, D.C.: Age of Enlightenment Publications.

[24] Maharishi Mahesh Yogi (1973) *Love and God.* Los Angeles: MIU Press.

[25] Wallace, R. K. (1986). *The Maharishi Technology of the Unified Field: The neurophysiology of enlightenment.* Fairfield, IA: Maharishi International University Press.

[26] *Creating an Ideal Society: A global undertaking.* (1977). Rheinweiler, W. Germany: MERU Press.

[27] Anonymous. (2005, May) Personal communication. Fairfield, IA.

[28] Fichte, J. G. (1800/1956). *The vocation of man.* Chisholm, R. M. (Ed.). Indianapolis: Bobbs-Merrill.

[29] Raine, K. (1975). *The land unknown.* New York: George Braziller.

[30] Edwards, J. (1986). Personal Narrative. In *Norton Anthology of American Literature,* 2nd Edition. New York: W. W. Norton.

[31] Tagore, R. (1929). *Letters to a friend.* Andrews, C. F. (Trans.). New York: Macmillan.

[32] Anonymous (2005, May). Personal communication. Fairfield, IA.

[33] Hegel, G. W. F. (1910/1949). *Phenomenology of mind.* Braille, J. B. (Trans.). (2nd Ed. Rev.). London: George Allen & Unwin.

[34] Hegel, G. W. F. (1892). *The logic of Hegel.* Wallace, W. (Trans.). (2nd Ed. Rev.). London: Oxford University Press. [Trans. from the *Encyclopedia of the Philosophical Sciences.* 1st Ed. publ. 1873].

[35] Jephcott, E. F. N. *Proust and Rilke: The literature of expanded consciousness.* London: Chatto & Windus.

[36] Carpenter, E. (1916). *My days and dreams.*

[37] Merrel-Wolff, F. (1973). *Pathways through to space.* New York: Julian Press.

[38] *Celebrating perfection in education* (1997). India: Age of Enlightenment Publications (Printers).

[39] Maharishi Mahesh Yogi (1988, February). Personal communication. Maharishi Nagar, India.

[40] *Maharishi Vedic University: Introduction* (1994). Holland: Maharishi Vedic University Press.

[41] Maharishi Mahesh Yogi (2004). Maharishi's Global News Conference, June 16.

[42] Nisargadatta Maharaj (1980). *Consciousness and the Absolute: The final talks of Sri Nisargadatta Maharaj.* Durham, NC: The Acorn Press.

[43] Maharishi Mahesh Yogi (2006). Maharishi's Global News Conference, March 8.

[44] Maharishi Mahesh Yogi (2008). 12 of January Celebration. MOU Channel, Vlodrop, Holland.

[45] *Inaugurating Maharishi Vedic University* (1996). India: Age of Enlightenment Publications.

Chapter 21

Systematic Technologies for Full Human Development I: The TM and TM-Sidhi Program

In this chapter, we will overview the mechanics of the TM technique and the advanced TM-Sidhi program including 'yogic flying.' The full range and evolutionary course of the transcendental, subtle, and gross levels of nature is applied in a comprehensive technology of human development. With growing clarity of the holistic view of nature, the unique and profound basis for Transcendental Meditation and the other technologies in Maharishi Vedic Science and Technology can be appreciated more fully. The overall main point of this chapter is that systematic empirically verifiable developmental technologies are now available with the power to unfold the full potential of individual life, alleviate suffering in our civilization, and create a genuine Age of Enlightenment based on fulfilled, spontaneously self-governing, peaceful individuals.

Hopefully by now it will seem at least plausible that Maharishi Vedic Science and Technology encompasses a much more comprehensive and integrated view of nature than has developed in modern science. Because reductive modern science historically fixated on the outer gross relative domain and had been closed to a non-material, nonlocal understanding of nature, such holistic relationships were not viewed as *real*. Nonetheless cutting edge scientific theories are now consistent with them, and this holistic understanding is increasingly appearing on the horizon of modern science, as summarized in Chapters 5-10. It is strongly supported by the well-established experimental finding of nonlocality (Bell's theorem), the theories of abstract geometries of information space associated with the concepts of quantum *reality* and nonlocal mind underlying physical matter, and the logical implications of unified field theory.

Maharishi Vedic Science and Technology applies this holistic understanding in specific technologies to align with the total potential of natural law and facilitate higher development in individual and collective life. These Vedic technologies involve precise and systematic applications of knowledge about the relationships between human behavior and universal laws that govern the expression of intelligent order throughout nature. Because these technologies recognize deterministic relationships in nature subtler than the physical, some of these technologies may be difficult to appreciate for those still locked into the untenable belief in a closed local space-time model of the universe. A discussion of these technologies makes more practical and tangible the planks on the bridge to unity across which modern science is now rapidly progressing. In the same way that objectivity can be understood as a special case of subjectivity, as discussed in Chapter 1, the fragmenting objective approach of modern science can be understood to be a special case within a holistic science of total knowledge—Vedic science.

Great credit can be given to modern science and its objective means of gaining knowledge for technological advances that have given more power over nature—perhaps the most significant of which are applications of the laws of electromagnetism, but also now quantum theory. It is certainly laudable for modern scientists to pursue vigorously the ability to gain even more control of nature—especially because modern scientific technologies have not yet alleviated much of the suffering still common on Earth. But the strategy of dismantling and manipulating nature based on an objectified reductive physicalism has entered extremely dangerous territory—and urgently needs to be reevaluated and gone beyond. As somewhat deeper levels of natural physical structures are dismantled and altered, it is an increasingly perilous course of action. This outdated strategy doesn't even incorporate the more holistic knowledge that has been established over the past century in modern science.

Now that modern science has established theoretically and experimentally the nonlocal interconnectedness of nature, however, it is much less of a leap to accept the feasibility of technologies that apply it. The holistic approach of ancient Vedic science includes systematic means to apply subtle nonlocal interdependencies based on the complete holism of nature in the unified field. Now that this holistic approach is available for practical application, there is no need to push so hard and fast with the disintegrating approach based on partial fragmented understanding of natural law. Fortunately there is a logical, consistent, and integrated scientific alternative to tearing apart physical matter as the means to gain control over nature. Beginning to appreciate the ultimate holism and orderliness of nature at all levels—gross, subtle, and transcendental—a strategy of alignment with nature rather than dismantling it is finally starting to be recognized. This fundamentally different understanding of the relationship of human life to natural law recognizes and applies the ultimate seamless unity of nature. This strategy is based on ancient holistic technologies that develop consciousness and refine mind and body for the *natural* unfoldment of full human potential.

As the unified source of everything, the unified field is also the unification of individual and cosmic life. That ultimate holistic basis of nature can be known and mastered through attunement with it. But mastery of natural law is not based on the human intellect alone. Rather it is based on full enlivenment in the individual of the total potential of natural law. This comes about naturally through *being* that totality, having the totality fully lively in one's own consciousness. In that state individuals develop toward being spontaneously healthy and fulfilled, the basis for a '*natural law-abiding civilization*' [1] that enjoys life free from suffering. This chapter describes in more detail the developmental technologies in Maharishi Vedic Science and Technology that can foster this natural integrated unfoldment of full human potential. Although numerous other technologies are available in ancient Vedic science, this and the next chapter cover the most prominent ones.

The Transcendental Meditation (TM)™ technique

The ordinary habit of subjective experience in the waking state of human consciousness is active attention directed outward toward objects of sense in the objective environment. This can be viewed as in the opposite direction of settling down the mind to less excited states, transcending mental activity, and directly experiencing the inner silence of transcendental consciousness.

In *trying* to settle down and still the mind, however, the typical experience is that it is fickle and constantly shifting from one thing to another—sometimes likened to a monkey jumping from tree to tree. Although it is quite difficult to catch the monkey in the trees, it will naturally sit still in order to eat a banana. This simple example embodies profound principles of the functioning of the mind that are basic to the effortlessness and distinctiveness of the Transcendental Meditation technique. This technology is a simple, systematic, scientific procedure that is based on holistic knowledge and experience of the totality of the laws of nature. It reflects a profound scientific understanding of the human mind and its relationship to the laws of nature that is far more advanced and integrated than in the current psychological, physiological, biological, and physical sciences. Not appreciating its subtlety and holistic basis, some well-intentioned researchers have attempted to dismantle it and tear it apart, applying the same reductive mentality that has led to the fragmented and partial understanding of nature that has been greatly restricting scientific progress and has contributed to the current 'meaningless' materialistic madness.

The TM technique is practiced 15-20 minutes twice daily while sitting comfortably with eyes closed. Instruction in the technique provides systematic personal training in how to turn attention inward and allow the inner process of transcending to occur naturally. In the many years of teaching this systematic approach Maharishi has emphasized that transcending is a simple mechanical process—like a wave settling back into the ocean.

Action is based on thinking, thinking is based on being, and being is based on the inner silence of pure Being or consciousness itself. In the same way that anyone who can walk can stand still and anyone who can talk can be silent, anyone who can think can effortlessly settle down to the transcendental source or ground state of thought. Given the considerable degree of misunderstanding about meditation for many centuries—continuing in recent decades, including in many scientific reports—Maharishi's restoration of effortless transcending can be considered the most significant technological innovation in the history of knowledge development.

> One common misconception, often found in popular books...and even sometimes in textbooks and research articles, is that all meditation procedures are more or less "the same." But this is simply incorrect, for major meditation procedures often differ in important ways... [T]raditional meditation procedures can differ with regard to the mental *faculties* they use (attention, feeling, reasoning, visualization, memory, bodily awareness, etc.), the *way* these faculties are used (effortlessly, forcefully, actively, passively), and the *objects* they are directed to (thoughts, images, concepts, internal energy, breath, subtle aspects of the body, love, God). They also often differ strongly with regard to how they relate to questions of belief... Recognizing these differences is thus essential to understanding the procedures themselves. It is also necessary for understanding the significance of the considerable body of research on meditation. This research now clearly shows that different procedures often have very different effects."—*Jonathan Shear, philosopher (pp. 9-10)* [2]

The TM technique is a standardized, systematic, reliable technology that is not based on a particular belief system or lifestyle. Individuals frequently report positive experiences from the beginning session, as well as positive outcomes even in the first few days or weeks of regular practice. Thus compliance rate for the technique tends to be high. [3][4][5][6][7] It is the most widely researched meditation technique, and it is supported by comprehensive principles of psychophysical laws of nature and systematic development of higher states of consciousness. [8][9]

Effortless transcending. Transcending during TM practice is based on two fundamental principles regarding the nature of the mind—introduced in Chapter 18. The first principle is that *the natural tendency of the mind is to go to a field of greater happiness*. The mind constantly shifts, but it does so in pursuit of greater happiness and enjoyment. The second principle is that *subtler, more refined levels of experience are inherently more enjoyable*. Based on these two principles, the mind naturally settles down to less excited, more refined states when given the opportunity, spontaneously drawn or 'charmed' to more enjoyable levels of experience. Ancient Vedic records, logical consistency, and the empirical experience of millions of individuals practicing this technology attest to the validity of these principles.

Effortlessness in meditation is not a commercial gimmick. Lack of understanding the principles of effortless meditation has resulted in a long history of confusion in mental and spiritual development. Although the TM technique is easy to learn and practice, it is a delicate and subtle process that is based on a holistic understanding of natural processes in mind and body. Its effectiveness depends on its simplicity and naturalness. If expectations enter into the process—such as for a particular experience, a specific result, or trying to 'simplify' or expedite it in some way— these expectations will interfere with the natural process of transcending and eliminate its effectiveness. Such strategies are based on superficial understanding of the nature of the mind and the subtlety of knowledge embodied in ancient Vedic science upon which this technology is based.

Because the common experience is that it is difficult to control the mind, many religious and spiritual seekers resigned themselves to tremendous dedication and effort. After fleeting success trying to control the mind, it was concluded that the outer sensory world of everyday life must be renounced in favor of a life of quietude and inner mental effort, in order to have a chance to calm or still the mind and achieve a state of inner silence. This led to an unfortunate division between a so-called impractical reclusive life of sacrifice, contemplation, or concentration, and a practical life of action in the world. Maharishi has explained that this has undermined not only the spiritual path but also weakened motivation for success in society. Respecting both the reclusive and active householder styles of life, he consistently has been teaching to clear up these ingrained misunderstandings and reestablish the universal value of effortless transcending. [10]

Maharishi first identified the technique of effortless transcending by the phrase 'deep meditation.' Because 'meditation' was generally interpreted to mean contemplation or thinking about something, however, he began using the phrase *transcendental meditation* to clarify that the process involves going beyond the common interpretation of meditation. By transcending the ordinary surface level of thought, the more powerful deeper levels of nature are activated in individual awareness. When thought is transcended altogether, the individual gains full contact with the total potential of natural law—the unified field. The power of softer, more refined and subtle levels of thought is much greater, and is at its maximum at the transcendental source of thought—in a similar manner that the deeper nuclear level of nature is many times more powerful than the molecular level, and the underlying transcendent unified field can be said to be infinitely powerful. To again refer back to the quote by Maharishi in the Prologue:

"Transcending thought is infinitely more powerful than thinking." (p. 444) [11]

Although perhaps difficult to fathom by those with the habit and intent of *trying* to control the mind, the process of transcending is not accomplished effectively through mental effort. It involves less and less mental activity, softer and softer thinking, to settle back to the ground state of the mind. Many prominent approaches to meditation have not recognized and applied these fundamental but quite subtle principles about the nature of the mind. For the most part other approaches define meditation in terms of *contemplation*—reflective thinking about something—or *concentration*—effortful focus on a particular object of attention—in trying to make the fickle mind settle down to inner silence and peace.

The TM technique is not contemplation or concentration. In the general context of mental and spiritual development, contemplation has come to mean reflecting on concepts such as peace, inner stillness, and the glory, compassion, or grace of God or the Godhead. It is frequently associated with different forms of prayer. A key aspect of contemplation is attention to the meaning of the concept or idea being considered, and exploring its connotations through quiet reflection. Although frequently a positive and constructive experience, the mind typically remains on the ordinary surface level of thought, sometimes expanding into deeper appreciation that can extend into emotions and feelings of devotion.

Contemplation is sometimes modified in attempts to deepen the experience by adding more emotional intensity—as if squeezing more sincerity into it. These strategies typically increase rather than decrease mental excitation or effort—bringing the mind outward rather than settling to deeper, more silent, refined, and powerful levels.

Because the mind tends to wander, contemplation sometimes is modified in attempts to reduce or eliminate intrusive thoughts and feelings—such as excitability or laxity (imbalanced rajas or tamas)—that distract the mind from being one-pointed, resolute, and still. This strategy involves different forms of concentration. These include such approaches as rote mental or verbal repetition of a concept, holding the senses on a particular external object such a candle, picture, or the breath, or attempting to maintain an internal representation such as a visual image from memory. One form of concentrative strategy is associated with the term mindfulness. This is said to involve attending to some object or process—such as the breath—and acknowledging or simply observing it without evaluating it, letting go of the thoughts and emotions that may arise, repeatedly reevaluating or introspecting about the mental state, and redirecting attention back to the object of focus.

In these strategies, for the most part the mind continues to be engaged actively on the levels of semantic meaning or focused attention—involving feelings in the heart, concepts in the intellect, or sensory-perceptual experiences such as body awareness. Under these active and effortful mental conditions the mind sometimes may tire due to excessive straining. It then may even momentarily relax enough to begin transcending toward a deeper, less excited level. But transcendent experiences have long been reported by people applying these strategies to be rare, unsystematic, and difficult to replicate—and thus requiring extensive disciplined practice.

"For thousands of years Indian civilization has given unusual attention to systematic explorations of the realm of consciousness... The predominant meditation practices in much of Asia, from Tibet to Japan, derive from Indian knowledge and traditions. Sometimes the Indian traditions evolved into new forms when they encountered indigenous knowledge, as in China where Buddhism and Taoism together yielded Zen... At other times knowledge derived by the Indian traditions became incorporated into practices of other Eastern traditions... This process is continuing in the West. Christian Centering Prayer is a major example. It spread...from St. Joseph's Abbey, where the monks had many years of experience with the TM technique, and many details of this Prayer resemble the Indian-derived TM much more than the earlier, highly concentrative practices associated with the ancient Christian *Cloud of Unknowing...* On the other hand, it is also clearly different from TM in many ways, in its use of relationship to God, indifference to the mental "sound" central to TM, and explicit Christian metaphysics.—*Jonathan Shear, philosopher (p. 14)* [2]

Holding the mind on the active surface level of experience frequently prevents transcending to deeper levels of the mind and to transcendental consciousness. On the surface level, thinking is much less powerful, and especially it doesn't involve direct contact with the underlying universal level that is the total potential of natural law. Even for centuries this simple profound point unfortunately has been overlooked in approaches to meditation and spiritual development.

In contrast, the TM technique is a systematic procedure designed to avoid mental effort and avoid engaging the mind in sensory, intellectual, or emotional processing that hold the mind on the surface level of thought or keep the individual mindful of some object of experience. It applies the natural tendency of the mind to settle down and spontaneously withdraw the senses from their outer objects of experience. Inner mental attention is turned back on itself by employing the natural tendencies of the mind. The outer-directed senses are naturally and effortlessly withdrawn from their objects and collected together, as they follow the mind inward to less excited, more settled, more refined and enjoyable states of inner experience. Any form of mental effort or control interferes with this natural settling process.

In effortless transcending, attention naturally is turned inward to experience impulses of thought coming from the source of thought—consciousness itself. Given the strong outward objectifying habit of the direction of attention in the ordinary waking state, frequently there is no clear experience or understanding that there is a *source of thought*. Thoughts just seem to appear in the mind as if from nowhere. Turning attention inward naturally allows the mind to begin to experience thoughts at earlier, less expressed, softer, deeper, subtler levels. Because experience is more enjoyable at these subtler levels, the mind is charmed to pick up the thought at an earlier, less expressed stage of the thought impulse. As this settling process continues—systematically involving experience of less excited, less expressed, and sometimes less distinct or more vague stages of thought impulses in their earlier stages of expression—the mind settles back into the source of thought. Correspondingly the body settles into deep physiologic rest and relaxation. This natural process, based on natural tendencies of the mind, is automatic and increasingly enjoyable, comfortable, and pleasant if it is not interrupted or prevented.

The mantra and its use in TM practice. Impulses of thought originate deep within the mind, in the transcendental field of consciousness itself, the source of thought as well as the source of everything. As described in Chapter 17, these impulses of intelligence and energy, or subtle vibratory forms of thought, become increasingly expressed and more concrete as they rise up—so to speak—through the levels of mind from their subtlest to grossest expression in verbal speech or other behavior. Taking the mind inward to subtler, softer, more refined levels of internal speech or sound vibrations in the form of increasingly subtle thoughts is a process that naturally draws attention away from outer sensory experience. This natural process—which the mind is capable of by its inherent nature—is facilitated by the particular object of thought used.

The TM technique is described as experiencing finer and finer levels of the object of experience until the object of experience and process of experience are transcended. A particular thought is used as the object of experience—the *mantra* or Vedic sound—that has specific features. One important and commonly misunderstood feature of the mantra is that it is *not* experienced in terms of its semantic value or meaning in the inward process of TM practice. It is not contemplating on the meaning of the thought—and not focusing on repeating the mantra—this frees up the mind from its habit of mental activity on the level of meaning. The TM technique is not an intellectual, reflective contemplative, or concentrative process. It is the *sound property* of the thought—and not the semantic value or meaning—that is relevant in effortless transcending. The sound property of the mantra, and not its meaning nor other aspects of it, clearly distinguishes the TM technique from contemplative practices; and its natural, automatic effortlessness clearly distinguishes it from concentrative practices. These differences have great significance for the naturalness and thereby the efficacy of meditation practice.

Another important aspect of the mantra is that its sound properties are known from the long history of the ancient Vedic tradition to facilitate transcending. (The sound value of Vedic language will be discussed in an upcoming chapter). The specific sounds or mantras have consistent effects at subtler levels of thought where they are more powerful. Other sounds can have effects that interfere with the transcending process, as well as other unintended effects. This important feature also is not recognized in other approaches to meditation. Typically only the surface levels of the process, and the surface meaning values of the concepts or words, are considered in other practices; and they are not based on a long time-tested history of direct experience of subtler and transcendental levels. Unfortunately this has resulted in deep experiences of transcending becoming unsystematic and rare, even in the field of spiritual practices.

Knowledge of the effects of sounds is a profound feature of Vedic language. It is based on a comprehensive understanding of the effects at each level from the unmanifest self-interacting dynamics of the unified field of consciousness to its phenomenal subtle and gross levels. The subtle effects of mantra sounds have been established by ancient Vedic rishis who directly cognized them in the structure of the Veda, the source of natural law. Assignment of the mantra during TM instruction, however, is a simple process that is part of the training to become certified as a TM teacher—along with how to provide guidance in using the mantra correctly in TM practice.

The TM technique is a holistic, integrated procedure. This has been demonstrated as a result of dismantling studies by various researchers attempting to break it down and isolate its core features, based on rather crude understanding of the dynamics of mental functioning. Reductive dismantling research to isolate and utilize a particular aspect of the technique has repeatedly obtained marginal or insignificant experimental outcomes compared to empirical outcomes based on proper instruction and practice of the TM program. The other practices also are frequently associated with higher rates of attrition, as well as reports that the practices aren't enjoyable.[12] These dismantling strategies reflect lack of knowledge of the holistic basis of TM practice. Nonetheless it is not uncommon in scientific research reports for the TM technique to be grouped into a general category along with these other strategies. Also research findings specifically on TM practice have been used as evidence to support generic approaches to meditation that don't produce equivalent results. Unfortunately this has blurred the distinctiveness and scientific understanding of the TM technique, making it more difficult to recognize its particular subtlety and effectiveness. It has reduced interest in and recognition of the value of meditation, which has decreased the accessibility of effortless transcending that efficiently facilitates higher development. It has led to the superficial objectification of meditation, as well as criticism of it as commercialized spirituality that has restricted its popularization and thereby restricted its benefits for society. Even the description of TM practice as easy and effortless has been viewed as an advertising gimmick, with no recognition of this instruction as being essential to the efficacy of meditation practice.

Personal instruction. The TM technique requires personal instruction by an instructor that has been certified through extensive training to teach the technique in a standardized procedure. Because instructions are personalized depending on *in vivo* experiences—including responses to questions to guide the delicate experiences during instruction—it cannot be taught through means such as a textbook.

The technique is taught in a sequence of standardized steps, including preparatory lectures, a personal interview with a TM teacher, private individual instruction including assignment of a mantra and how to practice the technique effortlessly; and practical feedback regarding experiences to ensure correctness of the practice. The instructional process includes a simple ceremony that establishes for the teacher the framework to teach the technique properly. The ceremony is a well-established cultural tradition for beginning instruction in new knowledge. It recognizes the ancient tradition of Vedic scientist-teachers and the importance of maintaining the time-tested effectiveness of the teaching. Also a systematic procedure to ensure correct, effortless practice after instruction—called *checking*—is included as part of the regular life-long follow-up program.

The process of normalization. The state of transcendental consciousness is described as an experience of inner bliss—in which the mind naturally wants to remain. But maintaining that state requires culturing the mind-body system to purify it of deep-rooted imbalances due to poor habits such as of eating and sleeping, stress, past trauma, and strain which coarsen its functioning. When the mind-body system contains deep-rooted imbalances, its functioning is not refined enough to maintain the inner silence of transcendental consciousness except for brief episodes.

When the mind-body system settles down to this simplest and deeply relaxed state, it gains deep rest that is the basis for purifying itself. The mind-body system has natural healing mechanisms that remove obstacles to healthy functioning, activated through deep rest in the same way that sleep and dreaming rejuvenate mind and body from fatigue. Rest is a natural antidote to stress and disease, and is the first prescription for most ailments. As a general rule the deeper the rest, the deeper the healing. Research has shown that TM practice produces deeper rest than simply resting with eyes-closed. Studies also have shown indicators of deeper rest during TM practice than during deep sleep. For example, as mentioned in Chapter 2, sub-periods during TM practice in which subjects report having experienced transcendental consciousness by pressing a button are positively correlated with breath quiescence or virtual breath suspension, suggestive of a profound state of physiological rest. The button presses were also positively correlated with simultaneous increases in skin conductance orienting and peak alpha power in the EEG. These and other indicators of deep physiological rest also have been shown to be associated with increased alertness.

The contrast of the inner silent, restful state of transcendental consciousness and active mental states associated with other forms of meditation is becoming clearer based on recent EEG findings. For example, mindfulness training has been associated with increased gamma synchrony, proposed as the best neural correlate of consciousness [13][14]—that is, of the ordinary waking state of consciousness but not the inner silence of the psychophysiologically distinct fourth state of transcendental consciousness. [15][16]

As mentioned in the Prologue, the process of settling down the mind as an efficient means to purify mind and body of stress and strain can be compared to physical processes associated with the 3rd law of thermodynamics. On the gross physical level this law or principle of nature refers to the process of decreasing entropy as temperature is reduced, which reduces activity in the physical system. For example, freezing salt water automatically purifies out the salt from the water. Although the principle is the same, herein its application is on the subtle inner mental level, referring to settling down and refining mental activity to establish more coherence in mind-body functioning. In the same way that impurities are naturally removed by reducing the temperature or activity in a physical system, mind and body naturally are purified when mental activity—analogous to mental temperature—is reduced to its least excited state of deep rest. This fundamental physical principle of nature can be applied to the nature of the mind.

Because of the one-to-one correspondence of mental and physical activity—discussed in prior chapters—when the mind settles into less excited states, the body follows into deeper rest. As mental activity is transcended, the body attains a profound state of rest. This is referred to as the *inward* stroke of TM practice. In this framework, when there is a sufficient degree of deep rest, biochemical and structural imbalances in the nervous system are naturally dissolved—termed the *process of normalization*. Correspondingly the activity of dissolving imbalances in the body increases mental activity, bringing the mind outward into more activity, in the form of thoughts during meditation—called the *outward* stroke. Mental activity arising in this way results from the physical activity of dissolving of incoherent biochemical and structural imbalances in the body.

Typically the content of the thoughts during this process reflect a conglomeration of different levels and types of imbalance released at that moment, and thus don't relate directly to the nature of the imbalances that have been eliminated. Because thoughts that arise in this way occur subsequent to the physical process of normalization, they don't directly facilitate the process. In this framework, their significance is that they are by-products of the purification and normalization that have taken place. Attention is not placed on them, and no time is spent considering or analyzing them. On occasion some of the thoughts associated with purification and normalization can be somewhat distressing or uncomfortable. The mind is expanded and stronger when experiencing refined states of mental activity and is the strongest when mental activity is transcended at the underlying unbounded ground state of the mind. The expanded alertness strengthens the mind and protects it during the normalization process.

Correctness of TM practice is in allowing the natural inward stroke of transcending conducted by the mind and the natural outward stroke conducted by the body to occur without the interference of mental effort, and without analyzing these natural automatic processes. Unnatural straining that interrupts their effortlessness can produce discomfort, such as a headache. This reduces enjoyment of the experience, as well as its effectiveness. The standard protocol of checking is used to verify and maintain the smooth effortlessness of the practice. Maharishi explains further:

"The inward stroke of Transcendental Meditation, in which the awareness experiences increasing charm, provides a soothing influence for the whole body. However, there may be a very tight, stressed area of the body that is unable to accept the first influence of increasing harmony, so it starts to cry. This crying is the process of release of stress, which will give that area the ability to participate in the increasing harmony in its surroundings—it is a natural process." (p.202) [17]

As a technology of healing, the TM technique is holistic and efficient. With regular experience of transcendental consciousness the mind-body system is cultured to function simultaneously in a more stable, settled state of relaxation along with heightened alertness—discussed in Chapter 20. This is supported by numerous studies showing positive correlations of these experiences with number of months of TM practice and also reported frequency of transcendental consciousness. Both subjective and objective empirical evidence support the efficacy of the program.

The TM technique has been taught to a wide range of individuals with different levels of intelligence and educational, religious, and cultural backgrounds. It has been applied successfully in diverse conditions including special venues such as maximum security prisons, psychiatric wards, substance abuse treatment programs, and nursing homes. In addition, insurance statistics indicate utilization rates for many chronic medical and psychiatric conditions are lower in regular TM practitioners compared to matched samples drawn from the general population, which includes a broad range of psychological and physiological conditions and states. These findings support the wide applicability and safety of regular TM practice.

As with about anything, however, proper use of the TM technique according to instructions is necessary. In the same way that pure water is important for health but excessive consumption can have undesirable effects, the TM technique is a powerful technology that needs to be applied in proper balance. After adding TM practice as a twice-daily routine, instructions emphasize the importance of alternating the deep rest achieved during TM practice with regular daily activity. The focus is on both regular TM practice and dynamic activity that does not involve straining or excessive intensity on either the mental or physical level. A moderate regular daily routine is recommended that promotes balance and refinement toward optimal brain functioning and integration of consciousness, mind, and body.

The necessity of eliminating the restrictive effects of accumulated stress, strain, and fatigue in order to develop a stress-free nervous system is an important part of this developmental technology. But in order to avoid any form of effort—such as interfering expectations or mental concentration—instructions emphasize the cumulative benefits in daily life outside of TM practice with gradual progress toward permanent higher states of consciousness. The emphasis is on developing a positive habit or routine of regular practice that supports success and enjoyment of increased mental potential outside of TM practice in daily life, rather than the mechanics of the program or the processes of purification and normalization.

In this respect, in sharp contrast to many common psychological approaches, TM practice does not involve bringing pathogenic beliefs or repressed material into attention and analyzing, re-enacting, or working through them consciously—neither during TM practice nor afterward. Many contemporary problem-focused therapies such as in applied psychiatry and psychology foster analytic habits that can have *iatrogenic* effects. These approaches can complicate the mind. They are based on a completely different understanding of the consciousness-mind-body relationship constrained by the reductive physicalist paradigm. Whether behavioral or psychological, a general feature of these therapies is that they attempt to objectify feelings, thoughts, and mental content. Indeed the *objectification* of experience is a fundamental tenet of the modern scientific paradigm in which most contemporary therapeutic approaches are conceptually embedded. In one form or another they attempt to bring traumatic memories outward, toward more expressed levels of experience—which potentially can have retraumatizing effects. This fundamental point about the subtle workings of the human mind is almost completely missing in therapeutic psychological practices. Maharishi makes it clear that approaches based on bringing past traumatic experiences to conscious attention are:

"...severely limited in their effectiveness by the fact that they do not include any means of simultaneously strengthening the mind and nervous system. These experimental attempts...are laudable in intent; however, they are proceeding on the basis of present weaknesses of the system and the results are therefore limited at best and can be undesirable." (pp. 40-41) [18]

"Analyzing an individual's way of thinking and bringing to the conscious level the buried misery of the past, even for the purpose of enabling him to see the cause of the stress and suffering, is highly deplorable; for it helps to strengthen directly the impressions of the miserable past and serves to suppress his consciousness in the present." (pp. 258-259) [10]

In contrast, the TM technique can be said to allow the mind to *expand inward* to deeper levels of experience. Maharishi explains:

"The mind is stronger when it is experiencing very refined states of mental activity. At such times it is very powerful and does not experience the impact of the release of stress with the same intensity with which it experienced the original impact; thus the influence is greatly diluted." (p. 41) [18]

Direct experience of transcendental consciousness, and the related expansion and refinement of mental and physical functioning, is the key missing element in many modern approaches to healing. As referred to in the Prologue, natural, direct experience of transcendental consciousness is the *active ingredient* in TM practice and the basis for its effectiveness.

The above points exemplify the uniqueness of the TM technique with respect to its naturalness, subtlety, and efficacy. This certainly is not to suggest that it is the only way to transcend or the only means for growth to higher states of consciousness. However, it is to recognize that it is the most extensively researched and validated, widely applicable, and systematically taught personal development technology. About 700 scientific research reports have been published in over one hundred different institutions around the world that uniquely document its wide ranging benefits. Importantly it also is drawn from and supported in theory and practice by the most comprehensive and enduring tradition of knowledge—Vedic science. This approach emphasizes holistic developmental technologies, which cover all levels of consciousness, mind, and body. Whether the particular technology primarily emphasizes physical exercise, daily routine, diet, social behavior, environmental structure, or mental practices, each Maharishi Vedic technology is based on and consistent with a completely integrated, holistic developmental approach. Each part or area of specialization is designed to be consistent with the totality of knowledge; each part embodies its connection to the integrated wholeness of knowledge based in the unified field of natural law.

Maharishi Vedic technologies are not based on isolated insights into a particular level of body or mind—typical of the numerous approaches that have been developed in recent years based on partial understanding of the mind-body relationship. This ensures that each Maharishi Vedic technology is consistent with its other approaches and thus effectively contributes to the balanced integration of life as a whole. Hopefully the discussions in this book thus far help clarify the subtle and profound knowledge that underpins these developmental Vedic technologies. It is a vastly more integrated approach than is found in the many personal development approaches emerging since Maharishi made the TM program available 50 years ago.

> "As a scientist and particularly as a student of physics, I find it quite reasonable that the existing laws of nature have a level which comes into direct contact with human consciousness and which opens entirely new possibilities for the direct interaction of the mental and the physical realms. Everything we know about scientific theory, especially in the past fifty years, points in this direction... The new and higher technology Maharishi is teaching today simply goes to a deeper level where consciousness and human physiology are involved directly rather than indirectly. I am confident that when the classic siddhis or supernormal powers are fully analysed, they will be seen to form a continuous extension of science rather than a contradiction to it."—*Lawrence H. Domash, experimental physicist (p.3)* [19]

Once the mind and body have become more familiar with the process of transcending, and the effortless of the process has been established as a natural habit, then advanced developmental technologies can be added. After establishing a regular daily routine of TM practice, Maharishi Vedic Science and Technology makes available additional technologies to accelerate the development of higher states of consciousness, including the advanced *TM-Sidhi program*. This program is based on the *Yoga Sutras* aspect of Vedic literature. It applies the expanded understanding of gross, subtle, and transcendental levels of nature that has been described throughout this book.

The TM-Sidhi™ Program

The practice of yoga has become increasingly popular in recent years, with growing interest in healthier lifestyles and appreciation of the interconnectedness of mind and body in holistic health and preventative medicine. Frequently the emphasis is on the part of yoga associated with light physical exercise and body postures, *hatha* yoga, a beneficial physical approach. Somewhat less emphasized, but much more significant for personal growth and evolution, is the deeper aspect of yoga—which is primarily a mental technology for full human development. The mental technology associated with yoga is detailed in the Yoga Sutras.

Yoga refers to *union*, and sutra means *stitch*. The Yoga Sutras are a collection of Vedic technologies that *stitch together* the *union* of individual self and universal Self. The Yoga Sutras describe mental practices to attain perfection of the consciousness-mind-body relationship, called *sidhis*. They have been revived from ancient Vedic science into a systematic developmental technology by Maharishi as the *TM-Sidhi™ program*. Maharishi explains:

> "The TM-Sidhi Programme is an advanced aspect of Transcendental Meditation. It trains the individual to think and act from the level of Transcendental Consciousness, greatly enhancing the co-ordination between mind and body, and developing the ability to enliven Natural Law to support all avenues of life to fulfill one's desires." (p. 435) [20]

The TM-Sidhi program unfolds the natural ability to engage in mental activity while also maintaining the underlying direct experience of pure transcendental consciousness—which is fundamental to the fifth state, cosmic consciousness, as well as higher states (refer to Chapter 20). It can be described as facilitating the natural integration of individual attention, or awareness of some object of experience, with universal consciousness.

In a scientific context the Yoga Sutras serve not only as practical means to develop higher states of consciousness, but also importantly as *empirical tests* of the degree to which higher states have been stabilized. The Yoga Sutras relate to verification that the individual's state of consciousness is developed enough to command different laws of nature, demonstrated in the ability to produce tangible physical effects by mere mental intention through practice of the mental techniques.

In other words the Yoga Sutras develop experiences that empirically demonstrate mind over matter—and ultimately consciousness over mind. They provide systematic empirical means to validate that mental *reality* is more fundamental than physical *reality*, and that both are nothing other than the phenomenal fluctuations of consciousness itself, the unified field of natural law, the core of one's own inner self, the universal Self.

Relationship of consciousness and attention in TM-Sidhis practice. The Yoga Sutras describe three components involved in attainment of the intended result of a particular sidhi: *Dharana*, *Dhyana*, and *Samadhi*. Briefly, dharana refers to naturally attending to a particular object of experience (in this case a particular sutra); dhyana refers to naturally experiencing finer and finer levels of the object of experience (the sutra); and samadhi refers to the perfect evenness of the simplest state of awareness, transcendental consciousness. When these three components occur effortlessly in unison—called *Samyama*—then the intention or object of the sutra, the sidhi, is automatically fulfilled. This is said to validate perfection in the connection of individual attention to universal consciousness. It demonstrates the power of attention in the state of samadhi—the field of all possibilities—to materialize intentions. A correspondence can be made between these three mental processes integrated in samyama and the object of experience, process of experiencing, and experiencer in the three-in-one self-referral dynamics of the unified field.

The Yoga Sutras foster development of the ability for individual attention to project a specific intention from samadhi. With regular practice they unfold the ability to experience all of the subtle levels of nature that underlie the gross objective domain and to gain mastery of them. Through repeatedly experiencing all the levels of nature that emerge from the unified field, individual development gradually progresses toward ultimate unity in the highest state—unity consciousness.

The Yoga Sutras describe eight areas or aspects, sometimes called the *eight limbs of yoga (Astanga Yoga)*, which cover the entire range of individual life: [21]

1. The aspect of life outside the individual that is influenced by individual thoughts and actions. The limb of yoga associated with this aspect or level of life, called *yama*, or the "five qualities of observance," (p. 474) [21] is established when the individual spontaneously upholds such behaviors as harmlessness and non-possessiveness. These behaviors are natural products of freedom from the boundaries of individual life, based on unbounded bliss awareness.

2. The physical structure of the individual nervous system and body. This limb of yoga, called *niyama*, or the "five rules of life," (p. 474) [21] is established when the individual spontaneously expresses inner purity, contentment, and devotion to the highest levels of knowledge and experience that comes about as the natural product of the repeated experience of samadhi or transcendental consciousness.

3. The aspect of physical posture, called *asana*, or the "sphere of posture." (p. 474) [21] This limb is established when the individual naturally maintains purity, flexibility, and softness of mind and body that results in physical coordination and steady posture.

4. The aspect of individual breath, associated with *pranayama*, or the "sphere of individual breath." (p. 474) [21] This limb is established when the individual naturally maintains efficiency and effortlessness of breathing. This limb of yoga is naturally developed by repeated experience of refinement and softness of breath through increasingly settled states of mind and body to the most settled, ground state of the mind in samadhi.

464

5. The aspect of life between the senses and the objects of sense. Associated with *pratyahara*, or "turning away the senses from their objects," (p. 474) [21] this limb of yoga is established when the senses and their objects function automatically—naturally non-attached from and witnessed by the individual in samadhi.

6. The aspect of life between the senses and the mind. This is associated with *dharana*, or "steadiness of mind," (p. 474) [21]—attention to the mind and withdrawal of the mind from its association with the excited states of outer-directed sensory activity.

7. The aspect of life between the mind and pure Being, the ground state of the mind. This limb of yoga, associated with *dhyana*, or "meditation," (p. 474) [21] involves increasing refinement of mental activity to the state of samadhi, transcendental consciousness, the field of all possibilities. It is directly related to the process of effortless transcending.

8. The aspect of pure Being, the field of all possibilities, the totality of natural law, *samadhi*, or "Transcendental Consciousness." (p. 474) [21]

Consistent with the common belief that controlling the mind requires effort, the first seven limbs sometimes have been interpreted as preparation for the eighth limb—samadhi. Maharishi has made the important clarification that through regular experience of samadhi—and the accompanying purification and refinement of mind and body—spontaneous action in accord with natural law develops. This fosters development of all of the limbs of yoga simultaneously. Otherwise, trying to perfect each limb in preparation for the next limb toward the ability to attain samadhi can be a long, arduous, and effortful struggle. Maharishi further explains:

> "For hundreds of years these different limbs of Yoga have been mistakenly regarded as different steps in the development of the state of Yoga, whereas in truth each limb is designed to create the state of Yoga in the level of life to which it relates. With the continuous practice of all these limbs, or means simultaneously, the state of Yoga grows simultaneously in all the eight levels of life, eventually to become permanent." (p. 486) [11]

By utilizing the natural tendency of the mind to settle down to less excited states, the body naturally follows. When the mind repeatedly settles into the simplest state, the state of samadhi, all the various levels of mind, senses, body, behavior, and their relationships—all the 'limbs of yoga'—are developed and enlivened simultaneously.

Practice of the TM-Sidhis involves allowing attention to experience subtler aspects of the sutra until the thought of the sutra settles into the source of the thought—while maintaining the state of samadhi. This means experiencing the sutra at more refined, more delicate, softer levels of thinking to the ultimate settled state of thought—the field of all possibilities—while at the same time maintaining the state of samadhi or pure consciousness. It is a very refined and delicate inner experience of feeling in the mind at the finest levels of individual attention and mental activity, integrating individual attention and pure consciousness. When the intention of the sutra merges into the unbounded wakefulness of samadhi, the laws of nature residing in the unified field—the cosmic computer—compute and fulfill the object of the sutra automatically in the individual's experience. If the underlying pure consciousness or samadhi is not established, or if the process of attention is not functioning on all the increasingly subtle levels of thought, then the sutra will not be fully actualized. Practice of this developmental technology is said to foster refinement of mind and body, eventually resulting in perfection in the process and full actualization of the intention in the sutra and automatic performance of the particular sidhi.

The Yoga Sutras apply the same universal mechanics of nature through which the unified field manifests everything in phenomenal creation. This delicate, highly refined process also can be understood in terms of the mechanics of the four stages of the gap described in Chapter 18—the fundamental mechanics of change throughout nature in the infinite dynamism of infinite silence. Attention to the sutra settling down to the unified field of consciousness—merging into the state of samadhi through the processes of dharana and dhyana—relates to the first stage. In this process the initial stage (Pradhwansa-Abhava) is reduced to infinite silence (Anyonya-Abhava), the second stage, which is the state of samadhi. Inherent in infinite silence is its enlivenment as infinite dynamism, the total potential of natural law (Atyanta-Abhava), the third stage. This is the field of all possibilities, which is the unmanifest self-interacting source of all the laws of nature that conducts all change. These self-interacting dynamics compute and manifest the new state (Prag-Abhava), the fourth stage. The automatic result is performance of the sidhi. The automatic complete performance of the sidhi is said to demonstrate perfect attunement with and mastery of the mechanics of change or transformation throughout nature by the individual—being fully awake inside to these most unmanifest dynamics of manifestation of all change in nature. Practice of the TM-Sidhis attunes individual experience and functioning to the natural processes and dynamics through which Nature manifests phenomenal creation.

This process can be described as well in terms of the mechanics of memory or *Smriti*—the cosmic computer—also discussed in Chapter 18. From this perspective, experiencing subtler levels takes the object of attention to its holistic basis in the unified field of consciousness. This field of all possibilities is the storehouse of memory and the switchboard of nature—Smriti—from which all change in nature is computed on the basis of the infinite correlation of the field of all possibilities. The specific activation of Smriti involves a vibration or sound at the unmanifest universal level of nature—Shruti—which then expresses itself in manifest form in space and time as the objective empirical result of the sidhi. In this process, infinite variety of phenomenal states of mental activity is transcended into the unmanifest wholeness or unity of the unified field of consciousness. From this unified wholeness in the field of all possibilities emerges the specific performance of the sidhi as a holistic experience. This process encompasses the entire range of the phenomenal object from its infinite value to its point value in individual consciousness, mind, and body—computed automatically by Smriti, the cosmic computer. It is the individual activation by mere intention of the mechanics of how Nature manifests itself. This is possible because individual consciousness and universal consciousness are the same at the transcendental level—but it requires perfection in the consciousness-mind-body relationship in the individual.

> "[B]rain wave (EEG) studies show that actually the brain becomes more "coherent" in meditation. This means that the firings of the billions of neurons in the brain occur more "in step" with one another. The electrical impulses from all the cells become less random, less independent. This state of mind has enormous power to affect reality, based on the synchronized 4-D holographic model... The DNA in our cells can naturally produce ..."phase-conjugate" waves... If this is the case, then each cell is a tiny radio transmitter which is capable of sending phase conjugate waves... The real power of DNA and the use of phase conjugate waves...occurs when millions or billions of cells transmit in phase. When this happens, the strength of the pattern increases as the square of the number of cells which are acting in unison... It is likely that the DNA molecules of each cell can be brought into coherence... This would enable the brain, when it is quiet and coherent, to combine together the signals of many DNA molecules. In this way, the desired image...can be brought into being."—*Claude Swanson, applied physicist (pp. 269-271)* [22]

The Yoga Sutras cover a vast range of empirical outcomes, from more cognitive and affective experiences to tangible objective results in the gross material domain. For example one of the Yoga Sutras develops the quality of compassion. Another sutra relates to refinement of the distinction between the intellect (buddhi) and consciousness itself (Purusha), fundamental to the discrimination of temporality and timelessness, relative and absolute. Another sutra develops 'divine hearing,' sometimes described as perception of the subtle essence of sound (tanmatra of sound), as well as clairaudience. Maharishi also describes it in a simple and easily understandable way as the practical ability to hear what is useful and respond for maximum advantage for oneself and the environment.

Another sutra is said to develop the ability in an individual to recall memories from past life-times. It is interesting that this controversial phenomenon may under certain conditions be able to be tested empirically and rigorously, though not yet done, given sufficiently detailed historical records and specificity of the supposed memories. Although historical attempts to do these types of experiments have not been able to eliminate major plausible alternative explanations to authenticate the reports, the research is becoming stronger. If this ability could be validated using well-controlled research, it of course would be strong evidence that memory is not just in the material brain—an important understanding in Maharishi Vedic Science and Technology discussed in Chapter 18.

Another sutra is said to refine the five sensory modalities and enliven their connection with the intuitive sense, the deepest value of intellect and (psychological) heart. Called '*Ritam-bhara pragya,*' this ability, involving the finest level of mental activity, is described as awareness that provides complete, accurate experiences. It is a state in which consciousness materializes thoughts. Maharishi states that at the most refined level where impulses emerge from consciousness:

> "...the difference between the intellect and the senses is minimal because that is the level which is closest to Transcendental Consciousness, where no differences exist. At this level, knowledge derived from perception and knowledge derived from understanding will be in perfect accordance with each other." (p. 110) [17]

As this ability develops naturally, the thought of an object—a pear, for example—can become so clear and precise that its visual form as well as all its other sensory qualia can be experienced fully on the screen of the mind. In ordinary experience the thought or memory of an object carries with it some degree of the qualities of an object, but typically not precise or complete. In more advanced development of Ritam-bhara pragya, thoughts of objects can be held in the mind and examined in detail, giving knowledge of the nature of the object by mere mental intention or attention. This is possible because the thought of an object is a vibratory pattern that is the same as the manifest object, only in subtle seed form (thought-form) rather than gross objective form. Here are several anecdotal reports of this experience on the level of thought and perception from TM and TM-Sidhi practitioners: (pp. 5-7) [19]

> "...I heard a bird singing outside and wondered what kind it was. All at once, I saw it within my mind, in full colour and liveliness, with its detailed markings and bold, perky posture clearly visible. I had never noted such markings before. Later while taking a walk I saw that these precise markings were to be found on the small local song birds."

> "A friend had lost a pen. He said it was in the hotel kitchen that he had lost it and when I did the technique I saw it in my mind under some boards by the stove. When I went to look for it later, I found it exactly where I had seen it while practicing the technique."

> "While experiencing a very refined level of awareness during the TM technique, I had the thought, 'A Mexican dinner certainly would be good.' Instantly dinner appeared in the field of all possibilities. It was very distinct, three-dimensional and in colour. The plate was hot to the touch, the taste of enchiladas, arroz, frijoles and chopped black olives was easy to perceive and the aroma was quite tempting. It was real as 'real' life. Afterwards the desire for food was gone. I felt satisfied."

"I experienced a wonderful feeling of inner freedom and intense bliss within my un-bounded inner space; the glowing heavenly bodies (stars, moon, etc.) each had a distinct taste. The taste of space itself was that of pure bliss; that of the pole star—a crisp, tingling sensation on the tip of the tongue, and tasting the moon was like putting my tongue into a pool of clear nectar."

"There are many examples of growing intuition, but basically what is experienced is foreknowledge of an even or sequence of events, of people's thoughts or desires, and the increased ability to fulfil the needs of the environment in a spontaneous manner. I seem to say and do the right thing at the right time spontaneously, without even thinking what is right or wrong. This, I think may be the most refined level of intuition. And there is bubbling bliss flowing and enlivening everything which I perceive. Beaming, radiant bliss pours forth from my Being. Sometimes there is a desire on the mental level, which drops into the heart, like a drop into an ocean, turning into waves of bliss and giving rise to the feeling 'I can do anything."

Experiences of Ritam-bhara pragya relate to the deep relationship between an object or vibra-tory form, its vibratory name in Vedic language, and the nature of thought as a mental vibration—all of these being vibratory patterns or fluctuations emerging within the unified field. It reflects a much more integrated understanding and experience of nature, associated with how the gross level manifests from the subtle manifest and unmanifest transcendental levels of the field of all possibilities. This will be gone into further in the upcoming discussion of Vedic language. It re-fers to a much more profound form of intuition and cognition than described in Chapters 1 and 9. The level of knowledge and experience that is developed through practice of the Yoga Sutras is far beyond the less refined and fragmented understanding and experience that is the product of abstract reasoning and ordinary sensory experience emphasized in modern science. However, it can be understood as the natural extension of ordinary perceptual and cognitive abilities that are unnaturally overshadowed by accumulated stress and incoherent mental and physical function-ing in lesser developed stages. This higher level of knowledge and experience has been referred to throughout history by revered individuals who achieved these higher levels of development but didn't provide well-articulated, systematic, and accessible means for others to develop them. For example Plato described similar ideas and experiences in his writings on Platonic forms.

The TM-Sidhi practice of 'yogic flying.' Perhaps the most tangible sidhi practice is the TM-Sidhi called 'yogic flying.' The Yoga Sutras state that by practice of *samyama* on the relation-ship of the body and *akasha,* generally interpreted as the property of space (refer to Chapters 11-12), the inherent ability eventually becomes perfected to levitate the gross physical body and float through the air. This ability certainly would be a tangible demonstration and rigorous empirical test of mind over matter.

Sometimes thought to be an impossible violation of the force of gravity, it is rather that more fundamental laws at the completely unified level of nature may be activated that supersede classi-cal gravitational effects—analogous to quantum mechanical phenomena such as superfluidity or superconductivity that might appear to violate classical laws. Historically Newton's' laws of grav-ity were superseded by Einstein's theories, which in turn were quickly superseded by supergravity and quantum gravity theories.[23] Freedom from the limitations of gravity through practice of 'yogic flying' reflects further development of understanding and experience of the nature of space-time. In our modern civilization such demonstrations of mind over matter seemed to be only imaginary concepts from science fiction, or phenomena of illusion and trickery. However, as summarized in earlier chapters, the theoretical and experimental basis for such physical phenomena is now established. In considering the possibility of this mental effect on the physical body, it is helpful to keep in mind that the state of samadhi or transcendental consciousness is described as direct contact with the unified field, the field of all possibilities. The unified field is theorized to be the source of all the laws of nature that structure and maintain the entire creation. Even within rigorous scientific theory, at the level of the unified field anything possible can happen.

As mind-body coordination is developed through practice of the 'yogic flying' sidhi, the process of beginning to be able to coordinate laws of nature more fundamental than classical gravity in order to free oneself from its limitations on the body produces waves of inner freedom and blissful energy. Described as the experience of *'bubbling bliss,'* it is frequently reported by TM-Sidhi practitioners of 'yogic flying' within the first few sessions of starting the practice. It is described as a highly exhilarating experience of freedom from limitations, expansion of mind, and lightness of body. Electrophysiological recordings of the specific period of time in which this experience begins during TM-Sidhi practice have shown that it is accompanied by the highest levels of global EEG coherence in the brain. In this unique psychophysiological state the brain displays the highest levels of coherence.

In the first stage of the development of yogic flying the subjective experience of 'bubbling bliss' is accompanied by waves of uplifting energy in the body, during which the body is impelled spontaneously to move upward in a jumping or hopping motion, sometimes described as 'hopping like a frog.' Along with this outer activity in the body—involving a thrusting upward in a hop—is an even more engaging inner experience of expansion and bliss. With increasing refinement of the sutra practice there are said to be brief episodes of the gross physical body levitating in the air. According to the Yoga Sutras, eventually in the final stage the adept sidhi practitioner is able to fly through the air at will—based on perfection of mind-body coordination with respect to the relationship of the body and space. Maharishi points out:

"The body lifting up in the air by virtue of a thought...demonstrates the command of individual awareness over gravity...proof of our ability to function on the level which is completely free... The science is there...the technology is there..." [24]

A TM practitioner describes an initial experience of the TM-Sidhi technique of 'yogic flying:'

"Just as I was starting the practice of yogic flying, I began to feel more expanded and very light. It felt as if a big magnet in the ceiling was turned on that began to lift me up." [25]

A TM-Sidhi practitioner describes experiences of 'yogic flying' in a large group of practitioners in Washington, D.C. in 1993:

"Sometimes during yogic flying in a group, I feel like I am sitting in an undulating field of pure energy, like a gull sitting on ocean swells that are impelling my body to lift upward. At other times, it feels like my entire body is a vertical tube of pure, uplifting energy, sort of like a wind tunnel of pure energy rising up through me, which is purifying my body, making it more refined and lighter, and pulling it upward. At other times, I sometimes feel unbounded and perfectly still inside; and in this perfect stillness, it feels like I am on the verge of lifting up into the air, like a feather by the slightest breeze." [26]

Here is another account of experience during practice of the 'yogic flying' sutra as part of the TM-Sidhi program:

"One particular day my heart was serene...and the dynamism of the light during the sidhis was very strong. When I began yogic flying...the sutra was like a liquid flowing through me... And then suddenly my body started to hop in place... The body felt like a wave or puff of air, and it would go up and then come down, and there was no sound when I came down. Then it would go up and each time stay a little longer. There was no heaviness, and then an explosion of light through my spine... Having tasted that lightness, I only wanted to experience it again." [27]

Here are several other anecdotal reports of experiences that spontaneously occurred during yogic flying: (pp. 6-7) [19]

"I was sitting on a couch meditating at the time. I felt a tremendous amount of energy go through me and simultaneously I had a vision of my spine and my chest being just white light and a form in the air some place and then my body moved up and down on the couch about three times. I thought, 'Oh, what is this?' and the next experience I had was hearing my body touch the floor. I say 'hearing' because I didn't feel it until after I heard it. It touched down, very, very softly. There was very little feeling of contact. I moved about a six foot distance at that time."

"The first time it was totally unexpected and I had the sensation that I only had time to grab a hold of my knees before I took off [from the lotus position]. The first time I hopped I wasn't aware of my surroundings or anything else. I had the sensation that I was hopping through black outer space and nothing else was there but me. The longer I do it the more I am aware of the environment as well as myself. I don't experience the moment of take-off. I experience being in the air and I experience coming down. At the end of it it's as if my body is a clear tube and its' as if I've blown a lot of things out so that I'm way more clear inside—as if I'm a column filled with light and with a special kind of very clear air and also a feeling of bubbling joy."

"I feel a very fine generating of energy within myself just before I bounce. And then very easily without any heaving or effort I take off. The body is becoming lighter and lighter, turning more into an ethereal substance."

"The energy started to flow through me really smoothly, kind of lifting me up in an upwards direction and I was rocking back and forth and this one time I remember, I was in the lotus position and I rocked forward up to my knees and instead of stopping there I just continued to lift up off the ground maybe a couple of inches, I don't know how much, and then set down. I didn't move any distance forward. I just kind of went up and came down in the same place."

"It's the greatest feeling of freedom I ever had. I'm free from today and I'm free from myself and yet I'm completely full in my individual self and it's very, very blissful."

To reiterate, electrophysiological research shows that such reports of 'yogic flying' experiences correlate with a brain state marked by the highest periods of global EEG coherence. Both the subjective reports and the brain physiology document a unique psychophysiological state. However, the above descriptions of experience don't include verified reports of actual levitation or flying through the air. In recent years there appears to be no formally documented reports of such advanced experiences—although undocumented anecdotal reports occur occasionally. This can be said to be indicative of the quite high degree of stabilization of higher states of consciousness and mind-body coordination necessary to perform fully this sidhi. On the other hand there are extensive historical reports of spontaneous experiences of levitation, in Vedic literature as well as in Western religious traditions, especially in the archival records of Catholic saints. [28] Given the deeper understanding of subtler levels or fields of nature associated with nonlocal, nonconventional space described in detail in Part 1 of this book, such anecdotal records now have a strong theoretical basis. Summarizing historical literature on the topic of levitation and 'yogic flying,' Gelderloos and van den Berg state: [29]

"In *Butler's Lives of the Saints,* it is reported that levitation—the raising of the human body from the ground by no apparent physical force—is recorded for over 200 saints and holy persons (Thurston & Attwater, 1962). In the Vedic texts very detailed descriptions of the progression of the phenomenon are given. The development of Yogic Flying is described to take place in three stages: "hopping," "floating," and "flying." For instance, in the *Yogatattvopanishad* these stages are described as follows:

Then, from more practice, *Darduri* is born to him. Just as the frog (Dardura) moves, continually hopping, so does the Yogi, sitting in lotus, move along the ground. Then, from more practice, he leaves the earth. [That Yogi], sitting in lotus position, leaving the ground, departs. (*Yogatattvopanishad,* 53-55, quoted in Mahashabde, 1977 translation William Sands)

This description authenticates the age-old understanding of the phenomenon of Yogic Flying and underscores the universal availability of such abilities. The TM-Sidhi program cultivates systematically this optimal mind-body coordination, so that over time the field of all possibilities becomes a living reality in the daily lives of the people..."

"A meta-analysis was published in *American Psychologist* comparing all studies on the physiological effect of the Transcendental Meditation technique and simply sitting with eyes closed (Dillbeck & Orme-Johnson, 1987)... [S]ubjects practicing Transcendental Meditation showed significantly greater physiological relaxation on...increased basal skin resistance, decreased respiratory rate, and decreases in the stress-related hormone plasma lactate... [T]he effect sizes were relatively large: .826, -.461, and -.617, respectively (measured in standard deviation units), indicating strong effects that are generally consistent across studies... The deep state of rest to mind and body during Transcendental Meditation also results in psychological trait changes indicative of enhanced peace. A recent meta-analysis, based on 146 independent outcomes, showed that the effects of the Transcendental Meditation technique on trait anxiety or chronic stress, is significantly greater than the average effect produced by other forms of meditation and relaxation (Eppley, Abrams, & Shear, 1989). All meditation and relaxation techniques for which trait anxiety had been empirically studied were compared. Except for concentration techniques, which produced virtually no effect on reducing anxiety, the other techniques (including placebos) had similar modest effect sizes, while the Transcendental Meditation technique produced a large statistical effect... Eppley et al. (1989) investigated the effects on anxiety of such possible confounding variables as type of population, age, gender, experimental design, duration and hours of treatment, pretest anxiety, demand characteristics, experimenter attitude, type of publication, and attrition... Statistically controlling for these variables via regression analysis did not alter the overall conclusions... Random assignment of studies by authors with neutral or negative attitudes towards the Transcendental Meditation technique showed an even larger effects size for this technique. Thus large effects associated with the Transcendental Meditation cannot be attributed simply to "pro-TM allegiance" on the part of researchers... With the use of large, permanent groups of coherence-creating groups for each country, it therefore becomes possible for the first time in history to create not only a peaceful mind and body, but a truly peaceful world"—*Charles Alexander, developmental psychologist and Vedic researcher (pp. 150-164)*[30]

Experiences of 'bubbling bliss' and 'hopping' during 'yogic flying,' correlated with increased brain coherence, are described as based on projecting an impulse of thought from the level of transcendental consciousness, the unified field of natural law characterized by infinite correlation, perfect order, and all possibilities (see Chapter 10).[31] This unique state is not only described as a thoroughly enjoyable experience for the individual. It also has been shown in numerous research studies to have significant environmental effects through a 'field effect' of coherence. This field effect is further magnified by the simultaneous group practice of the TM-Sidhi program and yogic flying. In the following section research is summarized that incorporates points from prior chapters on gross, subtle, and transcendental domains of nature toward a rational scientific explanation for the highly significant empirical findings documenting this unprecedented sociological effect.

"Five groups of elite members of the Middle East policy community—peer reviewers, newspaper reporters, Congress people, non-governmental influentials, and US diplomats—assessed a research study that explored a strategy for reducing conflict in the Middle East. That study was published in the *Journal of Conflict Resolution* (International Peace Project in the Middle East: The Effects of the Maharishi Technology of the Unified Field, or IPPME) and found that when a critical mass of people used the Transcendental Meditation technique, social stresses (e.g., crime and war intensity) were reduced in the surrounding population. Over half of each group reviewing the research rejected IPPME immediately without examining its scientific merit. Stereotyping and prejudice were evident. Others, who assessed scientific quality independently of their organizational philosophies and practices and exhibited greater curiosity, were likelier to consider IPPME further."—*Carla Linton Brown, education researcher* [32]

Collective Consciousness and the Maharishi Effect

Regular direct experience of the least excited state of the mind, the unified field of infinite correlation and perfect order, gradually purifies individual mind and body. This is said to foster attunement or alignment of individual behavior with the totality of natural law. Individual thought, speech, and action naturally become more coherent, orderly, and life-supporting—associated with increased coherence of brain functioning. Adding the TM-Sidhi program develops further the ability to think and act from the basis of full contact with the perfect orderliness of the unified field, amplifying the influence of coherent mind and brain functioning. This enhanced orderly influence is said to produce increased coherence through the subtle and gross levels of ego, intellect, mind, senses, body, and behavior. A growing body of research supports the view that it fosters coherent life-supporting behavior and contributes positively to the social environment.

Elimination of accumulated stress, strain and fatigue, increased coordination in mind-body functioning, and increased ability to maximize mental potential to fulfill personal desires in accord with others, naturally have a healthy influence in the surroundings. Individuals who are less stressed and function at higher levels of coherent brain functioning have a more orderly influence in their environment.

In other terms individuals at peace on the inside radiate a peaceful influence into their outer surroundings. It is commonly experienced that the immediate environment feels calmer around an individual who is in a deeply settled state of meditation. There is a subtle but sometimes noticeable influence in the environment associated with increased coherence in the mind and brain of individuals practicing the TM and TM-Sidhi programs. This influence is even more noticeable with a large group meditating together. Maharishi points out that this effect is well-documented in the Vedic literature:

"Tat sannidhau vairatyagah

Yog-Sutra 2.35

In the vicinity of Yogic influence—unifying influence, integrating influence, coherent and harmonious influence—conflicting tendencies do not arise." (p. 11) [20]

The quality of an entire society is a direct and sensitive reflection of the collective effects of the quality of the state of consciousness of its individual members. [34] In addition to the influence of individuals on others in the social environment, there is a reciprocal influence of a group on its individual members. In a similar manner that the state of consciousness of an individual determines the quality of thought and behavior, there is also a collective influence by a social group—family, city, state, region, nation, or human civilization as a whole—on each individual. This also is a common experience, such as the influence of an enthusiastic crowd at a championship sports event, or when entering a private home where the ambiance created by the family is noticeable.

The collective influence of a social group can be positive when social stress is low and harmony and coherence are high, or just the opposite when stress and incoherence are high. Positive coherent social environments are life-supporting and healthy. In Maharishi Vedic Science and Technology this influence is associated with the concept of *collective consciousness*. A positive coherent influence that is strong enough can influence the trends in an entire society toward healthier behavior. This phenomenon is apparent, for example, in different college campuses, businesses, or institutional facilities, as well as neighborhoods of a community and geographic regions.

"The implications of the Maharishi effect are obviously enormous. It would be foolish to ignore the possibility that the voluntary entry by a group of persons into a specific state of consciousness can literally change the world for the better. While our politicians and statesmen struggle with the apparently impossible task of achieving world order through traditional means, we must acknowledge that there may be other ways to achieve harmony and order which are disarmingly unadorned and effective. These ways apparently depend on something as simple as sitting down, clearing the mind, and entering a unique state of awareness—no colossal expenditure for defense, no involvement in internecine politics..."—*Larry Dossey, physician and science writer (pp. 262-263)* [33]

"Acausal connections and nonlocal correlations may be a powerful new way of understanding the universe. While the correlations in Bell's theorem are specific to quantum systems, they may be manifestations of something even more general. Space itself may be built up in ways that lie beyond simple locality. The ideas of nonlocality may also be necessary for a proper understanding of the operation of the brain, body, ecology, and even of society itself... Indeed a new form of activity, which I called "gentle action," could form the basis of a more appropriate response to today's challenges in economy, ecology, society, and international relations... The traditional response to social, economic, and environmental threat is to analyze the situation, isolate the problem, propose a solution, and then implement it in an active way. But such solutions are generally confined to a particular region or, in a fragmentary way, to some particular aspect of the system. Experience suggests that such "solutions' can have unexpected effects and that, when it comes to highly complex and interrelated systems, the cure may even be worse than the problem itself!... An alternative approach is to operate throughout the whole system in a gentle, nonlocal way. Rather than attempting to change the direction of a process or actively oppose some particular effect, the key would be to operate in a subtle but global way and seek to restore harmony through gentle correlations... The key to the serious problems that face us today may lie not in a call for immediate action but in careful and sensitive observation and a gentle instinct for balance and harmony."—*F. David Peat, theoretical physicist (pp. 163-164)* [34]

Research on the Maharishi Effect. In 1960 Maharishi first predicted significant positive changes in societal trends when a sufficient number of individuals in an area or region practiced the TM program—called the *Maharishi Effect*.[35] In the same way that TM and TM-Sidhi practice reduces stress and increases coherence in the individual, research also has found that collective practice of this program generates coherence in collective consciousness and reduces collective stress. This positive coherent influence is held to be more powerful because it involves mental functioning at increasingly unified levels of nature—and most fundamentally the field of infinite correlation and perfect order. A considerable body of well-controlled studies has been published in refereed journals on the Maharishi Effect. These studies repeatedly have found significant improvements in health, economic growth, crime rate, and a variety of other positive social indicators of the quality of collective life on local, state, national, and international levels.[36]

The first scientific research on the Maharishi Effect was conducted in 1974.[37] In 11 U.S. cities where *1%* of the population had been instructed in the TM program, there was an average decrease in crime rate of 8.2% in 1973-4, compared to control cities matched for geographic region and population, in which crime increased 8.3% consistent with national crime rate statistics during the same period.[35] A study in 1982 using a random sample of 160 U.S. cities obtained smaller but similar results. This study demonstrated through a cross-lagged panel design that the correlation between TM practice and crime reduction very likely was not due to unmeasured variables. This finding supports the conclusion that the lower crime rates resulted from the Maharishi Effect.[38]

The Extended Maharishi Effect. After introducing the advanced TM-Sidhi program in 1976, Maharishi predicted an even more powerful effect of collective coherence. Subsequent research of coherence-creating groups have found positive societal trends when as few as the *square root of 1%* of a population (also called the N^2 effect) practiced the advanced TM-Sidhi program together in a group.

For example a study was reported of 2500 TM and TM-Sidhi practitioners attending a coherence-creating group during a 6-week period in 1979 conducted at Amherst, Massachusetts.[39] As predicted, statistically significant outcomes showed violent crime fell 10% in Massachusetts and 3.4% in the U.S. as a whole during the experimental period, compared to trends over the same time of year in the previous eight years. Other sociological trends also changed in a positive direction, including significant decreases in transportation fatalities and fatal accidents, and increases in stock market indexes.

Another study assessed the effects of coherence-creating groups of TM and TM-Sidhi practitioners on political violence in the Lebanon war, 1983-1985.[40] During this time period there were seven separate sub-periods in which coherence-creating groups were assembled in sufficient numbers to have the predicted effects. Highly significant decreases in war intensity (48%), fatalities (71%), and injuries (68%) were obtained, while indicators of cooperation significantly increased (66%), controlling for a variety of relevant alternative explanations.

The Global Maharishi Effect. Similar findings were obtained on a world-wide basis when the TM-Sidhi program of 'yogic flying' was collectively practiced by a group that was large enough to be the square root of the entire world population. This *Global Maharishi Effect* was empirically demonstrated in three studies involving 7000 or more practitioners at large World Peace Assemblies held at different times in the U.S., Holland, and India.[41 42]

The success of these research programs led to a major highly publicized demonstration project in Washington, D.C. in 1993 applying sophisticated time series analysis to assess the effects of a coherence-creating group on violent crime. In this prospective study specific predictions were submitted prior the experiment to an independent body of evaluators. Consistent with the predictions, during the time period of the study in which the largest coherence-creating group participated—a two-week period with about 4000 TM and TM-Sidhi practitioners—approximately a 21% decrease was obtained in violent crime rate in the Washington, D.C. area, controlling for a variety of covariates. [43] This research documents a powerful positive field effect that neutralized social stress and promoted coherence in collective consciousness, reducing crime rate as well as other indicators of negative trends including accidents, illnesses, and economic problems. Other well-controlled published studies show a robust effect for which numerous alternative explanations have been ruled out through careful design and statistical controls. This research constitutes solid evidence that the coherent sociological effect is a product of the coherent influence of groups of TM and TM-Sidhi 'yogic flying' practitioners.

"The Maharishi Effect shows that individuals meditating in one place can influence individuals in another place with no direct interaction between the parties. This phenomenon has been further substantiated by three separate neurophysiological studies. The first study was conducted in August 1979. During that time, about 2,500 experts gathered together in Amherst, Massachusetts, for collective practice of the TM and TM-Sidhi program. Half a continent away, in Fairfield, Iowa, researchers found that during the Amherst meditation times, the intersubject EEG brainwave activity between several volunteer subjects in MUM's lab became significantly more coherent or in phase. These subjects were unaware of the purpose of the tests and had no knowledge of the meditation times of the Amherst group 1,200 miles away. The effect was measured on six consecutive days during the course, but there was no such effect on the same days in the following month after the larger group meditations had ended... A second study specifically measured the influence of increased brain wave coherence of an individual practicing the TM-Sidhi program on a non-meditator in an adjacent room. In each of a series of trials, different pairs of TM-Sidhi participants (Sidhas) and non-meditators were hooked up to the same EEG machine. This arrangement allowed a transfer function analysis of the relationship between the brain wave patterns of the two individuals. In each case, the Sidha's brain wave coherence led the non-meditator's coherence by several seconds... [W]hen there was an increase in the Sidha's brain wave coherence, several seconds later...was an increase in the non-meditator's brain wave coherence. The finding was highly significant statistically... The third set of studies correlated changes in the level of serotonin, a neurotransmitter, in non-meditators living in Fairfield, Iowa, with the number of people collectively practicing the TM-Sidhi program at MUM. Low levels of serotonin in the brain are associated with behavior problems, such as increased aggression and depression, and high levels are associated with experiences of well-being... Previous research had established that individuals practicing the TM technique have higher levels of serotonin. In this later set of studies, nightly excretion rates of 5-HIAA (the chief metabolite of serotonin) were measured over periods of 50-91 days. These studies found that increased group size of TM-Sidhi practitioners significantly correlated with increased levels of serotonin in non-meditators... Taken together these laboratory studies indicate that individuals acting from the underlying unified field of consciousness during practice of the TM and TM-Sidhi program positively influence the psycho-physiological functioning of non-meditators. This psycho-physiological influence appears to be the basis of the collective behavioral changes in society produced with increased levels of serotonin in non-meditators... It should be emphasized that in these studies the subjects never interacted, either directly or indirectly. Therefore, the observed effects cannot be explained by classical theories of social interaction."—*A Methodological Review of Maharishi Effect Research, Maharishi University of Management website.* [44]

What mediates the Maharishi Effect? The accumulated body of studies supports the Maharishi Effect as perhaps the most documented and statistically significant empirical finding in sociological research. The implications of this research on the collective practice of the TM and TM-Sidhi program including 'yogic flying' are revolutionary. It offers a means to create an influence of global coherence that can be the practical basis for changing societal trends in a healthy positive direction, and even to create world peace—a consciousness-based scientific approach to world peace. This is in a completely new direction than the failed strategy of 'violence to end violence' that has plagued human thinking and behavior for millennia, with increasingly devastating results in the past century—as well as the strategy of negotiating peace treaties that repeatedly fail. What could be the mechanism for propagating such a coherent peaceful sociological influence?

In many physical systems such as magnetism, laser light, crystallization, and primary organizer cells in the embryo—there is a phase transition to a coherent state in an entire system when about 1% of the elements in the system begin to function coherently. A similar coherent field effect of collective consciousness has been proposed to explain the mechanics of the Maharishi Effect. [40] In this theory the principle of coherence and orderliness evident in these examples from physics is generalized from a physical system to a social system such as a community, nation, and entire world population.

Typically social group effects are explained in terms of the communication of ideas and feelings between people via direct contact interactions. For example a courteous response, compliment, or 'random act of kindness' may result in the recipient extending a similar courtesy to others, and in this way the positive influence is transmitted through a social network, much like laughter can spread through an audience. As another example, when the weather is particularly mild and pleasant, a widespread positive influence can be generated in a community or society that is amplified by mutually courteous interactions based on positive moods. In such examples the positive effect is explainable in terms of ordinary means of communication—the physical effect of sunlight on a nice day that can affect peoples' moods, verbal communication, or other influences due to the transmission of physical energy such as a radio or television broadcast. But these clearly do not seem to be the means that would account for the coherent influence by which the Maharishi Effect is transmitted through social networks.

It has been proposed that the Maharishi Effect is a subtler field effect on the level of the unified field of consciousness. [45] The research evidence indicates that the coherent effect depends on a threshold size of the group generating it, as well as the type of program the group is practicing—theorized based on physical models to be either TM (1%) or the TM-Sidhi program including yogic flying (square root of 1%). It appears to be propagated through a social population in a manner that does not depend only on physical distance from the generating source, but rather is related more to population density in a geographic area.

Also cross-lagged panel design research indicates that the effect is time-dependent, varying with the generating source above threshold levels at least within a day or so of both initiating and ending it. Research on the immediacy of the effect has not been conducted more specifically than designs which document the onset of the coherent effect within one day. However, it would be expected that the nature of outcome measures of social trends such as accidents, violent crimes, economic changes and so on might be a limiting factor in documenting its immediacy—even if the effect were nearly instantaneous, mediated by subtle nonlocal impulses of energy and intelligence at or near the level of the unified field. The sociological nature of the coherent effect, spreading through a social network but not via ordinary contact interactions or other physical means of propagation, suggests some other causal process mediates the effect.

A physical analogy for the Maharishi Effect is the quantum mechanical version of the *Meissner Effect*, associated with superconductivity and superfluidity. [35][45] In an electrical conductor such as lead or another metal at ordinary room temperature, incoherent electrons in the metal can be influenced by an outside magnetic field that makes the electronic activity in the metal even more chaotic. When the metal is cooled to near absolute zero, there is a phase transition of a quantum mechanical field effect in which the electrons begin to function in a pattern of collective coherence. In this superconducting state, the coherent interactions of the electrons in the metal spontaneously resist the effects of an outside magnetic field. The flow of electrons in the superconductor become superfluid, and coherently move to produce a magnetic field that cancels out an outside magnetic field as it begins to penetrate the superconductor. Based on the 3rd Law of Thermodynamics, a coherent state is established in the metal that expels disordering influences when temperature and random motion are reduced.

A similar field effect is proposed to take place in a social system that begins to function when a threshold level of coherence in collective consciousness is achieved. In the case of this sociological influence of the Maharishi Effect, however, the mediating force field would be subtler. If the coherent effect is initiated from individual and group enlivenment of the perfect orderliness at the level of the unified field, it would have nonlocal distributive properties that first influence the inner subjective domain, and then extend into the gross objective level of ordinary behavior where the social outcome variables documenting the effect in geographical regions are measured in the empirical studies. This is consistent with the research outcomes, which indicate that the distributional properties of the Maharishi Effect—how the effect radiates from its source—are not localized in a manner that fits the field properties of the known fundamental forces in nature.

The finding of social coherence appears to be an extended field effect. But the gravitational force is too weak to mediate the effect, the strong nuclear and weak forces function at too short of distances, and the electromagnetic force also is too weak—as well as being screened out by metal buildings. The distances and distributional patterns involved in the Maharishi Effect clearly seem not able to be accounted for through the four fundamental forces. The findings suggest a coherent influence that is more fundamental than the known physical forces that mediate change on the gross objective level of behavior. The Maharishi Effect thus has been proposed to be a unified field-based effect. [45]

At the level of the unified field, all transformations are possible. From that level the Maharishi Effect would influence first the subtle mental and then gross physical levels of nature. A specific mental intention (dharana), settling down (dharana) to the unified field of consciousness, the transcendental field of perfect order (samadhi), enlivens coherence on the subtle mental level, reflected in coherence in neural activity (EEG coherence) as well as in gross macroscopic behavior. This nonlocal coherent influence increases coherence in individual minds—as well as collective social coherence—also influencing gross macroscopic behavior in a more coherent and life-supporting direction, a negentropic society-wide influence.

> "The transcendental level of nature's functioning is the level of infinite correlation. When the group awareness is brought in attunement with that level, then a very intensified influence of coherence radiates... Infinite correlation is a quality of the transcendental level of nature's functioning from where orderliness governs the universe. (p. 75) [46]

Large groups of individual practitioners contributing to collective coherence have been shown to have an exponentially more powerful effect on a sociological scale. The increased order in mental activity due to direct contact by the individual of the field of perfect order is reflected in neural activity in the form of maximum EEG coherence—producing increased order in the collective social network. This top-down causal interaction of transcendental, subtle, and gross levels of nature has been described extensively in previous chapters. On a global societal scale, it provides a rational and scientifically consistent explanation for the sociological findings that support the Maharishi Effect as a means to decrease negative incoherent trends and create a globally peaceful civilization. It is based on an expanded understanding of scientific principles that includes the nonlocal holism of nature at the level of the unified field and the subtle mental level of nature.

"Many carefully controlled experiments on the *Maharishi Effect*, the *Extended Maharishi Effect*, and the *Global Maharishi Effect* have appeared in leading scientific journals. These studies have utilized the most rigorous research designs and statistical methodologies to precisely evaluate the effect of large coherence-creating groups on standard sociological measures of the quality of life in cities, provinces, nations, and the world. The findings are consistent with the existence of the unified field of natural law and the capability of the human mind to function from it in the unified state of individual consciousness." (p. 12) [21]

The empirical findings on the Maharishi Effect indicate that the threshold for a phase transition into a coherent sociological state is approximately the square root of one percent of a population practicing the TM-Sidhi program of 'yogic flying'—the technology that produces the most coherent brain state in the individual. The square root of one percent of the current world population is about 8000 individuals. The research suggests that the establishment of such a group can produce a globally coherent influence sufficient to change the incoherent international trends of conflict and violence toward world-wide coherence in collective consciousness. Consistent with demonstrations of the Maharishi Effect on a global scale, the predicted result is positive, peaceful, life-supporting social indicators and spontaneous coherent societal trends the reflect progress toward world peace.

Based on the accumulated body of published research on collective coherence, Maharishi has been calling for the establishment of a coherence-creating group or groups of 8000 TM and TM-Sidhi practitioners as a systematic alternative to counteract the urgent threats of terrorism in this nuclear age and actually create world peace. Only a tiny fraction of the military budget of an advanced country, for example, would be required to implement fully and empirically test the immense practical potential of this unique sociological technology. As Maharishi has pointed out repeatedly:

"It is the weakness and lack of coherent thinking in the civilian sector of a nation which produces the situation where military action and war become necessary. Lack of intelligence, sharpness, quick decisions, and discipline gives rise to misinterpretations, muddles, confusion, and wrong relationships, which eventually become the cause of war... It is not the weapons that will save the country, it is the coherence of its national consciousness, and the unbounded awareness of its military men... Military is that section of society which is most concerned with action... As everyone knows, action comes from thought and thought comes from silence. So why should we not act from the first level of action? Fight the war with silence!... This will produce that influence of coherence and harmony in the nation which will keep all the cultural values fully alert and enlivened. The quality of national consciousness will be so very coherent...that the nation...will radiate no destructive influences at all. Therefore, no enemies will be created... The nation's high level of consciousness will disallow the birth of an enemy... The logic promoted by the arms-producing nations—that if one has the ability to destroy, the enemy will not arise—is self-defeating... The military has to avert the enemy from the absolute level of total power, which will avert the danger before it arises...

Heyam duhkham anagatam.
(Yog-Sutra, 2.16)

Avert the danger that has not yet come." (pp. 59-65) [20]

Maharishi explains further how this unique approach fulfills the duty of the military:

"Stopping war is possible only if we can stop the basis of war. It has been found in hundreds of cities in all parts of the world that when one per cent of the population practices the Transcendental Meditation Programme, sharp drops occur in the city's rates of crime and illness, and in other negative trends. This is a statistically established fact, not a hope... By profession, the military has a duty to equip itself with all possible systems of defence available. It should use both procedures—the dynamic approach to defence and the silent approach to defence... The dynamic approach is through weaponry, and the silent approach is through prevention. Prevention is best achieved by creating the 'Field Effect,' the *Maharishi Effect*... How is the enemy born? The answer is very simple... 'As you sow, so shall you reap'...

A stressed man will always create, in his own mind, doubts about others. Because of stress, his comprehension is narrow, so he loses contact with the elements outside his perception and comes to a wrong decision. The more the stresses, the worse the situation and the more the number of enemies that will be created... Through Transcendental Meditation stresses disappear and only values blossom which increase cordiality, love, and friendship... The only effective defence is prevention; that means disallow the birth of an enemy; that means we must learn how to be friends with Nature, because Natural Law is the only invincible force... The introduction of Transcendental Meditation in all fields of national life will create coherence in national consciousness, and creating a PREVEN-TION WING in the military will create an invincible armour for the nation and will give the defence department the ability to fulfil its goals... My formula for prevention in the field of defence concerns itself only with the domain of consciousness; it is completely impartial; it offers the military in every nation the chance to rise to invincibility... At almost no cost the military of any nation can become forever victorious... My programme also enables the military to fulfil its national duty of ensuring that coherence within the nation does not decrease. As a result, victory will be maintained before war without any effort at all... Invincibility to every nation from every other nation in the world—a beautiful concept of balance of power belonging to every nation." [20]

Military personnel are certainly among the most courageous, honorable, and noble in society. But the traditional military strategy to establish security unfortunately is based on the physicalist worldview that assumes matter over mind. It is fundamentally a default approach, due to weakness in the minds of the general population. The advent of powerfully destructive weaponry calls on us to focus on developing more powerful minds that can integrate our thinking and avert the danger of further deployment of terrible weaponry. Like the physicalist worldview, the destructive approach becomes outmoded when effective means are available for mind over matter. In a simple and profound statement of the dynamics underlying war and peace, Maharishi emphasizes:

"Diversity is destructive only in the absence of unity." [47]

Maharishi has proposed the only theoretically coherent, empirically supported, and economically viable approach available to make practical empirical progress toward world peace. It is obvious that nothing else has worked. Given the current volatility in our world family—with terrorist violence basically impossible to protect against given the means currently being used—no defensive or offensive weapon system can keep us safe. For our heads of state, scientific leaders, and captains of business and industry not to test and apply this approach, the only viable approach to world peace available, is quite inconsistent with socially responsible ethics and pragmatism.

Whether or not one understands or agrees with its increasingly well established theoretical underpinnings, the empirical research supports the conclusion that a practical means to promote world peace is now available. While continuing with the old strategies, including negotiation of peace treaties that can be helpful for short periods of time but historically have been shown to last only a few years, as well as the strategic deployment of military arms which give some sense of assurance but also assures the continuance of enemies and further war, this new and much more viable strategy at the very least warrants rigorous full-scale empirical testing.

In the next chapter additional technologies of Maharishi Vedic Science and Technology are introduced. These technologies all apply the holistic approach of the consciousness-mind-body connection expressed in the transcendental, subtle, and gross domains of nature. They represent advanced scientific understanding Maharishi has revived from ancient Vedic science that had been lost due to overlooking the fundamental underlying unity of nature and its systematic means to utilize it to develop the full potential inherent in human life now carefully systematized and available to be applied.

Chapter 21 Notes

[1] Hagelin, J. (2004). Maharishi's Global News Conference, September 8.

[2] Shear, J. (2006). *The experience of meditation: Experts introduce the major traditions.* London: Paragon Press.

[3] Alexander, C. N., Davies, J. L,, Dixon, C. A., Dillbeck, M. C., Oetzel, R. M., Druker, S. M., Muehlman, J. M., & Orme-Johnson, D. W. (1990). Growth of Higher States of Consciousness: Maharishi's Vedic Psychology of Human Development. In Alexander, C. N. & Langer, E. J. (Eds.). *Higher stages of human development: Perspectives on adult growth.* New York: Oxford University Press, pp. 286-341.

[4] Schneider, R. H., Alexander, C. N., & Wallace, R. K. (1992). In Search of an Optimal Behavioral Treatment for Hypertension: A Review and Focus on Transcendental Meditation. In Johnson, E. H., Gentry, W. D., & Julius, S. (Eds.). *Personality, elevated blood pressure, and essential hypertension.* Washington, D. C.: Hemisphere Publishing Corp.

[5] Denniston, D. (1986). *The TM book: How to enjoy the rest of your life,* 2nd Edition. Fairfield, IA: Fairfield Press.

[6] Forman, R. K. C. (1990). *The problem of consciousness.* New York: Oxford University Press.

[7] Roth R. (1994). *Transcendental meditation,* 2nd Ed. New York: Donald I. Fine.

[8] Murphy, M. & Donovan, S. (1988). *The physical and psychological effects of meditation—A review of contemporary meditation research with a comprehensive bibliography, 1931-1988.* Esalen Institute Study of Exceptional Functioning, San Rafael, CA.

[9] Alexander, C. N., Boyer, R., & Alexander, V. (1987). Higher States of Consciousness in the Vedic Psychology of Maharishi Mahesh Yogi: A Theoretical Introduction and Research Review. *Modern Science and Vedic Science,* Vol. 1, No. 1, January, pp. 89-126.

[10] Maharishi Mahesh Yogi (1963). *Science of being and art of living.* Washington, D.C.: Age of Enlightenment Publications.

[11] Maharishi Mahesh Yogi (1967). *Maharishi Mahesh Yogi on the Bhagavad-Gita: A new translation and commentary, chapters 1-6.* London: Penguin Books.

[12] Benson, H. (1975). *The relaxation response.* HarperCollins.

[13] Hameroff, S. R. (2008). *The conscious pilot: Synchronized dendritic webs move through brain neuro-computational networks to mediate consciousness.* April 11 plenary session, Toward a Science of Consciousness Conference, April 8-12, Tucson, AZ.

[14] Stapp, H. P. (2007). *Mindful universe: Quantum mechanics and the participating observer.* Berlin Heidelberg New York: Springer-Verlag.

[15] Travis, F. & Arenander, A. (2006). Cross-sectional and longitudinal study of effects of transcendental meditation practice on interhemispheric frontal asymmetry and frontal coherence. *The International Journal of Neuroscience,* 116, 1519-38.

[16] Orme-Johnson, D. W, (2008). <www.truthabouttm.com>

[17] Maharishi Mahesh Yogi (1997). *Maharishi speaks to educators*, Vol. 4, Edition 2: India: Age of Enlightenment Publications.

[18] Kanellakos, D. S. & Lukas, J. S. (Eds.) (1974). *The psychobiology of transcendental meditation: A literature review.* Menlo Park, CA: W. A. Benjamin.

[19] *Enlightenment and the siddhis: A new breakthrough in human potential* (1977). West Germany: MERU Press.

[20] Maharishi Mahesh Yogi (1996). *Maharishi's Absolute Theory of Defence.* India: Age of Enlightenment Publications.

[21] Nader, T. (2000). *Human physiology: Expression of Veda and Vedic Literature,* 4th Edition. Vlodrop, The Netherlands: Maharishi Vedic University.

[22] Swanson, C. *The synchronized universe: New science of the paranormal.* Tucson, AZ: Poseidia Press.

[23] Hagelin, J. The Physics of Flying. *The video magazine,* Vol. 7, Tape 1. N- 38. Maharishi University of Management.

[24] Maharishi Mahesh Yogi (2004). Maharishi's Global News Conference, December 29.

[25] Anonymous (1977, December). Personal communication. Interlaken, Switzerland.

[26] Anonymous (1993, August). Personal communication. Washington, D. C.

[27] Anonymous (2005, May). Personal communication. Fairfield, IA.

[28] Pearson, C. A. (2002). *The supreme awakening: Maharishi's model of higher states of consciousness applied to the experiences of individuals through history.* Doctoral dissertation, Maharishi University of Management, Fairfield, IA. Ann Arbor, MI: UMI Dissertation Services (Printers).

[29] Gelderloos, P. & van den Berg, W. (1989, Winter). Maharishi's TM-Sidhi Program: Participating in the Infinite Creativity of Nature to Enliven the Totality of the Cosmic Psyche in All Aspects of Life. *Modern Science and Vedic Science,* Vol. 2, No. 4, 372-412.

[30] Alexander, C. N. (1992). Peaceful Body, Peaceful Mind, Peaceful World. *Modern Science and Vedic Science,* Vol. 5, Nos. 1-2, 150-164.

[31] *Creating an Ideal Society: A global undertaking* (1977). Rheinweiler, W. Germany: MERU Press.

[32] Brown, C. L. (2005) Observing the Assessment of Research Information by Peer Reviewers, Newspaper Reporters, and Potential Governmental and Non-Governmental Users: International Peace Project in the Middle East (Maharishi Effect). In Schmidt-Wilk, J., Orme-Johnson, D. W., Alexander, V. K., & Schneider, R. H. (Eds.). *Applications of Maharishi Vedic Science: Honoring the lifework of Charles N. Alexander, Ph.D. Journal of Social Behavior and Personality,* Vol. 17, Special Issue, No. 1. Select Press.

[33] Dossey, L. (1989). *Recovering the soul: A scientific and spiritual search.* New York: Bantam Books.

[34] Peat, F. D. (1990). *Einstein's moon: Bell's theorem and the curious quest for quantum reality.* Contemporary Books, Inc.

[35] *Maharishi Vedic University: Introduction.* (1994). Holland: Maharishi Vedic University Press.

[36] Oates, R. M. (2002). *Permanent peace.* Fairfield, IA: Institute of Science, Technology and Public Policy.

[37] Borland, C. & Landrith, G. III. (1977). Improved Quality of City Life Through the Transcendental Meditation Program: Decreased Crime Rate. In Orme-Johnson, D. W. & Farrow, J. T. (Eds.). *Scientific research on the Transcendental Meditation program: Collected papers,* Vol. 1. Rheinweiler, W. Germany: Maharishi European Research University Press, pp. 639-648.

[38] Dillbeck, M. C., Landrith, G. III, Polanzi, C., & Baker, S. R. (1982). The Transcendental Meditation Program and Crime Rate Change: A Causal Analysis. In Chalmers, R. A., Clements, G., Schenkluhn, H., & Weinless, M. (Eds). *Scientific research on Maharishi's Transcendental Meditation and TM-Sidhi Program: Collected papers,* Vol. 4. Vlodrop, The Netherlands: Maharishi Vedic University Press, pp. 2515-2520).

[39] Davies, J. L. & Alexander, C. N. (1989). The Maharishi Technology of the Unified Field and Improved Quality of Life in the United States: A Study of the First World Peace Assembly, Amherst, Massachusetts, 1979 (MERU Research Report No. 323). In Chalmers, R. A., Clements, G., Schenkluhn, H., & Weinless, M. (Eds). *Scientific research on Maharishi's Transcendental Meditation and TM-Sidhi program: Collected papers,* Vols. 2-4. Vlodrop, The Netherlands: Maharishi Vedic University Press.

[40] Davies, J. L. & Alexander, C. N. (1990). Alleviating Political Violence Through Enhancing Coherence in Collective Consciousness: Impact Assessment Analyses of the Lebanon War. Summary of a paper presented at the 85th Annual Meeting of the American Political Science Association, September, 1989. In Wallace, R. K., Orme-Johnson, D. W., & Dillbeck, M. C. (Eds.) *Scientific research on the Transcendental Meditation program: Collected papers,* Vol. 5. Fairfield, IA: Maharishi International University Press, pp. 3260-3262.

[41] Orme-Johnson, D. W., Cavanaugh, K. L., Alexander, C. N., Gelderloos, P., Dillbeck, M. C., Lanford, A. G. & Abou Nader, T. M. (1989). The Influence of the Maharishi Technology of the Unified Field on World Events and Global Social Indicators: The Effects of the Taste of Utopia Assembly. In Chalmers, R. A., Clements, G., Schenkluhn, H., & Weinless, M. (Eds). *Scientific Research on Maharishi's Transcendental Meditation and TM-Sidhi Program: Collected Papers,* Vol. 4, Vlodrop, The Netherlands: Maharishi Vedic University Press, pp. 2730-2762.

[42] Cavanaugh, K. L., Orme-Johnson, D. W., & Gelderloos, P. (1989). The Effect of the Taste of Utopia Assembly on the World Index of International Stock Prices. In Chalmers, R. A., Clements, G., Schenkluhn, H., & Weinless, M. (Eds). *Scientific Research on Maharishi's Transcendental Meditation and TM-Sidhi Program: Collected Papers,* Vol. 4. Vlodrop, The Netherlands: Maharishi Vedic University Press, pp. 2715-2729.

[43] Hagelin, J. S., Rainforth, M. V., Orme-Johnson, D. W., Cananaugh, K. L, Alexander, C. N., Shatkin, S. F., Davies, J. L., Hughes, A. O., & Ross, E. (1999). Effects of Group Practice of the Transcendental Meditation Program on Preventing Violent Crime in Washington DC: Results of the National Demonstration Project, June-July 1993. *Social Indicators Research, 47,* 153-201.

[44] *A methodological review of Maharishi Effect research.*< www.mum.edu./m_effect/methodology.html >

[45] Hagelin, J. S. (1987). Is Consciousness the Unified Field? A Field Theorist's Perspective. *Modern Science and Vedic Science,* 1, 1, 29-87.

[46] Maharishi Mahesh Yogi (1986). *Life supported by Natural Law.* Washington: Age of Enlightenment Press.

[47] Maharishi Mahesh Yogi (2003). Maharishi's Global News Conference, June 25.

Chapter 22

Systematic Technologies for Full Human Development II: Maharishi Ayur Veda, Sthapatya Veda, Jyotish, and Yagya

In this chapter, we will overview key principles of the developmental technologies of Maharishi Ayur Veda, Sthapatya Veda, Jyotish, and Yagya. The full range and evolutionary course of the transcendental, subtle, and gross levels of nature is applied in gaining an understanding of the subtle mechanics of nature reflected in these technologies. The overall main point of this chapter is that Maharishi Vedic Science and Technology makes available holistic technologies for full human development that can now be appreciated based on a scientific understanding of the nonlocal interconnectedness of nature.

The approach to full human development in Maharishi Vedic Science and Technology applies systematic technologies to all the specific levels of inner and outer life. The technologies comprise a holistic approach based on how the totality of natural law is expressed on nonlocal and local levels in a manner that maintains full connection with the totality. The systematic technologies are said to develop increasingly refined balance and coordination of the different parts of the mind and body with the integrated wholeness that is the essence of every part. They enliven or reenliven natural healing mechanisms through subtle purifying and refining processes that eliminate imbalances and allow the natural unfoldment of higher, more integrated levels of physiological functioning and states of consciousness. The holistic approach is a top-down system that emphasizes direct development of the state of consciousness, and also natural technologies on each level of mind, body, behavior, and environment that support their stabilization.

Maharishi Ayur Veda

After systematizing the teaching of the TM and TM-Sidhi program Maharishi also revived *Maharishi Ayur Veda*, another important developmental technology in ancient Vedic science. The term *Ayurveda* refers to the *science of lifespan*. Maharishi Ayur Veda applies the fundamental interconnectedness of nature in a comprehensive system of natural medicine that focuses on preventative health. It sets forth in great detail the consciousness-mind-body connection and how to promote optimal levels of mental and physical health through attunement with the totality of natural law.

In Maharishi Ayur Veda the individual and the cosmos are spun from the same cloth—the individual is a reflection of the entire cosmos. Maharishi Ayur Veda describes the phenomenal structure of creation in the same terms as in Sankhya, summarized in Chapter 11. In this system there is a fundamental correspondence between the composition of the individual mind and body, the senses, and the objects of sense that comprise phenomenal nature. This integrated structure is applied in a system of natural health care to restore and maintain balance in individual life. In Maharishi Ayur Veda the concept of health relates to the wholeness of life, involving balance of universal consciousness and its different expressions in different parts of human physiology. This section is a brief introduction to some of the basic principles in this system of natural health care and disease prevention. On this issue, Maharishi has stated:

> "Modern medicine treats the body on the level of the body. The Vedic approach treats the body from its cause—consciousness. Consciousness is the flow of intelligence, the flow of awareness. The body's own inner intelligence cures the body." [1]

Maharishi Ayur Veda emphasizes mental technologies for the development of consciousness as the foundation for full mental and physical health, especially the TM and TM-Sidhi program. It also applies physical and behavioral approaches, including therapeutics of healthy diet and daily routines, for emotional and behavioral balance. It further includes an extensive system of medicinal herbs and how to prepare and apply them, as well as knowledge of natural biological rhythms, seasonal patterns, and their effects on homeostatic balance. It is primarily a preventative approach that incorporates noninvasive diagnostic procedures and multi-modal treatments.

Although supported by extensive clinical research, the technologies in Maharishi Ayur did not originate from experimental laboratory findings or theoretical insights of health practitioners based on fragmented understanding of isolated laws of nature typical of object-based, disease-oriented modern allopathic medicine. In Maharishi Ayur Veda the basis of all disease and disorder is lack of enlivenment of the connection of individual consciousness to the unified field of consciousness. Disorder and disease are a product of the mind and body not functioning in a sufficiently balanced and refined enough manner to maintain full attunement with the totality of natural law—referred to in previous chapters in terms of *Pragya aparadha*, the mistake of the intellect. The purpose of Maharishi Ayur Veda is to establish the universal value of consciousness in individual consciousness and reenliven and develop the inner intelligence of body and mind that supports attunement with the totality of natural law in the consciousness-mind-body connection.[2] In this framework, perfect health refers to the spontaneous ability to maintain full connection of individual consciousness in universal consciousness under all conditions and experiences in life.

In the holistic understanding of individual nature and cosmic nature the five subjective senses are directly correlated with the five objective subtle and gross elements. The objects are built of the qualities of space and sound, air and touch, fire and sight, water and taste, and earth and smell. These five elements and qualities combine to produce the phenomenal objects and their corresponding qualia. Experiencing the different objects of sense—whether by ingesting the elements through consumption of food, through respiration, or through even visual and auditory perception, for example—affects the consciousness-mind-body system of the individual.

In Maharishi Ayur Veda the scientific principle of cause-effect relationships in nature is applied to the level of behavior for the promotion of health. Human actions naturally have lawful consequences. The results of behavioral choices are not arbitrary; they have degrees of beneficial, detrimental, or mixed effects that accumulate to produce increased health and happiness, or suffering and disease. Sensory experience is directly related to the effects in body and mind of the objects of sense in the form of types of smells, tastes, sights, touches, and sounds. For example, with respect to the gustatory sense there are six basic tastes—sweet, salty, sour, bitter, pungent, astringent—each with its own effect on the body. Ingesting different tastes of food can either promote or disturb natural psychophysiological balances.

This fundamental design correspondence in the universe ties together actions, the environment, and the effects of actions on the environment and on oneself. It provides an integrated system for health-promoting behavior in accord with lawful patterns throughout nature. It thus provides a framework for a science of moral and ethical behavior. In this framework morality and ethics concern the promotion of mental and physical health through increasing attunement with the totality of natural law toward permanent higher states of consciousness in individuals and society (refer to Chapter 2). Taking things into the body and mind through any of the senses that are not healthy has a cumulative effect that decreases optimal functioning. Mind and body function more coarsely, clogging up and deteriorating normal functioning and resulting in disorder and disease. Countering this negative effect involves establishing an integrated balance. Adding to the daily routine the direct experience of transcendental consciousness is the most fundamental and effective means to promote optimal mind-body functioning in an overall preventative approach.

It is quite unfortunate that in modern civilization morality has become divorced from a systematic scientific framework for health promotion and human development. The original intent of religious moral codes and injunctions of right and wrong behavior, for example, was to guide people toward higher levels of development. Sometimes couched in terms of union with or being in the grace of God, the injunctions were based on the deeper, subtler interconnectedness and cause-effect relationships of behavior and its natural consequences. Not understanding these subtler relationships, the moral injunctions were grossly misinterpreted as seemingly arbitrary rules that God commands humans to comply with in order to receive favor that should be followed as a matter of religious faith even if not validated by ordinary reason or experience (refer to Chapter 1).

Divorced from their practical underpinnings, interpreted crudely with respect to only short-term immediate effects, their relevance to daily life as guides for improved physical, mental, and social health reasonably became in doubt. This degenerative process was fostered by inconsistent interpretations of the injunctions, the difficulty of validating the subtle cause-effect relationships using objective methodologies, and also by assumptions in modern science that nature is fundamentally random, *value-less,* and meaningless. It has been a fundamental quandary in modern science how nature can be universally deterministic and lawful on the one hand, and on the other hand most fundamentally random. This quandary is due to overlooking the subtle levels of nature that are deterministic, but so subtle that the relationships governing these levels have not been apparent within the limitations of objective modern scientific investigation.

As modern science has progressed toward deeper and more integrated understanding, inklings of the empirical efficacy of some religious wisdom has developed. For example there is growing recognition that some of the moral injunctions in religious doctrines have practical validity in natural medicine and preventative health, individually, socially, and epidemiologically. Likewise the religious community also is broadening its understanding of the universal meaning of religious texts common to the various religious doctrines. Also hopefully the religious community in general will again recognize the subtler influences that what they consume have on the ability to comprehend and behave in accord with religious principles and doctrines. Both secular and religious approaches can be said to have the same goal of identifying invariant laws of nature—or the Will of God, or whatever descriptors one wishes to apply—in order to eliminate suffering. Attention to the holistic basis of natural law underlying both secular scientific and religious approaches will bridge the gap between them.

Body types in Maharishi Ayur Veda. Recall that in Chapter 11 the fundamental division of knower, process of knowing, and known was associated on the level of the subtle relative domain with the terms *sattva, rajas,* and *tamas.* On the gross relative domain the corresponding terms are *vata, pitta,* and *kapha.* These three fundamental qualities, called *doshas* in the physical body, exist in all objects composed of the five gross elements (mahabhutas). On a basic level vata is associated with dryness, pitta with hotness, and kapha with heaviness.

An important feature of this system of natural health care is the classification into body types based on the principles of vata, pitta, and kapha—introduced in Chapter 17. This system classifies the inherited genetic constitution of the individual as manifested physically and behaviorally. Body types react differentially to basic cycles in nature, such as diurnal or seasonal cycles. Illness and disease are understood to be the consequences of behavior that is not in tune with the natural balance for the individual body type in the time and place in which the individual is living.

Although all three doshas make up each individual bodily form, specific physiological and behavioral tendencies can be identified according to their predominance in the individual. *Vata* dosha is most prominent in the gross elements of space and air; it is associated with motion, and primarily concerns regulation of locomotor functions. Centered in the colon area, it concerns for example circulation throughout the body, intestinal peristalsis, muscle function, cell division, as well as neural activity. It is involved with homeostatic balance, and controls the other two fundamental qualities, pitta and kapha. Typical psychophysical tendencies of individuals with a predominance of vata include: light, thin build; cold hands and feet, and dry skin; bursts of energy but quickness to becoming tired; tendency for insomnia and irregular digestion; quick performance of actions, including grasping and forgetting new information; enthusiasm, imagination, excitability, restlessness, worry, and changeability in general.

Pitta is most prominent in the gross elements of fire and water; it primarily concerns regulation of metabolic functions. Centered in the small intestine and associated with digestion, it involves biochemical reactions and energy systems in the body. It regulates digestive and metabolic processes as well as many glandular and hormonal functions and respiratory processes. Typical psychophysical tendencies of individuals with a predominance of pitta include: medium build, strength, and endurance; sharp intellect and speech; tendency toward irritability or anger under stress; strong digestion, hunger, and thirst; warm, flushed, fair or ruddy, often freckled skin with heavy perspiration; determined, orderly, and sometimes commanding or joyful demeanor.

Kapha is most prominent in the qualities of water and earth. Centered in the chest area, it concerns primarily cohesion in the body and control of the bodily structure, including intracellular structure. Typical psychophysical tendencies of individuals with a predominance of kapha include: solid build, physical strength, and tendency to be overweight; steady, graceful action; slow digestion and heavy sleep; cool, pale, soft, often oily skin; slow to grasp and forget information; affectionate, possessive, tranquil, complacent, and earthy.

As an example of daily and seasonal routines that match biorhythms and other natural cycles in Maharishi Ayur Veda, different times of the day, seasons, and stages of the lifespan are associated with the three different doshas. For example vata is associated approximately with late autumn and winter, the periods of 2-6 AM. and 2-6 PM, as well as the later adult years. Pitta is associated with midsummer and early autumn, 10 AM-2 PM and 10 PM-2 AM, and the early and middle adult years. Kapha is associated with spring and early summer, 6-10 AM and 6-10 PM, and the childhood years. For example kapha-related disorders such as colds are common in children and in the spring, vata-related disorders such as constipation and interrupted sleep are more prevalent among older adults, and pitta-related disorders occur more frequently in the summer.

Each gross bodily structure—human, animal, insect, plant—is made up of the predominance of certain permutations of vata, pitta, and kapha throughout all levels of gross physiology. In humans there are 10 basic body types determined by which dosha is primary, secondary, and tertiary: vata, vata-pitta-kapha, vata-kapha-pitta, pitta, pitta-vata-kapha, pitta-kapha-vata, kapha, kapha-vata-pitta, kapha-pitta-vata, and approximately equal amounts of the three (called *tridosha*). Good health is the natural result of balance in these fundamental qualities and elements according to the basic body type. When these qualities or doshas are in proper balance, associated with the term *prakriti*, good health is the naturally promoted. When aggravated or imbalanced, called *vikriti*, disorder and disease can result.

Each body type has particular strengths as well as vulnerabilities to imbalance. The body type system allows identification of risk factors and specification of preventative approaches as well as rapid diagnosis of imbalances, through assessment of prakriti and vrikriti. Common analogies for understanding the underlying role of vata, pitta, and kapha in human physiology are the mixing of a small set of primary colors in painting or in a television set that displays a great variety of images through three basic optical beams of color images. When the viewed images on the screen become unclear, it is the underlying dots of color projected onto the screen that need adjustment. Creating health through Ayurvedic technologies involves maintaining dynamic balance between the three doshas according to the fundamental body type or constitution of the individual. One basic general principle that leads to treatment strategies is to create balance by increasing the effect that is opposite of the imbalance.

Maharishi Ayur Veda contains diagnostic procedures to identify imbalances due to behavior that doesn't sustain natural health. An expert Ayurvedic doctor, called a *Vaidya*, determines the body type for an individual and the current prakriti and vikriti. While Western allopathic medicine focuses on diagnosis largely through the detection of visible symptoms, Maharishi Ayur Veda identifies six stages of imbalance. The first three stages—involving aggravation of the normal functioning of the doshas—are subtle and generally are not recognized in the allopathic approach of Western medicine.

One important diagnostic procedure is to observe the individual carefully and sensitively, based on detailed knowledge of the system of body typology. Another important diagnostic procedure, called *Nadi Vigyan*, involves feeling the pulse.[2] This approach recognizes that all interactions in nature involve vibratory patterns. Being able to sense these patterns in the human body can be the source of detailed information about its functioning. In this method all activity in the body is understood to send information to the heart, such as through the bloodstream. This information is coded in the heartbeat and can be accessed through the carrier wave of the pulse. The pulse reflects the underlying prakriti or constitution of the individual, as well as the vikriti or state of the three doshas or imbalances throughout the body. It contains information on the condition of the whole person as well as the various parts of the physiology—the balance of wholeness with the specific parts.

Each of the three doshas have five sub-doshas related to particular areas and functions of the body, such as the liver, stomach, and spinal column. Information about these 15 sub-doshas can be directly accessed through the pulse. It involves feeling the pulse at the radial artery above the wrist, at different depths and points using three fingers, each assessing information on vata, pitta, and kapha. Through this subtle method of assessment, an expert pulse diagnostician quickly and efficiently can obtain considerable detail about the condition of the body, including specific organ systems.

After assessing the doshas and their prakriti and vikriti through Nadi Vigyan as well as other diagnostic methods, there is a comprehensive system of therapeutic approaches in Maharishi Ayur Veda to reestablish and maintain balance in mind and body. In addition to the mental techniques such as the TM and TM-Sidhi programs there are detailed individualized approaches such as dietetics based on pure foods, herbal therapies, body purification therapies, exercise, music therapy, Vedic sound therapy, as well as daily and seasonal routines and other behavioral approaches.[2] These points serve as an introductory glimpse (or taste, for example, depending on your body type) of the principles of Maharishi Ayur Veda. The most integrated and holistic aspect of this approach to health, *Maharishi Vedic Healthcare*, will be introduced in Chapter 23.

Maharishi Sthapatya Veda

Another fascinating developmental technology that Maharishi has revived in recent years from ancient Vedic science concerns basic principles in the construction of natural living environments that promote health. As an example of the principles related to this approach, a recent study of ill-planned suburban areas on public health relates to the importance of city planning. The study suggests that people living in communities that have built up sporadically and chaotically in sub-urban sprawl have increased risks of obesity, high blood pressure, and other related chronic ill-nesses. [3]

The relationship between building design and physical and behavioral health is addressed in the aspect of ancient Vedic science called *Sthapatya Veda*. Sthapatya means 'to establish.' Maharishi Sthapatya Veda is an important developmental technology of architecture based on Vedic mathematics and engineering. As Maharishi explains:

"This field of Vedic Mathematics has only one basic principle of structuring, and that is...*Purnat purnam udachyate (Brihad-Aranyak Upanishad, 5.1.1)*—from fullness emerges fullness—from fullness is structured fullness—from total Natural Law emerges total Natural Law... In the process of transformation or evolution, it is the Totality that is reborn again and again...*Navo-Navo bhavati jayamanah (Rk Veda, 10.85.19)*. This means that in the sequential flow of evolution of Natural Law, all its expressions, at every step of evolution, are sustained in the quality of WHOLENESS." (p. 103) [4]

Maharishi Sthapatya Veda integrates abstract function and tangible structure, based on principles in Vedic mathematics, in order to align more fully individual life with cosmic life and the holistic structure of natural law. It focuses on the important influence of the architectural design and construction of physical structures such as office buildings and homes, and also city planning. In ancient Vedic science there is a direct correspondence between the structure of the cosmos and the human body. When physical structures are built according to these universal relation-ships based on the integrated wholeness of nature, they support progress toward higher levels of development. The principles of Maharishi Sthapatya Veda are said to be for the purpose of em-bodying qualities of unbounded consciousness into the boundaries of physical structure. Maha-rishi describes the fundamental principle of Sthapatya Veda:

"The principle of Sthapatya Veda is to establish any building, any village, any city, and country in full alignment with the structuring dynamics of the whole universe, which maintain the connectedness of everything with everything else... Buildings that are constructed according to Sthapatya Veda are very soothing, uplifting, and evolutionary to everyone because every individual is essentially Cosmic in nature—the structure and function of the individual physiology is an exact replica of the Cosmic Physiology, the physiology of the universe... The truth is that the individual is Cosmic on both levels—on the level of intelligence, or consciousness, and also on the level of his body, which is the expression of his consciousness, or intelligence. Because of this cosmic status of the individual, in order for the individual to be in peace and harmony within himself, every-thing about him should be in harmony with the universe; it is necessary that everything with which he is concerned, or anything that is in his environment, is in full alliance and harmony with the Cosmic Structure and its basis in Cosmic Intelligence... Sthapatya Veda is that aspect of the cosmic knowledge of Natural Law that maintains the buildings in which the individual lives and works, and the environment, in which he moves, well set in cosmic harmony." (p. 104) [4]

The proper orientation with respect to the cardinal directions, relationship to bodies of water, approaching and surrounding roads, the free flow of energy including air and light through a building, the proportions and dimensions of rooms and their lay-out for purposes such as cooking, sleeping, and meditating; the shape and slope of the building site and relationship to the local ecosystem—as well as the health effects of the construction materials—are all considered important in Maharishi Sthapatya Veda. [5] The proper relationship to the environment in order to establish harmony with natural law, called *Vastu*, is said to be crucial to whether a building promotes positive, life-supporting, and 'fortune-building' influences or contributes incoherence and entropy to the owners and occupants and to the local community.

"On its path from east to west the Sun generates different influences, different qualities of energy. Knowledge of these influences is provided by Maharishi Sthapatya Veda, which explains that a house should be designed so that the energy of the Sun always supports each activity of daily life—the dining room should be located where digestion will be most healthy; the study should be located where the intellect is most lively; the living room should be located where social life will enjoy greatest support, etc... Wrongly placed entrances alone contribute to inauspicious and negative influences in the life of the residents, creating anger, aggression, fear, poverty, lack of vitality, and even chronic disease. From eight possible major directions [the four cardinal directions and their diagonals, Southwest, etc.] only two [East and North] are auspicious, and therefore 75% of all buildings have an inauspicious orientation, contributing to ill health and other problems of society. Modern architecture simply does not possess the knowledge of proper orientation." (p. 505) [2]

The emphasis on orientation in Maharishi Sthapatya Veda is beginning to be corroborated by systematic research. For example recent neuroscientific research indicates that firing patterns of thalamic neurons appear to change with the direction of orientation individuals are facing. An interesting recent study may be the first evidence in support of the health effects of direction on orientation. Individuals sleeping with their heads to the north had significantly lower scores on the Mental Health Inventory compared to those who slept in other directions, and those whose homes have south-facing entrances had significantly lower overall scores on the Mental Health Inventory, and also reported more financial problems, than those with north or east entrances. [6] Another correlational study showed that homes with a south entrance were associated with 75 percent more burglaries than homes with other orientations." [7]

"According to Sthapatya Veda, when there is proper orientation of structure...then at this time there is the activation and firing of specific neurons. The activation of aspects of the nervous system based upon orientation means that the initiation of specific cycles and modes of functioning occurs in conjunction with specific structural position and orientation. (pp. 193-194) [2]

Another key principle is that the center of the building is best designed as an open unrestricted area of silence, called the *Brahmasthan*. This establishes an inner core of silence to the building that is reflective of the structure of cosmic life, which has at its inner core the infinite silence of the unified field of consciousness. Also hallways and windows are placed such that light and energy have unrestricted flow through the Brahmasthan and the entire building in the four cardinal directions, reducing the sense of boundaries in the building that impede the flow of energy both psychologically and physically. The overall feeling of a building constructed according to these principles is frequently described as embodying qualities of silence, freedom from restrictions, and natural comfort.

Maharishi Jyotish

Maharishi Jyotish, another important approach in ancient Vedic science, articulates an elaborate system of knowledge of fundamental relationships governed by laws of nature and how to predict events based on cycles and patterns. The Vedic word *Jyotish* is related to the concept of 'light,' in the sense of to shine, reveal, appear, illuminate, or manifest It applies Vedic mathematics to an analysis of natural cycles, including patterns of cosmic and astrophysical objects and their relationships, and how they influence human life on the subtle level in addition to the familiar gross level of nature. Maharishi describes Vedic mathematics that underpins jyotish:

> "Vedic Mathematics...is the systems and procedures available in the structure of the Veda and Vedic Literature through which order emerges and sustains the orderly universe... Vedic Mathematics is inherent in the structure of pure knowledge, the Veda... The COURSE of Vedic Mathematics is the unbounded field of the ever-expanding universe within the self-referral consciousness (Atma) of everyone. (pp. 335-336)[8]

Briefly, Vedic mathematics includes a number system based on unmanifest wholeness or oneness expressed in the symbol of 'zero.' In this system all numbers emerge out of the nothingness of zero, which represents the totality of unmanifest wholeness. All numbers are written in Vedic mathematics as a modification of the symbol for zero. This system focuses on holism and simultaneity, based on an inclusive rather than exclusive approach to logic and mathematical principles. In simple terms it emphasizes *and* rather than *or*, such as plus *and* minus together, rather than plus *or* minus; addition *and* subtraction together, multiplication *and* division together, infinity *and* point, unity *and* diversity, and so on.[89] These principles translate into efficient methods for complex mathematical calculation.

> "Vedic Mathematics is the mathematics of the Absolute Number...the mathematics of the absolute, self-referral field of pure consciousness, where everything is simultaneous, where everything is simultaneously administered on the level of perfect order... The common basis of all numbers and mathematical expressions is the Absolute Number... The expression of the Absolute Number is [10 encased in a circle], the expression that presents one zeroed—Unity zeroed—Unity made unmanifest—state of Unity devoid of expression (*Purusha*)...the unified state of all numbers...the common basis of all numbers, from where all number systems can spontaneously be handled with absolute precision and without any limitations. Vedic Mathematics has its status in the transcendental area of modern mathematics." (pp. 371-382)[8]

In Maharishi Jyotish which applies Vedic mathematics the entire phenomenal universe is understood to be pulsations, vibrations, reverberations, or mathematically precise cyclic patterns of the unified field in a nesting of levels of manifestation of unfathomable complexity. These cyclic patterns are found in cosmic manifestations and dissolutions, life and death, day and night, and the waxing and waning of all phenomena—animate and inanimate—from the biggest to the smallest fluctuations. Whether subtle or gross, these patterns theoretically can be calculated and used to predict events.

A helpful analogy for understanding Maharishi Jyotish is the prediction of weather, thought to be deterministic but quite difficult to predict due to inherent complexity, including non-linear interactions. This aspect of ancient Vedic science, which concerns predicting future events on the basis of cosmic, astrophysical, and local cyclic patterns, is sometimes interpreted as Vedic astrology. The astrological concept of the influence of cosmic objects on human behavior has long been rejected on grounds of causal impossibility as well as absence of experimental validation. On the other hand throughout history astrology has been a persistent belief system in most of the world's major civilizations, with considerable detail in methodology. Is there now a rational basis in modern science for such beliefs—for example, based on the nonlocal interconnectedness of nature in quantum and unified field theories?

A positive response to this inquiry is supported by the developing understanding of the gross, subtle, and transcendental levels of nature discussed throughout this book. Basically it involves a finer-grained analysis of nature than is reflected in the view that all action is conducted via the four fundamental forces conceptualized within the framework of Einstein locality. There are of course known influences of astrophysical objects on terrestrial events, such as the effects of the Moon on the tides, or sunspots on electromagnetic activity on Earth. These astrophysical cycles affect numerous biological cycles. Indeed perhaps the most powerful influence of any physical object on terrestrial life as we know it is the Sun. Practically speaking these influences can all be explained in terms of the four fundamental physical forces of nature. But the finding of nonlocality as a fundamental characteristic of the physical universe necessitates a more expanded understanding of relationships in nature within the framework of modern science.

Maharishi Jyotish incorporates these subtler, finer-grained relationships. When a planet or star formation is posited as influencing the human mind and body, the mechanism for this influence involves the flow of energy and intelligence in the subtle levels of relative domain, in addition to the gross levels. Such influences are posited to be causal and deterministic—just as in the gross relative domain—but nonlocal. The fundamental principle of cause and effect—embedded in the Vedic principle or law of *karma*—is applied to all levels of phenomenal manifest creation, but not just in terms of local influences within light-speed and the light cone as in classical relativistic space-time theory.

This scientific principle also is reflected in the religious tenet that 'you reap what you sow,' which appears in various forms in most all major religious traditions. This principle is the basis for the injunction that is common to most all major belief systems, sometimes called the *Golden Rule*. For example in Christianity it is sometimes described as '...all things whatsoever you would that men should do to you, do even so to them; in Hinduism it is sometimes described as, 'do not to others, which if done to you, would cause you pain;' in Islamism it is described as, 'no one of you is a believer until he loves for his brother what he loves for himself;' for Hebraism, it is, 'what is hurtful to yourself, do not to your fellow man; in Buddhism, it is described as, 'by treating them as he treats himself;' by Taoism, it is, 'regard your neighbor's gain as your own gain and regard your neighbor's loss as your own loss.' [10]

However, the scientific principle of cause and effect reflected in this universal injunction is not restricted to the notion of a classical unbroken chain of proximate, local, physical events in time and space only on the gross level of nature; it also includes subtle causal influences. This means that on the subtle level of nature there are causal influences from celestial bodies such as the Sun, moon, planets, and stars that affect mental as well as physical processes on Earth, in a similar manner on the gross level that they affect Earth via the fundamental force fields of electromagnetism and gravitation.

As lawful patterns, at least theoretically these subtle influences can be classified and quantified in a similar manner as we have been able to do with gross objective physical relationships. But the complexity of identifying the mechanics of the patterns can be compared to the complexity of establishing psychological laws of nature, in contrast to the less complex patterns characteristic of inert physical systems associated with the laws of physics, chemistry, or engineering—which themselves are quite complex when attempting to address even a few non-linear components of a causal nexus. The unfathomable number of cycles and nested cycles at all levels of nature—from the cosmic, astrophysical, terrestrial, biological, atomic, sub-atomic, and quantum environments, as well as the inner psychological environment—makes precise prediction quite challenging.

Identifying these subtle lawful patterns also has been confounded by strong paradigmatic beliefs that such relationships don't exist—which have led to almost no investment in well-designed research to investigate them. These beliefs are fostered further by imprecise popular applications of principles associated with these subtle astrophysical and cosmological relationships—such as astrological guides in daily newspapers that are not precise enough for rigorous empirical validation and thus reasonably are seen to contribute to counter-evidence. Without confirming evidence from research conducted carefully enough to test them, these subtler relationships are still categorically dismissed in mainstream modern science, even when their theoretical basis is becoming established both logically and experimentally.

In Maharishi Jyotish a cosmic object such as the Sun not only has its influence via electromagnetic and gravitational forces but also via nonlocal interactions that are said to influence directly processes in mind and body. Such influences are theorized to be felt and observed with development of subtle perception based on increasing refinement of mind and body. Experiences in the direction of subtler perception can be understood in terms of psychological undertones of ordinary experience, such as of different qualities of light and color brought out in poetic, allegorical, and other artistic works. In some instances these depictions may be based on veridical experiences of subtle levels of nature—exemplified in Chapter 21 on higher states of consciousness.

In this ancient knowledge system the subtle attributes of the astrophysical and cosmological objects that reflect light visible from Earth have been identified. In general terms their influences can be likened to the effects on daily behavior of changes in cyclic weather and climatic conditions—more or less sunlight, wind, moisture, dryness, and so on—related to the fundamental elements (tanmatras and mahabhutas) associated with the concepts of space, air, fire, water, and earth. These influences affect all levels of life on Earth from humans to animals, insects, plants, as well as the entire physical environment.

"A one-to-one Cosmic Counterpart correlation exists between the number of estimated stars within our galaxy and the estimated number of neurons in the human brain ($1\text{-}2 \times 10$ to the 11th) for each galactically and neurally."— *Thomas J. Routt, computer scientist (p. 193)* [11]

Maharishi Jyotish describes specific relationships between the structure of the cosmic physiology surrounding Earth and functional structures in human physiology. A remarkable correspondence has been identified between functional structures in the human body and these cosmic objects—called *cosmic counterparts*. As the cosmic counterparts move cyclically in space and time relative to Earth—the physical home of human physiology—their changing relationships have varying degrees of influence on feeling, thinking, and action, individually as well as collectively. In Maharishi Jyotish cosmic counterparts have been identified for the major functional structures in the human brain, including the deep-seated nuclei of the basal ganglia, cranial nerves, cortical and brainstem areas; as well as constituents of the DNA molecule including nucleotides; and groups of nucleic acids on the cellular level. For example the primary influences come from nine astrophysical objects or centers called *Grahas*, generally referred to as planets but also the Sun, Moon, and other centers or loci. As described in Maharishi Jyotish the influence of the Grahas are said to have a direct functional correspondence—as well as a general anatomical correspondence—with the major neural systems in the basal ganglia of the brain. According to Maharishi Jyotish, the cosmic counterparts for the major systems in the basal ganglia are as follows: [8]

Sun—Thalamus (central structure controlling sensory and motor input processes)

Mercury—Subthalamus (modulates output based on input from thalamus, other basal ganglia, and cortex related to cognitive functions of discrimination)

Venus—Substantia Nigra (involved in modulating putamen and caudate nuclei, and smoothness of movement)

Moon—Hypothalamus (involved in emotions, daily, monthly, and seasonal cycles such as feeding, body temperature, reproduction, hormonal function)

The Ascending Lunar Node—Nucleus Caudatus, head (associated with putamen, saccadic eye movement, spatial memory, cognitive and behavioral sets)

The Descending Lunar Node—Nucleus Caudatus, tail (associated with amygdala, emotions, learning, comprehension of emotional content of language)

Mars—Amygdala (the red nucleus, involved in defensive fight and flight reactions)

Jupiter—Globus Pallidus (associated with limbic system, involved in higher order planning and execution of complex behavior)

Saturn—Putamen (involved in motor activity and modulation of input to other basal ganglia)

The subtle influences of the cosmic counterparts on behavioral tendencies can be predicted from natural cosmological and astrophysical cycles. The birth date, time, and geographic location for an individual or for an event establish the phase of the cycles, from which the various influences can be calculated. The birth event can be thought of as a precise summary of the innumerable factors of the local and nonlocal laws of nature that go into the calculation of individual events by the cosmic computer. An individual's birth-time reflects specific influences in the moment in local space-time that match the entire history of the universe and the particular individual. The genetic inheritance of an individual is the record of many past factors contributing to the individual differences and the general tendencies of an individual's life. This can be understood as the crystallization of many natural processes and cycles, computed within the context of karma, the universal law of cause-effect relationships throughout nature. Not only immediate proximate physical factors contribute to the genetic inheritance but also nonlocal factors.

In the expanded framework of Maharishi Jyotish genetic inheritance is an important means through which past karmic effects that will be influencing the individual's current life are concretized—via specific attributes of the gross physical body. This genetic inheritance establishes the context for many tendencies that are likely to manifest in the individual's life, including environmental circumstances, health tendencies, proclivities for work, types of relationships, and numerous other individual biobehavioral patterns that contribute to individual personality. As an obvious example an individual who is 7' 3" in height has a greater likelihood of being a professional basketball player than a professional race horse jockey. When the cosmic relationships are calculated precisely enough, the information can be used to predict the likelihood and character of upcoming events that are shaped by natural cyclic patterns. This information can be used to 'avert the danger that has not yet come'[9]—so to speak, analogous to putting the top up on a convertible to avoid '*being rained on*.' The information also can be used to apply technologies that enhance positive and mitigate negative influences. This is an important aspect of the related field of Vedic engineering in Maharishi Vedic Science and Technology called *Maharishi Yagya*.

Maharishi Yagya

The ancient Vedic science of *Yagya* reflects an especially subtle application of the interdependence of all levels of nature at their unmanifest source. The term yagya, containing 'ya' and 'gyan,' refers to 'that which is knowledge'. It is a system of behavioral formulas to utilize the laws of nature that govern specific subtle impulses of intelligence and energy to actualize positive life-supporting intentions. The word *yagya* also refers generally to any action that is spontaneously in accord with the totality of natural law.[12] These actions include procedures to activate laws of nature that support and fulfill the evolutionary intentions of the individual.

The *Maharishi Yagya* program involves the practice of TM and its advanced developmental technologies including the TM-Sidhi program to develop the ability to function from the level of the unified field of natural law in transcendental consciousness—in samadhi. That level is described as direct access to the cosmic computer that structures and computes all change in nature. It is the unity of the silent witness of pure consciousness (Purusha) and the totality of the impulses of energy and intelligence that manifest all phenomena (Prakriti). In a similar way that the information that shapes all aspects of a tree is contained in the seed of the tree, or all aspects of the physical body are contained in the DNA code, the total intelligence that structures the entire manifest cosmos is contained in its unmanifest source—Veda, Prakriti, Nature, the home of all the laws of nature. Vedic yagyas are said to access the underlying structure of natural law at the level of the Veda—Prakriti or Nature—from which all phenomenal effects in nature emerge.

Although much more subtle, the technology of Vedic yagyas can be compared to and contrasted with creating certain effects in an experimental laboratory based on the laws of chemistry. The Veda contains procedures to influence patterns of thought and behavior, much like mixing chemicals to produce compounds with specific properties—but inclusive of subtle nonlocal relationships. The procedures are precisely detailed in the Veda, and are carefully implemented by Vedic scientists—*Vedic pundits*—extensively trained in how to utilize subtle nonlocal connections in nature at the level of the field of all possibilities.

In addition to technologies such as the TM-Sidhi program that develop the ability to fulfill mental intentions based on the integration of individual attention and samadhi, Maharishi Yagya includes the power of Vedic sounds and related behavioral formulas and procedures to produce specific effects. Vedic pundits perform highly structured and sequenced procedures along with recitation of corresponding Vedic sounds—from their most settled level of awareness. These performances enliven the connection between the intention of the individual for whom the yagya is carried out and fundamental laws that produce effects to fulfill specific intentions.

Introducing a specific intention at the deepest unmanifest level of the unified field, the field of all possibilities, is said to produce specific effects on the expressed manifest levels, similar to the mechanics of the Yoga Sutras. Because the unified field is the simultaneity of infinity and point, it is both nonlocal *and* local. Local information can be transmitted to a nonlocal level of the unified field and back, because embedded in locality is nonlocality, and embedded in nonlocality is the infinitely correlated level of the unified field which contains wholeness at every point—the level of the cosmic computer. The unified field is the level at which the entirety of natural law resides in its perfectly integrated, infinitely correlated state. Intentions at that level are spontaneously computed based on the totality of natural law—which are impossible to compute through the limited computational systems of the human intellect. This provides built-in protection from fragmented, partial manipulations at grosser levels of nature. Specific intentions introduced at the underlying holistic level of nature produce effects based on the totality of natural law that are described as spontaneously integrating, evolutionary, and life-supporting. The unified field sustains everything in nature; it is eternally maintaining and wholly life-supporting.

This can be contrasted with the procedures used such as in genetic engineering of the DNA molecule. In the case of genetic engineering there is a dismantling of the intelligence of nature embedded in the DNA molecule. Manipulations introduced at this somewhat deeper microscopic physical level cannot take into account the totality of natural law—they are based on incomplete understanding of the innumerable laws of nature, restricted to the gross level of nature.

Even with good intentions the tearing apart of naturally ordered genetic structures inevitably produces significant unintended side effects that are fundamentally disintegrating. The DNA molecule, whether in plants or animals, was not built by the human intellect, and its integrated complexity is far beyond current understanding in modern science. It is based on the total integrated functioning of all the laws of nature. As Maharishi succinctly and profoundly has pointed out:

"Man himself is not man-made." [13]

In genetic engineering it is not that the laws of nature are violated in the sense of violating some obscure and seemingly arbitrary moral code. It is rather that manipulation of nature based only on partial understanding at a somewhat deeper level—completely ignoring the subtler and holistic self-interacting levels—inevitably has disintegrating effects. It is a product of the crudeness of the level of research—based on the physicalist worldview—that functions 'as if there were no deeper *reality* and no *subtler* consequences' (refer to Prologue). The entire genetic history of life on Earth is beginning to be altered via this fragmented technology, with consequences that cannot be even comprehended for years at this stage of evaluation of effects. It can be likened to the designing of the automobile that created increased freedom and efficiency of transportation, but had the unintended consequence of polluting the atmosphere—only much riskier due to the deeper and more powerful level of nature being dismantled. It is a well-intentioned, misfortunate application of fragmented science that could have much more unfortunate consequences than smog to human physiology, potentially a devastating effect to the essential nature of human life, reflected in the recent intellectual theme of the post-human era. [14] [15]

From the perspective of an outside observer who is not familiar with the subtler mechanics of nature, Maharishi Yagyas may appear to be merely elaborate rituals for purposes of religious devotion. Because of restricted cultural, religious as well as other paradigmatic perspectives, the technologies are not viewed in a scientific context. They are commonly misclassified as ritualistic practices based on faith, due to lack of knowledge of the holistic structure of natural law and of the subtle nonlocal interdependencies in nature. This is a general issue in reactions to many of the developmental technologies in Maharishi Vedic Science and Technology. It also has been due to some practitioners of these ancient Vedic technologies themselves having lost the holistic basis of the technologies through loss of direct experience of the totality of natural law in consciousness itself.

Although ancient Vedic science and Maharishi Vedic Science and Technology in some ways may appear on their surface to be similar to some religious traditions, they don't represent a faith-based system of knowledge typical of religion. In the Western intellectual tradition science is generally distinguished from religion in that its fundamental tenets are based on logical consistency and public empirical validation according to the ordinary waking state of consciousness. As demonstrated in the course of this book, Maharishi Vedic Science and Technology is consistent with these core scientific principles. However, it also recognizes that empirical validation of the most integrated universal knowledge requires development of the knower, the observer of nature. It extends beyond contemporary knowledge in modern science by developing the state of consciousness of the knower and ability to experience the full range of nature—gross, subtle, and transcendental—not just the ordinary gross sensory level. It contains systematic technologies to make use of the unified field by aligning with the totality of natural law within the simplest state of consciousness itself, which are empirically testable.

Beyond Intellectual Analysis into the Discrimination of Wholeness

This chapter introduced some of the Maharishi Vedic technologies in a rational scientific context that otherwise may be challenging to comprehend because of the highly advanced principles of holistic knowledge upon which they are based. These technologies can evoke even incredulous reactions among those who may not be informed about the latest progress in quantum gravity and unified field theories, and their practical implications. Chapters 23-24 will go even deeper into the Vedic understanding of holistic relationships between the structure of the human body and the cosmic structure of nature. It will address in more detail the relationship between the microcosm of the individual and the macrocosm of the entire cosmos—the complete integration of individuality and universality. How human physiology is a copy of the structure of the entire local and nonlocal cosmos will be described. In preparation for this discussion the principle of holism in the fabric of nature will be located in the structure of Vedic language and Vedic literature.

The final part of this book may be much more challenging intellectually and experientially for some readers. In tangible forms it introduces the most abstract unmanifest mechanics of nature that are beyond the world of the senses and that even transcend the tremendous imaginative capacity of the human mind. Our patterns of thought are so structured and conditioned by the experience of locality and individuality that the notion of our inner core and the entire cosmos as infinite and eternal—that the individual is cosmic—is beyond the ability of the intellectual mind to appreciate fully. To think that life is infinite, with no beginning and no end, that it already has been going on forever and will continue to do so forever, and that every part of existence is infinitely correlated with every other part, is simply beyond human comprehension and imagination.

Getting a sense of the unmanifest mechanics of nature described in this next section of the book can be especially challenging for the intellectual mind that has developed the strong habit of reductive thinking. It requires integration of both fine focus of discriminative attention and wide-angle awareness to encompass the simultaneity of contraction and expansion, collapsed and uncollapsed states, part and whole, point value and infinite value, individuality and universality, impersonal and personal, unmanifest and manifest, infinite silence and infinite dynamism. This simultaneity of the coexistence of opposites ultimately requires going beyond the logical intellect and the power of imagination to the ultimate underlying holistic nature of consciousness itself.

By transcending the mind, heart, and intellect entirely, the individual directly experiences the infinite eternal unity of the universal Self. Repeated experience of that unity allows the mind to incorporate more holistic conceptions of nature—integrating heart and mind in the intellect in order to 'discriminate' the unbounded wholeness of nature, rather than just its parts. In that indivisible self-referral wholeness of complete unity, eventually all the diverse parts of nature are discriminated sequentially as the mechanics of natural law within one's own consciousness, as the structure of the Veda.

The upcoming introduction to the unmanifest mechanics of natural law—how nature is expressed phenomenally into tangible form—points to Maharishi's remarkable contribution in describing these abstract processes simply and clearly. He has revealed underlying principles of dynamic order in nature that had been unrecognized at least for millennia, and he has made available systematic technologies—some of which have been summarized in this chapter—to verify them directly for the full enrichment of human life based on systematic, scientific developmental technologies.

Chapter 22 Notes

[1] Maharishi Mahesh Yogi (2005). Maharishi's Global News Conference, April 27.

[2] Nader, T. (2000). *Human physiology: Expression of Veda and Vedic Literature,* 4th Edition. Vlodrop, The Netherlands: Maharishi Vedic University.

[3] Geller, A. (2003). Smart growth: A Prescription for Livable Cities. *American Journal of Public Health,* Sep 2003, 93, 1410-1415.

[4] *Celebrating perfection in education: Dawn of Total Knowledge* (1997). India: Age of Enlightenment Publications (Printers).

[5] *Inaugurating Maharishi Vedic University.* (1996). India: Age of Enlightenment Publications (Printers).

[6] Travis, F. T., Bonshek, A., Butler, V., Rainforth, M., Alexander, C. N., & Lipman, J. (2005). Can a Building's Orientation Affect the Quality of Life of the People Within? Testing Principles of Maharishi Sthapatya Veda. *Journal of Social Behavior and Personality,* 17, 553-564.

[7] Sprawl May Harm Health, Study Finds (2004). *The Washington Post,* September 27, p. A.03.

[8] Maharishi Mahesh Yogi (1996). *Maharishi's Absolute Theory of Defence: Sovereignty in Invincibility.* India: Age of Enlightenment Publications (Printers).

[9] Maharishi Mahesh Yogi (2004). Maharishi's Global News Conference, September 8.

[10] *What shall man live by?* Ft. Worth, TX: Parker Chiropractric Research Foundation.

[11] Routt, T. J. (2005). *Quantum computing: The Vedic fabric of the digital universe.* Fairfield, IA: !st World Publishing.

[12] Maharishi Mahesh Yogi (1967). *Maharishi Mahesh Yogi on the Bhagavad-Gita: A new translation and commentary, chapters 1 to 6.* London: Penguin Books.

[13] *Inaugurating Maharishi Vedic University.* (1996). India: Age of Enlightenment Press (Printers).

[14] Maharishi Mahesh Yogi (2004). Maharishi's Global News Conference, September 22.

[15] Haney II, W. S. (2006). *Cyberculture, cyborgs and science fiction: Consciousness and the posthuman.* New York: Rodopi.

Part III

INTRODUCTION TO VEDA AND VEDIC LITERATURE: HERE IS INFINITY

This third and final part of the book introduces the self-referral structure of Veda and Vedic literature as the Language of Nature and the deep inner mechanics of the manifestation and evolution of the entire phenomenal creation. It identifies the infinite self-referral dynamics of natural law as the dynamics of consciousness itself. It describes how it is sequentially elaborated in the Language of Nature at all levels of nature—from the structure of the cosmos to the structure of individual human physiology. These chapters introduce the completely integrated and holistic scientific framework of ancient Vedic science revived in Maharishi Vedic Science and Technology. This represents understanding and experience of nature that is far beyond the limited boundaries in modern thought and modern science, and it includes systematic, empirically verifiable technologies to develop fully the cosmic status of each individual.

Chapter 23

Veda and Vedic Literature in Vedic Language and Human Physiology

In this chapter, we will consider the structure of Veda and Vedic Literature and their expression in human physiology Veda as total knowledge—the totality of knower, process of knowing, and known—appears through its infinitely self-interacting dynamics to express itself in sequential levels of manifest structure. The mechanics of this phenomenal manifestation are experienceable as the finest fabric of consciousness in the vibrations or sounds of Vedic language—the Language of Nature. Cosmic natural law is fully expressed in individual human physiology. This is the basis for the practical health technologies of Maharishi Vedic Healthcare. The overall main point of this chapter is that the totality of natural law, the Veda, is the fine fabric of individual and universal consciousness—the universal Self, the ultimate unified field—and is expressed in the structure and function of human physiology.

In modern science there has been a basic conceptual distinction between objective forms of energy and matter that express the laws of nature and subjective theories about them. This reflects the distinction of objectivity and subjectivity, the long-standing mind-body problem, and the belief that the natural world exists independent of our subjective understanding and experience of it. Fundamentally this distinction is a product of the fragmented and limited experiences characteristic of the ordinary waking state of consciousness. Hopefully becoming clearer based on planks laid down throughout this book is a holistic understanding that objectivity and subjectivity are directly connected to each other in the unified field of natural law. The laws of nature are neither constructions by the human intellect as theoretical approximations of what exists in nature, nor do they exist entirely independent of human subjectivity.

Ancient Vedic science describes in detail the phenomenal levels of nature in terms of a seamless continuity, in which the objective domain is underlain by the subjective domain, both underlain by and nothing other than the completely unified field. The unified basis of objectivity and subjectivity can be located as the source of our own individual self—the universal Self. The laws of nature structure subjectivity from inside us, as well as structuring the entire cosmos that appears in some states of consciousness to be outside us, but that also ultimately can be experienced as the fabric of our own universal consciousness. As described in detail in Maharishi Vedic Science and Technology, we can develop the ability through increased refinement, balance, and purity to live spontaneously the cosmic totality of nature in our individual experience.

Living in accord with the totality of natural law doesn't mean knowing all the innumerable laws of nature on the level of intellectual understanding. It means spontaneously getting their full support for maximum benefit in daily life—enjoying the 'fruit of all knowledge.'[1] In limited states of consciousness we have various degrees of ability to live the full value of natural law. But ultimately we *are* it—it is our essential nature. It is only within the boundaries of underdeveloped states of consciousness that we live and understand it only partially, and that we habitually view outer objectivity as independent of inner subjectivity.

In that we have free will, the degree to which we align with and live total natural law is determined by the choices and actions we take—given our level of development, our state of consciousness. We evolve to increasing attunement with the total potential of natural law because it is what we are determined to do, and it is what we want most. It is natural and inherent in us that we want to be all that we can be. That ultimately is the universal Self, Atma, the infinite, eternal totality of nature, fully awake in itself—Brahm. In the state of Brahm—unity consciousness—natural law automatically computes the fulfillment of impulses of intelligence and energy in the form of desire emerging from fully awake self-referral consciousness in accord with all the laws of nature. When the mind is free from distortions due to deep rooted stress and strain that limit its full potential, the unity of objective and subjective nature as fluctuations of our universal Self is the natural primary experience. Our thoughts and actions are spontaneously in accord with the full value of natural law—we gain mastery of natural law. Maharishi identifies that state as the birthright of every human being. [2]

> "Every man is born with the potential to become the master of Natural Law and fulfil all his desires. When the total potential of Natural Law is enlivened, no negative tendencies can arise, because the individual is always in alliance with that state of Natural Law which is at the basis of all positive and evolutionary trends in life. (p. 67) [3]

Veda is Total Knowledge

Appreciation of natural law as inherent in our own consciousness is crucial for understanding the term *Veda*. In Maharishi Vedic Science and Technology, it is in the completely unified, holistic understanding and experience of nature that the significance of Veda and Vedic literature—and its vibratory forms or sounds as Vedic language—is fully appreciated. When our individuality is fully awake to our universality, the mechanics of how the laws of nature structure or engineer creation are open to direct experience as the finest fabric of our own self-referral consciousness.

Maharishi emphasizes that Veda is not a humanly conceived theory, model, representational language, worldview, or perspective about nature. Veda refers to that ultimate totality that is self-existent, self-created, self-sustaining. It is the eternally existing source of nature itself, including our own nature. It is not created by the human mind—it is the uncreated source of all minds, the cosmic mind or *cosmic psyche*. It is the home of all the laws of nature, the unlimited unbounded field of total knowledge. In it resides the entire collection of abstract laws of nature or impulses of creative intelligence that govern and administer creation. As noted in prior chapters Maharishi sometimes refers to it as the *Constitution of the Universe*, or the *blueprint of nature*. Emphasizing its dynamic role in structuring creation, he also describes it as the *cosmic computer* that computes all change via the infinitely correlated, three-in one self-referral dynamics of Nature.

The Veda is Nature itself, the totality of infinite silence and infinite dynamism, infinity and point, wholeness and part, cosmic and individual, simultaneous and sequential—reverberating within itself. As the field of all possibilities the Veda contains the self-referral awareness of itself both in terms of wholeness and parts, point and infinity. At the level of the Veda the part is the whole and vice versa—infinite self-referral. Being the wholeness and the parts simultaneously, it appears to express itself into a sequence of parts, each part continually referring back to and containing within it the wholeness as it also expresses progressive limitations from the subtlest to the grossest levels of phenomenal creation.

> "Vedic' includes the whole path of knowledge from the knower to the known—the whole field of subjectivity, objectivity, and their relationship; the whole field of life, unmanifest and manifest; the whole field of 'Being' and 'Becoming'; the whole range of knowledge from its source to its goal—the eternal source, course, and goal of all knowledge." (p. 5) [1]

In ancient Vedic science, Veda is delineated in terms of its *mantra* and *brahmana* aspects:

"*Mantra Brahmanayor-Veda nama dheyam.*
(Apassamha Shrautasutram, 24.1.31)

Mantras are the structures of pure knowledge, the sounds of the Veda; Brahmanas are the internal dynamics of the structure of pure knowledge, the organizing power of the Mantras, the intelligence that structures the Mantras—the structuring dynamics of the mantras... Because Mantras and Brahmanas both together constitute the Veda, the word 'Vedic' is meaningful for both aspects of Veda—Mantra and Brahmana." (p. 3-4) [1]

The mantra aspect refers more to structure, and the brahmana aspect refers more to function. Function can be related to the science or theory of the Veda, and structure to the technology or practice of the Veda. Inherent in the Veda is all the orderliness or intelligence and the dynamism or energy through which natural law shapes and governs creation. This is represented in the dynamic sequence of sound and silence—syllable and gap—that sequentially expresses the Veda.

The unmanifest inner fabric of the Veda is the mechanics of manifestation into phenomenal nature. The manifestations of intelligence and energy, form and substance, function and structure, are vibrations of the Veda embedded in the unmanifest structure of our own pure consciousness. Eventually with increasing clarity of experience within the transcendental field of consciousness, all the mechanics of nature, the totality of natural law, can be experienced as the nature of one's own Self. This can be said to be both the subject and the object of the process of gaining knowledge in Maharishi Vedic Science and Technology.

Settling down the mind to its source in self-referral consciousness and attending to consciousness itself enlivens it fully in individual awareness. The knower is said to find inside himself or herself the totality of the process of knowing and the known. All knowledge emerges from that level of fully awake consciousness: "Knowledge is structured in consciousness." [4] Being or knowing fully that self-referral consciousness through direct experience, we have access to unbounded unlimited total knowledge, because everything that exists is contained in it, there is nothing other than it, and it is our own self-referral consciousness.

"*Kasminnu bhagavo vighate sarvam*
Idam vigyatam bhavatiti
...(Mundaka Upanishad, 1, 1.3)

Know That, by which nothing else remains to be known."(p. 85) [3]

In Chapter 10 the 'nature' of the unified field was described as completely undifferentiated, indivisible wholeness or unity, and also as the total potential of natural law that manifests all diversity in phenomenal creation. It is completely undifferentiated and at the same time contains all possible differentiations—completely inseparable, and also containing infinite specificity into separate objects in diverse creation. When differences appear to predominate, we talk of phenomenal manifest objects of experience that are independent of each other. In phenomenal manifest creation, there are levels of separation, the finer-grained subtler levels characterized by nonlocal interactions and interdependencies and the grossest levels characterized by local interactions and independence.

In this context the concept of *manifest* can be associated with phenomenal degrees of independence of objects, and the concept of *unmanifest* can be associated with infinite self-interaction in which every distinction or point value always is fully lively in the wholeness value—infinity in every point, infinite correlation. It is in this sense that we can talk about the unified field as being completely undifferentiated or unified and also infinitely diverse. It is an additional way to express the distinction of infinite silence—Purusha—and infinite dynamism—Prakriti. Phenomenal creation manifests within the unmanifest level of infinite dynamism that is the home of all the laws of nature, the Veda itself.

With respect to first experiencing the undifferentiated silence of transcendental consciousness and then unfolding dynamism within it, Maharishi uses the analogy of hearing the undifferentiated hum of a marketplace from a distance, and then distinguishing more detail—individual sounds, voices, and so on—as one approaches and can hear the fine fabric in the hum. Likewise, within the complete silence of transcendental consciousness the phenomenal dynamics of its finest fabrics eventually unfold—within the unmanifest flow, reverberation, or unified sound of the Veda—without changing the infinite silence at all. In this fully developed state it can be said that the entire relative creation is experienced as the surface fluctuations or vibrations of the eternally silent unchanging universal Self.

The first stage toward phenomenal expression is the unmanifest reverberation of *Shruti*. Shruti can be described as abstract impulses of pure intelligence that can be heard, which then precipitate further into the tangible expression of phenomenal forms that comprise the subtle wave-like and gross particle-like domains of nature. The cosmic principle of Shruti is associated with the cosmic principles of Smriti or memory, and *Puran* which means ancient, eternal. In terms of the three-in-one dynamics of natural law, these can be associated with the heard (Shruti), the process of hearing (Smriti), and the hearer (Puran)—all within the self-interacting dynamics of the Veda.

Remaining absolute, eternal, non-changing silence, the Veda appears to manifest into the relative ever-changing flow of nature. In that eternally non-changing silence, its ever-changing appearance in manifest form is said to be *phenomenal*. Its expressions are *real* as relative conditional *realities* that are most fundamentally the absolute unchanging eternal Reality. As described in Chapter 20 on full development in unity consciousness, eventually the infinite, eternal, non-changing wholeness of the Self is fully appreciated in the individual as the one Reality.

Vedic Language: The Language of Nature

In first discriminating and distinguishing the underlying mechanics of the Veda manifesting into phenomenal creation, it is referred to as the three-in-one self-interacting dynamics of the unified field—the knower, process of knowing, and known. These self-interacting dynamics can be described as vibrations between infinity and point at infinite frequency. In this process of phenomenal creation different frequencies of vibration become expressed as limitations of the infinite frequency. The different frequencies are the vibratory forms or sounds of the Veda contained within the unmanifest infinite frequency.

In ancient Vedic science the simultaneity and sequence of infinite vibration is reflected in the structure of the Veda as expressed in the sequence of Vedic sounds—the language of the Veda. *Vedic language* refers to the essential processes of engineering or mechanics of nature as they become phenomenally expressed in terms of sequential vibrations or sounds. It is how Nature expresses itself—it is the *Language of Nature*. The Veda, as the Language of Nature, is the uncreated story Nature tells itself of how phenomenal creation manifests.

"When the quiet state of consciousness functions within itself, it knows its own fabrics. These expressions of Natural Law in the unmanifest are the speech of Nature, the language of pure knowledge, the Veda... The story of the Veda is not spoken by anyone; it is the eternal speech of Nature at the basis of all creation... When Nature speaks, it speaks in the form of creation: from finer expressions of creation to the grossest expression of creation. This is the Language of Nature, which is called...*Shruti*—'that which is heard." (pp. 107-113) [3]

Vedic language as the Language of Nature is not a humanly-conceived language such as English, Arabic, Chinese, Latin, or Sanskrit (in its popularized form). It refers to the reverberations or vibrations that are the

> "In the beginning was the Word, and the Word was with God, and the Word was God."—*John 1:1, Holy Bible (p. 710)* [5]

essential mechanics of nature. Some sense of this meaning of language is contained, for example, in the concept of the language or alphabet used to identify dynamic structures of the DNA molecule. It is different from this meaning of language, however, in that Vedic language is not a culturally based mode of symbolic expression. It is not an arbitrary representational system typical of human languages, as well as typical of how natural processes are labeled in modern science—such as the meaningless identification of theoretical phenomena in quantum physics using terms such as *flavor*, *color*, or *charm*. Any humanly conceived language is not the Language of Nature in the ultimate sense of Vedic language. Although, for example, mathematics is sometimes thought of as the language of nature, [6] it is better conceived as a language of the intellect—like French, for example, is considered to be the language of romance. The differences between the Language of Nature and other languages will be discussed further shortly.

Four levels of speech. In ancient Vedic science and Maharishi Vedic Science and Technology four levels of speech are identified, corresponding to levels of phenomenal creation. [8] The level of speech expression that we are familiar with on the ordinary gross level, the most manifest level, is called *Baikhari*. The next subtler level, called *Madhyama*, refers to the value of speech embedded in the subtler level of thinking, the level of mind. Thinking can be considered internalized speech; verbal speech can be considered externalized thought. These are the only levels of speech people are familiar with in the ordinary waking state of consciousness. The most subtle manifest level of speech is called *Pashyanti*, associated with the finest level of the intellect, the level of intuition, the finest level of manifestation. The unmanifest, transcendental, infinitely self-referral level of speech is called *Para*, the level described as the Language of Nature, the field of Vedic language, on the level of pure Being. [7][8]

These levels from gross to subtle to transcendental reflect increasing degrees of interrelationship between the form of the object and the form of the language expression, the sound vibration. At the level of Para the interrelationship is infinitely self-interacting and self-referral. The Language of Nature as the transcendental unmanifest level of speech is characterized by infinite correlation.

Within the sound of our own voice is the subtle sound or vibratory form of our thought. Within the sound of our thought is the subtlest, finest, softest sound or vibratory form of inner speech. Within the sound of the subtlest form of inner speech is the unmanifest self-referral level of speech, the Para level of speech. That level of speech is the infinite vibration of the sound of dynamic silence, the transcendental field that contains all the vibratory forms that compose the entire phenomenal creation—total potential of natural law, the Language of Nature. This is the meaning of the statement in the Prologue that the overlooked knowledge is *within the sound of our own voice*. Total knowledge is within us; it is our essence, and it makes up all parts of us.

Name and form. In Vedic language the sound of the eternal transformation of singularity into diversity, heard in fully awake consciousness, is the Vedic sound—Shruti. This sound, the eternal infinite vibration or continuous hum of nature, evolves and transforms itself sequentially into the structure and flow of the Veda. It also constitutes the Vedic alphabet, the vowels, consonants, syllables, words, phrases, and sentences of Vedic language—based on the eternal memory of Smriti. Smriti is the unmanifest eternal record or memory of the laws of nature and Shruti is their unmanifest infinitely self-referral vibratory form that can be experienced or heard within consciousness itself.

As the unmanifest infinitely self-interacting level of self-referral consciousness—the Para level of speech—every point is infinitely correlated with every other point and the wholeness. In the process of manifestation these self-referral expressions become quantified into self-referral loops—the circular form of a *mandala*—of infinite frequency structured in terms of sequences of syllable and gap, flow and stop, sound and silence, point and infinity in the Vedic alphabet. The syllables—*Akshara*—relate to dynamism, and the gaps—*Sandhi*—relate to silence as the coexistence of opposites, the 'nature' of the unified field of consciousness.

These abstract self-interacting vibratory structures or forms in the unified field of consciousness are further expressed into impulses of energy and intelligence that make up all the subtle and gross phenomenal forms of relative physical creation in manifest time and space. From within the non-material, self-referral, transcendental Vedic sound, different frequencies of vibration that express energetic and material forms throughout phenomenal creation are generated.

"*Prakritim swamavashtabhya visrijami punah punah*,

Taking recourse to My own self-referral Nature, I create again and again." p. 356) [9]

The entire manifest creation is fluctuations of consciousness expressed in sound vibrations that concretize into corresponding phenomenal forms. In Vedic language the word that names an object is composed of the same vibratory structures as its phenomenal sensory form. In Vedic language sound and meaning are the same—name and form are equivalent. They are self-referral, infinitely correlated—one referring to the other and to the totality, and vice versa. It is due to this relationship that Vedic language as the Language of Nature most fundamentally cannot be translated, because the meaning changes when the sound changes. [3] This deep relationship of name and form can be experienced directly, however, when the individual consciousness-mind-body system is awake within itself to the Veda, the home of all the laws of nature—in fully awake *self-referral* consciousness. This is another way of describing the state referred to in prior chapters as *Ritam-bhara-pragya*.

In the process of manifestation the separation of name and form appears to become more prominent. The infinitely correlated self-referral nature of Vedic sounds appears to take on more limited independent expressions in relative vibratory forms and speech that reflect various degrees of the total value of speech. The three deeper levels of speech, and deeper relationships between speech sounds as vibratory patterns and their inherent meaning and effects, are generally not recognized in modern understanding of speech

Vedic language and humanly-conceived languages. Maharishi points out that the entire range of speech—manifest to unmanifest—can be experienced directly through the process of transcending, and also can be verified in the Vedic records. [3] He further makes the distinction between the Vedic records which are the Vedic language itself, the speech of Nature at the unmanifest level of Veda, and *Sanskrit*, the spoken human language that most closely approximates it:

"The Vedic Records verify these experiences, but one's awareness alone unfolds its own value from within itself... These authentic Vedic Records are of two kinds: the Vedic language itself, which is an eternal reality at the basis of creation and the speech of Nature in the Transcendent, and the human language that most nearly expresses the completeness and exactness of the Language of Nature. This human language is Sanskrit, which means 'purified speech'—that speech which purifies human speech so that it becomes the speech of Nature... The Vedic Expressions are the eternal, absolute, non-changing expressions of the mechanics of Nature. These expressions are translated for their maximum value in the Sanskrit language, whose breath is derived from the Vedic Grammar. The Vedic language is uncreated, while the Sanskrit language is a human creation." (pp. 109-110) [3]

The total value of natural law exists at every point and level of creation, including every expression of language, sometimes phenomenally expressed and sometimes not. Each impulse of energy and intelligence emerges from the total potential of natural law in the unified field of consciousness, and rises from that level of infinite silence into dynamic phenomenal expression in thought, speech, and action. Ultimately all expressions in nature are the expressions of the Language of Nature. In the unified state all the five senses are fully awake in consciousness perceiving its own structure and flow. When the total value—the wholeness value—becomes overshadowed and lost in the expression of limiting values, the expressions reflect only partial values of their power, meaning, and significance. Then speech can be said to contain less equivalence of name and form.

Different languages embody different degrees of the total potential of nature. The sounds used in the various languages become divorced from the meaning—name and form appear to become separated and lose their interrelationship as infinitely correlated. In some languages their relationship is arbitrary, because the subtle knowledge of the vibratory relationships of name and form are completely overlooked. The particular language then has less power to convey the full value of feeling and meaning. Written language also has this limitation, and typically conveys only a small fraction of the total potential of natural law embedded in every expression in nature. (Hence we end up with complicated books with many words to explain the most simple Reality.)

It is worth clarifying further the relationship of name and form in the Language of Nature, inasmuch as the general academic understanding seems to be in the direction of the view that: "[There can be no unique mapping between words and the world." (p. 233) [10] According to this relativistic understanding of language, there are no extrinsic, mind-independent objects in the world to which representational language refers. Objects are dependent on the conceptual scheme of description, through which we shape and 'create' the objects of experience. So to speak, there is no thing such as an 'object-in-itself.' While this is a useful advance in understanding over naïve realism, it considers the distinction between objective objects and subjective subjects—as well as the circularity of their independence and dependence—only in terms of individual mind or individual self, leaving out entirely the universal mind or universal Self. It thus is a partial understanding of natural laws and the nature of language and meaning that contributes to the common fragmented experience of nature.

Taking into consideration the simultaneity of individuality and universality—described at length in prior chapters—there can be said to be both mind-independent objects and mind-dependent objects of experience. The 'unique mapping' of words and the world is the natural relationship of name and form. This relationship is beyond any particular individual mind, and beyond any humanly-conceived and constructed language or system of semantics. The Language of Nature means that there is a natural relationship between the form of an object, the name that represents it, and the thoughts and feelings in the subjective mind. These are different levels of vibratory patterns that are associated with the same manifestation of phenomenal experience. At deeper levels of nature the vibratory pattern of the name of an object and the vibratory pattern of the object have direct correspondence, and ultimately are equivalent at the completely unified level of Shruti in the self-interacting dynamics of the unified field.

When development is not complete enough to experience directly and fully appreciate their vibratory equivalence, languages with arbitrary representations are constructed. In such humanly-conceived languages, the equivalence of name and form is lost to ordinary experience. The 'groundlessness' or relativistic nature of language, as well as all individual experience, is from the perspective of individual selves. [10] However, this individual 'groundlessness' has its ultimate grounding in the universal 'ground of Being,' the unified field of consciousness, the totality of natural law beyond all relative manifestations.

As the totality of natural law is expressed in degrees of differentiation and individuation, different values of natural law manifest as the phenomenal diversity of nature. These different values of natural law shape different cultures in different regions of the world. The language of a region develops based on the specific values of the local laws of nature as they are expressed in the particular land. Along with the different cultures, different languages emerge, and their natural holistic relationships of name and form become less evident and less accessible to ordinary phenomenal experience.

Maharishi points out that in a particular region or locality associated with its own cultural traditions, the local language—the *mother tongue*—reflects the local laws of nature through which the language is structured. [12] The local laws of nature also structure the specific values of the physiology of the people from the particular region or land.

All languages embody the Language of Nature at their basis, but express it in different degrees. For a particular region there is a somewhat closer relationship between the mother tongue or local language and the Language of Nature that is the universal basis of all languages. Thus teaching and learning in the mother tongue of the people of a region has a tendency to facilitate deeper values of knowledge. However, the Language of Nature as the universal basis of all local languages contains the universal value of language in any local region, wholeness in each part. [1]

Direct Cognition of the Veda by Vedic Rishis

Maharishi provides an interpretation of a verse in the Rk Veda (1.164.39) that summarizes key points in this and the next chapter:

"Richo Ak-kshare parame vyoman
Yasmin Deva adhi vishwe nisheduh
Yastanna Veda kim richa karishyati
Ya it tad vidus to ime samasate.

The verses of the Veda exist in the collapse of fullness... in the transcendental field, in which reside all the...(Devatas), the impulses of Creative Intelligence, the Laws of Nature, responsible for the whole manifest universe. He whose awareness is not open to this field, what can the verses accomplish for him? Those who know this level of reality are established in evenness, wholeness of life." (p. 18) [3]

From time to time along the long corridor of history, Vedic rishis developed the exceptionally high degree of refinement of body, mind, and inner wakefulness to cognize directly the verses of the Veda in the silent dynamism of their own self-referral consciousness—the most profound form of direct intuition, called *Vedic cognition*. In India long-standing oral traditions have been established to preserve these Vedic cognitions or direct expressions of the Language of Nature. Maharishi elaborates on the verse above to explain the process of Vedic cognition:

"The Seers of...Rk Veda cognized the Laws of Nature structured in their own consciousness. In the expression of cognition, consciousness reverberated in the values of sound and form simultaneously... *Richa,* the Hymns of the Veda, are structured in...*Ak-kshara,* the imperishable—in the transcendental value of...*Vyoman,* the absolute, unmanifest, unbounded field of immortality... The characteristic of this field of immortality is described by...*Rk Veda* as...*Yasmin Deva adhi vishwe nisheduh,* 'All the impulses of Creative Intelligence, the Laws of Nature, abide in the reverberating consciousness of the Seer, the knower.' The expressions of knowledge that the Seer cognizes are the expressions of his own consciousness... Complete knowledge of the entirety of creation and evolution is contained in each...*Richa.* Every Natural Law expressed in a cognition contains the wholeness of knowledge—knowledge of the finite and knowledge of the infinite. Cognition is the transcendental expression of all knowledge, unfolded within the knower and therefore known in its entirety...*Yastanna Veda kim richa karishyati,* 'He whose awareness is not open to the infinite, unbounded, immortal field of the Transcendent, what will the...*Richa* accomplish for him?' For knowledge to be lively it must be structured in consciousness... *Ya it tad vidus ta ime samasate,* 'He who knows it is evenly established in that wholeness of the total expression of Creative Intelligence." (pp. 132-136) [11]

When individual consciousness is fully awake to universal consciousness, the individual can experience and comprehend the specific mechanics of natural law in the flow of the verses of the Veda.

"The human nervous system is able to promote that level of consciousness which is active and silent at the same time. From this all-powerful level of thought and speech, the potential of action is infinite... That awareness which is both silence and dynamism is fully awake in itself. The...Rk Veda (5.44.14) describes this fully awakened state of consciousness...

Yo jagar tam richah kamayante,

'He who is awake, the Richas of the Veda seek him out'...
The...Richas are the Hymns of the Veda, the initial impulses of Natural Law..." (p. 129) [3]

Direct experience of the hymns of the Veda is through Shruti, the Vedic sound of total knowledge 'heard' as the self-interacting dynamics of self-referral consciousness. Physician and Vedic scientist Tony Nader explains:

"Cognition of the Veda is neither a direct sensory experience, nor an act of imagination, nor an intellectual treatise. Only a very clear, refined, fully awake individual who is firmly established in pure consciousness, and in whom total Natural Law is fully lively, can evoke the memory of Veda. Only a living, awake Veda can experience Veda." (p. 563) [8]

At this level of experience, also sometimes called *Jyotish-mati pragya,* there is full wakefulness of all the possible structures of total knowledge—Veda—within consciousness itself. Every Vedic cognition is lively in the self-interacting totality of natural law. However, it is also lively in a particular value of natural law. The particular value of the cognition is determined by the specific physiology of the cognizer—universal and individual values, point and infinity, at the same time.

All Vedic cognitions are fully lively in the same universal value, but also in the specific values of the physiology of the particular Vedic rishi. [12] These specific values or qualities of the cognition of specific Vedic verses relate to the various branches of the Veda—the *Shakhas.* The classification into Shakhas is associated with family traditions over thousands of years in India—the Land of the Veda—that help maintain the genetic ancestry of the specific physiologies. Different Vedic richas or verses are associated with different Vedic rishis or seers that have cognized that branch of the Veda. The family traditions maintain the specific physiology associated with the particular branch through genetic inheritance as well as through memorization and recitation in their respective oral traditions. Genetic inheritance is an important contribution to the preservation of a particular branch of the Veda and particular Vedic richas or verses.

The Uncreated Commentary of the Veda: Apaurusheya Bhashya

Maharishi has revealed that the structure of the Veda contains within it its own commentary, referred to in the phrase *Apaurusheya Bhashya* (uncreated commentary):

"Veda is its own commentary." (p. 52) [8]

He has cognized this structure in detail in the first Veda, Rk Veda. He also has shown how it is elaborated in the entire Vedic literature. The Apaurusheya Bhasya in Maharishi Vedic Science and Technology demonstrates how the sequential sounds of the Veda maintain their simultaneous self-referral nature by each sound incorporating the wholeness, elaborating the wholeness in terms of parts, and each part referring back to the wholeness in sequential manifestation and evolution. It brings out the self-referral nature of the unified field of consciousness itself as the inner fabric of the Language of Nature.

"Seen sequentially, the structure of Veda itself contains the mechanics of its unfoldment; it is the different values of the structuring dynamics of Veda that are displayed in different aspects of the Vedic Literature. The sequential unfoldment of Veda itself constitutes the process of creation and, at the same time, reveals the fundamental principles underlying the mechanics of evolution." (p. 53) [8]

"According to the Vedic tradition, the verses of the Rig Veda are structured to comprise a sequential progression of elaboration. Each part of the Rig Veda is an elaborated commentary on the section which immediately preceded it. What this means is that the knowledge contained in the Rig Veda is developed and elaborated through a nested series of sequentially larger or expanded units. Maharishi has referred to this unique structural characteristic as the "Apaurusheya Bhashya"... The phrase "Apaurusheya Bhashya" is simply a Sanskrit expression which translates as the "uncreated commentary" or "self-commentary." Rig Veda is described as a "self-commentary" because every segment of the Rig Veda is a commentary or elaboration of the part which preceded it. This commmentary is inherent within the actual verses and is not something created externally."—*Nirmal Pugh and Derek Pugh, Vedic Researchers (p. 20)* [12]

'A,' the first sound of nature. The uncreated commentary of the structure and flow of the Veda is contained in the sound of the first syllable of the Veda—the expression 'A'. This is the first sound in the first syllable (*Akshara*) in the first word (*Shabda*)—the word *Agnim*—in the first verse (*Richa*), in the first phrase (*Pada*) in the first paragraph or hymn (*Sukta*) of the first chapter (*Mandala*) of the first book of the Veda (*Rk Veda*). It is the sound of the entire universe in its unmanifest completely unified state, the entire Veda—a one-syllable expression of total knowledge. All the subsequent sequences of sounds and words in the language of Veda contain 'A' and are progressive self-referral elaborations of it. It is the sound of *Atma*, the totality, the unified field of consciousness, the universal Self.

Contained in the one syllable of 'A,' within the holistic sound of unity, is the engineering dynamics and infinite organizing power of consciousness that diversifies itself into different vowels and consonants of Vedic expressions. 'A' is contained in all sounds or vibrations in nature, and thus also all syllables of Vedic language. Maharishi details how 'A' expands into the sequential progression of sounds in the Rk Veda—analogous to the wholeness of a seed that develops into the whole tree:

> "All knowledge through the first letter of Rk Veda: [Vedic expression] (A);
> All knowledge through the first syllable of Rk Veda: [Vedic expression] (Ak);
> All knowledge through the first word of Rk Veda: [Vedic expression] Agnim;
> All knowledge through the first Richa (verse) of Rk Veda..."

> *Agnimile purohitam*
> *Yagyasya devan ritvijam,*
> *Hotaram ratn dhatamam.*

> All knowledge through the first Sukta (collection of 9 Richas) of Rk Veda;
> All knowledge through the first Mandala of Rk Veda (192 Suktas arranged in a Mandala—indestructible continuum);
> All knowledge through the ten Mandalas of Rk Veda..." (p. 126) [1]

This sequential elaboration also is reflected in the other three aspects of Veda—*Sama, Yajur,* and *Atharva Samhitas*—as well as the entirety of the Vedic literature, outlined later in this chapter. In unmanifest form the Veda is the *oneness* of the three-in-one self-interacting dynamics of consciousness itself, also called the *samhita* of rishi, devata, and chhandas—introduced in Chapter 11. The samhita or unity of rishi, devata, and chhandas is 'A,' the eternal flow of self-referral consciousness, the seat of total knowledge in our own self-referral consciousness. Our own consciousness is the coexistence of the two contradictory, opposite qualities of oneness or singularity of samhita—the experience of samadhi—and the trinity of rishi, devata, and chhandas—the three-in-one self-interacting dynamics of the unified field.

All the diverse aspects of Vedic literature emerge as the self-referral interactions of the three fundamental qualities of rishi, devata, and chhandas within the samhita—the trinity of knower, process of knowing, and known—a fundamental unifying theme throughout this book. As referred to in earlier chapters, the knower is associated with the term *rishi*, the process of knowing with the term *devata*, and the known with the term *chhandas*. They have a particular significance and role in each verse of the Rk Veda: *rishi* with the knower, experiencer, or seer of the verse; *devata* with the dynamic laws of nature the verse expresses; and *chhandas* with the structure, form, or metre of the verse.[1]

Maharishi further explains that it is due to the rishi quality—knower or intelligence—within the wholeness of samhita that it cognizes itself. The devata quality—dynamism or energy—within the samhita initiates the flow as the process of cognition. The chhandas quality—structure— within the samhita shapes the process or flow into metric form. The rise and fall—metric form—of this flow of infinite vibration or unmanifest frequency is the self-interacting dynamics of rishi, devata, and chhandas. Variations in the flow—chhandas—structure specific values that are the relationships between the laws of nature. It can be said that the fundamental metric of nature is the value of the metre in Vedic sounds, associated with chhandas, and heard in terms of the unmanifest reverberations within consciousness of Shruti.

'A' is a continuum of infinite frequency that contains all other sounds. As infinite and beyond all discrete values it is eternally silent—infinite silence. In infinite silence is the potential for all dynamic expression—infinite dynamism, the total potential of natural law. Maharishi explains in more depth:

"The absolute structure of pure knowledge is available in Rk Veda as a continuum of silence expressed by...(A), the first letter of Rk Veda. The dynamism latent in the silent nature of consciousness, the dynamism latent within...(A), is expressed by the syllable...(Ak)... Wholeness of...(A), being fully awake, eternally maintains the memory of its point values, and in this what we see is the dynamism of infinity collapsing onto its own point—infinite dynamism prevailing within infinite silence... This is how we see...(Ak) represents infinite eternal dynamism, and...(A) represents infinite silence... Thus we find the first letter of Rk Veda, (A), the expression of eternal silence, is the potential of eternal dynamism... Silence and dynamism, the two qualities constituting the nature of consciousness, are expressed by... (Ak).

'This expression of silence and dynamism in the nature of consciousness, in the nature of self-referral consciousness, the Self, as expressed by...(Ak), is further elaborated as... Aknim [also Agnim], the first word of Rk Veda, which is the seed of all the four Vedas—Rk, Sama, Yajur, Atharva.

...(Ak), the seed (first syllable) of Rk Veda:
...(N), the seed (first letter) of Sama Veda;
...(I), the seed (first letter) of Yajur-Veda;
...(M), the seed (first letter) of Atharva Veda;

'The sound...Aknim is the seed expression of total knowledge, the Veda. The expression of complete knowledge is so perfect, orderly, and sequential, because it expresses the silent, evolutionary nature of self-referral consciousness in terms of the four Vedas... The process of three Vedas emerging from Rk Veda is the process of triggering diversification from within the structure of unity... The relationship of unity and diversity within the nature of unity is self-generated, self-perpetuated, and self-sustained by the nature of unity itself... This is the reality of the Veda and Vedic Literature present within the self-referral field of consciousness... The entire evolutionary process, the entire mechanics of transformation present within the field of consciousness—the entire material universe present within the self-referral field of consciousness—the entire diversity lively in unity—this reality can be experienced and intellectually understood in the state of one's own self-referral consciousness—Transcendental Consciousness." (pp. 134-138) [1]

The eternal continuous flow of the Veda can be described as shaping itself into a whirlpool of flow and obstruction within the wholeness of samhita, within the infinite, unbounded, non-changing, silent ocean of consciousness that is eternally in self-interacting dynamic waves of motion within itself. This whirlpool of flow and obstruction produces a whirling sound of *rrrrrr*, the rolling sound within the name of *Rk*. The sound of the 'R' in the word *Rk* expresses the dynamism of infinity to a point. This reverberation of 'R' expresses dynamism, and the 'K' expresses the stop of dynamism—the point value. This reflects the collapse of dynamism to a point—increasing silence as dynamism decreases, to complete silence. The word *Rk* refers to the collapse of dynamism from infinity to a point. [1]

"Thus it is clear that Rk Veda is Veda of all possible transformations of the COLLAPSE of the dynamism aspect of the Ultimate Reality and also of all possible transformations of the COLLAPSE of the silent aspect of the Ultimate Reality. Rk presents dynamic silence...the knowledge of collapse of silence into dynamism and dynamism into silence... As 'silence' stands for the 'unmanifest', dynamic silence means dynamic unmanifest—it means that the unmanifest is dynamic. Rk Veda is the knowledge of the dynamism of the silent, unmanifest reality, the field of Transcendental Consciousness." (pp. 339-340) [1]

The collapse of dynamism into silence and silence into dynamism also can be described in terms of the mechanics of the four stages of the gap, introduced in prior chapters as being the fundamental mechanics of all transformation or change in nature. Here it is applied to transformation of consciousness itself into sound, or vibration. All expressions of sound manifest through the unmanifest mechanics of the four stages of the gap—the unmanifest silence that is the basis of speech, consciousness itself. An impulse of sound dissolves (Pradhwamsa-Abhava) into infinite silence, self-referral consciousness ('A') (Atyanta-Abhava), which is simultaneously infinite dynamism (Anyonya-Abhava), and then emerges (Prag-Abhava) into a new sound.

This is the mechanics of transformation between syllables and gaps—*Akshara* and *Sandhi*— that express the pure knowledge and infinite organizing power sequentially in the self-referral structure of the Veda. All expressions of speech are contained in the collapse of 'A' to 'K,' as well as all the mechanics of one sound transforming into the next sound in the sequence of syllable and gap, flow and stop.

Maharishi further points out that the dynamics of syllable and gap, flow and stop, in 'A' and 'K' also have their significance with respect to human phonology. Studies of Vedic language extensively describe the relationships between the unmanifest mechanics of natural law in the flow of the Vedic richas or verses, the structure of relative creation, the structure of human physiology, and the manifestation of Vedic sounds in their phonological expression through speech.

"The pronunciation of...(A) requires full opening of the mouth, indicating that...(A) is the total value of speech... (A) presents unbounded totality...(A) is the total potential of speech. Pronunciation of...(K) requires complete closing of the channels of speech (the throat)... Full opening followed by full closing displays the phenomenon of collapse of the unbounded field (of speech) to the point value (of speech)... Similar is the situation of Rk, the name of the Veda. The collapse of...(Ak) is the collapse of absolute silence, and the collapse of Rk is the collapse of absolute dynamism." (pp. 354-355) [1]

In the following quote Maharishi summarizes the profound significance of the uncreated commentary of Apaurusheya Bhashya:

"So far, whatever commentaries are available on Veda, all commentators have commented on the *Akshara*, or syllables; the *Shabda*, or words; the *Pada*, or phrases; the *Richa*, or verses; the *Sukta*, or hymns; etc. These commentaries do not bring to light the value and meaning of the gaps. The gaps actually contain the mechanics of transformation.

Creation is a phenomenon of constant transformation. Transformation or evolution is the reality of existence. The mechanics of transformation takes place in the unmanifest field; that is why when this field of transformation, within the reality of the gap, was not brought to light by the commentators, the whole field of pure knowledge and its infinite organizing power remained <u>out of sight</u>; insight into the mechanics of the sequential progression of pure knowledge and the significance of its structuring dynamics remained <u>out of sight</u>; Veda and its utility remained <u>out of sight</u>; Law, Natural Law, and its ordering intelligence remained <u>out of sight</u>; how creation emerged from Veda—how Veda structures itself into *Vishwa* (creation)—remained <u>out of sight</u>; the relationship of the unmanifest with the manifest, and how unmanifest consciousness, self-referral consciousness, Transcendental Consciousness, structures itself into the structure of Veda remained <u>out of sight</u>; how Veda is the whole universe remained <u>out of sight</u>; how the Atma—Transcendental

Consciousness—has the whole universe within it remained <u>out of sight</u>; how the part is the whole remained <u>out of sight</u>; how point is infinity remained <u>out of sight</u>; how mortality is essentially immortality remained <u>out of sight</u>; how mortality expresses immortality remained <u>out of sight</u>; how the infinite, unbounded nature of life can become the living reality of daily life remained <u>out of sight</u>. Total potential, freedom, and bliss were lost—ignorance and suffering became real.

Total Knowledge of Natural Law Available to Everyone

Now, with the cognition of the reality of the gap, all that was out of sight becomes a concrete vision. This is the time when full enlightenment is available to everyone, and now the total potential of Natural Law is at home with everyone. With Maharishi's Vedic Science and Technology, the Age of Enlightenment is available to everyone, everywhere. This unique cognition identifies the structuring dynamics of Veda to be the structuring dynamics of consciousness, of the physiology, and of the entire creation. It explains that total knowledge (the Samhita of Rishi, Devata, and Chhandas) and its infinite organizing power are completely contained, expressed, and demonstrated in the sequential unfoldment of the structure of Rk Veda. This orderly, sequential unfoldment of the structure of Rk Veda is available to anyone at any time, intellectually in Maharishi's *Apaurusheya Bhashya*, and experientially in one's own Transcendental Consciousness through Maharishi's Transcendental Meditation." (pp. 50-51) [8]

Sequential Elaboration of Rk Veda

The structure of Rk Veda revealed in the Apaurusheya Bhashya is unfolded in the sequential elaboration of the Rk Veda. In Maharishi Vedic Science and Technology this is organized as the 40 aspects of Veda and Vedic literature. [8] The most concentrated—and at the same time unbounded—expression of totality, the Vedic sound 'A,' is elaborated in the 192 verses of the samhita or totality of Rk Veda. The one samhita of Rk Veda also is elaborated into the sequence of the three other samhitas of Veda: *Sama, Yajur, Atharva*. The next level of sequential elaboration is into the 36 aspects of Vedic Literature. These include six sets of six aspects each:

Vedanga: Shiksha, Kalp, Vyakaran, Jyotish, Nirukt, Chhand;
Upanga (the Darshanas referenced in earlier chapters): *Nyaya, Vaisheshik, Sankhya, Vedant, Karma Mimansa, Yoga;*
Upa-veda: Gandharva Veda, Dhanur-Veda, Sthapatya Veda, Bhel Samhita, Harita Samhita, Kashyap Samhita;
Ayur-Veda: Vagbhatt Samhita, Sushrut Samhita, Charak Samhita, Bhava-Prakash Samhita, Sharngadhar Samhita, Madhav Nidan Samhita;
Pratishakhya: Rk Veda, Shaki-Yajur Veda, Atharva Veda, Sama Veda (Pushpa Sutram), Krishn-Yajur-Veda (Tattriya), Atharva Veda (Chaturadhyayi).

Each of these 40 aspects of Veda and Vedic Literature (4 of Vedas and 36 of Vedic Literature), while each containing the totality of natural law, unfold specific values or qualities of natural law and also elaborate the relationship between the specific value and the total value—knower, known, and process of knowing. A most amazing scientific finding is that these correspondences of Vedic literature and specific qualities of natural law have their direct correspondence in the major functional systems of human physiology. These correspondences are shown in Nader's book entitled *Human Physiology: Expression of Veda and Vedic Literature.*[8] As an example of the direct relationship of aspects of human physiology and Vedic Literature, the one-to-one correspondence between the cerebral cortex and the Yoga Sutras aspect of Vedic Literature is summarized in the following quote from Vedic researcher Craig Pearson:

> "The Yoga Sutras are divided into *four chapters.* The cortex is divided into *four lobes* (occipital, frontal, parietal, and temporal)... In the Yoga Sutras, there are 51 verses (sutras) in the first chapter, 55 in the second, 55 in the third, and 34 in the fourth—for a total of 195 verses. The cortex's many folds, or gyri, are connected by "association fibers." There are 51 association fibers in the occipital lobe, 55 in the frontal, 55 in the parietal, and 34 in the temporal—for a total of 195 sets... Each sutra in the Yoga Sutras can be matched with a corresponding association fiber according to its respective length and size. In Chapter 1, for example, sutras 1, 2, 3, 4, and 23 are short, and they correspond to short folds on the gyri of the brain. Sutras 14, 15, 24, 30, and 41 are long, and they correspond to long folds... The cerebral cortex also serves the same *function* in the brain that the Yoga Sutras serve in the Vedic literature. The cortex performs a *unifying* function. We are constantly bombarded by millions of sensory inputs; our physiology continuously performs millions of tasks to maintain balance and integrity and to integrate differences. This unifying function is managed by the cortex, in particular by the association fibers...which are woven together into a structure resembling a fine fabric... So to in the Vedic Literature, Maharishi explains. The Yoga Sutras serve to unify all experience and thus cultivate the highest level of human development, unity consciousness. Yoga represents the unified, unifying quality of pure self-referral consciousness, the experience of Atma—the Self... Thus we find precise and detailed correspondences between both the structure and the function of the cerebral cortex and the ancient Yoga Sutras." (p. 390) [13]

Maharishi Vedic Healthcare. In recent years Maharishi's research into the Apaurusheya Bhashya—the uncreated commentary of the sequence of the 40 aspects of the Veda and Vedic Literature—as well as their specific values of natural law has revealed more detailed relationships. This research demonstrates how the self-interacting dynamics of the unified field are expressed repeatedly in the structure of the different levels of nature, including the gross relative domain associated with the human body. These universal and specific values of Veda and Vedic Literature have a direct correspondence to functional structures in human physiology.

Specific knowledge of structure and function in the physiology of the human body in the fields of neuroscience, biology, and medicine has progressed to the point at which it now can be matched up with specific aspects of the Veda and Vedic literature. Each aspect of natural law in the Veda and Vedic literature has a direct correspondence to specific structures and functions in human physiology. Total natural law is expressed inside each of us, including in the structure of our own body. These correspondences constitute a tangible demonstration that each individual human being is cosmic.

The remarkable correspondence between the structure of the Veda and Vedic Literature and human physiology means that there is a direct relationship between parts of the physiology and the sounds, words, and verses of the Veda, the Language of Nature. This is the basis for the ultimate system of Vedic health technology that applies directly these fundamental relationships— *Maharishi Vedic Healthcare.* It is challenging to convey how profound this Vedic healthcare technology is, and how vastly more integrated and holistic it is compared to the fragmented approaches in modern medicine. Eventually this completely holistic technology will be recognized and applied—the sooner the better.

Recall that earlier in the chapter we discussed the unique feature of Vedic language that the name of an object and its sensory form are the same vibratory impulses at the most abstract, subtlest level of nature. At this most unified level each specific impulse also contains the totality of natural law. This not only means that each Vedic sound is lively in the total potential of natural law, but also that each Vedic sound relates to the impulses or laws of nature that structure the corresponding specific parts of human physiology. It further means that specific Vedic sounds can be used to attune its corresponding vibratory form expressed as a specific part of the physiology to the perfect orderliness of the total potential of natural law. They can be used to refer back, reconnect, and realign any disconnected part with the integrated wholeness to restore and promote healthy functioning of the consciousness-mind-body system.

As mentioned in the brief introduction to Maharishi Ayur Veda in Chapter 20, the fundamental cause of disease and disorder is lack of enlivenment of the connection of individual consciousness to the unified field, the field of infinite correlation and perfect order. In Maharishi Vedic Healthcare this reconnection or reenlivenment can be promoted directly through the specific Vedic sounds that relate to the aspect of physiology exhibiting disorderly, disintegrated functioning.

In this holistic technology Vedic sounds and verses are used to reorder disordered areas of physiology at their deepest level—through accessing the level of nature where individual point values and the universal infinite value simultaneously reverberate with each other. Symptoms are assessed, and the corresponding Vedic sounds associated with the aspects of physiology involved are identified and presented to the individual. The Vedic sounds resonate in the subtle level of physiology of the individual to re-order the sequence of vibrations in tune with the total potential of natural law to promote naturally healthy functioning. [8]

Maharishi has identified this knowledge of Veda and Vedic Literature in human physiology as the most advanced finding in the history of science, leading to a profound, elegant, efficient, and practical holistic health technology. In this completely holistic approach, "Modern science and ancient Vedic science discover the fabrics of immortality in the human physiology." (p. 58)[8]

Physician and Vedic scientist Tony Nader, who has been working directly with Maharishi to unfold the knowledge of the Veda and Vedic Literature in human physiology, describes the steps in unfolding this most advanced health technology:

> "The various functions of the branches of the Vedic Literature fit beautifully with the functions of different organs of the physiology, particularly the different parts of the nervous system. Furthermore, the dynamics of the relationships between different aspects within the Vedic Literature correspond to the intricate dynamics of the relationships between the different aspects of the physiology... Great confidence in the identity of the functional design of these two aspects of pure knowledge (intelligence and matter—Veda and physiology) emerged from the first phase of research... The next step was to compare the actual structures of the components that were discovered to be similar in function. It was a great revelation to discover that the structure of the 40 aspects of Veda and the Vedic Literature have an exact one-to-one correspondence with the structures of the 40 aspects of the physiology... For example, wherever there are five, ten, or thirteen chapters in a specific branch of the Vedic Literature, exactly five, ten, or thirteen divisions of the homologous organ or structure in the physiology were found. It was even more astounding to discover that the subdivisions of the chapters or books correspond to the subdivisions of the related structures in the physiology. In some cases it has been possible to find correspondence at even the fourth or fifth level of subdivision. The third phase led to greater insights into the details of the relationships of Veda and the Vedic Literature

and the physiology, as well as the Devata aspect... Research is continuing to determine all the levels of correlation. Since the number of components involved at the finer levels of consideration are in the millions, we feel that this discovery opens the door for an almost endless field of research. Such research will require the involvement of many scholars inspired to uncover the structuring dynamics of life and creation based on this new avenue of research." (pp. 561-562) [8]

The final chapter of this book will bring to light how this remarkable discovery of the Veda and Vedic literature in human physiology has been applied to unfold new levels of integrated knowledge. It summarizes research extending the self-interacting relationship of name and form in Vedic language and human physiology to expressions of the *devata* value of natural law. Vedic devatas represent the tangible expression of universal natural laws that administrator nature in its totality. This research is discussed in the context of the new modern scientific discipline of neurotheology.

Chapter 23 Notes

[1] *Maharishi Vedic University: Introduction* (1994). Holland: Maharishi Vedic University Press.

[2] Maharishi Mahesh Yogi (2003). Maharishi's Global News Conference, December 10.

[3] Maharishi Mahesh Yogi (1997). *Maharishi speaks to educators: Mastery over Natural Law,* Vol. 2. India: Age of Enlightenment Publications (Printers).

[4] *Maharishi International University Catalogue,* 1974/75. Los Angeles: MIU Press.

[5] *Holy Bible.* King James Version (1973). Nashville, Tennessee: Dove Bible Publishers. John 1:1.

[6] Penrose, R. (2005). *The road to reality: A complete guide to the laws of the universe.* New York: Alfred A. Knopf.

[7] *Inaugurating Maharishi Vedic University* (1996). India: Age of Enlightenment Publications.

[8] Nader, T. (2000). *Human physiology: Expression of Veda and Vedic Literature,* 4th Edition. Vlodrop, The Netherlands: Maharishi Vedic University.

[9] Maharishi Mahesh Yogi (1996). *Maharishi's Absolute Theory of Defence: Sovereignty in Invincibility.* India: Age of Enlightenment Publications (Printers).

[10] Varela, F. T., Thompson, E. T., & Rosch, E. (1993). *The embodied mind: Cognitive science and human experience.* Cambridge, MA: The MIT Press.

[11] Maharishi Mahesh Yogi (1997). *Maharishi speaks to educators,* Vol. 4. India: Age of Enlightenment Publications.

[12] Pugh, N. D. & Pugh, D. C. (1999). *Unveiling creation: Eight is the key.* Fairfield, IA: Sunstar Publishing, Ltd.

[13] Pearson, C. A. (2002). *The supreme awakening: Maharishi's model of higher states of consciousness applied to the experiences of individuals through history.* Doctoral dissertation, Maharishi University of Management, Fairfield, IA. Ann Arbor, MI: UMI Dissertation Services (Printers).

Chapter 24

The Cosmic Individual: Macrocosm in the Microcosm

In this final chapter, we will unfold more fully the deep correspondence between the cosmic structure of the laws of nature and the structure of individual human physiology. We will review how Vedic Devata, the laws of nature or cosmic administrators of natural law, correspond to functional structures in human physiology. This remarkable knowledge is introduced in relation to the new field in modern science of neurotheology, which studies the relationship between human physiology and religious and spiritual experience. A key issue we will address is couched in terms of whether 'man is made in the image of God,' or God is made in the image of man. The overall main point of this chapter is that on all levels of nature—including gross physiology—every individual human being is cosmic.

In the past few years the new field of neurotheology has emerged. This field utilizes advanced technologies such as in genetics and neuroscience for systematic investigation of the biological basis of religious or spiritual feelings and experiences. For example brain imaging techniques including fMRI, CT, SPECT, and PET have been employed to measure changes in blood flow, metabolism, and neurotransmitter activity correlated with reported religious and spiritual states.

One strategy of neurotheological research is exemplified by a recent study correlating specific gene activity with reported 'self-transcendence,' dubbed research on the 'God gene.' [1] This study used a self-report questionnaire in which respondents gave yes/no answers to questions such as whether they have intuitive insights and ESP, have experiences of being absent-minded and so engrossed in activities that time and surroundings are forgotten, have feelings of unboundedness and of connectedness to others and to the universe; have openness to phenomena not scientifically validated; and have helped animals and plants avoid extinction. Higher scores on the self-report questionnaire positively correlated with the presence of cytosine and low scores with adenine in the VMATZ (vesicular monoamine transporter) gene. The study indicates that reported religious and spiritual feelings correlate with specific patterns of genetic activity.

Another research strategy uses in vivo measures to investigate neural activity in the brain during reported religious and spiritual experiences. As an example of this type of research, a SPECT study of Tibetan Buddhists involved intravenous injection of cerebral blood flow tracers during baseline resting versus self-reported 'peak' experiences while engaging in their mental practices. Increases in cerebral blood flow in bilateral frontal cortices, cingulate gyrus, and thalami were noted, with decreases in the superior parietal lobes bilaterally. [2] This and related research demonstrates differences in brain activity associated with different types of mental practices and experiences.

Neurotheological research also has found correlations between verbal attributions of religiosity and transient electrical activity in the temporal lobes. These correlations have been reported to be exhibited in normal subjects such as during practice of meditative states, as well as in patients with temporal lobe disorders ranging from mild seizures to epilepsy. [3] The association of epileptic seizures and religious experiences has a long history in Western culture. Related research has demonstrated that exposure of the cerebral cortex to weak complex magnetic fields is associated with reported experiences interpreted by subjects to be of a religious character:

"Most people reported a "sensed presence" of a Sentient Being. The interpretation of the 'presence' was a function of: 1) the person's temporal lobe sensitivity that can be inferred by both electroencephalographic activity and self reports from questionnaires, 2) available cultural attributions of beliefs for the experience, and 3) the shape of the magnetic fields. The experiences were emotional and personally significant. The reconstructions of these ephemeral experiences as a component of autobiographical memory are strongly influenced by the explanation the person attached to the experiences within a few seconds of their occurrence." (p. 280) [4]

The finding of seizure-like activity in temporal lobe areas of the brain, however, has not been found in the extensive published research on correlations between transcendent experiences and electrophysiological patterns during TM practice. These brain wave patterns—such as frontal coherence, globally increased alpha and decreased beta, and increased task-appropriate CNV activity—appear to be quite distinct from the findings of temporal lobe disorders as well as other research using magnetic fields to induce reports of religious experiences. Importantly these findings on TM practitioners also are accompanied by an extensive body of research documenting positive affective, cognitive, behavioral and social behaviors indicative of improved psychological functioning and advanced development. [5]

In this fascinating new field of neurotheology fundamental differences in scientific and religious perspectives are attempting to be integrated into a consistent and empirically verifiable theoretical framework. A core issue relevant to this research in human physiology and religious experience is the nature of the causal direction of the correlations. It is another approach to the perennial mind-body problem and the related issues of whether the mind is just in the brain, whether the physical is all that exists, whether nonlocal connections in nature are experienceable and have causal directionality, and more generally whether there is a scientific basis for morality as expressed in religious injunctions. These issues can be couched in terms of whether man is made in the image of God, or God is made in the image of man.

Although numerous opinions and arguments have been put forth for both positions on this fundamental issue, the field of neurotheology doesn't necessarily make conclusions about the causal direction of the correlations. From the physicalist view still prominent in the mainstream of modern science, however, the evidence is interpreted as support for the conclusion that spiritual feelings and experiences are caused by neurobiological activity. The 'God gene' research, for example, is sometimes interpreted as support for the view that religious beliefs and feelings are determined by genetic make-up. [1] The research showing that reports of religious and spiritual feelings can be generated by electromagnetic stimulation of the brain or by psychoactive drugs is interpreted as even stronger evidence of a causal direction consistent with the physicalist view. [4]

These research findings are consistent with the principles of the one-to-one correspondence of mind and body, [6] as well as the related principle of structural coherence [7]—discussed in Chapters 13-14. However, a deeper understanding of levels of nature is needed to address the issue of the fundamental causal direction implied by the correlations relating neuroscientific findings and theology. The issue of causal direction will be referred to again toward the end of this chapter, after other research uncovering remarkable correspondences have been summarized that embody the issue much more profoundly.

Neurotheological research can be viewed as a small first step in the direction of exploring issues relevant to the relationship between the abstract laws of nature as forms of Vedic Devata and their corresponding structures in human physiology—the focus of this chapter. As discussed in the prior chapter Maharishi Vedic Healthcare represents a much more advanced approach to these issues that already provides a holistic system of practical applications. It recognizes and applies the knowledge that Nature expresses itself according to the same patterns, cycles, and laws at all levels of phenomenal manifestation. In this holistic view the macrocosm of the entire cosmos—*Vishwa*—is embodied in the *microcosm* of individual human physiology—*Sharir*.

The Underlying Unity of Modern Science, Religion, and Spirituality

Systematic knowledge of the cosmic nature of the individual summarized in this chapter is so integrative that it concerns nothing less than the fundamental unification of science and religious and spiritual experiences. It demonstrates such a tangible connection between ordinary human physiology and extraordinary cosmic levels of nature that it initially may be very challenging for many to accept. The connections stretch wide and deep the restricted worldviews in our disintegrated modern civilization. It may be quite challenging for those steeped in Western secular conceptions that assume gross physical creation is the only level of existence. It may be even more challenging for those with long-held culturally specific conceptions of God and what *man in the image of God* might mean. To establish planks for understanding this mind-expanding holistic understanding of nature, we first refer back to the major knowledge systems of religion and science briefly discussed in the Introduction, and now emphasize more strongly their underlying unity.

Modern science has provided systematic means of verification that have raised the standard for reliable knowledge, eliminated many mistaken beliefs, and provided numerous practical applications. On the other hand it is appropriate to appreciate the profound contribution of religion in providing practical guidance for many in how to live daily life with commitment, deep meaning, and moral values—issues modern science largely had concluded it cannot or should not address, in part based on belief in modern science that nature is fundamentally random and meaningless.

For centuries there has been an antagonism between religion and science, thought by some to be irreconcilable. This is directly related to the deep tear in the psychosocial fabric of society that has plagued modern civilization—briefly mentioned in the Prologue. However, this historical picture is on the verge of profound transformation based on the holistic understanding of nature, an important aspect of which is the development of unified field theory in modern science. Now religion and science can join hands to unfold a completely unified understanding of creation, supporting universal knowledge and practical experience of how to live in fulfillment individually and collectively that honors both knowledge systems. A universal knowledge system, a system of total knowledge, would honor and effectively incorporate the useful aspects of different knowledge approaches, including language and perspectives from religious traditions as well as modern science. At the same time it would provide practical means to unfold their common universal basis.

To develop universal knowledge it is necessary to get to *universality*. The division of science and religion is transcended in a completely holistic universal knowledge system—in a *mature* science. It is from the holistic, unified perspective that Maharishi Vedic Science and Technology is identifying deep correspondences between the structure of individual human physiology and the structure of cosmic devata. It is important that this not lead to confusion or misunderstanding about the fundamental scientific nature of Maharishi Vedic Science and Technology.

In order to consider the plausibility of these correspondences it is perhaps useful again to overview the direction in this book toward an expanded understanding of the means of gaining reliable knowledge, and the implications of these expanded means with respect to empirical validation of Vedic Devata as embodiments of natural law. These key planks of understanding will be helpful as we proceed to the extraordinary core content of this chapter that tangibly addresses the relationship between universal cosmic law and individual physiology.

As described in the first section of this book, the objective means of gaining knowledge in modern science has gone beyond its founding assumption of the independence of subject and object—via quantum and unified field theories and findings—to the underlying unity of all existence. Modern scientific theories are beginning to corroborate ancient religious and spiritual intuitions of the ultimate unity of nature in a unified field theory of everything. In unified field theory the limits of objective investigation are being reached. Further progress requires more integrated perspectives that incorporate mind and consciousness into the investigation, not only just theoretically but also empirically, or experientially. Chapters 7-17 brought out the necessity of addressing the subjectivity of the investigator—the knower—in building coherent theories of even physical *reality*.

Fortunately we have arrived at the time in which the impersonal objective experimental approach in modern science is turning back upon itself to incorporate a systematic, personal, subjective, experiential approach. This is bringing the attention of the scientific investigator full circle, referring back to the investigator's own subjective experience, and to systematic means for directly developing higher states of consciousness. The totality of investigator, process of investigating, and object of investigation is beginning to be considered as essential to a coherent investigation into the nature of nature. This holistic empirical approach is the specialty of the subjective means of gaining knowledge in ancient Vedic science, revived and fully integrated in Maharishi Vedic Science and Technology.

The subjective means of gaining knowledge involves direct experience of the ultimate source of all experience—the unified field. That ultimate source is understood to be both the home of all the laws of nature and the transcendental core of one's own individual self. Through repeated direct experience of consciousness itself, the mind and body of the investigator is purified of accumulated deep-seated stress associated with distorted incomplete understanding, and is systematically developed to function at more refined, integrated levels of experience. This naturally unfreezes human development and unfolds the inherent capability to validate empirically the full range of nature, including those 'out of sight' levels that link the ordinary gross sensory level to the source of all the laws of nature that govern phenomenal creation. With increasing refinement of mind and body through the direct experience of consciousness itself, the total potential of natural law is said to open to direct experience in the fine fabric of the investigator's own consciousness. When the individual is fully awake inside—when the black box is fully illumined—the unmanifest structuring dynamics of nature, the unmanifest laws of nature themselves, become available to direct empirical validation. This is the *devata* value of nature.

Universal principles of natural law express themselves repetitively throughout phenomenal creation. When our mind and body are functioning with a sufficient degree of coherence and consistency, the universal principles can be uncovered directly within our own consciousness. With the progress of modern science using its objective approach to identify universal natural laws, we can now corroborate their consistency with principles uncovered by the more direct and developmentally efficacious subjective approach of ancient Vedic science.

These universal principles are at the core of the major religious traditions as well—especially such principles as unity, duality, and trinity—expressed in various cultural styles of language and symbology. Diversity of the various traditions enriches our experience as long as the underlying unity is not missed and overlooked. All the diverse traditions have at their core the universal principles of natural law, and are useful paths to the ultimate unity. They can become antagonistic to each other if the underlying unity has been obscured by inability to go beyond the surface diversity, dominated by reduction into parts and loss of wholeness or unity—Pragya aparadha.

The goal of modern science is accurate knowledge of the universal laws of nature, and the goal of its technological applications can be said to be the betterment of life. Religion has corresponding goals, only in a different language—such as in terms of knowledge of *God's Law*, eternal union with *God*, or living in accord with the *Will of God*. Both science and religion seek reliable knowledge of the absolute, invariant principles of life and existence—natural law. Both generally apply the practical view that nature is orderly, and that knowledge and action applying that order can benefit human life.

Conflicts between religions have been due to different interpretations of natural law, due to different cultural traditions, and also based on partial understanding of the universal knowledge embedded in the core of all religious traditions. Modern science has attempted to avoid these conflicts by establishing an unbiased, 'objective' approach to validate theoretical knowledge of natural law. But it has not incorporated empirical validation through direct experience of the unified totality of natural law. At this stage of its development, the most fundamental understanding in modern science of the laws of nature is represented in relativity theory and quantum theory, attempting to be reconciled in unified field theory. A core principle of relativity theory is that there is no absolute background or frame of reference of space and time. A core principle of many interpretations of quantum theory is that nature is fundamentally random. These principles correspond in society to secularist beliefs in moral relativism and the meaninglessness and groundlessness of life. Importantly and fortunately, unified field theory is leading beyond these partial views and fragmenting beliefs to a deeper, more integrated understanding.

The distinction between religious and secular remains relatively strong in Western society, due to the success of modern science in accounting for aspects of the natural world that religion seemed unable to do. Its sociopolitical application in the principle of the separation of church and state also is more prominent in the West. This distinction is now frequently couched in terms of faith-based versus empirically-based approaches. This has its roots in dark historical periods of Western civilization—and Eastern civilization as well—in which religion as well as anti-religion were misused to persecute and subjugate different groups in society. In both cases, unfortunately, life has been lived for the most part on the basis of only the gross level of empirical experience.

Traditionally in Eastern societies there has not been a sharp distinction between church and state within a particular sociopolitical system, such as a national government. Governments tend to either directly associate with a particular religion, or ban all religions. Issues related to church versus state, and religion versus religion, directly relate to much of the uncivilized violent conflict in our present-day world. These issues are a consequence of lack of effective means to unfold direct empirical experience of the ultimate unity of nature and integrate mind and heart into more holistic knowledge and experience—whether through religion or secularized modern science.

In our modern civilization the scientific tradition has been the most widely accepted and revered approach to reliable knowledge—this is an Age of Science. It is in this context that scientific descriptions of nature can be understood to be most prominent in Maharishi Vedic Science and Technology. The scientific descriptions help avoid misinterpretations of ancient Vedic science that sometimes have led to its misclassification as another religion. More practically, Maharishi Vedic Science and Technology focuses on experience of the ultimate unity intellectually uncovered in modern science and historically uncovered through direct intuition in religion. It honors the validity of both approaches by providing *systematic means to experience* the underlying unity toward which both modern science and ancient religion aspire and have been slowly progressing. That underlying unity is the same, in whatever various ways it is conceptualized, named, and espoused in the lives of scientists or religionists.

As noted in the Prologue, in Maharishi Vedic Science and Technology the essence of spirituality is the direct experience of consciousness in its pure state—the unified field of consciousness. It can be described as a systematic spiritual science and technology that incorporates both the impersonal values of natural law in the perspective and language of modern science, while also honoring the more personal perspectives and corresponding language in religion. The wide diversity of approaches is honored, while also emphasizing that the various approaches are all ultimately embedded in and founded on the same universal principles of the fundamental laws of nature.

Maharishi has repeatedly emphasized the importance of personal development within one's own cultural tradition, which frequently includes personal religious histories and beliefs. If there is an ultimate unity of life, all paths would eventually lead to it, and it would encompass all paths in it, whether less or more direct—scientific, religious, and spiritual. These paths merge together as they encompass deeper, more comprehensive levels in approaching the ultimate goal. With empirical experience in higher states of consciousness, the nonlocal universality of total knowledge becomes fully lively in all of its local diversity.

Because India happens to be the land where the tradition of Vedic science has been maintained—the Land of the Veda—religious traditions emerging in this geographic region would be expected to have similarities with it. One major religious tradition with cultural proximity to ancient Vedic science is the popular religion in India called *Hinduism*. This term appears to be a product of the British influence in India over the past couple centuries, which reflects the imposition of Western conceptions of religion onto Indian society. Unlike many Western conceptions of religion, however, there is no system in the tradition of Vedic science of doctrinal conversion; and it is not a system of faith-based beliefs unable to be subjected to systematic empirical validation.

In Maharishi Vedic Science and Technology the focus is on effective systematic means to unfold total knowledge and directly validate it within one's own individual experience. This bridges the gap between knowledge and experience, intuitive belief or faith and its empirical validation. The focus is not on distinctions of impersonal secularism and personal religion. It emphasizes systematic development of the ability to validate universal knowledge already contained in one's own cultural, intellectual, or religious traditions. Simple, systematic, effective means to experience unity is the missing ingredient that fosters development of the full value of one's own cultural background—whether religious, secular scientific, neither, or both.

To be clear, this doesn't mean that one's particular religious or secular beliefs need to be abandoned. Those who are atheistic or agnostic don't need to start believing in God. Those who believe in the oneness of life don't also need to believe in a personal God. Those who believe that the path to God is through works, those who believe it is through faith, and those who believe it is through God's grace, don't need to change their beliefs. Those who believe, for example, that *Jesus Christ* is the only son of the living God and all must pass through *Christ* to get to *God* don't need to change these beliefs; and neither do those who believe in the name of Allah, Buddha, Yahweh, any other name of the God of all creation, the Godhead, or superstring or unified field theories.

It does mean that the deeper significance of these beliefs and clarity of the underlying laws of nature unfold as one grows in understanding and direct experience of unity. Our level of understanding of religion and of science is affected by the functioning of our mind and body. As we evolve to higher stages of development the meaning attributed to scientific or religious texts and concepts also naturally evolves, becoming more abstract, all-encompassing, and universal. There are systematic means to develop optimal functioning of mind and body that naturally support holistic understanding and experience of the ultimate unity permeating all of nature's diverse expressions. Eventually the differences are reconciled and fully appreciated in terms of the underlying unity, harmony, and perfect order of nature—the unified field of nature.

From the holistic perspective it can be said that ultimately there is a unified religious approach, in the sense of the universal knowledge in every religion—the trunk of all the diverse branches of religion—and that it is scientifically valid. Likewise there is one unified science, and it has full religious meaning and moral significance. Both religion and science essentially contain the authority of total natural law. The different perspectives of religion and science find their fulfillment in pure spirituality—direct experience of total natural law, full development of the universal value of the knower. Any approach leads to the ultimate knowledge when the mind is allowed to unfold its holistic basis—the unified field of natural law, the Will of God, or however one wants to label It. It is in the direct experience of unity that deeper understanding unfolds, and it is with increasing purity of functioning of mind and body that it is permanently stabilized.

As to how long it takes for individuals and society to become established in the harmony and perfect order of the totality of natural law, Maharishi explains:

> "As long as it takes [the] Muslim world to follow Mohammed, and as long as it takes the Christian world to follow Christ, and as long as it takes the Buddhist world to follow Buddha...and as long as it takes the Vedic world to follow the Veda. All that...Mohammed has said, all that Christ meant...Buddha the same, as long as it takes...religious people to follow their ideals, so long it will take for the world to be heaven on earth. As long as it takes a scientist to follow the science, which is a systematic investigation into the reality—and systematic investigation into the reality, because reality is everywhere omnipresent, it should not take long. But if it takes long, they are misguided people, they are roaming in darkness, they have lost their path... If any religious man is suffering...he is not following his religion, because following his religion means he is breathing that reality which is bliss... The whole thing is a muddle of reality. Reality is so simple... Simpler than simplest is to dive within oneself." [8]

Unifying the impersonal and the personal. At some point in the development of reliable and personally fulfilling scientific knowledge, the impersonal, third-person, experimental, objective approach needs to expand deeper and become more comprehensive by referring back to personal, first-person, subjective experience. The impersonal objective approach in modern science that views the world as outside and independent of one's own subjectivity—the known—needs to be integrated with the personal inside approach—the knower—historically more associated with religion and spirituality. More broadly the universal and the individual need to be integrated, the formless and the form, unmanifest and manifest, need to be recognized and experienced in their ultimate unity. Wholeness needs to be experienced in every part, infinity in every point. In the research in Maharishi Vedic Science and Technology on the devata or processing of knowing value of nature, these two major aspects of natural law are tangibly integrated.

The discovery introduced in Chapter 23 that the detailed structure of Veda and Vedic Literature matches neural, DNA, and cellular structures in human physiology—expressions of *chhandas*, the known—has led to further investigation into the *devata* aspect, the organizing power of nature, the process of knowing. Devata is a Vedic word referring to the laws that maintain the perfectly orchestrated flow of nature, the administrators responsible for initiating and maintaining the evolutionary direction of all phenomenal change, the structuring dynamics of nature that phenomenally express pure intelligence (knower or *rishi* value) in specific forms and evolutionary flow (*chhandas* value). Not only the correspondence of the chhandas (structure, flow, or metre) of Veda and Vedic Literature with human physiology is being uncovered, but also the correspondence of the devata (laws or administering intelligence) with human physiology.

"The profound insights into ancient Vedic Literature, brought to light by His Holiness Maharishi Mahesh Yogi over the past 40 years, have guided the discovery that the laws that construct the human mind and body are the same as those that give structure to the syllables, verses, chapters, and books of the Vedic Literature, and to the administering intelligence of Natural Law described in the Vedic Literature as Vedic Devatas... The human physiology (including the DNA at its core) has the same structure and function as the holistic, self sufficient, self-referral reality expressed in Rk Veda. The specialized components, organs, and organ systems of the human physiology, including all the various parts of the nervous system, match the 40 branches of Veda and the Vedic Literature one-to-one, both in structure and function. This structure and function of Natural Law in its 40 aspects is also available in every cell of the body and in the DNA of every cell. Various groups of components of the human body have also been found to correspond in structure and function to the administering intelligence of Natural Law described in the Vedic Literature as Vedic Devata... Study of physiology in terms of the structure of Veda is that revelation of our scientific age which raises the individual dignity of human beings to the cosmic dignity of the universe." (p. VIII) [9]

As described in the Physics Unbound part of this book, the unified field is the field of all possibilities. It can be described as the omnipresent, omnipotent, omniscient level of nature that is the basis for all finite expressions of energy and intelligence. The unified field itself can *be, know, and do anything by itself*, within itself. It is self-referral, self-interacting, self-sufficient—the source of all intelligent, dynamic, self-directed behavior. It is conscious of itself—consciousness itself—and it is the source of life—life itself. It doesn't have to take in anything or do anything other than be itself in order to be conscious and alive. It is infinite eternal Being; it is its own self-referral physiology, and contained within its fine fabric is the specific substances, forms, and physiologies of all manifest objects and beings in phenomenal creation.

The cycles that constitute the phenomenal phases of living and non-living existence are fluctuations of that level of pure Being—the unified field of consciousness. According to its phenomenal structure, its grossest levels appear to be inert particles or atoms such as that compose water and earth which reflect the least its inherent self-interacting dynamism and self-referral consciousness. On the gross level of nature the inherent intelligence and energy of nature appears to be the least integrated—consciousness and matter appear to be the most separate.

Gross matter coheres into biological organismic selves that negentropically sustain their own individual lives by virtue of their connection with the subtle level of nature that coheres and guides it. At the subtler level energy and intelligence are naturally more integrated. This level has more the character of mental *reality* and thought forms than gross physical *reality* and material forms. Even in ordinary behavior in gross space and time it is the subtle mind that initiates and directs the gross body to move, carried out through subtle, nonlocal field effects that are automatically expressed in the gross relative domain via classical mechanics of physical motion.

The mind also initiates and directs movement of subtle forms, but subtle motion can be understood to be through impulses of thought directly, rather than via classical transduction mechanisms associated with the motion of material particles and expressed in neural activity in the gross body. Mind and body interact with each other and reciprocally influence each other. However, as the deepest levels of mind and body become highly developed, the super-ordinate control of mind over matter becomes increasingly evident.

The subtler mental level involves less separation between the sensory form of an object of experience and the impulse of thought. Their fundamental relationship as the same vibratory form is more prominent. The vibratory equivalence of name and form—thought and sensory form—that is characteristic of Vedic language is embodied more fully on the more interconnected subtle levels of nature compared to the more discrete, local, independent gross levels of nature .

This doesn't mean, however, that objects in the subtle relative domain of nature are less precise or less individualistic. They reflect increased refinement, subtlety, and precise locality while at the same time they are nonlocal. They are more reflective of the coexistence of point value and infinite value. We could also say they are simultaneously more personal and impersonal, individual and cosmic. At the infinitely self-referral transcendental level of nature, personal and impersonal are equivalent—the part *is* the whole, point *is* infinity, individuality *is* universality, oneness *is* Oneness. As stated in the Prologue, 'the point is the whole, and the whole is the point.'

On the subtle level of thought and feeling not bound to gross physical limitations, all the senses are involved and can sense both the subtle and the gross levels of nature. However, subtle levels are—practically speaking—opaque and sometimes even completely unavailable to gross sensory functioning. But it is not that the subtle aspects of bodily form cannot influence the gross levels of nature. There is a relatively continuous, seamless causal connection between gross and subtle levels of nature, and ultimately the transcendental level of nature—the top-down superordinate control or hierarchy in nature. Gross is embedded in subtle, and subtle is embedded in the transcendental, unbounded, infinitely correlated unified field. Whatever can be done on the gross level can also be done on the subtle and transcendental levels—including influencing the gross level. But not completely vice versa, because more expressed levels involve more limitation, as discussed in Chapter 11. On the subtle level transduction processes are so integrated and efficient that the flow of energy that initiates and directs motion can be undetectable from the vantage point of the gross level.

There is a rich and vast variety of subtle phenomena to be experienced that haven't been envisioned within the fragmented, highly limited framework of objective modern science and materialistic modern life. But none of them approach the enrichment and fulfillment that naturally unfolds with direct experience of the simultaneous totality of natural law in unity consciousness. That ultimate unity is said to be *infinitely fulfilling*.

As described in the past chapter, the laws of nature reside in the unmanifest level of Nature (Prakriti)—the Veda—that underlies and permeates all subtle and gross phenomenal expressions. The laws of nature are the abstract impulses of creative intelligence that are the structuring dynamics of nature. They constitute the administering intelligence that maintains the perfect orderliness of phenomenal creation—the governing Constitution of the Universe—unmanifest with respect to subtle and gross domains.

In ancient Vedic science and Vedic language, the laws of nature—*devata*—can be understood to embody both impersonal and personal senses of the notion of administrators of orderly patterns and cycles throughout creation. They are said to be unmanifest in the sense that their 'individuality' or specific qualities are fully lively in the infinitely self-interacting, self-referral dynamics of the totality of natural law, the universal, transcendental field of Nature or Prakriti underlying local and nonlocal relative manifest creation.

As *one* evolves through higher states of consciousness to *One*, universal values of life become more apparent, accessible, and lively in daily life. More and more, individual life embodies universal values and more perfect order in nature, and ultimately there is complete enlivenment of universal cosmic life in individual life. Increasing refinement of body and mind unlocks and enlivens deeper experience of the full range of creation, local, nonlocal, and transcendental; the '*doors of perception*' naturally open wide, the '*window to the soul*' eventually becomes completely transparent, and the universal value of life fully shines through all phenomenal experience.

As perception expands with this natural development, the inherent ability unfolds to perceive the subtle relative domain of nature that permeates the gross relative domain. As alluded to in Chapter 19 on higher states of consciousness, these subtle levels are said to include what are sometimes called celestial levels of life. According to the Veda, as the individual human nervous system develops the subtlety and flexibility to experience subtler phenomena in nature and the individual becomes fully awake inside, the laws of nature "seek out that individual." (p. 129) [iv]

Throughout recorded history inspired religious and spiritual artistic expressions have been replete with visual and verbal imagery attempting to depict celestial life. A rich celestial hierarchy also is described in many religious texts, some of which contain considerable detail about various levels of heaven and celestial beings with specific roles as attendants of *God*. Reports of such visionary experiences have been more frequent among individuals who have devoted their lives to religious and spiritual development. Anecdotal reports by some of these individuals suggest that the experiences arise spontaneously with increasing purity of mind and body. Based on growing acknowledgement in recent years of their potential relevance to models of higher human development, research such as in neurotheology that examines phenomenal reports in the direction of these sometimes rare and elusive experiences are appearing in reputable scientific journals.

Different geographical regions with their local laws and patterns of nature influence cultural patterns—local languages, lifestyles, music, art, social systems, and so on. In the same manner, religious depictions of God, heaven, and celestial life also frequently have culturally specific features. Religious imagery—whether imagined or directly observed—tends to reflect local and regional cultural values. Frequently what is imagined or observed is interpreted in the context of one's own cherished cultural background and history of what is comfortable, dear, and valued. Beauty—and *reality*—are in the eye of the beholder. This means that although a universal principle of nature applies universally and impersonally, it is reasonable to expect that the specific way in which it is understood, perceived, described, and depicted would reflect specific personally significant values and culturally specific qualitative features. This is an important principle with respect to the study of Vedic devata as laws of nature.

Vedic Devata as the Embodiments of Natural Law

In ancient Vedic science the concept of devata associated with the unmanifest level of the laws of nature integrates name and form, individual and universal, personal and impersonal, part and whole, infinite value and point value. In the same way that at the level of pure transcendental consciousness the individual value of consciousness is simultaneous with the universal value of consciousness, at the level of Vedic devata the universal value of the laws of nature are embodied in individual natural laws or collections of natural laws. The abstract, universal, self-interacting unity of natural law is embodied in individual laws of nature that govern a specific range of the phenomenal expression of the totality of nature. This is the same self-referral quality described in the past chapter on Vedic language, where each specific expression contains the totality.

This can be likened to the administration of a national or state government, in which each office is connected to and represents the entire government but also at the same time is responsible for a specific area of administration. In its jurisdiction of a specific area of the government, the office also is responsible for upholding all the laws that make up the entire government.

A further analogy can be drawn between the office of administration and the individual officer or administrator who heads the office. The governmental officer also simultaneously represents the office and the entire government, overseeing specific actions of the government in accordance with the entire set of laws of its constitution. The officer administers a particular set of tasks defined by the organizational laws or rules, which are consistent with the constitution of the whole organization.

Similarly the administering intelligence associated with specific laws of nature can be viewed as particular jurisdictions embedded in and consistent with the total potential of natural law—the Constitution of the Universe. It reflects a specific aspect of natural law, and it also can be understood in terms of the embodiment of a specific individual administrative officer who holds the office. In some cultural and religious traditions, the impersonal aspect of Oneness or the totality of Nature is emphasized; in others a more personal representation is emphasized. Such representations are depicted using different imagery and symbology that can help the very abstract principles of nature become more tangible, concrete, and personally meaningful to the populace.

In ancient Vedic science laws of nature are depicted in all of these ways. The more abstract, impersonal sense of devata as a universal law of nature also can be depicted more tangibly and personally in terms of its embodiment as an administrative officer of a specific law or collection of laws of nature. The devata value can be represented as a specific law within the totality of natural law, and further as a specific officer of the totality of natural law primarily responsible for a defined range of natural law.

Maharishi Vedic Science and Technology emphasizes the scientific nature of the devata value, in the form of the laws of nature, the impulses of creative intelligence that administer phenomenal creation. These impulses of creative intelligence are specific vibratory patterns of natural law that contain both name and form. The name relates to specific sounds and syllables in Vedic language, and the form can be depicted in terms of a specific individual embodiment.

The highest administrator of a government is frequently called the president—or sometimes the king—who embodies and presides over the entire government or kingdom and entire constitution of governmental laws. With respect to the government of Nature and the Constitution of the Universe, this unitary embodiment, conceptualized in various forms in various religions according to cultural context, is frequently associated with some conception or image of Go*d*. The form or image of God—as the omnipresent, omnipotent, omniscient totality of creation, the creator and administrator of the entire creation—is frequently depicted with personal and culturally significant qualities in the various traditions.

Modern science focuses on the impersonal secular concept of totality as the unified field of nature—the source of everything. Perhaps the closest religious concept is of the Godhead. More personally, the unified field of natural law also can be understood as pure Being, the universal Self, and even more personally as the one infinite, eternal God that creates an infinite variety of manifestations. All and each are nothing other than *God* in ultimate wholeness and infinite oneness, but also reflecting specific qualities and attributes. This recognizes the omnipotence, omniscience, and omnipresence of God, whether understood personally or impersonally.

But if God, or the Godhead, or the unified field, is omnipresent, then that omnipresence is everywhere, including permeating you and me. Recognizing the cosmic nature of a specific individual human being doesn't mean that 'man' is God, but it does mean that each individual is necessarily directly connected to the totality of nature that can be called God, or the One, or Infinity. The omnipresent totality of all things necessarily encompass the totality of 'man'—inside as well as outside. That direct connection is said to be naturally and spontaneously experienced in higher states of consciousness, when the individual self settles down and opens to the universal Self in the most refined inner awareness.

The more impersonal secular approach to understanding the unified field (of the Godhead) is a valid perspective; and the more personal religious approach in terms of God also is valid. If there is a unified field of nature, then in this broad sense there is also God. It is important to appreciate and respect that both science and religion approach the same universal understanding of natural law from different traditional perspectives. Again, there is only one completely unified field.

The simultaneous totality of all things, the unified field of natural law, or the Godhead, or God, would have no restrictions in the ability to be, know, sense, and do anything. 'Man' has the ability to experience the totality of God, in that man's essence must be God if God is everything—but limited to the channel of an individual human body. The individual human being can experience totality on the level of the infinite unbounded wholeness of consciousness itself. But with respect to finite individuality, the individual is limited to the experience of one thing at a time. As described in Chapter 18 in the section on individual attention, the human nervous system and body individualize experience into a single integrated array—the channel or spotlight of attention. Omniscient God, the unified field, or whatever we wish to call It, would have no such restriction, and would be and know all individual and universal expressions of creation simultaneously.

In the increasing unfoldment of unity into diversity in phenomenal manifest creation, it can be said that omnipresence, omnipotence, and omniscience are expressed in various specific degrees of power. The specific expressions are structured in a hierarchy of near infinite totality of natural law to lesser and lesser phenomenal expressions of totality. One aspect of understanding this is in terms of the spontaneous symmetry breaking of the unified field into the fundamental quantum particle and force fields, and their localized expression in various lawful phenomena throughout the levels of nature.

In religious traditions generally these are depicted in terms of various levels of God's creation, and are embodied in various levels of attendants of God as the administrators of the different aspects of God's Kingdom. These can be associated with specific laws of nature as well as levels or dominions in the natural world. The subtler levels of nature are sometimes depicted in terms of various levels of gods, angels, or other celestial attendants of the one God or Godhead according to religious and cultural traditions, in whose likeness 'man' is said to have been imaged. For example the first book in the Holy Bible, Genesis, has the interesting and relevant passage:

"And God said, "Let us make man in our image, after our likeness..." (p. 9) [11]

Vedic Devata in Human Physiology

Maharishi brings to light that Vedic Devata is another term for the various aspects of natural law that organize the entire universe and maintain it in perfect order. Different terms and languages reflect different ways to conceptualize the principles and phenomenal structures of natural law. Vedic Devata are laws of nature, or collections of laws of nature, with specific administrative functions that provide for the creation, maintenance, and dissolution of the entire universe. They are the creative powers of cosmic dimension, permeating the whole creation. All the devata, all the impulses of creative intelligence of natural law, are present in every point of creation.

"The Vedic Devata are the administrators of every aspect of creation. In the same way that the law of gravity exercises its 'rulership' over the attraction between different masses, the Vedic Devata exercise their administrative role over every form and function. For example, the Devata that administers silence is called Shiva, and the Devata that administers dynamism is Vishnu.

"The Vedic Devata are not separate from the ultimate reality of the Self—Atma—the Unified Field. Every point in creation, manifest or unmanifest, animate or inanimate, microscopic or cosmic in dimension, in its ultimate reality is nothing but the Unified Field. Every individual Law of Nature or any collection of the Laws of Nature, acting under any circumstances within time and space, is nothing but the dynamics of the Unified Field. Every ant or elephant, every bird or tree, every human being...and the Devata themselves, ultimately are the Unified Field... [T]he Devata are integral aspects of our own human physiology. They are embodied in every human being with the same forms and functions described in the Vedic Literature. The Devata are present in the physiology of everyone, no matter what one's race, belief system, or religion, no matter which political party one associates with, or in which geographical area one is born... The Devata, therefore, are presented here not as religious, philosophical, or poetic concepts, but as a scientific reality." (pp. 333-334) [9]

As discussed in Chapter 11 the 'nature' of the indivisible indescribable unified field can be described in terms of the coexistence of the opposite qualities of infinite silence and infinite dynamism, and also the concepts of unity and diversity. In ancient Vedic science these concepts are associated with Vedic names representing universal laws and principles of nature. Sometimes infinite silence is associated with the devata name of *Shiva*, and infinite dynamism is associated with the devata name of *Vishnu*. [12] The approach is completely monistic, or non-dual, while also recognizing the simultaneous phenomenal unity and diversity of nature in its unified totality.

"Reality, therefore, has two aspects, Unity and diversity, or silence and dynamism. A truly enlightened perspective is the one that transcends both of these values, yet, at the same time, puts these two contradictory values together, allowing the coexistence of opposite values... The Vedic Literature repeatedly mentions that Shiva is in the heart of Vishnu, and Vishnu is in the heart of Shiva. According to Maharishi, this can be understood as the coexistence of silence and dynamism, or the true oneness of Shiva and Vishnu. In this perspective, Shiva and Vishnu can be seen as both one and the same, and yet as two different aspects of the Unified Field of Natural Law. It is in this manner that the Devata are different from each other, because they administer different aspects of reality, and yet remain one undifferentiated wholeness of Natural Law." (p. 333-334) [9]

'Man in the image of God.' Given that the universe is orderly according to the laws of nature, it makes sense that the laws of nature in their impersonal and personal forms also would be reflected in similar patterns of structure in human physiology—if '*man*' is made in the '*image of God.*' The indivisible Oneness of nature is expressed in many forms of phenomenal diversity.

For example one traditional representation of Shiva as the totality of infinite silence is in the form of a *lingum*. Its correspondence in human physiology is the entire brain, which controls the total activity and, functioning at its optimal state of coherence, supports the experience of unbounded inner silence. [12] In ancient Vedic science Shiva is sometimes also depicted in a more personal form, accompanied by various objects. Each of the objects not only has meaning in terms of natural law, but also corresponds to structures and their related functions in human physiology.

"The Vedic Literature gives a precise description of Shiva and the items surrounding Him. From Shiva's head, on which the crescent moon is placed, the Ganga river flows in three directions. On His forehead is a third eye. His hair has a knot tied by a string of Rudraksha beads. Around His neck is a great cobra. He has with Him a drum, a water pot, and a trident." (p. 335) [9]

These objects, traditionally associated with the principle of infinite silence of Shiva, can be understood to have anatomical and functional correspondence to neural structures in the brain. The trident corresponds to a radial section view of the ventricular system in the brain, which cushions the brain and helps maintain balance in movement; the crescent moon corresponds to the hypothalamus, which controls such functions as metabolism, hormones, emotion, and growth; the Rudraksha beads correspond to the choroid plexus, vessels resembling a string of beads which secretes the cerebrospinal fluid that fills the ventricles, flowing in three possible directions, which correspond to the Ganga river said to flow in three directions according to ancient Vedic literature; the third eye corresponds to the pineal gland in the center of the brain on the ventricular system, which is light-sensitive and secretes the hormone melatonin related to wakefulness, moods, and states of awareness; the drum corresponds to the septum pellicidium within the ventricles, which is related to emotion and motivation; the cobra corresponds to the caudate nucleus and fornix, which are involved in smooth movement, memory, and emotion; and the pot corresponds to the third ventricle.

As another example the devata or collection of laws of nature identified as the chief attendant of Shiva, embodied in the form of an elephant, is associated with the name *Ganesh*. Ganesh is described as sitting at the entrance to Shiva's cave—and all that goes in or out of the cave passes by. Correspondingly the set of neural structures called the pons, medulla, and cerebellum, which function as the gateway to the brain, have similar anatomical structures. The ears of Ganesh correspond to the cerebellum, which governs balance, eye movement, equilibrium, and intentional action. The face of Ganesh corresponds to the pons, which mediates awareness, behavior and input-output processes. The trunk of Ganesh corresponds to the medulla, the pathway of the functioning of the pons controlling visceral functions such as breathing and heart rate. The eyes of Ganesh correspond to the trigeminal nerves, which mediate skin, muscle, and joints in the face and mouth and are involved in sensory input about the body. The tusks of Ganesh correspond to the pontine nerves, which relate to balance, eye movement, and posture. [9]

Similarly correspondences are being identified with all the Vedic Devata, including *Brahma,* the *Creator* aspect or principle of natural law. In ancient Vedic science the three-in-one self-interacting dynamics of natural law is depicted in terms of the Vedic names of Vishnu, Brahma, and Shiva, as well as *Saraswati, Lakshmi* and *Durga*, names of the corresponding female depictions of Nature, or Mother Divine—all with specific correspondences in human physiology. [9]

Further the levels of creation described in ancient Vedic literature have their correspondence to levels of skin tissue enclosing the body, as well as the six cortical levels. Other correspondences have been identified in the DNA and throughout the nervous system and bodily structure. (Again, refer to Nader's book, *Human physiology: Expression of the Veda and Vedic Literature* for a much more expanded exposition including graphic depictions of the relationship between Vedic Devata, the Veda and Vedic literature, the cosmic counterparts, and the functions and structures in human physiology). [9]

A final series of examples relates to the correspondences between human physiology and the principle of infinite dynamism embodied in the name *Vishnu*:

> "The Vedic Literature describes the appearance of Vishnu, as well as the objects that surround Him, in great detail... Vishnu is described as lying at the bottom of the ocean, sheltered under a serpent with a big head and a long slender tail. Vishnu's hands hold a lotus, a discus, a mace, and a conch. He is described as having 10 main incarnations, in which He appears in the form of a fish, a tortoise, a boar, a half-lion, a dwarf, Parashuram, Buddha, and Kalki. In the Ramayan He appears as Ram, and in the Mahabharat as Krishna.

> In general, Vishnu resides in the central hollow structures of the nervous system and physiology. In the physiology, the serpent covering Vishnu is the central nervous system (CNS). The brain, with its various lobes, represents the head or heads of the serpent, and the spinal cord its tail. The limbic system, with the caudate nucleus, fornix, and other basal brain components, form C- shaped structures that look like the cobra sitting behind Vishnu's head...

This central hollow structure in the brain is called the ventricular system, and in the spinal cord it is called the central canal. The central hollow canal and ventricles, as well as all the gaps between cells, contain a fluid that represents the ocean in which Vishnu is said to reside. This is the cerebrospinal fluid. It is limpid and crystal-clear, but venous blood is reflected in it with a faint bluish taint... Vishnu is also depicted as having a bluish colour. The cerebrospinal fluid contains nourishment for the entire physiology in the form of sugar, amino acids, energy sources, vitamins, etc. It also contains the ions and molecules that are at the basis of all the dynamic processing, communication, and exchange features of the entire physiology.

The four arms of Vishnu correspond to the four cerebral peduncles, two upward and two downward; the lotus corresponds to the basal nuclei, which form a lotus-like shape in the brain; the discus corresponds to the functions of the cranial nerves; the conch corresponds to the organs of speech and hearing; the mace corresponds to the medulla and pons." (pp. 348-351). [9]

As described in ancient Vedic literature there are 10 incarnations of Vishnu, each of which also has its correspondence in brain physiology: the fish (*Matsya*) with the hormones circulating in the body fluids; the tortoise (*Kurma*) with the portal system in the hypothysis sitting on the sella turcica of the brain; the boar (*Varaha*) with the brainstem; the half-lion/ half-man (*Narasimha*) with the diencephalon in the brain center; the dwarf (*Vaman*) with the homunculus of the motor cortex; *Parashuram* with the corpus callosum; *Ram* with the somatosensory and sensorimotor cortex; *Krishna* with the parietal lobe and sensory cortex; *Buddha* with the diencephalon and brainstem; and *Kalki* and his horse with the hippocampus. [9]

Here is an extended citation that exemplifies the correspondence of the Vedic Devata of *Krishna* and brain structures. It draws from two well-known aspects of Vedic literature—the *Mahabharat*, the story of a major war, and a pivotal sub-section, the *Bhagavad-Gita*, the story of *Krishna* teaching the hero of the war, the great warrior Arjun, the key to skill in action—through maintenance of transcendental consciousness. *Krishna*, as the full embodiment of infinite dynamism of *Vishnu*, corresponds in the nervous system to:

"...the ventricular system and central canal of the spinal cord and brain... In the nervous system, there are two directions of the flow of information besides the association fibres of the great intermediate net—inward (the sensory system) and outward (the motor system). Both are under Vishnu's control in his incarnation as either Krishna or Ram. The sensory system is under Krishna's rulership and the motor system under Ram's rulership. Krishna's main seat of action in the physiology involves the entire nervous system, and more specifically the parietal lobe and sensory cortex of the brain...

The senses draw the attention outward. Sensory input, however, travels in an inward direction from the object via the sensory apparatus to the subject. Krishna's teaching takes the inward direction to its ultimate value in the transcendental field of consciousness. It is only from and within the transcendental field that perfect action can be undertaken. Mastery over the senses, and what they seem to highlight in the field of diversity, comes through transcendence. This is one of the essential aspects of Vishnu's teaching in his incarnation as Krishna.

Krishna, therefore, has control over the inward stroke of life and all physiological inputs. These inputs are carried through the fibres, nerves, and tracts in the physiology that carry sensory impulses from the right and left sides of the body to the left and right sides of the brain respectively. All of these impulses cross at the midline around the central canal, representing Krishna.

The right and left sides of the brain have, respectively, synthetic and analytic qualities. These qualities can conflict, and this is the origin of the conflict in the mind of Arjun, the great hero of the Mahabharat battle. At the beginning of the battle, Krishna brought Arjun to the middle of the field, between the two opposing armies, to survey the field. At this critical moment, as the battle was about to begin, Arjun found himself unable to act because of the conflict within himself. As a warrior, he was unable to reconcile his *Dharma*, his duty to fight and destroy his kinsmen, with his overwhelming love for them. Duty requires a very sharp, analytical quality of thinking, with the ability to act in a balanced manner. This is behaviour that originates in the left side of the brain. The unifying quality of love, on the other hand, is mostly dominated by the right brain.

Arjun was unable to balance these two principle styles of brain functioning, and it became Krishna's role (representing Transcendence) to guide him to a style of functioning that balances the two, allowing the most perfect and effective action.

As Arjun wavered between the two armies, Krishna taught him how to act from the Transcendent. This is the level of the central canal—the point of crossing of all the fibres in the physiology, particularly the sensory fibres that ascend within the spinal cord and brainstem to the brain. This is one reason why Krishna is located in this junction area, as well as in the sensory cortex, where He has the maximum balancing effect over the senses.

In the absence of Krishna's teaching of proper balance and proportion in action based on the Transcendent, the emotional aspects of the physiology can clash with rational thinking and the proper sense of duty. Basic emotional drives are embodied in the limbic system and its related structures. These number 100, and correspond to the 100 Kauravas, the 100 sons of King Dritarashtra who, although members of the same family, initiated the unjust war against the Pandavas [Note: The Pandavas are five brothers, one of whom is Arjun; see the Maharishi Mahesh Yogi on the Bhagavad-Gita: A new translation and commentary, Chapters 1-6].

The 5 Pandavas are embodied in the 5 senses. In this case we can take the senses to be the 5 channels of conscious experience, fine pathways of experience in accord with Natural Law. They are based upon a wakeful, alert state of consciousness, and are contrasted here to action that is based upon more basic, instinctual defence mechanisms—action that may be guided or controlled by past experience, desires, or even stresses, resulting from previous memories or impressions.

The conflict in Arjun's mind between love and duty, between the synthetic and the analytic, can occur in any mind, including those involved in the war of the Mahabharat. Any time the ability to fully appreciate and balance both values, synthetic and analytic, is disrupted in any of its aspects, problems and indecision can arise. Both values, fully enlivened, are needed for successful action. This is why the Mahabharat is a living reality for everyone at all times—it represents the internal dynamics of the functioning of our human brain physiology." (pp. 368-370) [9]

One more example will be presented, showing the correspondence between the incarnation of *Vishnu*, the principle of infinite dynamism in nature in the embodiment of *Buddha*, and structures in the human nervous system.

"Vishnu's incarnation as Buddha can be visualized in the physiology in the diencephalon and brainstem areas... Buddha can be seen in the physiology sitting in front of the main shaft of the spinal cord, which extends into the brainstem. This shaft is the tree with its branches and leaves extending into the brain. The head of Buddha coincides with the thalamus, His hands and feet correspond to right and left groups of cranial nerves. His abdomen is the pons.

[T]he thalamus is connected with both the kingship or rulership quality and the balance between inner and outer—the relative and the Absolute. These are main characteristics of Buddha, who was born in a royal family and who embodied the quality of enlightenment in Cosmic Consciousness—*Nirvana*. In His teaching of enlightenment Buddha expounded upon the difference between inner and outer—the Absolute and the relative. Buddha is light, the effulgence of light and enlightenment, and the thalamus in the brain represents the Sun as well as the radiating fibres that carry all conscious sensory experience. All these are aspects of Buddha's effulgent qualities." (pp. 371-372) [9]

"Through the preceding stories and examples, we see that all the Devata, the *Rishis,* in fact all the aspects of Natural Law available in the universe, are embodied in our human physiology. All the events and stories that occur in the Itihas, Puran, Upanishad, and in the whole of the Vedic Literature, are an all-time reality, taking place at every moment in our physiology. Even such details as the trees have their expression in the physiology. In this case, they are embodied in the form of neurons and the many tree-like structures formed by the branches of nerves, blood vessels, and other structures. The tree-like structure in the brain formed by the corona radiate that connects to the cortex is a wish yielding tree (*Kalp Vrikshit*) since it connects to the area of the brain that represents Yoga, where all the *Siddhis* are formulated and represented in the physiology... Many illustrations can be cited; we have only mentioned some of the more prominent... The reality is that every individual is really Cosmic." (p. 419) [9]

Eternal simultaneity and temporal sequence. The stories of Vedic Devata in Vedic literature have various phenomenal expressions of name and form that emphasize particular qualities. They serve as depictions of the mechanics of universal laws of nature in action that are occurring simultaneously at every moment in time, as well as of the structures and functions in human physiology. They also sometimes serve as accounts of historical events. They represent a vast reservoir of knowledge of universal patterns of nature in specific levels and embodiments.

The principle of infinity and point, whole and part, universality and individuality, simultaneity and sequence in the Veda and Vedic literature also can be understood in terms of both the ever-present, simultaneous mechanics of natural law that govern all phenomenal change *and* the historical sequence of phenomenal expressions in nature. In ancient Vedic science natural law expresses itself in cyclic patterns, and these cyclic patterns also are expressed in the form of historical embodiments of natural law.

As described in Chapter 22 the notion of sequence as cyclic patterns through time is prominent in Maharishi Jyotish. This discipline describes vast epochs of time as the self-repeating phenomenal display of Nature. It echoes the Old Testament adage that 'there is nothing new under the sun.' This relates to the dynamics of all change in nature as manifestations of the principle of *Smriti* or memory. This principle is also referred to as the cosmic computer that computes all change holistically. According to this holistic principle any change anywhere and anytime in nature incorporates the entire history of the universe, simultaneously consistent with the unified totality and diversified locality of the laws of nature. It is the ever-new expression of the eternal repeating reverberation of the self-interacting dynamics of the eternal non-changing *Smriti* or memory of natural law. Ancient Vedic science incorporates a much more expanded conception of history than is embodied in the consensus theory of cosmological origins in modern science. A richer sense of the cycles of time within eternity associated with the holistic principle of Smriti may be obtained from the following excerpt of Maharishi's commentary on the *Bhagavad-Gita*:

> "Time is a conception to measure eternity... To arrive at some conception of the eternal, the best measure will be the life-span of something that has the greatest longevity in the relative field of creation... The eternity of the eternal life of absolute Being is conceived in terms of innumerable lives of the Divine Mother, a single one of whose lives encompasses a thousand life-spans of Lord Shiva. One life of Lord Shiva covers the time of a thousand life-spans of Lord Vishnu. One life of Lord Vishnu equals the duration of a thousand life-spans of Brahma, the Creator. A single life-span of Brahma is conceived in terms of one hundred years of Brahma; each year of Brahma comprises 12 months of Brahma, and each month comprises thirty days of Brahma. One day of Brahma is called a Kalpa. One Kalpa is equal to the time of fourteen Manus. The time of one Manu is called a Manvantara. One Manvantara equals seventy-one Chaturyugis. One Chaturyugi comprises the total sum of four Yugas i.e. Sat-yuga, Treta-yuga, Dvapara-yuga and Kali-yuga. The span of the Yugas is conceived in terms of the duration of Sat-yuga. Thus the span of Treta-yuga is equal to three quarters of that of Sat-yuga; the span of Dvapara-yuga is half that of Sat-yuga, and the span of Kali-yuga one quarter that of Sat-yuga. The span of Kali-yuga equals 432,000 years of man's life." (pp. 253-254) [13]

Within the comparatively limited framework of recorded history, key events depicted in Vedic literature include the two accounts referred to above of the incarnations of *Vishnu* as *Krishna* and as *Buddha*. The life of *Krishna* is chronicled in descriptions of the *Mahabharata,* such as in the Bhagavad-Gita, which has a basis in historical authenticity. In his commentary on the Bhagavad-Gita, Maharishi places the events into both a timeless universal perspective as well as a historical chronology:

> "The Bhagavad-Gita is the highest expression of divine intelligence understandable by man. Dealing with the unseen aspects of life, it also touches on the past and present of the world of our daily life. Furthermore, the Bhagavad-Gita, while expounding universal Truth, is itself a historical record and relates incidents that took place five thousand years ago." (p. 252) [13]

Maharishi also refers to historical accounts that date the incarnation of *Buddha* around 3000 years ago. According to the sequence of the 10 incarnations of *Vishnu* described above, the next incarnation of *Vishnu* as *Kalki* will take place toward the end of Kali-yuga. Because Kali-yuga is said to have begun right after the historical period of *Krishna,* about 5000 years ago, this next incarnation would be approximately 427,000 years into the future.

These cycles of time encompass periods much vaster than the time since the origin of the universe calculated according to recent theories of the big bang—about 13.7 billion years. The concept of the big bang (or big condensation) thus might represent a particular cycle within the larger context of cycles of manifestation and dissolution associated with the days and nights or lifespans of Vedic Devata described in Maharishi Jyotish. Such cycles of natural law within the infinite eternity would necessarily encompass time-spans far beyond the theorized birth of the Sun, our solar system, and human life as we know it now. This is included as another example of the integration of abstract impersonal principles of universal timeless laws of nature and their personal historical expressions in time and space. This example provides an additional glimpse at the expansiveness of the Veda and Vedic literature as total knowledge.

God and Man: The Causal Direction

It may now be apparent that ancient Vedic science and Maharishi Vedic Science and Technology represent a vastly more holistic and integrated account of nature than has been achieved through the fragmented and restricted objective means of gaining knowledge. Of primary practical significance, it includes systematic scientific means to validate holistic knowledge empirically in direct experience. Theoretical knowledge isn't fully satisfying until it is directly experienced to be accurate; theories are most meaningful on the basis of the authority of direct empirical experience. It is sometimes said that the essence of science is tangible evidence, not just good theory. As Maharishi repeatedly emphasizes, however, total knowledge requires total development of the ability to experience the evidence. With respect to total knowledge, one needs to *be* the knowledge, not just intellectually know it. Ultimately, reason and experience merge together: total knowledge must include total experience, in order to be total.

Direct empirical validation of total knowledge involves development of each individual investigator in the inner laboratory of his or her own non-local mind through systematic empirical research, such as regular practice of the developmental technologies for full human development described in this book. On the basis of full development, full knowledge can be gained. Until then the knowledge is not fully validated, and the knower is not fully competent to contribute to the consensual validation of total knowledge.

> "It doesn't make much sense to ask a blind man to confirm what you see."—Jonathan Shear, philosopher [14]

At least one understanding we have gained from modern science is that things are not always what they appear to be. [15] For example the Sun doesn't go around the Earth, as it appears to do in naive sensory observation. Gross sensory experience is not capable of being the final arbiter of *reality*. Full objective knowledge is based on full subjective ability to experience the total range of knowledge. The principle of empirical verification is crucial to knowledge, but cannot be based only on gross sensory functioning. For the empirical verification to be *fully valid*, it needs to be based on the inner development of full consistent subjectivity. This establishes the competence to investigate fully and gain control of the natural world—mastery over natural law—through alignment with the totality of natural law.

Generally it has not been appreciated in modern scientific circles that 'empirical evidence' is a product of the level of ability to experience. The objectified approach to knowledge in modern science has overlooked the state of the knower and his or her ability to experience the totality of nature—constrained only to ordinary gross empirical experience, which even within the limitations of modern science we now know from indirect evidence is quite incomplete and far from the total picture of nature. Gaining an intellectual understanding of the full range of levels of nature is helpful and inspiring, but it will never be sufficient and fully satisfying validation. Mathematical formulation of the unified field also won't yield satisfying validation. Hopefully this recognition will result in practical action to develop direct experience of the full range of gross, subtle, and transcendental domains and their ultimate unity—the means for complete validation. This requires each individual to refine and purify his or her own mind and body to be able to experience total knowledge directly. Essentially what has been overlooked in modern science and modern thought is systematic means to experience all the levels of nature from gross to subtle to transcendental. Maharishi has revived systematic technologies to facilitate this experience based on simple effortless transcending. This developmental process typically takes some time and regular practice—along with healthy daily routines, pure foods, regular exercise, and other developmental technologies.

Even understanding these fundamental points, however, it nonetheless may be helpful to conclude by restating the current progress in validating the holistic view described in this book within the limitations of the objective, reductive physicalist view. The view common among investigators who contribute to the general consensus at this stage of modern scientific knowledge is the view from the window of the ordinary waking state of consciousness. Within these limitations, what factual support is there for the holistic view of nature? Most anything can be explained by theory. What is the proof that can be understood within the generally accepted means of validation, current state of understanding, and restricted range of experience in the physicalist paradigm still common among modern scientists, religionists, and scholars?

Clearly the implications of the holistic view are life-transforming, both individually and sociologically. If it is as significant as it seems to be, wouldn't it be unequivocally demonstrable in the natural world of our ordinary everyday experience? If according to the holistic view the causal direction is consciousness over mind and mind over matter—transcendental to subtle to gross—wouldn't it be verifiable in the ordinary gross level of nature?

The correspondence between the 'cosmic counterparts' described in the past chapter and the 'cosmic devata' in this chapter to the structures of human physiology at several layers of detail can be understood to be an extraordinary objective indicator of the fundamental top-down structure of natural order in the universe. As the full significance of this research unfolds, hopefully it will be recognized as the most profound and integrative scientific discovery that reflects a comprehensive understanding of natural laws in Vedic science, far exceeding previous knowledge discovered though modern science. However, one might take the position that these detailed anatomical and functional correlations, while quite remarkable and compelling, are just correlations—and as we say, correlation doesn't prove causation. Where is the tangible proof? Key planks of 'objective' evidence will now be summarized that support the holistic view described in this book, a fundamental aspect of which is the causal direction of mind over matter.

In summarizing these planks of knowledge it is interesting to note that in an important sense the ancient Vedic records themselves can be thought of as obvious empirical verification of the holistic consciousness-mind-matter ontology in nature. These ancient records extensively described the unified field long before modern science developed to the point where it could probe nature to its non-material foundation, leading to unified field theory. This is suggestive of direct intuitive means of gaining knowledge that is beyond matter and beyond ordinary experience and reason. Because the unified field is beyond sensory experience, knowledge of it must have come from within the mind somehow in a top-down process, long before any ordinary bottom-up sensory experience could provide data from which to reason about a completely abstract unified field that cannot be seen, touched, or tasted. Notwithstanding this quite significant point, however, one may ask what evidence is there in the physical sciences that is consistent with the causal principle of mind over matter?

With respect to physics, 'obvious' empirical verification would include the major collection of experimental findings that strongly support the quantum wave-particle duality of nature beyond ordinary classical matter, as well as the numerous quantum mechanical applications in the ordinary macroscopic world. Additional 'obvious' empirical evidence includes the experimental tests of Bell's theorem that strongly support the fundamental nonlocality of nature. At least these findings clearly demonstrate that matter is not fundamental in nature, which takes us a bit closer.'

It also is tempting to include the orthodox interpretations of quantum theory that conscious mind collapses the quantum wave function into classical matter, and that quantum gravity theories—such as loop quantum gravity, black hole thermodynamics, and holographic models—propose an abstract information space underlying and generating physical space. Although relevant to the issue, they may not seem to fit into the category of 'obvious' objective evidence, however, in that they depend on theory more than empirical outcomes. As well, theories that associate an underlying information space with nonlocal quantum mind are still not widely understood and appreciated in the mainstream of modern science. Models in chaos theory that macroscopic effects can come about from quite small influences are in the right direction, but don't extend deep enough to go beyond the ordinary understanding of the physical causal nexus. Unified field theories and quantum gravity theories do extend beyond the physical, but don't yet obviously incorporate how the underlying non-material substrate demonstrates differential causal influences by mental intention—by the proposed nonlocal field of mind. There are major experimental attempts to investigate such influences, but the findings may not seem to constitute 'obvious' empirical evidence—yet.

In the psychological sciences it may be tempting to include related findings from paranormal research, which in recent years reflect considerably more experimental sophistication. Certainly unequivocal out-of-body, near-death, and psychic ability experiments would constitute strong evidence for mind over matter. However, again the findings seem to be considered not yet reliable, straightforward, or credible enough to fit the strict—if naive—criterion of 'obvious' empirical validation.'

The body of research on the TM program summarized in this book is quite relevant, especially because it demonstrates that this mental technology must be fundamental in order to produce such a wide range of benefits on the macroscopic level of ordinary daily life. These include improved performance on a variety of measures of intelligence, cognitive and memory processes, sensorimotor coordination, as well as social skills. This research also shows that the wide-ranging psychological and behavioral benefits are correlated with improved physical health, such as reductions in healthcare utilization rates as indicated by insurance statistics. Widespread improvements in performance measures and in health are positively correlated with number of years of TM practice, as well as reported clarity of transcendental consciousness. This constitutes strong 'obvious' empirical evidence that a mental practice can lead to developmental gains in many areas of ordinary human behavior. It also is corroborated generally by the voluminous amount of general research in the fields of mind-body medicine and preventative health in recent years.

Further, experimental research on the TM program shows that this mental technology produces reliable neurophysiological, biochemical, and experiential reports that distinguish a major fourth state of human consciousness—such as, for example, the unique findings of global EEG coherence and virtual breath suspensions. These changes positively correlate with reports of other developmentally advanced experiences, and with reports of higher human development consistent with theories in developmental psychology. Unique psychophysiological states that positively correlate with reports of higher consciousness and historical descriptions of enlightenment consistently result from regular practice of this mental technology. This research documents reliable physical changes accompanying a mental practice. The mental practices precede the physical changes, and otherwise are not observed; obviously, physical changes have not been observed as preceding and impelling individuals to practice the developmental technologies.

Experimental research on the advanced TM-Sidhi program also has identified unique psychophysiological patterns. For example research on 'yogic flying' indicates that the moment of the strongest subjective experience of lightness and impulse to 'lift off' is correlated with global EEG coherence across the brain cortex. This electroencephalographic pattern has not been reported in studies of religious and spiritual feelings produced by psychoactive drugs or electromagnetic manipulation of neural activity in the brain. Moreover, no physically induced neural changes have been shown to produce the wide range and depth of positive psychological and biobehavioral changes that accompany regular practice of the TM technique. The body of research on the TM and TM-Sidhi program constitutes strong evidence that these mental techniques cause consistent holistic psychophysiological and behavioral effects on the gross physical level.

Still, one might take the position that these empirical results are not unequivocal demonstrations of mind over matter from our criterion of 'obvious' empirical evidence. Although there are many historical reports of levitation, for example, there seem to be no current demonstrations of levitation reliable enough to be tested under rigorous laboratory conditions. Also it might be considered that this research doesn't yet constitute unequivocal evidence that these numerous physical effects are not caused by underlying neural activity in the brain that accompany them.

Unequivocal evidence would show a physical effect that is not explainable by any physical process, such as neural activity. The evidence that seems to do the best job of satisfying this criterion is the sociological research on the Maharishi Effect, summarized in Chapter 21. Recall that this research investigates the effect on social indicators such as war violence, crime rate, accident rate, and economic trends preceded by TM and TM-Sidhi practice including 'yogic flying' by large groups of practitioners. The highly significant statistical data in this well-controlled research indicates that the distributional range of the effects don't fit any model based on the four fundamental physical forces. Thus the case cannot be made that this sociological effect is due only to neural activity in the brain. No neural activity connects humans in this collective manner, and no physical force field radiates from its generating source in the manner that fits the empirical findings. The findings cannot be accounted for only in terms of the four fundamental force fields—and they are the only choices in a physicalist explanation. If the findings are indeed *real*, they constitute strong empirical verification of a coherent collective influence of mind over matter.

But in the same way one might say you can explain anything with theories and ad hoc supplements, one might also say you can do all the studies you want but they still may not be believable. It might be suggested that the research still has not controlled *all* plausible confounds, or that it relies on complicated statistics that could have errors, or that it has not been replicated enough, or that it is not accompanied by an adequate theoretical model, or that it must involve yet unknown experimenter biases, or that it does not match generally accepted theoretical principles in mainstream science, and so on. As mentioned in Chapter 1 there are no established criteria in modern science about the amount of validation or consensus necessary to accept a particular explanation of a pattern of phenomena as indicative of a natural law. When well-established paradigms are challenged, the bottom line might be whether it fits the currently popular intuitive-like sensibilities and beliefs and range of empirical experience, as discussed in Chapter 1—even when there is strong empirical evidence for a new and more inclusive paradigm.

If the empirical evidence for an expanded paradigm is strong and valid, however, then at some point it becomes useful to consider seriously what the *plausible* alternative explanations might be. The theoretical explanation for the Maharishi Effect research is based on the holistic consciousness-mind-matter view of nature. This holistic view emphasizes top-down super-ordinate causal influence of consciousness over mind, and mind over matter. This also is consistent with the deep intuitive sense that we are not robots with no free will, that we have a causal effect on the world, and that our minds and consciousness actually exist. The main popular alternative to the holistic view seems to be the reductive physicalist view. But the four fundamental forces of nature that exhaust the physical possibilities in this contemporary view don't provide a model to account for the experimental findings. More fundamentally, however, does the physicalist view offer a rational and logically consistent *plausible* alternative?

In the reductive physicalist view the only direction to go in pursuit of the basis of top-down super-ordinate causal control of a biological organism is to the underpinnings of neural activity in the brain—because the mind is thought not to exist outside of the brain. This takes us to underlying biophysical processes in the body and even deeper to inert physical processes that show no evidence of self-control or negentropic behavior but rather just random motion associated with inherent quantum indeterminism. The general understanding is that inherently dynamic quantum fields reflect fundamentally purposeless and meaningless behavior not differentially influenced by any mind or intelligent controlling agency.

In this view the laws of nature spontaneously and instantaneously emerged from randomness or disorder. Random disorder has inherent dynamism, but not inherent order, intelligence, life, or consciousness. Order in the form of the laws of nature is thought somehow to have emerged at the instant of the big bang, without any precedents—from literally *nothing*. Life later emerged from the dynamism and spontaneous order through random mutation and biological selectivity. Biological selectivity is suggestive of intelligence, emerging from physical processes that are theorized to exhibit no intelligence. Consciousness emerges from complex systems, whether living or non-living. Conscious mind then somehow gains super-ordinate control over the physical, while not being anything other than the physical. It seems quite reasonable to question the plausibility of these explanations, which posit that all the order—all the correspondences and correlations we have found in scientific research and that we observe throughout nature—is fundamentally due to random disorder. Even the concept of a plausible alternative implies causal determinism, not randomness. Doesn't the physicalist explanation seem fragmented, fundamentally inconsistent, and incomplete? Given contemporary theory and research, it is difficult not to question this approach as a plausible alternative explanation. Modern science has already gone beyond the physical. Isn't it also time to go beyond random disorder as the bottom line of nature—as unified field theory, the 2nd law of thermodynamics, the principle of decoherence, and the arrow of time logically imply? Isn't it logically consistent that parts come from the whole, than that the whole comes from the parts and then control them, and the parts ultimately come from *nothing*?

In pursuing further the issue of mind over matter, let's now return to neurotheology and the question of the causal direction in terms of whether 'man is made in the image of *God*, or *God* is made in the image of man. Both perspectives are based on assumptions of orderly deterministic laws of nature. In the physicalist view of biological organisms, mental activity and consciousness come from neural activity. With sufficient neural complexity, consciousness and mind emerge from or are the same as the physical neural activity—and don't exist without it. Although this does not prove the causal direction one way or the other, it frequently is interpreted as supporting the conclusion that religious and spiritual feelings are caused by certain patterns of neural activity in the brain, through which ideas such as of God are projected. (On the other hand one might say that if neural activity and conscious intention are the very same thing—the functional identity hypothesis—then the question of causal direction would become moot, in that one could not cause the other if they are the same thing).

In the physicalist view religious feelings and ideas are *real* as imagined phenomena—*real* thoughts and feelings in the brain, but only the neural activity of *unreal* illusory experiences. One neurotheological researcher states this perspective quite bluntly: "God "...is an artifact of the brain." (p. 68) [16] In this view the object of the ideas and feelings, such as God, is a mental projection that has no external form or existence outside of the physical brain of the individual having the illusory or hallucinatory experience. Agreement about God across individuals—outside of individual brains—means that the brains have similar neural patterns and project similar illusory ideas and feelings. But there is not believed to be an external God apart from the imaginations or hallucinations. Some people happen to have similarities of neural patterns associated with spontaneous religious and spiritual feelings, even perhaps genetically based; but the patterns also can be evoked through certain patterns of neural stimulation. In other words God is a figment of the imagination of man, a product of certain neural firing patterns. The experiences may have an adaptive role such as to reduce psychological angst, but God has no *real* existence as an object in nature outside the experiencer. So to speak, God is made in the image of man.

Progress in modern science beyond the boundaries of the physicalist view is reflected in quantum gravity and unified field theories which posit non-material sources of everything in the physical. As described in Chapters 10-11 the unified field is not only inherently dynamic but also inherently intelligent—at least inherently orderly— as the source of all order in nature. Carrying this perspective further, the unified field as omnipresent, omnipotent, and omniscient can be associated directly with God, or the Godhead, inasmuch as these universal attributes also are fundamental attributions of God. Whether using the impersonal secular scientific language of the unified field or the personal religious language of *God*, the causal direction is that God—or at least that meaning of God that can be described in terms of the unified field—must be the source of 'man.' The unified field is typically theorized to be the source of everything in nature. No one seems to be proposing that certain neural patterns evoke the illusory idea that there is a unified field. A theory that neural processes are the cause of the unified field would likely not have many proponents, even among physicalists.

On the other hand the holistic view discussed throughout this book proposes that matter and mind depend on the unified field. In Maharishi Vedic Science and Technology the unified field is inherently dynamic, intelligent, living consciousness. Gross physical matter reflects these attributes the least. Neural complexity allows them to be expressed in the gross physical body, but the physical body does not express them when the subtle manifestations of energy and intelligence depart from it. In this case *real* religious and spiritual ideas and feelings certainly can be imagined projections in the brain. But they also can be veridical perceptions of *real* objects in phenomenal nature outside of individual brains.

Both the impersonal scientific and the personal religious descriptions have their validity in expressing the self-interacting dynamics of natural law at different levels of creation, including human physiology. The personal religious image of God is based on genetic history and cultural experiences, grounded in human physiology. In this sense the image of God that 'man' has is influenced by and comes from 'man.' The image of God can be said to be made in the image of 'man' in the sense that they are influenced by culturally specific features projected by 'man.' But more fundamentally 'man' is made of the same energy, intelligence, and laws as the entire cosmos, which reside in the unified field. At least in this sense 'man' is made in the image of *God*. 'Man in the image of God' refers to a profound statement in ancient holy scriptures about the nature of self-referral—the Self referring back to Itself at all phenomenal levels of nature. It corroborates the expression of the self-referral mechanics of natural law described in ancient Vedic science (*Brihad-Aranyak Upanishad 1.4.10*). Maharishi describes Vedic expressions as reflective of the correspondence of the structure of individual physiology and the structure of the entire cosmos:

"*Yatah pende, yatah brahmande*

The individual is cosmic." [17]

This again brings us to one of the core issues running through this entire book that needs to be clear and explicit: Is there an *objective reality* with objects that are independent of the subjectivity of the observer? Perhaps by now the answer also will be obvious. With respect to individual consciousness the answer is *yes*; and with respect to universal consciousness the answer is *no*. It depends on the state of consciousness of the observer, whether individual or universal consciousness predominates—whether partial or holistic knowledge and experience predominates. Ultimately, however, everything in nature is connected, and most fundamentally is one—the unified field. In ancient Vedic science and Maharishi Vedic Science and Technology the unified field is universal consciousness itself. Its 'nature', *my* nature, *your* nature, *our* nature is the simultaneous coexistence of infinite value and point value, part and whole, individuality and universality. At that level the wholeness of the unitary experience of individual consciousness is the wholeness of universal consciousness—relative oneness of the self is the absolute Oneness of the Self. It is the unified experience of finite relative creation as nothing other than the infinite absolute unity of one's Self, again completely summarized in the Upanishadic statements:

"I am That, Thou art That, all this is That, That alone is, and there is nothing else but That." (p. 34) [18]

It is wholly appropriate at this point to recognize the depth, breadth, and clarity in Maharishi's revival of ancient Vedic science—of which this book respectfully has attempted to provide a glimpse. In the unfoldment of total knowledge he has systematically addressed perennial questions in modern and historical thought by clarifying the ultimate seamless unity of nature. The singular power of his articulation of total knowledge itself constitutes profound validation of the consciousness-mind-matter understanding of nature. Further he has provided systematic means to validate that understanding through direct experience of self-referral consciousness, and he has provided it in simple holistic developmental technologies—especially the TM and TM-Sidhi program including 'yogic flying.' In the long corridor of time our civilization is indeed fortunate that total knowledge is now within our easy grasp, in our own self-referral consciousness. Through systematic application of Maharishi's Vedic Science and Technology an Age of Enlightenment is dawning. *That* is the *whole point* of the opening statement of this book in the Prologue, hopefully now more clearly unfolded, and referred to once again here:

*A*ll the research in modern science has led to the doorstep of the ultimate unity of nature. *The next step is direct empirical validation of unity. That requires going beyond sensory experience and reasoning, the basic means of gaining knowledge in modern science. It is most fortunate that we now have systematic means to unify completely our theoretical understanding and empirical experience of nature based on a holistic science of consciousness. Consciousness is the 'lamp at the door,' illuminating both the outer diversified field of knowledge and the inner unified field of knowledge. Through this more inclusive science, we are stepping into a genuine Age of Enlightenment.*

Chapter 24 Notes

[1] Hamer, D. H. (2004). *The God gene: How faith is hardwired into our genes.* Doubleday.

[2] Newberg, A. B. & Iverson, J. (2002). On the "Neuro" in Neurotheology. In Joseph, R. (Ed.). *NeuroTheology: Brain, science, spirituality, religious experience.* San Jose, California: University Press, pp. 252-253.

[3] Persinger, M. A. (2002). The Temporal Lobe: The Biological Basis of the God Experience. In Joseph, R. (Ed.). *NeuroTheology: Brain, science, spirituality, religious experience.* San Jose, California: University Press, pp. 273-278.

[4] Persinger, M. A. (2002). Experimental Simulation of the God experience: Implications for Religious Beliefs and the Future of the Human Species. In Joseph, R. (Ed.). *NeuroTheology: Brain, science, spirituality, religious experience.* San Jose, California: University Press, pp. 279-292.

[5] *Scientific research on Maharishi's Transcendental Mediation and TM-Sidhi programme—Collected papers,* Vols. 1-5. Fairfield, IA: Maharishi University of Management Press.

[6] Maharishi Mahesh Yogi. (1972). *The Science of Creative Intelligence: Knowledge and Experience.* (Videotaped course). Los Angeles: MIU Press.

[7] Chalmers, D. J. (1996). *The conscious mind: In search of a fundamental theory.* New York: Oxford University Press.

[8] Maharishi Mahesh Yogi (2004). Maharishi's Global News Conference, December 22.

[9] Nader, T. (2000). *Human physiology: Expression of Veda and Vedic Literature,* 4th Edition. Vlodrop, The Netherlands: Maharishi Vedic University.

[10] Maharishi Mahesh Yogi (1997). *Maharishi speaks to educators,* Vol. 2. India: Age of Enlightenment Publications.

[11] *Holy Bible,* King James Version (1973). Nashville, Tennessee: Dove Bible Publishers, Genesis 1:26.

[12] Maharishi Mahesh Yogi. (2003). Maharishi's Global News Conference, April 7.

[13] Maharishi Mahesh Yogi (1967). *Maharishi Mahesh Yogi on the Bhagavad-Gita: A new translation and commentary, chapters 1-6.* London: Penguin Books.

[14] Shear, J. (2004, October). Personal phone communication, Fairfield, IA.

[15] Greene, B. (2004). *The fabric of the cosmos: Space, time, and the texture of reality.* New York: Alfred A. Knopf.

[16] Persinger, M. (2004). As quoted in, Is God in Our Genes? *Time,* October 25, Vol. 164, No. 17. New York: Time, Inc.

[17] Maharishi Mahesh Yogi (2003). Maharishi's Global News Conference, December 10.

[18] Maharishi Mahesh Yogi, (1963). *Science of being and art of living.* Washington, D.C.: Age of Enlightenment Publications.

DEDICATION AND ACKNOWLEDGEMENTS

It is my great honor to dedicate this book to His Holiness Maharishi Mahesh Yogi and the Vedic Tradition of Masters who maintain knowledge and experience of the integration of life. All the useful knowledge in this book is from carefully listening to Maharishi. *Jai Guru Dev*

It also is with my deepest love and appreciation that I acknowledge the patient assistance and support of my wife, Connie, our families, Chip Charleston, Park Hensley, Jerry Jarvis, David Scharf, Sam Boothby; Max Winters and Liz Howard who helped format the manuscript; and Denyce Rusch for design work.

1526146